A STUDY OF
GLOBAL SAND SEAS

COMPLEX LINEAR RIDGE with reversing and star dunes superimposed; broad interdunes on each side. Aerial photograph, taken southwest of the Kuiseb River, Namib Desert, South-West Africa, by E. Tad Nichols. (Frontispiece.)

A Study of Global Sand Seas

Edwin D. McKee, *Editor*

GEOLOGICAL SURVEY PROFESSIONAL PAPER 1052

*Prepared in cooperation with the
National Aeronautics and Space Administration*

University Press of the Pacific
Honolulu, Hawaii

A Study of Global Sand Seas

Edited by
Edwin D. McKee

ISBN: 1-4102-1457-5

Reprinted from the 1979 edition

University Press of the Pacific
Honolulu, Hawaii
http://www.universitypressofthepacific.com

FOREWORD

ONE HUNDRED YEARS AGO, the U.S. Geological Survey had its inception in Washington, D.C., when Congress established a unified geological survey under the Department of the Interior, combining the duties of and replacing four earlier government surveys. The Survey's program included "classification of the public land, and examination of the geological structure, mineral resources, and products of the national domain." It seems entirely appropriate, therefore, that a professional paper summarizing the progress of investigations and synthesizing our present state of knowledge in the important field of eolian deposition be one of those issued to commemorate the Geological Survey's 100th anniversary.

Early descriptions and considerations of the characteristics of modern eolian sand deposits, mostly in the great sand seas of the world, date back to the 1880's and 1890's, when pioneer geologists and explorers, including Walther, von Zittel, and Sven Hedin, wrote their classic papers on desert dunes. One or two decades later, interest in eolian processes greatly increased when wind-formed deposits were recognized in ancient sandstones in many parts of the world and in rocks of many ages.

Throughout the 20th century, as the science of geology has expanded and the programs of the U.S. Geological Survey have proliferated to keep pace, dune studies have had a similar growth. Work was initially concentrated mostly on the description of dune forms or morphology and on analysis of textural features; by midcentury, however, major contributions had been made to the physics of eolian sand, as exemplified by the classic work of Bagnold, by detailed studies and interpretations of minor eolian structures, by statistical analyses of cross-strata dip directions, and by the development of systems for dune classification. Most recently, interest has been renewed in detailed grain studies, in the study of cross-strata, and in interpretation of dune patterns by means of aerial photographs and Landsat imagery.

A major feature of this report on global sand seas is the compilation and comparison of available data based on many different methods of investigation. Evidence at one extreme is obtained from the detailed studies of minute particles and from analysis of individual grains. At the opposite extreme is evidence obtained from remote sensing, in which dune patterns, recorded from approximately 500 miles in space, are compared from one sand sea to another. Furthermore, comparisons are made between ancient and modern deposits with respect to textures, structures, and other characteristics. Criteria for recognizing the eolian origin of various ancient deposits are discussed. The application of these studies to economic problems, which is described in one chapter, clearly illustrates the importance of eolian deposits to our present culture and to human welfare.

H. William Menard

H. William Menard
Director, U.S. Geological Survey

Contents

[For convenience in cross-referencing, each chapter of this publication has a letter designation]

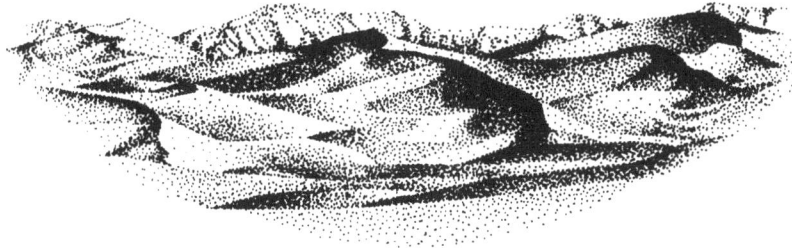

A STUDY OF GLOBAL SAND SEAS

INTRODUCTION TO
A STUDY OF GLOBAL SAND SEAS

Chapter A

By EDWIN D. McKEE

Contents

1

Illustrations

Table

Subject Matter

𝕿 HE BIRTH of the idea that led to this publication on "Global Sand Seas" dates back to the late 1920's. At that time I was engaged in a study of the Coconino Sandstone of Arizona's Grand Canyon. Considerable controversy existed then as to whether this sandstone was a subaqueous deposit or was composed of wind-formed dunes. It became apparent that definitive literature was sparse or lacking on types of dunes, global distribution of these types, the mechanics of their development, the precise nature of their internal structure of cross-stratificiation, and the relation of wind systems to these sand forms. Especially lacking were data on criteria that could confidently be used in the recognition of ancient dunes.

The common denominator in this publication is eolian sand bodies. Although the book is concerned primarily with desert sand seas, the subject matter is not restricted to deserts; it includes many references to deposits of coastal sand and to sand bodies in humid climates. Nor does the book deal exclusively with dunes, which, according to most definitions, involve mounds or hills. Many references are made to sand sheets, sand stringers, and other types of sand deposits that have no prominent topographic expression. All sand bodies accumulated by the action of wind are discussed.

Chapters A–J of this publication are primarily topical. Chapters cover the grain texture, the color, and the structure of modern dunes and other eolian sands. Special treatment is given to the relation of wind data to dune interpretation, the evolution of form in current-deposited sand bodies as determined from experimental studies, and the discriminant analysis technique for differentiating between coastal and inland desert sands. This topical part of the publication also includes an analysis of criteria used in ancient deposits to interpret their eolian genesis and a consideration of economic application of the principles described, including a discussion of potentials and problems associated with eolian hydrocarbon reservoirs. The final chapters present a discussion of the morphology and distribution of dunes as determined largely from Landsat images.

Chapter K of the publication is devoted to descriptions of major sand seas based largely on thematic maps derived from Landsat (ERTS) mosaics. Although inclusion herein of the actual mosaics proved to be impractical, the maps derived from them do show the distribution and abundance of various dune types and the relations of these types to certain associated features, such as bedrock, water bodies, and juxtaposed dunes. Furthermore, sand roses included with each of these maps enable the user to draw conclusions on the probable relations of wind strength and direction to dune type in a particular area.

Regional studies (chapter K) were a team effort. Analysis of the Landsat (ERTS) mosaics and mapping boundaries of individual dune types were by Carol Breed. Synthesis of the rather voluminous literature and preparation of abstracts covering it was by Camilla MacCauley. Actual preparation of maps was by Franci Lennartz and later by Sarah Andrews. The gathering of data on wind, the calculation of wind roses, and the interpretation of their relations to sand bodies were by Steven Fryberger, assisted by Gary Dean.

Facets of Dune Study

DURING THE COURSE of several decades, eolian dune studies have been made in ever-increasing numbers and from many different approaches. As might be expected most of the early studies were largely descriptive, with observations focused especially on external form and movement of sand bodies. Attempts to classify dune types advanced slowly, but basic forms were early recognized in relation to such factors as wind regime and physical barriers.

Principal approaches to dune study in the United States, and probably elsewhere to a considerable extent, have been as follows: (1) Classification of dune forms and sand patterns, first from ground studies, later from aerial photography, and finally from remote sensing; (2) textural studies involving characteristics of grain surfaces and shapes and comparisons with sand textures representing other environments; (3) studies of the physical processes involved in wind transport and deposition of sand; (4) statistical studies of cross-strata attitudes or dip directions and determination of resultant vectors; (5) investigations, both in the laboratory and in the field, of environments favorable to the preservation of surface structures and animal tracks or trails in sand; (6) analysis in three dimensions of eolian cross-strata including statistical studies of basic types; (7) quantitative morphology studies made from products of remote sensing in which constant scale allows direct comparisons between sand seas; and (8) studies, based primarily on the scanning electron microscope, of the factors controlling the red color that is characteristic of many dunes.

Dune Classification

Most geologists involved in the study of sand bodies have been faced with the problem of describing the type or types of dunes in one area and of assigning a name — either formalized from local terminology, adapted from some other sand sea, or newly designated. Thus, for the most part, local terminologies have proliferated, but formal classifications that apply to many regions have been neglected. Lack of means to make direct comparisons, at scale, between dunes of different regions has been a major deterrent in this field.

In North America two principal attempts — both in the 1940's — were made to classify dune types. "A Tentative Classification of Sand Dunes: Its Application to Dune History in the Southern High Plains," by Melton (1940), distinguished into classes those dunes on bare surfaces or loose sand and those formed in conflict with vegetation. "Dunes of the Western Navajo Country," by Hack (1941), indicated three basic controls in development; wind regime, available sand, and plant growth. The present publication describes a general classification with two variants, one based on ground observation of dune forms and internal structures, the other based largely on dune patterns recognized on images from remote sensing.

Sand-Texture Studies

The texture of eolian sand grains has received the attention of investigators for a very long time (Udden, 1898; Dake, 1921; Ries and Conant, 1931; Wentworth, 1932). Interest in textures came about partly because, on theoretical grounds, wind-abraded sand grains were believed to show certain features that are rare or absent among waterborne sand grains and partly because laboratory experiments disclosed major differences, especially in the rate of developing certain textures. Further knowledge of textures resulted from studies com-

paring modern sands of dune, marine, and fluvial types, which differ markedly in various textural features.

The principal attributes that are involved in textural considerations are grain size (mean diameter), sorting, sphericity, roundness, pitting, and frosting. Numerous scales and techniques for measuring and recording these properties have been developed through the years, ranging from rough visual estimates to highly sophisticated measurements and from simple tabulations to elaborate histograms and curves.

Despite all the studies that have been made and the voluminous literature on the subject, results from the textural approach in recognizing and defining eolian deposits have remained highly debatable for two inherent reasons. First, the grain properties that can be described are primarily the results of transportation, not deposition. They may represent the effects of intensive movement by water, although deposition was by wind, or the reverse. Second, the grains of a particular wind deposit may be and commonly are the product of a second or third generation, reflecting the properties inherited from some earlier sediment. Furthermore, some surface features, such as frosting, may not be primary; rather, they are the result of diagenesis and, therefore, unrelated to the origin of the deposit.

The application of textural parameters to eolian sands is discussed in chapter B by Ahlbrandt. Discriminant analysis, which is used to differentiate beach, river, and eolian sands, is described by Moiola and Spencer (chapter C).

Physical Processes Involved

Sand physics, focusing especially on the movement of sand in air, is critical to an understanding of the growth and movement of dunes. This field of investigation, which necessarily involves both observational data from active dunes and the results of wind-tunnel experiments on sand transport, was the subject of a classic study by Bagnold (1941) entitled "The Physics of Blown Sand." Bagnold's treatise clarified many features of the mechanics of dune development and gave much impetus to the general study of sand seas.

The investigations of Bagnold constituted primarily a study of the behavior of sand grains in a stream of wind and consisted of "precise measurement and experimental verification." The resulting information explained qualitatively many dune forms and, with relatively few modifications, has stood the test of checking since publication. Additional important contributions to our knowledge of the interaction between wind and sand have subsequently been described by Sharp (1963, 1966).

Statistical Studies of Mean Current Directions

The value of measuring cross-strata dip directions in order to determine regional trends in dune movement has been recognized since before the start of the 20th century. This approach has been used in studies both of modern dunes and of ancient sandstones believed to be eolian in origin. A model for eolian paleocurrents has been described by Selley (1968, p. 103).

Results of directional vector measurements commonly have been presented in either of two forms: (1) a circular histogram or rose diagram, as illustrated by the early work of Knight (1929) on the Casper Formation of Wyoming, U.S.A., and by Shotton (1937) on the Bunter Mottled Sandstone of England; and (2) a stereonet as introduced by Reiche (1938) for analysis of the Coconino Sandstone in Arizona.

Both methods of illustrating vectors serve well to illustrate the spread in wind current directions and to focus attention on mean or dominant azimuth directions. The stereonet has the added advantage of showing the degree of dip, which is useful in differentiating between dune slipfaces and windward-side deposits and in indicating the weight to be given various readings. Many statistical investigations of this type, in both modern and ancient eolian cross-strata, have been made within recent years; they not only help in determining the direction or directions of sand transport but also, in some situations, indicate the type of dune involved.

Preservation of Surface Structures, Tracks, and Trails

The environmental conditions necessary for the preservation of various markings in eolian sand deposits have been investigated in considerable detail by numerous geologists. These markings consist of primary sedimentary structures, varieties of contorted bedding, and the tracks and trails of animals. Such studies are especially important in the recognition of ancient eolian sands because they furnish some compelling evidence of genesis.

Minor sedimentary structures that are moderately common and that have distinctive features when formed in windblown sand include ripple marks, rain pits, and slump markings. These structures have been studied both in the field and through laboratory experiments during and after the 1940's (McKee 1945; Poole, 1962; Tanner, 1966) and compared with corresponding structures formed under aqueous conditions.

Studies of penecontemporaneous folding (contorted bedding) in eolian sands were conducted in the laboratory (McKee, Douglass, and Rittenhouse, 1971) and later in the field (McKee and Bigarella, 1972). A principal result of these studies is a means of distinguishing in ancient eolian sandstones whether they were formed in wet sand, saturated sand, damp crusty sand, or dry sand. Also, the studies supplied a means for distinguishing between sand accumulated in the zone of tension (upper slope) and sand in the compressional zone at the base of foresets. Other structural features served to differentiate between the products of different processes, such as avalanching, loading, and surface dragging (McKee, Reynolds, and Baker, 1962a, b).

Investigations of the conditions under which the tracks and trails of various animals were formed and preserved were made during the 1940's, both in the laboratory (McKee, 1944, 1947) and through field observations (Peabody, 1940). These studies established criteria for determining the degree of cohesiveness of track-bearing sand at the time prints were made and for explaining selective orientation of tracks as commonly observed in eolian cross-strata.

Analysis of Cross-Strata

In the United States the method of using a direct approach in determing structure in various types of dunes started with the trenching of a double barchan dune near Leupp, Ariz., in the summer of 1951 (McKee, 1957). The dune selected was about

270 feet (80 m) high; dips on the windward slope ranged from 5° to 7°, but most dips on the lee side were slightly more than 30°. The method used was to wet the dune by pouring water into it near the crest, allowing the water to spread outward and downward along cross-strata planes, and then to section the dune in selected places by digging trenches and pits with a shovel. The result recorded in three dimensions the structure within the dune.

Methods used in the experiment at Leupp, though primitive, proved so successful that, with minor modifications, they have been used in many places, and by many geologists since, and a considerable fund of information on dune structure has resulted. During trenching near Sabhā, Libya, in 1961, a pump was available for lifting water to the dune crest (McKee and Tibbitts, 1964); this was far easier than the earlier method of transporting water in buckets. At White Sands National Monument, New Mexico, U. S. A., in 1964 –65 trenches were dug by bulldozer — a tremendous improvement over handshovel work as done elsewhere (McKee, 1966). However, in most places, it has been necessary to use whatever equipment was already on hand at the scene.

Many important results from the trenching of coastal dunes in Brazil have been obtained in recent years by Bigarella and associates (Bigarella, Becker, and Duarte, 1969; Bigarella, 1972, 1975a, b). In these dunes internal moisture resulting from the humid climate has made digging possible without first pouring water. Likewise, at Killpecker dune field in Wyoming, U.S.A., the moisture within some dunes was sufficient to permit trenching at various places (Ahlbrandt, 1973).

Following exposure of the cross-strata pattern, a record normally is made by photography or by measuring and sketching. The most exact and most permanent method of recording details of structure, however, has been to make peels with latex or some other material.

Application of Remote Sensing (ERTS)

With the launching on July 27, 1972, of the unmanned Earth Resources Technology Satellite (ERTS – 1), later renamed Landsat – 1, a new and important approach to the investigation of sand seas became available. In an 18-day earth-orbiting cycle, images were obtained from an altitude of about 570 miles (915 km) by a multispectral scanner (MSS) aboard the satellite. These images cover broad bands or swaths around the globe between lat 80° N. and lat 80° S.

A project designated as "The Morphology, Provenance, and Movement of Desert Sand Seas" was developed with the acquisition of ERTS imagery that covered 15 major desert areas around the globe (fig. 1). These images, though representing widely spaced regions, were all at the same scale, thus permitting direct comparisons of dune patterns and meaningful descriptions of form and distribution. The principal investigator on this project was E. D. McKee and the coinvestigator was Carol Breed. Four to six persons assisted in the program at various stages.

Work was done originally using single-band black-and-white positive transparencies and paper prints at a scale of 1:1,000,000; subsequently, false-color ERTS images at the same scale became available and proved to have many advantages. These false-color images were assembled into mosiacs of selected sand seas and formed the base for thematic maps portraying patterns of dune types and their regional distribution. Through use of these maps, surface morphologies could be directly compared.

The study of sand seas from false-color Landsat (ERTS) imagery has contributed to an interpretation of the depositional environment, to an understanding of dune movement, and to the prediction of future sand sea development. Advances were made through the description and comparison of surface morphologies. For a complete understanding of the sand bodies, however, ground-based studies of climate, field studies of structures determined through trenching, and laboratory studies of texture were also necessary. The structural and textural studies contributed to an analysis of factors too small to establish from space; the meteorological data, especially wind vectors when superimposed on images, helped in understanding the complex relations between dune forms and wind activity.

Data from Skylab 4

During the winter and spring of 1974 when Skylab 4 was in orbit a considerable number of photographs were taken with a handheld camera, by request, of desert features in the major areas being studied on this project. These photographs

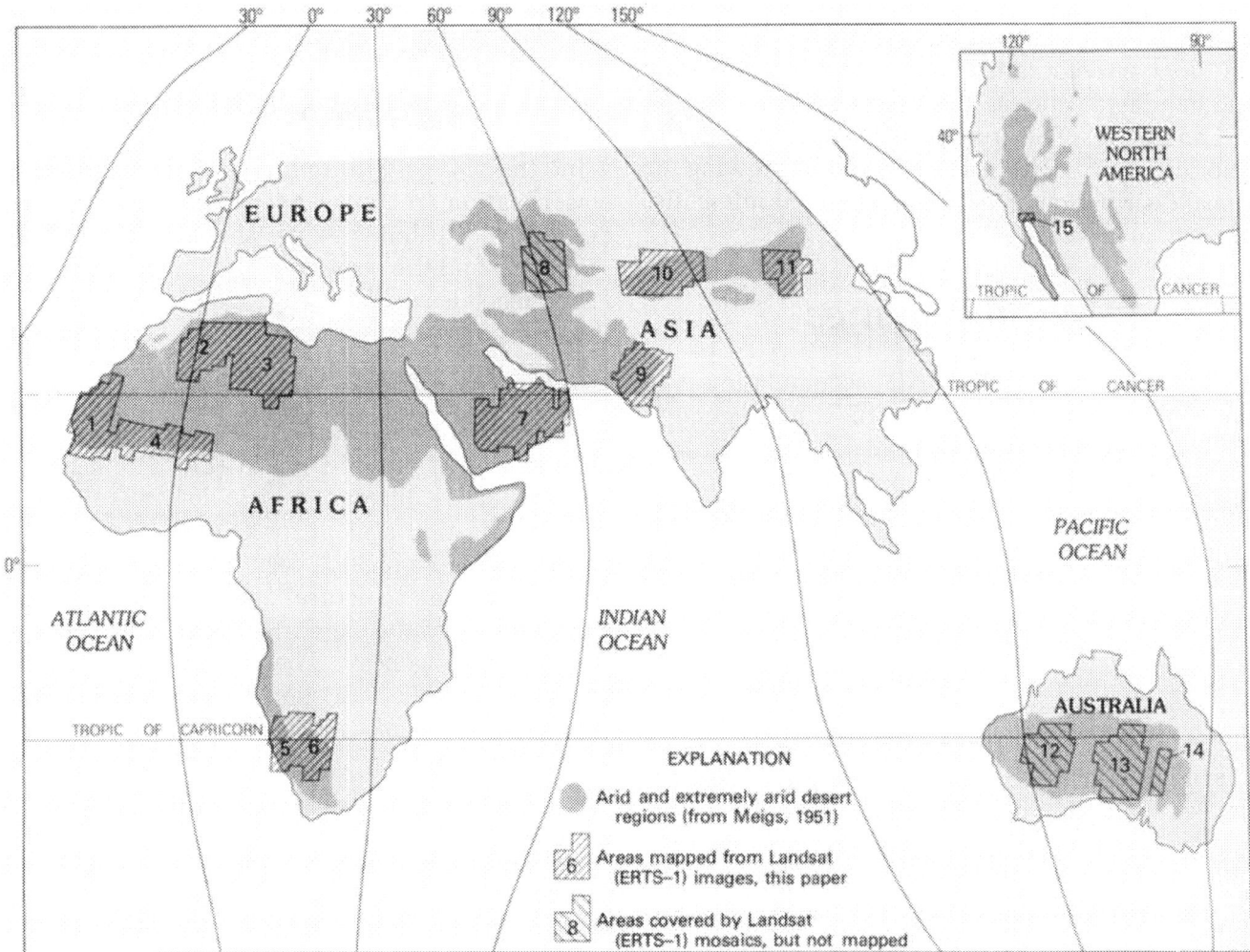

PRINCIPAL DESERT AND SEMIDESERT REGIONS and selected test areas: 1, Western Sahara, Mauritania and Senegal; 2, Northern Sahara (Grand Erg Occidental and Grand Erg Oriental), Algeria; 3, Northern Sahara, Libya; 4, Southern Sahara and Sahel, Mauritania, Mali, and Niger; 5, Namib Desert, South-West Africa; 6, Kalahari Desert, South-West Africa, Botswana, and South Africa; 7, Rub' al Khali, Saudi Arabia, United Arab Emirates, Oman, and Yemen; 8, Peski Karakumy and Peski Kyzylkum, U.S.S.R.; 9, Thar Desert, India and Pakistan; 10, Takla Makan, People's Republic of China; 11, Ala Shan Desert, People's Republic of China; 12, Great Sandy Desert, Australia; 13, Great Victoria Desert, Australia; 14, Simpson Desert, Australia; 15, Algodones Desert and Gran Desierto, U.S.A. and Mexico. (Fig. 1.)

were taken from distances much closer than those of the ERTS images and showed more detail in many places. Also, they had the advantage of being produced in true color. Sites chosen for taking these photographs were northern Africa, southern Africa, Saudi Arabia, India, China, U. S. S. R., and Australia.

Studies of Dune Color

A distinctive feature of most sand seas is the brilliant coloring of the sand — largely red with various degrees of intensity. Color has long been recognized as a function of iron oxide pigments associated with individual grains of sand and in various stages of oxidation, but how and when the iron has been introduced and when the oxidation has taken place are controversial problems with no single, simple answer.

Recent detailed studies of the red coloring in dune sands were made by T. R. Walker, using a scanning electron microscope, as well as a petrographic microscope, for analysis. Chapter D describes some of his results, Especially significant is the conclusion that at least seven separate and independent factors control the degree (variations in intensity) of redness in a sand deposit. This illustrates the complexity of the problem.

Colors as determined from satellite observation and handheld cameras and from false-color images of Landsat (ERTS) are useful in locating areas of color difference within a region and in showing the global distribution of various shades and intensities of particular colors.

Wind Regimes

BECAUSE EOLIAN SAND accumulation and dune migration are functions of wind regime, a study of available meteorological data for each sand sea was considered to be a necessary part of the global program as designed for ERTS imagery. A comparison of wind regimes with dune types was made for various localities around the world and is included in the parts of this Professional Paper dealing with regional analyses.

A type of sand rose, as described by Fryberger in chapter F was developed to show from accumulated wind data the direction and relative amount of sand drift at a given locality. By use of a "weighting equation" the relatively much greater ability of winds of high velocity to cause sand drift is taken into account. Resulting sand roses have been superimposed on the various regional maps in this publication to allow direct comparison with associated dune types as mapped.

Types of Dunes

A PRINCIPAL AIM of this investigation has been to construct an objective and descriptive classification which segregates into types the various distinctive dune forms and relates these types to specific environments of deposition. Two innovations in methods of study contributed substantially to the construction of this classification.

The first factor in establishing a sound basis for classifying eolian sand bodies is improved knowledge of process of construction, obtained from direct observation of their structure in pits and trenches. The second factor is availability of data from remote sensing programs, especially in the form of Landsat (ERTS) imagery, in which a uniform scale permits direct comparisons between various parts of the world. These two approaches

have greatly stimulated efforts to better describe and classify dune forms.

Basis for Classifying

In the investigation of eolian sands, the classification recommended is based on two descriptive attributes — the shape or form of the sand body and the position and number of slipfaces (steep lee-side surfaces) (fig. 2; table 1). Generally, these two attributes constitute recognizable features both in closeup ground views and at the great distances represented by ERTS images. Exceptions are those sand features that are too small to be identified on the Landsat (ERTS) images.

Two slightly different terminologies are used in the classification system for this publication. One version (table 1, col. 3) is derived primarily from an analysis of dune structures and so constitutes both a description of the sand dune forms and a partial record of their genesis. The other version (table 1, col. 6) of the classification is developed largely from the patterns of Landsat (ERTS) imagery and is used both in the thematic maps prepared from the ERTS mosaics of the principal sand seas and in ERTS-based studies of distribution and morphology. One version of the classification stresses factors that record the growth and development of individual dune bodies; the other version emphasizes the gross features of dune patterns where details cannot be observed.

An examination of sand seas from various approaches — including studies of Landsat (ERTS) images, air photos, ground records, and other material — suggests that the number of basic or simple dune forms is fairly limited (table 1). However, a large number of combinations of dune types and an almost infinite number of varieties occur. Dune combinations are of two categories: (1) compound dunes made up of two or more dunes of the same type, coalescing or overlapping, and (2) complex dunes in which two different basic types are combined or superimposed.

Basic (Simple) Dune Types

The simple dune types, because they are classified partly on overall form and partly on number of slipfaces, generally reflect factors of the depositional environments in which they devel-

STAR DUNES, showing three curving arms with steep slipfaces oriented in various directions. Tsondab Vlei, northeastern part of Namib Desert, South-West Africa. Photograph by E. Tad Nichols. (Fig. 2.)

oped, especially the wind (strength and direction), the sand supply, vegetation, physical barriers (rock outcrops and water bodies), and distance from source. Various combinations of these factors are responsible for each distinctive dune type and permit some generalizations concerning their genesis.

Probably the best known and commonest basic dune forms result from winds having a single dominant direction and are oriented with their axes at right angles to wind direction. Such dunes range from the small crescent-shaped type commonly known as barchan (fig. 3), through parallel rows of coalesced barchans, here referred to as barchanoid ridges (fig. 4), to essentially parallel straight ridges known as transverse dunes (fig. 5). All three types, which may occur in gradational sequence, are referred to as "crescentic" on the regional sandseas maps presented in this publication. Most barchan types are too small to be recognized on Landsat (ERTS) images, and at that scale the transverse dunes are difficult to distinguish from some of the smaller barchanoid ridges and from many linear dunes.

The genetic relationship between the three dune types previously referred to seems to be reasonably well established through the observations of many geologists and has been described in numerous papers. These dunes are all characterized by slipfaces in one direction and therefore represent unidirectional wind movement. The three types commonly occur in a definite sequence (fig. 6) downwind as a result of diminishing sand supply. In general, the type represented in a particular area seems to be a function of the amount of available sand. Barchan dunes clearly result from sparse sand and in many areas are scattered over bare rock surfaces. Transverse dunes, the other extreme, develop where sand is abundant, and they attain maximum height where there is a balance between strength of wind and available sand.

Because most barchanoid-type dunes — barchans, barchanoid ridges, and transverse dunes — form patterns of parallel wavy lines on ERTS images, they were referred to collectively as "parallel wavy dunes" in the classification of dune patterns used during ERTS studies (McKee and Breed, 1974a, b).

TABLE 1. — *Terminology of basic dune types and other eolian deposits*

[Examples may show simple or compound dunes; however, they are designated by basic dune type only]

Form	Number of slipfaces	Name used in ground study of form, slipface, and internal structure	Block diagram figure numbers (Chapter A)	Name used in space-imagery and air-photo study of pattern and morphology (Chapters J, K)	Examples from Landsat imagery (0–16 MILES; 0–25 KILOMETRES)
Sheetlike with broad, flat surface	None	Sheet	None, flat surface	Sheet[1]	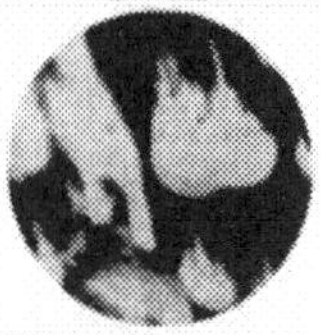
Thin, elongate strip	None	Stringer	None, flat surface	Streak[1]	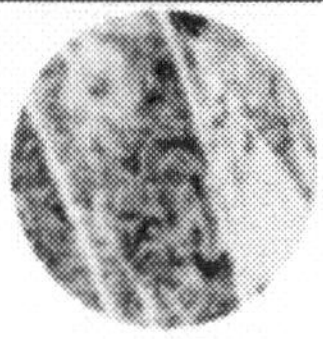
Circular or elliptical mound	None[2]	Dome	For detail see fig. 7	Dome-shaped	[3]
Crescent in plan view	1	Barchan	For detail see fig. 3	Crescentic: Barchan	[3]
Row of connected crescents in plan view	1	Barchanoid ridge	For detail see fig. 4	Crescentic: Barchanoid ridge	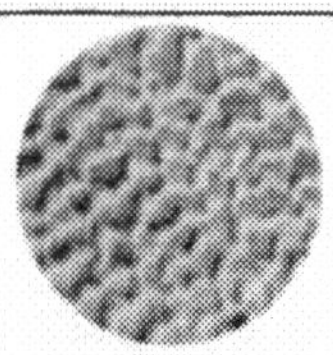
Asymmetrical ridge	1	Transverse ridge	For detail see fig. 5	Crescentic: Barchanoid ridge	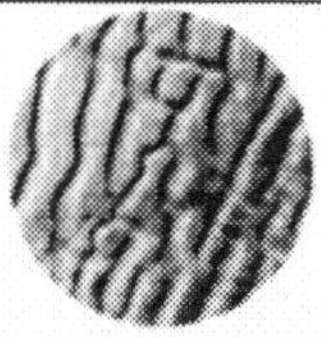
Circular rim of depression	1 or more	Blowout[4]	For detail see fig. 8	Not recognized	[3]
"U" shape in plan view	1 or more	Parabolic[4]	For detail see fig. 9	Parabolic	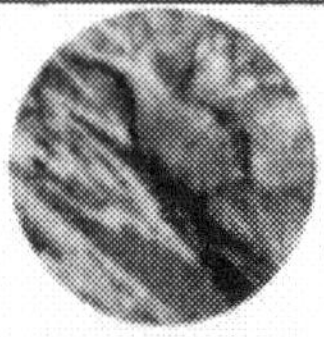
Symmetrical ridge	2	Linear (seif)	For detail see fig. 10	Linear	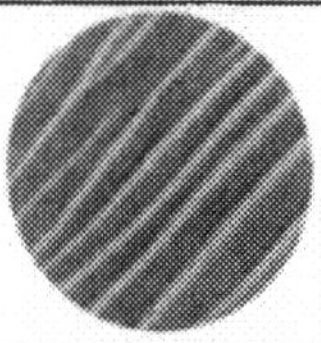
Asymmetrical ridge	2	Reversing	For detail see fig. 12	Reversing	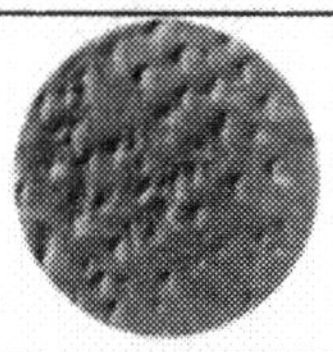
Central peak with 3 or more arms	3 or more	Star	For detail see fig. 11	Star	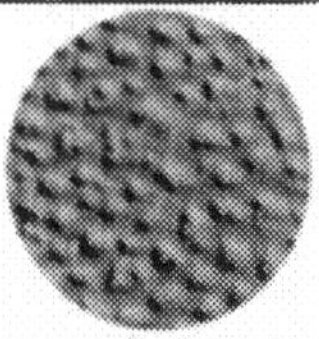

[1]May include mounds and other features too small to appear on Landsat imagery.
[2]Internal structures may show embryo barchan-type with one slipface.
[3]Because of their small size, dome dunes, blowout dunes and individual barchan dunes are difficult to distinguish on Landsat imagery.
[4]Dunes controlled by vegetation.

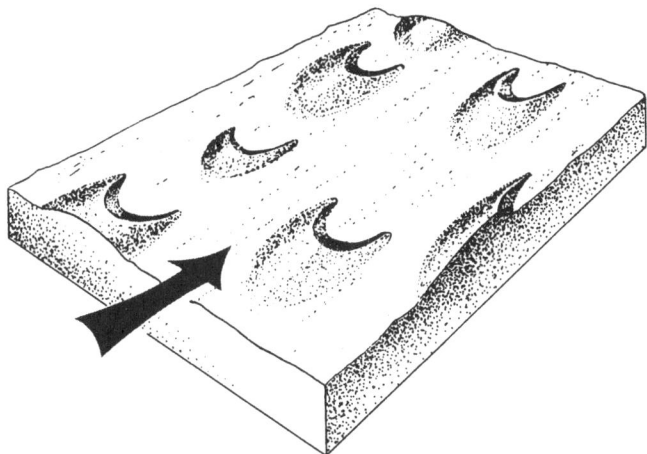

BARCHAN DUNES. Arrow shows prevailing wind direction. (Fig. 3.)

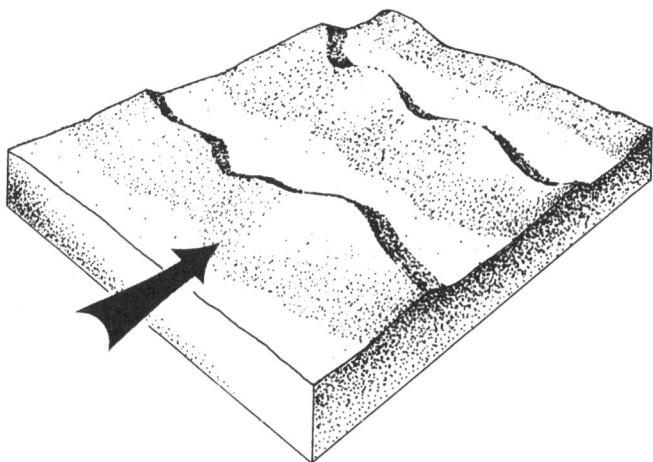

BARCHANOID RIDGE. Arrow shows prevailing wind direction. (Fig. 4.)

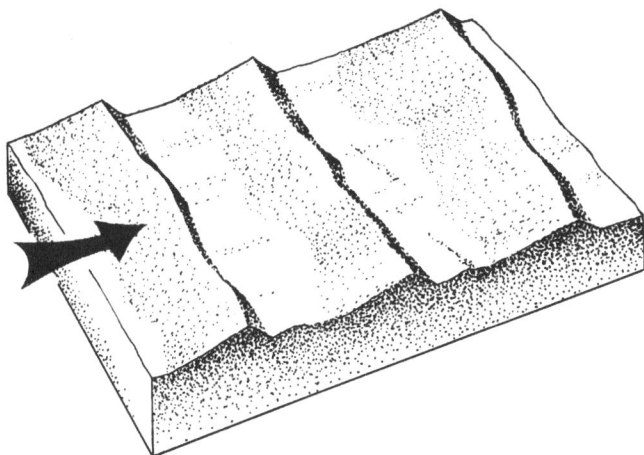

TRANSVERSE DUNE. Arrow shows prevailing wind direction. (Fig. 5.)

Some sand bodies display no slipfaces. These bodies are commonly referred to as sheets and stringers, using the geologic terminology usually applied to bodies of rock or sediment having such forms. These sand bodies are not considered to be dunes, because the term "dune," according to most definitions, refers to a mound or ridge. A stringer may be considered to be a variety of sand sheet having a stringlike shape, and, commonly, though not exclusively, it develops along the margins or at the downwind extremity of a sheet. The term "streak" is used instead of "stringer" on thematic maps in this publication. Streaks identified on the basis of Landsat (ERTS) mosaics may, therefore, include barchans and other small dunes which are not visible on the image.

Dome dunes (fig. 7), which probably consist of more than one type (as discussed in chapter E on dune structure), normally show no external slipfaces and are circular or elliptical in plan view. However, some dunes of this type internally disclose slipfaces dipping in one direction, suggesting that they began development as barchans, and implying that the dome form is the result of strong winds that truncated the top and flattened the lee slopes of a barchan. Many dome dunes of coastal areas probably have their shapes controlled more by moisture and vegetation than by strength of wind, as described later.

A class of dunes in which the forms apparently are controlled more by partial stabilization from vegetation or moisture or both, than by wind strength and direction, is represented by the blowout (fig. 8) and the parabolic dunes (fig. 9). Such dunes may have slipfaces sloping in one or many directions depending on what parts of their rims are free to migrate. They are, to a considerable extent, features of deflation by wind, but deposition occurs wherever margins are free to migrate as on other dunes. The blowout type is normally a circular bowl, whereas the parabolic develops a U-shape or V-shape from arms that are anchored and nose that migrates downwind. A coastal dune of ridge form which reflects control by vegetation is referred to by some authors as a retention ridge; it may have small blowouts developed on it.

Perhaps the most controversial type of dune is the linear dune (fig. 10) commonly known as the

SEQUENCE OF DUNE TYPES with unidirectional wind and (from left to right) diminishing sand supply; transverse, barchanoid ridge, and barchan. Arrow shows prevailing wind direction. (Fig. 6.)

seif dune in Africa and Saudi Arabia and as the longitudinal dune in many other regions. Whatever the mechanics of its genesis and number of wind directions involved, it is, in general, a straight ridge with slipfaces on both sides; groups of these ridges, as seen in ERTS images or air photos, appear as parallel straight lines.

Dunes of multiple slipfaces are the result of winds from several directions; they commonly have a high central peak and three or more arms extending radially. Such forms are called star dunes (fig. 11) and have an infinite variety of shapes. They mostly grow vertically, rather than migrating laterally.

Intermediate in character between the star dune and the transverse ridge is the reversing dune (fig. 12). It characteristically forms where two winds

from nearly opposite directions are balanced with respect to strength and duration. Such dunes may have the general form of transverse ridges, but a

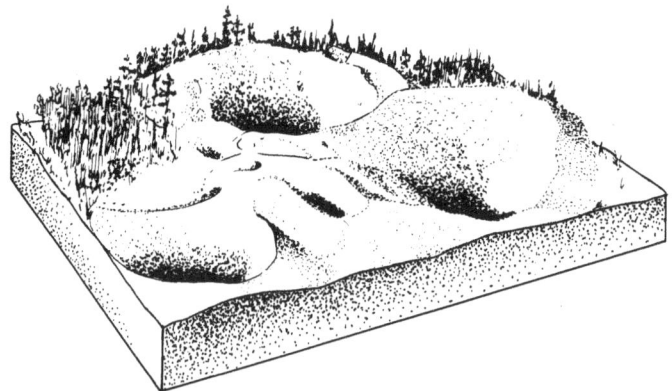

BLOWOUT DUNES. (Fig. 8.)

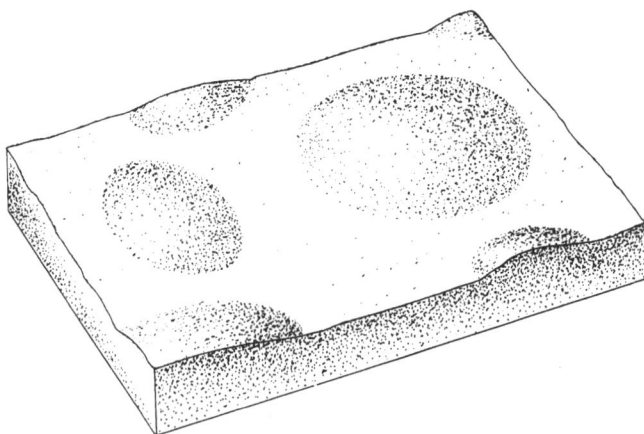

DOME DUNES. (Fig. 7.)

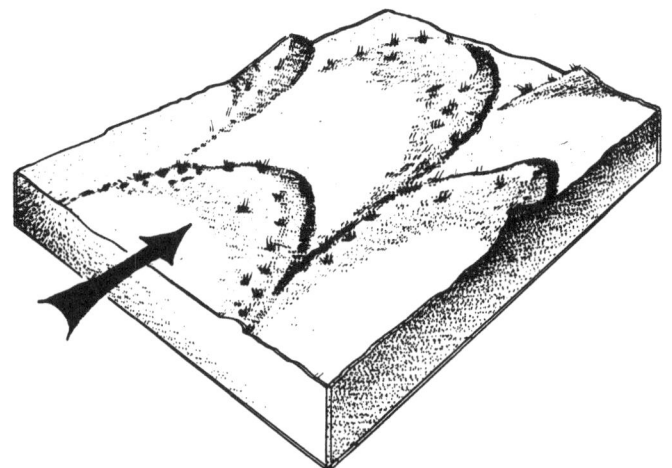

PARABOLIC DUNES. Arrow shows prevailing wind direction. (Fig. 9.)

second slipface, oriented in a direction nearly opposite that of the primary slipface, develops periodically.

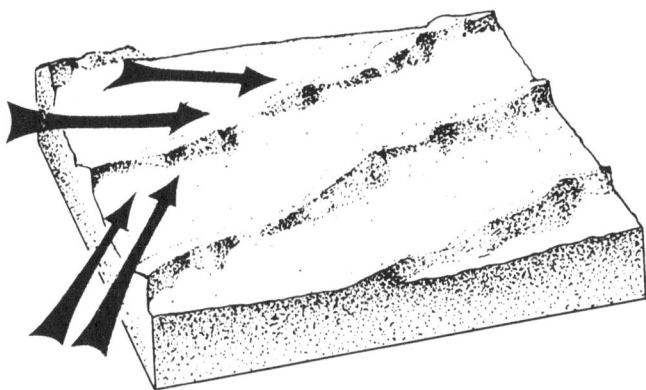

LINEAR DUNES. Arrows show probable dominant winds. (Fig. 10.)

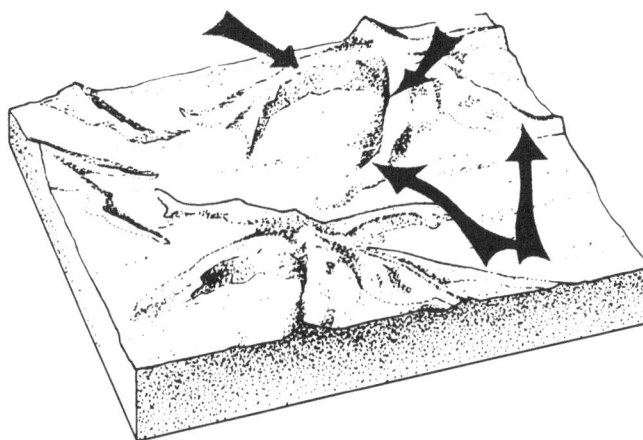

STAR DUNES. Arrows show effective wind directions. (Fig. 11.)

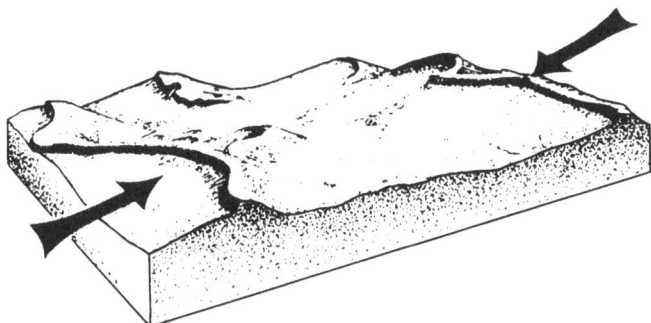

REVERSING DUNES. Arrows show wind directions. (Fig. 12.)

Compound Dune Types

In addition to the basic or simple dune forms just described, compound forms are abundant in most sand seas. These consist of two or more of the same type combined by overlapping or by being superimposed (fig. 13) on one another. Examples of compound dunes are (1) barchanoid ridges coalescing, (2) star dunes coalescing to form an intricate pattern of arms and peaks, (3) little barchan dunes on the windward flanks of large ones, (4) small parabolic dunes between the arms of a larger one, and (5) major ridges covered by many smaller linear dunes.

Complex Dune Types

Complex dunes in which two different basic types have coalesced or grown together are represented in most sand seas by various combinations (fig. 14). These combinations include linear dunes in parallel rows with star dunes on their crests; these are called "chains of stars" and are well developed in several major sand seas. Complex dunes also are illustrated by small barchan dunes in the corridors of linear dunes, by blowout dunes on transverse dunes, and by large star dunes with superimposed barchanoid dunes.

Varieties of Dunes

For all dune types — simple, compound, and complex — an almost infinite number of varieties occurs. The varieties are too numerous to describe or even classify by name herein, for they probably represent transitions from one basic type to another and result from differences in the direction and strength of wind, the amount of available sand, the physical obstructions, and other factors that control dune types. Some varieties that are especially conspicuous and have been much discussed in the literature are barchans with one horn greatly extended, linear dunes with many branches diverging from one end suggesting the name feather dunes, and parabolic dunes that are V-shaped rather than U-shaped (fig. 15).

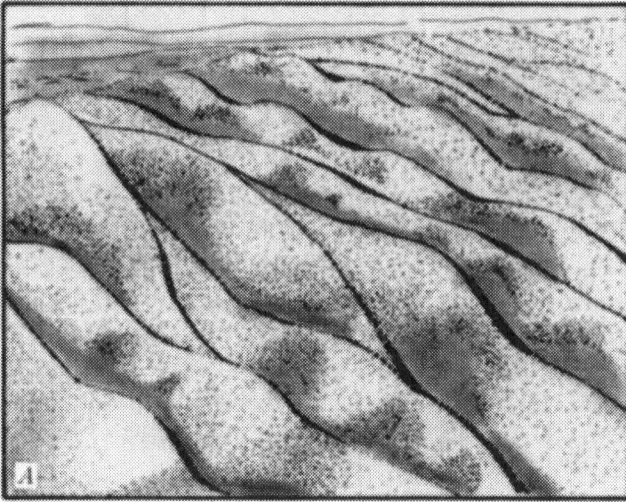

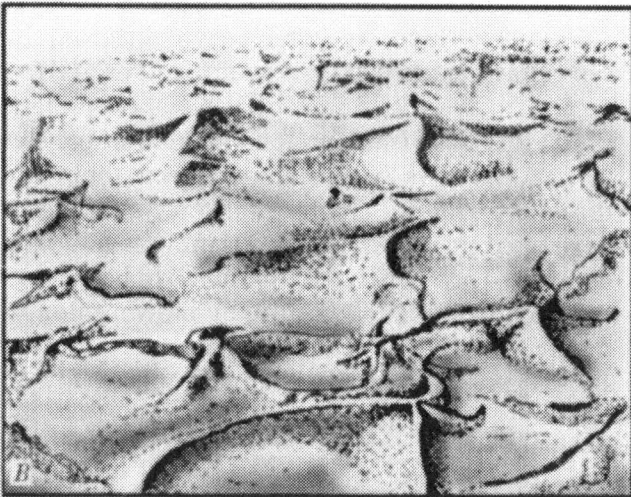

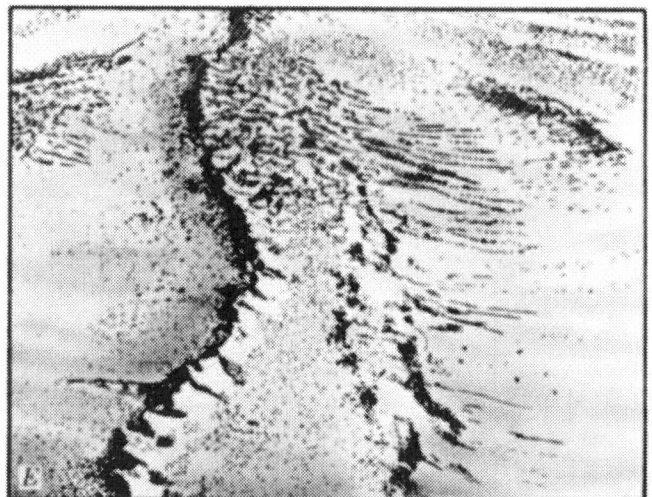

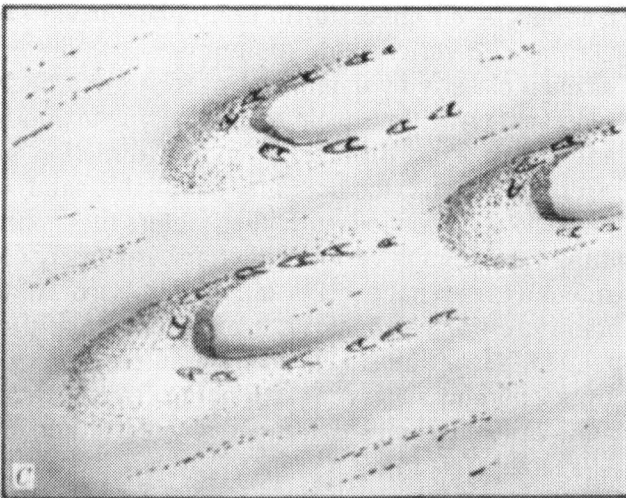

COMPOUND DUNES: *A*, Barchanoid ridges coalescing. *B*, Star dunes coalescing. *C*, Small barchans on large barchan. *D*, Parabolic dunes within a large parabolic dune. *E*, Linear dunes on large linear ridge. (Fig. 13.)

Interpretation of Ancient Eolian Sandstones

A MAJOR EFFORT of this investigation was to establish criteria for the recognition of ancient eolian-type sandstones (chapter H). Recognition of wind-deposited sand in ancient sandstones was recorded nearly 100 years ago by Walther (1888, p. 253) when he stated that a detailed study of the Nubian Sandstone in Sinai and adjacent parts of the Arabian Desert "has led me to the conviction that these sandstone deposits are an eolian dune development * * * and that North Africa contained deserts of Paleozoic and Mesozoic age."

Subsequently, the eolian origin of certain sandstones in the Southwestern United States was postulated by Huntington (1907, p. 380) because of textural and structural features "frequently seen among modern sand dunes of Persia, Transcaspia, and Chinese Turkestan." Other early references to sandstones believed to have been wind-deposited are Fourtau (1902), Sherzer and Grabau (1909), Twenhofel (1926, p. 174, 557), and Cuvillier (1930). In general, however, compelling evidence for the recognition of eolian sandstones had to await the fairly sophisticated analyses of cross-strata characteristics and direction vectors, studies of minor structures and their preservation, and other approaches that were introduced about 1937 and later, as described in chapter E on modern dune structures.

Terminology and Geographic Names

TERMS USED in this professional paper are defined in the glossary. Most definitions are from "A Desert Glossary," by R. O. Stone (1967), and from "Glossary of Geology," published by the American Geological Institute (1972), although various other glossaries were also consulted. A few terms introduced for the first time in this paper are also defined.

The names of geographic localities outside the United States are according to the gazetteer for each of the countries concerned. The spellings have been approved by the U. S. Board on Geographic Names

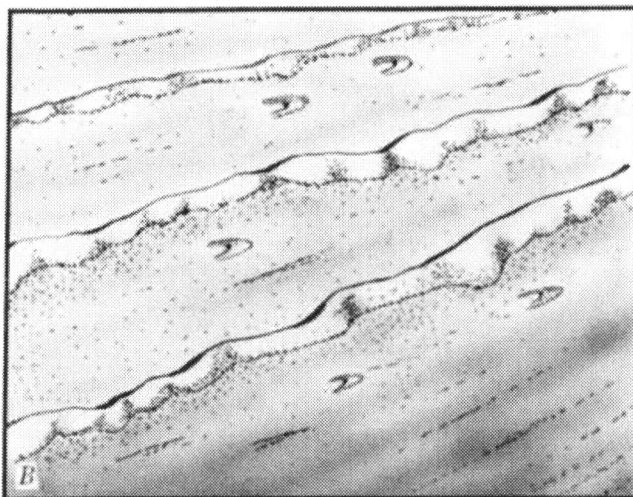

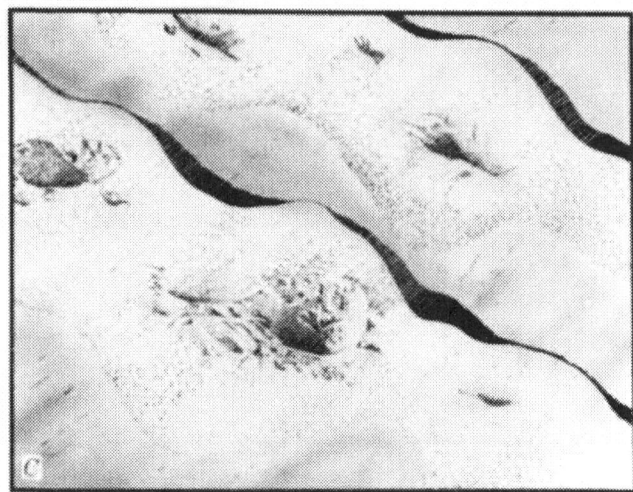

COMPLEX DUNES: A, Star dunes on linear dune. B, Linear dunes with barchans in interdunes. C, Blowouts on transverse dune. (Fig. 14.)

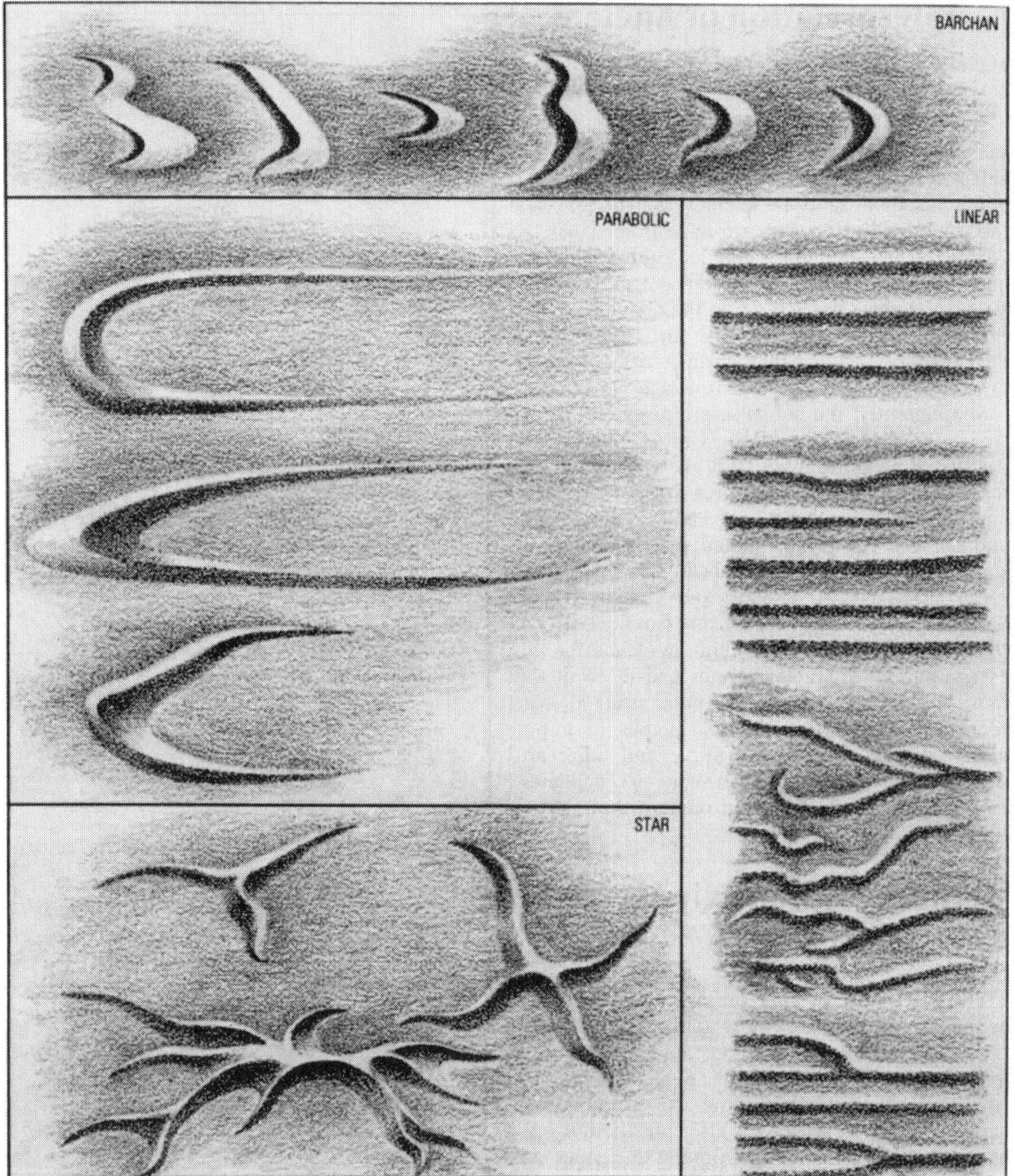

SOME DISTINCTIVE VARIETIES of basic dune types. (Fig. 15.)

and the U. S. Department of Defense, Defense Mapping Agency. International boundaries and geographic names and their spellings do not necessarily reflect recognition of political status of an area.

Systems of Measurement

THIS TREATMENT OF SAND SEAS on a global scale includes measurements of many kinds and from many sources. Resulting figures should preserve as nearly as possible the original measurements as derived, but at the same time should be readily visualized by people of various backgrounds and training. Thus, where an original measurement was made in terms of the metric system, it is expressed in that system, but for the convenience of some readers, its converted equivalent in the English system follows in parentheses. Likewise, where an object was measured in inches, feet, or miles, that measurement as recorded is presented in the text, succeeded by its metric equivalent in parentheses.

The following English-metric conversion scales are given for the convenience of the reader. They are for visual comparison only and do not represent actual units of measurement. Abbreviations shown in parentheses are those used in this publication.

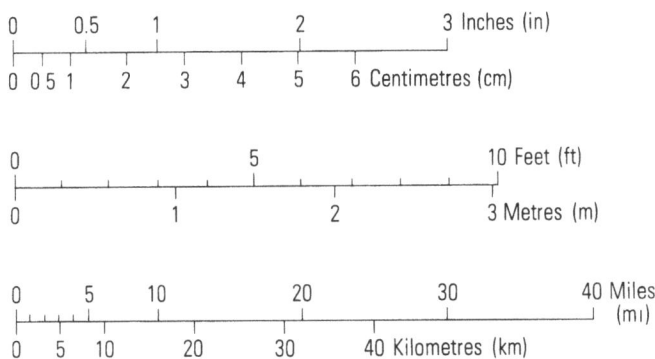

| 0 | 0.5 | 1 | | 2 | | 3 Inches (in) |
| 0 | 05 1 | 2 | 3 | 4 | 5 | 6 Centimetres (cm) |

| 0 | | 5 | | 10 Feet (ft) |
| 0 | 1 | | 2 | 3 Metres (m) |

| 0 | 5 | 10 | | 20 | | 30 | | 40 Miles (mi) |
| 0 | 5 | 10 | 20 | 30 | | 40 Kilometres (km) |

Comparative English and metric scales of distance.

Most of the conversions from one system to the other were made by Patricia Melrose of the U.S. Geological Survey, using a conversion calculator. Except where a high degree of precision was available and required by a particular situation, references throughout the text were converted only approximately.

Acknowledgments

A VERY LARGE NUMBER of organizations and individuals have contributed ideas, facts, and materials, as well as time and effort, to assist in the preparation of this treatise. To all of these we extend our gratitude. Contributions that pertain to particular chapters are, for the most part, acknowledged within the appropriate chapters; contributions that affect numerous chapters or the entire Professional Paper are acknowledged here.

Geologic and Geographic Data

For assistance in interpreting local features of various sand seas:

Dr. G. Lennis Berlin, Northern Arizona University, Flagstaff, Arizona, U. S. A.

Dr. Arthur L. Bowsher, Arabian American Oil Co., Dhahran, Saudi Arabia.

Dr. I. H. Hamrabaev, Institute of Geology and Geophysics, Tashkent, U. S. S. R.

Mr. Donald Holm, formerly with Arabian American Oil Co., Saudi Arabia.

Mr. Brad Musick, Office of Arid Lands Studies, Tucson, Arizona, U. S. A.

Dr. Shumil E. Roy, Indian Agricultural Research Institute, New Delhi, India.

Dr. M. K. Seely, Desert Ecological Research Institute, Walvis Bay, South-West Africa.

Remote Sensing Program

For guidance, support, and assistance in those phases of sand sea study based on remote sensing programs, we are indebeted to the following people in the United States:

National Aeronautics and Space Administration (NASA):
G. Richard Stonesifer, Technical Monitor.
Paul D. Lowman, Scientific Monitor.

EROS Program, Office of U.S. Geological Survey;
J.M. DeNoyer, Director.
Richard S. Williams, Jr., Geologist.
W. Douglas Carter, Geologist.
Priscilla Woll. EROS Technical Publications Office.

Landsat (ERTS) Office of U.S. Geological Survey:

> W. R. Hemphill, Chief, Resources Working Group.
>
> Lawrence C. Rowan, Staff Geologist and Coordinator for Remote Sensing.
>
> Allan N. Kover, Staff Geologist, Remote Sensing.
>
> Robert D. Regan, Staff Geologist, Remote Sensing.

Skylab, NASA:

> V. R. Wilmarth, Program Scientist.
>
> John Kaltenbach, Project Scientist.
>
> Michael McEwen, Geologist.

Wind Record and Other Climate Data

The following persons and institutions assisted in the gathering and interpreting of meteorologic records used in compiling wind roses;

> Arabian American Oil Co., Dhahran, Saudi Arabia — Mr. J. P. Mandaville.
>
> Central Arid Zone Research Institute, India — Dr. H. S. Mann.
>
> Desert Ecological Research Institute, Walvis Bay, South-West Africa — Dr. M. K. Seely.
>
> Indian Meteorological Department, Poona, India — Dr. K. Krishnamurthy.
>
> Meteorological Department, Ministry of Communications, Tripoli, Libyan Arab Republic — Mr. M. M. Sasy.
>
> National Climatic Center, Ashville, N. C., U.S.A. — Dr. Harold L. Crutcher.

Manuscript Review

For helpful suggestions and criticism, after their technical review of drafts for various chapters we thank the following:

> John Dingler, University of California at San Diego, California U. S. A.
>
> Gerald Friedman, Rensselaer Polytechnic Institute, Troy, New York, U. S. A.
>
> Dr. A. S. Goudie, Oxford University, England.
>
> Dr. R. L. Maby and colleagues, Arabian American Oil Co., Dhahran, Saudi Arabia.
>
> Dr. John Rogers, University of Cape Town, Rondebosch, South Africa.
>
> Dr. D. B. Thompson, Manchester University, England.

Volunteer Assistance for Work on Landsat (ERTS-1) Program

Many phases of the remote sensing program for Sand Seas Analysis, conducted at Flagstaff, Arizona, U. S. A., were made possible by the volunteer services of geology students, mostly from the Northern Arizona University. The following persons are given special thanks for their contributions to this program:

> Marion Blancett: photographic work.
>
> Dana Gebel: Cataloguing imagery acquisitions, work on chart of dune terminology, rough draft maps.
>
> Kevin Horstman: Measuring dune and interdune areas with density slicer.
>
> Katherine Nation: Preparing mosaics from ERTS imagery.
>
> Iris Sanchez: Vegetation studies, dune measurement studies.

Photographic and Other Illustrative Materials

Photographs for use in illustrating sand bodies of various types and localities were kindly furnished by the following in the United States:

> Dr. Stanley Froden, NASA, Ames Research Laboratory, Iowa.
>
> George Morrison, National Park Service, White Sands National Monument, New Mexico.
>
> E. Tad Nichols, Tucson, Arizona.
>
> J. Philip Schafer, Flagstaff, Arizona.
>
> Joseph F. Schreiber, University of Arizona, Tucson, Arizona.
>
> John Shelton, La Jolla, California.
>
> H. T. U. Smith (now deceased), Department of Geology, University of Massachusetts, Amherst, Massachusetts.

Editing and Preparation of Text and Illustrations

Because of the considerable size and numerous ramifications involved in this publication, far more than normal time and effort was required by some people of the U.S. Geological Survey to bring the text and illustrations to their final form. As editors, Patricia Melrose, and Shirley Person did much valuable work on preliminary drafts of the manuscript. For work on the completed text during early stages, much credit is due Priscilla Patton; for later painstaking review and for seeing the manuscript to completion, credit is due Diane Schnabel; and an excellent job of illustration editing was done by Sigrid Asher-Bolinder. Final editing and preparation for printing was by Carol Hurr; graphics are by Arthur Isom.

Library and Literature Review Research

The extensive program of literature review that was required in support of the ERTS program was greatly facilitated by the help of the following librarians and reference-service personnel here in the United States:

Katharine Bartlett, Museum of Northern Arizona library, Flagstaff, Arizona.

Elizabeth Behrendt and Helen King, U. S. Geological Survey, Denver, Colorado.

Mary Haymes and Pamela Noaecker, National Center for Atmospheric Research, Boulder, Colorado.

James Nation, U.S. Geological Survey, Flagstaff, Arizona.

Patricia Paylore, University of Arizona library, Tucson, Arizona.

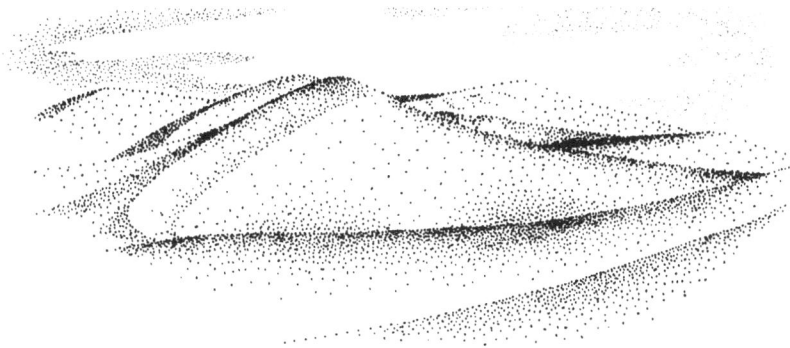

A STUDY OF GLOBAL SAND SEAS

TEXTURAL PARAMETERS OF EOLIAN DEPOSITS

Chapter B

By THOMAS S. AHLBRANDT

Contents

Illustrations

Tables

Summary of Conclusions

𝔄 SUITE OF 506 eolian sand samples was divided into 291 coastal dune samples, 175 inland dune samples, and 40 interdune and serir samples. Textural parameters for each sample are listed. These three subenvironments of eolian deposits have different textural parameters: (1) well-sorted to very well sorted fine coastal dune sand. (2) moderately sorted to well-sorted fine to medium inland dune sand, and (3) poorly sorted interdune and serir sand. The dune-sand samples, including those from coastal and inland dunes, range in mean grain size from −0.68 ϕ (1.6 millimetres) to 3.4 ϕ (0.1 millimetres). Most interdune and serir samples are bimodal in the sand fraction and also have higher silt and clay contents compared with adjacent dune-sand samples. Textural contrasts occur among samples from different positions on a given dune type, and among samples from dunes formed under different wind regimes.

In the absence of sedimentary structures as primary evidence of eolian deposits, textural criteria become important. Alluvial, littoral, and eolian sands differ in their porosity and permeability, which are functions of mean grain size, sorting, and packing arrangements. Impermeable or low permeability deposits, such as interdune or fluvial deposits adjacent to more permeable deposits, particularly well-sorted coastal dune sand, may form a barrier to fluid migration — oil or water. Therefore, differentiation of depositional environments in eolianites can enhance fluid recovery operations. Skewness and kurtosis values for eolian deposits are highly variable and were not found to be diagnostic. Sophisticated approaches, such as discriminant analysis or moment measures, rather than graphical measures of textural parameters may be a means to differentiate alluvial, littoral, and eolian sands on a textural basis as discussed by Moiola and Spencer (chapter C, this paper).

Introduction

THE DEGREE to which textural parameters may be used to identify eolian deposits is highly controversial. Folk and Ward (1957), Mason and Folk (1958), Friedman (1961), Folk (1962), Mabesoone (1963), Visher (1971), and others believe that textural parameters, such as excellent sorting, positive skewness, mean grain size, or rounded and frosted grains uniquely define eolian deposits. In contrast, Udden (1914), Shepard and Young (1961), Schlee, Uchupi, and Trumbull (1964), Moiola and Weiser (1968), Bigarella (1972), Ahlbrandt (1974b), and others consider these parameters to be considerably less diagnostic of eolian deposits.

In this chapter and in chapter C, particle-size diameter is measured in the phi (ϕ) grade scale followed by the equivalent in millimetres (mm) (Wentworth grade scale). The progression is as follows:

$$
\begin{aligned}
\ldots\; -2\; \phi &= 4 \text{ mm} \\
-1\; \phi &= 2 \text{ mm} \\
0\; \phi &= 1 \text{ mm} \\
1\; \phi &= 0.5 \text{ mm} \\
2\; \phi &= 0.25 \text{ mm} \\
3\; \phi &= 0.125 \text{ mm} \\
4\; \phi &= 0.0625 \text{ mm} \ldots
\end{aligned}
$$

Grains of sand size −1.0 to 4.0 ϕ (2.0−0.062 mm) compose most sand sea deposits (Cooke and Warren, 1973, p. 255−267); thus, the sand fraction of eolian deposits is discussed in this chapter. Eolian silts (loess), eolian clays, clay coatings on grains (Walker, chapter D, this paper) and sand size aggregates of silt and clay are not discussed.

Deficiencies of certain sizes of clastic material are considered to be eolian indicators by Udden (1914) and Kuenen (1960). Deficiencies of eolian material larger than 0.0 ϕ (1 mm) and smaller than 3.3 ϕ (0.1 mm) were noted by Kuenen (1960), and deficiencies of clastics smaller than 4.3 ϕ (0.05 mm) were observed by Bagnold (1941). In general, deficiencies of clastic particles in the very coarse sand to granule range, 0.0 to −2.0 ϕ (1−4 mm), and the very fine sand to silt range, 3.0 to 4.0 ϕ (0.125−0.062 mm), have been observed in a number of depositional environments (Shea, 1974). On the basis of analyses of 11,212 samples, Shea (1974) concluded that these clastic deficiencies do not exist. Although clastic particles larger than very coarse sand, < −0.68 ϕ (1.6 mm), and smaller than very fine sand, >3.4 ϕ (0.1 mm), are absent in the dune-sand samples used in this study, the interdune samples from the sample suite contain material that ranges in size from granule to clay. Because of the problems with the deficiencies concept discussed by Shea (1974) and the variety of material that occurs in the interdune and serir samples, this chapter will focus on the clastic material in the samples.

Several other types of textural studies will be briefly discussed. Upturned plates of quartz (chevrons) formed during the frosting process can be detected with the scanning electron microscope and, thus, is a potential indicator of eolian processes (Kuenen, 1960; Margolis and Krinsley, 1971; Rogers and Tankard, 1974). In indurated rocks, defrosting by the diagenetic effects of pressure solution and grain alteration reduce the reliability of this indicator (Kuenen and Perdok, 1962; Margolis and Krinsley, 1971). Despite these problems, Le Ribault (1974) suggested that original surficial textures may persist in the geologic record.

Differential settling velocities of light and heavy minerals have been used to document eolianites and their traction, saltation, and suspension components (Friedman, 1961; Hand, 1967; Steidtmann, 1974). Provided that all grain sizes are available and stable heavy minerals are used, this technique should be diagnostic because the relative sizes of hydraulically equivalent light and heavy grains deposited in air are different from those deposited in water.

The purpose of this discussion is to present and synthesize data for a broad-based collection of eolian deposits throughout the world and the graphical measures of their textural parameters, and to provide impetus for future textural studies of eolian deposits. Data used in this chapter are not the result of systematic worldwide sampling networks to prove or disprove a particular hypothesis, such as I. G. Wilson's (1973) hypothesis of granulometric control of eolian bedforms. There are few systematic textural studies of entire dune fields; however, the overall large and varied nature of the population studied in this report (506 samples) may lead to significant conclusions. This population comprises 191 sand samples primarily from inland dune fields and 315 samples mostly from coastal dunes in Brazil (fig. 16). The forms sampled include star, seif, dome, transverse, barchan, parabolic, hanging (falling), climbing, produne, and coastal dunes. Moreover, the samples came from various positions on the dunes, such as the crest, windward and leeward or slipface, from ripples, sand ridges, interdune and serir deposits. The composition of the samples includes carbonate sands (Bahama Islands) and clastic sands; the latter including gyp-

sum (White Sands National Monument, New Mexico), olivene (Hawaii), quartzose and feldspathic sands.

The textural parameters, derived from quarter phi sieve data, of mean grain size (M_z), inclusive graphic standard deviation (σ_I), inclusive graphic skewness (Sk_I), and transgraphic kurtosis (K'_G) as defined by Folk (1968b) will be used in this chapter.

Acknowledgments

SAMPLES WERE SIEVED on quarter phi intervals by R. J. Moiola and A. B. Spencer, both of Mobil Research and Development Corp., Dallas, Texas, and their data are at the end of this chapter. Edwin D. McKee of the U.S. Geological Survey provided 191 sand samples for this study. J. J. Bigarella of the Geological Institute, Federal University of Paraná, Brazil, and L. R. Martins of the Ceco-Instituto de Geaciencias-UFRGS, Porto Alegre, Brazil, provided the data for 315 samples from Brazilian coastal dunes. Textural data from the literature were obtained by Camilla McCauley of the U.S. Geological Survey. I thank them all.

Textural Parameters of Costal Dune; Inland Dune, and Interdune and Serir Deposits

A COMPARISON of published information of mean grain size and sorting values for selected dune sands (as opposed to interdune) from five continents indicates that the dune sands are generally in the fine sand range (0.125–0.25 mm) and are moderately well sorted (σ_I values from 0.50–0.71, Trask values from 1.4–2.0). (See table 2.) These data represent a variable number of samples, a multitude of sampling and laboratory techniques, many different dune types, and different locations of samples on the dunes. Because of the general confusion in the literature regarding textures of eolian deposits, generalizations based upon textures as discussed in the literature seem inappropriate. The textural parameters for the 506 sand samples used in this chapter and by Moiola and Spencer (chapter C, this paper) were all derived from quarter phi sieve data, thus allowing textural comparisons.

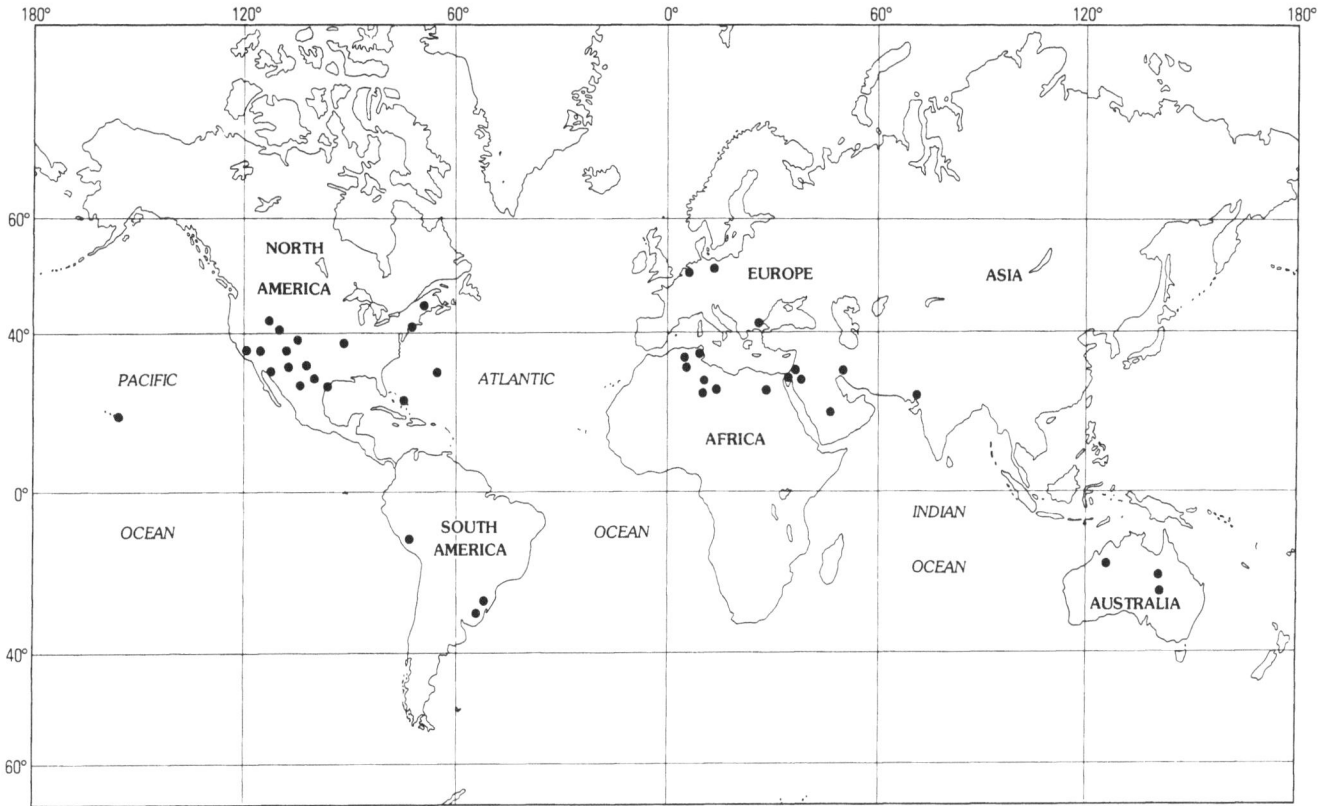

GENERAL LOCATIONS of the 506 eolian sand samples studied. (fig. 16.)

The majority of the 191 samples collected by Edwin D. McKee throughout the world represent inland dune and interdune environments (fig. 17 and frontispiece), and demonstrate the maximum range of textural parameters for eolian deposits. The dominant mean grain sizes for these samples are in the medium and fine sand ranges (fig. 18). Sorting values show a wide range with a considerable portion occurring in the poor sorting category (fig. 18). The histogram of the skewness values shows that the majority of the samples are positively skewed, although at least 20 percent of the samples are negatively skewed (fig. 18). The kurtosis histogram shows considerable scatter (fig. 18).

Data for another 315 samples, mostly from Brazilian coastal dunes, were combined with McKee's data and then divided into three major categories (where n is number of samples): (1) coastal dune sand ($n=291$), (2) inland dune sand ($n=175$), and (3) interdune and serir sand ($n=40$). The data at the end of this chapter (tables 5–7) are presented according to these categories and are grouped alphabetically by geographic area within

these categories. The statistical differentiation, using discriminant analysis, of coastal dune sand samples from inland dune samples, and of inland dune sand samples from interdune samples is presented by Moiola and Spencer (chapter C, this paper). The textural contrasts among coastal, inland, and interdune samples are illustrated by plots of their textural parameters.

Comparison of 291 coastal dune sand samples with 175 inland dune sand samples shows several differences. The coastal dunes are composed almost totally of very well sorted fine sand (fig. 19). Furthermore, they are symmetrical to positively (fine) skewed and are platykurtic to mesokurtic (fig. 19). Inland dune sand samples, by contrast, show a much greater range in mean grain size and sorting values (fig. 20). Skewness values show wide scatter, from positive to negative, and the kurtosis values tend to be mostly platykurtic (fig. 20).

The most significant textural differences occur between dune sands and interdune and serir sands. The mean grain size of 40 interdune and serir sand samples is highly variable, due to their poor sorting

TABLE 2. — *Comparison of textural parameters of sand from selected dune fields and sand seas*

[Leaders (. . .) indicate no data]

Selected area	Mean grain size (millimetres)	Sorting		Reference
		σ^1	Trask[2]	
Africa				
Grand Erg Occidental, Grand Erg Oriental, Algeria.	[3]0.10 − 0.30	. . .	. . .	Wilson (1973, p. 104).
Mauritania, Mali	[3]0.25 − 0.40	. . .	. . .	Tricart and Brochu (1955, p. 161).
Libya, Niger, Chad	0.20	. . .	. . .	Warren (1972, p. 42).
. . . . . . do	0.34	0.55	. . .	Hoque (1975, p. 395).
Kalahari Desert, South Africa	0.21	. . .	. . .	Goudie, Allchin, and Hedge (1973, p. 248).
Namib Desert, South-West Africa . .	0.23	. . .	. . .	Do.
Asia				
Ala Shan Desert, China	[4]0.19	. . .	. . .	Yu and others (1962, p. 276).
Takla Makan Desert, China	[4]0.17	. . .	. . .	Chu and others (1962, p. 23).
Thar Desert, India	[3]0.10 − 0.15	. . .	. . .	Goudie, Allchin, and Hedge (1973, p. 248).
Rub' al Khali	0.25	0.66	. . .	Ahlbrandt (this chapter).
Saudi Arabia (Oman)	0.19	[3]0.22 − 0.54	. . .	Glennie (1970, p. 82).
Peski Kyzylkum, U.S.S.R.	[3]0.13 − 0.25	. . .	. . .	Yakubov and Bespalova (1961, p. 654).
North America and South America				
Algodones dunes, California, U.S.A.	0.28	. . .	[3]1.22 − 3.47	Beheiry (1967, p. 29).
	0.23	. . .	1.13	Norris (1966, p. 298).
	[3]0.09 − 0.30	. . .	. . .	Van de Kamp (1973, p. 846).
Great Sand Dunes National Monument, Colorado, U.S.A.	[5]0.73 (reversing dunes[6]) [3] 0.24 − 1.51	[3]0.15 − 1.25	. . .	Ahlbrandt (this chapter).
Killpecker dunes, Wyoming, U.S.A.	[3]0.13 − 0.43 (transverse dunes[6])	[3]0.18 − 0.93	. . .	Ahlbrandt (1974b, p. 43).
Navajo Indian Reservation, Arizona, U.S.A.	[4]0.20	[5]0.52	. . .	Ahlbrandt (this chapter).
	[3] 0.15 − 1.10	[3] 0.35 − 0.96	. . .	Do.
White Sands National Monument, New Mexico, U.S.A.	0.25 − 0.55 (transverse dunes[6])	0.55	. . .	McKee and Moiola (1975, p. 61); Ahlbrandt (this chapter).
Gran Desierto, Mexico	[3]0.13 − 0.25	. . .	1.27	Merriam (1969, p. 531).
Coastal dunes, Brazil	[3]0.15 − 0.24	[3]0.25 − 0.49	. . .	Bigarella, Alessi, Becker, and Duarte (1969, p. 75 − 76).
Australia				
Great Sandy Desert	0.28 (sheet deposits[6])	0.53	. . .	Bettenay (1962, p. 15).
	0.53 (parabolic dunes[6])	. . .	. . .	Do.
Great Victoria Desert	0.50 (crests[6]) [3]0.26 − 0.30	. . .	1.40	Wopfner and Twidale (1967, p. 132).
Simpson Desert	0.16 [3]0.14 − 0.19	0.43 (crests[6]) 0.57 (flanks[6])	. . . . . .	Folk (1971a, p. 27). Do.

[1] Textural parameters as defined by Folk and Ward (1957).

[2] Textural parameters as defined by Trask (1931).

[3] Range of values for this parameter.

[4] Values calculated from data in original manuscript.

[5] Average of values for this parameter.

[6] Details of sample location or dune type(s) sampled, when known.

(fig. 21). The bimodality of interdune and serir sands accounts for the poor sorting values of these deposits (fig. 22), as has been noted by authors in deserts around the world (Beheiry, 1967; Folk, 1968a; Warren, 1972; Cooke and Warren, 1973). Skewness values generally show a tendency toward negative skewness with decreasing grain size (fig. 21), a phenomenon also observed in inland dunes of Wyoming, using settling tube data (Ahlbrandt, 1974b). Kurtosis values show considerable scatter but are predominantly platykurtic (fig. 21).

Recognition of Depositional Environments

Traditionally, three depositional environments have been differentiated on a textural basis — alluvial, littoral, and eolian. Their delineation based solely on graphical measures of textural parameters has not been satisfactory, as summarized by Friedman (1973, p. 626, 627): "Hence after many years of empirical measurements, experimentation and compilation of statistical parameters based on size-frequency distribution, textural parameters as clues to depositional environment have been a source of disenchantment."

Although the three eolian subenvironments, coastal dune, inland dune, and interdune and serir, have widely differing textural parameters, they reflect their source which is usually alluvial or littoral sand.

Inland dunes have variable mean grain size and sorting values, because their source materials are normally from alluvial fan, glaciofluviatile, fluviatile, lacustrine, or playa deposits.

The textural contrasts among inland dunes are illustrated by comparing samples from different dune types and by comparing samples from the base and crest of a given dune type (fig. 22). Textural contrasts occur both among dunes formed in different wind regimes (fig. 22) and among dunes formed in a given wind regime. For example, in inland dune fields with predominantly unidirectional wind regimes, both McKee (1966) and Ahlbrandt (1974b, 1975) observed improved sorting and finer

DUNE AND INTERDUNE DEPOSITS in the northwestern Namib Desert, South-West Africa. The star dune in the background is approximately 300 feet (100 m) high. Photograph by E. Tad Nichols. (Fig. 17.)

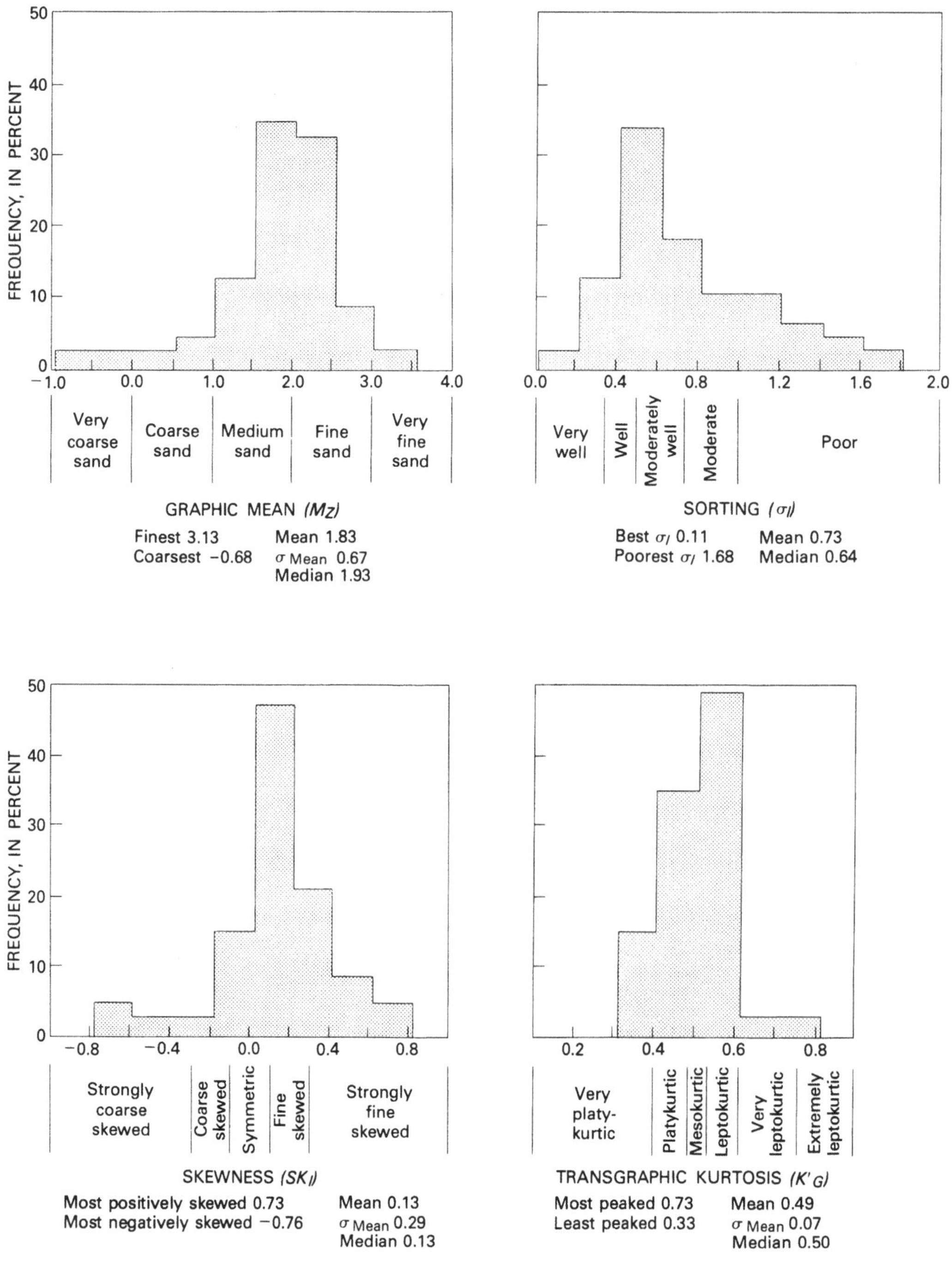

GRAPHICAL MEASURES for 191 eolian sand samples. (Fig. 18.)

mean grain size downwind from the source of sand in a sequence of dome, transverse, barchan, and parabolic dunes, respectively. Dome dunes, which occur on the upwind margins of dune fields, are noticeably coarser grained and more poorly sorted than are downwind dune types in the dunes at White Sands National Monument, New Mexico, and also in the Killpecker dune field in southwestern Wyoming. The textural differences among dune types are apparently not related to compositional differences as gypsum sand composes the dunes at White Sands National Monument (McKee,

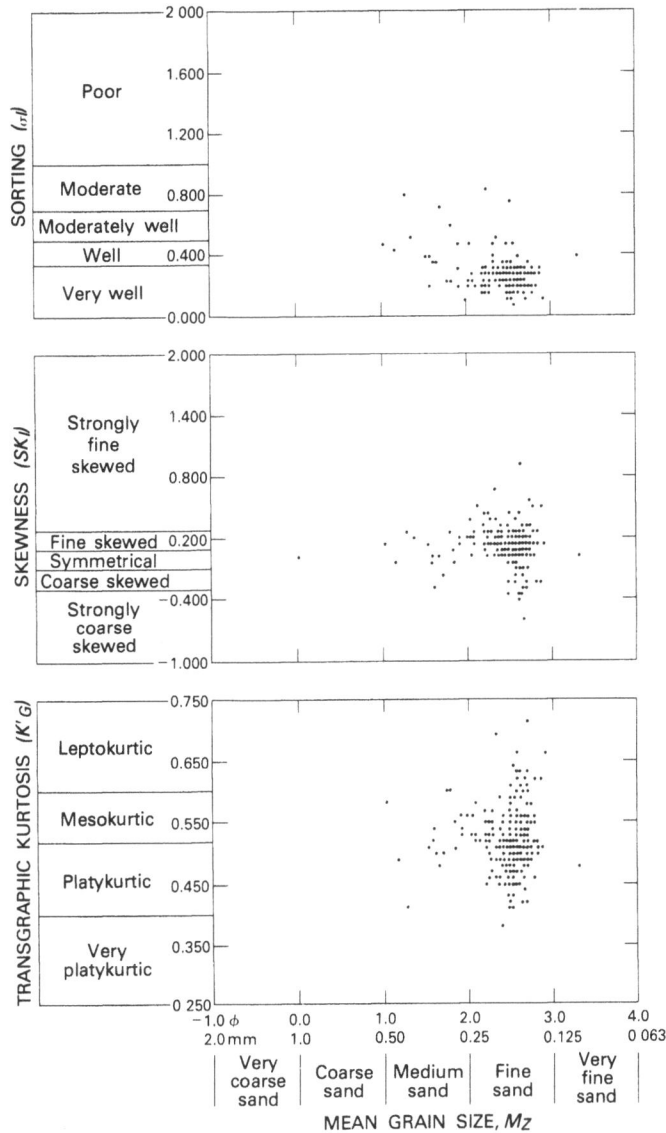

GRAPHICAL MEASURES of 291 coastal dune sand samples. (Fig. 19.)

1966), whereas quartzose and feldspathic sand compose the Killpecker dunes (Ahlbrandt, 1974b).

In reversing and multidirectional wind regimes, sand accumulates in dunes that may have little net lateral migration. Reversing dunes and star dunes are examples of dunes that grow vertically rather than migrate laterally. (See fig. 17.) The clastic material within these dunes represents a combination of available source material rather than the segregation and winnowing of source material downwind observed in dunes formed under unidirectional wind regimes. (Compare figs. 22C and D with fig. 22A.) For example, the Great Sand Dunes in Colorado have reversing wind flow —

that is, winds flow from nearly opposite directions at different times of the year (Merk, 1960). Streams emerge from the Sangre de Cristo Range onto alluvial fans east of the dune field providing poorly sorted coarse sand to the dunes. Periodically, wind flow reverses and unconsolidated alluvial and lacustrine or playa sediments in the San Luis Valley to the west of the dune field provide well-sorted fine sand (E. D. McKee, oral commun., 1975). The resultant eolian deposits are a combination of the two sands (fig. 22C).

Thus, considering the textural heterogeneity of inland dune sands, it is understandable why few textural studies using graphical measures can uni-

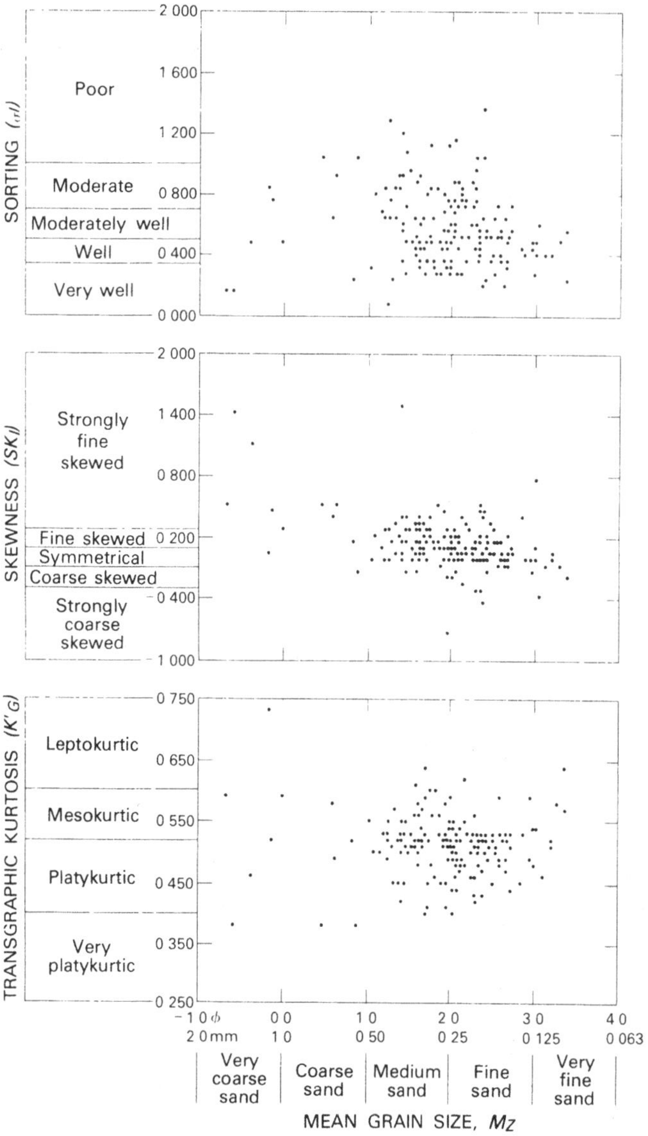
SORTING (σI)
Poor
Moderate
Moderately well
Well
Very well
2 000
1 600
1 200
0 800
0 400
0 000
SKEWNESS (SKI)
Strongly fine skewed
Fine skewed
Symmetrical
Coarse skewed
Strongly coarse skewed
2 000
1 400
0 800
0 200
−0 400
1 000
TRANSGRAPHIC KURTOSIS (K'G)
Leptokurtic
Mesokurtic
Platykurtic
Very platykurtic
0 750
0 650
0 550
0 450
0 350
0 250
−1 0φ 2 0mm
0 0 1 0
1 0 0 50
2 0 0 25
3 0 0 125
4 0 0 063
Very coarse sand
Coarse sand
Medium sand
Fine sand
Very fine sand
MEAN GRAIN SIZE, Mz

GRAPHICAL MEASURES of 175 inland dune sand samples. (Fig. 20.)

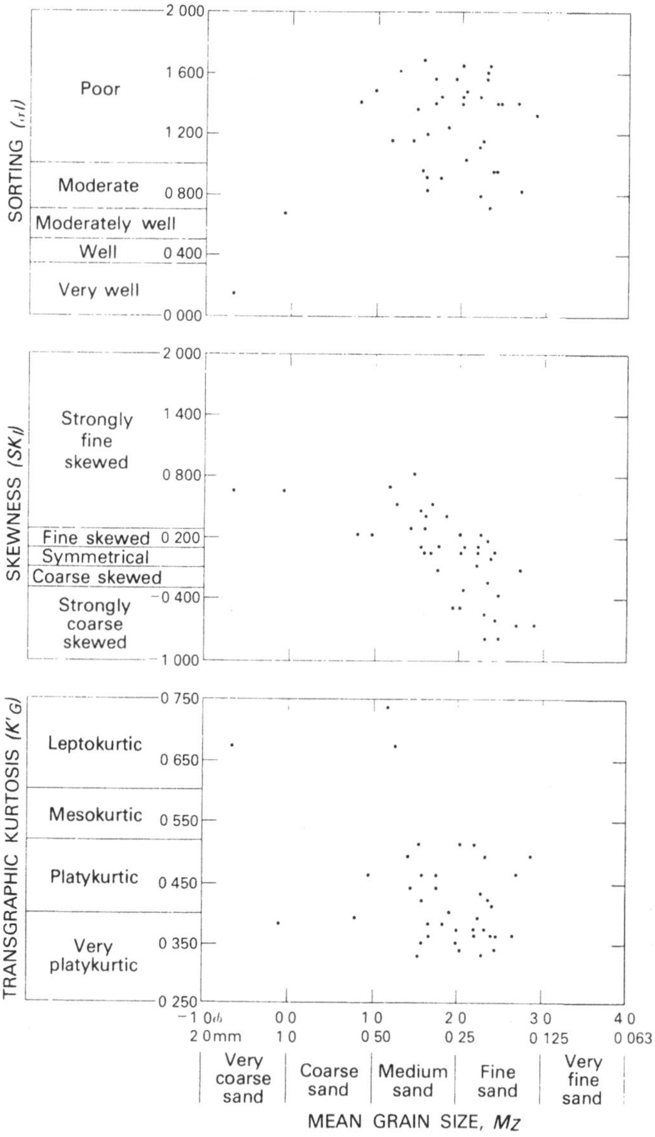
SORTING (σI)
Poor
Moderate
Moderately well
Well
Very well
2 000
1 600
1 200
0 800
0 400
0 000
SKEWNESS (SKI)
Strongly fine skewed
Fine skewed
Symmetrical
Coarse skewed
Strongly coarse skewed
2 000
1 400
0 800
0 200
−0 400
1 000
TRANSGRAPHIC KURTOSIS (K'G)
Leptokurtic
Mesokurtic
Platykurtic
Very platykurtic
0 750
0 650
0 550
0 450
0 350
0 250
−1 0φ 2 0mm
0 0 1 0
1 0 0 50
2 0 0 25
3 0 0 125
4 0 0 063
Very coarse sand
Coarse sand
Medium sand
Fine sand
Very fine sand
MEAN GRAIN SIZE, Mz

GRAPHICAL MEASURES of 40 interdune and serir sand samples. (Fig. 21.)

quely differentiate inland dune sand from its source (Moiola and Weiser, 1968). The compositional similarity of eolian sand and its fluvial sand source, the Colorado River delta, in the northwestern part of the Sonoran Desert has been shown by Merriam (1969). However, Friedman (1973, p. 627) suggested that moment measures may be more discriminative, such as the use of sorting of inland dune and river sand versus swash-zone sand or mean cubed deviation of inland dune versus river sand.

Coastal dunes are generally composed of well-sorted fine sand derived from beach deposits (fig. 23). Much recent detailed work with graphical measures of textural parameters attempting to differentiate coastal dune sand from littoral sand has met with limited success (Bigarella, Alessí, Becker, and Duarte, 1969; Bigarella, 1972; Friedman, 1966, 1973; Martins, 1967). Bigarella, Alessí, Becker, and Duarte (1969, p. 16) found that beach and dune sands in Brazil could not be distinguished using mean grain size and sorting when the source is a fine sand; however, medium- to coarse-grained sources showed dune sands to be better sorted and finer grained than beach deposits. Bigarella (1972, p. 14) questioned the significance of positive skew-

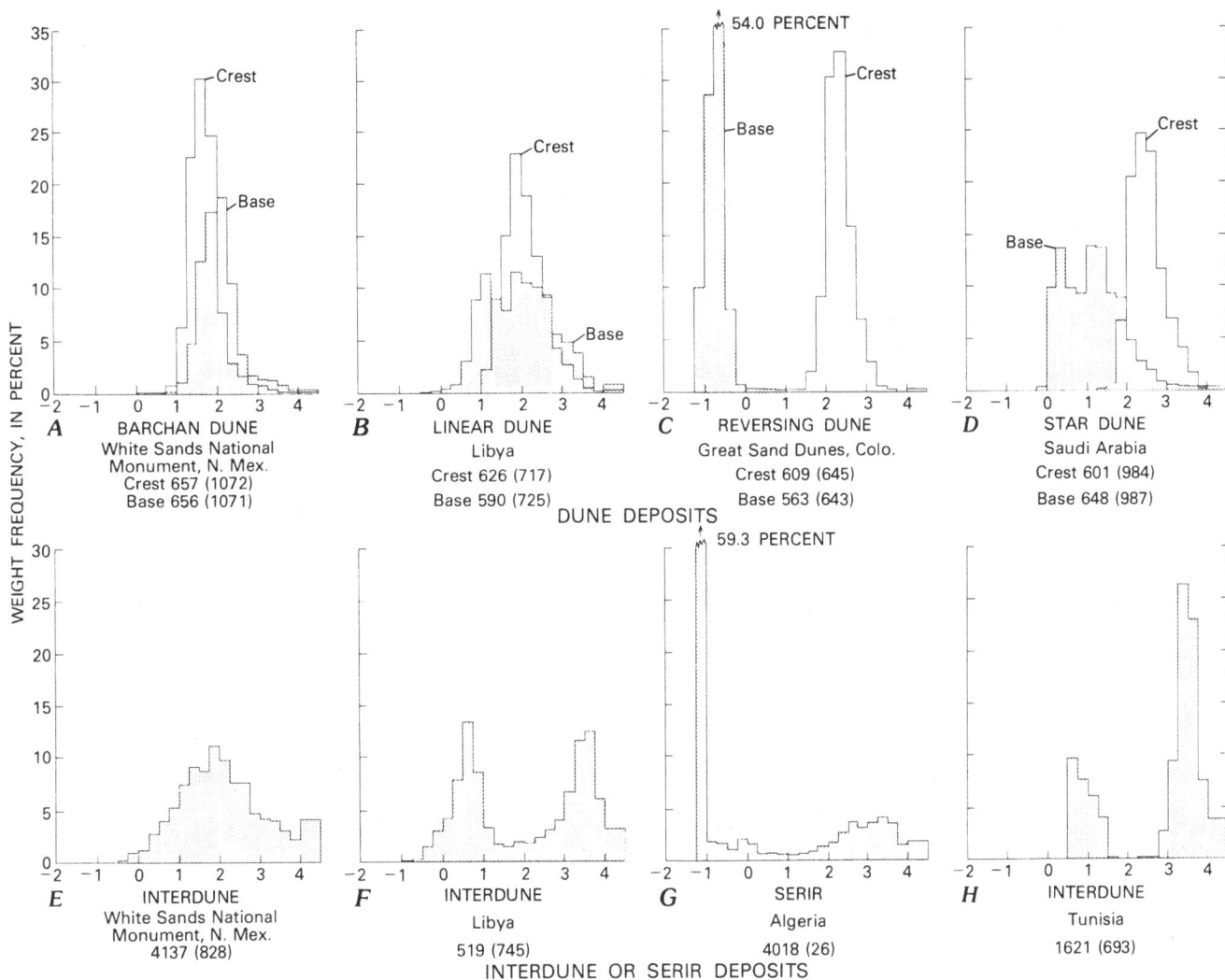

HISTOGRAMS SHOWING TEXTURAL CONTRASTS among dune and interdune samples. Values along abscissa are in whole phi; histogram class intervals are quarter phi. Numbers below each histogram are the computer index number (outside parentheses) and McKee index number (inside parentheses) of samples listed in tables 5–7. Silt- and clay-size fraction to the right of 4 ϕ (Fig. 22.)

ness as an eolian indicator because some dune sands have negative skewness values. Although dune sand is usually more rounded than beach sand, Bigarella (1972, p. 14) found the most dependable eolian indicator to be light- and heavy-mineral separations.

In the absence of sedimentary structures as primary evidence of eolian deposits, other criteria must be used. Discriminant analysis (Moiola and Spencer, chapter C), certain moment measures, and perhaps light- and heavy-mineral separations may be able to differentiate eolian sands from alluvial and littoral sands. Frosted grains, as viewed with the scanning electron microscope, have also been used as an eolian indicator; however, that technique may be limited by the degree of diagenesis grains have undergone. Graphical measures are certainly less discriminating and of more limited use than these latter techniques. Sorting and mean grain size values do have application as will be shown. The application of skewness and kurtosis are, at best, tenuous because they show wide scatter in eolian deposits (figs. 18–21) and because skewness is dependent on mean grain size (Ahlbrandt, 1975; this chapter, fig. 21).

LARGE COMPLEX LINEAR DUNES which derive sand from adjacent coastal deposits. Slipfaces on the dunes face inland at the time the photograph was taken. Sandfisch Bay, western part of Namib Desert, South-West Africa. Photograph by E. Tad Nichols. (Fig. 23.)

Porosity and Permeability Relationship to Textural Parameters

THE MEAN GRAIN SIZE, sorting, and packing arrangement of grains affect porosity and permeability in alluvial, littoral, and eolian sands. Porosity in modern sediments decreases with progressively more poorly sorted material regardless of grain size (fig. 24). Figure 24 indicates that the initial porosity (from 28 to 43 percent) is as much as 15 percent lower in very poorly sorted sand than in extremely well sorted sand. The corresponding pore-volume loss for a given grain-size range is 35 percent. This range of values demonstrates the difference between very well sorted

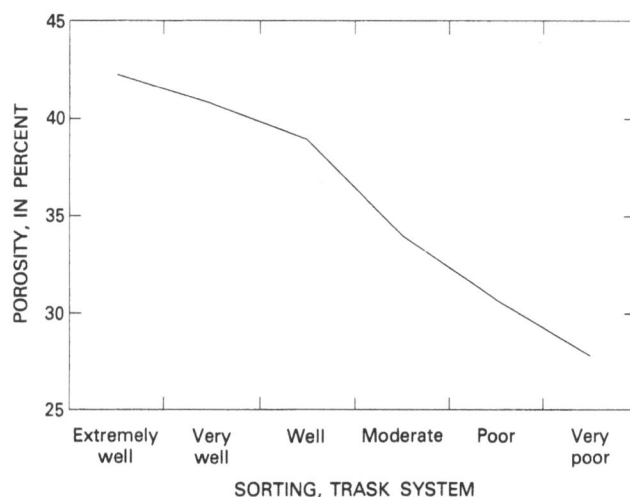

RELATIONSHIP BETWEEN initial porosity and sorting. The curve was drawn for wet, packed sand from data by Beard and Weyl (1973). (Fig. 24.)

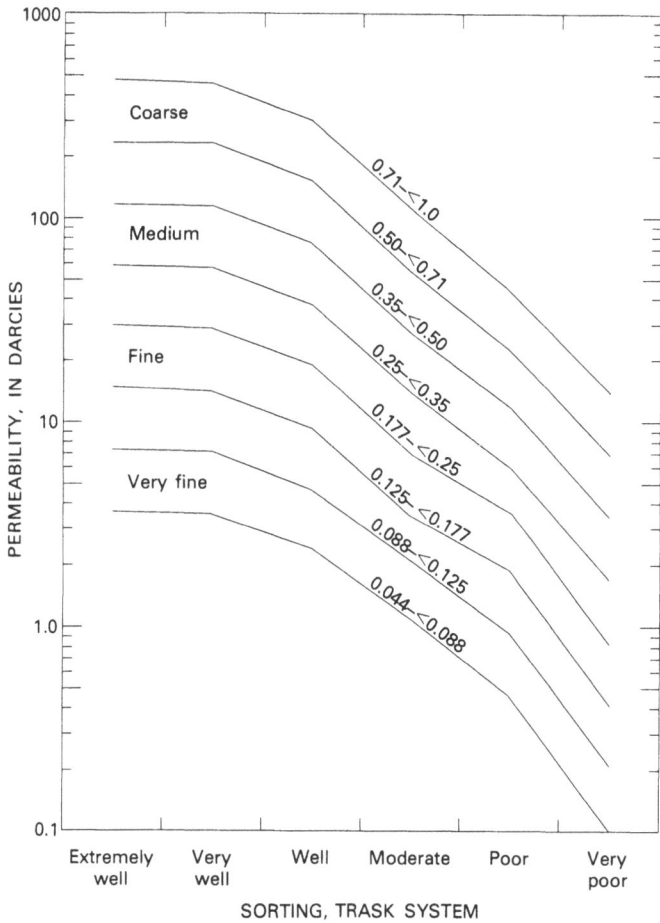

INTERRELATIONSHIP OF MEAN GRAIN SIZE (in millimetres) and sorting on initial permeability for wet, packed sand from data by Beard and Weyl (1973). (fig. 25.)

dune sand and poorly to very poorly sorted interdune and serir sands. Permeability decreases with poorer sorting and finer mean grain size (fig. 25) but is also affected by packing arrangements (Rittenhouse, 1971; Pryor 1973). Less symmetric packing — for example, orthorhombic packing (rotated 30°) — results in greater porosity loss due to solution than do cubicly packed grains (Rittenhouse, 1971, p. 82.) Porosity loss due to solution is maximized in very poorly sorted, extremely well sorted or very angular sand and minimized in very well rounded, well-sorted sand (Rittenhouse, 1971, p. 86). The well-rounded and well-sorted character of many eolian deposits favor them to have higher initial and diagenetically altered porosities (considering solution only) than do alluvial, and perhaps littoral, deposits.

Pryor (1973, p. 162) found that within a dune sand body there is little variability in porosity and permeability, whereas river bar and beach sands

have directional permeabilities that are parallel to the long dimension of the deposit. The heterogeneity of sediments and more irregular packing arrangement of grains in the river bar sand "packets" (Pryor, 1973, p. 187) resulted in lower effective permeabilities for them as compared with beach and dune sands (table 3). Effective permeability measures the flow across bedding units or "packets" of contrasting permeability.

TABLE 3. — *Porosity and permeability measurements in alluvial, littoral, and coastal dune sand*

[Modified from Pryor, 1973, p. 162. n, number of samples; D, darcies; mD, millidarcies]

	Type of sand		
	Alluvial (point bar) (n = 348)	Littoral (n = 480)	Coastal dune (n = 164)
Average porosity .. (percent)..	41	49	49
Range of porosity .. (percent)..	17 − 52	39 − 56	42 − 54
Average permeability	93 D	68 D	54 D
Range of permeability	4 mD − 500 D	3.6 − 166 D	5 − 104 D

The contrast in textures among well-sorted to very well sorted coastal dune sands, moderately sorted to well-sorted inland dune sands, and poorly sorted interdune and serir sands should show progressively lower porosity and permeability. In order to test the inland dune-interdune porosity and permeability relationship, cores from four drill holes in the White Sands National Monument, New Mexico, U.S.A., were studied. The differences in sorting for the dune and interdune environments at White Sands and elsewhere are shown in figure 22.

The cores penetrate dune and interdune deposits, both of which are weakly cemented owing to the gypsiferous composition of the sand. The 14 samples taken for porosity and permeability analyses (fig. 26) show a general reduction in permeability for the interdune deposits (fig. 27). The high porosity yet low permeability of the interdune

TABLE 4. — *Porosity and permeability measurements in eolian subenvironments: Coastal dune, inland dune, and interdune sand*

[n, number of samples; D, darcies; mD, millidarcies]

	Type of sand		
	Coastal dune (n = 164)[1]	Inland dune (n = 6)	Interdune (n = 6)
Average porosity .. (percent)..	49	26	31
Range of porosity .. (percent)..	42 − 54	23 − 30	25 − 38
Average permeability	54 D	137 mD	120 mD
Range of permeability	5 − 104 D	28 − 450 mD	33 − 270 mD

[1] Pryor, 1973, p. 162.

DRILLING SITE NUMBER

EXPLANATION

Dune
deposits

Interdune
deposits

Fluvial
deposits

Playa
deposits

No
recovery

TD

Total
depth

*

Plug taken for porosity and
permeability determination

EXPLANATION

③ Location of
drilling site

Sand

Playa

Mountain
base

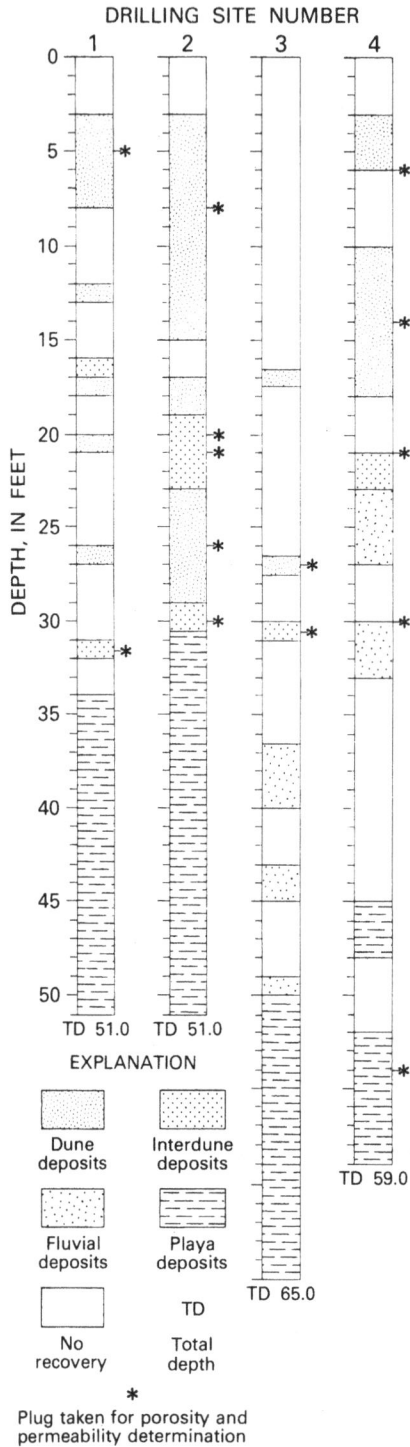

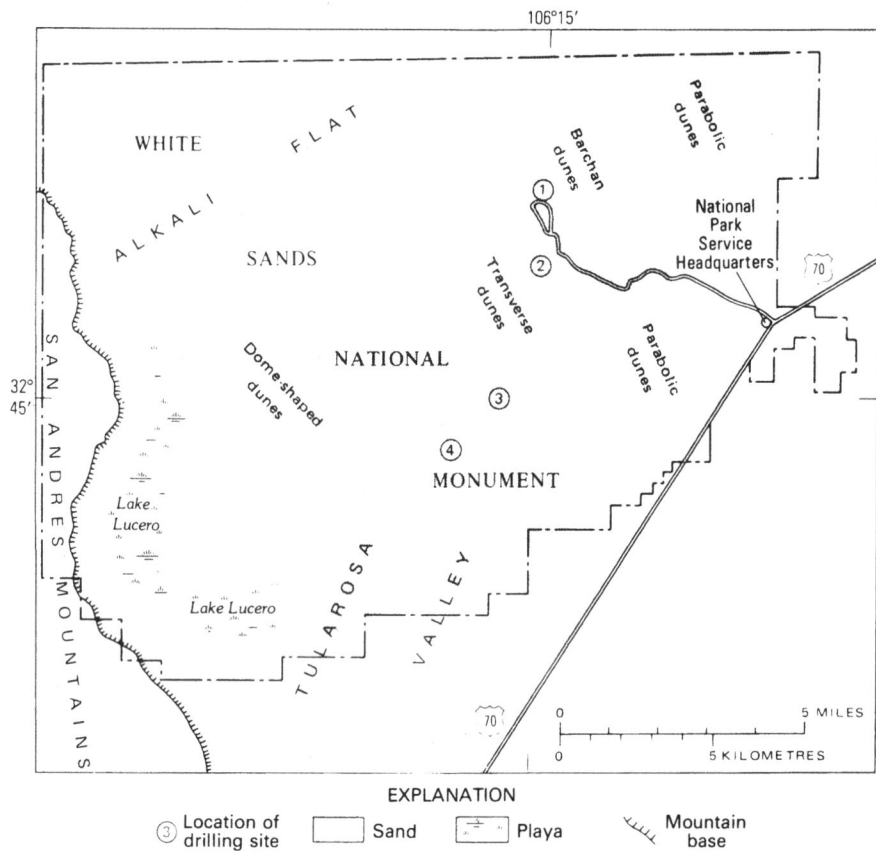

LOCATIONS OF CORES and depth of plug samples taken for porosity and permeability determinations, White Sands National Monument, New Mexico, U.S.A. Core descriptions and modified index map from McKee and Moiola (1975, p. 60, 65). (Fig. 26.)

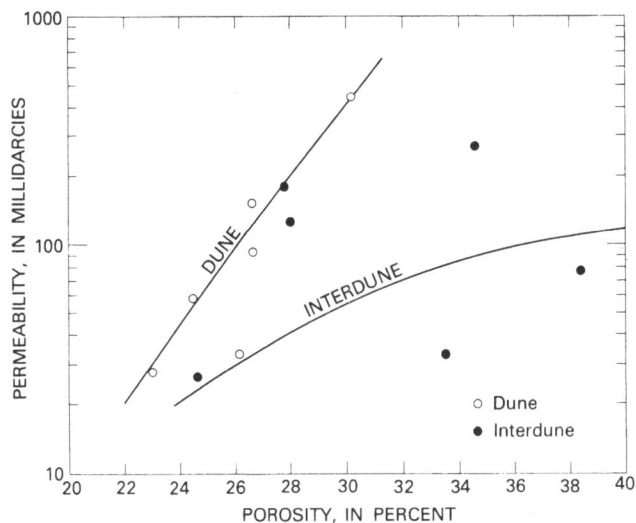

POROSITY-PERMEABILITY DISTRIBUTION of dune and inter-dune sediments from cores, White Sands National Monument, New Mexico, U.S.A. Analyses by Core Laboratories, Inc., Denver, Colorado, U.S.A. Location and description of cores are shown in figure 26. (fig. 27.)

deposits suggests that much of the porosity is in-effective and probably is incorporated into the fine-grained matrix.

It appears that the textural differences in eolian subenvironments also bear significantly on initial porosity and permeability (table 4).

The more poorly sorted interdune deposits may form effective permeability barriers after burial owing to greater grain-to-grain contacts which in-crease porosity loss by pressure solution and ce-mentation. Tightly cemented, poorly sorted inter-dune deposits separate porous well-sorted dune sandstones in the Tensleep and Weber Sandstones (Pennsylvanian and Permian) in the Wertz and Brady oil fields, Wyoming, U.S.A., forming local barriers to oil migration. The porosity and per-meability contrasts in the two deposits must relate in part to the textural differences of the dune-inter-dune sediments that compose part of the Tensleep and Weber Sandstones (chapter I).

TABLE 5. — *Graphical measures of sand samples from coastal dunes used as a data base*

Principal county or area	Specific area	Computer Index No.	McKee's Index No.	Mean grain size φ	Mean grain size mm	Median grain size φ	Median grain size mm	Sorting ρ_I	Skewness Sk_I	Kurtosis K'_G
Bahama Islands	Bimini	1622	580	1.916	0.2661	1.870	0.2736	0.495	0.221	0.536
		531	580	1.186	0.4383	1.200	0.4353	0.425	0.020	0.486
Bermuda	Elbow Beach	1619	586	1.710	0.3057	1.800	0.2872	0.704	0.137	0.499
Brazil	Garopaba Sirin (State of Santa Catarina).	1556	22	2.360	0.1948	2.280	0.2059	0.537	0.272	0.544
		1510	402	2.630	0.1615	2.610	0.1639	0.299	0.025	0.445
		1511	403	2.550	0.1708	2.550	0.1708	0.258	0.014	0.459
		1512	404	2.440	0.1843	2.400	0.1895	0.356	0.128	0.468
		1513	405	2.480	0.1792	2.480	0.1792	0.300	0.002	0.452
		1514	406	2.330	0.1989	2.270	0.2073	0.363	0.166	0.460
		1515	407	2.240	0.2117	2.120	0.2300	0.309	0.456	0.503
Brazil	Island of Santa Catarina	1516	408	2.660	0.1582	2.650	0.1593	0.233	0.117	0.513
		1517	409	2.630	0.1615	2.600	0.1649	0.313	0.166	0.449
		772	7	1.800	0.2872	1.810	0.2852	0.255	0.248	0.597
		773	6	2.200	0.2176	2.203	0.2172	0.318	0.053	0.514
		774	5	1.900	0.2679	1.920	0.2643	0.302	0.182	0.525
		775	4	1.920	0.2643	1.926	0.2632	0.211	0.127	0.559
		776	3	2.050	0.2415	2.076	0.2372	0.223	0.171	0.534
		777	2	2.100	0.2333	2.140	0.2269	0.234	0.261	0.521
		778	1	2.060	0.2398	2.096	0.2339	0.195	0.368	0.557
		779	4	2.190	0.2192	2.203	0.2172	0.232	0.167	0.510
		780	4	2.100	0.2333	2.130	0.2285	0.252	0.184	0.575
		781	4	2.020	0.2466	2.030	0.2449	0.119	0.223	0.561
		782	3	2.120	0.2300	2.173	0.2217	0.190	0.487	0.516
		783	3	2.100	0.2333	2.133	0.2280	0.294	0.132	0.530
		784	3	2.100	0.2333	2.126	0.2291	0.244	0.263	0.583
		785	2	2.290	0.2045	2.296	0.2036	0.176	0.130	0.546
		786	2	2.250	0.2102	2.276	0.2065	0.265	0.195	0.528
		787	2	2.300	0.2031	2.310	0.2017	0.285	0.157	0.494
		788	1	2.200	0.2176	2.230	0.2132	0.174	0.308	0.531
		789	1	2.200	0.2176	2.216	0.2152	0.185	0.263	0.571
		790	1	2.200	0.2176	2.210	0.2161	0.193	0.233	0.550
		791	8	2.200	0.2176	2.230	0.2132	0.178	0.230	0.520
		792	9	2.300	0.2031	2.220	0.2146	0.184	0.261	0.530
		793	10	2.250	0.2102	2.210	0.2161	0.183	0.250	0.524
		794	11	2.300	0.2031	2.210	0.2161	0.178	0.260	0.525
		795	12	2.280	0.2059	2.200	0.2176	0.177	0.250	0.530
		796	13	2.250	0.2102	2.210	0.2161	0.176	0.230	0.550
		797	14	2.300	0.2031	2.250	0.2102	0.178	0.260	0.573

Location									
Brazil .... Laguna (State of Santa Catarina) ..	798	15	2.290	0.2045	2.250	0.2102	0.180	0.261	0.571
	799	16	2.300	0.2031	2.210	0.2161	0.183	0.230	0.570
	1500	241	2.620	0.1627	2.700	0.1539	0.276	0.420	0.532
	1501	242	2.590	0.1661	2.600	0.1649	0.195	0.001	0.558
	1502	243	2.450	0.1830	2.450	0.1830	0.328	0.068	0.452
	1503	244	2.520	0.1743	2.550	0.1708	0.276	0.146	0.411
	1504	245	2.640	0.1604	2.700	0.1539	0.243	0.436	0.564
	1505	246	2.530	0.1731	2.580	0.1672	0.271	0.207	0.432
	1506	247	2.800	0.1436	2.800	0.1436	0.211	0.023	0.464
	1507	248	2.660	0.1582	2.700	0.1539	0.223	0.308	0.524
	1508	249	2.670	0.1571	2.700	0.1539	0.248	0.169	0.560
	1509	250	2.600	0.1649	2.610	0.1638	0.237	0.129	0.534
	1538	434	2.320	0.2003	2.400	0.1895	0.387	0.045	0.460
Brazil .... Morro Conventos (Iate Clube)	1539	435	2.550	0.1708	2.550	0.1708	0.326	0.050	0.450
	1540	436	2.530	0.1731	2.500	0.1768	0.337	0.131	0.498
	1541	437	2.600	0.1649	2.590	0.1661	0.266	0.077	0.490
	1542	438	2.540	0.1719	2.530	0.1731	0.296	0.027	0.411
	1543	439	2.520	0.1743	2.500	0.1768	0.299	0.148	0.424
	1544	440	2.540	0.1719	2.540	0.1719	0.344	0.047	0.467
	1545	441	2.550	0.1708	2.490	0.1780	0.331	0.235	0.460
	1546	442	2.490	0.1780	2.450	0.1830	0.351	0.158	0.425
	1548	444	2.440	0.1843	2.450	0.1830	0.329	0.007	0.475
Brazil .... Planicie Costeira (State of Rio Grande do Sol).	663	1	2.800	0.1436	2.833	0.1406	0.206	0.267	0.524
	664	3	2.900	0.1340	2.903	0.1340	0.125	0.148	0.662
	665	4	2.600	0.1649	2.603	0.1649	0.251	0.089	0.513
	666	8	2.760	0.1476	2.770	0.1466	0.209	0.124	0.677
	667	10	2.800	0.1436	2.886	0.1350	0.331	0.509	0.616
	668	11	2.890	0.1350	2.840	0.1397	0.297	0.233	0.509
	669	14	2.600	0.1649	2.603	0.1649	0.181	0.021	0.547
	670	15	2.500	0.1768	2.500	0.1768	0.221	ND	0.587
	671	16	2.590	0.1661	2.593	0.1661	0.186	0.013	0.535
	672	17	2.500	0.1768	2.526	0.1731	0.130	0.208	0.505
	673	18	1.590	0.3322	1.596	0.3299	0.203	0.040	0.537
	674	9	2.800	0.1436	2.806	0.1426	0.288	0.044	0.461
	675	19	2.560	0.1696	2.580	0.1672	0.212	0.120	0.500
	676	20	2.600	0.1649	2.600	0.1649	0.097	0.045	0.529
	677	21	2.570	0.1684	2.570	0.1684	0.170	0.069	0.542
	678	24	2.550	0.1708	2.550	0.1708	0.760	ND	0.487
	679	26	2.600	0.1649	2.610	0.1638	0.178	0.071	0.494
	680	28	2.500	0.1768	2.513	0.1756	0.180	0.129	0.543
	681	30	2.580	0.1672	2.580	0.1672	0.167	0.053	0.530
	682	34	2.500	0.1768	2.513	0.1756	0.200	0.156	0.519
	683	36	2.320	0.2003	2.360	0.1948	0.284	0.217	0.505

TABLE 5. — *Graphical measures of sand samples from coastal dunes used as a data base* — Continued

Principal county or area	Locality Specific area	Computer Index No.	McKee's Index No.	Mean grain size φ	Mean grain size mm	Median grain size φ	Median grain size mm	Sorting ρ_I	Skewness Sk_I	Kurtosis K_G
Brazil--Con.	Planicie Costeira (State of Rio Grande do Sol)--Con.	684	38	2.700	0.1539	2.680	0.1560	0.185	0.079	0.528
		685	39	2.250	0.2102	2.253	0.2102	0.143	0.018	0.518
		686	40	1.050	0.4841	1.093	0.4697	0.471	0.139	0.581
		687	41	2.390	0.1908	2.413	0.1882	0.279	0.158	0.520
		688	42	2.570	0.1684	2.590	0.1661	0.216	0.198	0.557
		689	43	2.480	0.1792	2.486	0.1780	0.251	0.098	0.520
		690	45	2.520	0.1743	2.530	0.1731	0.254	0.047	0.511
		691	46	2.750	0.1487	2.750	0.1487	0.206	ND	0.601
		692	44	2.500	0.1768	2.510	0.1736	0.253	0.086	0.512
		693	47	2.720	0.1518	2.760	0.1476	0.196	0.208	0.505
		694	48	2.500	0.1768	2.516	0.1743	0.249	0.134	0.482
		695	49	2.510	0.1756	2.530	0.1731	0.238	0.113	0.484
		696	50	2.480	0.1792	2.476	0.1792	0.230	0.024	0.524
		697	51	2.600	0.1649	2.636	0.1604	0.290	0.119	0.493
		698	53	2.500	0.1768	2.503	0.1768	0.303	0.148	0.534
		699	57	2.490	0.1780	2.506	0.1756	0.288	0.107	0.520
		700	59	2.480	0.1792	2.466	0.1805	0.291	0.020	0.560
		701	62	2.580	0.1672	2.596	0.1649	0.176	0.112	0.510
		702	63	2.460	0.1817	2.483	0.1792	0.280	0.154	0.477
		703	65	2.500	0.1768	2.530	0.1731	0.245	0.200	0.520
		704	67	2.580	0.1672	2.530	0.1731	0.298	0.107	0.471
		705	70	2.490	0.1780	2.503	0.1768	0.237	0.060	0.563
		706	71	2.410	0.1882	2.420	0.1869	0.271	0.322	0.477
		707	72	2.600	0.1649	2.600	0.1649	0.206	ND	0.550
		708	73	2.690	0.1550	2.636	0.1604	0.213	0.286	0.541
		709	78	2.510	0.1756	2.503	0.1768	0.276	0.018	0.521
		710	79	2.520	0.1743	2.533	0.1731	0.191	0.100	0.515
		711	80	2.400	0.1895	2.420	0.1869	0.243	0.196	0.533
		712	81	2.390	0.1908	2.403	0.1895	0.249	0.189	0.590
		713	83	2.410	0.1882	2.453	0.1830	0.271	0.258	0.511
		714	87	2.370	0.1934	2.391	0.1908	0.293	0.214	0.494
		715	89	2.600	0.1649	2.610	0.1638	0.256	0.111	0.599
		716	93	2.390	0.1908	2.393	0.1908	0.270	0.077	0.487
		717	97	2.520	0.1743	2.546	0.1708	0.257	0.172	0.482
		718	99	2.500	0.1768	2.520	0.1743	0.296	0.148	0.613

719	102	2.480	0.1792	2.506	0.1756	0.266	0.154	0.520
720	109	2.220	0.2146	2.306	0.2017	0.263	0.417	0.450
721	112	2.390	0.1908	2.366	0.1934	0.240	0.024	0.509
722	113	2.310	0.2017	2.343	0.1975	0.261	0.355	0.542
723	114	2.280	0.2059	2.283	0.2059	0.258	0.269	0.469
724	115	2.420	0.1869	2.466	0.1805	0.289	0.220	0.487
725	116	2.360	0.1948	2.386	0.1908	0.273	0.144	0.460
726	117	2.390	0.1908	2.401	0.1895	0.235	0.099	0.453
727	123	2.510	0.1756	2.540	0.1719	0.247	0.225	0.559
728	124	2.320	0.2003	2.393	0.1908	0.311	0.351	0.493
729	125	2.600	0.1749	2.603	0.1649	0.225	0.028	0.491
730	126	2.600	0.1749	2.630	0.1615	0.276	0.149	0.558
731	129	2.500	0.1768	2.516	0.1756	0.282	0.123	0.505
732	132	2.600	0.1649	2.596	0.1661	0.229	0.011	0.596
733	134	2.500	0.1768	2.510	0.1756	0.275	0.094	0.473
734	138	2.610	0.1638	2.650	0.1593	0.180	0.894	0.529
735	140	2.600	0.1649	2.606	0.1649	0.241	0.056	0.534
736	141	2.600	0.1649	2.566	0.1696	0.236	0.175	0.616
737	142	2.400	0.1895	2.420	0.1869	0.273	0.134	0.507
738	143	2.410	0.1882	2.436	0.1856	0.249	0.177	0.497
739	144	2.450	0.1830	2.483	0.1792	0.294	0.153	0.505
740	145	2.400	0.1895	2.406	0.1895	0.213	0.125	0.551
741	147	1.290	0.4090	1.466	0.3635	0.791	0.262	0.410
742	149	2.400	0.1895	2.403	0.1895	0.215	0.088	0.561
743	153	2.500	0.1768	2.523	0.1743	0.264	0.123	0.481
744	155	2.610	0.1638	2.623	0.1627	0.247	0.058	0.501
745	156	2.060	0.2398	2.066	0.2398	0.486	0.049	0.527
746	157	2.690	0.1550	2.703	0.1539	0.141	0.291	0.578
747	158	2.630	0.1615	2.646	0.1604	0.147	0.252	0.608
748	159	2.600	0.1649	2.626	0.1627	0.178	0.240	0.571
749	160	2.690	0.1559	2.693	0.1559	0.127	0.143	0.630
750	161	2.510	0.1756	2.536	0.1731	0.158	0.251	0.585
751	162	2.650	0.1593	2.680	0.1560	0.206	0.240	0.541
752	163	2.480	0.1792	2.520	0.1743	0.237	0.245	0.491
753	164	2.580	0.1672	2.576	0.1684	0.157	0.114	0.632
754	165	2.510	0.1756	2.543	0.1719	0.239	0.211	0.514
755	166	2.580	0.1672	2.576	0.1684	0.197	0.034	0.545
756	167	2.500	0.1768	2.516	0.1756	0.193	0.213	0.566
757	168	2.490	0.1780	2.496	0.1780	0.221	0.162	0.609
758	169	2.500	0.1768	2.500	0.1768	0.214	0.166	0.551
759	170	2.490	0.1780	2.470	0.1805	0.281	0.069	0.489
760	171	2.610	0.1638	2.626	0.1627	0.189	0.166	0.548
761	172	2.550	0.1708	2.566	0.1696	0.225	0.128	0.505
762	173	2.600	0.1649	2.606	0.1649	0.212	0.086	0.500
763	174	2.490	0.1780	2.533	0.1731	0.240	0.260	0.515

TABLE 5. — Graphical measures of sand samples from coastal dunes used as a data base — Continued

Principal county or area	Locality Specific area	Computer Index No.	McKee's Index No.	Mean grain size φ	Mean grain size mm	Median grain size φ	Median grain size mm	Sorting ρI	Skewness SkI	Kurtosis KG
Brazil—Con.	Planicie Costeira (State of Rio Grande do Sol)—Con.	764	175	2.600	0.1649	2.636	0.1615	0.228	0.183	0.488
		765	176	2.600	0.1649	2.626	0.1627	0.259	0.164	0.472
		766	177	2.600	0.1649	2.630	0.1615	0.238	0.219	0.522
		767	178	2.550	0.1708	2.556	0.1708	0.238	0.120	0.489
		768	179	2.600	0.1649	2.633	0.1615	0.206	0.196	0.544
		769	180	2.500	0.1768	2.513	0.1756	0.228	0.143	0.482
		770	181	2.510	0.1756	2.567	0.1696	0.210	0.305	0.481
		771	182	2.550	0.1708	2.570	0.1684	0.256	0.117	0.483
		1567	280	2.693	0.1550	2.700	0.1539	0.201	0.097	0.554
		1568	310	2.693	0.1550	2.690	0.1550	0.215	0.058	0.523
		1569	311	2.506	0.1768	2.600	0.1649	0.278	0.360	0.527
		1570	311	2.640	0.1604	2.610	0.1638	0.239	0.199	0.542
		1571	312	2.493	0.1780	2.510	0.1756	0.253	0.035	0.483
		1572	313	2.616	0.1638	2.600	0.1649	0.307	0.058	0.466
		1573	315	2.660	0.1582	2.650	0.1593	0.280	0.058	0.534
		1574	316	2.643	0.1604	2.630	0.1615	0.268	0.087	0.590
		1575	317	2.706	0.1539	2.700	0.1539	0.217	0.090	0.502
		1576	318	2.573	0.1684	2.550	0.1708	0.283	0.185	0.473
		1577	320	2.693	0.1550	2.690	0.1550	0.226	0.029	0.510
		1578	321	2.500	0.1768	2.500	0.1768	0.300	0.095	0.473
		1579	322	2.710	0.1528	2.700	0.1539	0.318	0.022	0.476
		1580	324	2.496	0.1780	2.430	0.1856	0.236	0.429	0.483
		1581	341	2.696	0.1550	2.700	0.1539	0.208	0.083	0.534
		1582	342	2.700	0.1539	2.700	0.1539	0.221	ND	0.609
		1583	343	2.706	0.1539	2.700	0.1539	0.118	0.068	0.534
		1584	344	2.753	0.1487	2.750	0.1487	0.190	0.049	0.582
		1585	345	2.623	0.1627	2.600	0.1649	0.228	0.206	0.539
		1586	346	2.640	0.1604	2.620	0.1627	0.178	0.200	0.514
		1587	347	2.606	0.1649	2.600	0.1649	0.204	0.090	0.551
		1588	348	2.500	0.1768	2.500	0.1768	0.280	0.023	0.456
		1589	350	2.573	0.1684	2.550	0.1708	0.311	0.128	0.518
		1590	351	2.553	0.1708	2.550	0.1708	0.275	0.035	0.493
		1591	352	2.653	0.1593	2.680	0.1560	0.264	0.352	0.438
		1592	353	2.636	0.1615	2.640	0.1604	0.170	0.096	0.559
		1593	356	2.450	0.1830	2.450	0.1830	0.286	ND	0.473

1594	362	2.740	0.1497	2.700	0.1539	0.198	0.342	0.551
1595	364	2.806	0.1436	2.800	0.1436	0.198	0.114	0.547
1596	365	2.643	0.1604	2.700	0.1539	0.253	0.346	0.586
1597	366	2.726	0.1518	2.700	0.1539	0.218	0.215	0.603
1598	367	2.660	0.1582	2.620	0.1627	0.246	0.169	0.595
1599	368	2.700	0.1539	2.710	0.1528	0.210	0.218	0.561
1600	368	2.703	0.1539	2.700	0.1539	0.213	0.006	0.611
1601	369	2.683	0.1560	2.700	0.1539	0.205	0.590	0.615
1602	370	2.686	0.1560	2.680	0.1560	0.218	0.008	0.457
1603	371	2.706	0.1539	2.680	0.1560	0.260	0.111	0.581
1604	386	2.703	0.1539	2.690	0.1550	0.219	0.119	0.713
1605	387	2.653	0.1593	2.650	0.1593	0.250	0.039	0.521
1606	388	2.676	0.1571	2.680	0.1560	0.228	0.014	0.530
1607	389	2.656	0.1593	2.630	0.1615	0.229	0.184	0.518
1608	390	2.510	0.1756	2.500	0.1768	0.215	0.069	0.523
1609	391	2.616	0.1638	2.600	0.1649	0.211	0.115	0.504
1610	392	2.700	0.1593	2.700	0.1593	0.206	0.071	0.524
1611	393	2.576	0.1684	2.680	0.1672	0.318	0.381	0.470
1612	398	2.500	0.1768	2.500	0.1768	0.277	ND	0.450
1613	399	2.400	0.1895	2.350	0.1961	0.276	0.342	0.497
1624	303	2.670	0.1571	2.740	0.1497	0.303	0.351	0.423
1625	304	2.730	0.1528	2.700	0.1539	0.257	0.073	0.495
1626	305	2.680	0.1507	2.700	0.1539	0.267	0.094	0.501
1627	306	2.680	0.1560	2.670	0.1571	0.220	0.131	0.505
1628	307	2.660	0.1582	2.660	0.1582	0.246	0.050	0.505
1629	308	2.690	0.1550	2.680	0.1560	0.215	0.084	0.513
1630	309	2.690	0.1550	2.690	0.1550	0.213	0.071	0.541
1518	414	2.830	0.1406	2.810	0.1426	0.186	0.145	0.501
1519	415	2.550	0.1708	2.500	0.1768	0.361	0.060	0.591
1520	416	2.680	0.1560	2.620	0.1627	0.239	0.302	0.489
1521	417	2.670	0.1571	2.650	0.1593	0.255	0.087	0.478
1522	418	2.720	0.1518	2.700	0.1539	0.255	0.094	0.492
1523	419	2.620	0.1627	2.560	0.1696	0.391	0.160	0.589
1524	420	2.490	0.1780	2.450	0.1830	0.485	0.060	0.540
1525	421	2.630	0.1615	2.600	0.1649	0.282	0.081	0.630
1526	422	2.580	0.1672	2.600	0.1649	0.482	0.244	0.661
1527	423	2.570	0.1684	2.560	0.1696	0.309	0.047	0.472
1528	424	2.600	0.1649	2.550	0.1708	0.287	0.284	0.495
1529	425	2.660	0.1582	2.660	0.1582	0.329	0.028	0.460
1530	426	2.580	0.1672	2.570	0.1684	0.300	0.028	0.452
1531	427	2.620	0.1627	2.600	0.1649	0.296	0.079	0.450
1532	428	2.720	0.1518	2.700	0.1539	0.221	0.157	0.488
1533	429	2.440	0.1843	2.420	0.1869	0.323	0.130	0.819
1534	430	2.640	0.1604	2.630	0.1615	0.218	0.017	0.589
1535	431	2.810	0.1426	2.800	0.1436	0.166	0.124	0.559

TABLE 5. — *Graphical measures of sand samples from coastal dunes used as a data base* — *Continued*

Principal county or area	Specific area	Computer Index No.	McKee's Index No.	Mean grain size φ	Mean grain size mm	Median grain size φ	Median grain size mm	Sorting ρ_I	Skewness Sk_I	Kurtosis K_G
Brazil	State of Rio Grande do Sul	1536	432	2.410	0.1882	2.420	0.1869	0.313	0.012	0.384
		1537	433	2.450	0.1830	2.420	0.1869	0.330	0.161	0.452
		500	14	2.603	0.1649	2.600	0.1649	0.181	0.021	0.547
		501	23	2.556	0.1696	2.550	0.1708	0.170	0.056	0.642
		502	17	2.526	0.1731	2.500	0.1708	0.130	0.208	0.505
		503	28	2.513	0.1756	2.500	0.1708	0.180	0.129	0.543
		504	34	2.513	0.1756	2.500	0.1708	0.200	0.156	0.519
		505	19	2.580	0.1672	2.560	0.1696	0.212	0.120	0.500
		506	21	2.570	0.1684	2.570	0.1684	0.170	0.069	0.542
		507	15	2.500	0.1708	2.500	0.1708	0.221	ND	0.578
		508	18	1.596	0.3299	1.590	0.3322	0.203	0.040	0.537
		509	16	2.593	0.1661	2.590	0.1661	0.186	0.013	0.535
		510	20	2.600	0.1649	2.600	0.1649	0.097	0.045	0.529
		511	30	2.580	0.1672	2.580	0.1672	0.167	0.053	0.530
		512	10	2.886	0.1350	2.800	0.1436	0.331	0.509	0.616
		513	8	2.770	0.1466	2.760	0.1476	0.209	0.124	0.677
		514	3	2.903	0.1340	2.900	0.1340	0.125	0.148	0.662
		515	11	2.840	0.1397	2.890	0.1350	0.297	0.233	0.509
		516	4	2.600	0.1649	2.603	0.1649	0.251	0.039	0.513
		517	9	2.806	0.1426	2.800	0.1436	0.288	0.044	0.461
		518	1	2.833	0.1406	2.800	0.1436	0.206	0.267	0.524
		533	51	2.636	0.1604	2.600	0.1649	0.290	0.119	0.493
		534	56	2.403	0.1895	2.400	0.1895	0.215	0.088	0.561
		535	42	2.590	0.1661	2.570	0.1684	0.216	0.198	0.557
		536	41	2.413	0.1882	2.390	0.1908	0.279	0.158	0.520
		537	43	2.486	0.1780	2.480	0.1792	0.251	0.098	0.520
		538	46	2.750	0.1487	2.750	0.1487	0.206	ND	0.601
		539	45	2.530	0.1731	2.520	0.1743	0.254	0.047	0.511
		540	50	2.476	0.1792	2.480	0.1792	0.230	0.024	0.524
		541	49	2.530	0.1731	2.510	0.1756	0.238	0.113	0.484
		542	47	2.760	0.1476	2.720	0.1518	0.196	0.207	0.505
		543	39	2.530	0.1731	2.250	0.2102	0.143	0.018	0.518
		544	44	2.510	0.1756	2.500	0.1768	0.253	0.086	0.512
		545	26	2.610	0.1638	2.600	0.1649	0.178	0.071	0.494
		546	24	2.550	0.1708	2.550	0.1708	0.760	ND	0.487
		547	36	2.360	0.1948	2.320	0.2003	0.284	0.217	0.505
		548	586	1.393	0.3896	1.310	0.4033	0.525	0.178	1.147
		552	38	2.680	0.1560	2.700	0.1539	0.185	0.079	0.528

Country	Locality									
California (U.S.A.)	Pismo Beach	530	548	3.353	0.0981	3.350	0.0981	0.388	0.007	0.484
		635	548	2.730	0.1507	2.710	0.1528	0.355	0.580	0.494
Canada	Northumberland Coast	1620	623	2.233	0.2132	2.200	0.2176	0.826	0.132	0.464
Libya	Ghat	569	623	1.823	0.2832	1.820	0.2832	0.233	0.014	0.513
Mexico	Cholla Bay	1560	31	1.833	0.2813	1.800	0.2872	0.617	0.082	0.549
New Jersey (U.S.A.)	Long Beach Island	1565	659	1.586	0.3345	1.600	0.3299	0.386	0.064	0.515
		610	658	1.680	0.3121	1.680	0.3121	0.365	0.008	0.476
		522	659	1.613	0.3276	1.680	0.3121	0.366	0.285	0.495
		613	660	1.560	0.3392	1.530	0.3463	0.400	0.122	0.514

TABLE 6. — *Graphical measures of sand samples from inland dunes used as a data base*

[Localities are listed alphabetically by country, and specific area where known. The computer index number and McKee's index number are also listed. Specific sampling information relating to dune type or position on a dune is listed under "remarks". ND indicates no data]

Principal country or area	Locality / Specific area	Computer Index No.	McKee's Index No.	Mean grain size phi	Mean grain size mm	Median grain size phi	Median grain size mm	Sorting rho I	Skewness SkI	Kurtosis K'G	Remarks
	Inland dunes										
Texas (U.S.A.)	Padre Island	628	962	2.480	0.1792	2.800	0.1436	0.270	0.141	0.500	
		627	965	2.690	0.1550	2.680	0.1560	0.310	0.076	0.421	
		629	967	2.350	0.1961	2.360	0.1948	0.470	0.683	0.685	
Algeria	Bou Saada	4019	27	2.026	0.2455	ND	ND	0.549	0.247	0.536	Crest of dune.
	Ouargla	1558	28	2.190	0.2192	2.100	0.2333	0.504	0.311	0.531	Seif dune, slipface.
Arizona (U.S.A.) ..	Coal Mountain	1559	29	1.563	0.3392	1.500	0.3536	0.409	0.321	0.612	
	Carrizo Mountains	4173	1134	3.012	0.1240	ND	ND	0.474	0.116	0.481	
	Grand Canyon	1616	566	3.056	0.1207	3.210	0.1081	0.613	0.357	0.540	
		566	566	2.703	0.1539	2.690	0.1550	0.545	0.131	0.529	
	Grand Falls	4028	37	1.195	0.4368	ND	ND	0.638	0.291	0.533	
	Hotevilla	605	570	1.900	0.2679	1.900	0.2679	0.366	ND	0.511	
	Leupp	1551	9	2.686	0.1560	2.700	0.1539	0.530	0.020	0.526	Barchan dune, swale between crests.
		1552	10	2.296	0.2045	2.250	0.2102	0.427	0.220	0.527	Barchan dune, crest.
		1553	11	2.403	0.1895	2.380	0.1921	0.516	0.136	0.528	Do.
		4010	12	2.457	0.1821	ND	ND	0.516	0.091	0.507	Do.
		4011	13	2.402	0.1892	ND	ND	0.573	0.023	0.519	Barchan dune, windward slope.
		1554	14	2.233	0.2132	2.200	0.2176	0.608	0.090	0.521	Do.
		1555	16	2.336	0.1989	2.280	0.2059	0.433	0.193	0.515	Barchan dune, crest.
	Moenkopi	634	569	2.670	0.1571	2.610	0.1638	0.480	0.230	0.468	Hanging dune, slipface.
	Page	558	568	2.036	0.2432	1.980	0.2535	0.611	0.156	0.500	Seif dune, windward slope.

Region	Locality	No.									Remarks
	Red Lake	1617	568	2.273	0.2073	2.250	0.2102	0.753	0.023	0.432	Do.
		584	654	2.293	0.2045	2.300	0.2031	0.615	0.004	0.512	Slipface of dune, at 20-ft height.
	White Sands National Monument.	4022	30	2.675	0.1566	ND	ND	0.363	0.056	0.516	
	Tonalea	1561	33	2.300	0.2031	2.200	0.2176	0.961	0.161	0.427	Crest of dune.
		4025	34	2.612	0.1636	ND	ND	0.654	-0.006	0.515	Windward slope of dune.
		4016	23	2.698	0.1541	ND	ND	0.731	0.056	0.438	Crest of dune.
	Yuma	1562	35	2.490	0.1780	2.380	0.1921	0.627	0.296	0.520	Do.
		1563	36	2.363	0.1948	2.300	0.2031	1.021	0.080	0.427	
Australia	Betoota	4071	699	2.175	0.2214	ND	ND	0.364	0.364	0.552	Seif dune.
	Canning Basin	4041	625	2.058	0.2401	ND	ND	0.574	0.299	0.513	Do.
		4042	626	2.558	0.1698	ND	ND	0.518	0.192	0.511	Do.
		4043	627	1.665	0.3153	ND	ND	0.450	0.359	0.572	Do.
		4044	628	1.746	0.2981	ND	ND	1.119	0.422	0.450	Do.
		4045	629	2.402	0.1892	ND	ND	0.540	0.267	0.519	Do.
		4046	630	1.611	0.3274	ND	ND	0.450	0.383	0.560	Do.
		4047	631	2.236	0.2123	ND	ND	0.738	0.018	0.500	Do.
		4048	632	1.979	0.2537	ND	ND	0.474	0.317	0.579	Do.
		4049	633	1.882	0.2713	ND	ND	0.264	0.134	0.557	Do.
	Monkira	578	698	2.340	0.1975	2.300	0.2031	0.354	0.243	0.531	Seif dune, crest.
Belgium	Oostende	560	620	2.433	0.1856	2.440	0.1843	0.249	0.081	0.504	Do.
California(U.S.A.)	Fallon	568	556	1.813	0.2852	1.800	0.2872	0.283	0.200	0.602	Transverse dune, 300 ft high, base of slipface.
	Twentynine Palms	528	556	1.810	0.2852	1.790	0.2892	0.303	0.196	0.563	Do.
		562	653	2.320	0.2003	2.310	0.2017	0.479	0.009	0.513	Transverse dune, slipface.
Colorado(U.S.A.)	Fort Morgan	4167	1092	1.947	0.2594	ND	ND	0.566	0.033	0.510	Climbing dune.
	Great Sand Dunes National Monument.	606	18	1.810	0.2852	1.800	0.2872	0.364	0.134	0.475	Reversing dune.
		4015	19	1.939	0.2608	ND	ND	0.321	0.035	0.531	Reversing dune, high on slipface.
		572	21	-0.146	1.1096	0.250	1.1892	0.825	0.081	0.727	Reversing dune, base.

TABLE 6. — Graphical measures of sand samples from inland dunes used as a data base — Continued

Principal country or area	Specific area	Computer Index No.	McKee's Index No.	Mean grain size φ	Mean grain size mm	Median grain size φ	Median grain size mm	Sorting ρI	Skewness SkI	Kurtosis KG	Remarks
Colorado(U.S.A.)--Con.	Great Sand Dunes National Monument.--Con.	563	643	-0.586	1.5052	0.600	1.5157	0.146	1.485	0.376	Do.
		593	644	0.833	0.5625	0.800	0.5743	0.246	0.212	0.521	Reversing dune, 300 ft above base.
		609	645	2.090	0.2339	2.100	0.2333	0.262	0.050	0.530	Reversing dune, crest.
			1211	2.000	0.2500	ND	ND	0.484	-0.157	0.588	Ripple trough.
		4177	1212	1.977	0.2540	ND	ND	0.433	0.043	0.522	Do.
		4097	723	1.455	0.3648	ND	ND	1.077	-0.072	0.447	Do.
		4178	1213	1.892	0.2694	ND	ND	0.498	-0.025	0.533	Do.
		4179	1214	2.038	0.2435	ND	ND	0.507	-0.006	0.511	Do.
		4180	1215	1.917	0.2648	ND	ND	0.443	-0.009	0.514	Do.
		4181	1216	0.453	0.7305	ND	ND	1.029	0.535	0.380	Ripple crest.
		4182	1217	ND	1.0000	ND	ND	0.484	0.292	0.585	Do.
		4183	1218	0.872	0.5464	ND	ND	1.028	-0.106	0.376	Do.
	Lamar	1557	24	1.460	0.3635	1.400	0.3789	0.706	0.199	0.554	Windward slope.
Egypt	Ayn Amur, El Kharea	4131	768	2.040	0.2432	ND	ND	1.148	-0.090	0.398	Barchan dune.
	Al Wahat al Kharijah ..	4132	769	-0.112	1.0807	ND	ND	0.769	0.493	0.521	Climbing dune.
	Ayn Amur, El Kharea ...	4133	770	0.609	0.6557	ND	ND	0.934	0.584	0.491	Do.
Germany	Wadi Aou Arta, Sinai ..	4129	766	2.085	0.2357	ND	ND	0.444	0.040	0.480	Do.
Greece	Kremberg-Berlin	1566	692	1.403	0.3789	1.380	0.3842	0.569	0.108	0.499	
India	Macedonia	577	696	3.203	0.1088	3.200	0.1088	0.395	0.047	0.517	
	Hooghly River	651	1020	3.000	0.1233	3.020	0.1250	0.290	0.798	0.536	
		652	21	3.200	0.1073	3.220	0.1088	0.388	0.079	0.509	
		653	1022	3.380	0.0954	3.390	0.0960	0.550	0.155	0.638	
Iran	Alwaz	4162	1065	3.011	0.1241	ND	ND	0.396	0.048	0.480	Produne(?).
Israel	Tel Aviv	633	686	2.503	0.1768	2.500	0.1768	0.260	0.056	0.447	
Jordan	Ram Fort	4128	765	1.434	0.3701	ND	ND	0.599	0.206	0.518	Climbing dune.
Hawaii(U.S.A.) ...	Kilauea	607	17	1.530	0.3463	1.620	0.3253	0.489	0.143	0.507	
Libya	Fezzan	641	714	2.320	0.2003	2.320	0.2003	0.400	0.003	0.530	Seif dune, 45 ft high, crest.
		581	715	1.260	0.4176	1.000	0.5000	0.630	0.613	0.518	Do.
		636	716	2.420	0.1869	2.340	0.1975	0.520	0.223	0.519	Do.

Locality	No.	No.								Remarks
	626	717	2.010	0.2483	2.000	0.2500	0.340	0.117	0.541	Seif dune, 45 ft high, high on slipface.
	587	711	2.330	0.1989	2.140	0.2269	0.534	0.515	0.531	Do.
	632	712	2.500	0.1768	2.500	0.1768	0.394	0.204	0.530	Do.
	618	713	2.320	0.2003	2.400	0.1895	0.450	0.260	0.479	Do.
	588	718	2.176	0.2207	2.150	0.2253	0.729	0.121	0.621	Seif dune, 45 ft high, middle of slipface.
	589	719	1.696	0.3078	1.700	0.3078	0.807	0.221	0.402	Do.
	520	720	1.963	0.3099	2.180	0.2207	1.103	0.696	0.413	Do.
	4094	721	2.097	0.2337	ND	ND	0.700	0.013	0.499	Do.
	617	721	2.070	0.2382	2.050	0.2415	0.880	0.030	0.487	Do.
	615	722	1.950	0.2588	1.950	0.2588	0.458	0.116	0.512	Do.
	638	724	1.810	0.2852	1.750	0.2973	0.650	0.194	0.479	Do.
	4099	724	1.581	0.3343	ND	ND	0.646	0.303	0.519	Do.
	555	701	2.280	0.2059	2.250	0.2102	0.436	0.136	0.559	Do.
	4080	708	2.003	0.2495	ND	ND	0.592	0.069	0.516	Do.
	640	709	2.040	0.2432	2.000	0.2500	0.740	0.138	0.435	Do.
	621	710	2.360	0.1948	2.380	0.1921	0.190	0.029	0.425	Do.
Fezzan—Con.	590	725	2.060	0.2398	2.100	0.2333	0.861	0.266	0.491	Seif dune, 45 ft high, base of slipface.
	4101	725	1.929	0.2626	ND	ND	0.819	0.012	0.448	Do.
	619	726	2.100	0.2333	2.050	0.2415	0.879	0.075	0.441	Do.
	580	704	1.766	0.2932	1.600	0.3299	0.825	0.321	0.405	Do.
	602	705	1.380	0.3842	1.800	0.2872	0.935	1.497	0.445	Do.
	620	706	1.260	0.4175	1.080	0.4730	1.270	0.334	0.489	Do.
	570	707	2.290	0.2045	2.510	0.1756	1.024	0.259	0.421	Do.
	567	700	1.953	0.2588	2.080	0.2365	0.753	0.173	0.445	Do.
Libya—Con.	521	723	2.133	0.2285	2.250	0.2102	0.849	0.201	0.542	Seif dune, 45 ft high, swale, fill on slipface.
	4097	723	1.455	0.3648	ND	ND	1.077	-0.072	0.447	Seif dune, slipface.
Sabha	611	662	1.620	0.3253	1.600	0.3299	0.340	0.086	0.521	Seif dune, slipface.
	604	663	1.910	0.2661	1.810	0.2852	0.357	0.288	0.517	Seif dune, windward slope.
	612	665	1.820	0.2832	1.480	0.3585	0.820	0.540	0.456	Seif dune, 75 ft high, crest.
	579	666	0.596	0.6598	0.500	0.7071	0.635	0.445	0.584	Seif dune, base.
	529	668	2.383	0.1921	2.800	0.1436	1.344	0.384	0.468	Seif dune.

TABLE 6. — *Graphical measures of sand samples from inland dunes used as a data base — Continued*

Principal country or area	Locality Specific area	Computer Index No.	McKee's Index No.	Mean grain size φ	Mean grain size mm	Median grain size φ	Median grain size mm	Sorting ρ$_I$	Skewness Sk$_I$	Kurtosis K'$_G$	Remarks
	Sedales	608	673	2.070	0.2382	2.000	0.2500	0.890	0.281	0.459	Seif dune, slipface.
Marshall Islands .	Three Mountains	614	674	1.930	0.2624	1.900	0.2679	0.576	0.142	0.514	Do.
		4065	679	2.566	0.1689	ND	ND	0.560	0.019	0.481	Seif dune.
	Mejatto Islet	582	519	1.980	0.2535	1.980	0.2535	0.318	0.071	0.507	Seif dune.
Mexico	Chihuahua	1550	6	2.593	0.1661	2.570	0.1684	0.365	0.074	0.590	Crest of dune.
Missouri(U.S.A.) .	Waverly	4004	5	3.134	0.1139	ND	ND	0.403	-0.031	0.462	
New Mexico(U.S.A.) .	Santa Rosa	4003	4	1.357	0.3904	ND	ND	0.832	0.234	0.519	
	White Sands National Monument.	4023	32	1.047	0.4840	ND	ND	0.301	0.001	0.545	
		586	825	1.636	0.3209	1.600	0.3299	0.308	0.187	0.476	Crest of dune.
		574	826	1.763	0.2952	1.700	0.3078	0.495	0.262	0.520	Windward slope of dune.
		623	827	1.570	0.3368	1.500	0.3536	0.409	0.380	0.507	
		527	829	1.263	0.4175	1.150	0.4506	0.765	0.332	0.554	Dome dune.
		565	837	1.536	0.3439	1.540	0.3439	0.467	0.025	0.533	
		575	849	1.663	0.3164	1.600	0.3299	0.442	0.268	0.521	Parabolic dune.
		532	850	1.400	0.3789	1.400	0.3789	0.930	0.038	0.551	Vicinity of star dunes(?).
		603	851	1.610	0.3276	1.600	0.3299	0.900	0.041	0.476	Parabolic dune.
		549	971	1.270	0.4147	1.220	0.4293	0.083	0.008	0.532	Dome dune, near base.
		658	971	1.220	0.4147	1.270	0.4293	0.830	0.008	0.532	Do.
		659	972	2.120	0.2238	2.163	0.2300	0.777	0.093	0.482	Do.
		550	972	2.163	0.2238	2.120	0.2300	0.777	0.093	0.482	Do.
		660	973	1.330	0.3869	1.370	0.3978	0.823	0.102	0.451	Do.
		661	974	1.910	0.2643	1.920	0.2661	0.677	0.506	0.509	Do.
		645	975	1.100	0.4665	1.000	0.5000	0.780	0.240	0.497	Do.
		644	976	1.603	0.3299	1.580	0.3345	0.890	0.076	0.514	Do.
		643	977	2.700	0.1539	2.650	0.1593	0.720	0.130	0.512	Do.
		592	977	1.490	0.3560	1.320	0.4005	0.944	0.326	0.501	Do.
		591	979	1.766	0.2932	1.720	0.3035	0.515	0.306	0.604	Parabolic dune, near crest.
		642	979	2.820	0.1416	1.760	0.2952	0.440	0.420	0.447	Do.
		637	980	1.606	0.3276	1.480	0.3585	0.590	0.420	0.584	Barchan dune. crest.

Locality	Place	Sample no.	Sample no.							Description
		662	981	1.700	0.2952	1.760	0.3078	0.292	0.039	0.530	Transverse dune, crest.
		654	1069	1.310	0.3938	1.330	0.4033	0.255	0.190	0.507	Parabolic dune, base of nose.
		655	1070	1.400	0.3711	1.430	0.3789	0.350	0.221	0.530	Parabolic dune, crest.
		656	1071	1.630	0.4118	1.280	0.3231	0.470	0.294	0.520	Barchan dune, crest.
		657	1072	1.720	0.2932	1.770	0.3035	0.350	0.310	0.644	Barchan dune, base of slipface.
		4168	1099	1.719	0.3038	ND	ND	0.482	0.315	0.594	Barchan dune, crest.
		1623	1100	1.566	0.3392	1.520	0.3487	0.429	0.221	0.564	Do.
		557	0	1.663	0.3164	1.680	0.3121	0.633	0.055	0.514	Parabolic dune, crest.
		4187	0	1.462	0.3630	ND	ND	0.495	0.427	0.532	Barchan dune.
		4188	0	2.255	0.2095	ND	ND	0.870	0.164	0.463	Do.
		4189	0	2.278	0.2062	ND	ND	0.858	0.180	0.464	Do.
		4190	0	2.005	0.2491	ND	ND	0.717	0.170	0.493	Do.
Nevada (U.S.A.) ..	Winnemucca	1564	652	2.966	0.1285	3.000	0.1250	0.452	0.088	0.489	Parabolic dune, crest.
		630	652	2.870	0.1368	2.900	0.1340	0.380	0.005	0.526	Parabolic dune.
Peru	Lomo de Corvina	4185	1226	2.133	0.2280	ND	ND	0.524	0.054	0.518	
	Otuma	4033	249	-0.667	1.5878	ND	ND	0.153	0.585	0.591	Ripples.
	Pachacamac	4184	1224	1.314	0.4022	ND	ND	0.645	0.398	0.573	
Saudi Arabia	Irq as Subay	601	984	2.670	0.1571	2.680	0.1560	0.320	0.051	0.487	Star dune, crest.
		649	988	2.130	0.2285	1.910	0.2661	0.800	0.450	0.470	Do.
		646	985	1.430	0.3711	1.040	0.4863	1.180	0.450	0.415	Star dune, base.
		647	986	2.040	0.2432	1.950	0.2588	0.660	0.266	0.547	Do.
		648	987	1.170	0.4444	1.120	0.4601	0.670	0.141	0.497	Do.
		650	989	2.330	0.1989	2.300	0.2031	0.540	0.555	0.507	Star dune, middle of slipface.
		600	993	-0.376	1.3014	0.910	1.8790	0.495	1.144	0.464	Star dune, bench on slipface.
Texas (U.S.A.)	Midland	4006	8	2.583	0.1669	ND	ND	0.533	0.203	0.525	
	Monahans	4174	1156	2.115	0.2308	ND	ND	0.271	-0.008	0.486	Base of slipface.
	Orange	4001	1	1.584	0.3336	ND	ND	0.511	-0.096	0.507	
	Signal Peak	551	3	2.640	0.1604	2.650	0.1593	0.193	-0.057	0.520	
		1549	3	2.083	0.2365	2.030	0.2449	0.448	0.190	0.521	
Tunisia	Bir Sultan	554	693	3.386	0.0954	3.360	0.0974	0.225	0.226	0.566	Seif dune, middle of slipface.
Utah (U.S.A.)	Great Salt Lake	4175	1180	1.650	0.3186	ND	ND	0.422	0.116	0.536	
		4186	1234	1.629	0.3233	ND	ND	0.471	0.141	0.500	

TABLE 6. — *Graphical measures of sand samples from inland dunes used as a data base — Continued*

Locality		Computer Index No.	McKee's Index No.	Mean grain size		Median grain size		Sorting ρ_I	Skewness Sk_I	Kurtosis K_G	Remarks
Principal country or area	Specific area			φ	mm	φ	mm				
Kanab		624	846	2.420	0.1869	2.400	0.1895	0.240	0.193	0.502	Base of slipface.
		631	847	2.590	0.1661	2.560	0.1696	0.390	0.202	0.487	Crest of dune, 20 ft high.
		571	848	2.476	0.1792	2.390	0.1908	0.356	0.400	0.485	Windward slope of dune.
Lund		625	848	1.440	0.3686	1.390	0.3816	0.287	0.332	0.514	Do.
		1614	243	3.276	0.1037	3.280	0.1029	0.468	0.019	0.575	Do.
		1615	245	2.976	0.1276	2.980	0.1267	0.492	0.019	0.593	Slipface of dune.
Monument Valley		4172	1106	2.533	0.1728	ND	ND	0.727	-0.035	0.495	Hanging dune.
Mount Carmel Junction .		1618	571	2.316	0.2017	2.250	0.2102	0.555	0.238	0.495	Produne(?), slipface.
San Rafael Swell		583	571	1.993	0.2517	1.950	0.2588	0.283	0.272	0.545	Do.
		4170	1101	2.394	0.1903	ND	ND	0.644	0.444	0.521	Barchan dune, crest.
		4171	1102	2.390	0.1908	ND	ND	0.650	0.476	0.530	Barchan dune, base of slipface.

Interdune and serir sands

TABLE 7. — *Graphical measures of sand samples from interdune and serir deposits used as a data base*

[Localities are listed alphabetically by country, and specific area where known. The computer index number and McKee's index number are also listed. ND indicates no data]

Locality		Computer Index No.	McKee's Index No.	Mean grain size		Median grain size		Sorting rho I	Skewness SkI	Kurtosis K'G
Principal country or area	Specific area			phi	mm	phi	mm			
Algeria	Ouargla	4018	26	1.935	0.2615	ND	ND	1.560	-0.471	0.404
Egypt	Ayn Amur	4130	767	-0.683	1.6055	ND	ND	0.179	0.702	0.668
Jordan	Ram Fort	594	685	1.403	0.3789	1.200	0.4353	1.177	0.318	0.489

Country	Locality									
Lybia	Fezzan	622	702	2.190	0.2192	2.200	0.2176	0.812	0.015	0.507
		639	703	2.350	0.1961	2.490	0.1780	0.710	0.240	0.492
		556	727	2.370	0.1934	2.350	0.1961	0.971	0.006	0.422
		616	728	1.570	0.3368	1.290	0.4090	1.180	0.316	0.421
		4106	729	1.754	0.2965	ND	ND	0.913	0.117	0.439
		4107	729	1.578	0.3349	ND	ND	0.936	0.425	0.462
		595	730	2.333	0.1989	3.000	0.1250	1.630	0.208	0.371
		4109	730	2.279	0.2060	ND	ND	1.569	-0.758	0.334
		597	731	2.213	0.2161	2.000	0.2500	1.456	0.169	0.360
		4111	731	1.684	0.3112	ND	ND	1.392	0.541	0.361
		561	732	2.003	0.2500	1.700	0.3078	1.420	0.261	0.366
		573	733	1.839	0.2793	1.400	0.3789	1.253	0.435	0.382
		585	734	1.450	0.3660	0.700	0.6156	1.348	0.860	0.441
		4115	734	1.155	0.4491	ND	ND	1.170	0.725	0.731
		523	735	2.886	0.1350	3.680	0.1560	1.305	0.641	0.486
		4117	735	2.412	0.1879	ND	ND	1.408	-0.565	0.362
		564	736	1.563	0.3392	1.400	0.3789	0.848	0.087	0.346
		524	737	2.670	0.1571	3.400	0.0947	1.386	0.667	0.356
		598	738	1.230	0.4263	0.350	0.7846	1.616	0.540	0.668
		4121	739	2.468	0.1807	ND	ND	1.397	-0.748	0.363
		4122	739	2.044	0.2425	ND	ND	1.472	-0.294	0.343
		4123	740	1.677	0.3127	ND	ND	1.553	0.067	0.377
	Ghat	4124	743	0.961	0.5137	ND	ND	1.464	0.273	0.459
		599	744	1.560	0.3392	0.800	0.5743	1.671	0.497	0.333
		519	745	2.450	0.1830	2.850	0.1387	1.390	0.369	0.342
		576	746	2.000	0.2500	1.900	0.2679	1.417	0.082	0.351
		559	675	2.006	0.2483	2.620	0.1627	1.636	0.448	0.368
	Sabha	553	664	1.553	0.3415	1.550	0.3415	0.940	0.146	0.505
		525	667	2.300	0.2031	2.950	0.1294	1.588	0.515	0.434
	Sedales	4061	672	1.769	0.2934	ND	ND	1.431	-0.080	0.462
	Three Mountains	596	680	2.236	0.2117	2.000	0.2500	1.164	0.246	0.391
		526	681	2.210	0.2161	2.150	0.2253	1.113	0.068	0.370
New Mexico(U.S.A.)	White Sands National Monument	4137	828	2.047	0.2420	ND	ND	1.035	0.126	0.510
Peru		4191	0	2.702	0.1537	ND	ND	0.820	-0.113	0.460
	Pampe Corre Viento, Lomitas	4031	247	-0.093	1.0666	ND	ND	0.694	0.701	0.378
		4032	248	0.787	0.5795	ND	ND	1.411	0.251	0.387
Tunisia	Bir Sultan	1621	693	2.416	0.1882	2.350	0.1961	0.941	0.056	0.405

A STUDY OF GLOBAL SAND SEAS

DIFFERENTIATION OF EOLIAN DEPOSITS BY DISCRIMINANT ANALYSIS

Chapter C

By R. J. MOIOLA[1] and A. B. SPENCER[1]

Contents

[1] Mobil Research and Development Corp., P.O. Box 900, Dallas, Texas 75221.

Illustrations

Summary of Conclusions

𝒜 LTHOUGH IT WAS NECESSARY to use textural parameters (mean, standard deviation, skewness, kurtosis) –which in themselves are an abridgment of the grain-size distribution — as input variables, inland and coastal eolian deposits were effectively differentiated by discriminant analysis. Previous studies have demonstrated that the weight fractions of quarter phi grain-size analyses are the preferred input for discriminant analysis (Moiola, Spencer, and Weiser, 1974). Therefore, if quarter phi data were available, even better separation of these deposits could be expected. This conclusion is supported by the excellent separation obtained for the inland dune and interdune deposits of Fezzan, Libya, using quarter phi grain-size data.

These results confirm the contention of Moiola, Spencer, and McKee (1973) and Moiola, Spencer, and Weiser (1974) that discriminant analysis is perhaps the most effective method for differentiating modern eolian sands by their textural characteristics.

Introduction

VARIOUS GRAIN-SIZE TECHNIQUES are available for differentiating sand bodies (Pettijohn, Potter, and Siever, 1972). These techniques are primarily attempts to classify and assign sets of unknown samples to one of two or more populations. Mathematically, they have become progressively more sophisticated, beginning with simple visual comparison of histograms, to plotting textural parameters, to graphical analysis of cumulative curves, and finally to such techniques as discriminant analysis.

Perhaps the most popular approach to date has been the use of textural parameters plotted as variables on a two-dimensional coordinate system (Mason and Folk, 1958; Friedman, 1961, 1967). Unfortunately, this use of textural parameters has met with only partial success (Shepard and Young, 1961; Schlee, Uchupi, and Trumbull, 1964; Gees, 1965; Sevon, 1966; Moiola and Weiser, 1968). For example, Moiola and Weiser (1968) demonstrated that certain combinations of textural parameters, such as mean diameter and skewness, can be used to differentiate certain sand deposits, but textural

parameters could not differentiate beach and coastal dune sands and could only poorly differentiate river and inland dune sands.

Recent studies by Sahu (1964), Sevon (1966), Greenwood (1969), Tillman (1971), Moiola, Spencer, and McKee (1973), and Moiola, Spencer, and Weiser (1974) suggest that discriminant analysis is probably the most effective grain-size technique for differentiating modern sand bodies. In this study, the discriminant technique is used to differentiate inland dune, coastal dune, and interdune deposits.

Discriminant Analysis

DISCRIMINANT ANALYSIS is a classifying technique that differentiates distinct populations by using a number of independent attributes. It was developed by Fisher (1936) to solve a taxonomic problem. The method uses a classifying function to assign samples individually to one of two or more populations. A priori knowledge that the populations differ is a prerequisite for its use. The technique defines the coefficients of a function, which allows samples described by a number of independent variables to be projected onto a line so that the distance between projected populations is maximized relative to the scatter within each projected population.

The detailed mathematical treatment of the technique is not presented here but can be found in a number of standard references, such as Wilks (1962), Miller and Kahn (1962), and Krumbein and Graybill (1965). Its application to grain-size data has been amplified by Moiola, Spencer, and Weiser (1974). The independent variables can be the weight fractions of each size class, textural parameters, or other measurable attributes. Both weight fractions of quarter phi grain-size analyses and textural parameters calculated from sieve data according to the scheme of Folk and Ward (1957) are used in this study.

Application

THE EFFECTIVENESS of linear discriminant analysis for differentiating modern beach, river, inland dune, and coastal dune sands has been demonstrated by Moiola, Spencer, and Weiser

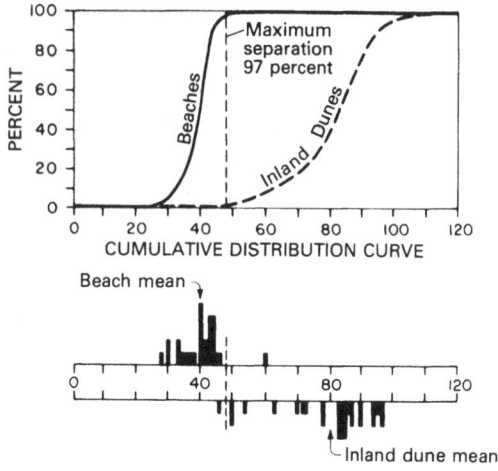

A

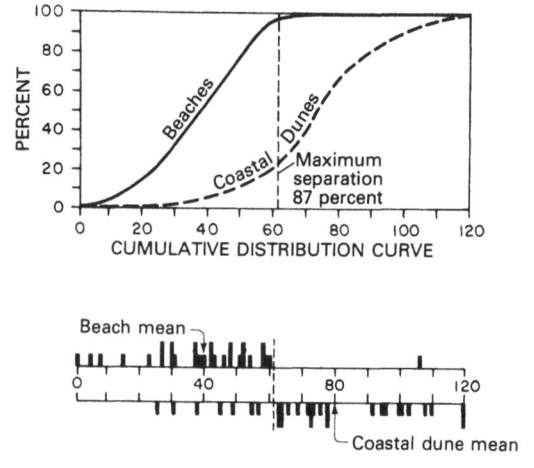

B

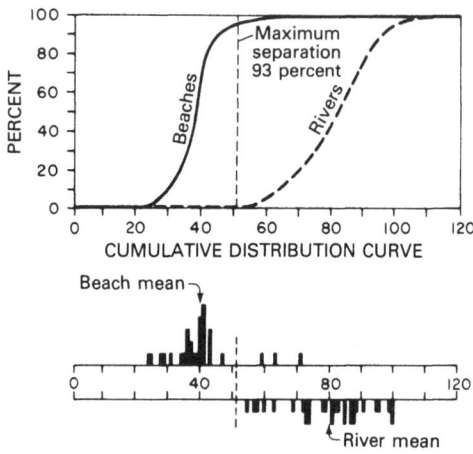

C

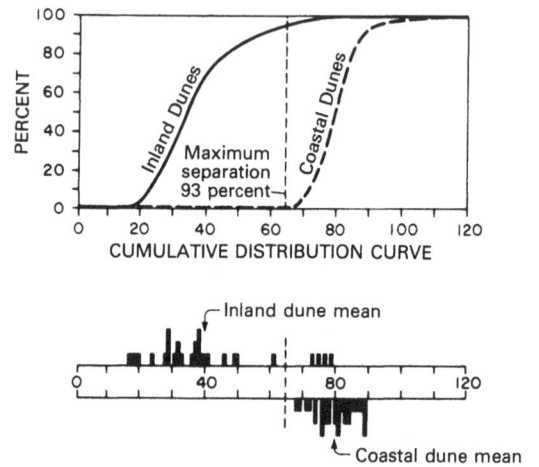

D

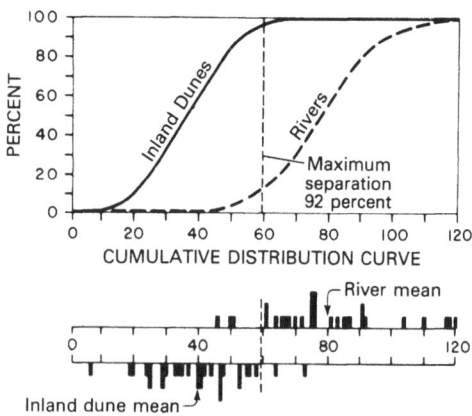

E

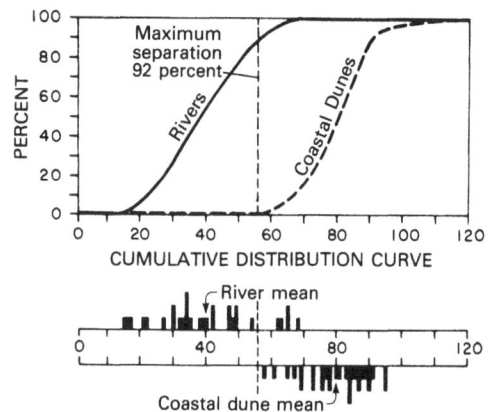

F

DISCRIMINANT FREQUENCY plots and cumulative distribution curves of beach, river, coastal dune, and inland dune samples (modified from Moiola, Spencer, and Weiser, 1974). Scale of 0 −120 is arbitrary. (Fig. 28.)

(1974). In their study, using whole phi analyses of sets of 30 samples from each environment, the differentiation obtained between pairs of sets was no less than 87 percent (fig. 28). They also demonstrated that linear discriminant analysis using whole phi data is at least as effective as the best combination of textural parameters based on quarter phi data in differentiating sands from these environments. In addition, the technique successfully differentiated beach and coastal sands, as well as river and inland dune sands, whereas textural parameter plots did not.

In this study various eolian deposits are differentiated. A selected suite of 152 inland dune, 291 coastal dune, and 38 interdune samples, compiled by Ahlbrandt (chapter B), is the data base. The discriminant function used is of the following form:

$$z = c_1x_1 + c_2x_2 + \ldots c_nx_n + b,$$

where z is the transformed sample value; $c_1, c_2 \ldots c_n$ are the coefficients of the discriminant function; $x_1, x_2 \ldots x_n$ are the descriptive variables; and b provides scaling so that the means of the two projected populations fall at 40 and 80 on an arbitrarily chosen scale of 0–120.

Because weight fractions of each size class, which are ordinarily used, were not available for the set of coastal dunes, it was necessary to find alternative variables for discriminant analysis. Textural parameters were selected. A sub-set of eolian samples from Fezzan, Libya, for which quarter phi weight fractions were available, was used to determine whether associated inland dune and interdune deposits could be differentiated.

Results

BIVARIANT TEXTURAL PARAMETER plots have not proven particularly effective in separating environments in general. The parameters themselves, when used as descriptive variables in discriminant analysis, have been found to be sensitive descriptors (Sevon, 1966). Results of the present study support this contention.

Using Folk and Ward's (1957) approximations of the mean, standard deviation, skewness, and kurtosis of the grain-size distribution as input for linear discriminant analysis, sets of inland dune and coastal dune samples can be easily differentiated (fig. 29). The population of projected coastal dune samples tightly clusters around its mean at 80 on the arbitrarily chosen scale of 0–120. Although the population of projected inland dunes is loosely dispersed about its mean at 40, maximum separation (91 percent) occurs at approximately 69 on the scale, with only 24 coastal dune samples and 13 inland dune samples from the total population of 443 falling outside their fields. Comparison of the relative dispersions of the coastal and inland dune populations suggests that the textural character of

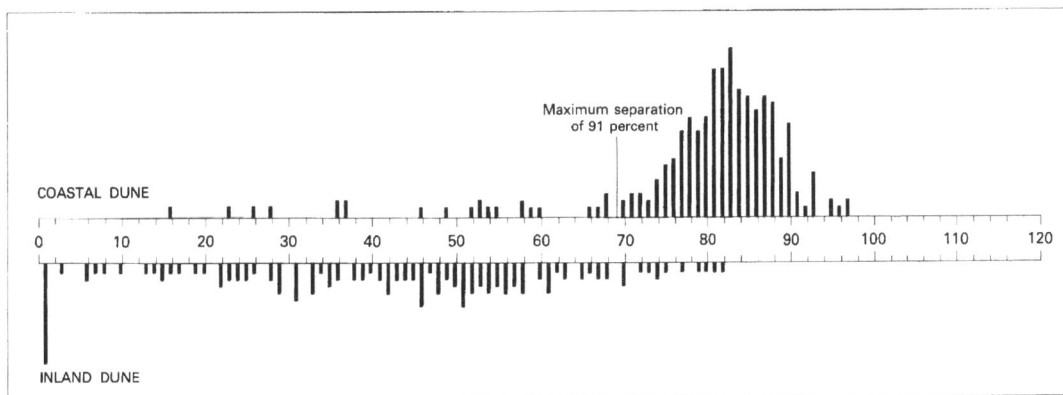

DISCRIMINANT FREQUENCY plot of 152 inland dune and 291 coastal dune samples. Inland dune samples indicated below abscissa; coastal dune samples indicated above abscissa. Maximum separation of 91 percent occurs at 69 on the arbitrarily chosen scale of 0–120.(Fig. 29.)

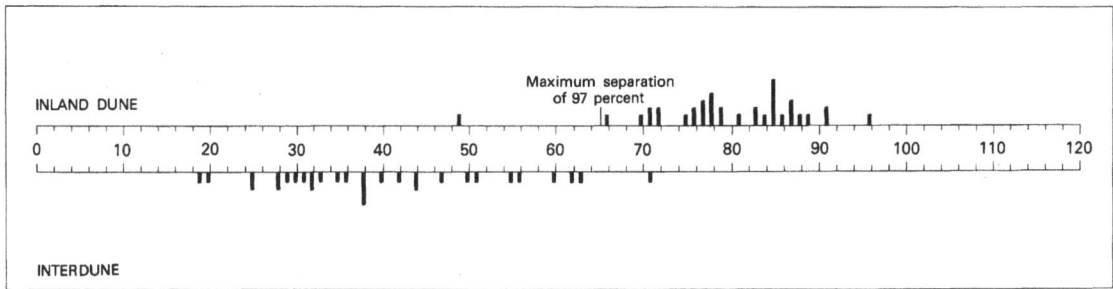

DISCRIMINANT FREQUENCY plot of 31 interdune and 38 inland dune samples. Interdune samples indicated below abscissa ; inland dune samples indicated above abscissa. Maximum separation of 97 percent occurs at 65 on the arbitrarily chosen scale of 0 −120. (Fig. 30.)

inland dune sands is more variable than that of coastal dunes. (See Ahlbrandt, chapter B). This may reflect the more varied source of inland eolian deposits.

The effectiveness of the linear discriminant technique in differentiating the intimately associated eolian subenvironments, such as dune and inter-dune, is shown in figure 30. Of the 69 samples in the combined set of dune and interdune samples from Fezzan, Libya, only one interdune sample out of 31 and one dune sample out of 38 fall outside their respective fields. Thus, 97 percent of the samples are successfully differentiated with maximum separation occuring at 65 on the scale.

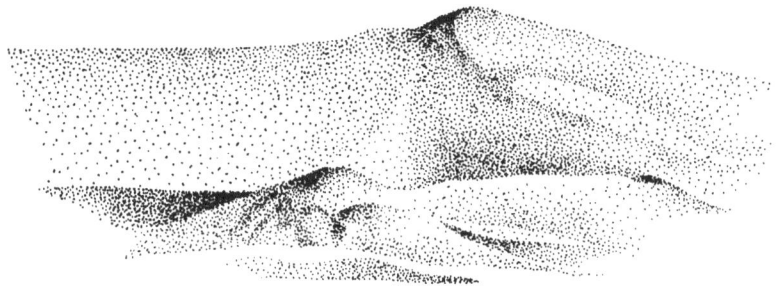

A STUDY OF GLOBAL SAND SEAS

RED COLOR IN DUNE SAND

Chapter D

By THEODORE R. WALKER[2]

Contents

[2] University of Colorado, Boulder, Colorado 80309.

Illustrations

Table

Summary of Conclusions

Œ OLIAN DEPOSITS in western Libya provide an outstanding example of modern dune sands that have reddened with age and with distance of transport. The pigment in the dunes is chemically precipitated iron oxide, which is derived in part from alteration of ferromagnesian silicate minerals and in part from iron-bearing clay minerals that adhere to the surfaces of sand grains. The clay minerals of the surface coatings are derived mainly from airborne dust mechanically infiltrated into the dunes and into other surficial deposits from which the dune sands are derived.

To the extent that iron-bearing minerals are not in equilibrium with meteoric water, they react with dew, rain, and interstitial water, and release iron, which, owing to the oxidizing and alkaline chemical environment, precipitates on the surfaces of the sand grains as yellowish ferric hydrate (limonite). Subsequently, as these grains are abraded during eolian transport, the limonite and limonite-stained clay are removed from exposed surfaces of coarser grains, but these materials are retained in indentations where they are protected from abrasion. The fine grains are abraded much more slowly than the coarse ones and therefore tend to have more indentations in which to preserve the pigment. Thus, the finer grains usually are more heavily stained than are the coarser grains. Conditions that favor the acquisition of additional iron oxide and clay coatings on the surface of grains are likely to occur whenever the grains are at rest.

The light-yellow ferric hydrate is metastable, and, upon aging under the chemical conditions existing in this desert region, it ultimately converts to hematite. The aging of ferric hydrate is accompanied by reddening; hence, the sands become redder with increasing age and distance of transport. Although aging is prerequisite to the reddening of the dune sands, the time interval involved is also affected by such factors as the amount of moisture available, availability of iron, and amount and type of clay minerals retained on grains. The effects of these factors are regionally variable; therefore, although the dunes redden with increasing age, the intensity of redness is not a reliable measure of the age of the deposits, and it is not a reliable criterion for correlation.

Each step in the sequence of events leading to the accumulation of the stain and its gradual reddening can be documented in deposits that are forming in the desert today. This important fact indicates that the reddening is a product of the modern desert environment, not the product of an alleged earlier moist climate.

Introduction

A COMMONLY OBSERVED feature of deserts is that dune sands have a reddish hue and become increasingly redder with distance of transport and with increasing age (Norris, 1969). The origin and climatic significance of the pigment in such dunes has been a subject of considerable controversy among geologists for many decades. What is the nature of the pigmenting material and how and when is it accumulated on the grains? Does the pigment reflect the reworking of sands which developed their color during earlier, much more moist climates than those existing now (as for example, during Pleistocene pluvial intervals), or is it the product of processes that are going on at present in deserts? Is the intensity of redness a reliable measure of the age of dunes, or are factors other than age equally important in the reddening? Hopefully, the evidence presented in this chapter will help to answer these questions.

Excellent examples of dune sands that show reddening in the direction of transport are the Ṣaḥrā' Awbārī and Ṣaḥrā' Marzūq of western Libya (fig. 31). In the author's opinion, these and other dune sands in Libya convincingly document successive stages in the development of red pigment, and they are believed to form models that illustrate the processes of dune-reddening active today in hot desert regions throughout the world.

The evidence presented here supports the conclusions of Norris (1969) that much of the iron in the pigment is derived internally from the dunes and that processes leading to the reddening are active in modern deserts. The evidence also supports the conclusion of Folk (1969) that airborne dust plays an important role in the reddening of dunes. The dunes studied by Folk, in the Simpson Desert of Australia, however, are considered by him to have attained redness from dust that initially was red and was derived from red lateritic soils

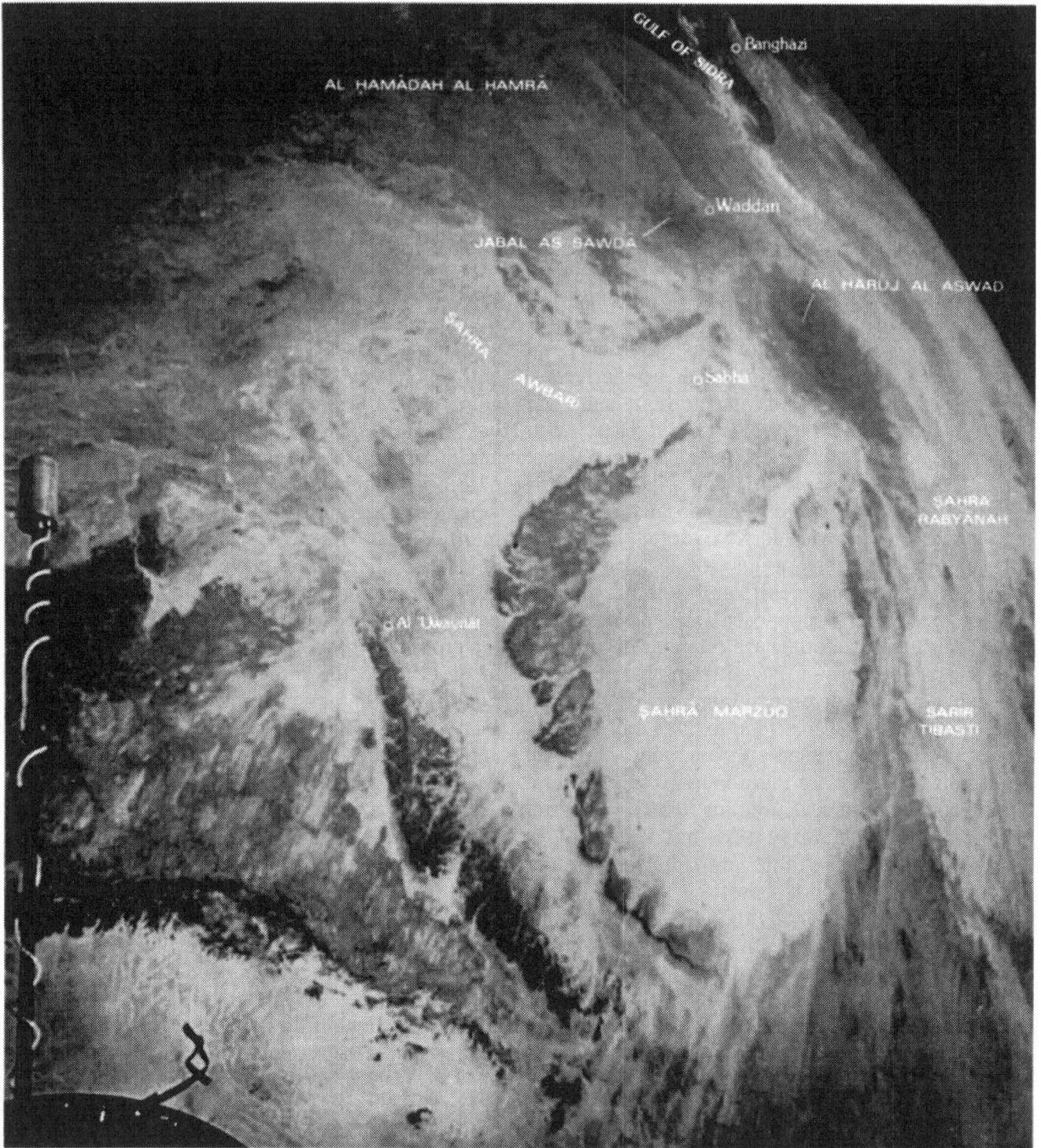

GEMINI XI SPACE PHOTOGRAPH of Libya. Northeastward view from above eastern Algeria. Prevailing wind is from northeast (upper right). Note marked increase in redness of Ṣaḥrā' Awbārī and Ṣaḥrā' Marzūq in direction of sand transport. (Fig. 31.)

developed under a former, more humid climate. In contrast, evidence presented here indicates that dust involved in the reddening of dunes in the Sahara initially was not red, but has become red in the existing arid climate.

Acknowledgments

This research was supported by National Science Foundation Grant GA –20738 and the University of

Colorado Council on Research and Creative Work. John C. Harms (Marathon Oil Co.) and James R. Steidtmann (University of Wyoming) critically reviewed the manuscript. A. J. Crone provided valuable laboratory assistance. All of this aid is gratefully acknowledged.

Regional Setting

Moisture

WITH THE EXCEPTION of a narrow belt of arid steppe climate along the Mediterranean coast, western Libya is mostly a region of severe desert (fig. 32). Average annual rainfall along the coast westward from the Gulf of Sidra is about 200 mm (8 in.), and the amount diminishes markedly within a short distance inland. Along the southern margin of the Gulf of Sidra the desert begins almost at the coastline (fig. 32). The central part of the country lies in the driest part of the Sahara, where, at times several years may pass with no rainfall whatever. However, nearly all areas — even those in the driest parts of the desert — receive some rainfall (Fürst, 1970), and even seemingly negligible amounts become important geologically because over long periods they can provide enough moisture to cause mineral alterations that result in reddening. In addition, diurnal temperature changes commonly are extreme in the Sahara, and they lead to frequent occurrences of heavy dew. Although apparently insignificant, such small amounts of moisture, when replenished over and over again, may play an important role in the reddening processes described herein. The importance of dew as an agent of mineral alteration in deserts has been stressed by many previous writers (among them, Engel and Sharp, 1958; Kuenen and Perdok, 1962; Margolis and Krinsley, 1971).

Winds and Eolian Transport

Throughout most of Libya prevailing winds blow from the northeast except near the coast, where westerlies prevail during the winter months. The major direction of eolian transport into the interior of the desert, therefore, is southwestward. Accordingly, the major sand seas that dominate the landscape in southwestern Libya (Şaḥrā' Awbārī and Şaḥrā' Marzūq) are composed of sand derived principally from sources lying northeastward toward the Gulf of Sidra and in northeastern Libya (figs. 31, 32). In addition, the Şaḥrā' Awbārī, particularly its northern arm, probably receives a significant contribution of sand blown southwestward from the coastal steppe region across the Al Ḥamādahal Ḥamrā plateau (figs. 31, 32).

Bedrock

Bedrock throughout Libya consists dominantly of non-red sedimentary rocks of Paleozoic to Tertiary age overlain locally by Tertiary volcanic rocks (fig. 32). All geologic periods are represented by the sedimentary rocks, but those rocks of Cretaceous and Tertiary age are the most important insofar as the present discussion is concerned. They are areally dominant, and they provide the major bedrock sources of the eolian deposits. The Cretaceous and Tertiary sedimentary rocks consist mainly of marine limestones and shales but locally contain interbedded continental deposits of sandstone and shale (Conant and Goudarzi, 1964).

The Cretaceous and Tertiary sandstones are mostly orthoquartzites and probably provide a major source of the quartz sand which dominates the dunes. Cretaceous and Tertiary limestone, shales, and volcanic rocks provide sources of sand-size rock fragments, but grains of these lithologies in most places constitute only a very small percentage of the dune sands.

Tertiary volcanic rocks are the probable major sources of unstable ferromagnesian silicate minerals, such as augite, hornblende, and epidote, which occur in the heavy-mineral fractions of dune sands at all the sampled localities. In addition, shales, sandstones, and, to a lesser degree, limestones are the ultimate sources of airborne dust, which, as will be shown in the subsequent discussion, is a significant factor in the development of red pigment in the dune sands. The bedrock formations include some red beds of Mesozoic age, but these beds are of limited occurrence and are relatively unimportant as sources of pigment in the dune sands.

Soils

In most places bedrock is overlain by a veneer of surficial deposits, which are mainly light red eolian

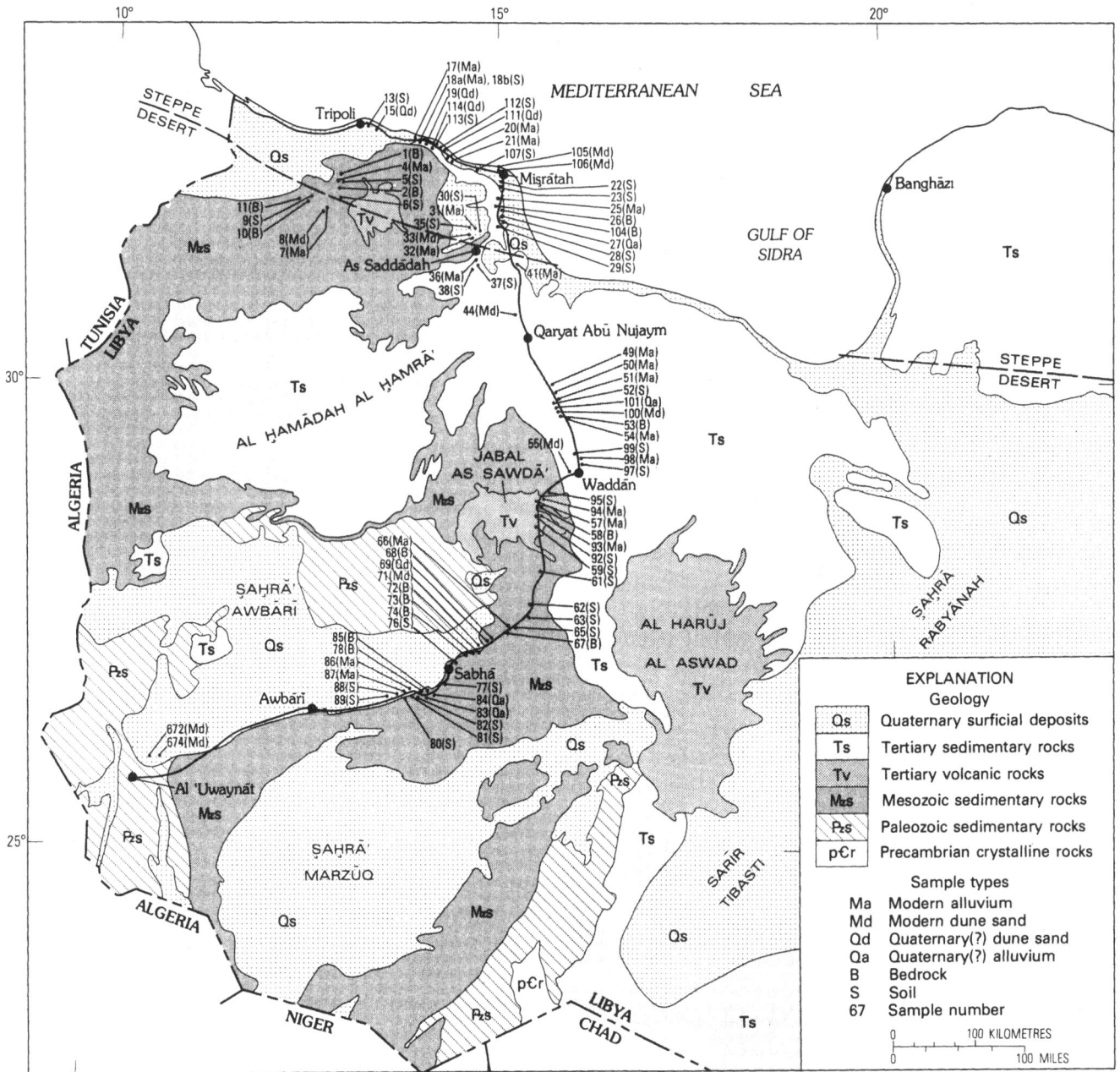

GENERALIZED GEOLOGIC MAP of western Libya, showing sample localities. Geology generalized from Conant and Goudarzi (1964). (Fig. 32.)

sands. Where migration of these deposits has been temporarily halted either by vegetation, as in many places in the coastal region, or by the development of protective gravel pavements, as on deflation surfaces in the interior of the desert, and on the upland surfaces of dissected alluvial fans, weakly developed but strikingly reddish (2.5 YR to 5 YR, Munsell Soil Color Chart, Munsell Color Co., 1954) soils have been formed. These soils belong to the

great soil group known as "Red Desert Soils" and, as evidenced by the following characteristics, they attest to long-term regional aridity.

The soils are calcareous throughout their profiles, and in places near the coast they contain prominent concentrations of carbonate deposits (caliches) which commonly are a metre (3 ft) or more thick. Somewhat farther inland, where aridity is greater (as in the area beginning about 75 km (47 mi) south of Miṣrātah and extending to Qaryat Abū Nujaym), caliche zones are composed mainly of gypsum. Gypsum caliches also occur nearby in the desert regions of Tunisia (Page, 1972). In most places the soils also contain sand- and pebble-sized rock fragments of limestone, and in the coastal region they typically contain grains of carbonate shell debris.

The occurrence of carbonate and gypsum caliche, and the presence of carbonate clasts, imply an absence of significant leaching during soil formation. Moreover, soils in the more arid regions commonly contain palygorskite as a major constituent of the clay fraction (table 8). This clay mineral requires an alkaline environment for development and survival. The importance of all these characteristics can hardly be over-emphasized because they indicate that the soils formed under nonleaching, alkaline conditions. The nature of the soils provides evidence that for at least as long as the age of these soils, which presumably extends from early Pleistocene to Holocene, the climate in this part of the desert has not been significantly more moist than it is at present. Accordingly, the characteristics of the soils provide strong evidence that the associated reddish pigment, is the product of an arid climate.

TABLE 8. — *Clay minerals identified in samples analyzed for this study*

[Sample localities are shown in figure 31. Clay minerals are listed in order of decreasing abundance. Capital letter symbols indicate major constituent; lower-case letter symbols indicate minor constituent; parentheses, trace amount. K, kaolinite; I, illite; ML-IM mixed-layer illite montmorillonite; P, palygorskite]

Sample	Clay minerals	Age and rock type
		Bedrock
1	I, ml-im	Triassic shale
2	K, I, ml-im	Cretaceous shale
10	I, K, ML-IM	Do.
11	K, I, ml-im	Do.
26	ML-IM, I, K	Miocene shale
53	I, K, ml-im	Tertiary clay
58	K, ML-IM, i	Cretaceous shale
67	K, i, ml-im	Cretaceous sandstone
68	K, P, I, ml-im	Tertiary sandstone
72	K, ml-im	Cretaceous claystone
73	K, I, ml-im	Cretaceous sandstone
74	K, I, ml-im	Cretaceous shale
78	K, i, ml-im	Cretaceous sandstone
85	K, i	Cretaceous conglomerate
104	I, K, ml-im	Tertiary(?) mudstone

TABLE 8 — Continued

Sample	Clay mineral	Sample	Clay mineral
		Soils	
5	I, K, ml-im	63	P, K, i
6	I, K, ml-im	65	P, K, I, ml-im
9	I, K, ml-im	76	K, I, ml-im
13	I, K, ml-im	77	K, I, ml-im
18b	I, K, ml-im	80	I, K, ml-im
22	I, K, ml-im	81	P, K, i, ml-im
23	I, K, ml-im	82	P, K, i, ml-im
28	I, K, ml-im, p	88	K, P, i, ml-im
29	P, K, i, ml-im	89	I, K, ml-im
30	P, K, i, ml-im	92	I, K, ml-im
35	P, I, K, ml-im	95	I, K, ml-im
37	P, I, K, ml-im	97	I, K, ml-im
38	P, K, I, ml-im	99	I, K, ml-im
52	I, K, ml-im	107	I, K, ml-im
59	I, K, ml-im, (p)	112	I, K, ml-im
61	I, K, ml-im, (p)	113	I, K, ml-im
62	I, K, ml-im		
		Modern alluvium	
4	I, K, ml-im	49	P, I, K, ml-im
7	I, K, ml-im	50	I, K, ML-IM
17	I, K, ml-im	51	I, K, ML-IM
18a	I, K, ml-im	54	I, K, ml-im
20	I, K, ml-im	57	I, K, ML-IM
21	I, K, ml-im	66	I, K, ml-im
25	I, K, ml-im	86	K, I, ml-im
31	I, K, ml-im, p	87	I, K, ML-IM
32	I, K, ml-im	93	ML-IM, I, K
36	I, K, ml-im, (p)	94	I, K, ML-IM
41	I, K, ml-im	98	I, K, ML-IM
		Modern dune sand	
8	K, I, ml-im	100	I, K, ml-im
33	I, P, K, ml-im	105	K, I, ml-im
44	I, K, ml-im	106	K, I, ml-im
55	I, K, ml-im	672	K, I, ML-IM
71	K, i, ml-im, (p)	674	K, I, ML-IM
		Quaternary(?) dune sand	
15	I, K, ml-im	111	I, K, ml-im
19	I, K, ml-im	114	I, K, ml-im
69	P, K, I, ML-IM		
		Quaternary(?) alluvium	
27	K, I, p, ml-im	84	K, I, ml-im
83	K, I, ML-IM	101	I, K, ml-im

Distribution of Red Sands

Eolian sands that show pronounced reddening in the direction of transport occur in two parts of western Libya. One is in the belt of arid steppe climate that extends inland for several tens of kilometres along the Mediterranean coast, west of the Gulf of Sidra (fig. 32). Here, sand is blown landward from the beach and piled into dunes which, through time, migrate further inland. The beach sands and dunes adjacent to the beach are not even slightly red, but the dunes become progressively redder inland, and within a few kilometres of the coast they have reddish hues ranging from 5YR to a maximum redness of about 2.5YR, values from 5 to 7, and chromas from 4 to 8, as determined by comparisons with the Munsell Soil Color Chart (Munsell Color Co., 1954). The most commonly occurring colors are designated as "reddish yellow" and "yellowish red" on that chart.

A second region of red dunes is that occupied by the sand seas called Şaḥrā᾽ Awbārī and Şaḥrā᾽ Marzūq in the intensely arid interior of the desert (figs. 31, 32). Most of the sand in these two dune fields has been transported from sources that lie several hundred kilometres to the northeast in the very arid coastal region south of the Gulf of Sidra. These sands, like those in the northern coastal region of western Libya, redden progressively in the direction of transport, but, because of the greater aridity of this region, the distance of transport required to achieve a comparable degree of redness is much greater here.

The full range of color change in the interior deposits was not studied because those sands were not traced to their source area. However, in the area between Waddān and Sabhā (figs. 31, 32), which lies at the upwind end of the sampled area, the color of the sands range from about 8YR 7/4 (pink) to about 6.5YR 6/6 (reddish yellow), whereas near Al 'Uwaynāt, which lies near the downwind end of the southern arm of the Şaḥrā Awbārī (fig. 32) the color is about 4YR 5/8 (yellowish red). The distance across which this amount of reddening occurs is nearly 700 km (435 mi), or about 10 times the distance necessary to achieve approximately the same degree of redness in the northern coastal region of western Libya.

Characteristics of the Pigment

WHEN SAMPLES of the stained dune sands from both of the regions studied are examined under a binocular microscope, the pigment is seen to occur, in part, as extremely thin films which generally form complete coatings on grains and in part as thicker concentrations which lie within indentations on grains, where the material has been protected from abrasion. The latter material charac-

teristically is the reddest. Moreover, fine and very fine grains commonly are more heavily stained and redder than are coarser grains.

The mineralogy, texture, and chemical composition of both the thin films and thicker concentrations of pigment were determined by combining X-ray-diffraction techniques with scanning electron microscope (SEM) and energy-dispersive X-ray analysis. Representative samples of dunes showing different degrees of redness were treated in the following manner. Between 10 and 20 grams of sand were split from the bulk sample, and this portion was washed and treated in an ultrasonic cleaner until enough of the pigmenting material was available for X-ray analysis. Untreated portions of the bulk sample were examined under a binocular microscope; several grains which showed typical staining characteristics were selected from each sample, and the grains were mounted on SEM stubs.

After mounting the grains on stubs — but prior to coating them with a conductive film of gold and paladium — each grain was again examined under the binocular microscope, and a sketch was drawn, showing the distribution of concentrations of the pigmenting material. These sketches were used during the SEM examination as maps for locating areas on the grains that would be particularly suitable for detailed study. All of the mounted grains from each sample were then examined using a Cambridge Stereoscan SEM.[3] One or more representative grains from each was photographed at magnifications ranging from about × 200 to about × 11,000. Finally, using a Kevex X-ray energy spectrometer attached to the SEM, semiquantitative chemical analyses of the pigmenting material were made at locations that could be identified on the photomicrographs. Representative results are illustrated in figures 33 and 34 and 37 – 40.

Pigment in the thin films apparently is composed of amorphous ferric hydrate ("limonite") because, although energy-dispersive X-ray analyses show that they contain iron (fig. 33, trace 2), they show no evidence of crystallinity when viewed under the SEM at magnifications of × 50,000 and higher.

---

[3] Trade names and company names are included for the benefit of the reader and do not imply endorsement of the product or company by the U.S. Geological Survey.

Moreover, few X-ray-diffraction analyses of concentrates of the pigmenting material reveal the presence of crystalline iron oxide minerals. Some of the red material concentrated in indentations on grains is composed entirely of essentially pure iron oxide (fig. 34), but most commonly it is composed of iron oxide stained platelets of clay minerals, the composition of which is confirmed by energy-dispersive X-ray analyses. Such analyses reveal that most of the platelets are aluminum silicates (figs. 38 – 40), although some platelets are composed of pure silica (fig. 33). The clay-mineral and silica-platelets occur in aggregates that have clastic textures similar to those displayed by clay mechanically infiltrated into sand (Crone, 1974, 1975; Walker and Crone, 1974; Walker, 1976). For reasons explained later, such platelets are interpreted by the author to be dominantly of that origin.

X-ray analyses of concentrates of clayey material, removed from grains by using an ultrasonic cleaner, show that throughout Libya the clay minerals in the coatings are mixtures of dominantly kaolinite and illite, with subordinate amounts of mixed-layer illite-montmorillonite (table 8). X-ray analyses also show that only the reddest dune sands, such as those occurring near Al ‘Uwaynāt, at the downwind end of the Şahrā’ Awbārī (samples 672 and 674), contain detectable hematite in the grain coatings. In the author’s opinion this hematite represents iron oxide, which initially was precipitated as ferric hydrate and which upon aging either in place or during transport has converted to hematite in the manner discussed by Berner (1969) and by Langmuir (1971).

Sources of Iron

THE SOURCE OF IRON in the pigment cannot be determined unequivocally from available data. However, the fact that the redness of the dunes increases with age and distance of transport, coupled with the likelihood that source areas of the sands have not changed, suggests that the iron is derived internally from the dunes. Two likely sources are the iron-bearing accessory minerals within the sands and the clay minerals in the clay coatings.

Studies of heavy-mineral concentrates show that grains of iron-bearing minerals, such as augite, hornblende, epidote, ilmenite, and magnetite, are present in dunes throughout the area examined. Because these minerals are not in equilibrium with meteoric water, they provide potentially important sources of iron in the pigment. That at least some of the more unstable of these minerals, such as augite and hornblende, have undergone alteration is indicated by the fact that grains of these materials commonly show evidence of pitting by dissolution (fig. 35), and they commonly display cockscomblike terminations blunted by abrasion (fig. 36).

Regional comparisons of the heavy-mineral fractions show a noticeable increase in the degree of alteration of both augite and hornblende with distance of travel. For example, near the upwind end of the Şahrā’ Awbārī, augite — the more unstable of these two silicates — typically is only weakly etched, and hornblende is essentially unaltered. In contrast, near Al ‘Uwaynāt, at the downwind end of that sand sea where dunes are the reddest, augite is nearly absent in the sands, and hornblende has become noticeably etched (fig. 35). These relationships support Norris’ (1969) contention that grains of detrital iron-bearing minerals are important sources of iron for staining dune sands, even in very dry deserts.

Energy-dispersive X-ray analyses of clay coatings on representative sand grains in samples collected from dunes throughout the region, show that the clays characteristically are iron-bearing. Some of the iron occurs as limonitic coatings on the clay, but some also occurs within the lattices of the clay minerals because energy-dispersive X-ray analyses show that iron is present in unstained clay. Inasmuch as these clays are not authigenic, they probably are not in complete chemical equilibrium with meteoric water, and, hence, upon being repeatedly moistened by surface and interstitial waters, they likely release additional iron. The importance of iron-bearing clay minerals as potential sources of iron for iron oxide pigment in other desert sediments has been stressed previously (Walker and Honea, 1969; Walker, 1976).

Origin of Clay Coatings on Eolian Sand Grains

INASMUCH AS RED pigment commonly is associated with clay that partially coats desert sand grains, the origin of the coatings and the circumstances that allow them to be preserved on grains

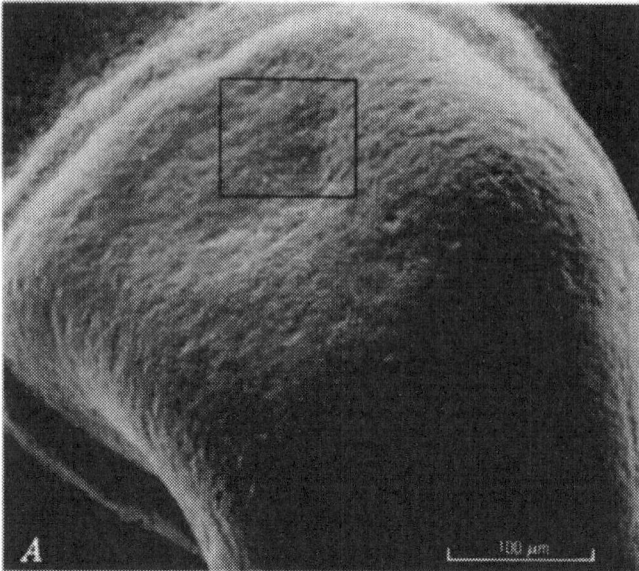

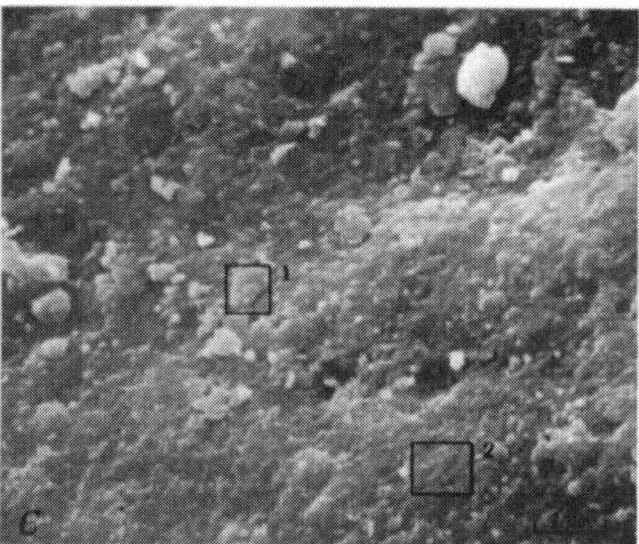

SCANNING ELECTRON MICROSCOPE (SEM) photo-micrographs (A–C) of a coarse rounded quartz grain which, when viewed under the binocular microscope, was completely coated by a film of reddish-yellow pigment (5YR6/6 on Munsell Soil Color Chart, Munsell Color Co., 1954); however, the grain showed no discernible evidence of concentrations of pigment in indentations on the surface. Rectangle in A is area of B; that in B is area of C. D, Energy-dispersive X-ray traces showing the elemental composition of the analyzed areas outlined in C (traces 1 and 2). The photomicrographs reveal that the shallow indentation on the grain contains scattered platelets that were not visible under the binocular microscope. The elemental data indicate that the platelets are composed of silica and a small amount of iron (trace 1 in C) The platelets that have a clastic texture and appear to be sitting on the surface of the grain (for example, the analyzed platelet) are interpreted to be silica dust acquired during transport and derived from eolian abrasion of quartz sand grains. Some of the more firmly attached platelets may be chips which are spalling from the surface as proposed by Margolis and Krinsley (1971). The iron is interpreted to be iron oxide which coats the entire surface of the grain (trace 2 in C), including the platelets (trace 1 in C). Gold and paladium recorded in the traces are contained in the conductive coating applied for SEM examination. (Fig. 33.)

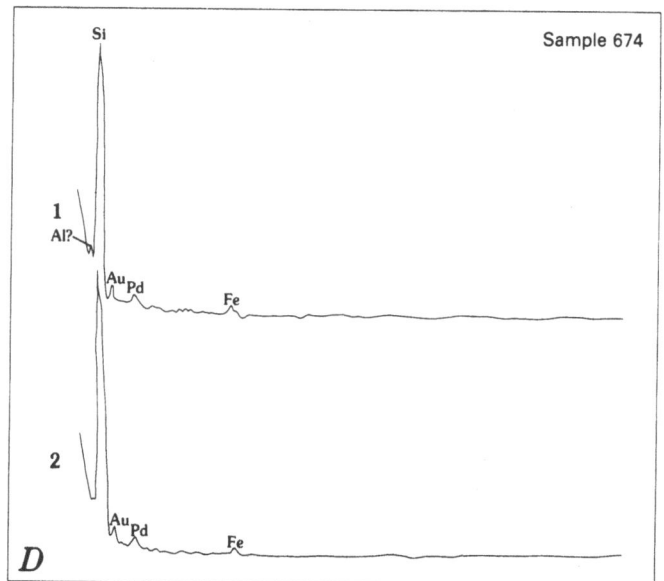

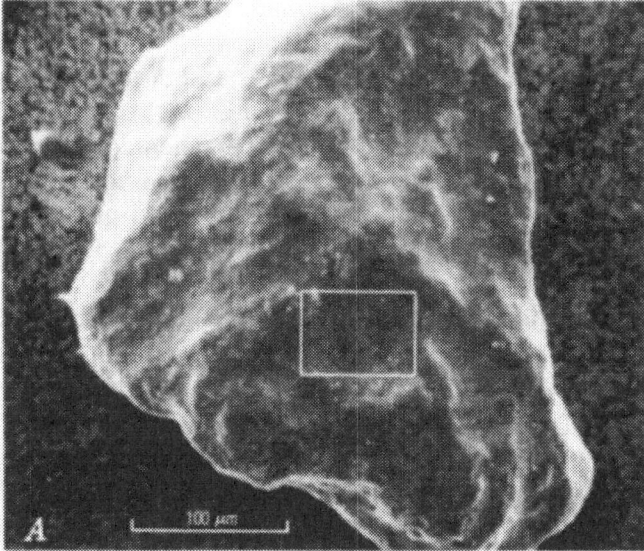

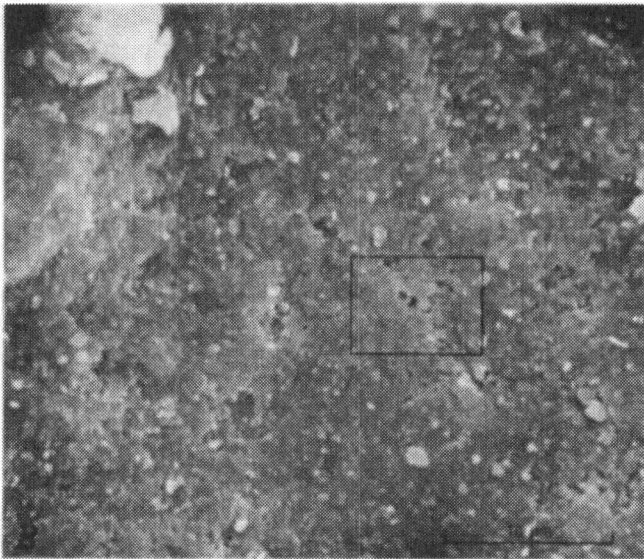

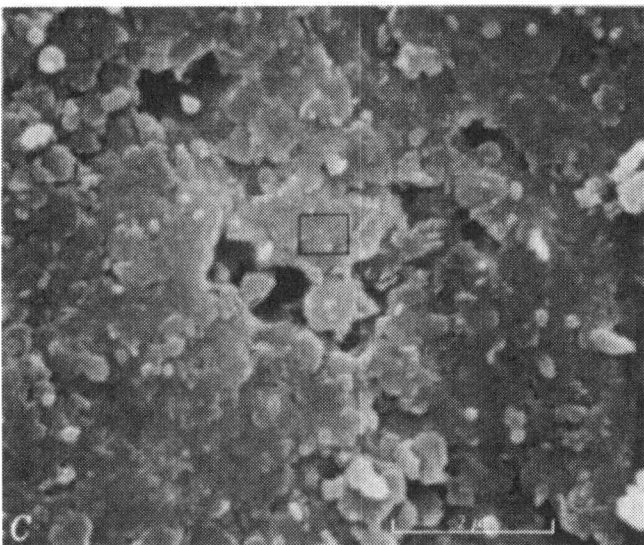

SCANNING ELECTRON MICROSCOPE (SEM) photo-micrographs (*A*–*C*) of a medium-sized subangular quartz grain which contains abundant red (4YR 5/6 on Munsell Soil Color Chart, Munsell Color Co., 1954) pigment concentrated in indentations on the grain surface. Rectangle in *A* is area of *B*; that in *B* is area of *C*. *D*, Energy-dispersive X-ray traces showing the elemental composition of the pigmenting material in the analyzed area outlined within rectangle in *C*. The photomicrographs reveal that the pigmenting material is composed of anhedral particles that are oriented parallel to the grain surface. The elemental data indicate that the particles are composed almost entirely of iron oxide. The particles are interpreted to be a poorly crystalline iron oxide mineral, possibly hematite. Gold and paladium recorded in the traces are contained in the conductive coating applied for SEM examination. (Fig. 34.)

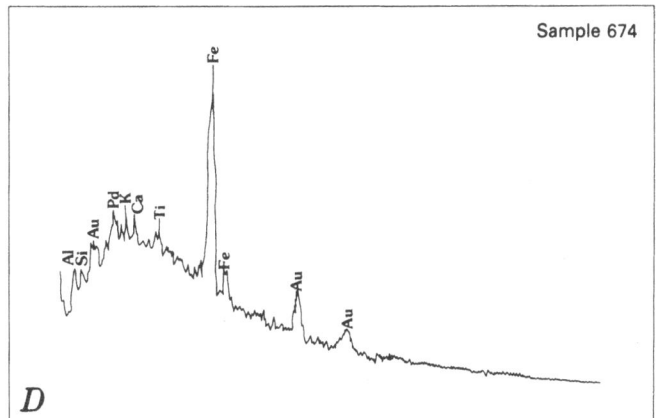

100 μm

Sample 672

ETCHED HORNBLENDE GRAIN from modern dune near Al 'Uwaynāt, Libya. (Fig. 35.)

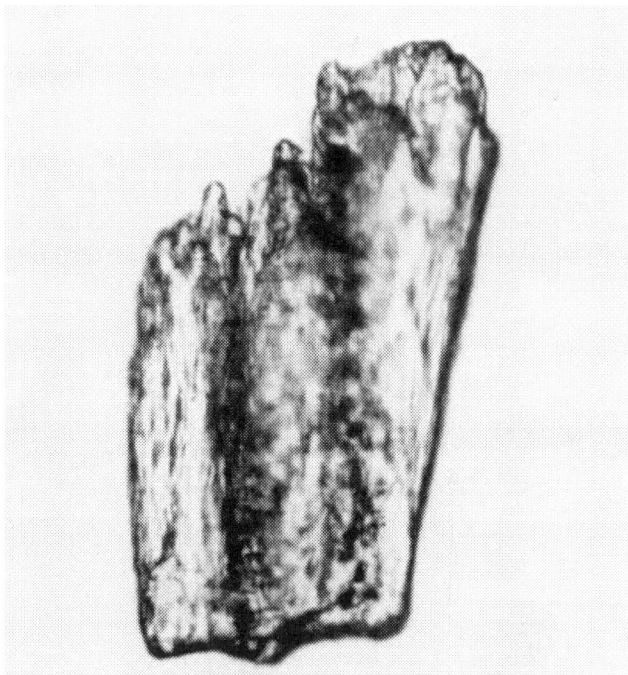

100 μm

Sample 33

ETCHED AUGITE GRAIN from modern dune near As Saddādah, Libya. (Fig. 36.)

despite long distances of eolian transport needs explanation. In part, the coatings may be inherited from matrix clays in sandstones and other bedrock lithologies from which the sand grains are derived. However, three lines of evidence, when considered together, indicate that clay coatings commonly consist of airborne dust emplaced on sand grains by mechanical infiltration of dust particles into the dunes or into other surficial deposits from which sand grains of the dunes are derived. This method of acquiring clay coatings is similar to that proposed by Folk (1969) to account for the formation of clay coatings on sand grains in Australian dunes.

The types of evidence are as follows:

1. Clay coatings are not present on sand grains of the beach from which coastal dune sands are derived, but they are common on grains in the dunes located a short distance inland from the beach. Thus, clay coatings must be acquired as the sand migrates inland.

2. Clay coatings on sand grains of dunes throughout the region characteristically are composed of aggregates of tiny platelets (figs. 37–40) which display clastic textures that are similar to those produced experimentally by infiltrating clay-bearing water into sand (Crone, 1974, 1975). These textures, in turn, are similar to those produced naturally by clay which has been infiltrated into sand by influent seepage of surface water (Walker and Crone, 1974; Crone, 1975; Walker, 1976). Such similarities indicate that the clay is composed of mechanically infiltrated particles, and influent rainwater is the most logical agent to accomplish the infiltration.

3. The mineralogy of the clay in the coatings is similar to that expected of the airborne dust, suggesting that dust is the probable source of the clay. This latter point needs further explanation.

Time limitations precluded the establishment of stations in Libya for the systematic collection of airborne dust samples. The mineralogical composition of airborne dust was predicted, however, by analyzing the clay fraction of representative samples of the major types of potential source materials. The types and number of samples analyzed were bedrock (15), Quaternary(?) alluvium (4), Quaternary(?) dune sand (5), soils (33), modern alluvium

(22). These are compared with the clay in coatings on sand grains in the modern dunes.

The locations of analyzed samples are shown in figure 32. Results of these analyses (table 8) show that essentially all types of materials which are exposed to the wind and which could provide sources of clay contain the same varieties of clay minerals in roughly the same proportions. In all types of source materials — if palygorskite, a likely authigenic clay, is exluded — kaolinite and illite are the predominant clay minerals, almost without exception, and mixed-layer illite-montmorillonite is the principal subordinate clay mineral. The same varieties of clay minerals in roughly the same proportions occur in the coatings on sand grains in the modern dunes (table 8). These mineralogical similarities and the clastic texture of the clay strongly suggest that the clay coatings are composed of infiltrated dust derived ultimately from the bedrock.

To what extent clay minerals, other than palygorskite, have formed by processes of surface weathering is not known, but in view of the striking similarities between the clay mineralogy of the soils and that of the bedrock (table 8), it seems likely that clays in the soils also are derived mainly from the bedrock, not from weathering. Indeed, all available evidence leads to the conclusion that there has been no significant authigenesis of clay minerals except for the palygorskite.

The preceding interpretation readily explains the origin of kaolinite which occurs as a dominant clay mineral in the alkaline desert soils, as it cannot reasonably be expected to have formed in place there. Furthermore, SEM studies of the soils show that clay in the argilic "B" horizons characteristically has a clastic texture similar to that shown in figure 37, and this texture is indicative of mechanically infiltrated material. The importance of infiltrated dust on the formation of desert soils has been stressed previously by Yaalon and Ganor (1973). The conclusion is reached therefore, that all the clay in all types of the surficial deposits has been mechanically infiltrated into the sediments. This may have been accomplished either by infiltration of airborne dust by occasional rain, as with dunes and soils, or by infiltration of clay suspended in the water of intermittent influent streams, as with alluvial deposits.

Presumably, part of the clay-size platelets of silica that occur on some sand grains (fig. 33) also are derived from airborne dust, but they consist of particles which have been abraded from quartz grains during eolian transport, rather than eroded from the bedrock. Some of the silica platelets may also represent chips that are in the process of spalling from the surface of the grains (fig. 33C) in the manner proposed by Margolis and Krinsley (1971, p. 3397).

The mechanism for producing clay coatings on sand grains by mechanical infiltration probably operates throughout the desert, but it is most effective in areas where rainfall is greatest because, in such areas, water penetrates more frequently and deeper into the deposits. Consequently, the amount of clay infiltrated into the surficial deposits is greatest and the rate of formation of clay coatings on the included sand grains is most rapid in the region of relatively high rainfall near the coast. The characteristic occurrence of mechanically infiltrated clay matrix in the soils and temporarily stabilized dunes in the coastal region (fig. 37) supports this interpretation.

Although the prevailing winds are onshore from the Mediterranean Sea, the airborne dust even in the coastal region is derived principally from North African sources, not from sources across the sea. The winter westerlies and other winds that diverge from the prevailing direction have ample energy to erode dust from local sources. The transportation of dust from the central Sahara northward into Europe and east-northeastward into the Middle East is well documented (Yaalon and Ganor, 1973).

The infiltrated clayey matrix tends to bind together the framework grains of initially unconsolidated surficial deposits; thus, the clay helps, at least temporarily, to anchor the deposits. However, owing in part to the friable nature of the surficial deposits and in part to the paucity of vegetation to anchor them, their stabilization in this dry climate is only temporary. Subsequent erosion releases the clay-coated sand grains to eolian transport.

As clay-coated grains are blown about by the wind, the clay is removed from exposed surfaces of the grains by abrasion, but it tends to remain in indentations, where it is protected from abrasion. Such clay-coated indentations are common on grains of all sizes but are most abundant on small

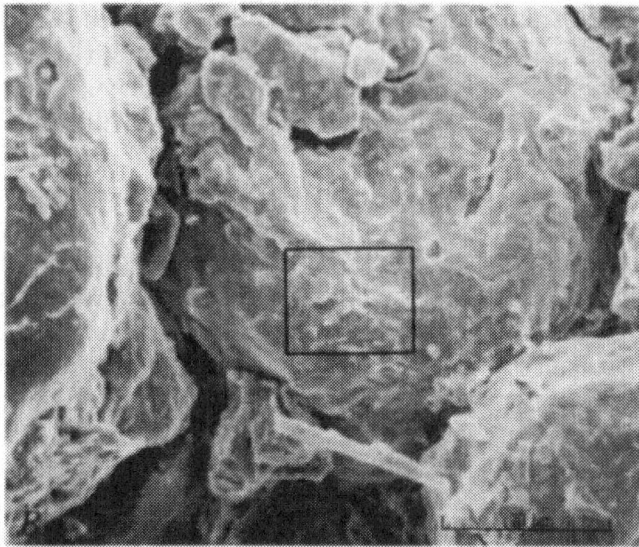

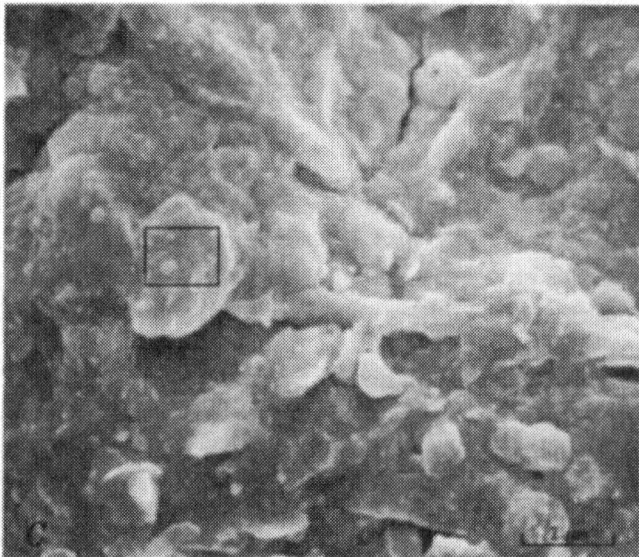

SCANNING ELECTRON MICROSCOPE (SEM) photomicrographs (A–C) of matrix-rich sand collected 2 metres below the surface of a weakly stabilized red dune in the northern coastal region near Miṣrātāh. The color of the clay as determined under a binocular microscope is reddish yellow (5YR 6/6 on Munsell Soil Color Chart, Munsell Color Co., 1954). Rectangle in A is area of B; that in B is area of C. D, Energy-dispersive X-ray traces showing the elemental composition of the matrix clay in the analyzed area outlined within rectangle in C. The photomicrographs reveal that the matrix is composed of an aggregate of clay-size platelets (C). The elemental data indicate that the platelets are composed mainly of aluminum silicates (clay minerals) and iron. The matrix is interpreted to be composed of airborne dust that has been mechanically infiltrated into the dune sand and has become reddened by iron oxide stain subsequent to infiltration. Gold and paladium recorded in the trace are contained in the conductive coating applied for SEM examination. (Fig. 37.)

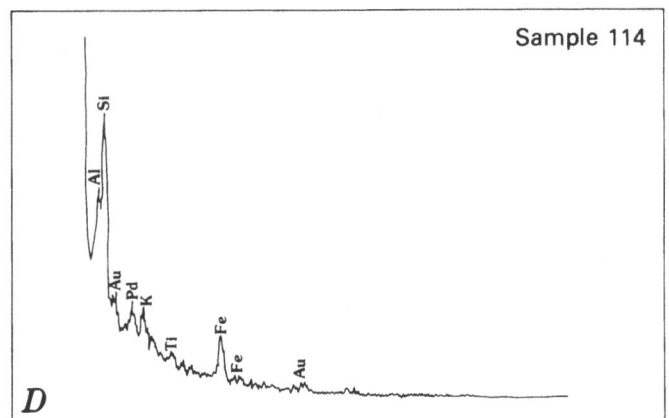

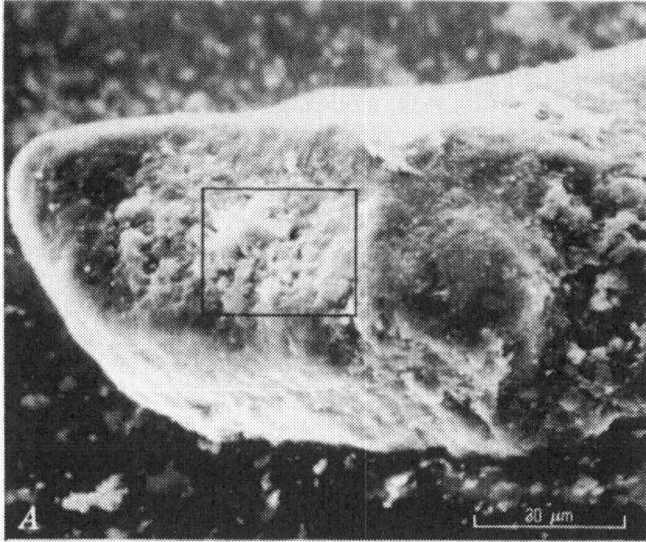

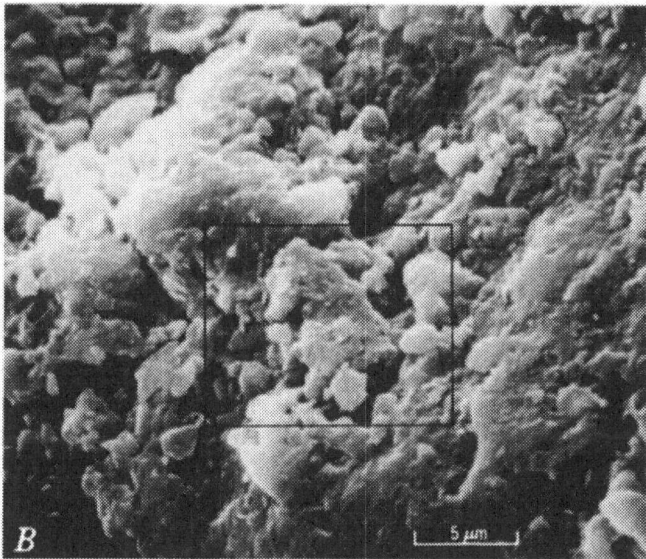

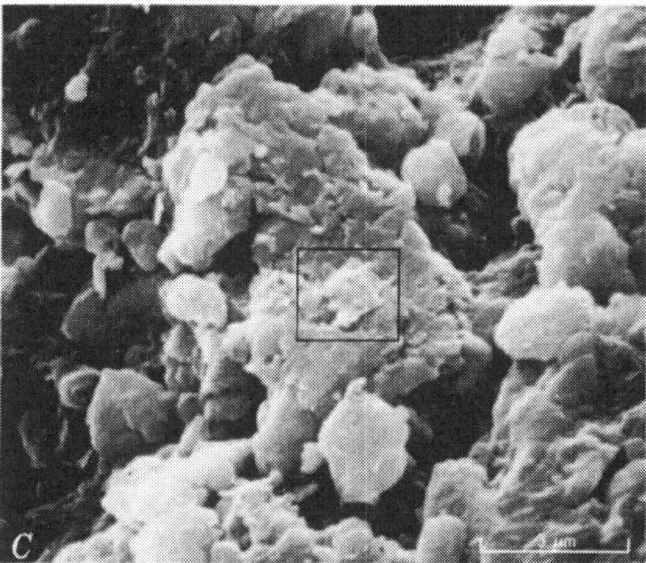

SCANNING ELECTRON MICROSCOPE (SEM) photo-micrographs (A–C) of a fine-grained subrounded quartz grain which contains abundant light-reddish-brown (5YR 6/4 on Munsell Soil Color Chart, Munsell Color Co., 1954) pigment concentrated in indentations. Rectangle in A is area of B; that in B is area of C. D, Energy-dispersive X-ray traces showing the elemental composition of the pigmenting material in the analyzed area within rectangle C. The photomicrographs reveal that the pigment is composed of aggregates of clay-size particles. The elemental data indicate that the particles are composed mainly of aluminum silicates (clay minerals) and iron. The sample from which this grain was selected was collected from the surface of a modern red dune about 100 km (62 mi) inland from the coast near Tripoli. The grain is typical of clay-coated sand grains which comprise a major portion of the dune. The sand in the dune has been reworked from nearby matrix-rich soils and dune sands that have characteristics similar to those shown in figure 37. The pigmenting material illustrated here is interpreted as mechanically infiltrated matrix clay which is analogous to but coarser than that shown in figure 37 and which during eolian transport has been retained in indentations on grains where it is protected from abrasion. Gold and paladium recorded in the trace are contained in the conductive coating applied for SEM examination. (Fig. 38.)

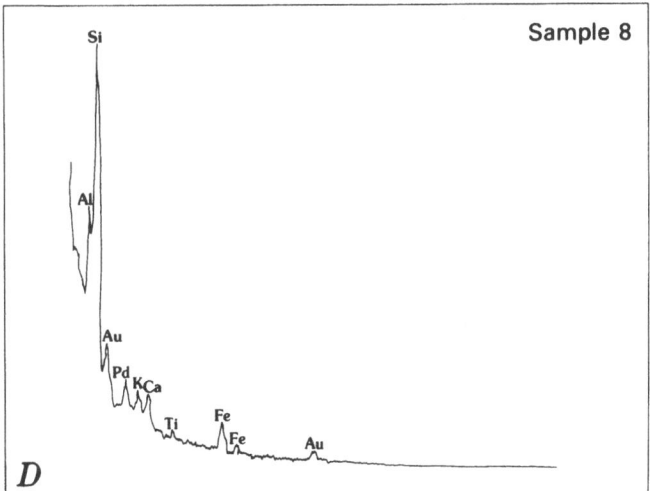

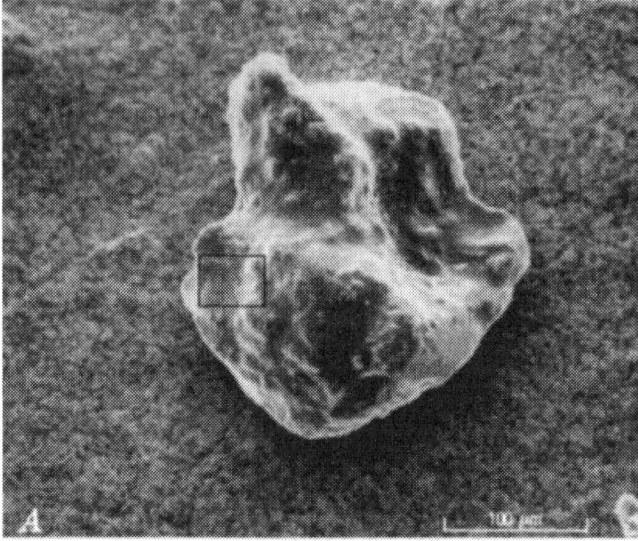

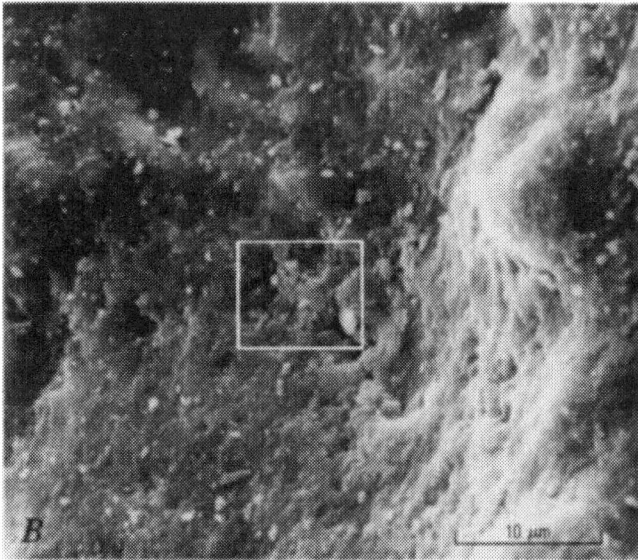

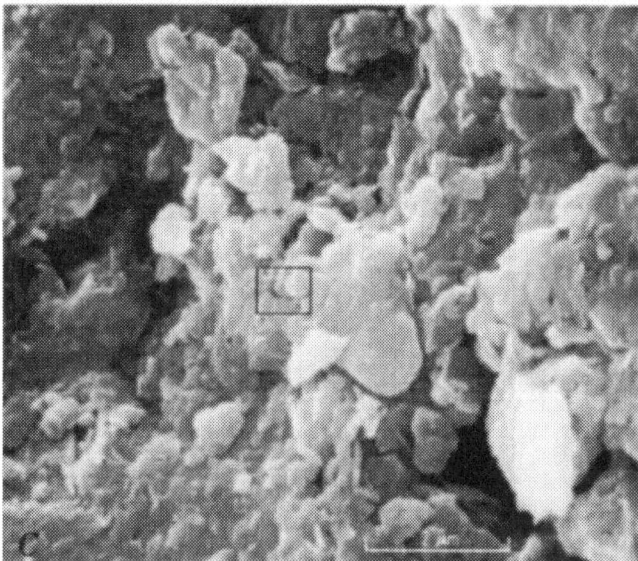

SCANNING ELECTRON MICROSCOPE (SEM) photo-micrographs (A–C) of a fine-grained subangular quartz grain collected from a pink (8YR 7/4 on Munsell Soil Color Chart, Munsell Color Co., 1954) modern dune near Qaryat Abū Nu-jaym. Rectangle in A is area of B; that in B is area of C. D, Energy-dispersive X-ray traces showing the elemental composition of the pigmenting material in the analyzed area within rectangle in C. This grain is typical of fine and very fine sand in dunes throughout the interior of the desert. Note low degree of round-ness and abundance of protected indentations containing con-centrations of clay platelets analogous to those illustrated in figure 38. Gold and paladium recorded in the trace are con-tained in the conductive coating applied for SEM examination. (Fig. 39.)

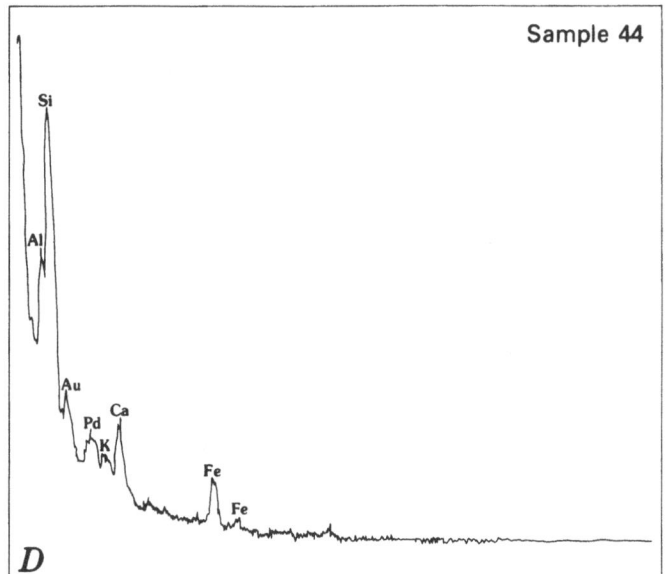

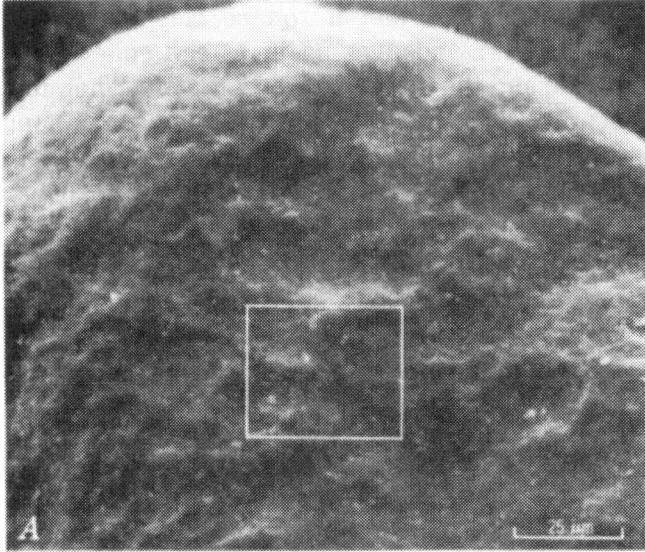

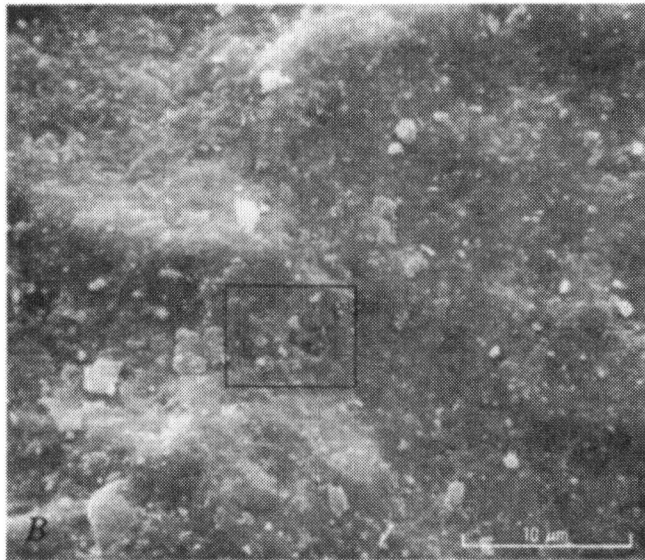

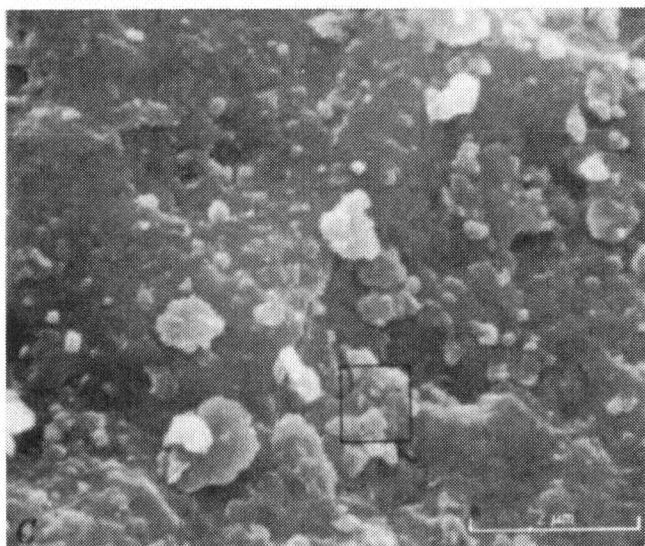

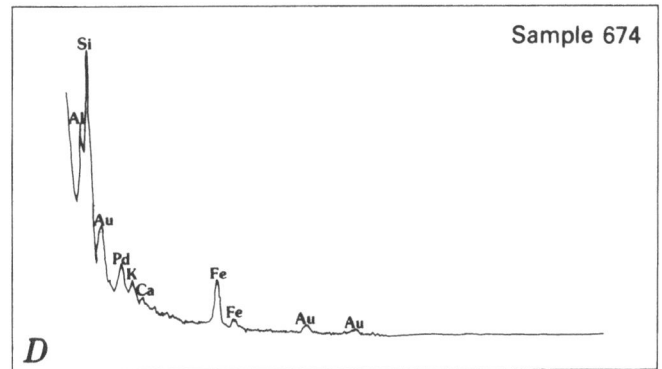

SCANNING ELECTRON MICROSCOPE (SEM) photomicrographs (A –C) of a medium-grained rounded quartz grain collected from a red (3YR 5/6 on Munsell Soil Color Chart, Munsell Color Co., 1954) dune near Al 'Uwaynāt. Rectangle in A is area of B; that in B is area of C. D, Energy-dispersive X-ray trace showing the elemental composition of the pigmenting material in the analyzed area within rectangle C. The photomicrographs, coupled with the elemental data, show that despite the rounded character of the grain, iron oxide stained clay platelets are abundant in shallow indentations. Such concentrations of clay and iron oxide are common on the surfaces of rounded and well-rounded sand grains throughout the desert. Gold and paladium recorded in the trace are contained in the conductive coating applied for SEM examination (Fig. 40.)

grains, such as those in the fine and very fine sand sizes. These small grains show little evidence of abrasion despite long distances of eolian transport. Characteristically, they tend to have angular to subangular shapes and to retain their original surface irregularities. Such observations support the experimental data of Kuenen (1960), who found that eolian abrasion decreases rapidly with decrease in grain size and becomes insignificant for sizes below 150 μm. A typical grain of fine sand is shown in figure 39. A typical coarser grain (medium sand) is shown in figure 40.

The development of coatings of infiltrated clay on sand grains of the Libyan dunes seems to be particularly favored by the readily available source of clay in the sedimentary rocks that dominate the bedrock source areas. Erosion of these rocks by wind, rain, and ephemeral streams provides an unending source of clay for airborne dust. In addition, the developing of the coatings is enhanced by prolonged residence of the sand grains in surficial deposits in the coastal steppe area, where conditions are especially favorable for the mechanical infiltration of clay. Sand grains that bypass this zone or migrate across it quickly are likely to reach the desert interior without having acquired the clay coatings, and acquisition of the coatings is much slower once the grains reach the very arid interior.

Unstained grains occur in all of the active dunes that were investigated in this study, both in the coastal region and in the interior of the desert, but in both regions they are progressively less common with increased distance of transport. This distribution suggests that the coatings are acquired during migration of the dunes, even in the intensely arid regions.

The mechanisms proposed by Folk (1969) to explain the acquisition of clay coating on sand grains in the Simpson Desert seems thoroughly adequate to accomplish the task here. That is, during calm periods between wind storms, airborne dust settles onto the dune surfaces where it adheres to sand grains, particularly if moisture from dew or rain is present. Moreover, during periods of rainfall additional dust is washed from the atmosphere and carried into the dunes by influent rain, where ultimately it also settles onto the surfaces of sand grains.

The processes that introduce clay presumably operate over and over again as the eolian sands migrate and the dune fields accumulate. Therefore, deep infiltration of the dust into the deposit is not required for the formation of coatings, and the coatings are acquired even where rainfall is negligible.

Conditions Favoring Formation of Iron Oxide Pigment

NO DATA ARE AVAILABLE concerning chemistry of the water in contact with the sand grains (for example, dew at the ground surface, pellicular films of water on grains within the vadose zone, and ground water below the water table), but the water can be assumed with confidence to be oxidizing and alkaline and, therefore, to lie in the Eh-pH stability field of ferric hydrates and hematite (Garrels and Christ, 1965, fig. 7.6). The paucity of vegetation — indeed, the complete lack of it throughout most of the desert —and the lack of any other reducing agents provide assurance that meteoric water, upon infiltration into the sand, retains dissolved oxygen obtained from the atmosphere. Accordingly, an oxidizing environment seems a certainty.

Moreover, the widespread occurrence of carbonate, gypsum, and palygorskite in surficial deposits of the desert indicates an alkaline weathering environment. Therefore, iron released by alteration of any of the above-mentioned iron-bearing minerals, including the clay minerals in clay coatings, precipitates as iron oxide, probably initially as amorphous ferric hydrate and (or) finely crystalline goethite, both of which are metastable and convert to hematite upon aging (Berner, 1969; Langmuir, 1971).

Stated another way, hot desert climates, such as those in north Africa, are ideally suited for the formation and preservation of iron oxide pigments. Indeed, precursor oxides of hematite and, ultimately, hematite should be expected to occur as stains in desert deposits wherever and whenever moisture, even in seemingly negligible amounts, comes into contact with any kind of iron-bearing minerals that are not in equilibrium with the water.

Factors Affecting Regional and Local Variations in Redness

REDDENING ACCOMPANIES aging of the initially precipitated precursor oxides of hematite; consequently, dunes tend to redden with increasing age and distance of transport. If aging of the ferric hydrates were the only factor involved in reddening, the intensity of redness of dunes would be a measure of their age, and their color probably could be used as a basis for correlation. The evidence that has been presented in this chapter, however, shows that redness also is affected by a variety of other factors which are not uniform everywhere. Consequently, local or regional variations in numerous environmental factors can be expected to cause variations in the color of the dunes. Following is a list summarizing these factors, many of them interrelated, and their effects on the color.

1. *Availability of unstable iron-bearing minerals in source rocks.*—Unstable iron-bearing minerals, particularly ferromagnesian silicates, such as augite and hornblende, are perhaps the most important sources of iron for pigment in dune sands. If other conditions are equal, dune sands derived from source rocks rich in these minerals probably will redden faster than sands derived from rocks deficient in these minerals. This point was stressed by Norris (1969).

The effect of differences in source-rock mineralogy may explain why the Ṣaḥrā' Awbārī and Ṣaḥrā' Marzūq are strikingly redder than the Ṣaḥrā' Rabyānah and Sarīr Tibasti (fig. 31), despite the fact that sand in the Ṣaḥrā' Rabyānah and Sarīr Tibasti has travelled at least as far as that in the other deposits. The Ṣaḥrā' Awbārī and Ṣaḥrā' Marzūq lie downwind from volcanic rocks exposed in the highlands of Al Harūj al Asway and Jabal as Sawdā' (figs. 31, 32), and both deposits almost certainly receive contributions of sand from these rocks. The Ṣaḥrā' Rabyānah and Sarīr Tibasti, on the other hand, have no comparable source of unstable iron-bearing minerals exposed upwind.

2. *Amount of clay retained on sand grains*. — Inasmuch as pigment in the dunes is carried in part by clay coatings on sand grains, dunes tend to become redder with the increase in the amount of clay that is acquired and retained on individual grains. Hence, a significant factor in determining the redness of dunes is the percentage of sand grains that have clay coatings, and also the percentage of the surface area of these grains that is coated.

The modern dune from which the sand grain in figure 38 was collected is strikingly red because it is composed mostly of grains which, like the one in the photograph, retain extensive coatings of stained clay. Sand in this dune was derived, in part, from nearby red soils of the desert type described earlier, and in part from older weakly stabilized red dunes, both of which contain abundant infiltrated clay matrix similar to that in the example shown in figure 37. Owing to the relatively short distance of transport, a large amount of clay is retained on the grains. Analogous red dunes of modern or near-modern age are common in the coastal region of northern Libya.

3. *Amount of moisture available.*—Water plays several important roles that enhance the reddening process: (a) It provides the medium for chemical hydrolysis of iron-bearing minerals (for example, ferromagnesian silicates and clay minerals in the grain coatings) and is responsible for the release of iron, which precipiates as oxide stain; (b) it provides a mechanism for mechanical infiltration of airborne clay into dune sands and into other surficial deposits from which dune sands are derived, thereby increasing the amount of clay available for coating and staining sand grains; (c) it assists in the growth of vegetation, which in turn tends to stabilize dunes and helps protect the desert soils from erosion, thus keeping grains longer in the environment where conditions are most favorable for reddening. Reddening of dune sands is much faster in the coastal region than in the interior of the desert because of the increased amount of moisture along the coast. The moisture may be rain, dew, or ground water.

4. *Time.*—Time is required for the alteration of both the ferromagnesian silicate minerals and the clay minerals in grain coatings. Together, they pro-

vide the major sources of iron in the pigment. Because these alterations are continuous and because reddening normally accompanies aging of iron oxides, dunes throughout western Libya have tended to become increasingly red with age. The time required to achieve a specific degree of redness, however, is decreased where the amount of moisture is increased, so reddening is faster in the steppe region along the coast than in the intensely arid interior of the desert.

5. *Grain size and grain shape.*—During eolian transport coarse grains are abraded at a faster rate and to a greater degree than are fine grains; hence, they become more rounded and the indentations caused by original irregularities in grain shape tend to be reduced in size and number. The coatings of iron oxide and iron oxide stained clay, which can persist only if protected from abrasion, are therefore less common on the surfaces of coarse sand than on fine sand. The finer grains, on the other hand, because they abrade very slowly, or not at all (Kuenen, 1960), tend to retain original irregularities, and iron oxide stain once acquired tends to persist. Fine sand grains in dunes, therefore, normally are coated with more clay, contain more pigment, and are redder than are the coarse grains.

The effect of grain size on color is striking in places where modern dunes show two distinctly different colors of sand. As an illustration, dunes located near As Saddādah (fig. 32) were examined by the author after a day of particularly strong winds. Surfaces of the slipfaces on the dunes were covered by a veneer of dominantly medium sand which had a color of pink to reddish yellow (7.5YR 7/5, Munsell Soil Color Chart, Munsell Color Co., 1954) and had been deposited by the strong winds of the previous day. Exposed in places on these slipfaces, beneath the veneer of medium sand, were distinctly redder (4YR 6/8), fine to very fine sands that had been deposited earlier by winds of lesser velocity. A likely analogous example has been illustrated by Fürst (1970, pl. 5) in a color photograph that clearly shows the dual color of the sand in some dunes in the Sahrā' Awbārī. However, the two different colors are attributed by Fürst to different ages of the sands rather than to different grains sizes. Other analogous relationships between color and grain size in dune sands have been recognized by Folk (1976), in the Simpson Desert in Australia.

6. *Distance of transport.*—This factor has the same effect as time. Increased distance of travel normally increases the time available for alterations that cause reddening; hence, reddening increases with distance of travel. One might expect the pigment to be removed from sand grains by erosion during prolonged eolian transport, but this is not the case, as demonstrated by the evidence presented already.

Two opposing and unequal processes act upon grains during their transport: (1) abrasion when the grains are in transit, and (2) accumulation of pigment when the grains are at rest. Abrasion is much more effective with coarse grains than with fine, because coarse grains, owing to greater mass, abrade more rapidly. Nevertheless, pigment characteristically is preserved even on coarse grains in pits, where it is protected from abrasion, and it commonly occurs as films that envelop coarse grains. Accumulation of pigment, on the other hand, is the dominant process with finer grains because, owing to their small mass, they abrade slowly, if at all.

Fine and very fine grains, which are the predominant sizes in the dunes studied in this investigation, tend to accumulate more pigment on their surfaces when they are at rest than they lose by abrasion when they are in transit. The accumulation of pigment in this manner, coupled with the fact that it reddens upon aging, causes the dunes to become increasingly redder with distance of transport. As is true of time, the distance required to achieve a specific degree of redness is shorter if the amount of moisture is increased. Consequently, in the intensely arid interior of the desert, the distance required to achieve a certain degree of redness is many times greater than it is in the moister coastal region.

7. *Types of clay minerals contained in coatings on sand grains.*—Iron is carried by clay minerals as a constituent of the clay-mineral lattice, as well as in coatings of iron oxides on the surfaces of clay particles (Carroll, 1958). Such iron can be released by weathering processes if the clay minerals are not in equilibrium with the meteoric water with which they come in contact. Because the Eh and pH of interstitial moisture are high in the desert environment, iron released by such weathering should precipitate almost immediately and increase the amount of iron oxide on the surface of clay particles. The amount of iron released

by such weathering depends in part on the type of clay minerals that are present in the coatings, because clays are not equally capable of carrying iron as a constituent of the crystal lattice (Carroll, 1958, table 1).

Among the clays occurring as coatings on sand grains in the dunes of western Libya, iron is likely to be more abundant in illite and mixed-layer illite-montmorillonite than in kaolinite. Grains with coatings containing high percentages of illite and (or) mixed-layer illite-montmorillonite, therefore, probably have greater potential for increasing redness faster than do those with low percentages of these minerals. The effect of such differences, however, may be insignificant in the sands studied in this investigation because the mixtures of clay minerals in the coatings are virtually the same throughout the region. In other parts of the Sahara or in other deserts, differences in the types of clay minerals occurring in clay coatings might conceivably make significant differences in the rate at which reddening occurs.

8. *Percentage of sand grains derived from older red beds.*—Red beds of Mesozoic age crop out locally in western Libya and red sand grains that apparently have been reworked from these deposits commonly can be identified in the dunes. These grains, in places, contribute in a minor way to the color of the dunes, but the author saw no red dunes in Libya that obviously owed their color mainly to concentrations of such grains. If such dunes occur there, they have only local significance. Grains of sand which have been reworked from the Mesozoic red beds usually can be distinguished readily from grains which have developed red pigment in the manner described earlier. For example, the reworked grains normally are much redder than the pigments forming today. They have hues ranging from 10R to 7.5R on the Munsell Soil Color Chart (Munsell Color Co., 1954) compared with hues of only 5YR to about 3YR in the younger pigments.

Most reworked sand includes rock fragments of red sandstone, red siltstone, and (or) red shale, all of which contain matrix with dark-red color, which attests to the derivation of the grains from red beds. Moreover, the red pigment in reworked quartz grains commonly penetrates the interior of the grains along incipient fractures and, hence, cannot easily be removed completely when the grains are treated with reducing agents. In contrast, pigments that are forming today are limited to the surfaces of grains, and they are easily and quickly removed by reducing agents. Although they are considered unimportant here, red sands that are reworked from older red beds may play important roles, at least locally, in producing red dunes elsewhere. Such dunes, in contrast to those discussed here, should show a decrease in redness with distance of transport away from the source of the sand because of dilution by non-red sands from other sources.

A STUDY OF GLOBAL SAND SEAS

SEDIMENTARY STRUCTURES IN DUNES

Chapter E

By EDWIN D. McKEE

With sections on THE LAGOA DUNE FIELD, BRAZIL

By JOÃO J. BIGARELLA [4]

[4]Geological Institute, Federal University of Paraná, Brazil

Contents

Illustrations

Tables

Summary of Conclusions

SEDIMENTARY STRUCTURES in dunes consist of cross-stratification and of minor surface features, such as ripple marks and rain pits. Pene-contemporaneous deformational features, including slump marks, also are present in a wide variety of forms. Interdune deposits differ from dunes in composition and texture, as well as in structure. This chapter is devoted largely to the description and analysis of the cross-strata that are characteristic of the various dune types. These structures provide the record of dune development, the means of recognizing and identifying dune types in the subsurface and in ancient rocks, and important controls of fluid migration and accumulation.

The study of dune cross-strata is still in a pioneer state and much more work must be done before meaningful generalizations can be made for many varieties of dunes. Most of the work done so far has consisted of a direct approach in which trenches or pits have been dug in dunes, exposing three-dimensional views of cross-strata. Additional studies have consisted of making statistical analyses of foreset dip directions, the vectors of which furnish information on the direction or directions of sand transport and the type of dune represented.

A structural classification based in part on the number of slipfaces, as represented by the principal directions of high-angle foreset planes, is useful. Thus, dunes having slipfaces inclined in one direction, two directions, or many directions are recognized, in addition to sheet sands and stringers and interdune deposits that have no slipfaces. Interdune deposits typically have flat or irregular stratification.

Descriptions of dune cross-strata are concerned primarily with the bounding surfaces of sets and the foreset planes within these sets. The cross-strata are mostly of two basic types — tabular planar and wedge planar. Trough-type cross-strata are rare in those dunes that have been examined internally. Although there are many exceptions, high-angle and large-scale foreset planes are typical of dune structures in general. The spread and pattern of dip directions of foresets give valuble information on dune types. Some types of dunes also can be differentiated by such features as the upward or downward concavity of laminae, the abundance of scour troughs, and the horizontality of apparent dips normal to wind direction.

Introduction

THE PRINCIPAL FACTORS that control dune development and are therefore responsible for dune form are an adequate source of sand and wind of sufficient strength and duration to transport and deposit it. The geomorphic form assumed by an accumulating sand mass is determined primarily by the direction and strength of the responsible wind but may be influenced locally by barriers or deflecting factors, such as bedrock, vegetation, water, or other sand masses. The internal structure or cross-stratification within each dune is a record of its depositional history.

Development of dune structures, as described by Bagnold (1941, p. 37), involves two principal processes: (1) transport of sand particles up the gentle windward (stoss) slope of the dune by saltation, or bouncing, of grains; and (2) avalanching of sand down the slipface on the steep lee side (fig. 41) either by slumping (mass movement) or by grain flowage. Where wind is strong, some sand commonly is carried outward from the dune crest and settles directly from suspension.

Sedimentary structures in eolian dunes commonly consist of sets of medium- to large-scale foresets that dip downwind and are separated by bounding surfaces that are horizontal or inclined at low angles. An exception to this basic pattern of cross-stratification develops where the available sand load exceeds the amount of sand that the wind can move up a dune slope, thereby causing deposition of low-angle strata dipping to windward. Other exceptions develop when the wind is unusually strong or changes direction, thus introducing partial erosion of earlier deposits to form either troughs or beveled surfaces that are subsequently buried and form part of the dune record.

Distinctive dune forms have been correlated with genetic factors by many students of sand seas, and numerous classifications of dune patterns (chapter J, tables 26–32) have resulted. A common method of subdividing dune forms is to recognize the characteristic shapes attributed to the direction or

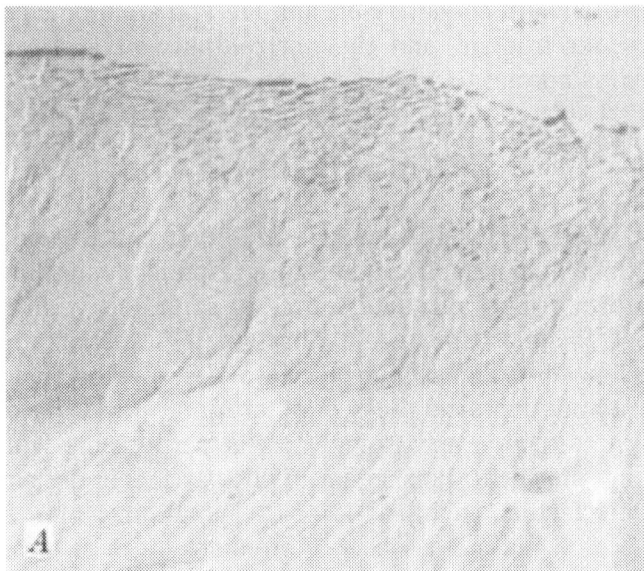

AVALANCHING SAND BY SLUMPING and grain flowage down slipface of dunes, White Sands National Monument, New Mexico, U.S.A. *A*, Small blocks of weakly cohesive sand formed from damp crust in upper part of slump masses. Length of base of ripple is 6 inches (15 cm). *B*, Curving crest of barchanoid dune delineated by top of avalanching sand. Ripple length in lower left is 4 inches (10 cm). *C*, Series of slumps partially burying ripple-marked slipface. Knife handle is 6 inches (15 cm) long. *D*, Scoop surface where slump mass broke loose near top of slipface on transverse dune. (Fig. 41.)

directions of the dominant wind or winds involved (table 9). Those dunes that are formed by a prevailing wind (or a single dominant wind) include types commonly referred to as barchan, barchanoid ridge, transverse, dome-shaped, and some parabolic. Dunes attributed to bidirectional winds may form linear ridges, locally known as longitudinal dunes or seif dunes, where the winds converge at low angles, or they may form reversing dunes if winds blow from nearly opposite directions. Where winds are shifting around the compass they form star dunes with three, four, or more arms extending outward from a central high point.

Although all eolian dunes are primarily controlled by wind activity and therefore share many structural features, some distinctive and probably unique structures characterize each type. The value of recognizing such distinguishing features should be apparent, for they make possible the interpretation of genesis and an understanding of subsurface trends in various sand bodies. Whether the sand has advanced in a constant direction to form a sheet

TABLE 9. — *Classification of principal dune types according to form and to number of slipfaces*

Number of slipfaces				
None (no mound)	None (mound)	One	Two	Multiple
Sheet.	Dome.[1]	Barchan.	Linear (seif or	Star.
Stringer.		Barchan ridge.	longitudinal).	Blowout.
Climbing.		Transverse ridge.	Reversing.	Parabolic.[2]
Falling.		Retention ridge.		

[1] Internal structure may show buried foresets.
[2] May have only one slipface developed.

DUNE TYPES

Simple: Dunes in the forms shown above.
Varieties: Dunes having differences in details of topographic form.
Compound: Dunes in which two or more of the same type are superimposed.
Complex: Dunes having two or more types combined.

or has built mostly vertically into a thick lens; whether it has advanced in one or in more than one direction; and whether the accretion of additional sediment has been rapid or slow, therefore resulting in relatively pure sand or in much-contaminated sand — these are questions that may be answered by structural analysis.

Dunes that have been studied in detail from the standpoint of structure are relatively few, but trenching and dissection have been undertaken in the United States at White Sands National Monument in New Mexico, at Great Sand Dunes National Monument in Colorado, and at Killpecker dune field in Wyoming. Similar studies have been made in Libya, in Saudi Arabia, and at several localities in Brazil. Some investigations have been made in solitary dunes, others in selected parts of large composite dunes, where extrapolation was necessary to reveal the complete structure. Many more studies of certain dune types in various geographical situations are needed before generalizations can be made with complete confidence, but some features seem very definitely established already.

Sedimentary Structures Common to Most Dunes

THE FOLLOWING STRUCTURES, on the basis of our limited present knowledge, are considered to be common to most types of eolian dunes:
1. Sets of medium- to large-scale cross-strata that typically consist of foresets dipping leeward at high angles (not uncommonly 30° –34°) and represent original angles of repose.
2. Sets of tabular-planar cross-strata that in vertical sequences in high dunes tend to be progressively thinned from the base upward; sloping laminae within these sets tend to repeat the pattern of foresets displayed in the thick basal set, but the angle of dip may be lower (figs. 43, 46, 48).
3. Bounding planes between individual sets of cross-strata that mostly are horizontal or dip leeward at low angles; in high dunes additional bounding surfaces commonly form in downwind parts, with dips at moderately high angles (20° –28°), and these planes truncate foresets that dip at still higher angles (28° –34°; fig. 48).

Various other structures apparently are characteristic of only one or two dune types. Such features will be discussed in the following pages under each of the principal dune types in which they occur.

Sedimentary Structures Characteristic of Certain Dune Types

Barchan and Barchanoid Ridge

THE BARCHAN DUNE, one of the best known and most elemental forms, is also probably the best-

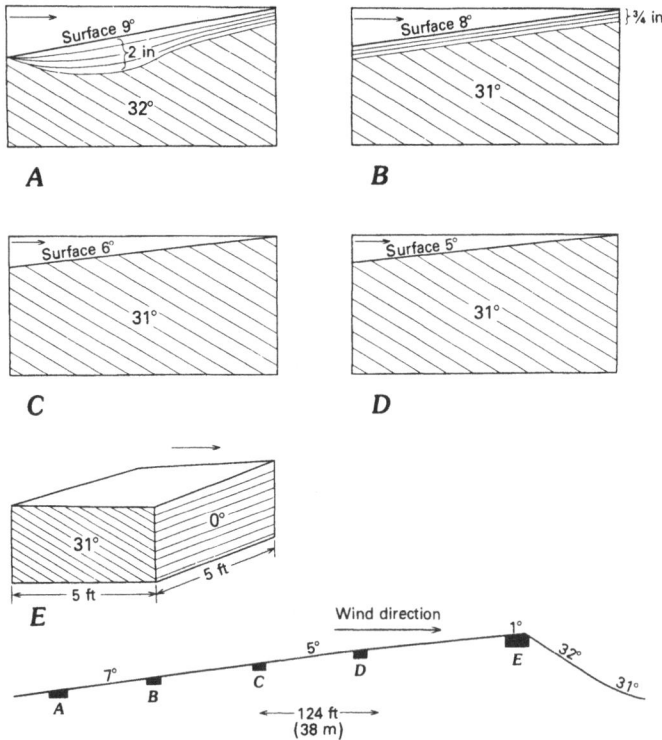

CROSS SECTIONS OF BARCHAN DUNE near Leupp, Coconino County, Arizona, U.S.A., taken through crest in direction of dominant wind. Modified from McKee (1957). *A*, 12 feet (3.7m) from base, 72 feet (22m) from crest; *B*, 24 feet (7.3m) from base, 60 feet (18.3m) from crest; *C*, 36 feet (11.0m) from base, 48 feet (14.6m) from crest; *D*, 48 feet (14.6m) from base, 36 feet (11.0m) from crest; *E*, crest, on east side. 100 feet = 30.5 metres. Scale of sections *A–D* = 1 ×2 feet. (Fig. 42.)

recorded dune in terms of sedimentary structure. The typical crescent shape, with horns pointing downwind, commonly develops as a single, isolated sand deposit; groups of barchans that coalesce to form barchanoid ridges in a parallel wavy pattern are also common.

A small isolated barchan dune located a few miles north of Leupp Trading Post, in eastern Coconino County, Arizona, was dissected and its structures were studied in detail in 1951 (McKee, 1957, p. 1721). The dune was a double crescent with a height of about 11 feet (8 m) and a width of 270 feet (80 m). Five test pits — three located from base to crest on the windward side and one on the forward part of each horn — provided good insight into the structure of a typical barchan.

Laminae in the middle part of the Arizona dune consisted of long, even layers dipping consistently 31°–32° downwind. In the lower part of the windward slope, the truncated tops of these laminae

were covered by a veneer of laminae dipping upwind at angles of 5°–7° (fig. 42). On the wings, variations from the normal trend of stratification occurred as changes in direction and amount of dip. The wings contained strata dipping with relatively low angles, resulting from growth in locations of low wind stability and at extremities of the slipface surface.

At White Sands National Monument, New Mexico, one segment of a barchanoid ridge dune was trenched and analyzed (McKee, 1966, p. 39). The crescent-shaped mound studied was about 170 feet (50 m) long, 290 feet (90 m) wide, and approximately 27 feet (9 m) high. Trenching by a bulldozer, in directions both parallel to and normal to the prevailing wind, exposed to view cross-lamination of the dune in three dimensions.

The pattern of the barchanoid ridge structure as recorded in the main trench consisted upwind of a series of nearly flat-lying tabular-planar sets of cross-strata, each about 3 or 4 feet (p.9–1.2 m) thick and containing foresets that dipped downwind at 26°–34° (figs. 43, 44). In the downwind part of the trench, the bounding surfaces of each set changed from nearly horizontal to steeply dipping (20°–28°) and the foresets between them attained dips up to 34° and lengths of 40 feet (12 m). A few low-angle (2°–5°) windward-dipping laminae capped the dune, probably formed as windward-side deposits at times when the wind was not competent to carry its entire load over the crest. Shallow, small-scale lenses of sand, scattered among sets of windward-dipping cross-strata, apparently represent scour-and-fill deposits of shifting winds.

Crossbedding, as seen in the walls of the trench, at right angles to wind direction (fig. 44*B*), was materially different from that exposed in the main trench. A section cut straight across the crescent exposed, near its middle, crossbed sets and their bounding planes that appear nearly horizontal. On the flanks both cross-strata and bounding surfaces dip outward at low angles as a result of curvature of the horns or wings; individual cross-strata sets form wedges that taper toward dune margins, and foresets show apparent dips of 12°–23°. Small trough structures, probably formed as blowouts during fluctuations of wind regime as the dune was forming, were conspicuous near the dune center.

Horizontal sections cut above the barchan dune floor showed the curving lines formed by trunca-

SERIES OF STEEPLY DIPPING CROSS-STRATA forming tabular-planar sets in barchanoid ridge dune. Strata tangential with basal bounding planes. White Sands National Monument, New Mexico, U.S.A. Photograph by E. Tad Nichols. (Fig. 43.)

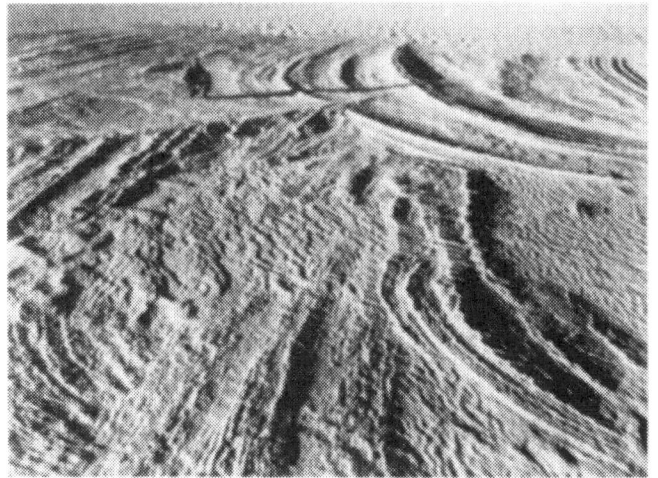

BEVELED REMNANT OF BARCHAN DUNES in interdune area, White Sands National Monument, New Mexico, U.S.A. From McKee (1966). (Fig. 45.)

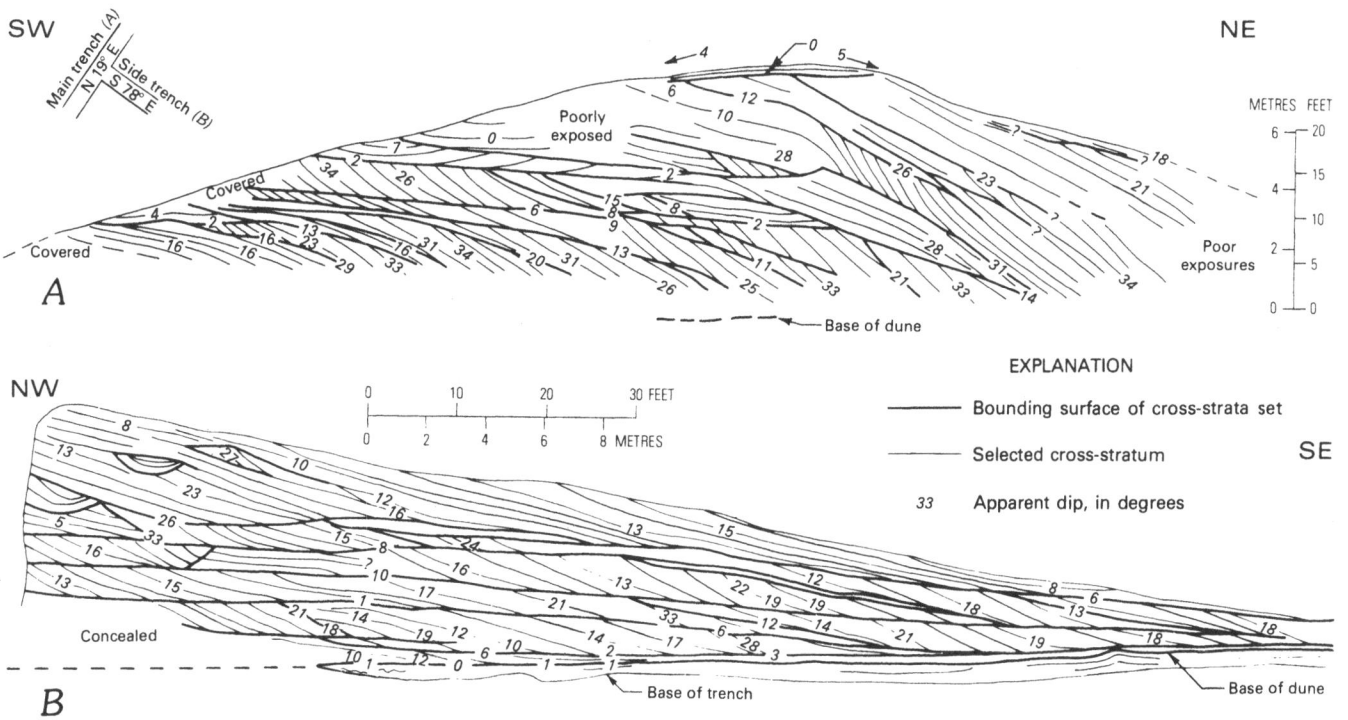

CROSS-STRATA IN BARCHANOID RIDGE DUNE, White Sands National Monument, New Mexico, U.S.A. A, Section parallel to dominant wind direction; B, section normal to dominant wind direction. Modified from McKee (1966). Lines queried where structure was obscure in field. (Fig. 44.)

tion of laminae that had formed on the slipfaces of the dune. The overall pattern was one of festoons, with one arc or crescent overlapping another (fig. 45).

Barchanoid ridge dunes of the Killpecker field in western Wyoming (Ahlbrandt, 1973, p. 38) showed virtually the same sedimentary structures (fig. 46) as those at White Sands. A trench on the lee side revealed foresets dipping at high angles (26° –34°) to lee and a bounding surface dipping 4° in the same direction. Deformational structures consisted of fadeout laminae and overthrusts.

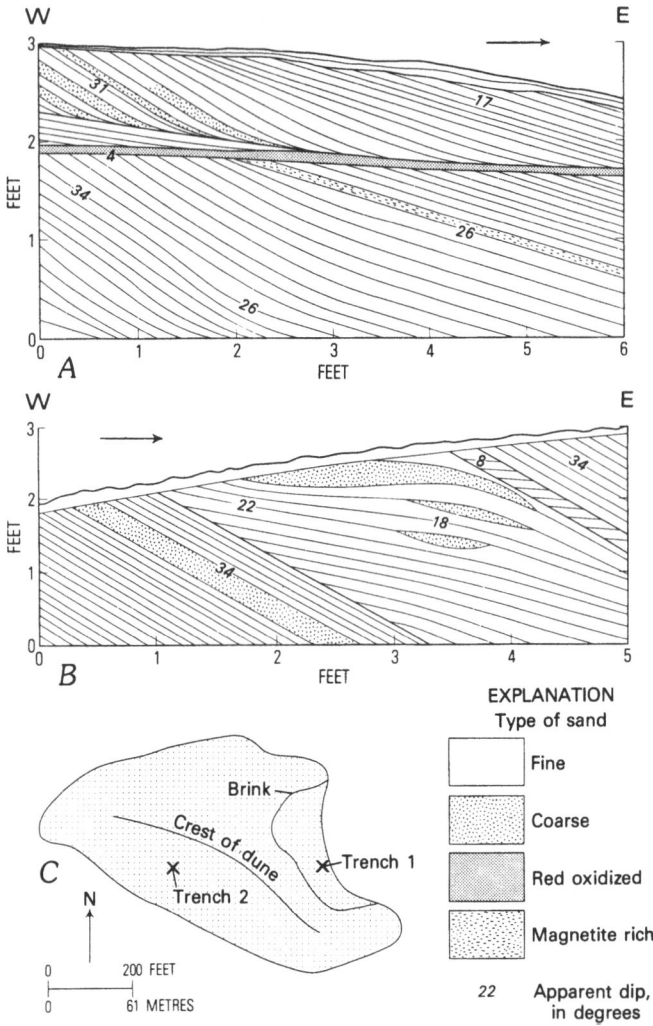

BARCHANOID RIDGE DUNE, showing stratification to dominant wind direction (arrow), Killpecker dune field, Wyoming, U.S.A. A, Cross section of lee side, trench 1; B, cross section of upwind side, trench 2; C, plan view showing trench locations. Modified from Ahlbrandt (1973), 1 foot = 0.3 m. (Fig. 46.)

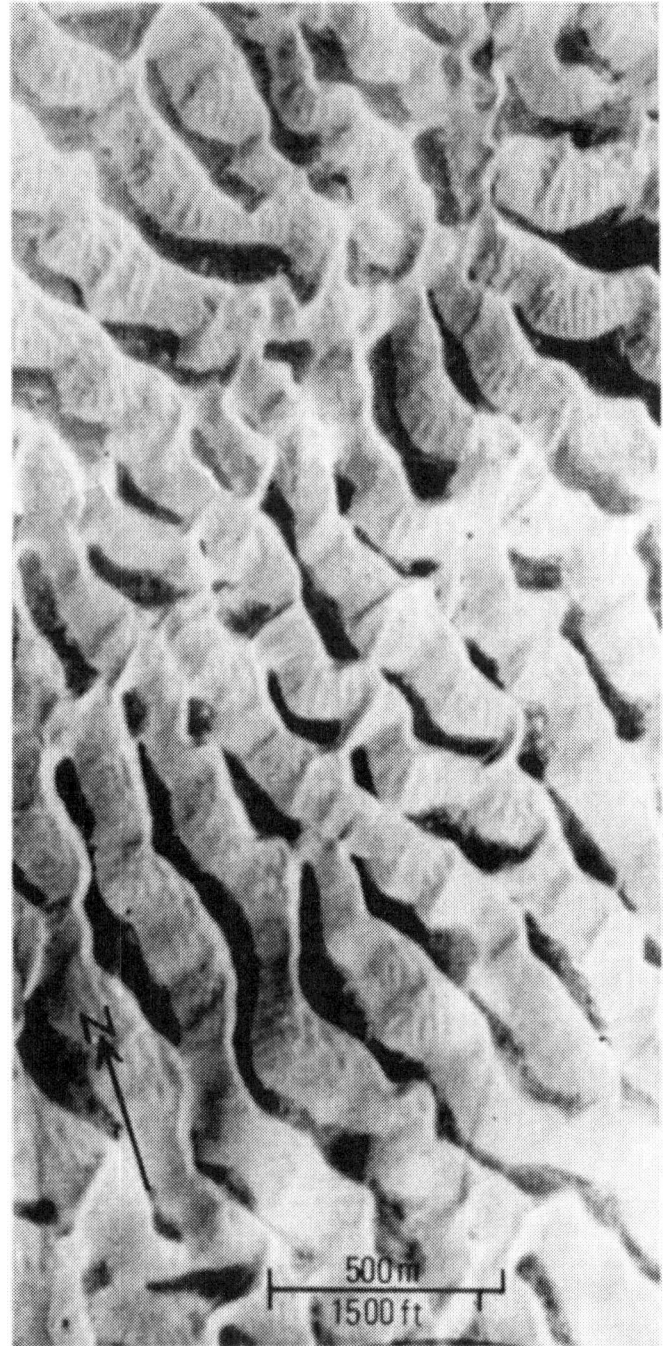

BARCHANOID RIDGE DUNES and interdunes, White Sands National Monument, New Mexico, U.S.A. Photograph by Holloman Air Force Base. (Fig. 47.)

A second trench in the Killpecker barchanoid ridge, dug on the upwind side, exposed four sets of cross-strata, all dipping leeward. In two sets the dips were high angle (34°), in a third they were medium angle (18°−22°), and, in a thin but persistent unit, they were low angle (8°) at one end but steepened downwind to the angle of repose. The low-angle strata contained small lenses with scour-and-fill structures.

One feature of the barchanoid ridge in the Killpecker dune field, not recorded in other barchans, was an alternation of coarse and fine laminae and a filling of lenses with relative coarse sand. Such contrasts in texture apparently resulted from poor sorting in the source sand and the proximity of this source. In terms of textural parameters (chapter B), interior desert dunes commonly are more poorly sorted than coastal dunes because of normal differences in source material.

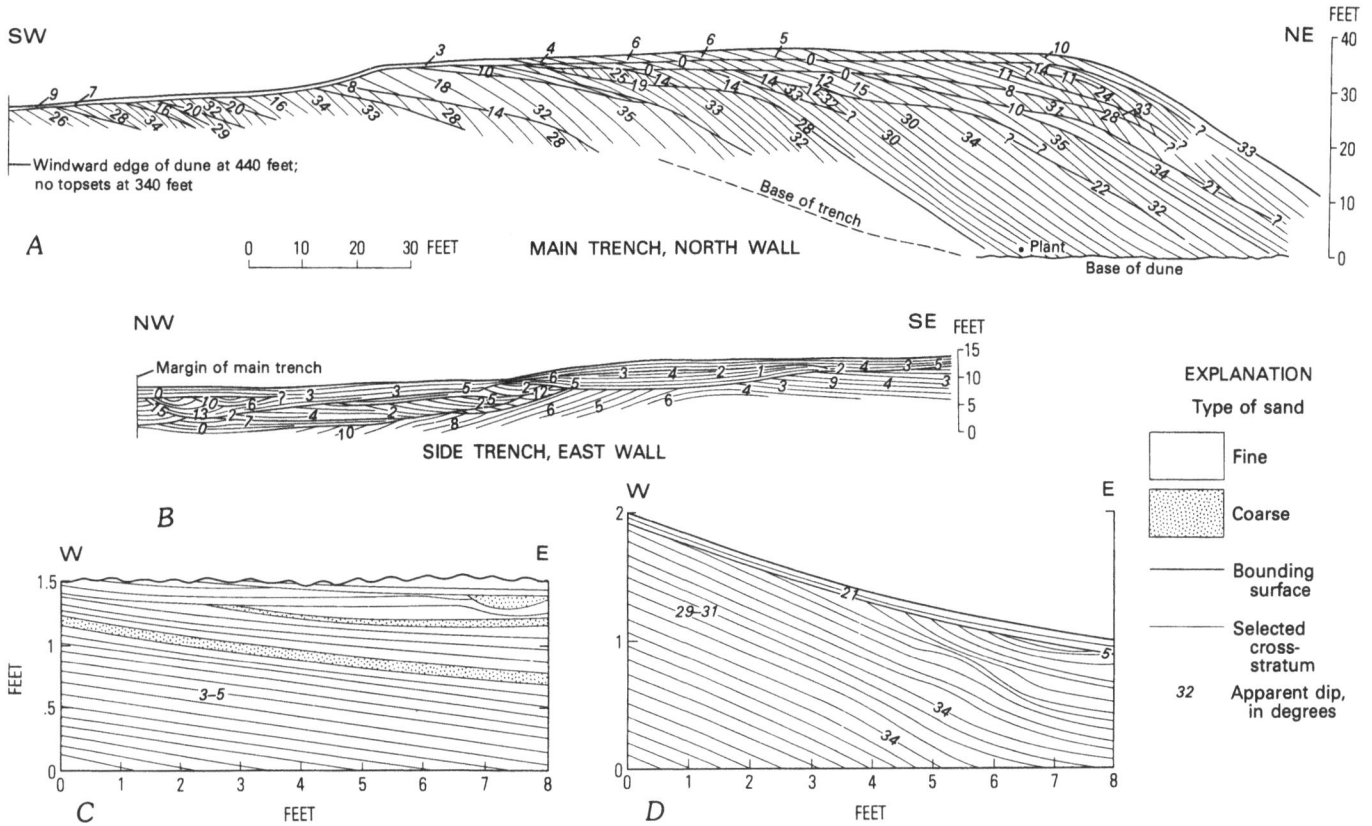

CROSS SECTIONS OF TWO TRANSVERSE DUNES in the United States. A, B, White Sands National Monument, New Mexico. From McKee (1966, fig. 7). C, D, Killpecker dune field, Wyoming. Modified from Ahlbrandt (1973). A, Parallel to dominant wind direction. B, Normal to dominant wind direction. C, Parallel to dominant wind direction on windward side. D, Parallel to dominant wind direction on lee side. 1 foot = 0.3 metre. (Fig. 48.)

Transverse

Transverse dunes resemble barchan dunes in being a product of unidirectional winds, with steep slipfaces oriented normal to wind direction. The transverse dune apparently evolves into the barchan downwind with an intervening intermediate form, here referred to as a barchanoid ridge. The barchanoid ridges consist of parallel rows of crescents or barchanlike forms that have combined to form parallel wavy sand ridges (fig. 47). Interdune areas, forming long corridors, occur between pairs of ridges. The barchanoid ridge and transverse types of dunes, although similar in general form, differ in internal structure.

Transverse dunes at White Sands National Monument, New Mexico, are the highest dunes in the area and form long parallel ridges, upwind from the smaller, much more irregular barchanoid types. A typical example selected for sectioning and study was about 40 feet (12 m) high, 400 feet (120 m) thick, and had a nearly straight crest extending for more than 800 feet (240 m) at right angles to the wind direction. The general straightness of this crestline was the feature that especially distinguished this dune type from adjacent barchanoid ridges and was reflected in the internal structures of transverse dunes.

Crossbedding in the main trench at the White Sands transverse dune, cut parallel to the dominant wind direction (fig. 48), consisted largely of tabular-planar sets of steeply dipping (30°–34°) strata within bounding planes that ranged from nearly horizontal to moderately high angle (20°–26°) downwind. Foresets in the lower part of the dune were very large, some strata extending for 55 feet

(16 m) downdip and terminating as tangential curves at the dune base. Crossbed sets in the upper part were relatively thin, mostly 2–3 feet (0.6–1.0 m) thick, with nearly parallel bounding planes.

Perhaps the most diagnostic feature of the transverse dune structure was the great extent of horizontal, or nearly horizontal, parallel laminae formed by the apparent dip of strata as seen in cross sections cut normal to wind direction (fig. 48). Some cross-strata in these sections dipped at low angles, and a few sets appeared as cut-and-fill (festoon) structures with axes parallel to wind direction, but high-angle foresets were absent.

A transverse dune at Killpecker dune field, Wyoming (fig. 48), which was trenched and analyzed by Ahlbrandt (1973, p. 36), showed most of the same features. Like the White Sands dunes, it was located downwind from dome dunes and to windward of barchans. It was formed largely of tabular sets of strata dipping steeply (29°–34°) downwind and bounding surfaces, that is, slipfaces, dipping as much as 21°. Water saturation in parts of the dune had made the sand cohesive, and overthrusts had developed. In dry areas fadeout laminae were abundant. Another feature of the dune was the presence of lenticular zones of coarse sand among laminae of fine sand, probably the result of scouring and later infilling among strata with low-angle dips on the windward side.

Blowout

In many coastal areas and others where sand becomes stabilized by vegetation cover, moisture content, or both, a common dune form is referred to as the "blowout" and, in general, appears like a crater (fig. 49). Blowouts differ widely in size and shape partly because their rims are anchored in so many places that even in regions of a dominant wind direction, prevailing winds are deflected in many directions, and local sand resistance to deflation and transport is extremely variable.

A circular blowout dune may evolve into a U-shaped parabolic form where a prevailing wind can cause the advance of an unstable section of the rim. In this manner, the nose of a parabolic dune may develop and migration of this dune type can be initiated. The internal structure of parabolic dunes has been examined and is described on succeeding pages, but the stratification of normal blowouts

does not seem to have been recorded as yet. Judging from the characteristic shapes and distribution of blowouts, one might expect a series of scour-and-fill structures to develop; however, this is yet to be determined by trenching techniques.

BLOWOUT DUNES in the United States. A, South shore of Lake Michigan; photograph by R. L. Gutschick. B, East of Salton Sea, California; photograph by John S. Shelton. (Fig. 49.)

Parabolic

A parabolic dune is U-shaped or V-shaped and represents a type of blowout in which the middle part has moved forward with respect to the sides, or arms (fig. 50). This kind of dune occurs in many parts of the world but is especially prevalent in coastal areas and along margins of desert regions where vegetation has obtained a foothold. A distinctive feature is the anchoring of the dune arms by plant growth, which causes the entire dune to be relatively stable. The middle, or blowout, part of

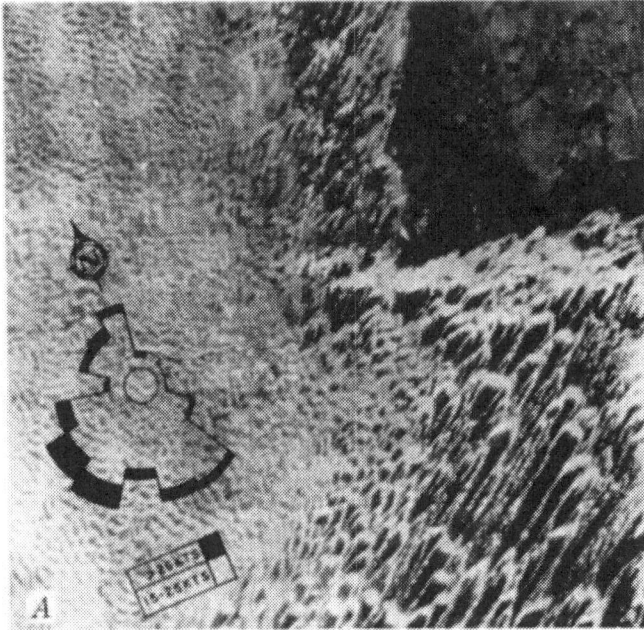

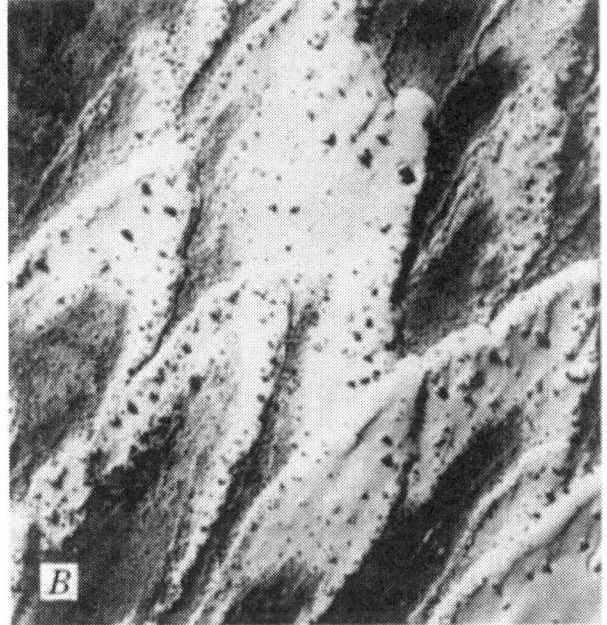

AERIAL VIEWS OF PARABOLIC DUNES at White Sands National Monument, New Mexico, U.S.A. *A*, Compare prevailing southwest winds (in knots) shown by wind rose with orientation of dunes. Photograph by Holloman Air Force Base. *B*, Vegetation anchoring arms of dunes. Orientation of dunes same as in *A*. Individual dunes, including the dragging arms, average about 700 feet (215 m) long in this area. (Fig. 50.)

the sand mass commonly migrates slowly, forming a rounded nose that points downwind.

Parabolic dunes vary greatly in size. In the Thar Desert of India and Pakistan, they commonly are very large, individual dunes extending several miles downwind, the U-shapes readily discernible in Landsat (ERTS) images. Elsewhere, for example, at Cabo Frio, on the Brazilian coast (McKee and Bigarella, 1972, p. 672), parabolic dunes are only a few hundred feet long, and among the Indiana dunes on the shore of Lake Michigan, they range in length "from a few tens of feet to several hundreds" (Cressey, 1928, p. 6).

Because parabolic dunes are largely the product of unidirectional winds, although secondary movements may be to both sides, their structures include many features common to all dunes that are dominated by one major wind direction. Their steep depositional faces (slipfaces), which form on the dune nose, have foresets consistently dipping downwind; however, strata in the arms dip normal to the dune axis, both right and left. Various combinations of parabolic dunes occur, such as series with one dune inserted inside another (common among dunes of the Thar Desert, India-Pakistan) and reversing types as at Lagoa, Brazil (McKee and

Bigarella, 1972, p. 672); these compound dunes are very complicated structurally.

A parabolic dune was trenched at White Sands National Monument, New Mexico (fig. 51). In its lower part this dune shows cross-strata with high-angle foresets as in barchanoid ridge and other dune types of the area, but in the upper part, especially on the upwind side, many of the strata are very low angle or are nearly horizontal and in thin sets. Deposition extended from a nearly flat top, over the crest, and down the slipface. Its dipping laminae, except near the top, are in sets mostly 2–5 feet (1–2 m) thick of tabular-planar or wedge-planar form. A few scattered sets of symmetrically filled trough cross-strata — about 25–30 feet (7–9 m) wide and 3–4 feet (1 m) deep — also occur near the top.

The White Sands parabolic dunes exhibit two structural features seemingly unique to this dune type: (1) many foresets are concave downward (fig. 51), probably resulting from crosswinds that undercut and oversteepened the bases of foresets on the dune nose, and (2) organic accumulations that locally warped strata, result from root growth, especially along the bedding surfaces.

The parabolic dune examined at White Sands showed an unusually wide directional spread of

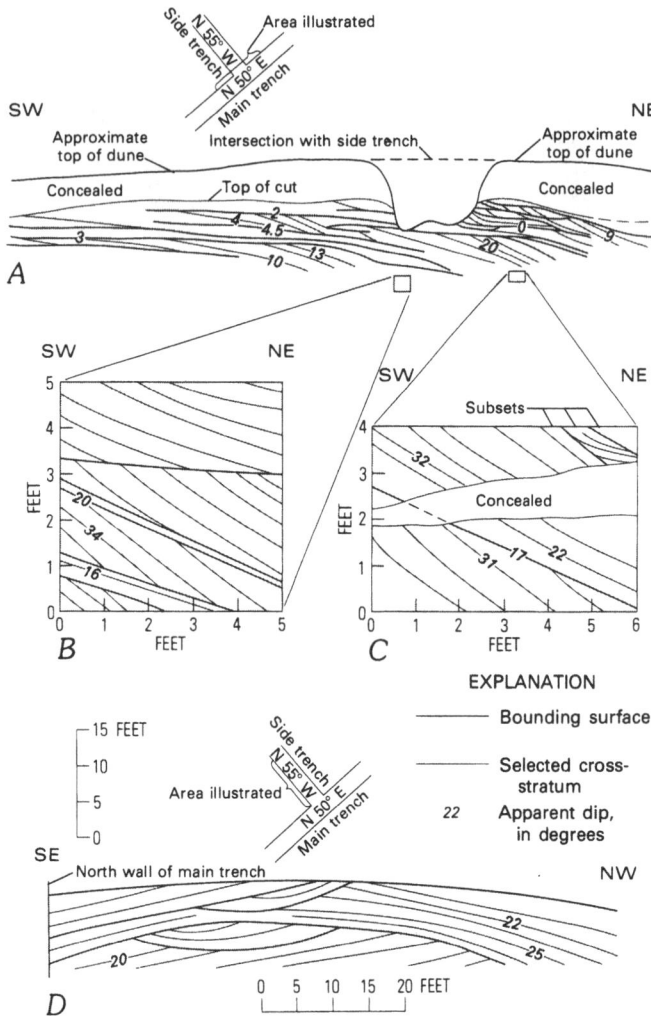

VARIOUSLY ORIENTED SECTIONS through a parabolic dune, White Sands National Monument, New Mexico, U.S.A. Modified from McKee (1966, fig. 9). *A*, Parallel to dominant wind; *B*, *C*, parallel to wind direction, showing high-angle foresets; *D*, normal to dominant wind. All measurements are in feet. 1 foot = 0.3 metre. (Fig. 51.)

crossbedding dips (McKee, 1966, table IV) — a distinctive feature of this dune type. Seventy-three readings of true dip in foresets showed a spread of 200°, and 161 readings of truncated foresets in adjacent interdune areas showed a spread of 210°. Barchanoid ridge, transverse, and other dunes upwind at White Sands showed much smaller spreads, ranging from 60° to 140°.

Structures of parabolic dunes on the coastal plain of Brazil at Cabo Frio (McKee and Bigarella, 1972, p. 672, fig. 4d) resemble those at White Sands. Recent extensive trenching, by Bigarella, of a large parabolic dune with smaller ones inside at the

Lagoa dune field in Santa Catarina, Brazil, has disclosed many additional features discussed later in this chapter.

Parabolic dunes, whether inland or coastal, seem to characteristically migrate very slowly (McKee and Douglass, 1971, p. D113). In most places they are anchored by vegetation and in coastal areas, especially, they are partly stabilized by moisture. In Brazilian retention dunes, which are in a similar coastal environment, "the supply of sand from the source area [beach] seems to be large enough to overcharge the gentle slope of the dune's surface * * * but the winds are not so intense nor the sand so dry that the complete removal of the previously deposited beds can be accomplished * * * [so] a new set of strata just barely covers the former one" (Bigarella, Becker, and Duarte, 1969, p. 38).

The partly stabilized coastal dunes of Brazil also show virtually concordant attitudes between slipface sets. The difference in angle between strata of various sets is small despite disturbances from slumping along slipface surfaces, according to Bigarella, Becker, and Duarte (1969, p. 38).

Dunes of normal parabolic shape are well developed at the high altitude and under the cool dry climatic conditions of western Wyoming. Their structures have been examined by Ahlbrandt (1973, p. 39 –41). One dune of this type is described as having foresets dipping 24° –34°, commonly 26° –30°. Internal structures are disturbed by plant roots, and laminae of the dune nose exhibit minor folding. A bulging of the dune nose is reflected in the convex-upward form of the laminae as seen in the test trench (fig. 52). Thus, the Wyoming dune displays the same principal structures observed at White Sands and in the Brazilian coastal dunes.

Retention Ridges

Retention ridges, common features of many coastal dune areas, characteristically form long narrow dunes parallel to the shore on the landward margins of backshore beach deposits. They develop where dune sand accumulates against vegetation barriers, forming a nearly stabilized type of dune that may gradually be enveloped by trees and other plants. This type of dune ridge is common along the west coast of the United States and in that area has been referred to by Cooper (1958, p. 55; 1967, p. 22) as "precipitation ridge."

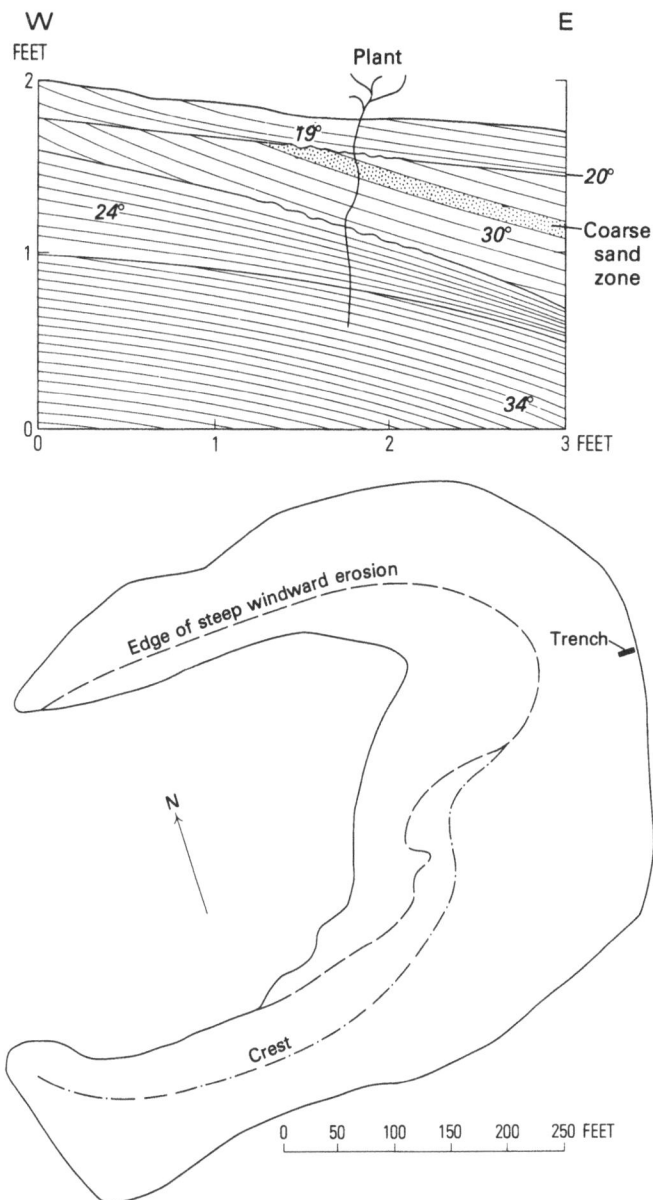

NORTH WALL OF A TRENCH on the nose of parabolic dune, Killpecker dune field, Wyoming, U.S.A. Strata are concave downward in lower two sets. From Ahlbrandt (1975, figs. 8, 14). Measurements in feet. 1 foot = 0.3 metre. (Fig. 52.)

Primary structures in retention ridges of Brazil were studied in detail by Bigarella, Becker, and Duarte (1969, p. 28–29) and by Bigarella (1972, p. 20–21). Trenches were dug and cross sections examined at Jardim São Pedro (fig. 53), Guairamar, and Pôrto Novo on the coast of Paraná in Brazil. These dune ridges formed under conditions of high humidity and considerable vegetation and included many "blowout" features. They apparently advanced very slowly and contained most of the

structural characteristics of parabolic dunes. "Convex-upward strata * * * are rather frequent in the dunes studied," and distinctive types of distortion on many of the slipface surfaces were common features (Bigarella, Becker, and Duarte, 1969, p. 37, 45).

A distinctive feature of Brazilian ridge dunes (beach dune ridges) with high moisture content is that "both the topset and foreset are made up by the same continuous set of strata, and not by different truncated sets" (Bigarella, Becker, and Duarte, 1969, p. 36). Probably the abundance of sand supply and the slow rate of movement (because of internal dampness) are responsible for this feature, suggesting that it results from the humid climate, rather than the type of wind. Sinuous sand layers, of medium scale, that occur in lower parts of the Brazilian retention ridges (fig. 53C) are similar to strata in small dome dunes that commonly occur to seaward. Above the wavy basal beds in the ridges examined were tabular-planar sets of cross-strata. In some of these sets the bounding surfaces dipped at low angles, and in others, at moderate to high angles (fig. 53). The tabular-planar sets included both nonerosional (simple) and erosional types; few trough-type cross-strata were associated.

Some cross-strata in the retention ridge dunes of Brazil dip downwind at very high angles (34°–39°) probably owing to high humidity at the time of deposition (Bigarella, 1972, p. 21). In sections normal to prevailing wind direction, however, most of the stratification is horizontal or nearly horizontal, except across protuberances in which strata dip at low angles toward the outer margins, giving a strongly curving appearance in section. Sets of cross-strata tended to be realtively thick (some more than 3 feet (1 m)) and to have relatively high angles of dip near the dune base, but in the upper part they were thinner (0.32–0.98 foot (0.1–0.3 m)) and had lower angles of dip.

The degree and direction of dip from several dune sections of the Guairamar dune are illustrated by stereonets (fig. 54). The spread of high-angle (>30°) foreset dip directions at Jardim São Pedro is more than 170° (Bigarella, Becker, and Duarte, 1969) which is very much greater than the spread of barchanoid and transverse dunes at White Sands, New Mexico, in the United States.

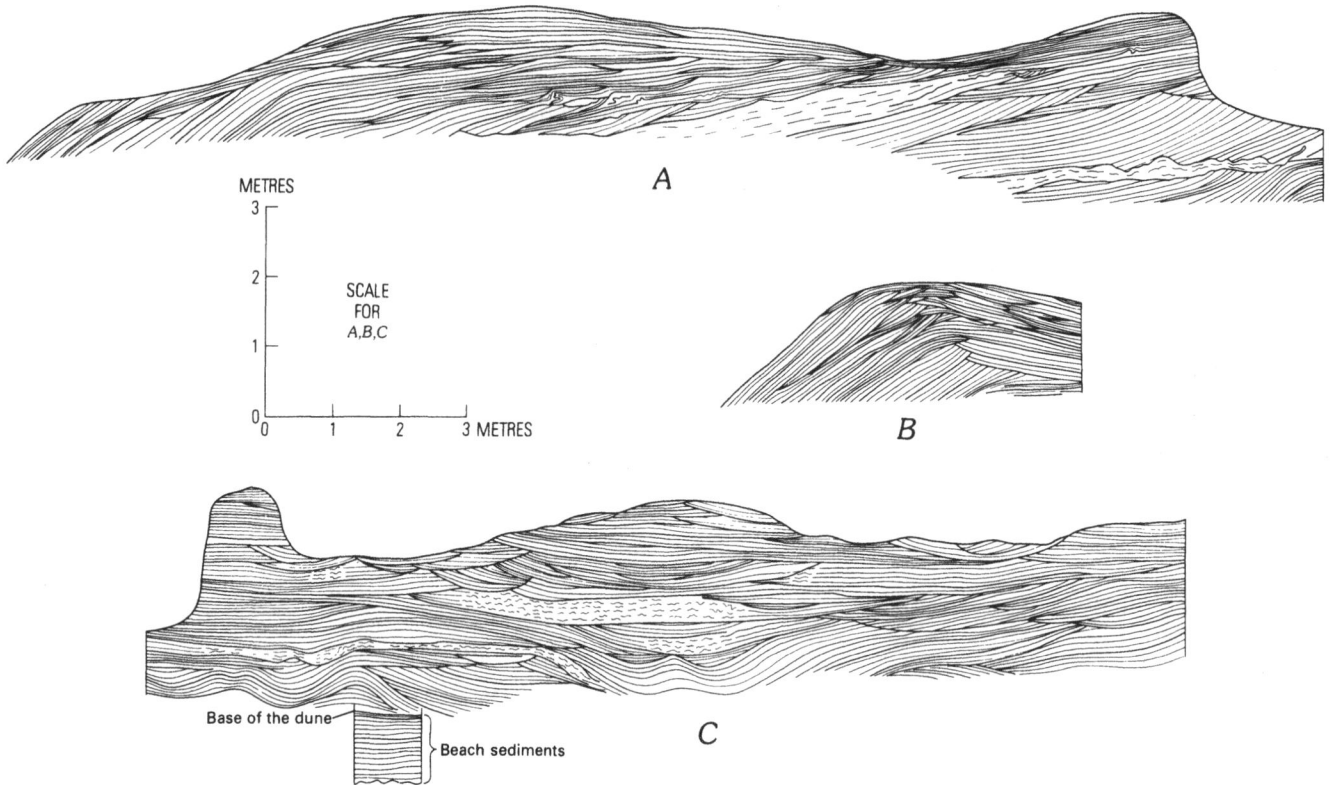

RETENTION DUNE CROSS SECTIONS showing details of stratification, Jardim São Pedro, Praia do Leste, Paraná, Brazil. A, Section parallel to prevailing wind; B, section normal to prevailing wind at linguoid protuberance where crest and slipface are curved; C, section normal to prevailing wind at middle part of dune. Modified from Bigarella, Becker, and Duarte (1969). 1 metre = 3.28 feet. (Fig. 53.)

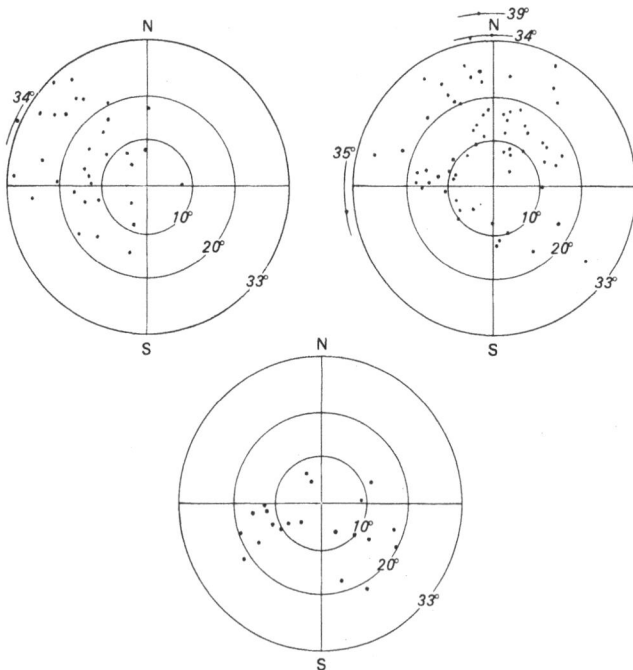

STEREONETS FOR THREE SECTIONS showing angle (distance from center) and direction (azimuth) of dip on foresets of retention ridge, Guairamar, Paraná, Brazil. Modified from Bigarella, Becker, and Duarte (1969). (Fig. 54.)

Dome

Dome dunes develop in the upwind parts of some dune fields, where winds are sufficiently strong to bevel and, thus, effectively retard a normal upward growth of dune crests. Three localities where dunes of this type are characteristically developed and where they have been trenched in order to determine the primary structure pattern are (1) White Sands National Monument, New Mexico, U.S.A., a desert environment, (2) the Killpecker dune field, Wyoming, U.S.A., a high-altitude, cold climate, and (3) along the southern coast of Brazil, a warm, humid environment.

The dome dunes at White Sands National Monument, New Mexico (described in detail by McKee, 1966, p. 26, 27), are round or oval in ground plan, have rounded tops, and do not display steep avalanche faces on their lee sides. One dune — 490 feet (150 m) wide, 460 feet (140 m) long, and 20 feet (6 m) high — considered to be typical of that area, was selected for trenching and study. Its texture showed poorer sorting and a higher percentage of

TRENCH THROUGH DOME DUNE, White Sands National Monument, New Mexico, U.S.A. Section parallel to dominant wind from left. Scale is in inches. 1 inch = 2.54 centimetres. (Fig. 55.)

coarse grains than did the other types of dunes located farther downwind.

Structures in the upwind part of the White Sands dome dune consisted largely of tabular-planar cross-strata (fig. 55) with foresets dipping downwind at high angles (28° –33°). In the downwind part, foresets dipped somewhat less (14° –27°), perhaps because much of the sand there had been deposited from suspension rather than mainly from slumping (fig. 56A). In a trench normal to the dominant wind direction, most of the stratification appeared to be horizontal or to have very low dip toward the dune margins (fig. 56B). Thus, in basic structure, this dune type resembles other unidirectional wind deposits at White Sands. The dome dunes probably began as isolated barchans, if shape and structure can be used as criteria.

Two features of the dome dune seem to be distinctive and are either absent or not well developed in other types of dunes in the area. First, cut-and-fill structures — probably the result of trough-forming deflation by strong, unchecked winds, followed by sand deposition under weaker winds — formed a series of sand-filled scours parallel to wind direction. Second, in the same section a layer of horizontal or nearly horizontal

laminae, capping the rounded dune top, formed topset beds that connected with dipping foresets downwind. Both features distinguish dome dune structures from those of other dune types in the area.

Dome dunes in the Killpecker dune field of western Wyoming, like those at White Sands, occur farthest upwind of all dunes, and are developed under the full force of approaching sand storms. Three trenches dug by Ahlbrandt (1973, p. 32) in a Killpecker dune revealed structures generally similar to those at White Sands. In sections parallel to wind movement, strata dipped consistently downwind at the angle of repose (about 34°) to form tabular-planar cross-strata, except on the lee side, where fractures (fig. 57) were developed from incipient slipface movement. These fractures defined slump blocks which locally were responsible for great oversteepening of beds, some of which dipped as much as 55°. The steepness of dips on such beds is attributed to the accumulation of snow layers within the deposit with subsequent melting and collapse of overlying cross-strata.

Other features of the dissected dome dune in the Killpecker field were (1) beds dipping 1° –4° to windward in the upper 1.5 feet (0.5 m) on the upwind side, apparently resulting from an overabundance of sand supply, and (2) slump structures on the downwind side consisting of many normal faults and a few fadeout laminae, typical of avalanches in many dune sands (McKee, Douglass, and Rittenhouse, 1971, p. 370 –372).

Dome dunes on the coasts of Paraná and Santa Catarina in Brazil occur as miniature foredunes, seaward of retention ridges that parallel the coast. They resemble dome dunes of the desert areas of the Western United States in that they also form in unobstructed upwind locations and develop with oval or circular shapes and with relatively flat tops. They differ from the desert dome dunes in their smaller size — typically less than 1 metre (3 ft) high and 12 –14 m (40 –45 ft) in diameter — and in the type of structures and large spread of dip directions of their stratification. The differences probably result mainly from contrasts in depositional environment. In coastal Brazil, the domes are formed under humid conditions and on a surface roughened by many grasses and small shrubs.

Structures in Brazilian dome dunes have been examined and described by Bigarella and Popp (1966, p. 143 –148) at Barra do Sul, by Bigarella, Becker,

MAIN TRENCH, NORTH WALL

A, continued

SIDE TRENCH, WEST WALL

B

EXPLANATION

—— Bounding surface

—— Selected cross-stratum

21 Apparent dip, in degrees

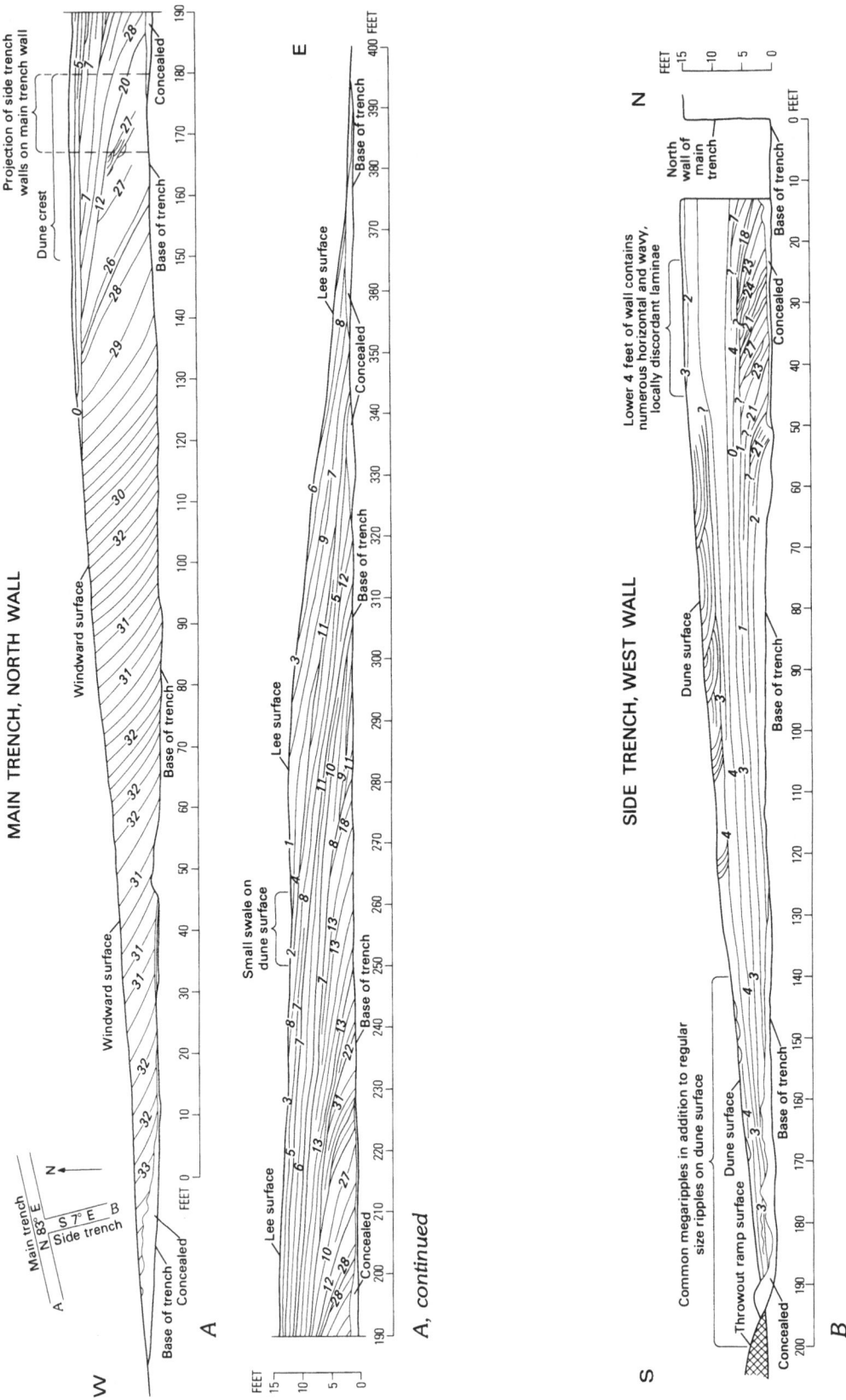

CROSS SECTIONS OF DOME DUNE, White Sands National Monument, New Mexico, U.S.A., (A) parallel to dominant wind direction and (B) normal to dominant wind direction. Modified from McKee (1966, fig. 6). All measurements are in feet. 100 feet=30.5 metres. Lines are queried where structure is obscure. (Fig. 56.)

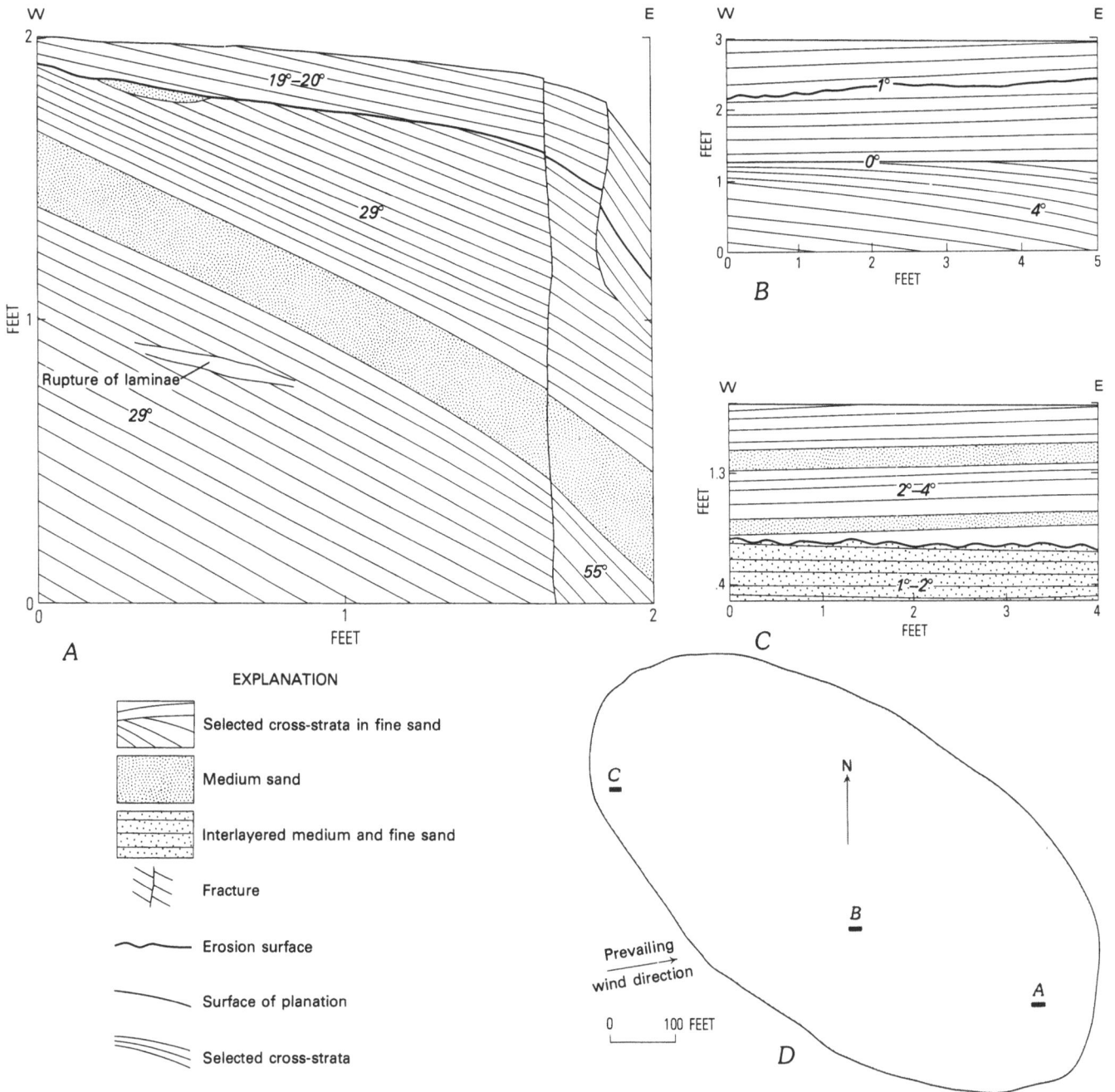

EXPLANATION

Selected cross-strata in fine sand

Medium sand

Interlayered medium and fine sand

Fracture

Erosion surface

Surface of planation

Selected cross-strata

DOME DUNE, Killpecker dune field, Wyoming, U.S.A. A, Stratification on steep lee side in section parallel to wind direction. Note fractures defining slump blocks caused by snowmelt and collapse of strata. B, Section near crest showing lee-dipping strata overlain by horizontal strata. C, Section upwind, showing low-angle capping beds dipping upwind. D, Ground plan of dome dune, showing location of sections A, B, and C. Modified from Ahlbrandt (1973). 1 foot = 0.3 metre. (Fig. 57.)

and Duarte (1969, p. 17) at Praia do Leste, and by Bigarella (1972, p. 24 –25) at Barra do Sul. They consist, in general, of numerous sets of superimposed, curving strata in which bounding planes and many laminae are convex upward (fig. 58). Uniformly dipping, high-angle foresets, such as form the cores

and lower parts of desert domes, are lacking, and the characteristic sets of low-angle strata, dipping in many directions, apparently represent windward-side deposits, topsets, and also foresets; few avalanche deposits are recognized. In brief, the domes consist of blanket deposits resulting from a

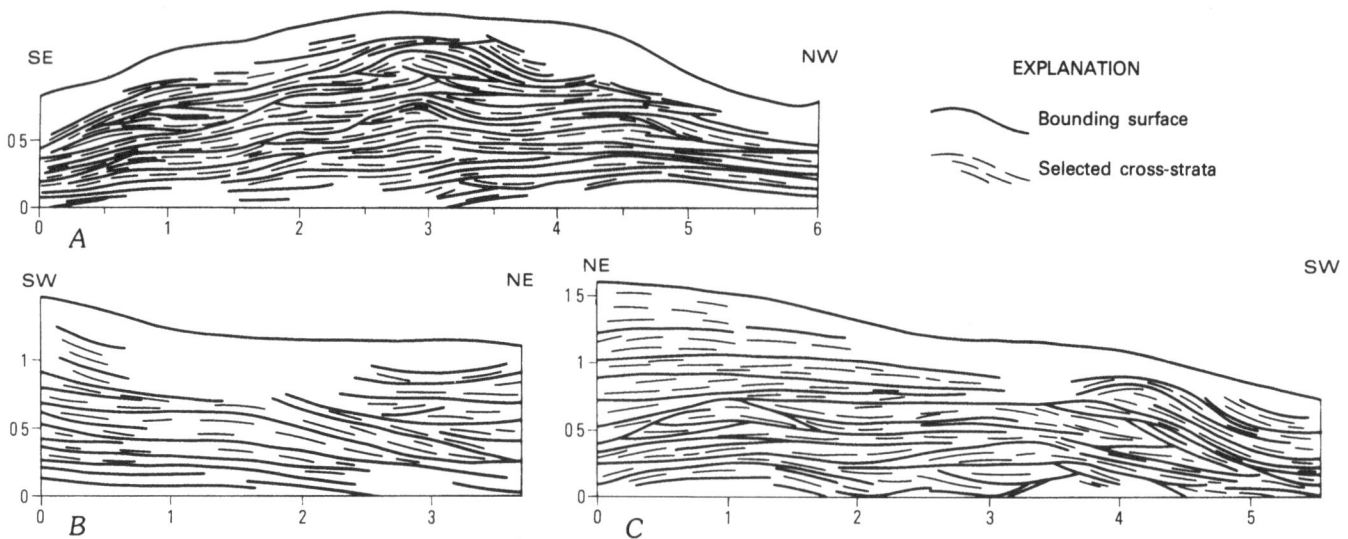

CROSS SECTIONS OF COASTAL DOME DUNE, Praia do Leste, Paraná, Brazil, showing (A) stratification in section parallel to dominant wind direction, and (B, C) stratification in sections normal to dominant wind directions. Modified from Bigarella (1972). All measurements are in metres. 1 metre = 3.28 feet. (Fig. 58.)

large source of material and a wide spread of winds that alternately deposit and erode sand on an irregular surface.

Brazilian coastal dome dunes are characterized by strata with "dip directions * * * distributed completely around the compass" (Bigarella, 1972, p. 25). This is partly the result of surface irregularities from vegetation and partly from retarding effects of moisture. Both upwind dips and nearly horizontal topset beds are also developed in domes of desert sands, but on a larger scale. They occur as blankets above normal foresets of tabular-planar sets. In both coastal and desert environments a large sand source is available, but in the humid environment with abundant vegetation, accumulation of sand is irregular, in time and direction, and associated blowout features are common.

Linear

One of the most common but least understood types of desert dune is the long straight ridge herein called a linear dune (fig. 59). It is commonly referred to as a seif dune in North Africa and Saudi Arabia, as a sand ridge in Australia, and as a longitudinal dune in many other places; its pattern, as observed on Landsat (ERTS) images (McKee and Breed, 1976, p. 81) has been termed "parallel straight." At least three subtypes of this dune form

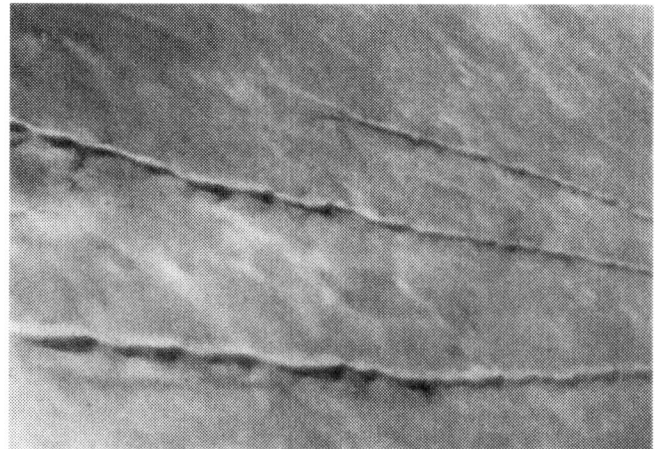

AERIAL VIEW OF SEIF, or linear, dunes, north of Sebha, Libya. Dune ridges are about 50 feet (15 m) high. Distance between dunes ranges from ¼ to ½ mile (0.4 to 0.8 km). (Fig. 59.)

may be distinguished, according to Price (1950, p. 462), which doubtless is one reason for the current confusion regarding its genesis. Differences in details of shape likewise occur in various sand seas, and many varieties are recorded, including the feather-type in Saudi Arabia, the zigzag type in Libya, and the converging, nonparallel type with junctures in Australia.

An environment favorable to development of linear dunes seems to be one in which there is high wind velocity (Glennie, 1970, p. 95; Bagnold, 1941,

p. 171; Twidale, 1972, p. 82). As pointed out by Bagnold (1941, p. 178), "a strong sand-laden wind with uniform drift over a uniform rough surface has transverse instability so sand tends to deposit in longitudinal strips." According to Glennie (1970, p. 95), "The higher wind velocity, the larger the seif dune and the greater the interdune spacing."

The origin of linear dunes has long been a matter of controversy; many geologists (Bagnold, 1941, p. 223; McKee and Tibbits, 1964; Twidale, 1972, p. 103; and others) believe them to be the product of two-directional winds, but others attribute them to a unidirectional or prevailing wind (Dubief, 1952; Glennie, 1970). Even among the geologists who adhere to one or the other concept of directional control by wind, there are many differences in the mechanisms by which this dune form is attained.

Some linear dunes are explained as forming from the trailing arms of parabolic dunes that have become separated from dune noses (Dubief, 1952; Capot-Rey, 1943; King, 1960; and others); other linear dunes are shown to have formed by elongation of one horn of a barchan by winds oblique to the prevailing transport direction (Bagnold, 1941, p. 223). An explanation that calls for the development of wind cells in interdune areas through pressure gradients along the dune sides was proposed by Glennie (1970, p. 90), and experiments with running a roller across a layer of grease suggested to Folk (1971b, p. 3464) a theory of helicoidal air flow. Recent work on Australian dunes led Twidale (1972, p. 103) to the conclusion that "longitudinal dunes are being initiated from leeside mounds bordering playas and alluvial plains" and that "these mounds cause changes in the airflow in a bi-directional regime."

Regardless of the mechanism by which linear dunes are formed, their structures make a distinctive pattern very different from those of barchanoid-type dunes. In linear dunes that have been examined thus far, slipfaces develop on both sides of the crest, and foreset dip directions form two clusters of points in nearly opposite directions, rather than a single grouping. Unfortunately, few detailed studies based on trenching have yet been made of these dunes, so examples in different sand seas that appear similar in size and form may not actually be alike structurally.

Probably the first reconstruction of linear dune structure was that by Bagnold (1941, p. 242) in which he showed a nearly symmetrical, hypothetical section based on "the few tests already made by the author." This section (fig. 60) shows the "arrangement of accretion and encroachment deposits," with the middle and upper parts of the dune composed of avalanched sand, and the low outer margins, or plinth, on both sides, formed of accretion deposits. The section shows growth of a perfectly regular type, but, as Bagnold pointed out, successive stages in nature doubtless are very irregular. In any event, deposition probably alternates from side to side by avalanching down the steep slipfaces.

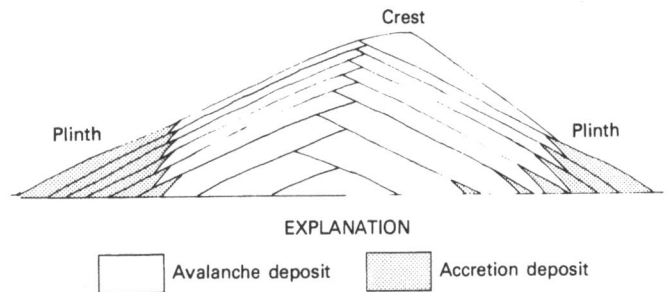

HYPOTHETICAL SECTION, showing structure of seif dune. Modified from Bagnold (1941, fig. 83). Note foreset dips in opposite directions. (Fig. 60.)

Confirmation of the validity of the basic structure pattern as determined by Bagnold is found in a series of trenches made by McKee and Tibbits (1964, p. 17) across a "seif dune" near Sabhā, Libya, and by laboratory experiments conducted in Denver, Colorado, U.S.A., in which a small linear dune was developed. In the section across the Libyan dune (fig. 61), high-angle cross-strata (26°–34°), presumably slipfaces, show at many places in the upper half of the dune and on both sides; low-angle (4°–14°) cross-strata occur on the lower flanks of the dune and probably represent accretion beds.

Reversing

The existence in central Asia of dunes formed as a result of reversing winds was noted by Sven Hedin in 1896 (p. 267–268). Likewise, dunes of this type were described and discussed by Cornish in 1897 (p. 286). Neither man had examined or

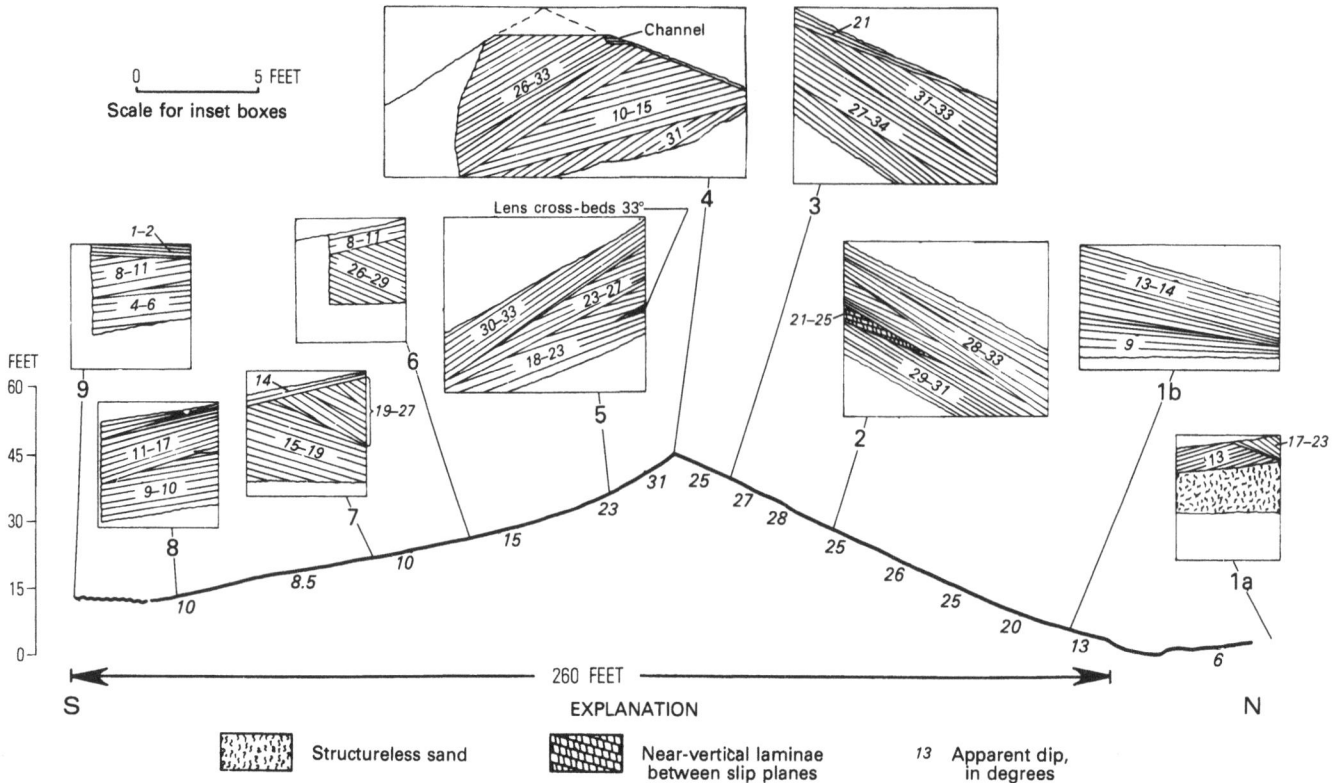

CROSS SECTION OF SEIF DUNE near Sebhā, Libya, based on series of test pits. From McKee and Tibbitts (1964, p. 17). All measurements are in feet. 1 foot = 0.3 metre. (Fig. 61.)

analyzed the internal structures of the dunes, but Cornish's sketches (1897, fig. 7) clearly indicate that steep slipfaces had developed on opposite sides of a sand hill with re-forming of the dune top during a wing reversal. Another early reference to reversing dunes is by Bagnold (1941, p. 242), who stated, in discussing dunes in North Africa: "Reverse slip-faces formed during bouts of wind from opposite directions can be distinguished at once by direction of the sloping shear-planes."

Reversing dunes at Great Sand Dunes National Monument in southern Colorado in the United States were examined by Merk (1960, p. 127) with regard to internal structures. From excavation of a few test pits, Merk showed that the bases and lower parts of the large dunes indicated deposition by a prevailing southwesterly wind, whereas the upper parts commonly re-formed with steep slipfaces toward the southwest, during periods of easterly storm winds. Thus, in a single dune of barchanoid or transverse type (fig. 62), two groups of steeply dipping foresets, facing nearly opposite directions, were developed (fig. 63). Further, in many large

dunes, a considerable number of low-angle laminae that probably formed as windward-slope deposits, have been buried in the lower parts during reversals of wind direction. Contorted structures commonly result from slumping during the frequent reversals.

Reversing dunes about 30 feet (9 m) high occur in Antarctica, where approximately 30 barchans of reversing type were reported from Victoria Valley by Lindsay (1973, p. 1799). His descriptions of these dunes indicate that they are much like reversing dunes elsewhere; however, they have frozen cores. Slopes developed on windward sides are said to dip an average of about 6° in their basal parts, but are much steeper (approximately 15°) in the upper parts; the zone of change apparently coinciding with the lower limit of surface moisture. Avalanche slopes are reported to range from 29.5° to 33°, which is normal for most dunes elsewhere; this steepness readily differentiates the slipfaces from other surfaces. Below the slipfaces are relatively low-angle (14°–16°) slopes referred to as "aprons."

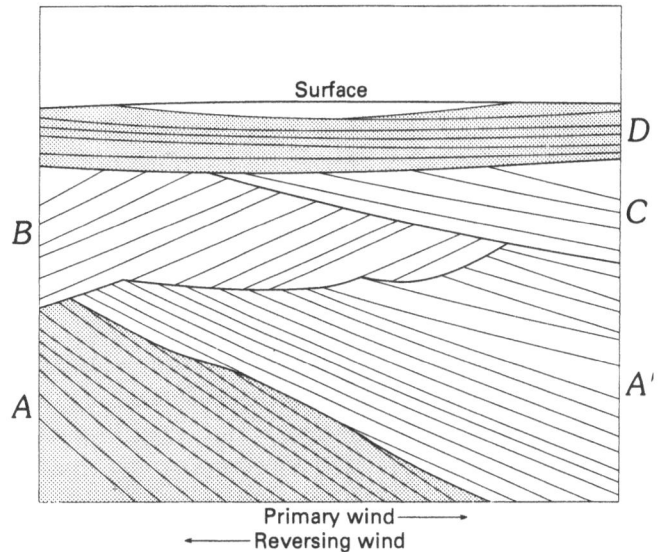

STRUCTURE IN REVERSING DUNE, parallel to dominant wind, Great Sand Dunes National Monument, Colorado, U.S.A. Based on photograph by Merk (1960). Primary wind from left to right; reversing wind from right to left. Set *A*, *A'*, primary dune foreset laminae; set *B*, modified dune windward laminae; set *C*, primary dune foreset laminae; set *D*, modified dune windward laminae. (Fig. 63.)

REVERSING DUNES, Great Sand Dunes National Monument, near Alamosa, Colorado, U.S.A. Higher dunes are as much as 700 feet (210 m). Aerial photographs by J. S. Shelton. (Fig. 62.)

The attitude of the cross-strata and the structure patterns of the reversing dunes in the Kelso dune field, California, U.S.A., as determined by Sharp (1966, p. 1062) illustrate several distinctive features: (1) High-angle foresets of lee-side deposits are relatively scarce, suggesting that the "steeply dipping (>30°) slipface deposits are rarely preserved within these dunes." (2) Lee-side bedding which underlies a "veneer of windward deposits" dipped leeward 10°–15° in many pits, indicating that most of these strata are from the lower part of foresets. (3) "Bedding relationships are more complex and thickness of units greater in the crestal areas" presumably because of frequent reversals in asymmetry and "back-and-forth movement of the crest."

At Lagoa, Ilha de Santa Catarina, on the southern coast of Brazil, a large reversing dune, caused by seasonal shifts in wind direction, was dissected and examined by McKee and Bigarella (1972, p. 677). In this dune, steep foresets, formed on slipfaces that dip in two nearly opposite directions, are conspicuous. Both stairstep folds and thrusts, characteristic of avalanches in wet, strongly cohesive sand, are abundant (McKee and Bigarella, 1972, fig. 3).

The elongate ridges of the Kelso dunes in the Mojave Desert of California probably have received more detailed study than any other reversing dunes. They lie in a valley nearly surrounded by desert ranges but are exposed on the northwest side, from which additional windblown sand is introduced. During an interval of 15 years, measurements were made and data kept at 10 stations (Sharp, 1966); sand movements were recorded, and changes both in dune form and in the facing direction of dunes were documented. Although no trenching was done, dune movements were measured precisely with respect to 10 series of fixed poles, and the nature of dune structures was determined by digging a large number of pits, 4–5 feet (1.0–1.5 m) deep, at various locations.

The most frequent winds in the Kelso dune field are from the west, but winds from other quadrants are believed (Sharp, 1966, p. 1049) to have higher velocity and, therefore, to be nearly equal in sand-moving capacity. In any event, a summary of many measurements shows that the dune ridges moved northeast at the slow rate of only 12 feet (3.6 m) in 12 years but that during that time "the crest shifted back and forth within a zone about 30 feet (9 m) wide." Thus, "scouring and truncation of dune crests occur when a sharp crest is subjected to a powerful reverse wind" (Sharp, 1966, p. 1050).

At White Sands National Monument, New Mexico, U.S.A., where dunes are mostly the result of unidirectional winds, cross-strata resulting from reversal in wind direction were preserved in some cross sections (fig. 64) parallel to the dominant wind direction (McKee and Douglass, 1971, p. D114, fig. 3). Weather records show that in this area the principal wind direction normally shifts from southwest to north for a short period each winter season, and this shift is recorded in the dune structure.

To summarize the relatively scant data available, reversing dunes are characteristically developed under many different climatic regimes. They occur in arid, low-altitude areas (Kelso dune fields, California, U.S.A.); in humid, low-altitude areas (Santa Catarina, Brazil); cold, high-altitude areas (Great Sand Dunes National Monument, Colorado, U.S.A.); and cold, low-altitude areas (Antarctica). Because their development seems to be basically a function of contrasting wind directions, regardless of other environmental factors, the topographic control (especially mountain ranges) of wind movements seems to dictate the type of dune pattern developed. In brief, most, if not all, of the reversing dunes cited are in areas where the wind pattern is locally but strongly affected by physical barriers.

Star

Star dunes (fig. 65), apparently the result of wind from several directions, are among the most complex of dunes, from the standpoint of primary structure. Two representative examples of such dunes, 100 feet (30 m) and 35 feet (11 m) high, in central Saudi Arabia, near Zalim, were examined and described (McKee, 1966, p. 65).

The larger star dune (fig. 65), with four arms or ridges extending out from a high central crest, indicates by its structure a shifting in the direction of wind movement. Each arm has a slipface on one side, with characteristic high-angle dips (31°–33°) and a more gentle slope (17°–21°) on the other (fig. 66). Thus, a seasonal rotating wind direction probably was necessary to initially establish the positions of these arms. Confirmation of a shift in direction of wind was found in cross sections exposed in trenches at the dune crest. These sections showed cross-laminae dipping at high angles in three principal directions which coincided with directions of seasonal wind movement in the area.

The smaller star dune that was trenched and studied near Zalim, Saudi Arabia, had three arms

SECTION IN A TRENCH parallel to dominant wind direction, White Sands National Monument, New Mexico, U.S.A., illustrating effects of wind reversal during winter season. Scale is in inches. 1 inch = 2.54 centimeters. (Fig. 64.)

STAR DUNES, southeast of Zalim, Saudi Arabia. Crests are 70–100 feet (20–30 m) high. Aerial photograph by E. Tad Nichols. (Fig. 65.)

TRENCH AT CREST of 100-foot-high (30.5-m-high) star dune, southeast of Zalim, Saudi Arabia. Wedge-planar cross-strata dip steeply in three directions. Scale is in inches. 1 inch = 2.54 centimetres. Photograph by E. Tad Nichols. (Fig. 66.)

and was structurally more complex and difficult to explain genetically than the large one. However, it also showed a complicated mixture of structures dipping in several directions (McKee, 1966, p. 68) and almost certainly was the result of alternation among several wind regimes.

Star dunes occur in many parts of the world and under a variety of conditions. They are especially numerous in the Rub' al Khali of Saudi Arabia and in the Grand Erg Oriental of Algeria but also are well developed in the Gran Desierto of Sonora, Mexico. In most places they appear as individual or separate mounds randomly scattered across a sand surface, but in other places two or more seem to have coalesced to form compound dunes. In several sand deserts (fig. 67), they are distributed along the crests of linear dunes, forming parallel rows of stars, sometimes referred to as "strings of beads." These are complex dunes, composed of two distinctive types.

So far as is known, neither complex nor compound star dunes have ever been trenched for structure analysis. Observations on the orientation of slipfaces suggest, however, that, as the product of multiwind development, their structures are complicated and show a wide spread in dip directions. Furthermore, the great heights attained by many star dunes indicate a dominant vertical, rather than horizontal, growth.

The rows of compound dunes suggest two generations of growth — the earlier one with a bimodal-wind environment, the later one with multiple-wind environment. The overall structure pattern represented in sand deserts of this type can only be determined with certainty when direct observations from trenching have been made. Entirely unrecorded are the structural characteristics of interdune areas either in fields of scattered star dunes or between parallel ridges of compound linear-star dunes.

Sedimentary Structures of Interdune Areas

INTERDUNE AREAS, which are the spaces between dunes, differ greatly in size and shape, depending largely on the types of dune that are separ-

STAR DUNES ALONG RIDGES of linear dunes, Namib Desert, 6 miles (10 km) south of Kuiseb River at Gobabeb, South-West Africa. Photograph by E. Tad Nichols. (Fig. 67.)

ated by them. Study of deposits that accumulate within interdune areas of barchanoid types show the sediment to be mostly (1) dust or silt-size particles that settle out of the atmosphere, (2) organic matter largely derived from vegetation, and (3) limited amounts of sand that have strayed from adjacent dunes (fig. 68). In the White Sands area that has been studied in detail these deposits are very thin, ranging in thickness from 1 to 4 feet (0.3 to 1.3 m), depending largely on how long the surface has been exposed before burial by advancing dune forsets.

When dunes migrate across a dune field with a unidirectional wind, the sand surface of the dune ahead of each interdune is constantly being beveled through truncation of its cross-strata (fig. 69), and, at the same time, the windward side of the interdune is being buried and preserved. If dune migration is very rapid, interdune deposits will be especially thin or even lacking. If migration is very slow, an appreciable amount of sediment may accumulate on the interdune surface before burial.

Regardless of the migration rate, the proportion of clean, well-sorted eolian sand to the impure, contaminated sediment of the interdunes is great, and the result is thick sand layers of crossbedded sets, separated by thin parting beds of mixed sediment with flat-lying or irregular strata. Such relations have been established through a drilling program for the White Sands dune field, New Mexico (McKee and Moiola, 1975), where three or four sequences of dune sand are separated by a corresponding number of interdune deposits.

In some dune fields — for example, in parts of Algeria and Australia — the sand supply is sparse and forms isolated dunes on a bedrock or other non-sand surface. In southern Algeria, sand dunes rest on a pediplane cutting Precambrian metamorphic rocks. Under such circumstances, sand is very transient and migrates until some sediment trap causes it to accumulate as a large body. In areas of considerable sand accumulation, the interdune surface may show beveled tops of earlier dunes that have recently been planed off. Such surfaces define on

INTERDUNE AREA showing miniature stringers and sheets of flat-bedded sand (wind-shadow deposits) advancing across interdune. White Sands National Monument, New Mexico, U.S.A. From McKee and Moiola (1975, fig. 6). (Fig. 68.)

horizontal planes the curving arcs of barchans (fig. 70), the long, relatively straight crest lines of transverse ridge dunes (fig. 71), or other patterns representing stages in dune migration. Where these interdune surfaces have been exposed for an appreciable time, the relict structures are obliterated by accumulating sediment.

Interdune deposits at White Sands National Monument, New Mexico, U.S.A., were examined in vertical sections by trenching in areas contiguous both to barchanoid ridges and to transverse dunes (McKee and Moiola, 1975, p. 61). Structures in these sections consisted of thin, flat or irregular beds of silts and sands with scattered fragments of plants and with laminae contorted by root growth (fig. 72). The interdune deposits examined at the surface, and also in cores, range in thickness from 1 to 4 feet (0.3 to 1.3 m). A few kilometres upwind in

the same dune field, where sand movement is much more rapid, the thickness probably is proportionately less. Interdune surfaces in many such places are represented merely by horizontal bedding planes.

The Killpecker dune field in Wyoming, also formed mainly by unidirectional winds, has been examined through trenching, and the nature of interdune structures has been determined (Ahlbrandt, 1973, p. 41, figs. 20, 66, 67). Trenches in this field showed mostly horizontal laminae disrupted in places by root tubules and consisting in part of marl developed in interdunal ponds. The deposits examined were formed in the upwind part of the field under the influence of very strong winds; the sand included some of the coarsest and most poorly sorted detritus in the entire dune field.

Structures of interdune areas among the coastal dunes of Brazil, like the just-described North American desert types, consist mostly of horizontal laminae that rest upon truncated cross-strata like those of nearby active dunes (Bigarella, 1972, p. 27). The beveled foresets are the remnants of earlier dunes that migrated across the area; they formed an overlapping pattern as "the result either of changes in wind direction or of irregular superposition of the curved foreset strata." Coastal-type interdunes in Brazil are further discussed by Bigarella later in this chapter.

Interdune surfaces in areas of parallel-straight or linear dunes, commonly referred to as seif or longitudinal, represent a depositional environment very different from that among transverse-type dunes. Between linear-type dunes, the interdune surfaces commonly are many times wider than the sand ridges that enclose them (fig. 59). Furthermore, these surfaces between the long, straight dunes extend in a direction generally parallel to the prevailing wind or winds. They have an open accessway upwind and few, if any, barriers downwind.

Detailed studies of a well-developed interdune deposit between linear (seif) dunes in the Fezzan of Libya were made in 1962 (McKee and Tibbits, 1964, p. 13). These studies indicated that effective winds moving down the length of a flat interdune surface deposited sand as flat-lying laminae to a thickness of at least 4 feet (1 m), as shown in test pits (fig. 73). No cross-strata were noted. The sand as determined by analysis was poorly sorted, ranging in size from

BEVELED WINDWARD SURFACE of transverse dune showing truncation of lee-side foresets, White Sands National Monument New Mexico, U.S.A. From McKee and Moiola (1975, fig. 6B). (Fig. 69.)

BEVELED REMNANT OF BARCHAN DUNES in interdune area, White Sands National Monument, New Mexico, U.S.A. Three truncated sets of cross-strata are visible. (Fig. 70.)

BEVELED REMNANT OF TRANSVERSE DUNE in interdune area, White Sands National Monument, New Mexico, U.S.A. From McKee (1966). (Fig. 71.)

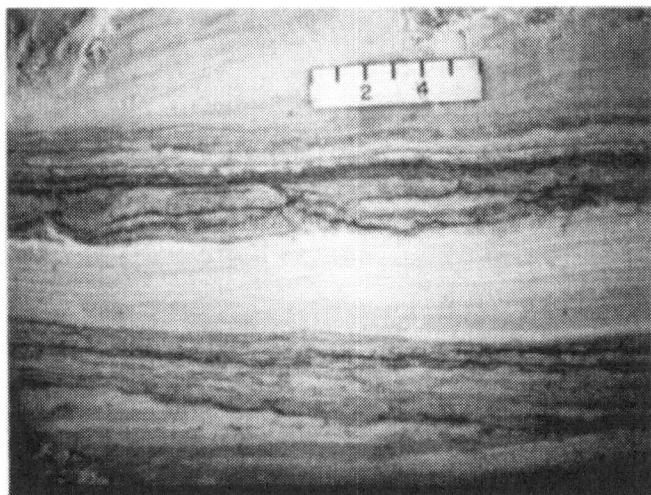

HORIZONTAL AND IRREGULARLY BEDDED STRATA in trench, interdune area between barchanoid ridges, White Sands National Monument, New Mexico, U.S.A. Scale is in inches. 1 inch = 2.54 centimetres. From McKee and Moiola (1975, fig. 6). (Fig. 72.)

very fine to coarse, in contrast to well-sorted sand in nearby dunes. Along the margins of the linear dunes, which have slipfaces on both sides dipping at high angles (as much as 35°), interdune deposits consisted partly of low-angle strata (10°–13°), dipping toward the dunes, and partly of very fine structureless sand. No information is available on the thickness of these interdune deposits.

The nature of interdune deposits between star dunes (fig. 74), which are the product of multidirectional wind regimes, has not been investigated, to the author's knowledge. Such dunes mainly build vertically, rather than migrating laterally. After attaining a maximum peak height permitted by the threshold velocity, they must spread outward from the base, so it seems probable that the interdunes develop few of the features characteristic of troughs (sediment traps) associated with the active migration of transverse-type dunes. The structural character of interdunes in fields of star dunes is yet to be determined.

Deformational Structures in Dunes

IN ADDITION TO sedimentary structures formed by deposition, numerous penecontemporaneous structures, caused by deformation of strata during or immediately following deposition, are characteristic of various dune types. Principal

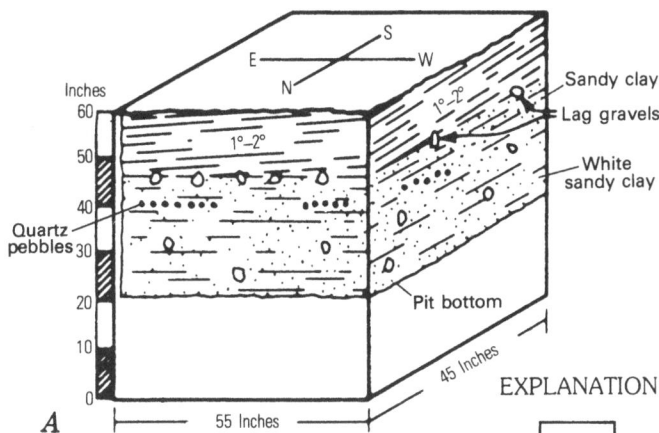

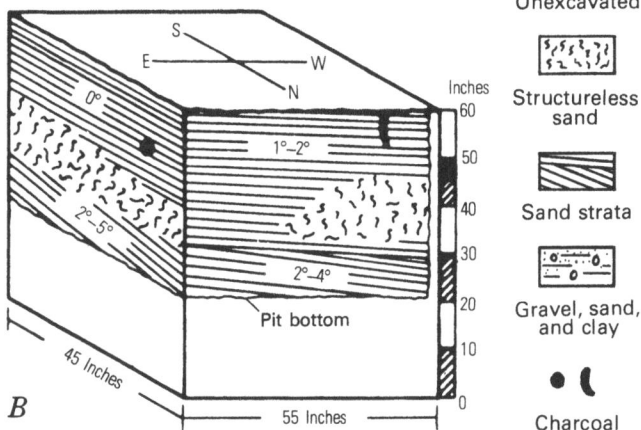

BLOCK DIAGRAMS (A, B) OF INTERDUNE STRUCTURES and photograph of test pit between seif dunes, northeast of Sebhā, Libya. A, pit 14; B, pit 12. From McKee and Tibbitts (1964, fig. 8a, c). C, Test pit in center of interdune area, shows horizontal stratification. Steel tape extends 4 feet (1 m) down. Photograph by Gordon C. Tibbitts, Jr. (Fig. 73.)

INTERDUNE SURFACE AMONG STAR DUNES near Tsondab Vlei, Namib Desert, South-West Africa. Photograph by E. Tad Nichols. (Fig. 74.)

forms of deformation in dune cross-stratification are folds, faults, and brecciated strata of which there are many varieties.

Penecontemporaneous deformation is caused in dunes primarily by avalanching, either through slumping (mass movement) or by grain flowage; less frequently it results from loading, plant growth, scour-and-fill processes, and sand drag. Virtually all forms are considered to be contorted bedding, and most are very irregular, especially when compared to similar-appearing structures formed in competent rock masses.

A classification of deformational structures in dunes, with suggested terminology for various forms that have been recognized, was proposed by McKee, Douglass, and Rittenhouse (1971). This classification was based partly on experimental work in the U.S. Geological Survey Sedimentary Structures Laboratory in Denver, Colorado, and partly on field investigations at White Sands National Monument, New Mexico. Subsequent field studies, testing the validity of this classification on coastal dunes of Brazil, were made by McKee and Bigarella (1972).

Principal types of deformation structures formed in avalanche deposits are referred to as rotated structures, warps or gentle folds, drag folds and flames, high-angle asymmetrical folds, overturned folds, overthrusts, break-aparts, breccias, and fadeout laminae (fig. 75). Some of these structures are characteristic of tensional stresses, common near the tops of dune slipfaces; others are principally the result of compressional forces that develop near the bottoms of dune foresets. Some are common in dry sand, others develop mostly in wetted sand, saturated sand, or sand with damp crusts.

Moisture greatly affects the type of contorted structure developed within a sand body because it largely controls the degree of cohesion of sand particles. In sand that is dry, cohesion of grains is slight, and the chief deformational structures are fadeout laminae, drag folds and flames, and warps or gentle folds. In sand that has been wetted, considerable cohesion develops, and common deforma-

tional structures are break-aparts and breccias, rotated plates and blocks, and high-angle asymmetrical folds. An analysis of these trends as determined in laboratory experiments and in a comparison of desert sands in New Mexico and coastal sands in Brazil was tabulated by McKee, Douglass, and Rittenhouse (1971, table 5).

In many dune sands, both coastal and desert, moisture may form wet crusts, as from early morning dew. In such deposits, certain layers have been dampened or wetted shortly after deposition, so wet and dry layers alternate.

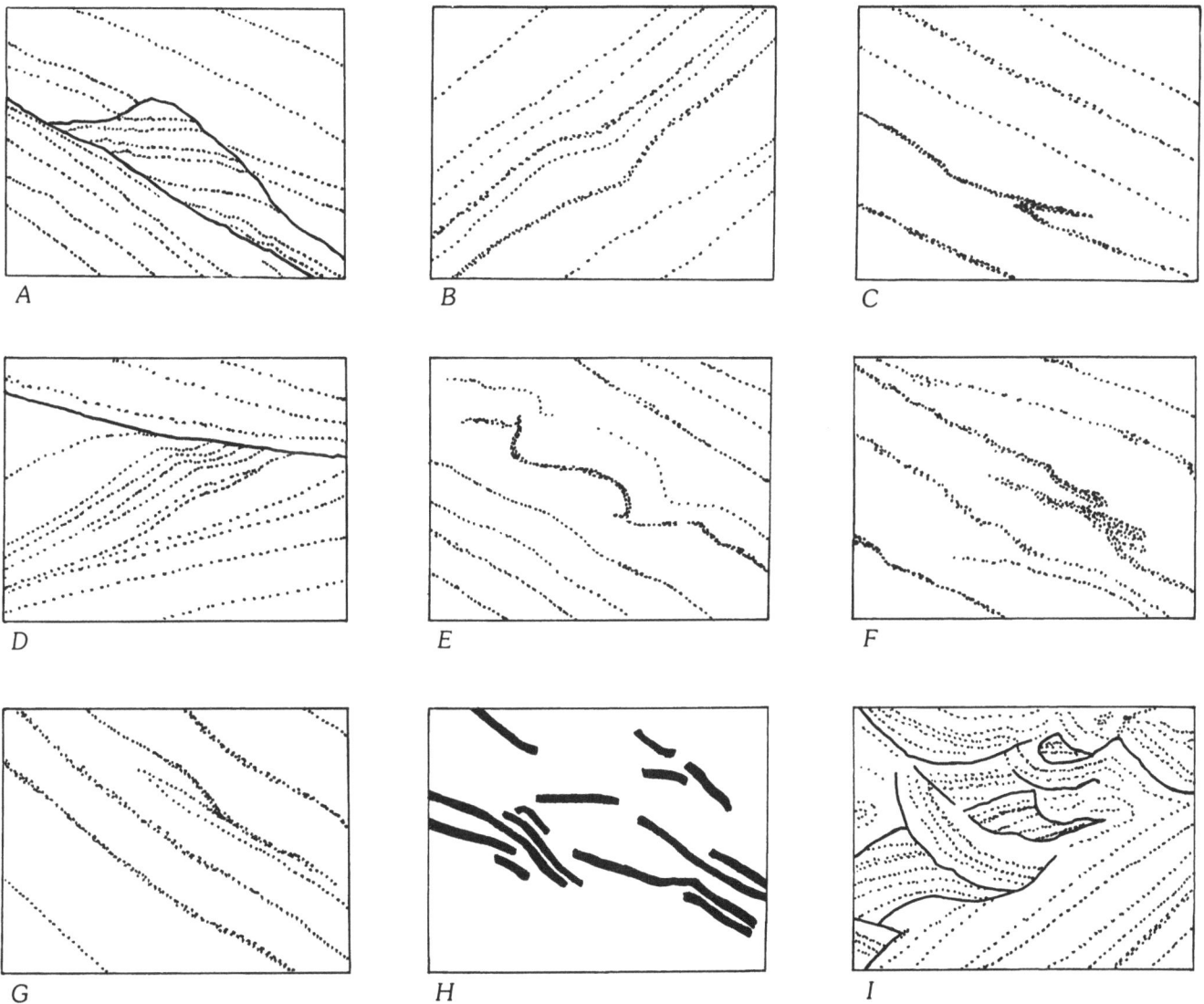

PRINCIPAL TYPES OF DEFORMATIONAL STRUCTURES in avalanche deposits of dunes: *A*, Rotated structures; *B*, warps or gentle folds; *C*, flame structure; *D*, drag fold; *E*, high-angle asymmetrical folds; *F*, overturned folds; *G*, overthrust; *H*, break-aparts; *I*, breccias. Each block represents an area approximately 6 × 4 inches (15 × 10 cm). (Fig. 75.)

Structural Features at Lagoa Dune Field, Brazil

By JOÃO J. BIGARELLA

Parabolic Dunes

THE LAGOA DUNE FIELD is in the east-central part of Ilha de Santa Catarina (State of Santa Catarina, Brazil). The dune field is bounded on the east by crystalline hills (mostly granite), on the south by the Campeche Beach (locally referred to as Joaquina or Lagoa Beach), and on the west and north by the Lagoa da Conceicão lagoon (fig. 76).

Prevailing winds in this area blow from the northeast, but short-term storm winds come from the south. The influence of the prevailing winds on the dune field is decreased considerably by granitic rock hills located upwind. The prevailing winds are deflected by the mountains to the west of the Lagoa da Conceicão and reach the dune area as northerly winds. Because the dune field is located in a shadow area, protected from the northeasterly prevailing winds, the deflected northerly winds cannot counteract the action of sand transport from the southern short-term storm winds. Thus, the main eolian sand movement is to the north (Bigarella, 1972).

The Joaquina Beach constitutes a main source of eolian sand. However, the amount of dry sand available for eolian transport on the beach is too small to counteract the deflation caused by the strength of the southern storm winds in the southern part of the dune field. Therefore great numbers of blowout features, including parabolic dunes, are well developed. In the northern part of the dune field, a series of reversing dunes has formed from the interaction of opposing winds. They are slowly moving northward.

Method of Study

The present study is of a compound parabolic dune located in the southern part of the Lagoa dune field about 480 metres (1,575 ft) from the beach. Most of the internal structures to be recorded and analyzed were exposed by digging a series of trenches and pits, and by cleaning two erosion scarps cut by deflation near the dune nose (fig. 77). Trenches C, D, I, and H are located in the most ac-tive part of the parabolic dune, trench I on the slipface, trench H on the windward side behind the crest, and trenches C and D oriented oblique to the wind direction. Trench E and pits E_1, E_2, and E_3 are alined roughly parallel to the wind direction and constitute a southern continuation of scarp B, extending down the windward slope into the interdune area. Trenches F and G, and pits 1–7 are oriented normal both to the dune's arms and to the dominant wind direction. These pits are located in the interdune area and dissect some of the small satellite parabolic dunes.

Cross-Stratification

Cross-strata in the parabolic dune are medium scale in the crest area but large scale in the nose and the arms. Many laminae dip at high angles (29°–34°) in the nose and also in the arms where they are normal to the dominant wind direction and to the dune axis. Both along the crest and in some parts of the arms, the bounding surfaces between sets of cross-strata are nearly horizontal. Elsewhere, as in the nose and parts of the arms, they dip at angles ranging from low to high. As verified for other dune types (McKee, 1966; Bigarella, Becker, and Duarte, 1969), individual sets of cross-strata tend to be thinner and the laminae flatter near the top than at the bottom of the dune (fig. 77).

Cross-strata measurements for all trenches, scarps, and pits in the various parts of the dune are summarized in tables 10 and 11, and in figure 78. Measurements in each part are controlled by its local morphological conditions (fig. 77).

Nose Area

In the nose, high-angle foresets are characteristic features in the walls of trenches C and I. These trenches were oriented oblique and parallel, respectively, to the southern storm winds (fig. 77). Trench C, excavated in the slipface, was oriented N. 52° W. at an angle of about 46° to the direction of the southern storm winds, but approximately in the average cross-strata dip direction as calculated from nine measurements. In trench I, excavated parallel to the direction of the storm winds, five measurements of cross-strata were made, and their dips were N. 2°–15° W. The average dip, N. 6° W., coincides with the main direction of the southern

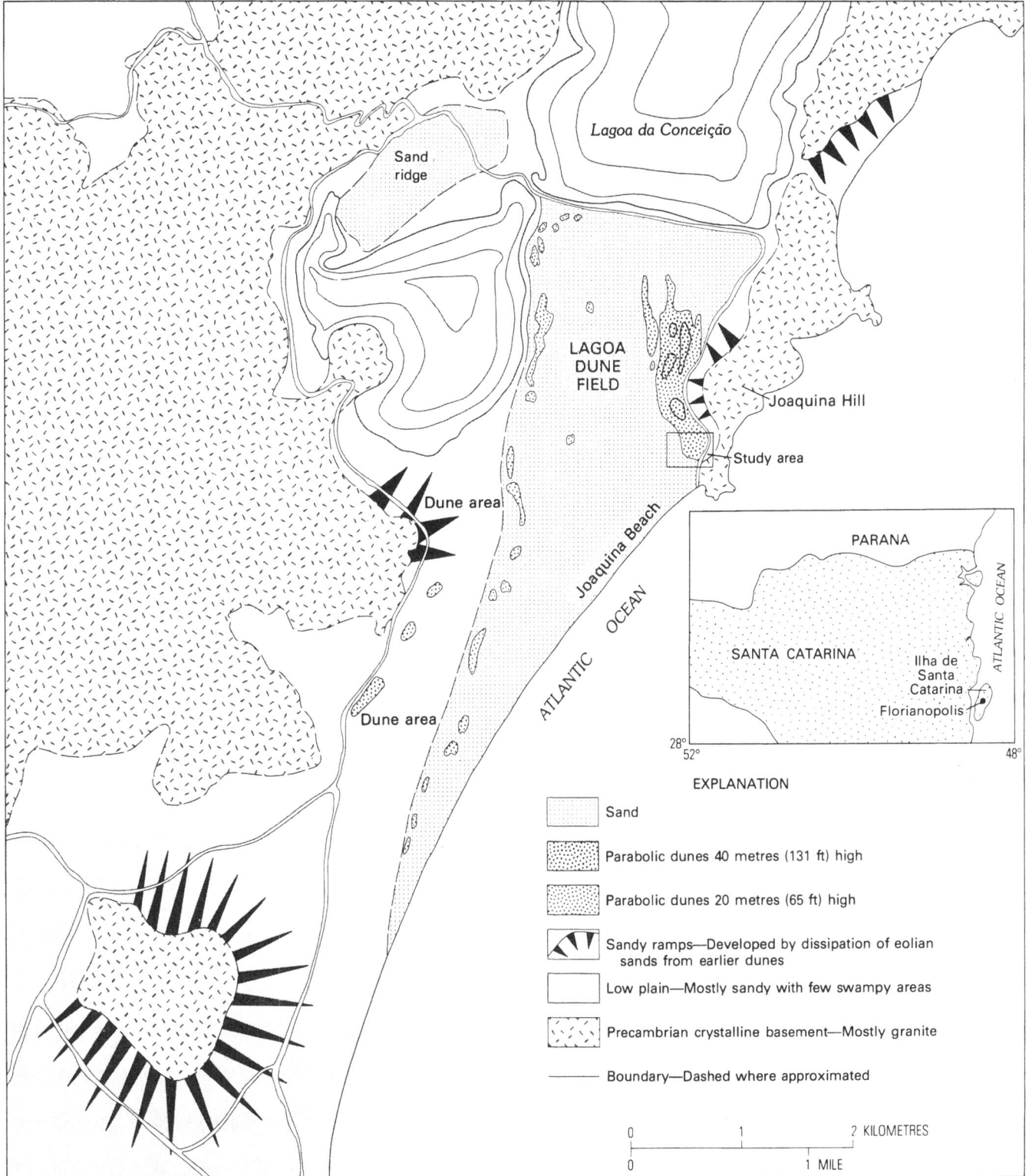

LOCATION OF LAGOA DUNE FIELD, Ilha de Santa Catarina, State of Santa Catarina, Brazil. Modified from Bigarella (1975b, fig. 1). (Fig. 76.)

EXPLANATION

Forward edge of dune slipface
Dune crest

Satellite dune

High area on dune

Trench
Scarp
× Pit
Cross-strata (thin lines) bounded by erosion surface (heavy line)

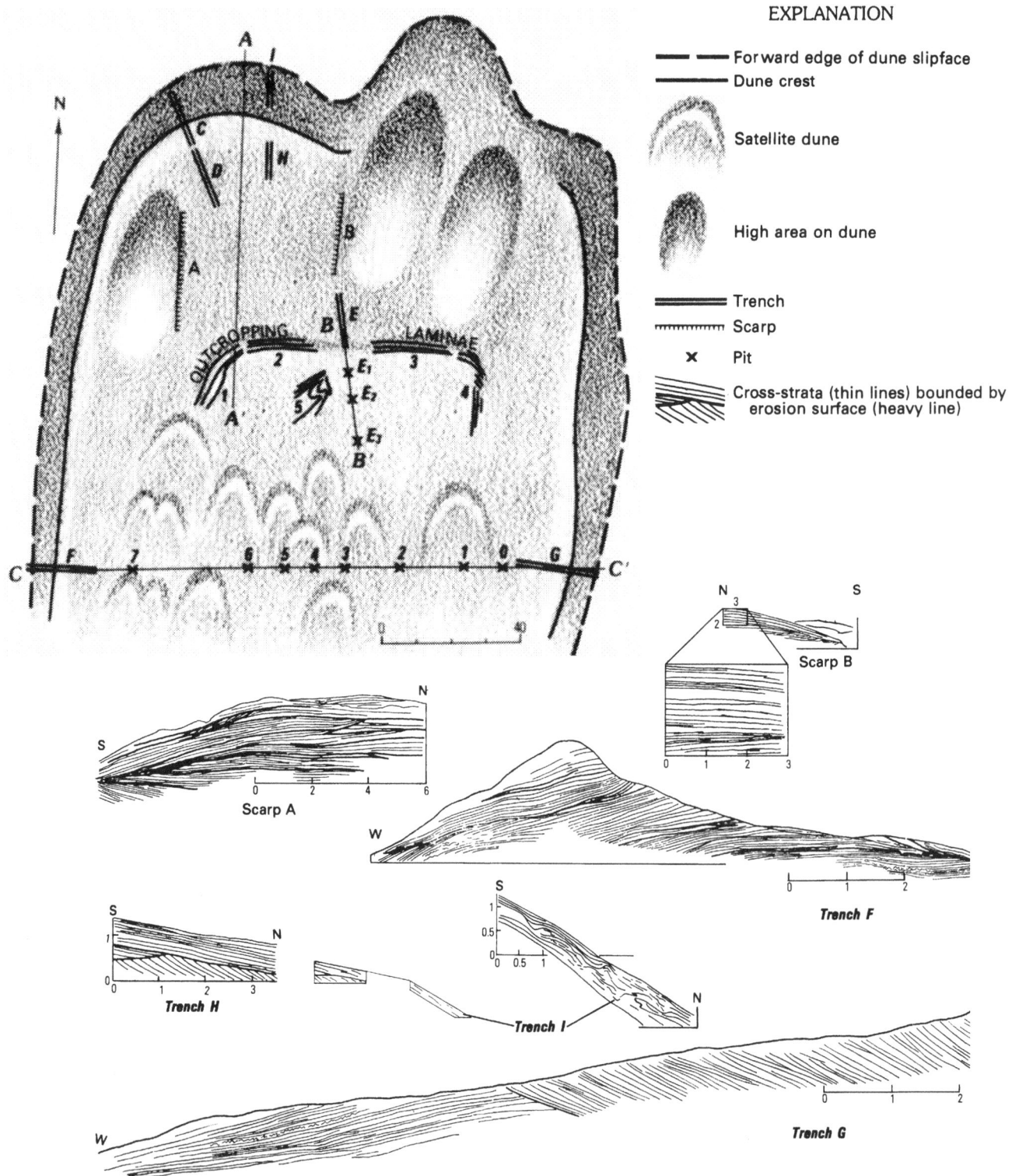

Scarp A

Scarp B

Trench F

Trench H

Trench I

Trench G

CROSS SECTIONS AND PLAN VIEW of parabolic dunes studied in the Lagoa dune field, Ilha de Santa Catarina, Brazil, showing A–A′, B–B′, and C–C′ show surface profiles across dune and interdune areas with locations of pits and trenches. All measure-

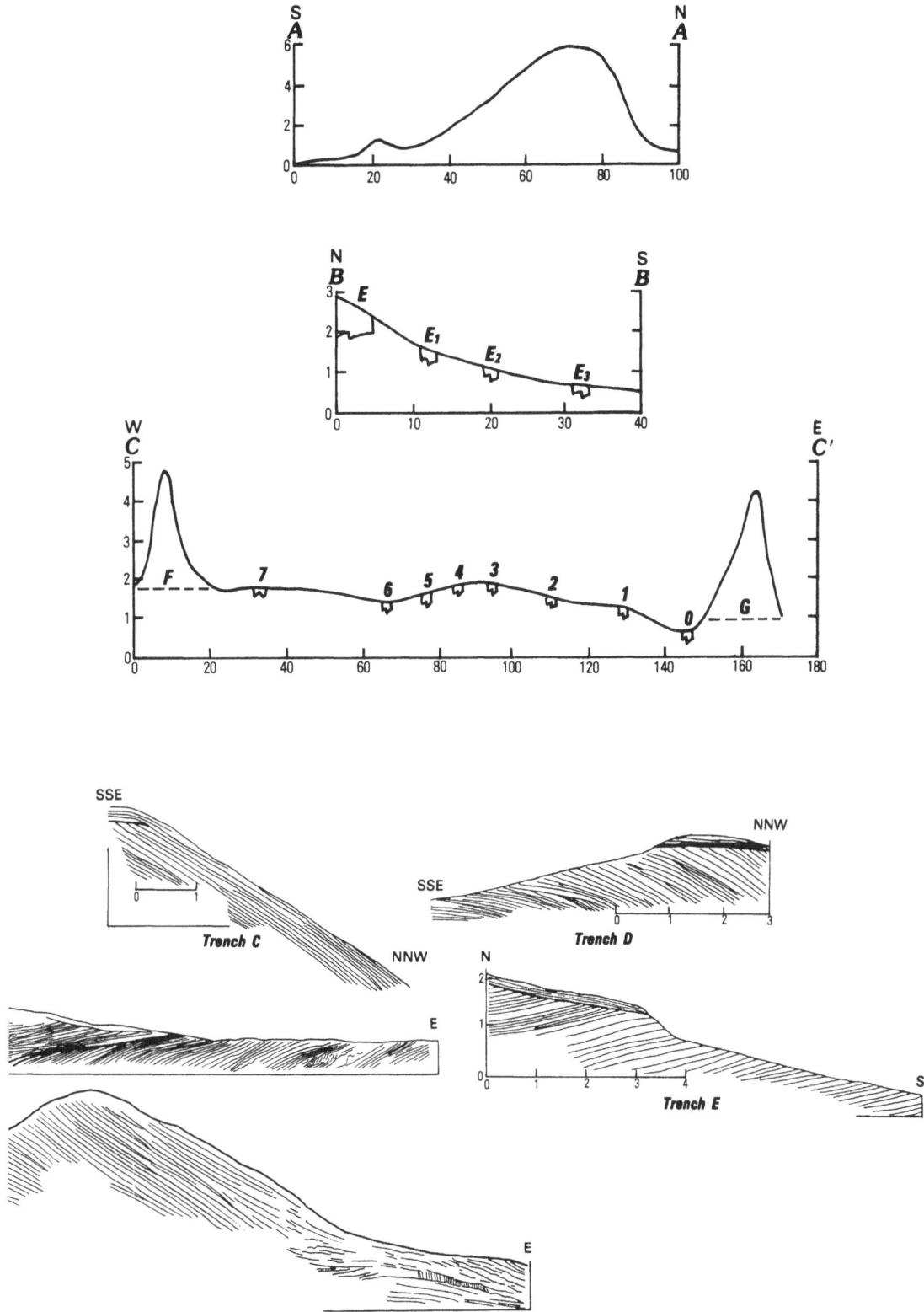

scarps, trenches, and pits. Groups of lines numbered 1–5 are approximate locations of outcropping laminae. Cross sections
ments in metres. 1 metre=3.28 feet. Modified from Bigarella (1975b, figs. 4, 5). (Fig. 77.)

storm winds (S. 6° E.) and with the axis orientation of blowout dunes inside the Lagoa dune field. In trench I, the average cross-strata dip direction coincides with the dip direction of the slipface surface.

The rose diagrams of figure 79 illustrate cross-strata dip directions from various trenches dug in the slipface and crest areas. The slipface and the crest just behind it have a similar cross-strata attitude, but differ in the average of the cross-strata dip angles, which are 10°–13° less in the crest beds.

Crest Area

In the present crest area, trenches D and H constitute southward continuations of trenches C and I in the slipface area (fig. 77). In trenches D and H, internal structures consist of two main parts: (1) a thick set of cross-strata with high-angle foresets, deposited as the nose of this parabolic dune at an earlier stage, and (2) thinner cross-strata sets with low- to moderate-angle dips deposited on the crest (fig. 77).

An erosion surface truncates the lower set in both trenches. On the west side of trench D (fig. 77), an early set of crest-strata dipping 6°–11° toward N. 37° W. is overlain by high-angle foresets, mostly dipping 29°–35° (average 28°) to N. 58° W. This succession suggests a former slipface on the western flank of the nose approximately where trench C is located (fig. 77). The truncated slipface is overlain by a sequence of strata dipping at low angles (1°–11°) toward the eastern quadrant N. 55° E.–S. 52° E.

High-angle cross-strata in the lower part of trench H (fig. 77) dip N. 13° E., whereas those from the upper part (crest beds) dip N. 27° E. (four measurements). These cross-strata of the crest continue northward to form strata of the slipface. In the area of transition between the crest and the slipface, strata are convex upward. Both parts of trench H have dip directions which diverge by 19°–33°, indicating that the former slipface dip direction was more north-northeastward than the slipface of trench I, which is alined with the wind direction.

Structures representing an earlier position of the crest are exposed scarps A and B. The former crests were cut in half by a deflation corridor and a new crest was built downwind. In the erosion scarps,[5] about 2.5 metres (8.2 ft) of sand sequences were ex-

posed—both windward strata and downwind strata (fig. 77). Sets of cross-strata are prominent in erosion scarps A and B, and bounding surfaces dip slightly to moderately windward. Some sets of strata dip downwind at low angles, other sets are approximately horizontal.

The whole crest seems to result from the coalescence of several low mounds of sand. Rose diagrams in figure 79 illustrate the peripherally dipping attitudes of cross-strata deposited in the windward and downwind parts of the former dune crest. High-angle dips (15°–30°) occur among lee-side sets of cross-strata on the west side of erosion scarp A (fig. 77). On this scarp, the average dip of cross-strata is 12.8°; many angles of dip are between 3°–11°. The resultant southeasterly vector and the low-angle strata indicate a predominance of windward-deposited strata (fig. 77), although about 20 percent of the strata dip peripherally in all directions around the compass, suggesting a very gentle dome structure.

Erosion scarp B, on the opposite side of the deflation corridor from scarp A, exposed strata dipping mostly less than 10° and averaging 8.4°. The strata dip consistently southwestward, suggesting that this part of the former crest of the parabolic dune had a dome shape.

Trench E extends southward from erosion scarp B (fig. 77) and has in its upper part strata that dip south and belong to the former dome-shaped crest. Internal structures indicate the presence of crest-type cross-strata, of which the average dip is 12.9°. The lower part of the trench is composed of low-angle strata dipping downwind 8°–16°, whereas the upper part is composed of windward-dipping strata with dips of 16°–18°. The sequence of erosion scarp B overlies strata of trench E. Trench E strata, in turn, overlie sets of high-angle cross-strata exposed in pits E_1, E_2, and E_3 in the present interdune area. The high-angle (25°–33°) strata in these pits are lower parts of slipface deposits of a parabolic dune nose which migrated northward and was truncated by deflation. A new crest was later developed (lower part of trench E deposits) but was eroded on its windward side, which formed the deposit downwind represented in the erosion scarp B sequence.

The crossbedding pattern exposed at the surface, upwind from the crest, at the transitional edge of the interdune area, is illustrated in figure 80 (outcropping laminae in fig. 77). The trace of high-angle cross-strata represents deposits from the nose and lower parts of the arms of this parabolic dune. Cross-stratification is periodically exposed by deflation in this area. The three main sets of nose deposits downwind suggest the way in which the crest and nose split and develop as wide and very low moundlike features.

Rose diagrams (fig. 78) show the attitude of cross-strata that are exposed behind (upwind of) the crest. Measurements were made on both the left (west) and right (east) sides of the dune axis. The average dip direction for both sides is within a spread of 33°. The resultant vector N. 13° E., is 18° to the east of the main direction of the southern storm winds.

Arms

Basic structures in the arms of the parabolic dune are illustrated in trenches F and G, which are nor-

---

[5] Erosion scarps refer to surfaces along the sides of deflated or "blowout" areas which have been smoothed by machete. Trench walls refer to elongate excavations of much longer dimensions than the pits.

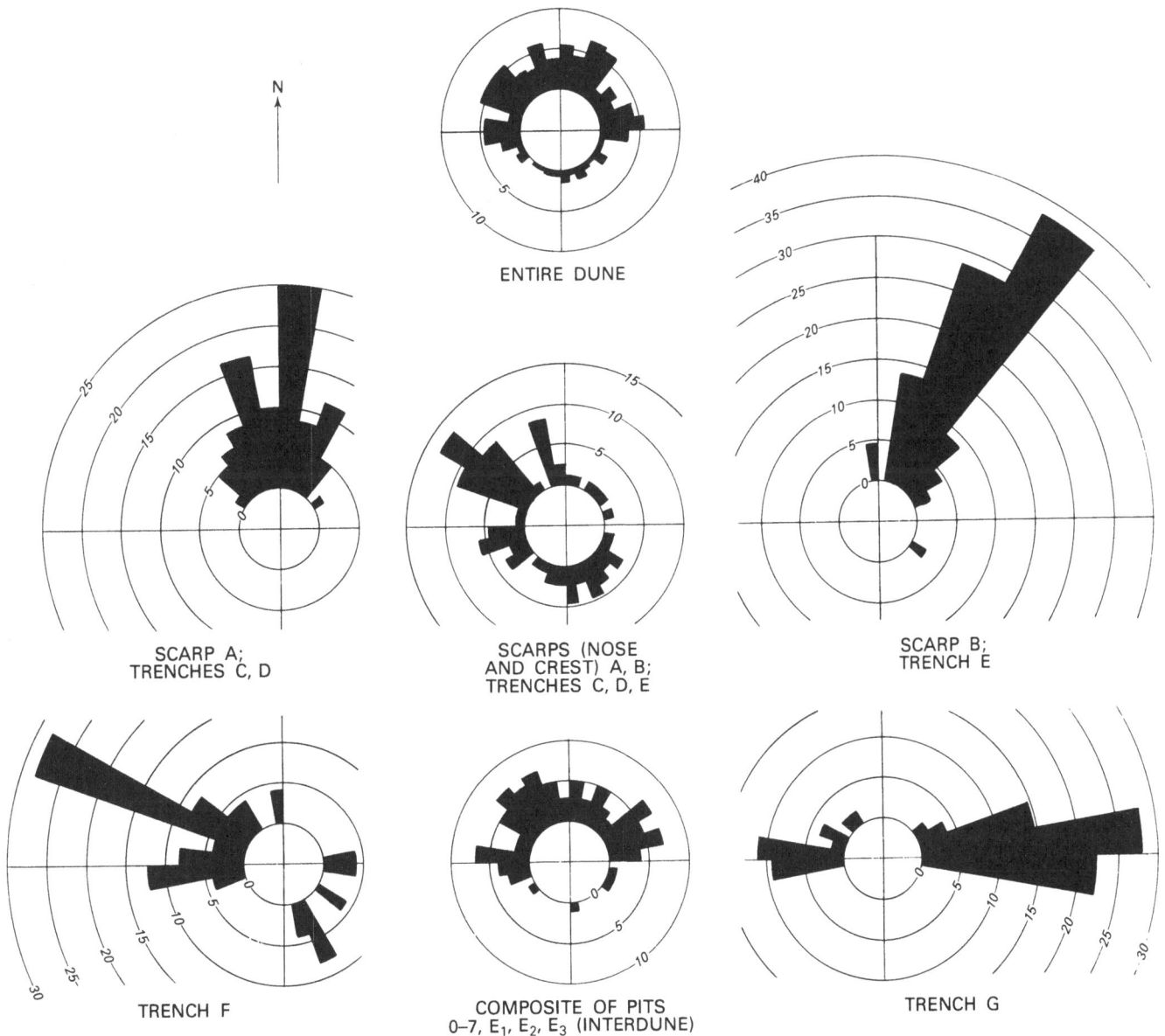

ROSE DIAGRAMS OF CROSS-STRATA DIP DIRECTIONS recorded for scarps, trenches, and pits representing principal parts of parabolic dune and interdune area. Values are in percent. Modified from Bigarella (1975b, fig. 6). (Fig. 78.)

TABLE 10. — *Frequency of dip-direction intervals for cross-strata in trenches, scarps, pits, and surface of a parabolic dune, Lagoa dune field, Brazil*

[Modified from Bigarella (1975b, fig. 6). Locations shown in fig. 77]

Interval (degrees)	Scarps and trenches								Pits											Surface			Entire dune
	A	B	C	D	E	F	G	Total	1	2	3	4	5	6	7	E₁	E₂	E₃	Total	West	East	Total	
0–30	1	⋯	⋯	⋯	1	⋯	⋯	2	⋯	5	2	1	2	1	1	1	⋯	⋯	13	27	19	46	61
30–60	2	⋯	⋯	1	⋯	⋯	1	4	1	⋯	1	2	5	⋯	2	⋯	1	⋯	12	4	22	26	42
60–90	1	⋯	⋯	⋯	⋯	1	25	27	4	⋯	⋯	1	7	⋯	4	⋯	⋯	⋯	16	⋯	1	1	44
90–120	1	⋯	⋯	1	⋯	1	12	15	⋯	⋯	⋯	⋯	⋯	⋯	2	⋯	⋯	⋯	2	⋯	⋯	⋯	17
120–150	6	⋯	⋯	2	⋯	1	⋯	9	1	⋯	1	⋯	⋯	⋯	⋯	⋯	⋯	⋯	2	⋯	1	1	12
150–180	10	⋯	⋯	⋯	⋯	3	⋯	13	1	⋯	⋯	⋯	⋯	⋯	⋯	⋯	⋯	⋯	1	⋯	⋯	⋯	14
180–210	3	⋯	⋯	⋯	2	⋯	⋯	5	⋯	⋯	⋯	⋯	⋯	⋯	⋯	⋯	⋯	⋯	⋯	⋯	⋯	⋯	5
210–240	1	3	⋯	⋯	⋯	⋯	⋯	4	1	⋯	⋯	⋯	⋯	⋯	⋯	⋯	⋯	⋯	1	⋯	⋯	⋯	5
240–270	1	10	⋯	⋯	⋯	4	5	20	4	⋯	⋯	⋯	⋯	1	⋯	2	⋯	⋯	7	⋯	⋯	⋯	27
270–300	1	1	2	6	⋯	10	9	29	5	2	3	⋯	⋯	1	⋯	⋯	⋯	⋯	11	⋯	⋯	⋯	40
300–330	⋯	⋯	7	11	1	4	2	25	1	5	5	⋯	⋯	⋯	⋯	⋯	2	5	18	8	⋯	8	51
330–360	1	⋯	⋯	2	6	1	⋯	10	⋯	5	5	⋯	⋯	⋯	2	2	1	⋯	15	21	2	23	48

TABLE 11. — *Direction and degree of dip for cross-strata in trenches, scarps, pits, and surface of a parabolic dune, Lagoa dune field, Brazil*

[Modified from Bigarella (1975b, fig. 6). Locations shown in fig. 77]

Locality	Number of measurements	Mean dip direction[1]	Consistency ratio[2]	Maximum dip	Average dip
Scarps A, B, trenches C, D, E	84	N. 83° W.	0.32	35°	16.1°
Trench F	25	N. 75° W.	.51	28°	14.1°
Trench G	54	N. 74° E.	.43	34°	22.2°
Total	163	N. 53° W.	.13	35°	17.8°
Total pits	98	N. 5° W.	0.15	37°	17.4°
Surface — East side	45	N. 31° E.	0.95	34°	25.5°
Surface — West side	60	N. 3° W.	.84	34°	23.6°
Total	105	N. 13° E.	0.44	34°	24.4°
Entire dune	366	N. 5° W.	0.42	37°	19.6°

[1] The frequency of dip-direction data from which the mean dip direction was derived are shown in table 10.

[2] Consistency ratio is obtained by dividing the length of the vector resultant of all observations at a given locality by the sum of the lengths of individual vectors. It is a weighted measure of degree of grouping of vectors. It ranges from 1 to 0.

mal to the direction of the southern storm winds (fig. 77). The cross-strata sets dip toward both the eastern and the western quadrants, almost at right angles, or oblique to those on the dune nose.

The profile across the arms of the parabolic dune may be divided into two parts — the inner and the outer. The inner side, less steep than the outer, commonly is an area of deflation. The steeper, outer profile has a slipface and is an area of deposition.

In trench G, on the right arm of the parabolic dune, the lower part of the sequence on the left side of the trench is composed of westerly dipping strata averaging 15.2° (fig. 77). The highest dips are 19°–22° and the lowest, 9°–13°, similar to the middle part of trench F below the crest, where dip angles are also 9°–13° (fig. 77). These cross-strata are not typical slipface deposits like those shown on the right (eastern) side of trench G (fig. 77). The left (western) part of trench G may represent a former outer part of the left arm of an earlier migrating parabolic dune, subsequently truncated by the present dune.

On the east side of trench G (fig. 77), foresets are mostly medium- to large-scale. The cross-strata dip eastward. Forty percent of them have high-angle dips (30°–34°); most of the others dip only 21°–29°. Bounding surfaces also have moderate dip angles. The average dip of cross-strata in trench G is 22.2°. High-angle strata from the outer part of the arm dissected by trench G seem to overlie low-angle interdune cross-strata, enriched with humate compounds. In some places, an incipient paleo A horizon is developed (nearly vertical lines, on east edge of trench G fig. 77).

The internal structure of trench F is far more complex than that of trench G. The right (east) side of trench F (fig 77) is composed in the lower part of truncated cross-strata dipping westward at moderate angles (18°–25°). They represent deposits of the left arm of a former parabolic dune which migrated across the area. Between this sequence and the one forming the crest are nearly flat lying strata (3°–5°) which probably represent interdune deposits. Above these interdune strata, the structures of the arm can be traced across a series of erosion surfaces, each series truncating slipface strata, up to the present position of the crest. Bounding surfaces in trench F dip moderately westward.

Low-angle cross-strata commonly are deposited on the windward sides of dunes, but in the arm of this parabolic dune some of the low-angle dipping strata seem to have been deposited on the inner side of the arm at almost right angles to the wind direction.

The opposing attitudes of cross-strata in trenches F and G are illustrated in a rose diagram (fig. 78; tables 12, 13).

TABLE 12. — Attitudes of 25 cross-strata sets in trench F of parabolic dune, Lagoa dune field, Brazil

[Modified from Bigarella (1975b, table 1)]

Set No.	Dip direction (°)	Amount of dip (°)	Set No.	Dip direction (°)	Amount of dip (°)
1	290	25	14	270	6
2	310	20	15	305	14
3	312	18	16	310	8
4	332	12	17	325	16
5	315	25	18	310	9
6	295	12	19	302	12
7	111	18	20	302	13
8	98	3	21	276	11
9	6	5	22	280	13
10	136	25	23	279	16
11	165	10	24	310	15
12	172	11	25	285	28
13	165	8			

Dip direction: For true north, subtract 12°.

Deformation

Penecontemporaneous deformation, caused by slumping of steep foresets, is common in most coastal dune structures (fig. 81). This gravity slumping occurs during or soon after deposition.

Deformational structures also occur in moderately to gently dipping strata. Deformed laminae may occur between undeformed beds (fig. 82A) or may affect adjoining sets of strata. The structures produced by deformation seem to be chaotic — folds, faults, and breccias which consist of broken, rolled, and crinkled masses of sand (fig. 81). Some contorted structures result from the activities of organisms (fig. 82C).

The response to stresses and the type of structures produced during deformation are controlled largely by the amount of moisture present in the sand (McKee, Douglass, and Rittenhouse, 1971; Bigarella, Becker, and Duarte, 1969; McKee and Bigarella, 1972).

TABLE 13. — *Attitudes of 54 cross-strata sets in trench G of parabolic dune, Lagoa dune field, Brazil*

[Modified from Bigarella (1975b, table 2)]

Set No.	Dip direction(°)	Amount of dip (°)	Set No.	Dip direction(°)	Amount of dip(°)
1	274	20	28	95	27
2	331	9	29	94	21
3	290	22	30	102	21
4	291	14	31	95	27
5	310	10	32	87	23
6	278	17	33	95	30
7	305	14	34	95	31
8	332	12	35	107	29
9	288	18	36	97	32
10	284	20	37	103	34
11	296	15	38	103	32
12	279	16	39	110	27
13	283	19	40	106	37
14	275	14	41	108	30
15	286	13	42	97	34
16	279	10	43	97	33
17	77	22	44	102	32
18	99	23	45	108	28
19	110	3	46	97	32
20	64	12	47	100	29
21	91	21	48	97	30
22	91	27	49	96	27
23	89	16	50	102	33
24	85	26	51	92	20
25	91	31	52	99	8
26	88	25	53	75	12
27	102	18	54	84	16

Dip direction: For true north, subtract 12°.

Interdune Areas Among Parabolic Dunes

In an interdune area, between the arms of a surveyed parabolic dune (fig. 77), two lines of pits were dug. The east-west line (pits 1 –7) provided information concerning the interdune area between the trenches G and F on the trailing arms of the parabolic dune. The north-south line (pits E_1 –E_3) forms a southward continuation of trench E. The pits are rectangular in cross section with sides approximately 1 –2 metres (3 –6 ft) long and depths of 1 –2 metres (3 –6 ft; figs. 83, 84).

Characteristic of the interdune area are the apparently structureless sands, enriched with humic compounds derived from incipient soil formation. In some places, paleosol A horizons alternate with stratified sand (pit 6, fig. 83).

The interdune area frequently becomes wet, sometimes swampy, after rains, and small ponds may form in low parts when the phreatic water level rises. A typical vegetation cover of grasses and shrubs grows in the interdune environment when the sand is wet, and many small plants, especially Gramineae and Cyperaceae germinate there. Humic compounds that originate from decay of the organic material stain the sandy substratum and are responsible for the brown color of both ground water and surface water.

The buried upper surface of the brown structureless sand is irregular. Above this surface are sequences of cross-stratified eolian sand deposited either by migration of the main parabolic dune or by growth of small satellite parabolic dunes (fig. 85).

The attitudes of 98 cross-strata were measured in the 10 pits. About 42 percent had dip angles greater than 20°, whereas 58 percent had dip angles less than 15°. The maximum dip recorded was 37°, and the average dip was 17°. The crossbedding pattern from the interdune area documents the migration or movement of former dune bodies across this area. The traces of several parabolic dunes, both nose and arms, can be recognized. A rose diagram (fig. 86) shows most strata in this area dipping toward the northern quadrants, with an average dip direction of N. 5° W.

Substantial changes in dune morphology along the Brazilian coast have been caused by climatic change (Bigarella, 1965, 1972). Recent action of concentrated rainfall has greatly modified the original sand body at Lagoa — the eolian sands were spread by floodwaters like a blanket, decreasing in thickness downstream.

Climatic changes in the past provided conditions under which extensive dune development alternated with dune modification. Early deposits of eolian sands have become mixed with colluvial material producing an unsorted sandy, muddy sediment. Mixing took place during colluvium movement caused by climatic change.

The dunes subjected to heavy rainfall became saturated with water. This rainfall caused wet sliding, sand flows, gullying, and the formation of many alluvial fans that coalesced to form dissipation ramps (fig. 76). Many faults were developed in the sand during avalanching. Sand-flow and rainwash deposits mostly form wavy or lens-shaped layers, and many of them are imbricated. Some deposits are largely derived from alteration of the dunes; others are mixtures of colluvium and modified eolian sands. In some places, clay was deposited in ephemeral interdune ponds. Times of soil formation alternated with times of erosion. In the same manner, dissipation deposits are intercalated with deposits having normal dune cross-stratification.

Interdune Areas Among Reversing Dunes

At Lagoa dune field the depositional environment of interdune areas among reversing dunes is much drier than the corresponding areas among parabolic dunes. Some ephemeral ponds are present, but most of the area is covered with Gramineae and Cyperaceae, and some parts of it consist of sand plains without vegetation. In the unvegetated area, four pits and two trenches were dug to expose the internal structures of an interdune (fig. 87). Trench VII shows a sequence of northward-dipping truncated foresets deposited by short-term storm winds from the south. In trenches VIII, XVII, XVIII, and XIX, high-angle foresets are covered by flat-lying beds, dipping less than 1°. Where these strata approach the slipface of the next dune upwind, however, the dip is as much as 10°. In trench IX the

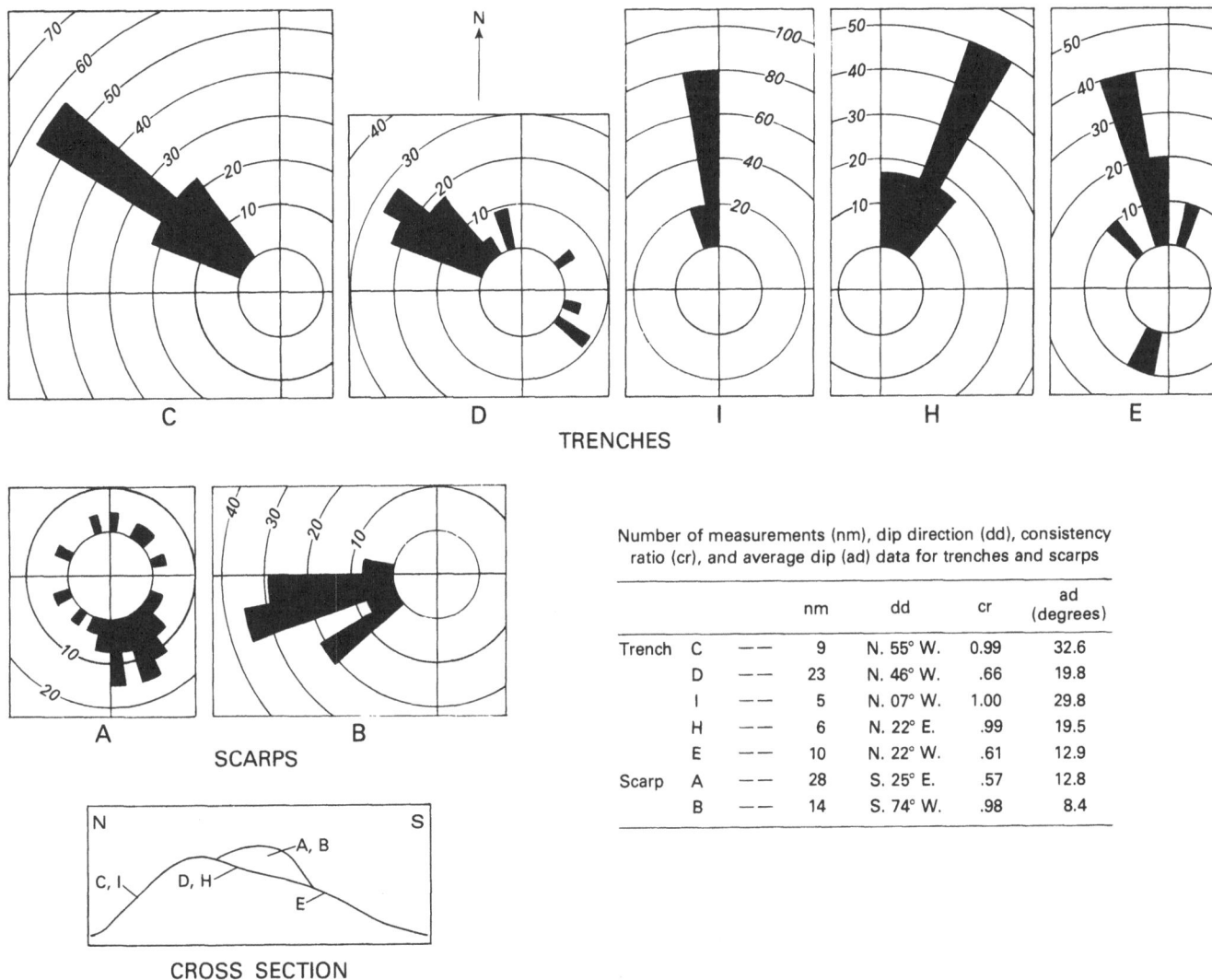

TRENCHES

SCARPS

CROSS SECTION

Number of measurements (nm), dip direction (dd), consistency ratio (cr), and average dip (ad) data for trenches and scarps

			nm	dd	cr	ad (degrees)
Trench	C	— —	9	N. 55° W.	0.99	32.6
	D	— —	23	N. 46° W.	.66	19.8
	I	— —	5	N. 07° W.	1.00	29.8
	H	— —	6	N. 22° E.	.99	19.5
	E	— —	10	N. 22° W.	.61	12.9
Scarp	A	— —	28	S. 25° E.	.57	12.8
	B	— —	14	S. 74° W.	.98	8.4

ROSE DIAGRAMS OF ATTITUDES of cross-strata deposited in an earlier dune crest (trenches D, H, E, and scarps A, B) and in the nose (trenches C, I) of a parabolic dune (fig. 77), Lagoa dune field, Brazil. Values are in percent. Generalized cross section through dune is parallel to dominant wind direction and shows location of trenches and scarps. Table gives statistical data. Modified from Bigarella (1975b, fig. 7). (Fig. 79.)

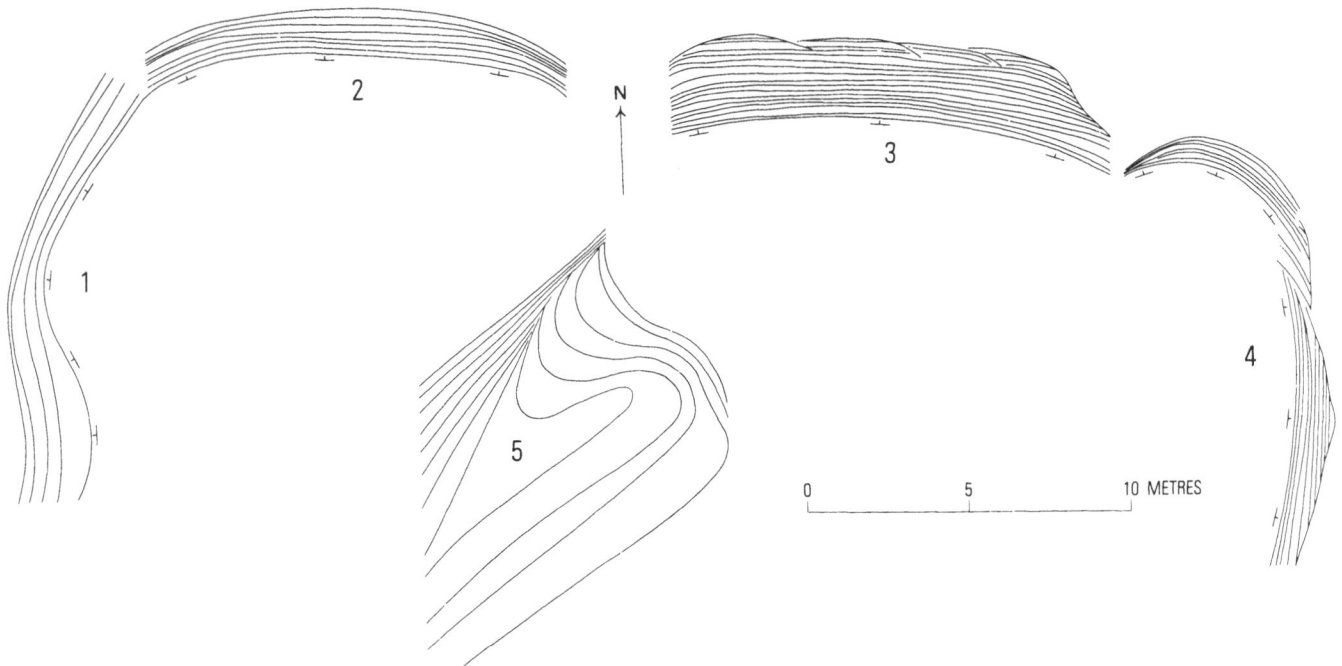

OUTCROPPING LAMINAE EXPOSED on beveled surface in interior of parabolic dune. Units 1 – 5 show details of traces of laminae as exposed by natural truncation in interdune area (fig. 77). Direction of strike and dip shown by symbols. Scale is approximate. Modified from Bigarella (1975b, fig. 6). (Fig. 80.)

high-angle foresets are covered by interdune deposits which range in dip from low angle to almost horizontal, and above these is another set of high-angle dune foresets.

Dissipation of Dunes, Lagoa, Brazil

By JOÃO J. BIGARELLA

General Features and Environment of Dune Dissipation

ALONG THE BRAZILIAN coastal plain flash floods have changed both internal pattern and morphology of the original dunes and beach-ridge deposits. This reworking (by dissipation) modifies or destroys original dune structures, and it concentrates colloidal materials in a wavy, irregular pattern known as dissipation structure.

In an area within the Lagoa dune field on the Ilha de Santa Catarina, Brazil (fig. 76), dissipation structures in parabolic dunes were studied. Flash flooding in this area forms the large sandy ramps and the small-scale structures that interrupt normal dune

bedding patterns. The processes seem to be cyclic and occur only under certain environmental conditions; accordingly, dissipation layers alternate with dune bedding and are recurrent in the stratigraphic section. The older dunes in the dune field have lost their original shape by dissipation. Some still show either a hummocky morphology or a rounded topography. Others have been changed to tongue-shaped layers that are not intercalated among other dune deposits.

Four principal sequences of eolian sands separated by paleosols (A horizons) occur as a dissipation ramp on the slopes of the granite hill (Joaquina Hill) just east of the dune field (fig. 88; study area in fig. 76). The contacts between the various sand sequences are irregular, owing to either erosion or the hilly topography of the dunes.

The oldest sequence, which is reddish brown, rests on a clayey lag deposit, composed of granite and quartz pebbles, which, in turn, overlies weathered granite. The eolian sand above the gravel pavement originally consisted of fine windblown grains. This sand subsequently became mixed with granules (primarily quartz and feldspar), clays, and silts from the weathered granitic mantle.

SCARP A

TRENCH I

TRENCH F

TRENCH G

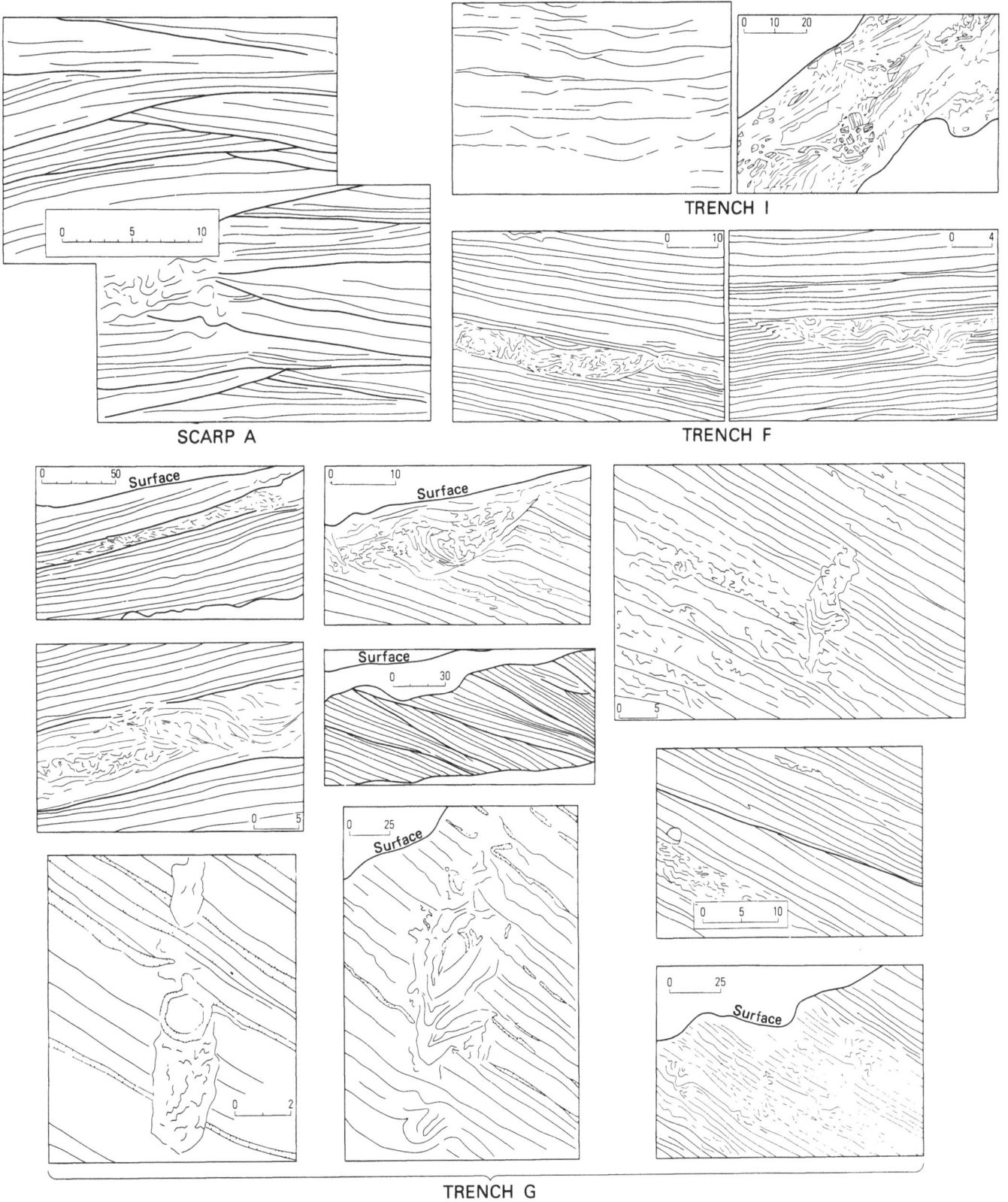

DEFORMATIONAL STRUCTURES OBSERVED in parabolic dune, Lagoa dune field, Brazil. All measurements are in centimetres. 1 centimetre = 0.39 inch. (Fig. 81.)

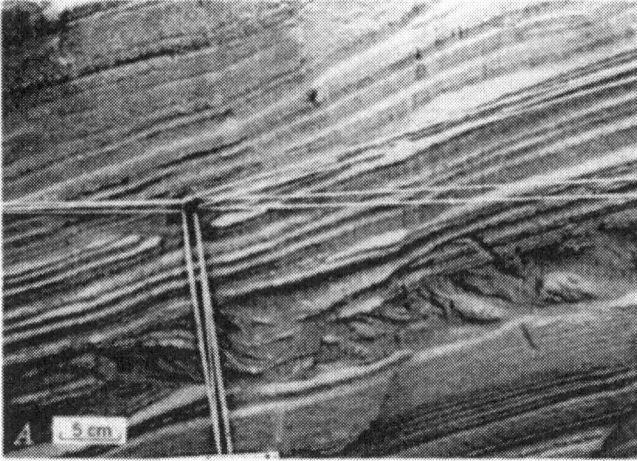

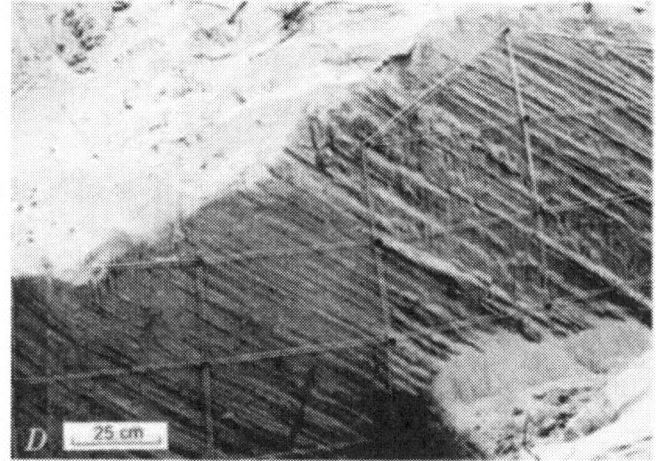

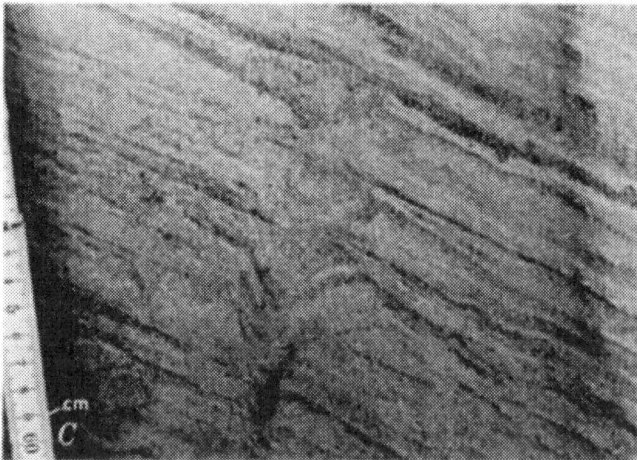

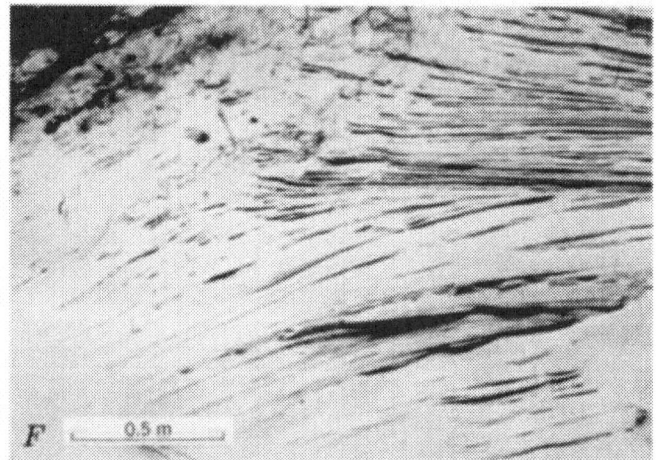

STRUCTURES IN A PARABOLIC DUNE, Lagoa dune field, Ilha de Santa Catarina, Brazil. A−D, Structures in trench G, eastern arm of parabolic dune. A, Contorted bedding in strata dipping toward the interdune area between the arms. B, Strata in trench showing a shift of the dune arm. The lower set dips toward the interdune area between the dune arms, and the upper set dips outward. C, Collapse structure caused by sand flow filling nearby burrowing crab tube. Strata are dipping eastward in trench. D, High-angle strata set dipping eastward in trench showing string-and-nail grid for recording structures. E, F, Structures in eastern arm of parabolic dune. E, Former crest area exposed by erosion (deflation), showing downwind strata dipping to the left, upwind strata dipping to the right and horizontal to subhorizontal strata corresponding to crest deposits. F, Former crest area showing transition from crest-area strata into foreset strata. (Fig. 82.)

Textural changes are numerous among the dissipation sands at Joaquina Hill. A mixing of textures occurred when the dune sands slid downslope, destroying original structures. Moving sands contributed detrital material from the underlying weathered mantle. Mud and sand flows containing granules, from upslope, mixed with and covered sliding eolian sands. This formed the final sediment layer of an unsorted light red sandy granule deposit.

An erosional unconformity, with local paleosol (A horizon) development, separates each sequence of eolian sand. Former dunes in the two intermediate sequences also underwent mass movement, causing structural and compositional changes similar to those previously discussed. As a result, each successive sequence has fewer granules.

Processes, Structures, and Soil Formation

Dunes

The original eolian deposits of the Lagoa dune field consist of reversing parabolic and captation dunes, such as stringers and hanging dunes. Sedimentary structures differentiate dune sands from beach sands of the neighboring source area and the patterns of eolian cross-strata allow differentiation of dune types.

At times large parts of dunes or entire dunes undergo subsequent structural changes, causing disappearance of the characteristic eolian cross-strata patterns. The original structures are replaced by much less clearly defined patterns or by structureless deposits — that is, dissipation structures.

Structures produced by reworking of eolian sands are not as easily visible as those of original dunes, but they constitute new sedimentary structures — that is, dissipation structures. Most of these structures resemble contorted bedding, but they are produced by heavy density flow or by loading. Some resemble cut-and-fill structures; others follow shearing planes and small intraformational faults produced by sand flowage. Tube-shaped features, reflecting root growth or the burrowing of animals, such as crabs, also occur in these deposits. Small circular or irregular patches may suggest root growth.

The new internal structure commonly is characterized by an irregular, wavy pattern accented by differences in concentrations of colloidal material composed predominantly of clays, hydrous iron oxides, and humic compounds (fig. 89). Previously these features generally were ascribed to soil-forming processes; now, they are known to be the result of reworking by dissipation processes. The wavy features are especially prominent because they are outlined with colloidal material along preferential lines. The concentration of colloidal material follows irregular directions that do not correspond to any previous stratification planes or dune structures. At Joaquina Hill, the colloidal material is composed predominantly of clay minerals and hydrous iron oxides, as well as minor amounts of humic compounds. Apparently, colloidal material tends to concentrate preferentially along structural patterns that developed during the dissipation of former dunes.

Flash floods cause the mineral colloidal material, such as from a weathered granitic mantle, to mix with the eolian sands. The amount of mineral matter reinforcing the structures is variable — in most places, less than 5 or 10 percent. After every sand flow, the colloidal material either concentrates at the uppermost section of the sedimentary layer as a "peel" or filters down into lower layers. In the lower layers, colloidal material tends to settle along structural lines, which makes the strata more readily visible. Structures later become vague as the colloidal enrichment progresses. With an increasing content of colloidal material, the structures lose identity, and the sediments develop a uniform, brown color.

Close to the source of weathered granite mantle, the usually thin colloidal peel increases in thickness and becomes a colluvial layer. Such layers alternate with reworked eolian sands. The development of a colluvial layer over a reworked eolian sand produces an irregular surface with tongues and other features that indicate inclusion of the sand in the colluvial layer. The overlying colluvial layer produces contorted structures in the underlying sediments.

In surface accumulations deposition of colloidal material may be contemporaneous with deposition of sand layers. Downward migration of colloidal material is a penecontemporaneous feature. The in-

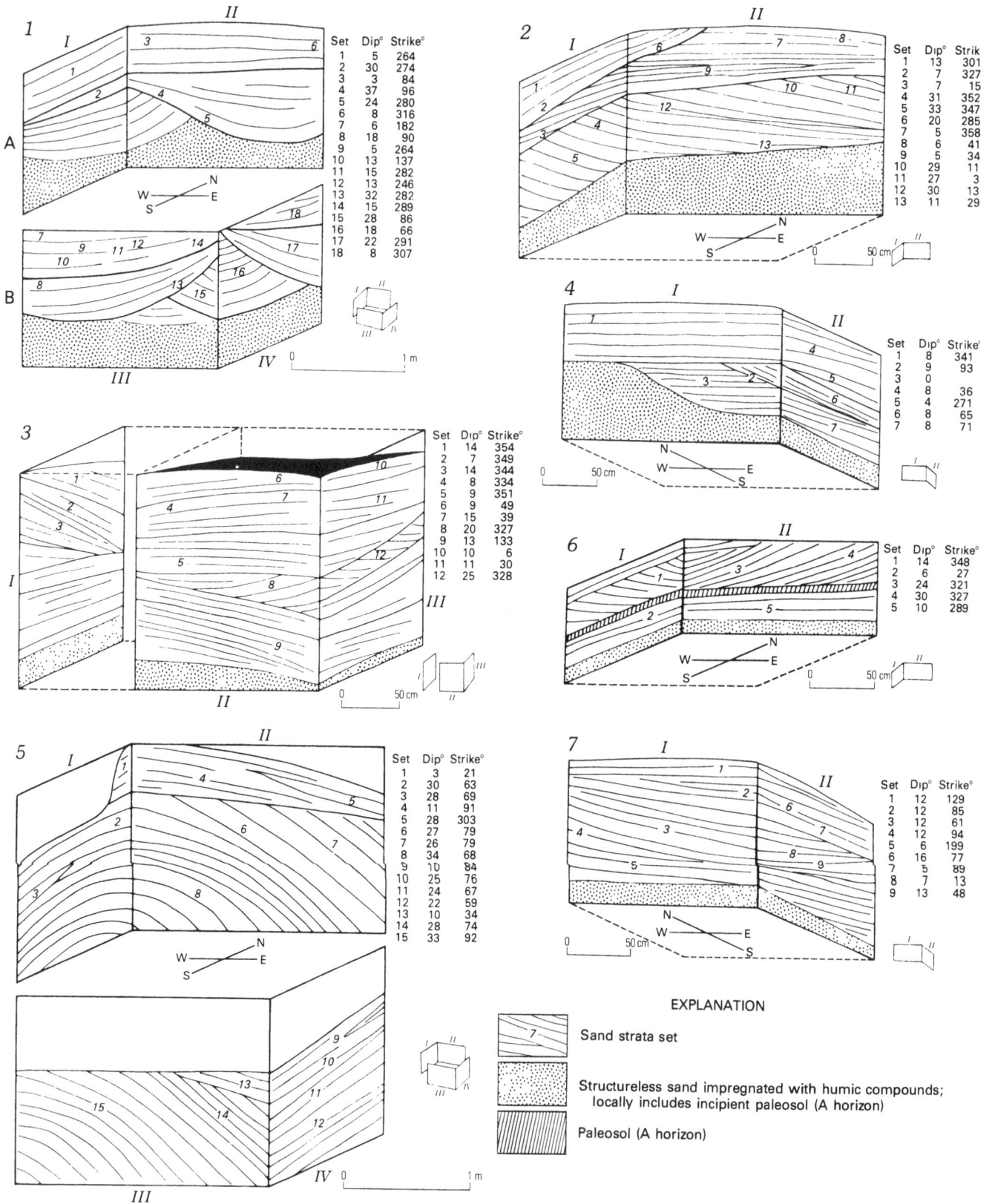

1

Set	Dip°	Strike°
1	5	264
2	30	274
3	3	84
4	37	96
5	24	280
6	8	316
7	6	182
8	18	90
9	5	264
10	13	137
11	15	282
12	13	246
13	32	282
14	15	289
15	28	86
16	18	66
17	22	291
18	8	307

2

Set	Dip°	Strik°
1	13	301
2	7	327
3	7	15
4	31	352
5	33	347
6	20	285
7	5	358
8	6	41
9	5	34
10	29	11
11	27	3
12	30	13
13	11	29

3

Set	Dip°	Strike°
1	14	354
2	7	349
3	14	344
4	8	334
5	9	351
6	9	49
7	15	39
8	20	327
9	13	133
10	10	6
11	11	30
12	25	328

4

Set	Dip°	Strike°
1	8	341
2	9	93
3	0	
4	8	36
5	4	271
6	8	65
7	8	71

6

Set	Dip°	Strike°
1	14	348
2	6	27
3	24	321
4	30	327
5	10	289

5

Set	Dip°	Strike°
1	3	21
2	30	63
3	28	69
4	11	91
5	28	303
6	27	79
7	26	79
8	34	68
9	10	84
10	25	76
11	24	67
12	22	59
13	10	34
14	28	74
15	33	92

7

Set	Dip°	Strike°
1	12	129
2	12	85
3	12	61
4	12	94
5	6	199
6	16	77
7	5	89
8	7	13
9	13	48

EXPLANATION

Sand strata set

Structureless sand impregnated with humic compounds; locally includes incipient paleosol (A horizon)

Paleosol (A horizon)

STRATIFICATION IN PITS 1–7 along east-west line in the interdune area between the arms of a parabolic dune in the Lagoa dune field, Ilha de Santa Catarina, Brazil. Dip and strike are given for each numbered cross-strata set within a pit. Pit locations are shown in figure 77. Modified from Bigarella (1975b, figs. 14–16). (Fig. 83.)

E_1

Set	Dip°	Strike°
1	28	332
2	23	01
3	26	356
4	31	21
5	26	335

0   50 cm

E_2

Set	Dip°	Strike°
1	15	286
2	25	327
3	23	340
4	24	331
5	31	342
6	11	346
7	13	43

0   1 m

E_3

Set	Dip°	Strike°
1	16	332
2	25	325
3	25	321
4	33	297
5	22	279
6	19	279
7	22	314
8	25	312

0   1 m

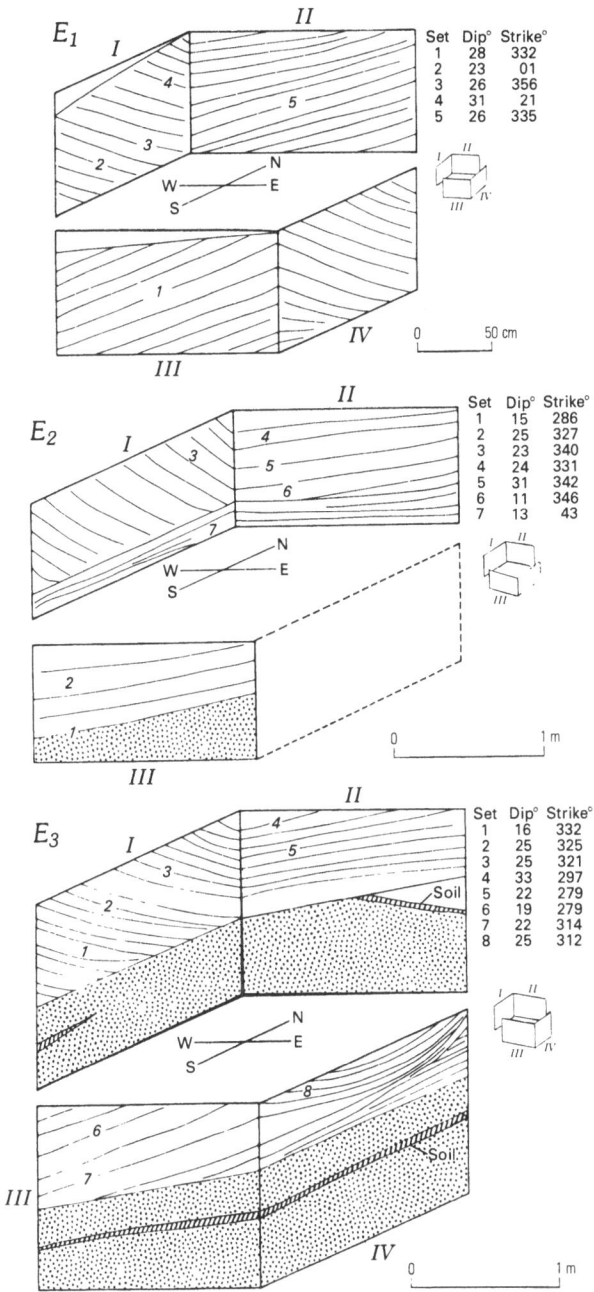

STRATIFICATION IN PITS E_1, E_2, and E_3 along north-south line of the interdune area between the arms of a parabolic dune, Lagoa dune field, Ilha de Santa Catarina, Brazil. Dip and strike are given for each numbered cross-strata set within a pit. Light stippled areas are nonstratified sand. Pit locations shown in figure 77. Modified from Bigarella (1975b, fig. 17). (Fig. 84.)

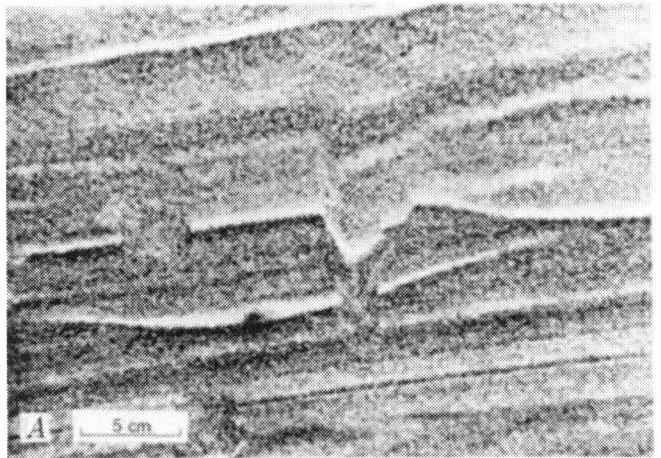

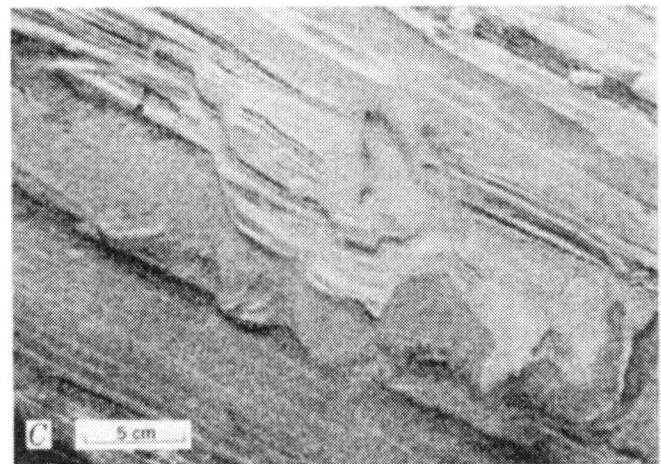

STRUCTURES OF INTERDUNE AREA between the arms of large parabolic dune at Lagoa dune field, Ilha de Santa Catarina, Brazil. A, Collapse structure of sand filling tube of burrowing animal; B, set of high-angle cross-strata of earlier dune between sets of interdune subhorizontal strata; C, high-angle faults in foresets of earlier dune preserved in interdune area. (Fig. 85.)

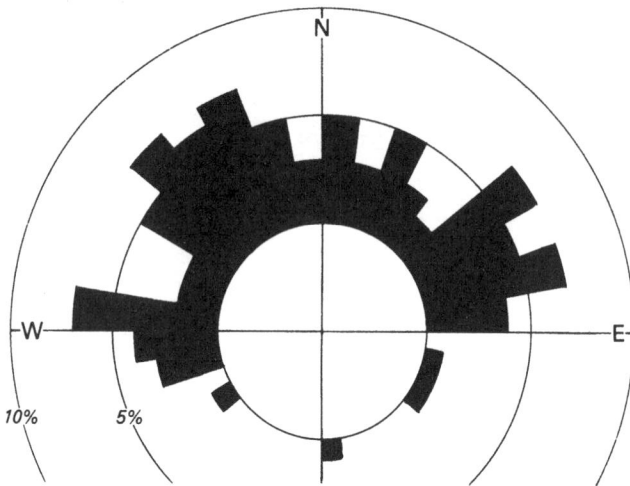

ROSE DIAGRAM OF CROSS-STRATA DIP DIRECTIONS, based on 98 measurements; interdune area, Lagoa dune field, Ilha de Santa Catarina, Brazil. (Fig. 86.)

troduced material follows and reinforces the primary structures developed by reworking of the eolian sands (dissipation of former dunes). In this way, the apparently structureless pattern that originated during the dissipation process becomes more evident.

In addition to the dissipation action of water, these dunes are subjected to soil-forming processes that create soil horizons. The red and brown colors of the apparently structureless eolian sands are generally attributed to soil formation; however, at the Lagoa dune field, most dunes investigated showed little red color to have been developed in situ by such processes. The amount of heavy minerals present in the dunes is not sufficient to provide the colloidal materials (clay and iron compounds) for staining the quartz grains under normal soil-forming conditions.

A surficial soil profile is not well developed on the dunes. Thus, the red and brown materials are not commonly a result of downward migration of colloidal material from soil formation. Large amounts of foreign materials, mostly colluvial, have been introduced in the dune field by mud and (or) sand flows, which probably originated during drier climates at times of regionally concentrated rainfall.

Foreign material obviously was introduced into the dune field, and the migration of this material definitely results from the action of percolating waters and soil formation. Evidences of migration are locally very clear in some of the trenches studied.

The highest concentration of organic matter is found in paleosol (A horizon). Humic compounds derived from the organic soil horizon and from the decay of plant debris migrate downward through the sand profile into the zone of saturation below the water table, causing the sand to become brown. In migrating, the humic compounds tend to impregnate first on the shear planes and sedimentary structures produced by the dissipation processes; the resulting near-black stain makes the structures easily visible. As the impregnation progresses laterally, the structures tend to be less visible and disappear into a dark-brown sandy area. In turn, the fluctuation of the phreatic water level, rich in humic colloids, is responsible for additional impregnation of the organically cemented dark-brown crusts of layers.

Interdune Areas

In the eastern part of the Lagoa dune field, not far from the granite hill, several trenches were dug in an area of parabolic dunes. The sediment in these trenches contained dissipation structures representing drier climate episodes, during which there was a regional loss of vegetation and only sporadic but intense rainfall. Flash floods had washed weathered material both from the side of the granite hill and from the dune into the interdune area. In the trench wall, sand- and granule-size stream deposits and thin clayey beds deposited in small ponds were exposed. These deposits alternate with sets of well-stratified dune sand. Thus, the dissipation structures represent times when the flash floods dominated in the development of dunes.

Deformation Results of Dune Dissipation

The response to stresses and the types of structure produced during deformation depend on the amount of moisture present in the sand (McKee, Douglass, and Rittenhouse, 1971). The cyclic sequences at the Lagoa dune field were developed under varying moisture conditions. Some units consist of typical eolian dune sediments; others are water-reworked dune deposits; some are mixed

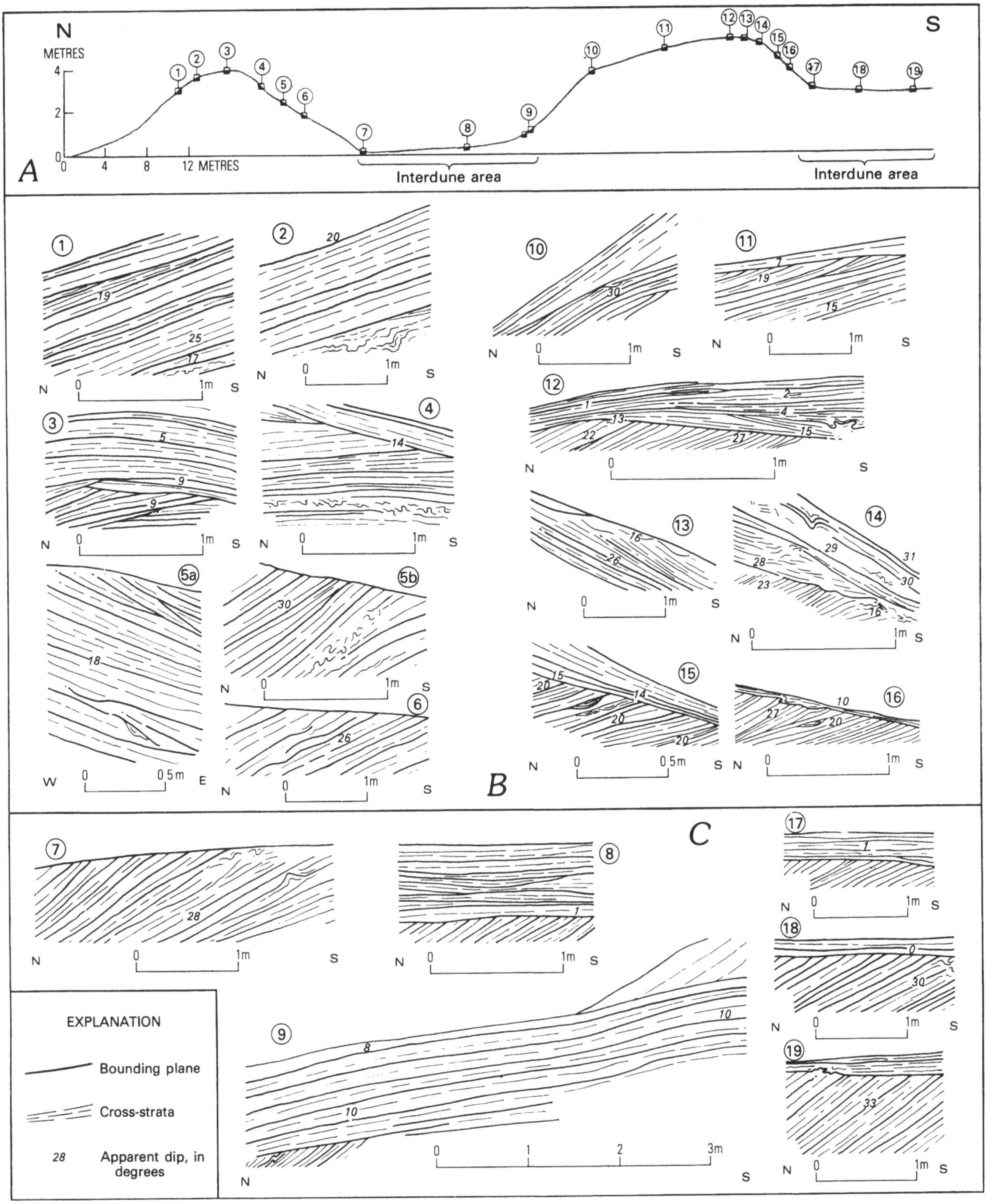

REVERSING DUNE AND INTERDUNE STRUCTURES, Lagoa dune field, Ilha de Santa Catarina, Brazil. A, Dune profile showing locations of trenches; B, trenches in reversing dunes; C, trenches in interdune areas. Dip of cross-strata is shown in degrees; no vertical exaggeration. Modified from Bigarella (1975b, fig. 13). (Fig. 87.)

A STUDY OF GLOBAL SAND SEAS

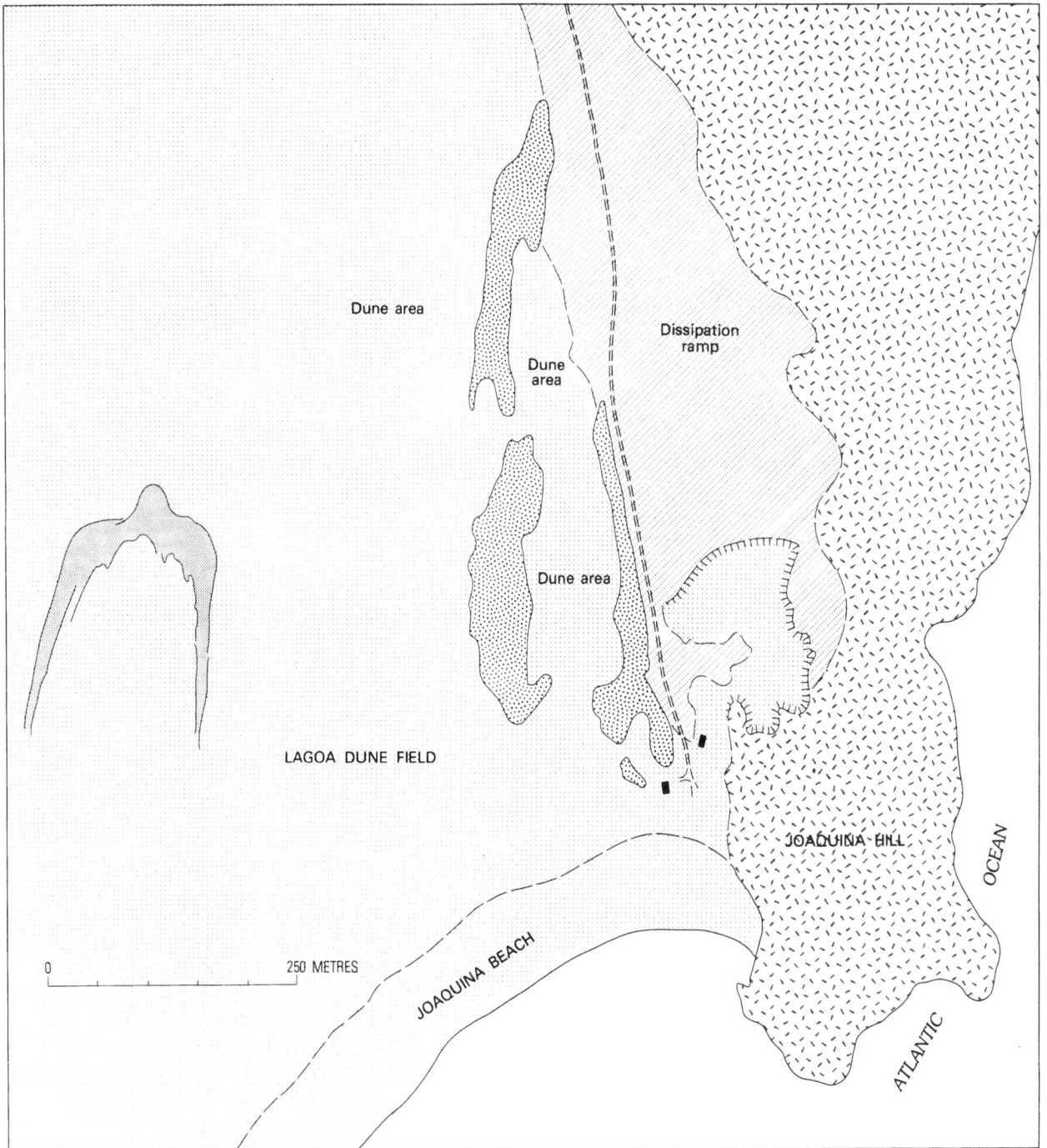

EXPLANATION

Sand	Stabilized parabolic dune	Precambrian crystalline basement; mostly granite
Dissipation ramp	Surveyed parabolic dune	Boundary; dashed where approximate

DISSIPATION RAMP LEADING DOWN from Joaquina Hill to Lagoa dune field, Ilha de Santa Catarina, Brazil. Hachures indicate inward slope. Map is study area of figure 76. Modified from Bigarella (1975b, fig. 2). (Fig. 88.)

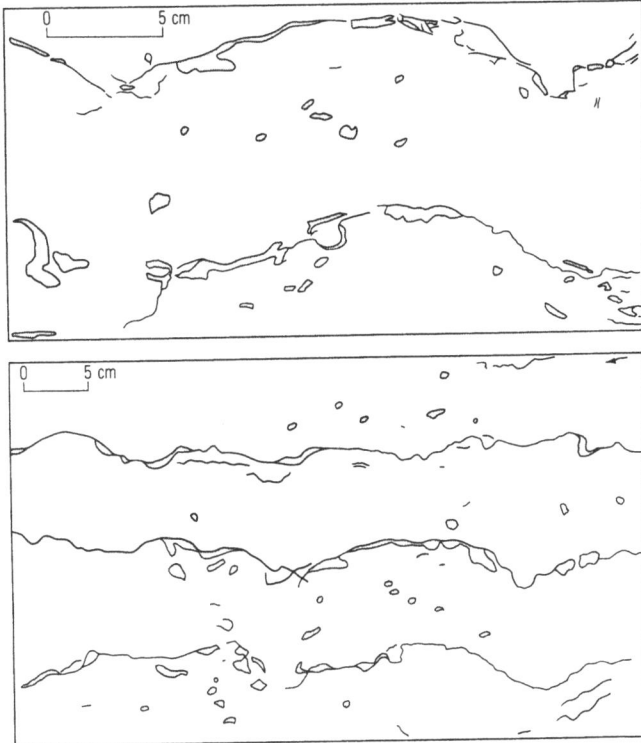

CONTORTED WAVY PATTERN IN DUNE SAND formed by colloidal material, mostly clay, separating individual sand-flow layers developed during dissipation processes. Cross-sections of dune normal to dip at Lagoa dune field, Ilha de Santa Catarina, Brazil. (Fig. 89.)

with sediments derived from the reworking of colluvial material from the nearby granite hills.

Under conditions normally present in an arid climate, the coastal dunes of Lagoa would have formed primarily by the accumulation of sands that avalanched down the steep lee faces. The most common structures from such avalanching, besides cross-stratification, are shear planes, thrust faults, high-angle asymmetrical folds, overturned folds, breccias, normal faults, fadeout laminae, and intertonguing sand lenses (McKee and Bigarella, 1972, p.

674). Dissipation conditions at Lagoa, however, were established by an increased amount of water. Periodic mass movements in these dunes (fig. 90), such as "wet avalanches," develop faults connected with systems of shear planes, brecciated sand masses, and folds. Downward movement of the saturated sediments differs from normal avalanching of dry or damp sand. In a more advanced stage, the action of sand flow and rainwash may spread across the whole sand body producing its dissipation, and the resulting structures display a wavy and lenticular pattern.

The avalanching of sand on the lee side of a dune produces shear planes in which heavy minerals are concentrated by downward migration through the voids between grains. Locally, avalanches in wet reworked dune sand resulted in wavy and lens-shaped structures. This movement may have been started in wet sand by undercutting at the base of the deposit. The avalanching of wet sand is a mass movement that develops more than one shear plane (fault) parallel or subparallel to the main movement. A secondary system of shear planes commonly forms roughly perpendicular to the main faults.

Normal faults and breccias are common to avalanching in wet sand (McKee and Bigarella, 1972, p. 674). The faulting due to avalanching is caused by heavy rainfall that soaks the dune. Associated interdune sediments indicate the presence of small ephemeral streams that may undercut the dune base to start sand movement.

Scour-and-fill structures may form as part of a single process in which the troughs are scoured by flowing water and subsequently filled with sand. Scouring and filling seem to be related to rainfall patterns and to the sediment carried in small ephemeral streams that flow in interdunal areas during the dissipation process.

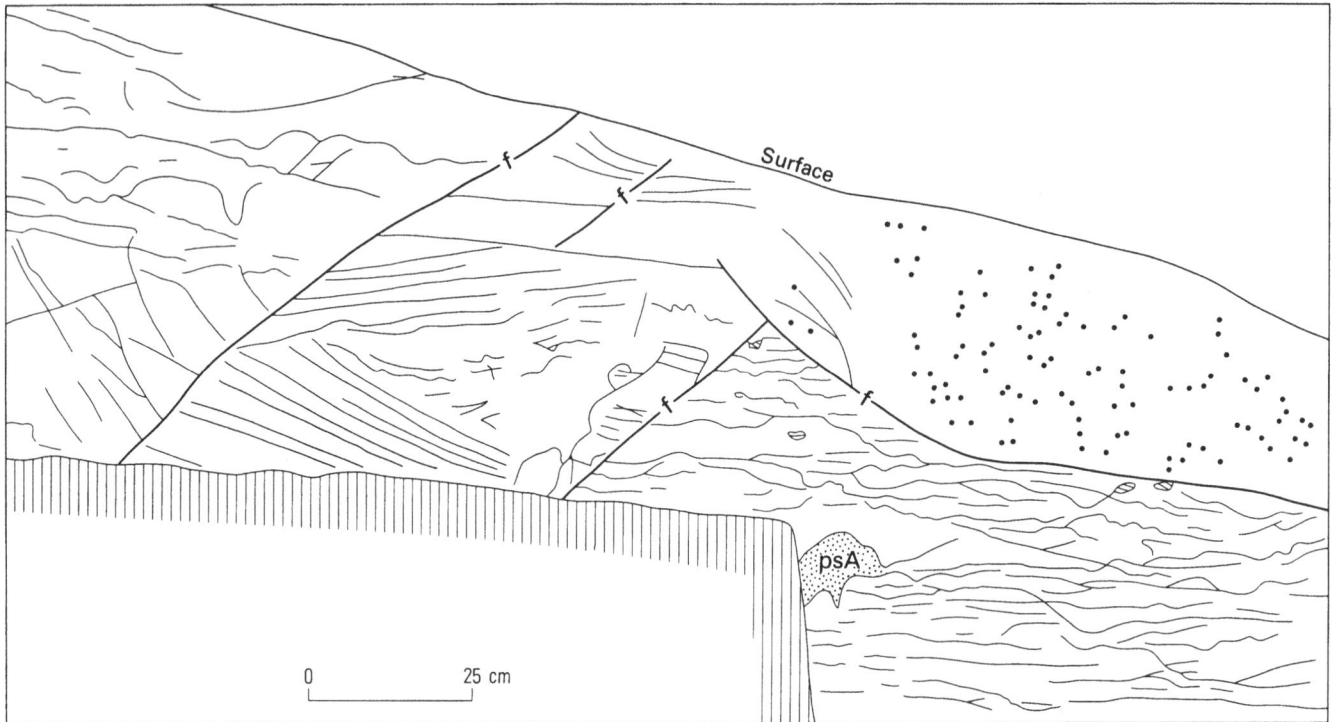

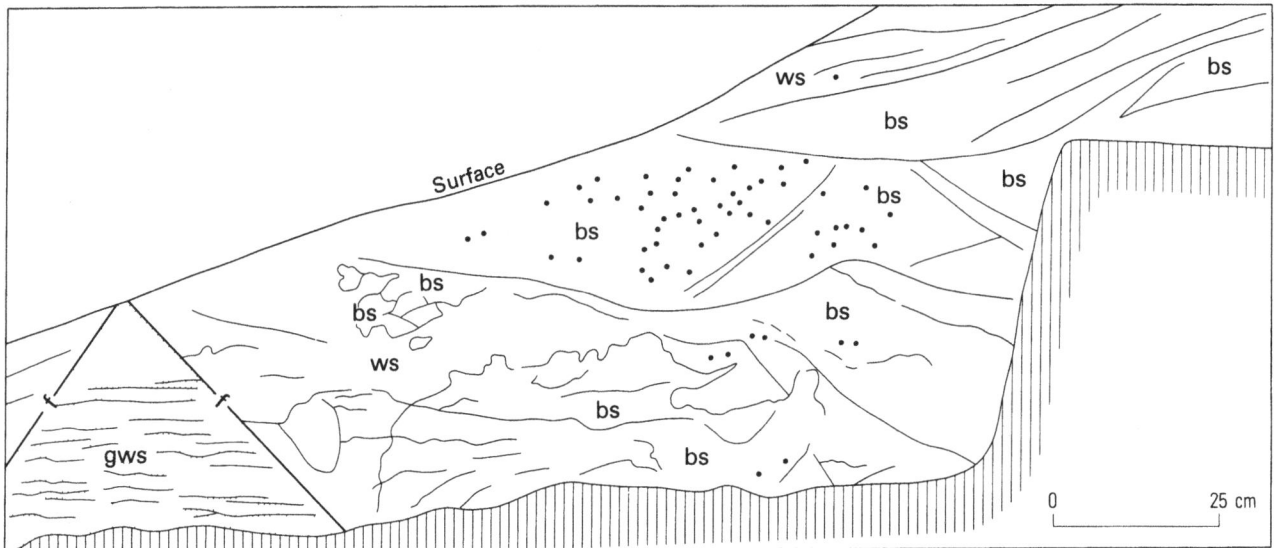

SCHEMATIC CROSS SECTIONS illustrating features such as shearing planes, faults, breccia-bed sands, wavy dissipation structures, and coarse sand layers that result from dissipation conditions. Black dots, colluvium pellets; f, fault; ws, white sand; gws, gray-whitish sand; bs, brown sand; psA, paleosol A, enriched with organic matter. Hachures indicate predune surface. Modified from Bigarella (1975b, fig. 20). 1 centimetre = 0.39 inch. (Fig. 90.)

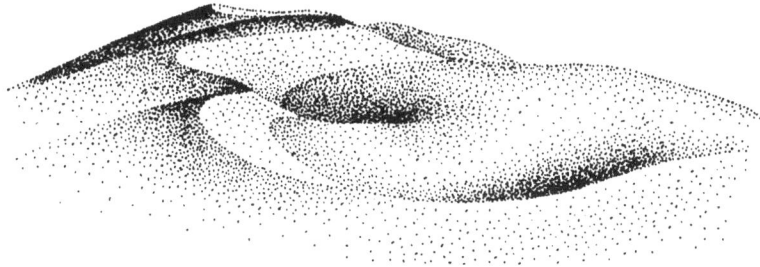

A STUDY OF GLOBAL SAND SEAS

DUNE FORMS AND WIND REGIME

Chapter F

By STEVEN G. FRYBERGER,
assisted by GARY DEAN

Contents

Illustrations

Tables

Summary of Conclusions

𝕿HE COMPARISON OF WIND REGIMES to dune forms on a worldwide basis has become possible through two types of data. The first is recently available Landsat imagery for deserts of the entire world, some parts of which are very difficult of access. The second is reliable surface wind information for these regions, available from the National Climatic Center, Asheville, North Carolina, U.S.A. In this report the morphologies of dunes both from well-studied and from relatively less studied parts of the world were compared with local and regional winds. Characteristic wind environments were discerned for barchanoid, linear, and star dunes.

By making several simplifying assumptions, surface winds were analyzed for direction and relative amount of sand migration (drift), using the Lettau[6] equation for sand drift:

$$q \cdot g / C'' \rho = V^{*2}(V^* - V_t^*).$$

This formula was modified to:

$$Q \propto V^2 (V - V_t) \cdot t,$$

Where Q is a proportionate amount of sand drift, V is average wind velocity at 10-metre (33-ft) height, V_t is impact threshold wind velocity, and t is time wind blew, expressed as a percentage in a wind summary.

Values derived from this formula by substitution reflect the sand-moving capacity of the wind for the time period of the wind summary and are herein known as drift potentials (DP), which are numerically expressed in vector units (VU). The direction and magnitude of the vector resultants of drift potentials from the 16 compass directions are herein known as the resultant drift direction (RDD) and the resultant drift potential (RDP), respectively. An index of the directional variability of the wind is the ratio of the resultant drift potential to the drift potential, herein known as RDP/DP. The greater the directional variability of the effective winds at a station, the lower its associated RDP/DP will be.

[6] K. and H. Lettau, Meteorology Department, University of Wisconsin, Madison, Wisconsin: unpublished data, 1975. Equation reproduced by permission of the authors.

The average annual amount of wind energy in the deserts studied ranges from approximately 80 to 489 VU. On the basis of rough groupings of deserts in terms of wind energy, the wind energy of desert environments is classified as follows: low energy, DP less than 200 VU; intermediate energy, DP, 200–399 VU; high energy, DP, 400 VU or greater.

Sand roses are circular histograms representing the potential effects of winds from the directions of the compass, as evaluated from the modified Lettau equation. Effective (sand-moving) winds tend to occur in directional groupings or modes, which are reflected in sand roses. The directional classification of effective wind regimes as represented on sand roses is based on the relationships of modes and includes the categories herein designated as "narrow unimodal," "wide unimodal," "acute bimodal," "obtuse bimodal," and "complex."

Barchanoid dunes, including barchans, domes, and transverse ridge dunes, have similar wind environments. Wind distributions associated with these dunes generally are narrow unimodal or wide unimodal but may be acute bimodal if one mode is much stronger than the other. Barchanoid dunes in environments with high drift potentials seem to be associated with a higher RDP/DP than those in environments with low drift potentials.

Barchanoid dunes are observed in present-day wind environments of low, intermediate, and high energy (drift potential). Large fields of barchanoid dunes occur in continental settings and in restricted high-energy settings along coasts. Of all the dune types observed, barchanoid dunes are associated with the least variability in effective wind direction for a given drift potential.

Linear dunes are associated with a variety of effective wind environments — from low to high energy. Effective wind distributions range from wide unimodal through complex. The wind environments of most linear dunes have a greater directional variability of effective winds than do those of barchanoid dunes for a given drift potential. Simple linear dunes are a nearly exclusive type over wide regions of the Peski Karakumy and the Kalahari, Simpson, Great Sandy, and Great Victoria Deserts. The unvegetated dunes in the southwest Kalahari and Simpson Deserts are associated with intermediate to high-energy bimodal or wide unimodal wind regimes. Partially vegetated dunes in the northwest Kalahari and Great Victoria Deserts are associated with low-energy bimodal or complex wind regimes.

Simple linear dunes can occur in the same high-energy wind environment as compound linear dunes. This suggests that factors other than wind, such as available sand and the age of the dunes, may determine the type of linear dune observed at a particular high-energy locality. Compound linear dunes are most frequently observed in intermediate- to high-energy wind environments, with wide unimodal or bimodal distributions of effective winds.

Star dunes have been observed in complex wind regimes of low, intermediate, and high energy. The complex wind regimes associated with star dunes can differ in several respects from those associated with linear dunes. Unlike linear dunes, star dunes have been observed in a complex wind environment of high drift potential, in which complex wind distributions prevail during the windiest months, and in which the dominant modes are directly opposed.

Star dunes in a high-energy wind regime near Ghudāmis, Libya, are isolated mounds, whereas star dunes in a low-energy wind regime near Beni Abbes, Algeria, seem to be connected by sinuous arms. Some star dunes are developed in linear patterns atop "sand sheets," linear dunes, or possibly crescentic dunes. Wind regimes associated with alined star dunes in the two examples available have characteristics intermediate between dune types in terms of *RDP/DP*.

Introduction

The following conclusions are theoretical because the author was not able to visit many of the localities studied from Landsat imagery during the course of the investigation. Early in the study, however, it was observed that a single dune type commonly occurred throughout wide stretches of desert. For example, barchanoid dunes characterize the eastern Takla Makan Desert of China; linear dunes, the Erg Maktëir of Mauritania; and star dunes, the Grand Erg Oriental of Algeria. Thus, if the winds of the various regions could be defined, it seemed possible to obtain insight into the wind regimes responsible for creating major dune types.

Following the suggestion of Bagnold (1941, p. 184), winds were evaluated and defined in terms of their potential sand-moving effectiveness through the use of a suitable weighting equation applied to a standardized set of data.

Parts of this chapter on barchanoid, linear, and star dunes are designed to give a summary of the author's observations, based on the included tables and on material in chapter K. In addition, examples are given of observed wind environments of certain dune types. These examples were chosen first to illustrate the most common wind environments in which the representative dune types were observed, and second to illustrate problems requiring further research. Finally, examples were chosen, where possible, to compare the wind environments of similar dune types in different regions. Examples were selected from localities where the author had most confidence that the winds measured at nearby meteorological stations were representative of the regional wind environment.

The classification of effective wind environments used in this paper will enable the reader to compare one effective wind environment with another but is not intended to be a formal system. Although the sand rose concept (discussed later) seems useful in delimiting the gross directional and energy properties of effective wind regimes, the more important problem of the effect on morphology of rare, high-speed winds versus winds of intermediate strength remains to be solved.

An important caution should be kept in mind by the reader in regard to the observations in this report. Although the available surface wind data provide a good picture of present-day effective winds, the maximum period of record seldom exceeds 15 years. In contrast, many of the observed dunes are very large and, therefore, possibly are very old. Thus, some of the large dunes observed, particularly in regions of low or intermediate drift potential, may in part have developed in response to older, different wind regimes. The most reliable comparisons, therefore, are between dunes and wind environments with high drift potentials.

Acknowledgments

Technical assistance and criticism for this chapter were generously given by Robert P. Sharp, California Institute of Technology; Heinz Lettau,

University of Wisconsin; Douglas L. Inman, Scripps Institute of Oceanography; Harold L. Crutcher, National Climatic Center; Harry Van Loon, National Center for Atmospheric Research; and Errol L. Montgomery, Northern Arizona University.

Methods of Study

Types of Wind Data Available

SAND DRIFT REFERS to the process of sand movement across the desert as a result of surface winds. Most estimates of sand drift given in this report were made from surface wind tabulations known as "N summaries." These were prepared by the Environmental Technical Applications Center of the U.S. Air Force. They are stored at the National Climatic Center, Asheville, North Carolina, U.S.A., and are available by station name or by World Meteorological Organization (W.M.O.) number. Each W.M.O. station summary normally contains both monthly and annual data. Wind speed is recorded in knots to the nearest 10° of direction at 3- to 6-hour intervals. The period of record for stations used in this chapter averages 10 years.

Two N summary formats are available (fig. 91). The first format (fig. 91A) is more useful because wind velocities are divided into 9 or 11 categories, whereas the second format (fig. 91B) has only 5 velocity categories. About 100 summaries using the first format were analyzed. The linear-regression technique was used to estimate relative sand drift from 34 summaries in the second format.

Detailed wind data, other than the N summaries, were also obtained from some government offices, desert research organizations, corporations, and libraries. These data were reduced to a form similar to that shown in figure 91A, then evaluated by the same methods.

Limitations of Wind Data

The surface-wind data used in this report generally was of good quality. However, methods of gathering data and summarization processes both introduced some systematic inaccuracies.

Inaccuracies introduced during data gathering occur primarily in two ways. The first, known as observer bias (Ratner, 1950, p. 185), is the tendency for weather observers to record wind occurrences from the prime directions rather than from intermediate directions of the compass. For example, when an observer is uncertain, he commonly records wind as coming from the northeast rather than from the east-northeast.

A second type of inaccuracy occurs during data gathering because of deviations from standard observing conditions as specified in the W.M.O. "Guide to Meteorological Practices." One common occurrence is the mounting of an anemometer at a height other than the standard 10 m (33 ft) specified by the World Meteorological Organization. For example, the anemometer height at El Golea, Algeria, was 4 m (13 ft) during 1949–56, 22 m (72 ft) during 1956–60, and 7 m (23 ft) from 1960–73. Additionally, a station may be sheltered from the wind by nearby trees, buildings, or high ground.

The mounting of anemometers at nonstandard heights may have slightly affected some calculations of drift potential, because threshold drag velocity assumed in calculations is based on wind velocity at a 10-m height. If anemometers are mounted lower than the standard 10 m, calculated drift potential will be slightly less than the true drift potential which would result from calculations based on a 10-m height. This is because wind velocities are lower near the ground; thus, the theoretical threshold velocity will be exceeded less often.

Inaccuracy also arises during summarization of data, and it usually develops in two principal ways. First, inaccuracy enters a summary when data is condensed from 36 to 16 compass directions. This is known as procedure error (Wallington, 1968, p. 293). The result is to create an apparent increase in observations from the prime compass directions at the expense of the intermediate directions. Second, summarizing of observations results in a coarsening of the resolution of the data in terms of velocity, direction, and percent occurrence. Most percentages on N summaries are expressed to the nearest 0.1 percent (fig. 91A). Depending on the number of observations, however, single occurrences may be represented in a summary as more than 0.1 percent. For example, an easterly maximum wind of 17 knots was recorded during October at T'ieh-kan-li-k'o, China, for which period only 137 observations

DATA PROCESSING DIVISION
ETAC/USAF
AIR WEATHER SERVICE/MAC

PERCENTAGE FREQUENCY OF WIND
DIRECTION AND SPEED
(FROM HOURLY OBSERVATIONS)

SURFACE WINDS

23195	YUMA ARIZONA IAP	48-71	ALL
STATION	STATION NAME	YEARS	MONTH

ALL WEATHER

CLASS   ALL HOURS (L S T)

CONDITION

SPEED (KNTS) DIR	1·3	4·6	7·10	11·16	17·21	22·27	28·33	34·40	41·47	48·55	≥56	%	MEAN WIND SPEED
N	1 0	3 4	3.1	1 8	4	1	0					9 7	8.0
NNE	1 0	3 4	2.6	7	1	0	0					7 7	6 6
NE	8	2 1	1 1	2	0	0	0					4 3	5 8
ENE	5	1 1	4	1	0	0						2 2	5 7
E	7	1 0	3	1	0	0	0					2 2	5 3
ESE	4	8	4	1	0	0						1.7	5 8
SE	6	1 4	1 9	1 6	.3	0	0					5.8	8 9
SSE	5	1 9	3 0	2 6	.4	0	0					8 5	9 5
S	8	2 6	3 2	1 6	1	0						8 4	7.8
SSW	6	1 7	1 8	7	.0	0	0					4 9	7 1
SW	8	2 6	1 9	3	0	0						5 6	6.2
WSW	6	2 6	2 1	4	0	0						5 7	6 5
W	7	3 0	2 5	9	2	1	0					7 3	7 5
WNW	6	2 5	2 2	1 3	3	1	0	0	0			6 9	8 2
NW	6	1 8	1 3	5	1	0	0					4 3	6 9
NNW	6	1 7	1 8	9	1	0	0					5 1	7 8
VARBL													
CALM	✕	✕	✕	✕	✕	✕	✕	✕	✕	✕	✕	9.6	
	10 8	33 8	29 6	13 8	2 1	3	0	0	0			100.0	6 7

TOTAL NUMBER OF OBSERVATIONS   160705

A

AIR WEATHER SERVICE

CLIMATIC CENTER

DATA PROCESSING DIVISION

PERCENTAGE FREQUENCY OF SURFACE WINDS

N SUMMARY #1

62002 NALUT LIBYA

31 52 N   010 59 E   2036 FT   MAR

YEARS 49,50,51,52,54,55,56,57,58,59,60,61,62,63

DIRECTION	WIND SPEED GROUPS IN KNOTS					TOTAL %	TOTAL OBS	MEAN WIND SPEED/KTS
	01-06	07-16	17-27	28-40	GTR 40			
N	5 8	10 9	2 1	0		18 8	468	10 2
NNE	2 6	3 8	1 0			7 4	184	10 0
NE	2 1	3.2	1 1			6 5	160	10 5
ENE	6	9	1			1 6	40	9.0
E	4	5	4	1		1 4	35	13 7
ESE	.2	2	1	1		.5	13	15 8
SE	2	8	3	2		1 5	37	16 6
SSE	6	1 2	1 4	3		3 6	87	16 5
S	2 4	5 1	4 0	6	1	12 2	303	15 2
SSW	1.0	2 2	1 2	3	0	4 8	118	14 2
SW	1 1	3 6	1 9	4	2	7 3	180	15 4
WSW	2 1	3 6	9	2	0	6 9	170	11.3
W	4 3	4 6	1 8	0		10 7	265	10.2
WNW	1 4	1 3	1			2 9	71	7 9
NW	1 3	1 3	1			2 7	66	8.0
NNW	1 0	2 0	1			3 1	78	9 2
VARIABLE								
CALM						8 2	203	
TOTALS	27 3	45 2	16 5	2 4	4	100 0	2478	10 8

MAXIMUM WIND SW 50 KTS

B

TWO TYPES OF N SUMMARIES containing surface wind data from which estimates of relative sand drift were made. *A*, 16 directional categories and 11 velocity categories. *B*, 16 directional categories and 5 velocity categories. (Fig. 91.)

are available. This single observation is represented as 0.7 percent of all observations. It is questionable whether this occurrence represents a group of winds which blew 0.7 percent of the time, or 5.2 hours of the 744 hours, in October.

Corrections were not made for the limitations just described because it was not known when and where they occurred. Furthermore, techniques available for correcting observer bias and procedure error (Wallington, 1968, p. 296; Ratner, 1950, p. 186) could not be applied uniformly to the different types of wind summaries used in this study. More detailed future work involving the collection and summarization of surface-wind data would be improved by taking the factors previously discussed into account. However, these factors probably did not affect the data enough to detract from the general conclusions of this study.

Evaluation of Wind Data

Selection of a Weighting Equation

A number of equations are available to compute rates of sand drift if the shear velocities are known. The most useful formulas, including those of Bagnold (1941), Kawamura (1951), and Zingg (1953) were tested by Belly (1964, p. 3–5) with data from his wind-tunnel studies. All formulas were found by Belly to describe the data well when suitably evaluated. A formula recently suggested by K. and H. Lettau (written commun., 1975) produces a theoretical curve that agrees very well with Belly's work for 0.44-mm-diameter sand (fig. 92). For this reason, the Lettau equation for the rate of sand drift was chosen in this study as the basis for the weighting equation for sand movement, although other formulas probably would have given similar results. Lettau's equation is as follows:

$$\frac{qg}{C''\rho} = V^{*2} \ (V^* - V_t^*) \ , \qquad (1)$$

where
- q = rate of sand drift;
- g = gravitational constant;
- C'' = empirical constant based on grain size;
- ρ = density of air;
- V^* = shear velocity; and
- V_t^* = impact threshold shear velocity, or the minimum shear velocity required to keep sand in saltation.

Additionally,

$$C'' = C' \ (\delta/\delta^*)^n \ ,$$

where
- C' = universal constant for sand (about 6.7),
- δ = grain diameter of sand moved,
- δ^* = 0.25 mm (standard diameter), and
- n = empirical constant $\simeq 0.5$.

All units are c.g.s. (centimetre-gram-second).

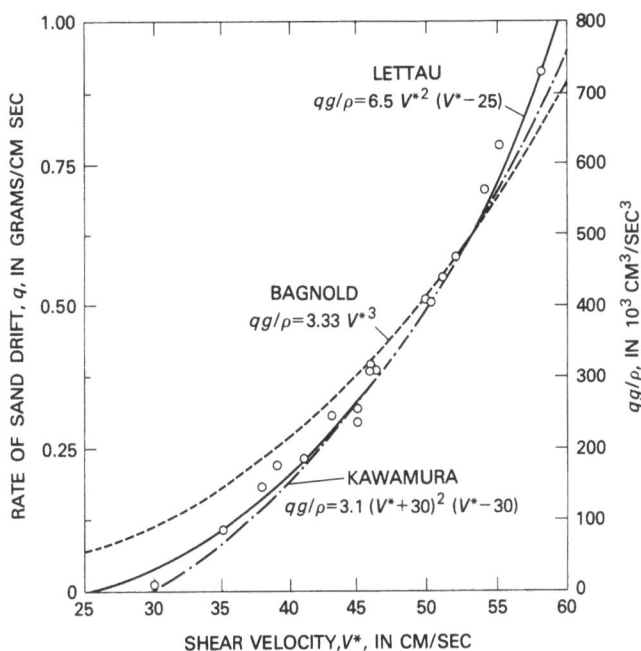

RATE OF SAND TRANSPORT (grams moving per second across a crosswind distance of 1 cm (0.40 in.)), q, versus shear velocity, V^*, of air motion, according to theoretical curves by Bagnold (1941), Kawamura (1951), and Heinz Lettau (1975, unpublished data) in comparison with wind tunnel measurements (circles) by Belly (1964, p. 17). Reproduced by permission of the authors, K. and H. Lettau, Meteorology Department, University of Wisconsin, Madison, Wisconsin, U.S.A. (fig. 92.)

Surface conditions, in addition to shear velocity, control the rate of sand drift. Four important factors are mean grain diameter of the sand (Bagnold, 1941, p. 67; Belly, 1964, p. 13), the degree of surface roughness (Chepil and Woodruff, 1963, p. 240; Bagnold, 1941, p. 71), the amount and kind of vegetative cover (Chepil and Woodruff, 1963, p. 221), and the amount of moisture in the sand (Belly, 1964, p. III 21). These parameters could not be evaluated for many of the localities studied. For this reason, localities were compared using relative quantities of *potential* sand drift.

The Lettau equation for the rate of sand transport was generalized as follows:

$$q \propto V^{*2}(V^* - V_t^*) \,. \qquad (2)$$

Drag velocity is proportional to wind velocity for a given height (Belly, 1964, p. 18). Therefore, as a first approximation, wind velocities at a 10-m (33-ft) height (the standard W.M.O. anemometer height) may be substituted for drag velocities. The Lettau equation then becomes:

$$q \propto V^2 (V - V_t) \,, \qquad (3)$$

where

V = wind velocity of at 10-m (33-ft) height, and

V_t = impact threshold wind velocity at 10 m (33 ft) (minimum velocity at 10 m (33 ft) to keep sand in saltation).

This relationship produces a number which expresses the relative amount of sand potentially moved by the wind during the time it is presumed to blow. When the factor of time is added to equation 3 (thereby creating equation 5), the resulting number is here referred to as the drift potential (DP) and is a measure of the relative amount of potential sand drift at a station for a stated period of time. For convenience, the units of drift potential are here called vector units (VU) because wind velocities are treated as vectors. The detailed method of computation of drift potentials is described next.

Calculation of Drift Potentials

Assumptions Required to Apply the Weighting Equation

In order to use a weighting equation to determine the effects of surface wind, the condition of the surface over which the wind blows must be assumed. For our purposes, this surface is assumed to consist of loose quartz sand grains with an average diameter of 0.25–0.30 mm. The surface is further assumed to be without bedforms larger than ripples, to be dry, and to be without vegetation. Similar surfaces have been used for most wind tunnel studies of sand drift, and 0.25–0.30 mm is the average diameter of many desert dune sands (Ahlbrandt, chapter B, fig. 21). The assumed surface might not serve to predict actual rates of sand drift in areas with very large dunes, but it is useful when comparing one area to another in terms of available wind energy.

A threshold wind velocity must be determined in order to use a weighting formula. For a sand surface of 0.30-mm average diameter quartz sand, the surface roughness factor (Z') as determined by Belly (1964, p. I1 – I2) during sand driving was 0.3048 cm (0.118 in.). The threshold wind velocity at height Z' (V_t') was 274 cm/s (107 in./s) and V_t^* was 16 cm/s (6 in./s). V_t' may be extrapolated to a 10-m (33-ft) height (the height at which most of the wind data were collected) using the equation (Bagnold, 1941, p. 104):

$$V_{(10\ m)} = 5.75\ V_t^* \log \frac{Z}{Z'} + V_t' \,. \qquad (4)$$

When this is done, a value of 11.6 knots is obtained for V_t.

This value indicates that for the conditions described, threshold wind velocity as measured at a 10-m (33-ft) height should be within the 11–16 knot velocity category on N summaries, such as that shown in figure 91A. For this study a value of 12 knots was chosen for $V_{t(10m)}$.

The assumption was made that a wind speed and direction component occurred in nature for an amount of time proportional to its percentage in the summary. With a few exceptions, such as that previously described for T'ieh-kan-li-k'o, China, this assumption seems reasonable because periods of record are long, observations are taken at different times of day and night, and each annual summary averages more than 1,000 observations.

Derivation of Weighting Factors and Application to Wind Data

Weighting factors as used here are numbers which represent the relative rates at which winds of differing average velocities can move sand. These numbers are derived by substitution of values for wind velocity (average wind speed of a velocity category) into the weighting equation of Lettau (equation 3) as shown in table 14.

TABLE 14. — *Derivation of weighting factors for relative rate of sand transport by substitution of average wind velocities into the generalized Lettau equation (equation 3 in text)*

[The value of $V^2 (V - V_t)$ is divided by 100 to reduce weighting factors to smaller sizes for convenience in plotting sand roses, as described in text. Velocities exceeding 40 knots are rare and are not computed]

N summary velocity category (knots)	Mean velocity of winds in category V	V^2	$(V - V_t)$	Weighting factor $V^2(V - V_t)/100$
11 – 16	13.5	182.3	1.5	2.7
17 – 21	19.0	361.0	7.0	25.3
22 – 27	24.5	600.3	12.5	75.0
28 – 33	30.5	930.3	18.5	172.1
34 – 40	37.0	1,369.0	25.0	342.3

The weighting factors represent rates of sand transport, and the percentages of wind occurrence in the summaries represent the length of time during which the winds blew. Therefore,

$$Q \propto V^2 (V - V_t) \, t \, ,$$

or $Q \propto$ (weighting factor) t ,   (5)

where

t = time wind blew, expressed as a percentage on N summary; and

Q = annual rate of sand drift.

To evaluate the relative amount of sand drift which potentially occurs at a station, the weighting factor derived from equation 3 for each velocity category is multiplied by the percentage occurrence of wind in that category for all 16 directions of the summary, and the results are totaled. This computation is shown for a single direction in table 15. Calculations of this sort are tedious; therefore, a programmable calculator was used for much of the work. The most convenient method of expression of the results is the sand rose, construction of which is described later.

TABLE 15. — *Computation, from N summary, of vector unit total from the west-northwest, Yuma, Arizona, U.S.A.*

[The drift potential at the station is the sum of the vector unit totals computed in the same manner from each of the 16 compass directions. Total vector units from west-northwest equals 18.6. Wind data shown in figure 91A]

	Velocity category (knots)				
	11 – 16	17 – 21	22 – 27	28 – 33	34 – 40
Weighting factor	2.7	25.3	75.0	172.0	342.3
Percent occurrence ...	1.3	.3	.1	0	0
Vector units	3.5	7.6	7.5	0	0

Calculation and Plotting of Sand Roses

A sand rose is a circular histogram which represents potential sand drift from the 16 directions of the compass (fig. 93). The arms of a sand rose are proportional in length to the potential sand drift from a given direction as computed in vector units. Thus, a sand rose expresses graphically both the amount of potential sand drift (drift potential) and its directional variability.

These diagrams differ from paleocurrent roses constructed from crossbedding dip directions, because the arms point toward the direction from which sediment moved, that is "into the wind." Sand roses are based on surface wind only and thus reflect only *potential* sand transport. The pattern suggested by the sand rose may be considerably modified by local conditions.

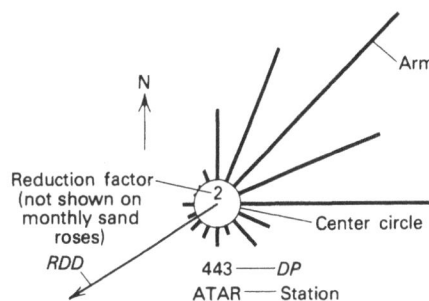

TYPE OF SAND ROSE USED in this chapter. Components are as follows: Arm (vector-unit totals, plotted in millimetres), proportional in length to potential amount of sand drift from a given direction toward center circle. Reduction factor, number by which vector-unit total of each sand rose arm was divided so the longest arm would plot at <50 mm (2 in.). *DP* (drift potential, in vector units), measure of relative sand-moving capability of wind; derived from reduction of surface-wind data through a weighting equation (equation 3 in text). *RDD* (resultant drift direction, in vector units), net trend of sand drift. Station, name of meteorological station at which wind data was recorded. (Fig. 93.)

Sand roses are plotted with a Hewlett-Packard 9810 A calculator and 9862 A plotter.[7] The vector unit totals from each direction are plotted in millimetres. When the vector unit total for any direction exceeds 50, all arms of that sand rose are divided by two until the longest arm of the sand rose plots at less than 50 mm (2 in.) in length. The number by which the arms are divided is known as

---

[7] Use of a specific brand name does not necessarily constitute endorsement of the product by the U.S. Geological Survey.

the reduction factor and is shown in the center circle of the sand rose (fig. 93).

Vector unit totals from the different directions may be vectorially resolved to a single resultant. The direction of this computed resultant is herein referred to as the resultant drift direction (RDD), as shown on figure 93. The RDD expresses the net trend of sand drift, or the direction in which sand would tend to drift under the influence of winds from the different directions. The magnitude of the resultant drift direction may be found from the same data, using the Pythagorean theorem, and is herein referred to as the resultant drift potential (RDP). The RDP expresses, in vector units, the net sand transport potential when winds from various directions interact.

Additional Computations

Drift potentials were estimated for a number of stations in India, Libya, and China, for which only those N summaries in the less detailed format (fig. 91B) were available. Percentages in each velocity category of the figure 91B summary were apportioned into two smaller velocity categories of the figure 91A summary, based on the ratios of adjacent percentages established by linear regression analysis, using data from 36 randomly selected

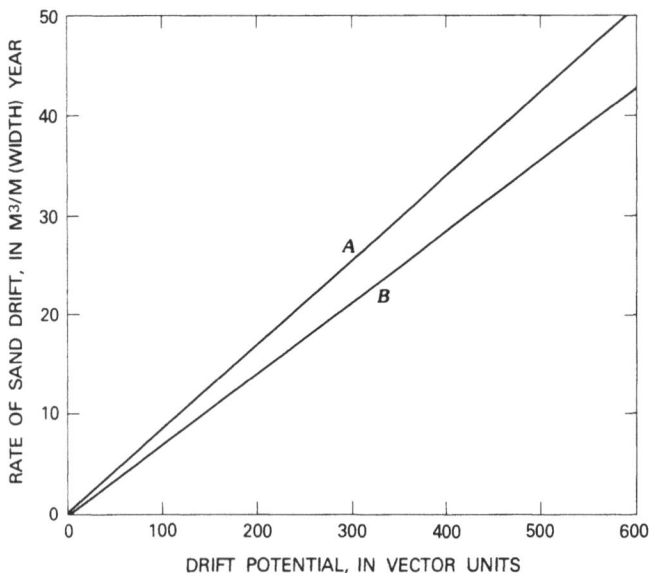

DRIFT POTENTIAL (ANNUAL) VERSUS rate of sand drift (m³/yr) across a 1-metre (3.3-ft) section normal to the drift direction for two threshold drag velocities, according to the Lettau equation (equation 1 in text) in which V_t^* = impact threshold shear velocity. Line A, V_t^* = 16 cm/s (0.5 ft/s) (Belly, 1964, p. 11); line B, V_t^* = 19 cm/s (0.58 ft/s) (Bagnold, 1941, p. 60). (Fig. 94.)   ·

detailed (fig. 91A) summaries for each computation.

The ratio of the resultant drift potential to the drift potential — the RDP/DP — was computed for all stations in the study. The RDP/DP is an index of the directional variability of the wind. Where the wind usually comes from the same direction, the RDP/DP approaches unity. In contrast, where the wind comes from many directions, the RDP/DP approaches zero because the resultant drift potential is very low.

Although this study required only qualitative estimates of rate of sand drift, quantitative rates can be predicted using equation 1. The relationship of drift potential to annual rate of sand drift predicted by this equation for two presumed threshold drag velocities is shown in figure 94.

Classification of Wind Environments
Direction of Surface Winds

MOST SETS OF SURFACE WIND observations, such as those used in this study, exhibit groupings, or distributions, in terms of both direction and speed (for example, fig. 91A, B). Some sets of surface-wind-direction distributions may be described as elliptical, or may be complicated because of mixed land and sea breezes, through mixtures of seasonal flows, or for other reasons (Crutcher and Baer, 1962, p. 522). For example, the 360-point directional wind rose for Juraid Island, Arabia, (fig. 95) suggests four groups of winds; a group from the west, northwest, northeast, and southeast. However, most directional observations at Juraid Island are encompassed within groups of winds from the west and northeast which make an obtuse angle with each other. Sand roses plotted from surface wind data reflect such directional groupings. Although surface wind distributions, such as that for Juraid Island, can be very complex in detail, experience indicates that, as a first approximation, five relationships of directional distributions occur frequently. These relationships form the basis for the classification of directional charcteristics of effective winds in terms of sand roses shown in figure 96. This scheme was adopted in order to simplify the discussion of wind regimes and associated dune forms that follows.

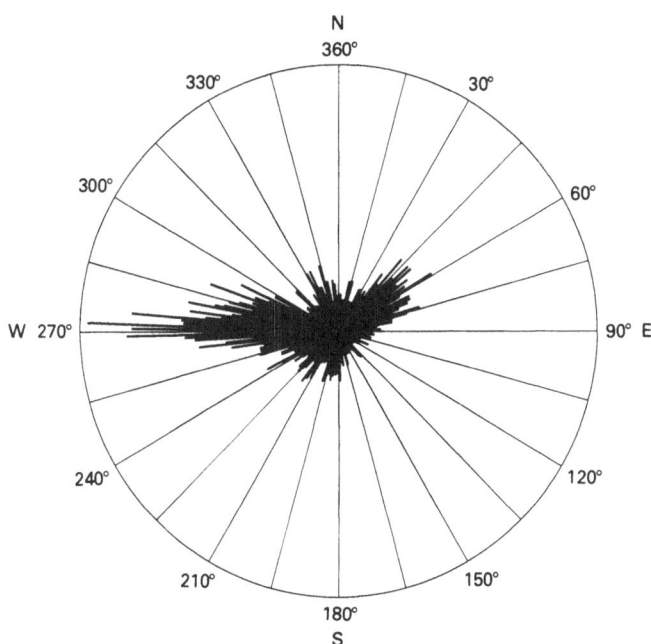

N
360°
330°   30°
300°   60°
W 270°   90° E
240°   120°
210°   150°
180°
S

WIND ROSE FOR JURAID ISLAND, ARABIA (lat 27°12'03" N., long 49°57'23" E.), showing relative numbers of wind observations for 360° of the compass. Lengths of arms of the rose are proportional to the number of times wind came from a given direction. Winds from the west and northeast constitute the dominant groups. Based on wind records during March 10–April 10, 1971, observations taken at 15-minute intervals. (Data from Arabian American Oil Co.). (Fig. 95.)

The five commonly occurring wind regimes (fig. 96) are:
1. Narrow unimodal. — 90 percent or more of the drift potential at a station falls within two adjacent directional categories, or within a 45° arc of the compass.

2. Wide unimodal. — any other directional distribution with a single peak or mode.
3. Acute bimodal. — a distribution with two modes, in which the peak directions of the distributions (longest arms on the sand rose) of the two modes form an acute angle (here arbitrarily including also the right angle, 90°).
4. Obtuse bimodal. — a distribution with two modes, in which the peak directions of the two modes form an obtuse angle.
5. Complex. — any distribution with more than two modes, or with poorly defined modes. The 16 point wind-direction data commonly will not clearly show more than three modes.

Modes observed on sand roses are not considered to be significant for purposes of classifying a wind regime if the modal direction and the two adjacent categories constitute less than about 15 percent of the drift potential at the station. This simplifies the classification of wind regimes by focusing on the dominant modes that are controlled by large pressure systems. All sand roses tend to reflect procedure and observer bias. Extreme bias results in a "sawtooth" pattern of arms (example, fig. 111B). This pattern can sometimes make unimodal or bimodal wind regimes seem complex.

The RDP/DP which is a measure of the directional variability of the wind, is arbitrarily classified as follows: 0.0 to less than 0.3, low; 0.3 to less than 0.8, intermediate; 0.8 or greater, high. Many low ratios (RDP/DP) are associated with complex or obtuse bimodal wind regimes, intermediate ratios with obtuse bimodal or acute bimo-

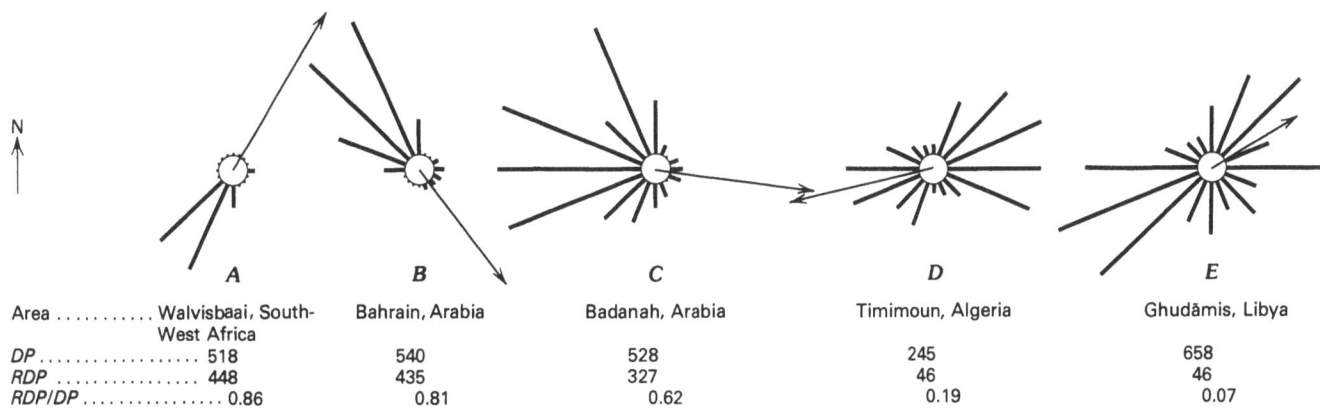

Area	Walvisbaai, South-West Africa	Bahrain, Arabia	Badanah, Arabia	Timimoun, Algeria	Ghudāmis, Libya
DP	518	540	528	245	658
RDP	448	435	327	46	46
RDP/DP	0.86	0.81	0.62	0.19	0.07

FIVE COMMONLY OCCURING RELATIONSHIPS of modes on sand roses. A, Narrow unimodal; B, wide unimodal; C, acute bimodal, D, obtuse bimodal, (a special example in which the modes are almost exactly opposed); and E, complex. DP (drift potential) and RDP (resultant drift potential), in vector units. Arrows indicate resultant drift direction (RDD). (Fig. 96.)

dal wind regimes, and high ratios with wide and narrow unimodal wind regimes.

Energy of Surface Winds

Drift potentials, which are measures of the energy of surface winds in terms of sand movement, are classified according to rough groupings of average annual drift potential for the desert regions shown in table 16. The deserts in China (127 *VU*; 81 *VU*) and India (82 *VU*) and the Kalahari Desert of South-West Africa (191 *VU*) constitute a low-energy group. The northern Saudi Arabian (489 *VU*) and Libyan (431 *VU*) deserts are a high-energy group. All other deserts studied are in the intermediate-energy group. On the basis of this classification, low-energy wind environments have drift potentials less than 200 *VU*; intermediate-energy wind environments have drift potentials of 200–399 *VU*; and high-energy wind environments have drift potentials of at least 400 *VU*. This grouping of regions by wind energy in terms of potential sand movement applies only to the generally arid regions surveyed during this study. Other regions probably have different average drift potentials (perhaps very much higher than desert regions, many of which are known to be relatively calm). However, within the regions studied, the differences in relative sand-moving power of wind are related to specific weather patterns. Aspects of

the wind-energy structure of various desert regions, with regard to sand movement only, are discussed in detail in chapter K.

Further, the relative wind energies (in terms of sand movement) shown in table 16 are rough approximations. In fact, most desert regions are strongly zoned in terms of wind energy. For example, high drift potentials occur in Northern Mauritania; intermediate to low *DP*'s occur farther south toward the Sénégal River. (See chapter K, fig. 197).

Dune Forms and Associated Wind Environments
Summary of Observed Wind Environments of the Various Dune Types

THE METHODS OF EVALUATING wind regimes discussed under "Methods of Study" in this chapter were used to compare wind environments with different dune types as observed on Landsat imagery and on aerial photographs. The dunes and associated wind regimes will be discussed in order of presumed increasing complexity of dune type — that is, barchanoid, linear and star. A synthesis of the more detailed discussions of

TABLE 16. — *Average monthly and annual drift potentials for 13 desert regions, based on data from selected stations*

[*, drift potentials estimated; leaders (. . .), no data]

Desert region	Number of stations	Jan.	Feb.	Mar.	Apr.	May	June	July	Aug.	Sept.	Oct.	Nov.	Dec.	Annual drift potential
High-energy wind environments														
Saudi Arabia and Kuwait (An Nafūd, north)	10	35	39	52	54	51	66	49	33	20	18	16	25	489
Libya (central, west)*	7	40	42	48	64	51	41	20	18	24	24	22	37	431
Intermediate-energy wind environments														
Australia (Simpson, south)	1	43	40	27	17	13	10	18	26	52	56	46	43	391
Mauritania	10	45	49	45	38	33	40	26	19	20	20	19	30	384
U.S.S.R. (Peski Karakumy, Peski Kyzylkum)	15	39	41	43	43	33	25	22	21	23	23	24	29	366
Algeria	21	21	27	37	48	32	27	18	13	15	16	16	23	293
South-West Africa (Namib)	5	8	2	6	17	13	50	19	22	27	44	17	12	237
Saudi Arabia (Rub' al Khali, north)	1	23	28	53	32	20	30	. . .	. . .	. . .	1	7	7	201
Low-energy wind environments														
South-West Africa (Kalahari)	7	14	11	8	10	9	11	18	24	26	26	17	18	191
Mali (Sahel, Niger River)	8	9	12	14	12	19	22	15	9	10	5	5	7	139
China (Gobi)*	5	9	11	16	23	20	11	7	5	5	5	7	8	127
India (Rājasthān, Thar)*	7	2	2	5	5	10	21	19	9	5	2	1	1	82
China (Takla Makan)*	11	3	2	9	16	16	9	9	5	4	5	2	1	81

each dune type and wind regime is given in figure 97 for comparative purposes. Figure 97 demonstrates the greater directional variability of effective winds associated with star dunes in contrast to the less variable wind regimes commonly associated with barchanoid dunes (in high-energy wind environments).

Narrow unimodal to wide unimodal winds are associated with transverse dunes, such as barchanoid ridges. Bimodal winds are commonly associated with linear dunes, and complex wind distributions, with star dunes.

Barchanoid (Transverse Dunes)

Subtypes and Sizes of Barchanoid Dunes Studied

Dunes referred to as crescentic in the morphological classification used for Landsat imag-

ery (chapter J), including the barchan, barchanoid ridge, and transverse ridge of the structural classification (chapter A) are included in the following discussion. Unique wind environments for each of the subtypes, as defined in chapter A, could not be distinguished with the regional data available. These dunes commonly occur in the same dune field (McKee, 1966, p. 4, fig. 1; Ahlbrandt, 1975, p. 61). Dome dunes and parabolic dunes may be found in association with barchans, barchanoid, and other transverse-type ridges. Dome dunes of the type studied by McKee at White Sands National Monument, New Mexico, U.S.A. (1966, p. 26), are not discussed in this chapter because they could not be identified by the author on Landsat imagery.

Small barchanoid dunes occur both singly and in environments where star dunes (Glennie, 1970, p. 87, figs. 70, 71) or linear dunes (Glennie, 1970, p. 94,

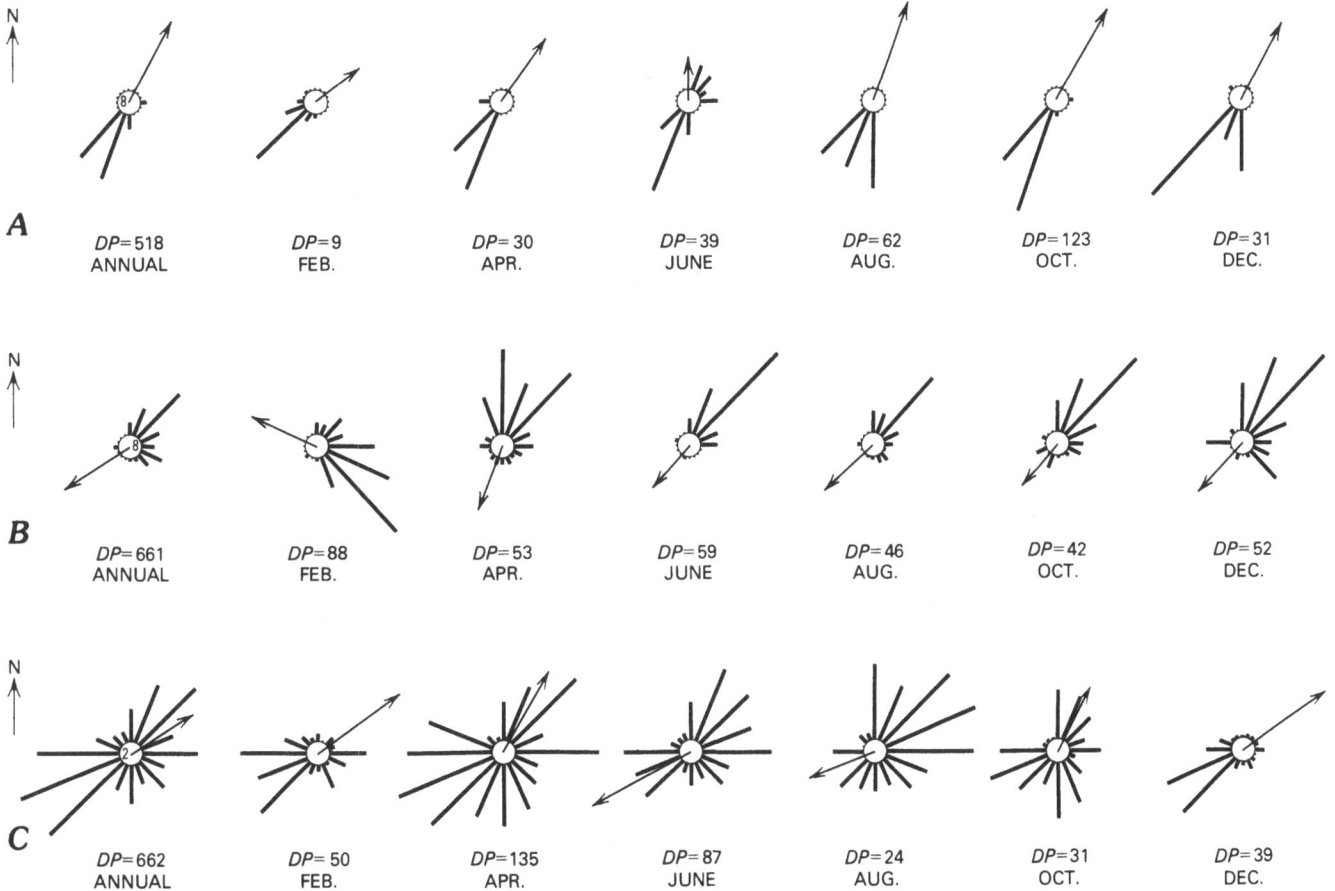

A						
DP=518 ANNUAL	DP=9 FEB.	DP=30 APR.	DP=39 JUNE	DP=62 AUG.	DP=123 OCT.	DP=31 DEC.

B						
DP=661 ANNUAL	DP=88 FEB.	DP=53 APR.	DP=59 JUNE	DP=46 AUG.	DP=42 OCT.	DP=52 DEC.

C						
DP=662 ANNUAL	DP=50 FEB.	DP=135 APR.	DP=87 JUNE	DP=24 AUG.	DP=31 OCT.	DP=39 DEC.

TYPICAL HIGH-ENERGY WIND REGIMES of three basic dune types. Annual and bimonthly sand roses depict distribution of effective winds at each station. A, Narrow unimodal; barchanoid dunes near Pelican Point, South-West Africa. B, Bimodal; linear dunes near Fort-Gouraud, Mauritania. C, Complex; star dunes near Ghudāmis, Libya. Number in center circle of each rose is reduction factor. DP (drift potential, in vector units) is given for each rose. Arrows indicate resultant drift direction (RDD). (Fig. 97.)

fig. 75) are the dominant type. In the latter places they apparently develop in response to a seasonal component of a wind regime (Glennie, 1970, p. 92). Often, relatively small dunes, such as those developed in response to a seasonal wind, are not visible on Landsat imagery. With a few exceptions (for which adequate field observations are available), dunes discussed in this paper are large enough to be clearly defined from Landsat imagery. These dunes apparently have developed in response to the annual wind regime, of which the annual sand rose is an approximate measure.

Summary of Observed Wind Environments of Barchanoid Dunes

The wind environments of barchanoid dunes are evaluated with data from the 14 stations listed in table 17, and from regional analyses in chapter K. Drift potentials range from low (51 *VU*, Pa-Ch'u, China) to extremely high (estimated 2,823 *VU*, Pomona, South-West Africa). The data in table 17 indicate that the barchanoid dune type is not restricted to either a low- or high-energy wind environment, assuming that dunes observed on the Landsat imagery are adjusted to present-day wind regimes. Active[8] barchanoid dunes seen on Landsat imagery seldom occur in a wind environment with an *RDP/DP* lower than 0.50 (fig. 98). They are associated with less directional variability of effective winds where they occur in environments of high energy (fig. 98). Barchanoid dunes develop in wind regimes with less directional variability than those of linear dunes or star dunes for a given energy environment. (Compare fig. 98 with figs. 104 and 116.) Annual wind regimes associated with barchanoid dunes may be unimodal or bimodal, although in bimodal regimes, one mode normally is much stronger than the other. Extensive fields of barchanoid dunes occur along coastlines swept by oceanic tradewinds or by monsoon winds, in both high- and low-energy settings. In addition, barchanoid dunes are the dominant type in some continental regions such as the Takla Makan and Ala Shan Deserts in China.

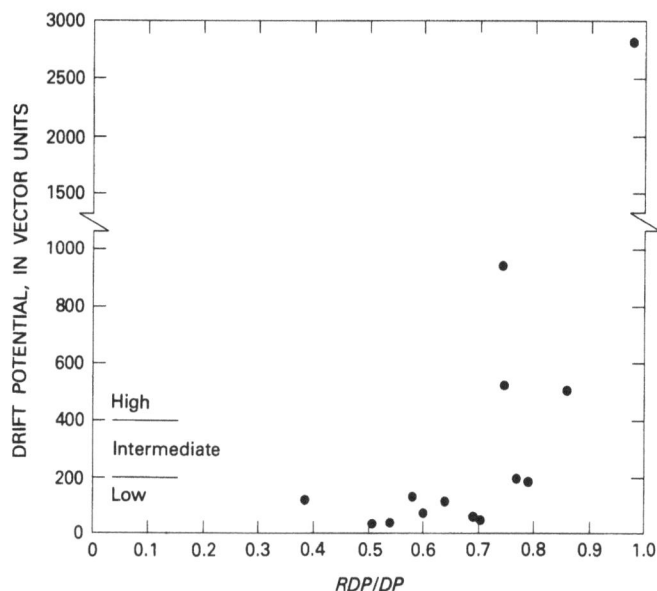

DRIFT POTENTIAL (*DP*) VERSUS *RDP/DP* for 14 stations near barchanoid dunes, which are represented by dots in the figure. All dunes except those near El Centro, California, U.S.A., and Pomona, South-West Africa, were large enough to be observed on Landsat imagery. Although based on few stations, the distribution of points indicates that barchanoid (transverse) dunes occur in wind environments with less directional variability than those of linear or star dunes. (Compare with figs. 104 and 116.) Barchanoid dunes in environments with high drift potentials (>399 *VU*) are associated with higher *RDP/DP*'s (less directional variability) than are barchanoid dunes in environments with low drift potentials (<200 *VU*). Based on data from table 17. (Fig. 98.)

Examples of Wind Environments of Barchanoid Dunes

White Sands National Monument, New Mexico, U.S.A.

Barchanoid dunes, dome dunes, and parabolic dunes can develop in low-energy wind environments having considerable directional variability of effective winds. These three types of dunes occur at White Sands National Monument, New Mexico (McKee, 1966, p. 4). The annual sand rose for Holloman Air Force Base, located downwind from the dunes at White Sands, has an obtuse bimodal wind distribution (fig. 99). Effective winds from the north-northwest and from the southeast occur during much of the year. However, these winds have less potential effect than do the southwesterly winds, which are strongest during April (fig. 99C). The drift potential at Holloman Air Force Base is 149 *VU* — a low value compared with most other desert regions around the world (table 16).

---

[8]Dunes observed on Landsat imagery are considered to be active where they occur in regions of less than 100 mm (3.93 in.) average annual rainfall, if vegetation (which appears red on false-color imagery) is not apparent, if they have distinct slipfaces and sharp outlines, or if evidence of movement is available from aerial photographs or published reports.

The alinement of parabolic dunes visible in figures 99A and 99B is parallel to the resultant drift direction at Holloman Air Force Base. The gross alinement of barchanoid dunes shown in figures 99A and 99B is roughly at right angles to the resultant drift direction. Individual slipfaces within the dune field may be oriented in various directions during the year (McKee, 1966, p. 14) in response to the effective winds from several directions (fig. 99C).

Namib Desert Coast, South-West Africa

Some barchanoid dunes occur in wind environments with a high drift potential and a narrow unimodal wind distribution. For example, transverse ridge, barchanoid ridge, and barchan dunes occur in such an environment along the coast of the Namib Desert, South-West Africa, near Pelican Point (figs. 97A, 100). Although barchan dunes at this locality are not visible on Landsat (ERTS) imagery, they were observed along the coast by M. K. Seely (oral commun., 1975). This coastal region is swept by onshore winds derived from the southeast tradewinds of the South Atlantic Ocean, which are strongest during October in the region of Pelican Point. Although the energy of effective wind — and, thus, the drift potential along the coastline — varies greatly during the year (chapter K, fig. 232), effective winds in the region as shown in figure 97A are mostly from the south or southwest and result in a narrow unimodal wind distribution. Barchanoid dunes occur in other coastal settings as well, particularly along shores swept by oceanic tradewinds or by monsoonal

TABLE 17. — *Localities used to evaluate the wind environments of barchanoid dunes*

[*DP*, drift potential; *RDP*, resultant drift potential]

Dune form(s)	Dune coordinates (lat, long)[1]	Nearest representative station (distance from dunes in km)	Station coordinates (lat, long)	Other data[2]	DP	RDP/DP	Wind regime classification, using sand rose
Barchan	31°50′ N., 115°50′ W.	El Centro, California, U.S.A. (40).	32°49′ N., 115°40′ W.	Yes	525	0.75	Narrow unimodal.
	Near station	Pomona, South-West Africa (5–10).	27°09′ S., 15°15′ E.	No	[3]2,823	.97	Do.
Barchanoid ridge	38°20′ N., 85°30′ E.	Cherchen (Charchan), China (22).	38°08′ N., 85°32′ E.	No	85	.60	Wide unimodal.
	37°10′ N., 79°30′ E.	Khotan, China (56)	37°07′ N., 79°55′ E.[4]	No	33	.70	Do.
	39°00′ N., 87°30′ E.	Charkhlik (Nochiang), China[5] (40).	38°40′ N., 88°03′ E.	No	185	.79	Do.
	38°00′ N., 78°00′ E.	So-ch′e (Yarkand), China (50).	38°25′ N. 77°15′ E.[4]	No	42	.69	Unclassified.[6]
	39°30′ N., 79°00′ E.	Pa-ch′u, China (30)	39°46′ N., 78°20′ E.[4]	No	51	.51	Obtuse bimodal.
	28°00′ N., 41°40′ E.	Hā′il, Arabia (40)	27°30′ N., 41°40′ E.[4]	No	(?)	.78	Wide unimodal.
Barchan, barchanoid ridge, parabolic.	Near station	Holloman Air Force Base, New Mexico, U.S.A.[5] (5).	32°18′ N., 106°55′ E.	Yes	149	.37	Obtuse bimodal.
Barchanoid ridge, barchan	. . .do	Walvisbaai, South-West Africa[5] (0).	22°53′ S., 14°26′ E.	Yes	518	.86	Narrow unimodal.
Compound and simple barchanoid ridges.	40°20′ N., 87°30′ E.	T′ieh-kan-li-k′o, China (20)	40°39′ N., 87°42′ E.	Yes	50	.54	Wide unimodal.
Compound barchanoid ridge, and barchanoid ridge.	40°15′ N., 100° – 100°30′ E.	Mao-mu (Ting-hsin), China[5] (15).	40°20′ N., 99°45′ E.	No	140	.57	Obtuse bimodal.
Compound barchanoid ridge . .	18°50′ N., 11°30′ W.	Tidjikdja, Mauritania (30).	18°33′ N., 11°26′ W.	No	133	.64	Wide unimodal.
Small barchanoid ridges, also linear dunes.	Near station	Bilma, Niger (5–10)	18°41′ N., 12°55′ E.	No	948	.74	Narrow unimodal.

[1] 10-km-radius circle centered on this point.

[2] "Yes" indicates airphoto or other data used to confirm identification of dune type.

[3] Estimated from Beaufort data, one year of record by John Rogers (written commun., 1975).

[4] Coordinates approximate.

[5] Discussed as example in text.

[6] Annual wind summary not available.

[7] Not computed.

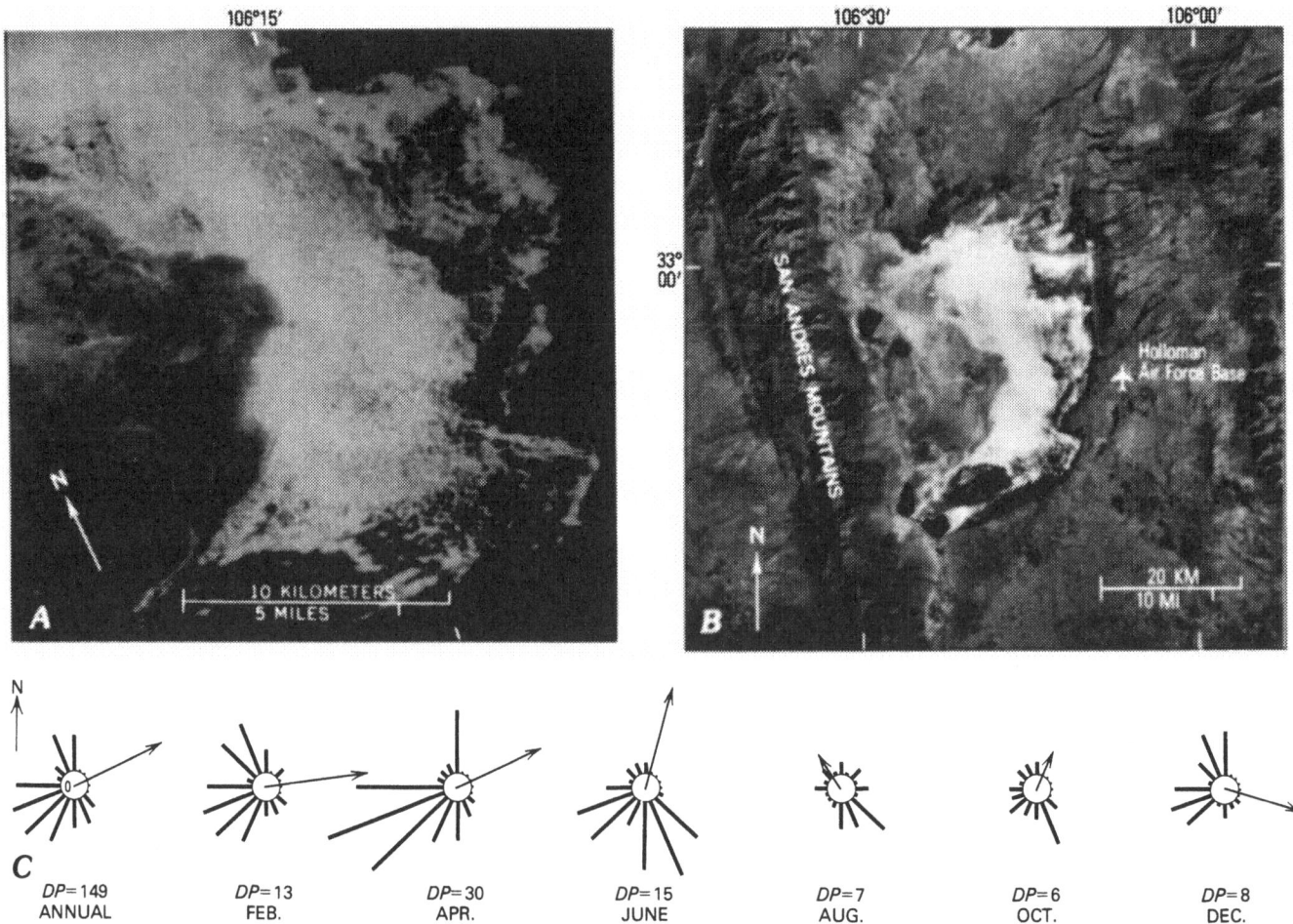

LANDSAT IMAGERY OF WHITE SANDS National Monument, New Mexico, U.S.A. (A, B), and sand roses for Holloman Air Force Base, New Mexico, near the downwind end of the dune field (C). A, Produced from computer-compatible tape at the Center of Astrogeology, Flagstaff, Arizona; scale approximately 1:290,000. Parabolic and barchanoid dunes are clearly visible. B, Produced by bulk-processing facility at Sioux Falls, South Dakota. At this scale (1:1,000,000) barchanoid dunes are barely visible, but parabolic dunes can still be clearly seen. A and B are included for comparative purposes to illustrate the resolution of the bulk-processed Landsat imagery used in this study. C, Annual sand rose and six monthly sand roses for Holloman Air Force Base, illustrating the directional variability of effective winds near the dunes. Annual wind regime is obtuse bimodal. Principal effective winds occur in April (DP = 30 VU) from the southwest. Number in center circle of rose is reduction factor. DP (drift potential, in vector units) is given for each rose. Arrows indicate resultant drift direction (RDD). (Fig. 99.)

winds, such as the Wahiba Sands of Arabia and the sands of the Thar Desert of eastern Pakistan and western India. The associated environments can have either high or low drift potentials.

Cherchen Desert, China

Several groups of barchanoid dunes may occur within a single dune field, each group characterized by a different size and orientation. In the eastern Cherchen Desert, China, three sets of barchanoid dunes have orientations differing from 45 to 90 degrees, depending on locality (fig. 101A). Effective surface winds in the Cherchen Desert, and in the eastern Takla Makan basin tend to be weak but very steady in direction (chapter K, fig. 206).

Most effective winds at Charkhlik (Nochiang), about 40 km (25 mi) southeast of the dunes shown in figure 101A, are weak (estimated DP at Nochiang is 185 VU) but very steady in direction during the year (fig. 101B). Thus, they do not provide a ready explanation for the different orientations of the transverse dunes. Possibly one set of dunes is older than the other and is inactive; however, no indications were noted from Landsat imagery, such as vegetation, that any of the dunes are stabilized. Another possible explanation for several orientations is that the dunes which are alined approximately north-northeast to south-southwest have developed transverse to the very strongest winds that occur at the station. These winds are from the

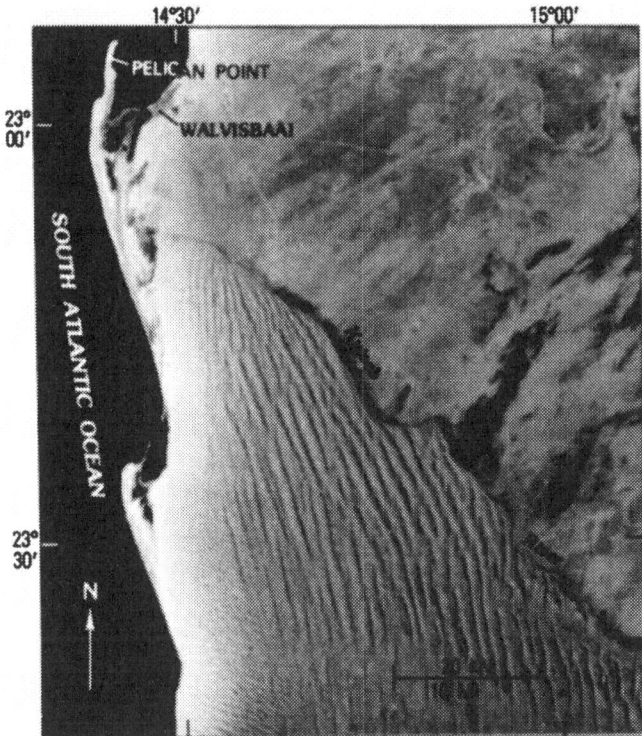

LANDSAT IMAGERY OF THE NORTHERN PART of the Namib Desert, South-West Africa, showing barchanoid dunes along the coastal margin of the desert. The wind regime which has produced the dunes is probably restricted to the coastal region. This assumption is based on the distribution of the dunes and other evidence (chapter K). Large linear dunes occur inland. Streaks north of the Kuiseb River, oriented northeast-southwest, are alined roughly with the resultant drift directions at the inland stations of Zwartbank, Rooi Bank, and Gobabeb during June and July, the windiest months at these stations. Sand roses for Pelican Point are shown in figure 97A. (Fig. 100.)

morphology from those of the Cherchen Desert to the west. In the Ala Shan region smaller transverse dunes grade westward into larger compound barchanoid ridge dunes. In contrast, the transverse dunes of the Cherchen Desert have developed as continuous ridges with relatively continuous slipfaces (fig. 101A). Large discontinuous slipfaces occur on the compound dunes of the Ala Shan region. These slipfaces are arranged to form a secondary alinement at an angle of approximately 22° to the annual resultant drift direction at Mao-Mu (Ting-hsin), the closest available wind station. Surface wind data are not presently available for the interior of the Ala Shan Desert. Regional analysis (chapter K) suggests that winds in the Ala Shan Desert are more variable in direction than those in the Takla Makan Desert. A spring sand-moving season occurs at both Charkhlik (Nochiang) and Mao-Mu (Ting-hsin). But winds during April, which is the month of highest drift potential at both stations, are about twice as effective at Charkhlik (Nochiang) as they are at Mao-Mu (Ting-hsin). The winds at Charkhlik (Nochiang) are also more directionally steady during this time (compare April sand roses for Charkhlik (Nochiang), fig. 101B, and Mao-Mu (Ting-hsin), fig. 103B).

Simple barchanoid ridge dunes near the station of Mao-Mu (Ting-hsin) occur in a wind environment similar to that of White Sands National Monument, New Mexico, U.S.A., previously described. The sand roses for each station are similar (compare annual sand roses, figs. 99C, 103B), and the estimated drift potential at Mao-Mu (Ting-hsin) (140 VU) is close to the drift potential at White Sands (Holloman Air Force Base, 149 VU).

east-northeast for 8 months of the year (figs. 101A, 102).

Occasional extreme gusts or winds, not sustained long, may have important morphological effects in such regions as the Cherchen Desert, where winds of intermediate strength occur infrequently and where the region as a whole has a low drift potential. A more detailed analysis of this problem with respect to the Cherchen Desert region, shown in figure 101A, is precluded because of limitations in available wind data. This example is presented, however, to illustrate the development in the same field of very large transverse dunes with orientations differing as much as 90°.

Ala Shan Desert, China

Barchanoid dunes of the western Ala Shan Desert of China (fig. 103A) differ greatly in

Linear Dunes

Summary of Observed Wind Environments of Linear Dunes

The active linear dunes observed on Landsat imagery occupy a wide range of wind environments in terms of both energy and directional variability. Drift potentials associated with linear dunes range from 45 VU (estimated) in an interdune corridor at Narabeb, South-West Africa, to 948 VU at Bilma, Niger (table 18). The wind environment of a linear dune seems to have a greater directional variability of effective winds (lower RDP/DP) than does the wind environment of a barchanoid dune, for a

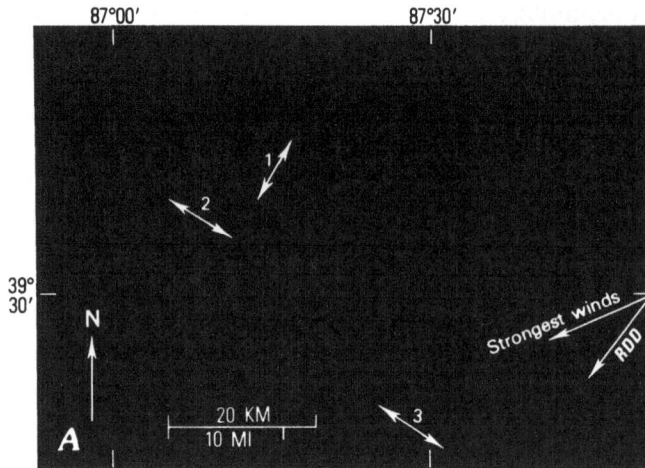

LANDSAT IMAGERY OF THE EASTERN CHERCHEN DESERT, China showing (A) dune field with three sets of transverse dunes. Two large sets of approximately equal size strike roughly northeast and northwest; a smaller set also strikes northwest. One set of dunes may be oriented transverse to strongest winds in the region indicated by arrow, and other sets may be transverse to more frequently occurring winds of lesser strength. Resultant of latter winds corresponds roughly to estimated annual resultant drift direction (RDD, indicated by arrow) at Nochiang. B, Annual and six monthly sand roses for Charkhlik (Nochiang), China, show that most of the effective winds at the station, and in the eastern Cherchen Desert, are from the northeast quadrant throughout the year. Number in center circle of rose is reduction factor. DP (drift potential, in vector units) is given for each rose. Arrows indicate resultant drift direction (RDD). (Fig. 101.)

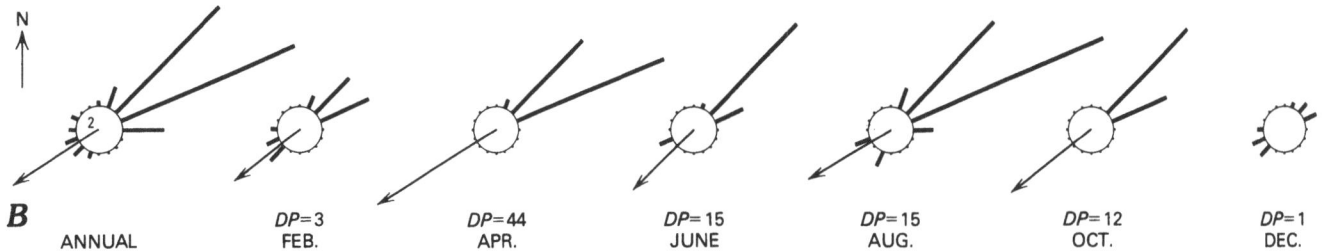

given drift potential (fig. 104). Also, a linear dune may require less directional variability of effective winds with increasing energy of wind environment (fig. 104). Wind distributions associated with active linear dunes range from wide unimodal through complex. Linear dunes are usually alined parallel to the resultant drift direction of effective winds in the surrounding environment. This is true whether the associated wind distribution is wide unimodal, bimodal, or complex. The alinement is closest at localities where wind data are most reliable and dunes most clearly active, as at Fort-Gouraud, Mauritania; Upington, South Africa; and Sabhā, Libya. Junctures ("tuning fork" or "Y") are a common feature of linear dunes and were observed at most localities. The junctures usually open into the effective wind (against the resultant drift direction) and are most common at localities with low-energy

complex wind regimes. Junctures are least common at localities with high-energy wide unimodal wind regimes.

Examples of Wind Environments of Linear Dunes

Southwest Kalahari Desert, South Africa

Simple linear dunes are the nearly exclusive type visible on Landsat imagery of the Kalahari Desert of South Africa (chapter K). Active linear dunes and sand sheets occur near Upington, in the southwestern part of the Kalahari Desert (fig. 105A). The linear dunes and the sand sheets are alined with the resultant drift direction of a high-energy obtuse bimodal wind regime (fig. 105B). Both the northwesterly and southwesterly components of the wind regime are persistent throughout the year.

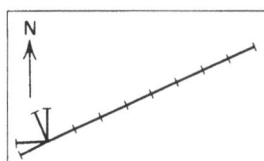

	Jan.	Feb.	Mar.	Apr.	May	June	July	Aug.	Sept.	Oct.	Nov.	Dec.
Direction from which maximum winds came	ENE.	ENE.	WSW.	ENE.	NNE.	W.	ENE.	ENE.	ENE.	NE.	ENE.	ENE.
Speed (knots)	35	27	27	35	35	31	27	31	31	31	27	19

WIND ROSE FOR MAXIMUM WINDS and table of average maximum monthly winds at Charkhlik (Nochiang), China, for 1956–61. Each tick mark on a wind-rose arm represents a maximum gust from the direction indicated by the arm. (Fig. 102.)

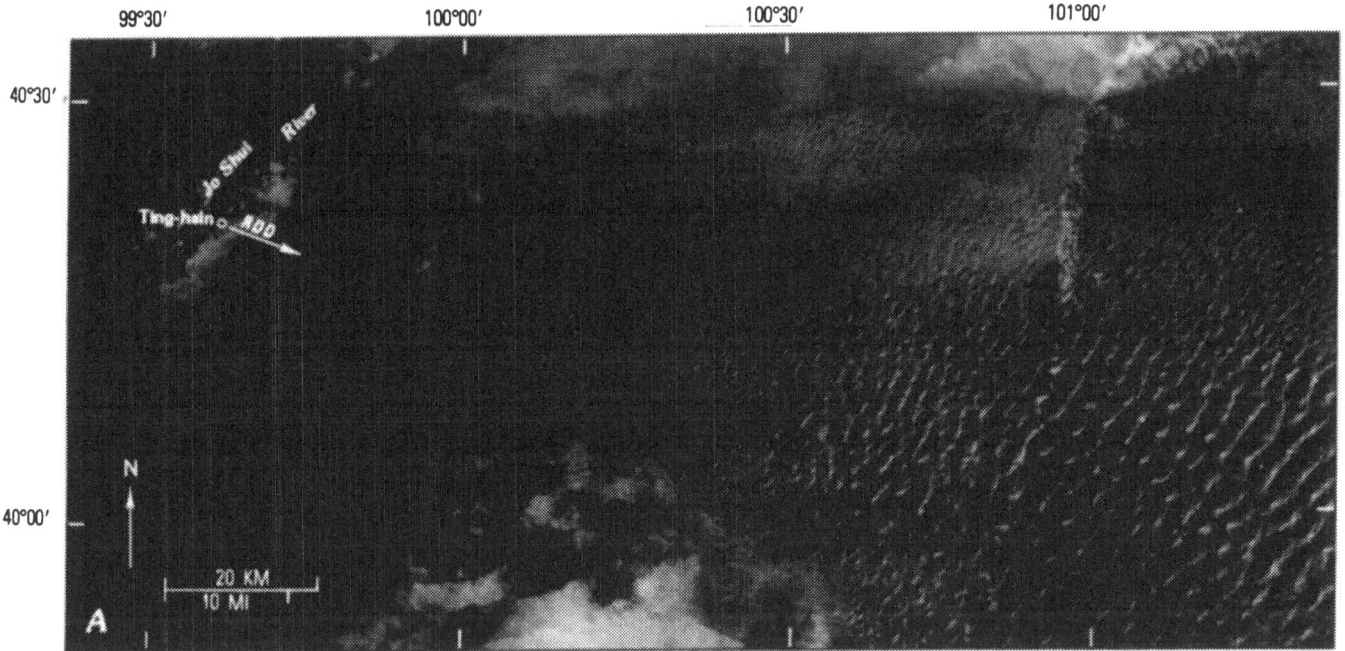

MOSAIC OF TWO LANDSAT IMAGES (A) of the western Ala Shan Desert, China. Small barchanoid ridge dunes grade eastward from the region of Mao-Mu (Ting-hsin) into larger compound barchanoid ridge dunes. Barchanoid dunes in this sand sea are peaked, unlike those in the Cherchen Desert (fig. 101A). Arrow on mosaic, and those on roses indicate resultant drift direction (RDD). B, Annual and six monthly sand roses for Mao-Mu (Ting-hsin), China, illustrating directional variability of low-energy effective winds near the westernmost part of the Ala Shan Desert. Number in center circle of rose is reduction factor. DP (drift potential, in vector units) is shown for each rose. (Fig. 103.)

The distribution in occurrence of maximum winds follows closely the pattern of effective winds of all velocities suggested by the annual sand rose (fig. 106). At this locality the resultant drift direction varies little throughout the year (fig. 105B).

Western Simpson Desert near Oodnadatta, Australia

Simple linear dunes are an exclusive type throughout much of the Simpson, Great Victoria, and Great Sandy Deserts of Australia. The driest and windiest region of the Simpson Desert may be characterized by a markedly different wind regime from that of the southwestern Kalahari Desert just discussed, although the dunes in both regions are of the same type (fig. 107A). The wind regime at Oodnadatta, Australia (located approximately 110 km (68 mi) from the dunes shown in fig. 107A), has a wide unimodal distribution of effective winds and an intermediate drift potential (fig. 107B). The northerly group (mode) of effective winds is here neglected; it has 12 percent of the total DP at the station and occurs during the period of weakest winds.

Instead of two distinct and relatively persistent components of the wind regime as at Upington, several components exist at Oodnadatta. These components change in relative intensity during the year, resulting in the wide unimodal annual distribution dominated by effective winds from the south. The systematic shift of effective winds from roughly southerly to southeasterly in January, to southwesterly in July, and back during the rest of the year results in a gradual swing of the resultant drift direction. It moves from a northwestward-

TABLE 18. — *Localities used to evaluate the wind environments of linear dunes*

[*DP*, drift potential; *RDP*, resultant drift potential]

Dune form(s)	Dune coordinates (lat, long)[1]	Nearest representative station (distance from dunes in km)	Station coordinates (lat, long)	Other data[2]	DP	RDP/DP	Wind regime classification, using sand rose
Simple linear	24°37′ S., 18°10′ E.	Mariental, South-West Africa[3] (25).	24°37′ S., 17°58′ E.	Yes	50	0.32	Complex.
	25°14′ S., 128°13′ E.	Giles, Australia[3] (25)	25°02′ S., 128°18′ E.	No	101	.23	Do.
	27°33′ S., 135° 27′ E.	Oodnadatta, Australia[3] (110) ..	27°33′ S., 135°27′ E.	No	388	.46	Wide unimodal.
	Same as station	Ekidze, U.S.S.R.[3] (0)	41°02′ N., 57°46′ E.[4]	No	111	.19	Complex.
	Near station	.. Sabhā, Libya[3] (5)	27°01′ N., 14°26′ E.	Yes	268	.20	Do.
Compound and simple linear.	Near station	.. Bilma, Niger[3] (5)	18°41′ N., 12°55′ E.	No	948	.74	Wide unimodal.
	22°30′ N., 12°16′ W.	Fort-Gouraud, Mauritania[3] (40 − 50).	22°41′ N., 12°42′ W.	No	661	.67	Acute bimodal.
	20°50′ N., 13°30′ W.	Atar, Mauritania (40)	20°31′ N., 13°04′ W.	No	443	.64	Wide unimodal.
	19°30′ N., 14°10′ W.	Akjoujt, Mauritania (20)	19°45′ N., 14°22′ W.	No	352	.68	Do.
Complex linear	Same as station	Narabeb, South-West Africa (in interdune corridor) (0).	23°52′ S., 14°52′ E.[4]	Yes	45	.24	Unclassified.[5]
	Near station	.. El Golea, Algeria (0)	30°30′ N., 02°52′ E.	No	270	.26	Complex.
Simple linear; also linear "sand sheets."	28°20′ S., 21°26′ E.	Upington, South Africa[3] (10 − 25).	28°24′ S., 21°16′ E.	No	520	.66	Obtuse bimodal.
Compound linear; also linear "sand sheets."	Near station	.. Ash Shāriqah (Sharjah), Arabia (5 − 10).	25°21′ N., 55°23′ E.	No	280	.60	Acute bimodal.

[1] 10-km-radius circle centered on this point.

[2] "Yes" indicates airphoto or other data used to confirm identification of dune type.

[3] Discussed as example in text.

[4] Coordinates approximate.

[5] Annual wind summary not available.

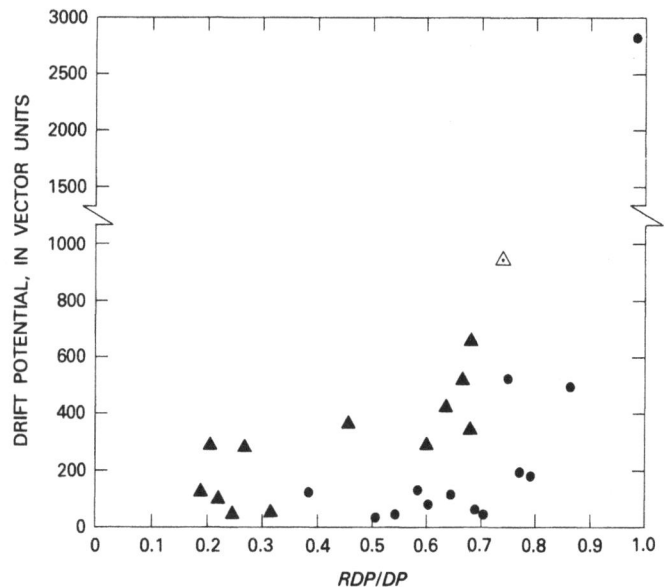

DRIFT POTENTIAL (*DP*) VERSUS *RDP/DP* for 13 stations near linear dunes visible on Landsat imagery. Linear dunes represented by triangles. Data from figure 98 also shown, barchanoid dunes represented by dots. Dot within triangle denotes a locality near which both dune types are well developed. The distribution of points indicates that wind environments of linear dunes have greater directional variability than do wind environments of barchanoid dunes for a particular drift potential. Linear dunes in wind environments with high drift potentials are associated with less directional variability of effective winds than are linear dunes in wind environments with low drift potentials. (Fig. 104.)

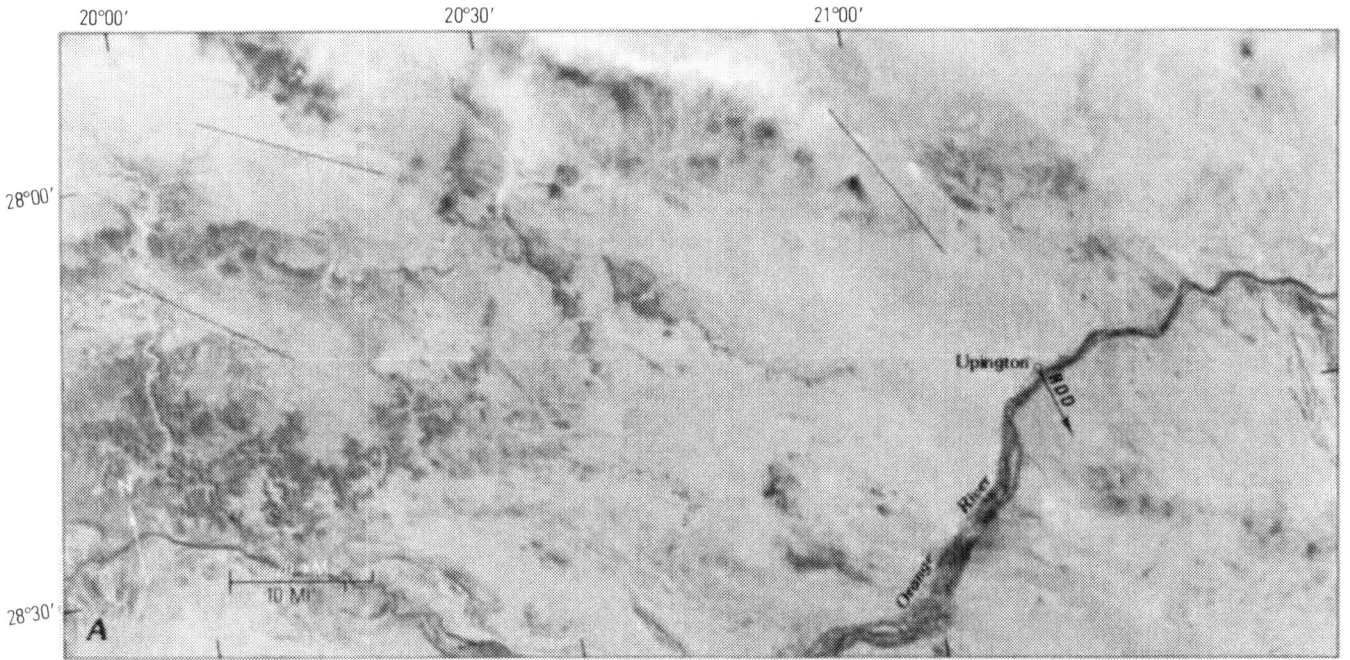

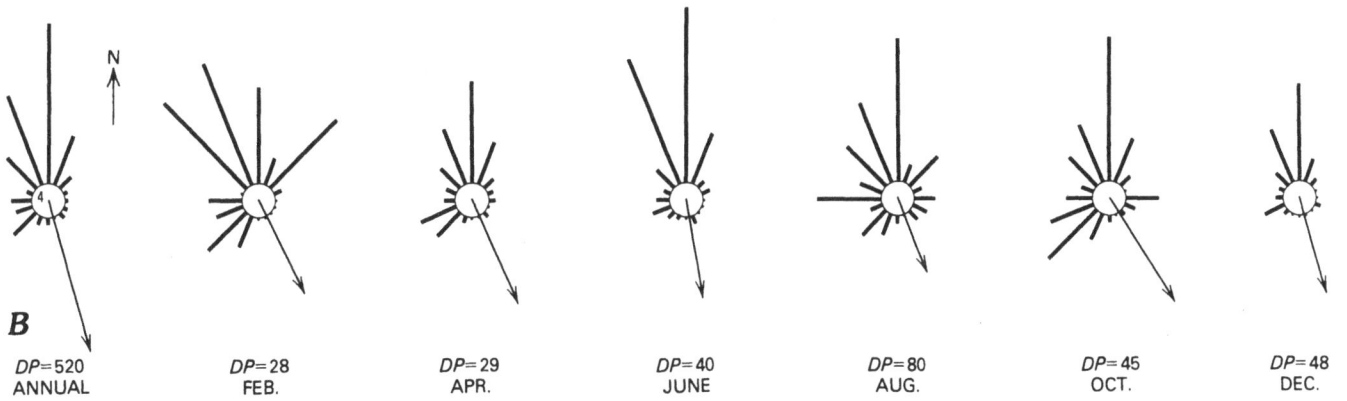

LANDSAT FALSE-COLOR IMAGERY (A) of part of the southwestern Kalahari Desert near Upington, South Africa. Red sand is yellow on Landsat false-color imagery. Linear dunes with sharp crests are visible near Upington and farther east near the Orange River (upper right corner of image). Wider linear streaks which lie across drainages and hills are marked by lines parallel to streaks. Although linear in gross aspect, they may contain transverse bedforms too small to be seen on Landsat imagery. (Similar streaks are visible on Landsat imagery of northern Arizona north of the Little Colorado River near Cameron, Arizona, in the United States and were examined in the field by the author. These streaks were extended sandy areas with barchanoid dunes.) Green vegetation along the Orange River shows as red on Landsat false-color imagery. Clouds are white. Arrow indicates annual resultant drift direction (RDD) at Upington. B, Annual and six monthly sand roses for Upington, South Africa, illustrating an obtuse bimodal wind regime. The resultant drift direction (arrows) remains steady and approximately parallel to the orientation of linear dunes in the region throughout the year. Number shown in center circle of rose is reduction factor. DP (drift potential, in vector units) is given for each rose (Fig. 105.)

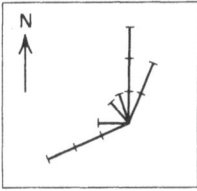

	Jan.	Feb.	Mar.	Apr.	May	June	July	Aug.	Sept.	Oct.	Nov.	Dec.
Direction from which maximum winds came	NNW.	NW.	N.	WSW.	WSW.	N.	N.	N.	NNE.	WSW.	NNE.	W.
Speed (knots)	33	40	35	31	32	29	36	37	52	35	33	37

WIND ROSE FOR MAXIMUM WINDS and table of average maximum monthly winds at Upington, South Africa, for 1956 – 67. Each tick mark on a wind rose arm represents a maximum gust from the direction indicated by the arm. Strongest winds at Upington closely reflect the obtuse bimodal distribution of all effective winds shown in the annual sand rose in figure 105B. (Fig. 106.)

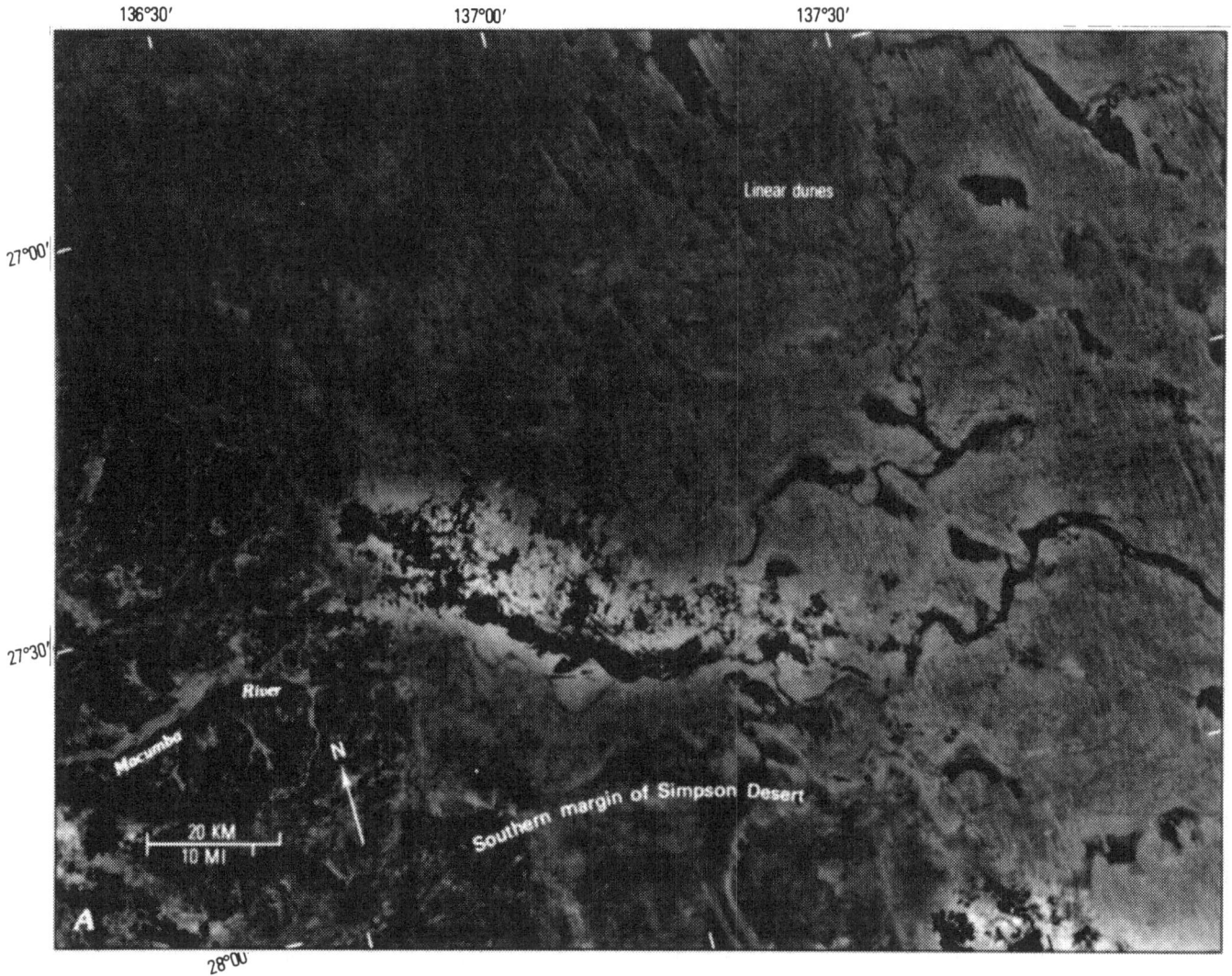

LANDSAT IMAGERY OF SOUTHWESTERN SIMPSON DESERT, Australia (A). Straight simple linear dunes are the only type visible on Landsat imagery throughout much of this desert and along its margins. B, Annual and twelve monthly sand roses for Oodnadatta, Australia, approximately 110 km (68 mi) from the dunes in A. Wide unimodal wind distribution of effective winds displayed by annual sand rose results from interaction of winds between the southwest and southeast during the year. Arrows indicate resultant drift direction. Number in center circle of rose is reduction factor. DP (drift potential, in vector units) is given for each rose. (Fig. 107.)

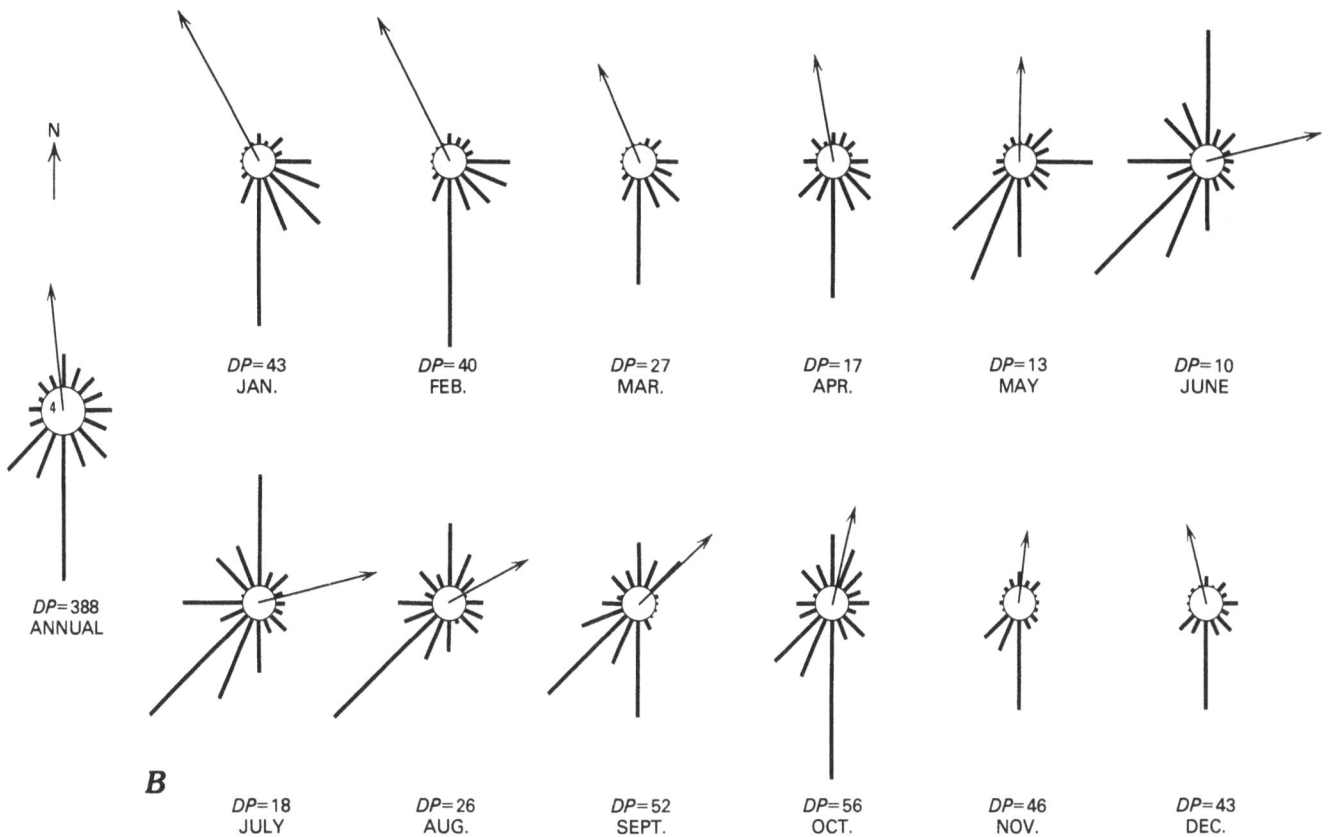

DP=43
JAN.

DP=40
FEB.

DP=27
MAR.

DP=17
APR.

DP=13
MAY

DP=10
JUNE

DP=388
ANNUAL

B

DP=18
JULY

DP=26
AUG.

DP=52
SEPT.

DP=56
OCT.

DP=46
NOV.

DP=43
DEC.

directed *RDD* in January, to an east-northeastward-directed *RDD* in July, and back. Another region characterized by a regular swing of resultant drift direction back and forth is western Mauritania (chapter K). This region is also characterized by linear dunes, although both wind regimes and dunes differ in detail from those near Oodnadatta.

Ramlat Zallāf sand sea near Sabhā, Libya, and Peski Karakumy, U. S. S. R.

Some simple linear dunes occur in regions of complex wind distributions of low or intermediate energy. Long, relatively narrow linear dunes, which are very gently curving on Landsat imagery (fig. 108A), occur west of Sabhā, Libya, in a complex intermediate-energy wind regime (fig. 108B). These dunes, in fact, have a zigzag pattern in plan when viewed from the ground or on aerial photographs (McKee and Tibbitts, 1964, p. 14–15, plates I, II). Wind regimes from month to month are complex, with the exception of May through August when easterly effective winds are most important. The occurrence of maximum winds suggests a com-

plex pattern (fig. 109). The trend of the dunes near Sabhā is exactly parallel to the annual resultant drift direction at the station (fig. 108B).

The parallelism of dune trend and effective wind direction at Sabhā suggests that each component of a complex wind regime may have a proportionate effect on the growth of dunes and their resulting orientation. On the other hand, the orientation of the dunes is also parallel to the resultant drift direction during May, the month of highest drift potential at the station. Therefore, it seems equally possible that most of the growth of the dunes occurs from May until September, the time of relatively steady easterly winds. Complex monthly wind regimes during the remainder of the year might be incapable of obliterating the linear form of the dunes established during the May to September season.

An example of linear dunes, probably active, in a complex wind regime of low energy occurs in the northern Peski Karakumy in the region of Ekidze, U.S.S.R. (fig. 110A). Near Ekidze, features that seem to be linear dunes — because they exhibit the

14°00'   14°30'

27°
30'

N

Trend of linear dunes

20 KM
10 MI

o Sabhā

A

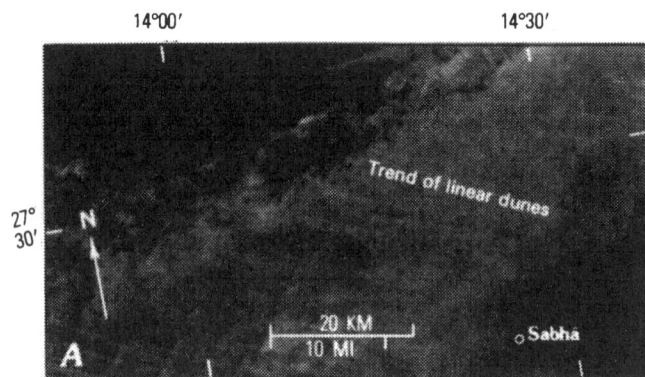

LANDSAT IMAGERY OF THE RAMLAT ZALLĀF SAND SEA near Sabhā, Libya (A). Linear dunes are gently curving, thin streaks extending from west-northwest to east-southeast across the image. B, Annual and twelve monthly sand roses for Sabhā, Libya, illustrating the complex pattern of effective surface winds. Dunes shown in A are alined with the annual resultant drift direction at the station and with the monthly resultant drift direction during May, when the highest drift potentials occur. Winds are relatively steady from May through August, during which about 35 percent of the average annual drift potential occurs. Number inside center circle is reduction factor. Arrows are resultant drift direction (RDD). DP (drift potential, in vector units) is given for each rose. (Fig. 108.)

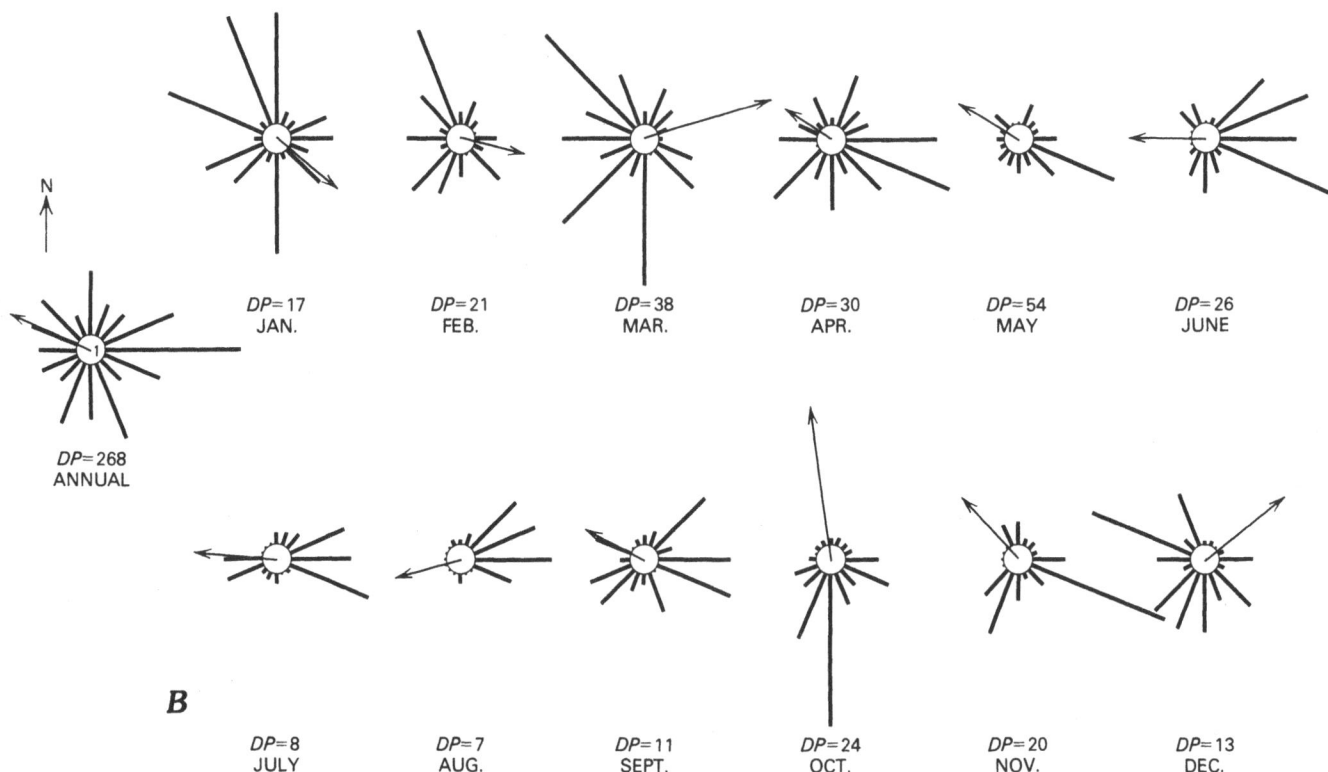

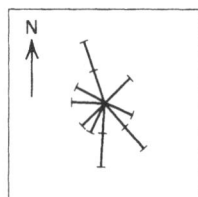

N

DP=17
JAN.

DP=21
FEB.

DP=38
MAR.

DP=30
APR.

DP=54
MAY

DP=26
JUNE

DP=268
ANNUAL

B

DP=8
JULY

DP=7
AUG.

DP=11
SEPT.

DP=24
OCT.

DP=20
NOV.

DP=13
DEC.

N

	Jan.	Feb.	Mar.	Apr.	May	June	July	Aug.	Sept.	Oct.	Nov.	Dec.
Direction from which maximum winds came	NNW.	NNW.	SW.	ESE.	SE.	S.	W.	NE.	WNW.	S.	SSW.	SE.
Speed (knots)	24	24	27	24	30	22	25	19	26	31	24	23

WIND ROSE FOR MAXIMUM WINDS and table of average maximum monthly winds at Sabhā, Libya, for 1949 and 1961−63. Each tick mark on a wind rose arm represents a maximum gust from the direction indicated by the arm. Strongest winds at Sabhā reflect the complex distribution of all effective winds shown on the annual sand rose in figure 108B. (Fig. 109.)

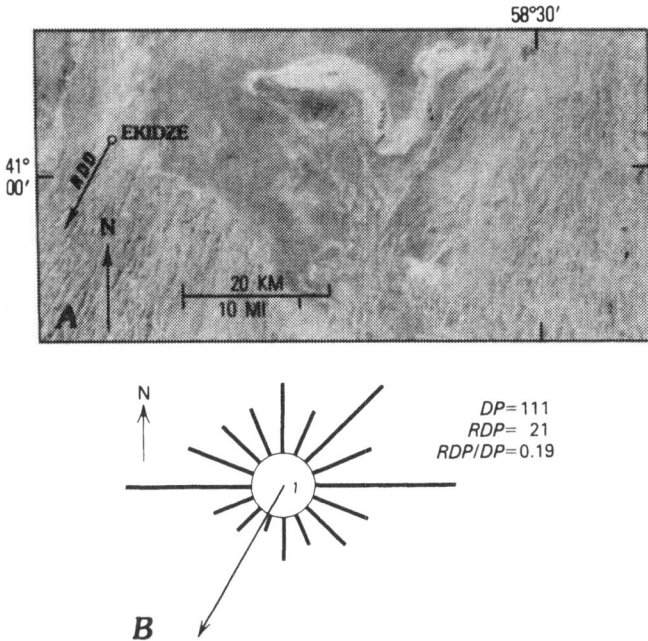

LANDSAT IMAGERY OF PRESUMED LINEAR DUNES (A) in the Peski Karakumy near Ekidze, U.S.S.R. Linear dunes have sharp crests and junctures which open upwind (against the resultant drift direction, RDD, indicated by arrow). B, Annual sand rose for Ekidze, illustrating a complex wind regime dominated by a mode from the east and a mode from the west. Linear dunes near Ekidze are alined with the annual RDD (arrow) at the station but exhibit more waviness in plan than do linear dunes near Upington, South Africa, and Oodnadatta, Australia, which have higher energy, bimodal effective wind distributions. DP, drift potential; RDP, resultant drift potential; both in vector units. (Fig. 110.)

windward-facing tuning-fork junctures characteristic of linear dunes — are perfectly alined with the resultant drift direction at Ekidze (arrows, figs. 110A,B). The complex wind regime at the station is dominated by two opposing modes of effective wind from roughly the west and the east.

Northwestern Kalahari Desert, South-West Africa, and Great Victoria Desert, Australia

Simple linear dunes are partially vegetated in wide regions of the Kalahari, Great Victoria, and Great Sandy Deserts. Near Mariental, South-West Africa (fig. 111A), and Giles, Australia (fig. 112A), partially vegetated linear dunes are associated with low-energy wind regimes (figs. 111B, 112B) compatible with the observed dune types.

The sand rose for Mariental reflects a high degree of observer or procedure bias. Nevertheless, as does

the distribution of maximum winds (fig. 113), this sand rose suggests a wide unimodal or perhaps acute bimodal distribution of effective winds. The wind regime at Giles is complex but composed primarily of wide modes from the north-northeast and south.

The degree to which the observed morphologies are a result of present day winds is difficult to determine from available data. For example, vegetation on dunes in the Sahel and southwestern Mauritania seems to be associated with climatic change which has led to the immobilization of the dunes (chapter K), although not all the observed dunes in these regions are out of adjustment, or alinement, with present-day winds.

In addition to vegetation, which tends to reduce sand movement, the wind regimes at Mariental and Giles have low drift potentials. For example, the drift potential of 50 VU at Mariental indicates negligible sand drift if the impact threshold shear velocity (V_t^*) is assumed to be 19 cm/s (0.58 ft/s) in figure 94. On the other hand, dunes near Mariental and possibly near Giles are only partially vegetated, indicating that they may be at least partly active.

The compatibility of the wind regimes with dune types at Giles and Mariental, and the continuity of dune trends from partially vegetated to unvegetated areas across considerable distances in both the Kalahari and Simpson Deserts (chapter K) indicate that the dune morphologies are at least partially the result of present-day winds. In addition, the degree of waviness in dune plan in the unvegetated and the vegetated regions of the Kalahari Desert and in the Australian deserts corresponds roughly to the directional variability of the wind regimes (compare dunes near Giles and Ekidze with those near Oodnadatta and Upington).

Even if present-day low-energy wind regimes have not created the dunes near Mariental and Giles, they may still serve to maintain the form of the dunes or have insufficient strength to obliterate them. More specific knowledge of the degree to which the dunes observed near Mariental and Giles are the product of present-day winds would facilitate interpretation of the paleoclimate of these regions.

18°30'

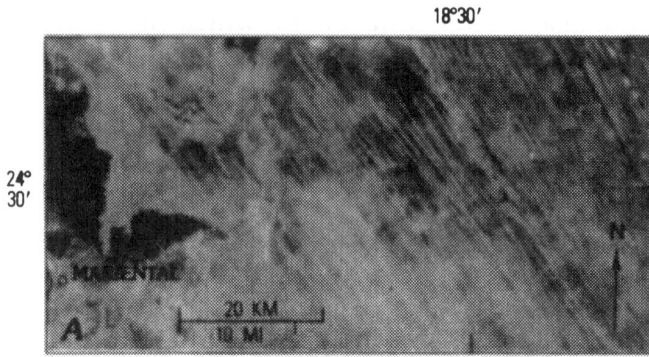

128°00'

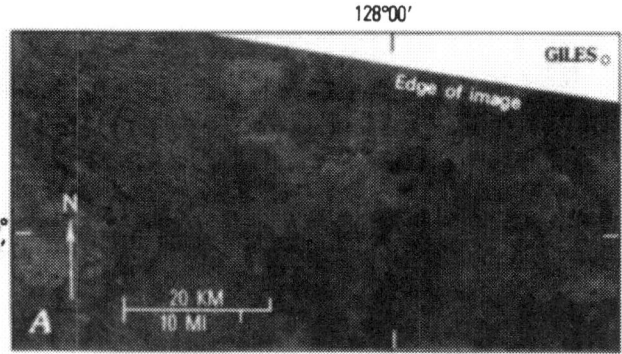

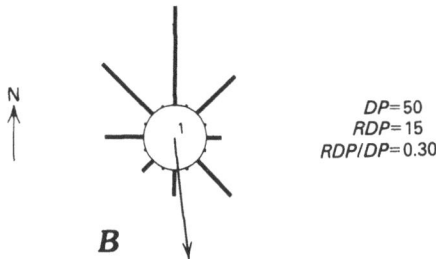

DP=50
RDP=15
RDP/DP=0.30

B

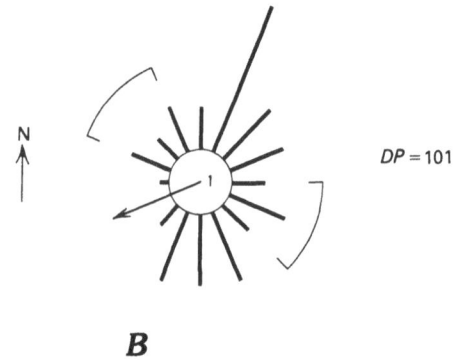

DP = 101

B

LANDSAT IMAGERY OF THE NORTHWESTERN KALAHARI DESERT (*A*) near Mariental, South-West Africa. Partially vegetated, straight linear dunes with junctures opening upwind, or against the resultant drift direction. Resultant drift direction at Mariental indicated by arrow. *B*, Annual sand rose for Mariental shows extreme observer or procedure bias (few observations between the prime points of the compass) but suggests wide unimodal, or perhaps acute bimodal, wind distribution. *DP*, drift potential; *RDP*, resultant drift potential; both in vector units. Arrow indicates resultant drift direction (*RDD*). (Fig. 111.)

LANDSAT IMAGERY OF LINEAR DUNES IN THE GREAT VICTORIA DESERT (*A*) near Giles, Australia. These dunes are more wavy in plan than those near Mariental, South-West Africa (fig. 111*A*). *B*, Annual sand rose for Giles illustrates not only the complex nature of effective winds but also the difficulties of classifying directional characteristics of effective wind regimes. The subsidiary groups of the distribution (from approximately the northwest and east-southeast, as bracketed) each comprise more than 15 percent of the annual drift potential at the station. Most of the effective winds at the station, however, occur in groups from the north-northwest and south. In this example, because maximum wind data are unavailable, the classification may be either complex or obtuse bimodal. The sand rose illustrates well, however, the greater directional variability of effective winds at Giles than at Mariental (fig. 111*B*). *DP*, drift potential, in vector units. Arrow indicates resultant drift direction (*RDD*). (Fig. 112.)

Erg Hammami Region, Mauritania, and Erg Bilma Region, Niger

Both compound and simple linear dunes occur in regions characterized by acute bimodal, or wide unimodal wind distributions and a high drift potential. Much of Mauritania northwest of a line from Akjoujt through Atar is characterized by acute bimodal winds, and by compound and simple linear dunes which extend subparallel to the resultant drift direction of present-day winds (fig. 114).

The long, straight, very large linear dunes near Fort-Gouraud, Mauritania (fig. 114), seem to have developed in an acute bimodal high-energy wind regime in which the two major components are mainly in different seasons (fig. 97*B*). The more dominant group of winds, from north through northeast, is active from spring through fall. This group results from anticyclonic tradewind circulation from the North Atlantic high-pressure cell,

which is strongest in midsummer (chapter K). The group of winds from the southwest may represent the effect of anticyclonic circulation from a wintertime high-pressure ridge over the northern Sahara.

Simple linear dunes and compound linear dunes, similar in appearance to those near Fort-Gouraud occur near Bilma, Niger (fig. 115*A*), in a region characterized by a wide unimodal high-energy wind regime (fig. 115*B*). In this region, simple linear dunes and compound linear dunes occur in

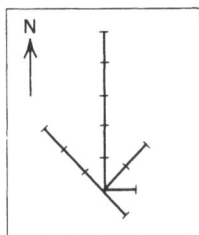

	Jan.	Feb.	Mar.	Apr.	May	June	July	Aug.	Sept.	Oct.	Nov.	Dec.
Direction from which maximum winds came	NW.	NE.	N.	SE.	NW.	N.	NE.	N.	N.	N.	NW.	E.
Speed (knots)	25	35	20	25	23	23	23	25	30	30	25	25

WIND ROSE FOR MAXIMUM WINDS and table of average monthly maximum winds at Mariental, South-West Africa, for 1960–67. Each tick mark on a wind rose arm represents a maximum gust from the direction indicated by the arm. Strongest winds at Mariental support the classification of effective winds in figure 111*B* as either wide unimodal or acute bimodal. (Fig. 113.)

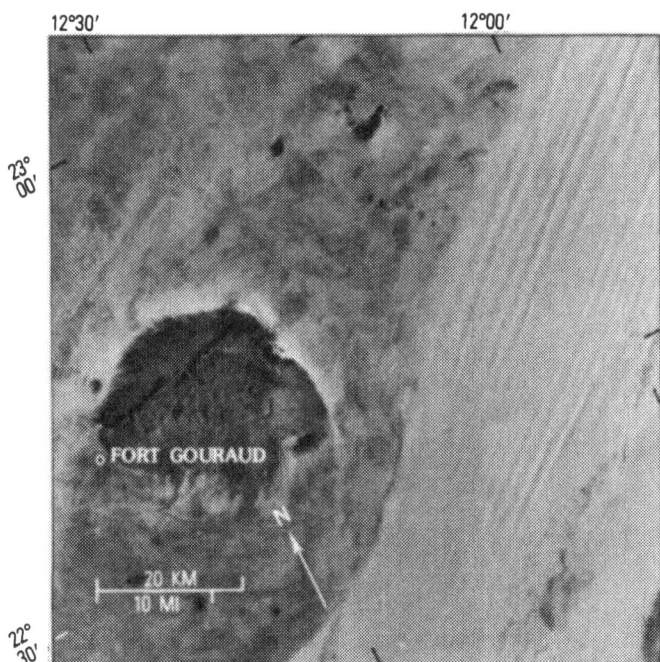

LANDSAT IMAGERY OF LINEAR DUNES in the Erg Hammami region, northern Mauritania. Long, large compound linear dunes in this region have developed in an acute bimodal high-energy wind regime (fig. 97*B*). (Fig. 114.)

two different areas. This evidence indicates that both simple linear and compound linear dunes can develop in high-energy wind regimes. Perhaps available sand supply and age of the dunes, as well as wind regime, may be important factors that determine whether simple linear or compound linear dunes will form at a given locality.

Star Dunes

Summary of Observed Wind Environments of Star Dunes

Star dunes observed in this study are in areas with complex wind regimes (table 19). Drift potentials range from low (73 *VU* at Illizi, Algeria) to high (658 *VU* at Ghudāmis, Libya), and *RDP/DP*'s range from intermediate to low. Isolated star dunes are associated with *RDP/DP* of less than 0.20, and star dunes in chains or atop linear features are associated with higher *RDP/DP* (fig. 116).

Examples of Wind Environments of Star Dunes
Eastern Grand Erg Oriental, Algeria-Libya

Isolated star dunes occur in the Grand Erg Oriental near Ghudāmis, Libya (fig. 117) in a high-energy complex wind regime (fig. 97*C*). This association supports the suggestion of McKee (1966, p. 68) that star dunes develop as a result of effective winds from several directions.

The complex wind regime at Ghudāmis differs from the complex wind regime associated with linear sand dunes at Sabhā (fig. 108*B*) in several significant ways. First, the drift potential at Ghudāmis is 2.5 times as great as at Sabhā; second, complex wind distributions prevail during the months of highest drift potential at Ghudāmis and become relatively narrow only during the seasons of low drift potential from December through February. At Sabhā, the reverse situation occurs. Finally, although two dominant modes may be recognized in the annual sand roses for both Ghudāmis and Sabhā, the dominant modes are perfectly opposed at Ghudāmis but are at approximately a right angle at Sahbā.

Grand Erg Occidental, Algeria

Some star dunes occur in low-energy complex wind regimes. Star dunes occur in such an environment along and atop an escarpment near Beni Abbes, Algeria (fig. 118A). The drift potential at Beni Abbes is low (185 VU), and effective winds during the month of highest drift potential are complex, as at Ghudāmis, Libya (fig. 118B). Star dunes near Beni Abbes, visible in an aerial photograph (Alimen and others, 1957, plate III), seem to be

composed of sinuous ridges rather than having formed isolated mounds, as at Ghudāmis.

Western Grand Erg Oriental, Algeria

Star dunes have been observed on Landsat imagery atop other sand features, many of which have a linear aspect in plan view. These linear features on

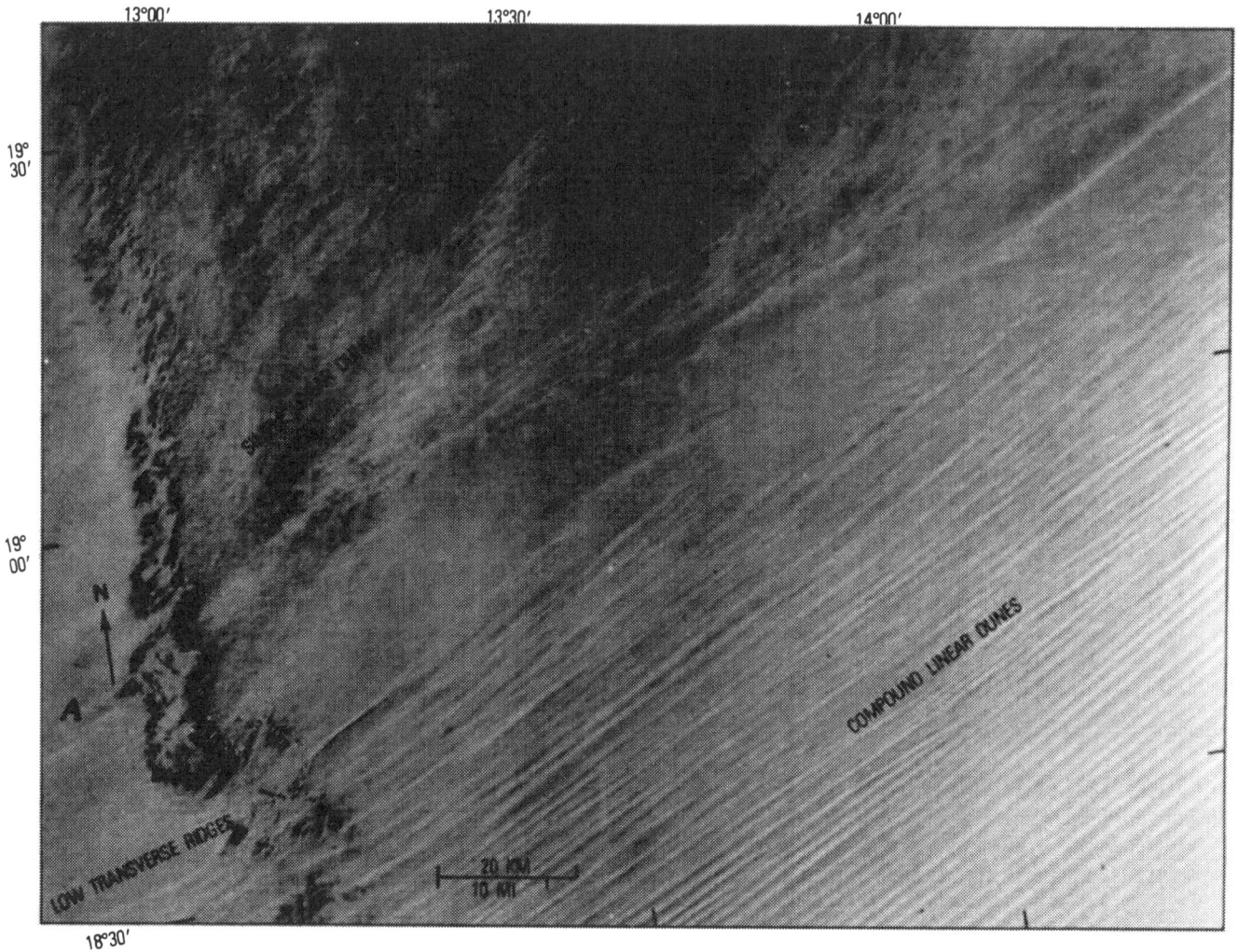

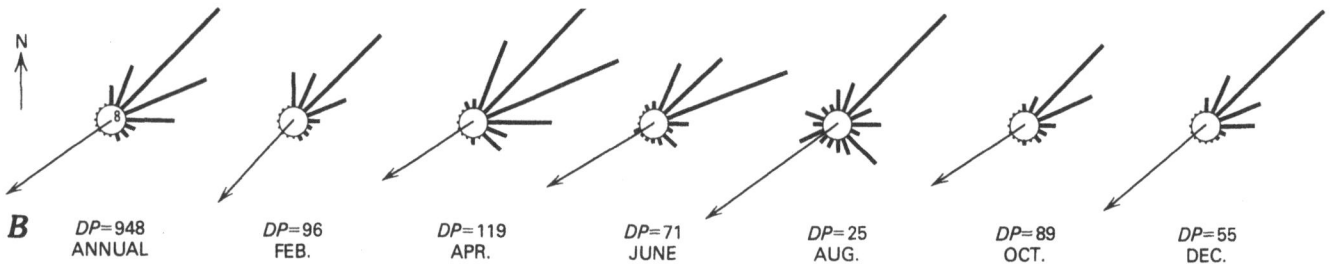

LANDSAT IMAGERY OF LINEAR DUNES (A) in the Erg Bilma. Both simple and compound linear dunes occur near Bilma in the same regional wind environment. B, Annual and six monthly sand roses for Bilma, Niger, illustrate a wide unimodal high-energy wind regime near the dunes. Arrow indicates resultant drift direction (RDD). Number in center circle of rose is reduction factor. DP (drift potential, in vector units) is given for each rose. (Fig. 115.)

TABLE 19. — *Localities used to evaluate the wind environments of star dunes*

[*DP*, drift potential; *RDP*, resultant drift potential]

Dune form(s)	Dune coordinates (lat, long)[1]	Nearest representative station (distance from dunes in km)	Station coordinates (lat, long)	Other data[2]	DP	RDP/DP	Wind regime classification, using sand rose
Star	Near station	Ghudāmis, Libya[3] (≈5)	30°08' N., 09°30' E.	Yes	658	0.09	Complex.
	Do.	Beni Abbes, Algeria[3] (0)	30°08' N., 02°10' W.	Yes	184	.11	Do.
	Do.	Illizi, Algeria (0)	26°30' N., 08°29' E.	No	73	.07	Do.
	26°22' S., 16°14' E.	Aus, South-West Africa (23) ..	26°42' S., 16°14' E.[4]	No	([5])	([5])	Wind rose only.
Star dunes in chains or on linear "sheets."	31°34' N., 06°15' E.	Hassi Messaoud, Algeria[3] (20) .	31°40' N., 06°09' E.	No	200	.31	Complex.
Star dunes atop linear or barchanoid features.	Near station	Bordj Omar Driss (Fort Flatters), Algeria (0).	28°08' N., 06°50' E.	No	112	.35	Do.

[1] 10-km-radius circle centered on this point.

[2] "Yes" indicates airphoto or other data used to confirm identification of dune types.

[3] Discussed as example in text.

[4] Station coordinates approximate.

[5] Drift potential unknown.

which the star dunes have developed may, in some places, represent older barchanoid or linear dunes, or extended linear "sand sheets." Star dunes near Hassi Messaoud, Algeria, have developed in chains atop extended linear sand features of indeterminate identity (fig. 119A). The annual effective wind regime at Hassi Messaoud is difficult to classify but is either obtuse bimodal or complex. Effective wind

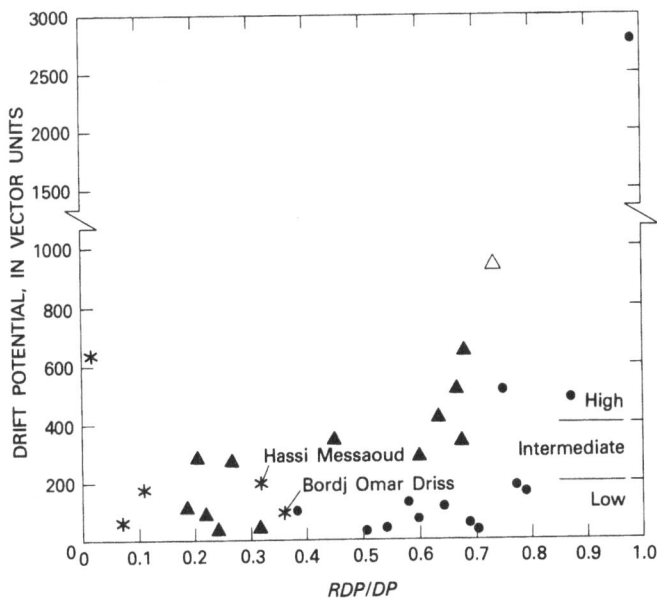

DRIFT POTENTIAL (*DP*) versus *RDP/DP* for five stations near star dunes, which are represented by stars. Data from figures 98 and 104 also shown; barchanoid dunes represented by dots, linear dunes represented by triangles. Dot within triangle indicates locality near which both barchanoid and linear types are well developed, Isolated star dunes were observed only in wind environments with low *RDP/DP*. Star dunes on broadly linear sand "sheets," or atop other linear features were observed in intermediate environments in terms of *RDP/DP*. Examples in Algeria are dunes near Hassi Messaoud and Bordj Omar Driss (Fort Flatters). (Fig. 116.)

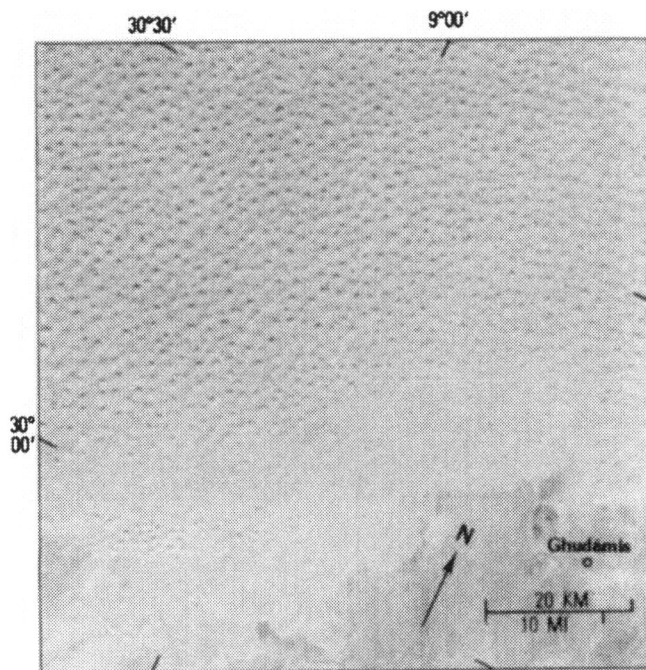

LANDSAT IMAGERY OF ISOLATED STAR DUNES in the Grand Erg Oriental near Ghudāmis, Libya. Annual and six monthly sand roses for Ghudāmis illustrating complex high-energy wind regime are shown in figure 97C. (Fig. 117.)

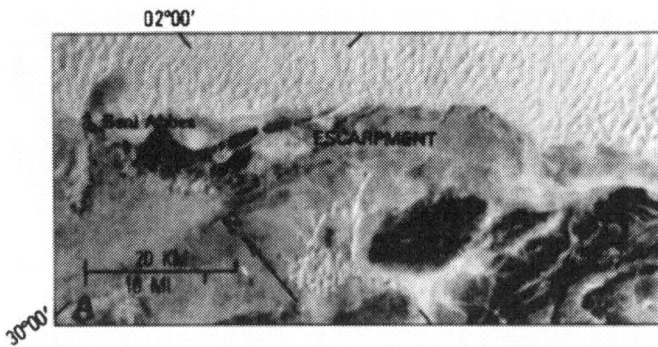

LANDSAT IMAGERY OF STAR DUNES (*A*) that occur along and atop an escarpment at the southern margin of the Grand Erg Occidental near Beni Abbes, Algeria. These star dunes are not as large as those near Ghudāmis, Libya. *B*, Annual and six monthly sand roses for Beni Abbes. Annual effective wind regime is complex, low energy. Effective wind regime is complex during April, the month of highest drift potential. Arrows indicate resultant drift direction (*RDD*). Number in center circle of rose is reduction factor. *DP* (drift potential, in vector units) is given for each rose. (Fig. 118.)

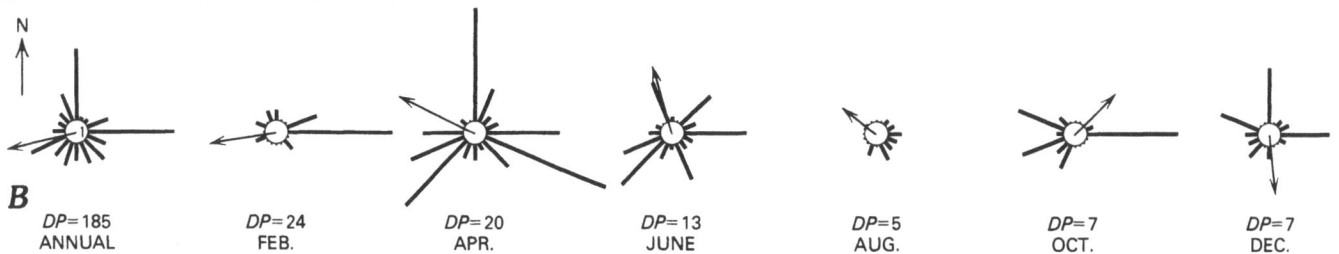

DP=185 ANNUAL	DP=24 FEB.	DP=20 APR.	DP=13 JUNE	DP=5 AUG.	DP=7 OCT.	DP=7 DEC.

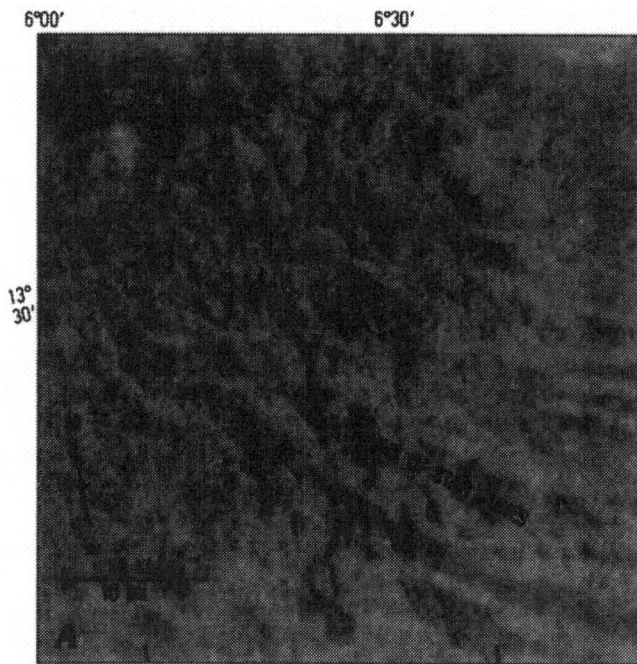

LANDSAT IMAGERY OF ALINED STAR DUNES (*A*) atop linear sand sheets near Hassi Messaoud, Grand Erg Oriental, Algeria. Exact nature of the sheets is unknown. *B*, Annual and six monthly sand roses for Hassi Messaoud, Algeria, illustrating complex (or possibly obtuse bimodal) intermediate-energy effective wind regime at the station. The present-day wind environment of the region seems intermediate between those of star and linear dunes (figs. 104, 116). The dune pattern may have developed from present-day winds or may represent the result of two or more stages of development involving different wind regimes. Arrows indicate resultant drift direction (*RDD*), in vector units. Number in center circle of rose is reduction factor. *DP* (drift potential, in vector units) is given for each rose. (Fig. 119.)

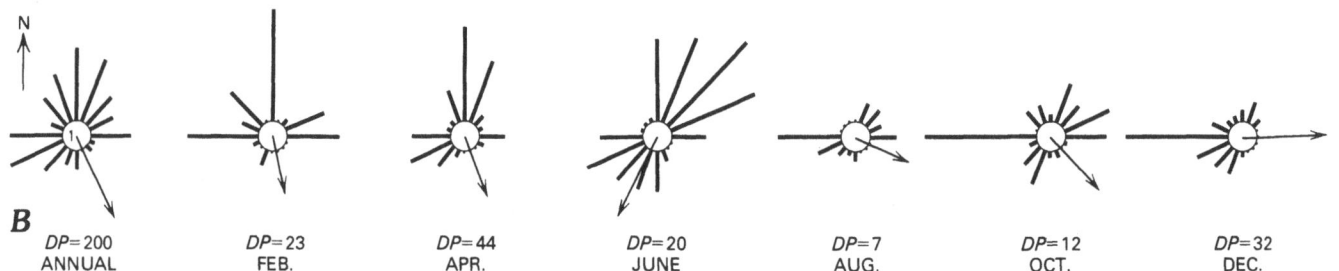

DP=200 ANNUAL	DP=23 FEB.	DP=44 APR.	DP=20 JUNE	DP=7 AUG.	DP=12 OCT.	DP=32 DEC.

distributions from month to month tend to be bimodal or complex (fig. 119B). This wind regime differs from other wind regimes associated with star dunes because the resultant drift direction at the station is relatively steady (toward the southeast) during the year. This wind environment seems to be intermediate between that associated with linear dunes and that with star dunes, in terms of RDP/DP (fig. 116).

An anomaly is presented by dunes in the region of Erg Irrarene near Bordj Omar Driss (Fort Flatters), Algeria. There, star dunes are atop what seem from Landsat imagery (chapter K) to be barchanoid dunes. If the sand bodies upon which the star dunes have developed were originally barchanoid dunes, a climatic change is indicated because the observed environments of star and barchanoid dune types differ widely (fig. 116). Present-day winds at Bordj Omar Driss are intermediate between those forming linear dunes and those forming star dunes, in terms of RDP/DP.

A STUDY OF GLOBAL SAND SEAS

LABORATORY STUDIES OF SAND PATTERNS RESULTING FROM CURRENT MOVEMENTS

Chapter G

By THEODORE F. TYLER

Contents

Illustrations

Tables

Summary of Conclusions

𝕿HIS STUDY IS A COMPARISON of laboratory-generated small-scale aqueous ripples, natural large-scale ripples (aqueous dunes), and eolian dunes to show similarities in their developmental sequences and resulting bedform patterns. Small-scale aqueous ripple development on smooth sloping sand beds and ripple development in the lee of barriers placed on smooth-leveled sand beds were studied in a laboratory wave channel by means of tracings and still photography. These tracings and photographs were compared with oblique aerial photographs and Landsat (ERTS) imagery to determine similarities between aqueous and eolian bedform patterns. The various bedforms show remarkable similarities although developed under different environmental conditions.

Bedforms can be categorized on the basis of their developmental sequences, as follows: (1) Isolated crescentic forms (isolated barchan), (2) coalescing crescentic form (coalescing barchan), (3) parallel-sinuous form (barchanoid ridge), and (4) parallel-straight form (transverse ridge).

Introduction

THIS STUDY WAS UNDERTAKEN in the U.S. Geological Survey sedimentary structures laboratory to determine the transition patterns of small-scale oscillation ripples on gently sloping surfaces, and on level surfaces when barriers were present to disrupt the normal wave motion. Tracings were made as the ripples developed under normal wave action, and a tracing was made every 10 minutes for a 50- to 60-minute period. Then the ripples were classified according to crest shape. Comparisons were made between the bedforms of these small-scale aqueous ripples and those of the large-scale ripples (aqueous dunes) developed under natural conditions. The aqueous forms were then compared with large-scale eolian dunes, by using aerial photographs and Landsat (ERTS) imagery. Emphasis was placed on the similarities between their plan views and transition patterns.

The following discussion falls into four categories: (1) hierarchy of bed roughness, (2) the wave tank and procedures used for this study, (3) ripple patterns developed during three laboratory experiments and a comparison of aqueous dunes with eolian dune development, and (4) effects of barriers on bedform development in both aqueous and eolian environments as evidenced by use of photographs of laboratory experiments, oblique aerial photographs, and Landsat (ERTS) imagery.

Background Information

BEDFORMS OCCUR on the surface of non-cohesive sediments in response to fluid motion. It is extremely difficult to determine the various parameters and mechanisms and the intricate interrelations which control bedform development in a natural situation. Under laboratory conditions one parameter at a time may be varied to study its effect on bedform development and to provide insight into this parameter's effect on bedform patterns observed in the field.

Hierarchy of Bed Roughness

A critical, or threshold, velocity, which is a function of grain size, must be reached to initiate motion in noncohesive sands. Before onset of grain motion on a flat bed, only grains themselves contribute to bed roughness. A second-order, and possibly a third-order, of bed roughness may develop once grain movement starts. This hierarchy of bed roughness is: (1) individual grains, (2) small-scale ripples, and (3) dunes. In general, the individual grain is the basic building block for this hierarchy of bed roughness, and for the development of a sedimentary structure. Bedform development is also dependent on other factors, such as the availability of sand, the water or wind velocity, water depth, and time.

Oscillation Ripple

A ripple is defined as any deviation from a flat bed that resembles or suggests a ripple of water and that is formed on the bedding surface of a sedimentary deposit. Aqueous ripples vary considerably in size and shape. Figure 120 shows a ripple profile at right angles to wave propagation and gives the

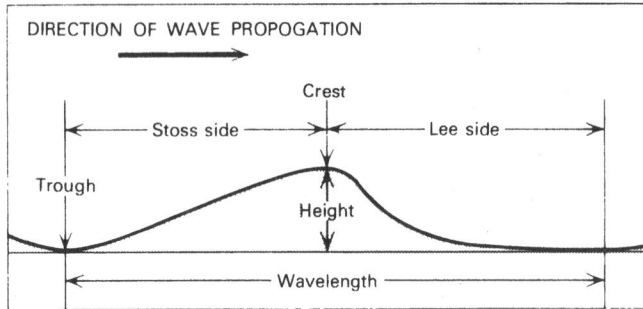

GENERALIZED RIPPLE PROFILE, showing gently sloping stoss side and steeply sloping lee side. (Fig. 120.)

terms commonly used to describe a ripple. In general, a ripple whose crest becomes more planar than those shown in figure 120 is defined as having a platform crest. Two ratios are useful in characterizing a ripple:

1. Ripple symmetry index (*RSI*):

$$RSI = \frac{\text{Horizontal projection of stoss side}}{\text{Horizontal projection of lee side}}.$$

For a symmetrical ripple formed under oscillatory flow conditions (wave ripple), this ratio should be equal to 1.0; for asymmetrical wave ripples this ratio ranges from 1.1 to 3.8 (Reineck and Singh, 1973, p. 25).

2. Ripple index (*RI*):

$$RI = \frac{\text{Wavelength}}{\text{Height}}$$

For both symmetrical and asymmetrical wave ripples the ripple index ranges from 5 to infinity, with most wavelengths being 6−8 (Dingler, 1975, p. 95; Reineck and Singh, 1973, p. 25). For unidirectional current ripples, the *RI* is always more than 5, with most values falling between 8−15 (Reineck and Singh, 1973, p. 29).

Small-scale oscillation ripples are defined as having wavelengths ranging from 0.4 to 78.8 inches (1 to 200 cm) and heights from 0.1 to 9.1 inches (0.3 to 23 cm) (Reineck and Singh, 1973, p. 24−25). Small-scale unidirectional current ripples are defined as having wavelengths of 1.6 to 23.6 inches (4 to 60 cm) and heights of 0.1 to 2.4 inches (0.3 to 6 cm) (Reineck and Singh, 1973, p. 29).

Large-scale ripples (aqueous dunes) are not as well described in the literature as are small-scale ripples, but some good examples can be found. A development of bedform patterns at the southeastern edge of the Tongue of the Ocean, Bahama Islands, in water 3−4 fathoms (5.5−7.3 m) deep is shown by Newell and Rigby (1957, pl. 10, fig. 1). Crescentic (barchanoid) forms occur as solitary forms which coalesce and eventually form parallel-sinuous and parallel-straight forms (fig. 121*A*). The horns of the simple crescentic forms are 50−200 feet (15−61 m) wide, and they coalesce to form great ripples (aqueous dunes) with wavelengths of 10−30 feet (3−9 m). Parallel-straight to parallel-sinuous transverse ripples (dunes) are developed in the vicinity of the Northwest Channel Light, Bahama Islands (fig. 121*B*). These forms have wavelengths of approximately 300 feet (91 m).

Oscillation megaripples and certain eolian dunes show similarities in their external profile and developmental patterns when viewed in plan. Thus, figure 120 and the associated terms can also be used to describe an eolian dune oriented transverse to the wind direction. The lee- and stoss-side dimensions and height of an eolian dune range from a few feet to tens of feet in magnitude, very similar to the megaripples (aqueous dunes) at the Tongue of the Ocean previously described.

Present Study

A DESCRIPTION FOLLOWS of the wave tank and of the procedures used for experiments conducted in the laboratory. Ripple development on gently sloping surfaces and the transition patterns of these ripples are compared for three experiments.

Equipment and Procedure

A 46-foot-long by 2-foot-deep by 1.5-foot-wide (14-m-long by 0.6-m-deep by 0.5-m-wide) wave tank was used for all experiments (fig. 122). Distances along the wave tank were measured from the downwave end of the tank. The tank has a piston-type paddle at the upwave end and a silica sand beach with a median grain size of 0.319 mm at the downwave end. The study area was the interval between 15 and 30 feet (4.6 and 9.1 m). This area in-

BEDFORM PATTERNS OF LARGE-SCALE RIPPLES (aqueous dunes) in the Bahama Islands. Sketched from photographs in Newell and Rigby (1957, pl. 10, fig. 1, and fig. 2) A, Southeastern edge of the Tongue of the Ocean. The water is 3 – 4 fathoms (5.5 – 7.3 m) deep in the area. B, Vicinity of the Northwest Channel Light. (Fig. 121.)

DIRECTION OF WAVE PROPOGATION

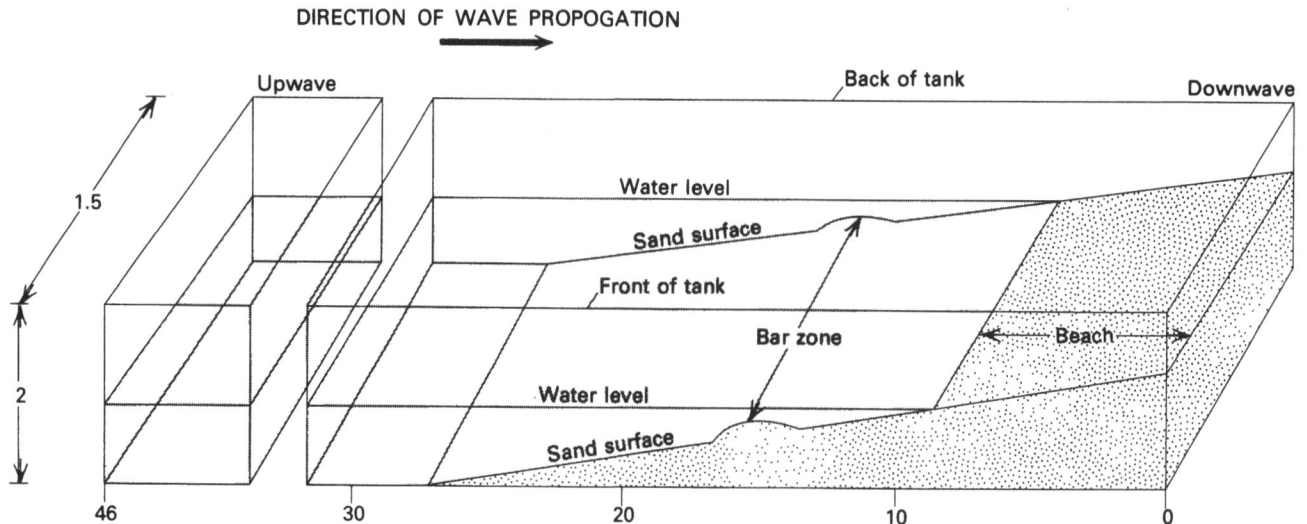

GENERALIZED DIAGRAM OF THE WAVE TANK used in the laboratory experiments. Dimensions of the tank are in feet (1 foot = 0.31 metres). (Fig. 122.)

cluded the sand bed from the breaker-related bar zone upwave to a point where the sand was about 0.4 inch (1.0 cm) thick. From 15 to 0 feet (4.6 to 0.0 m) the sand-water interphase was affected by the borelike motion of the water, which occurs after the wave breaks, and the interval was therefore excluded from this study.

The following parameters were held constant for the first three experiments: The grain size of the sand, D, was 0.319 mm; the water depth, h, at the 30-foot location (9.1 m) was 8.6 inches (21.8 cm); and wave period, T, was 2.5 seconds. Differences in these parameters, between experiments, reflected the initial variations in the sand thickness (fig. 123).

Another series of experiments were conducted to study the effect barriers had on ripple development. The sand in the tank from the first experiments was leveled, and the barrier was placed on this leveled sand surface.

The following procedures were used for the experiments:

A. Sloping sand bed without barriers:
  1. The sand was leveled to a predetermined profile.
  2. Waves were generated for a total time of 50–60 minutes per experiment.
  3. Wave generation was stopped at the end of each 10-minute interval, and tracings were made of the bottom (plan view).
  4. Ripple height was measured directly, using a metre stick.

DIRECTION OF WAVE PROPOGATION

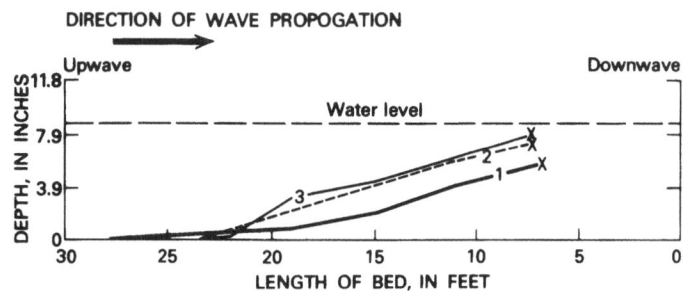

STARTING SAND PROFILE IN WAVE TANK, experiments 1, 2, and 3. 1 inch = 2.54 centimetres; 1 foot = 0.31 metres. Measurements for study stop at X's. (Fig. 123.)

  5. Potassium permanganate dye was injected into the water to permit visualization of the flow conditions.
  6. Magnetite grains were allowed to settle onto the sediment at various times to indicate surface changes as a function of time.
  7. The horizontal projections of the lee- and stoss-side dimensions were measured from the ripple tracings.
  8. The rate of ripple migration was calculated when possible by following the movement of specific ripples over a predetermined time interval.
  9. The ripple symmetry index (RSI) and the ripple index (RI) were calculated for groups of ripples with similar crest shapes.
10. Photographs of ripple development were taken and used for comparative purposes.

B. Flat sand bed with barriers:
1. The sand was leveled and a barrier was placed on it.
2. Waves were generated until ripple development began to propagate downwave in a uniform pattern.
3. Photographs of ripple development were taken and used for comparative purposes.

Wave-Tank Experiments

To avoid repetition, ripple development is discussed in detail for experiment 1 and only briefly for experiments 2 and 3. Illustrations, tables, and graphs are used to examine ripple growth from a gently sloping sand bed and their developmental sequences.

Experiment 1

Oscillatory waves were generated for 60 minutes during experiment 1. The initial sand profile is shown in figure 123, and the bed in plan at given time intervals is shown in figure 124. The wave height, H, ranged from 5 inches (12.7 cm) at 17 feet (5.2 m) to 3.2 inches (8.1 cm) at 24 feet (7.3 m). Assuming a horizontal bottom at 24 feet (7.3 m) and using a linear wave theory (Airy, 1845), the maximum near-bottom fluid velocity, U_m, is calculated to be 10.6 in./s (27 cm/s), where:

$$U_m = \pi H/T \sin h(kh),$$
$$k = 2\pi/L \text{ is the wave number, and}$$

L, which is the wavelength, is a function of h and T.

Initial ripple development took place during the first 5 minutes of wave action. Parallel-straight to slightly parallel-sinuous ripples developed between 15.5 and 16 feet (4.7 and 4.9 m) while nonuniformly spaced crescentic and coalescing crescentic ripples developed between 16 and 19.5 feet (4.9 and 5.9 m). After 10 minutes (fig. 124) bar development was located at 15.2 feet (4.6 m), and branching ripples and ripples that terminate in mid-tank formed between 15.2 and 18.5 feet (4.6 and 5.6 m). Parallel-sinuous ripples developed from 18.5 to 20.0 feet (5.6 to 6.1 m), and ripples which terminated in mid-tank developed between 20 and 21 feet (6.1 and 6.4 m) (fig. 124).

Branching ripple development took place early in the experiment, when ripples either joined (fig. 125, sets A, D) or separated to form individual ripples (fig. 125, sets B, C, E, F). Those ripples which did not completely cross the tank (ripple crest 3 of set B) seemed to initiate this branching process. The segment of the ripple that separated from the branching bedform migrated downwave and joined with the next ripple. This process could be repeated with the newly branched bedform acting as the nucleus.

The average ripple wavelength, height, stoss- and lee-side horizontal dimensions, and ripple symmetry index (RSI) were calculated for groups of similar ripple patterns (table 20).

After 30 minutes of wave action the bar zone was located at 15.3 feet (4.7 m) (fig. 124). Parallel-straight to slightly parallel-sinuous ripples were distinct crests developed throughout the wave tank (fig. 124).

One ripple in group E and one in group F were followed for 20 minutes more (fig. 124) to study rates of ripple migration. The ripple in group E migrated 0.74 foot (0.23 m) and the ripple in group F migrated 0.65 foot (0.20 m). This corresponds to rates of 0.43 inch (1.1 cm) per minute and 0.39 inch (1.0 cm) per minute, respectively.

The bar was located at 15.6 feet (4.8 m) after 50 minutes of wave action. Parallel-sinuous ripples occurred from 15.6 to 16.4 feet (4.8 to 5 m). Between 16.4 and 17.3 feet (5 and 5.3 m), the ripples did not extend across the full width of the tank. Instead, they terminated approximately 5.9 inches (15 cm) from the front side of the tank (fig. 124). These ripples still had a slightly parallel-sinuous pattern with platform crests and filled troughs. Crests lo-

TABLE 20. — *Summary of ripple dimensions (in centimetres) and ripple symmetry index for experiment 1 after 10, 30, and 50 minutes*

[Letters designating ripple groups correspond to those in figure 124]

Ripple group	Average wavelength	Average height	Average stoss	Average lee	Ripple symmetry index (RSI)
		After 10 minutes			
A	8.8	1.5	5.0	3.8	1.3
B	8.6	1.5	5.1	3.6	1.4
C	6.7	.9	4.1	2.6	1.6
		After 30 minutes			
D	9.1	1.4	6.3	3.0	2.1
E	11.0	1.7	8.3	2.7	3.1
F	8.8	1.4	6.6	2.8	2.3
G	9.9	1.5	7.0	2.8	2.5
		After 50 minutes			
H	10.9	1.7	7.4	3.5	2.2
I	10.1	1.8	7.0	3.1	2.3

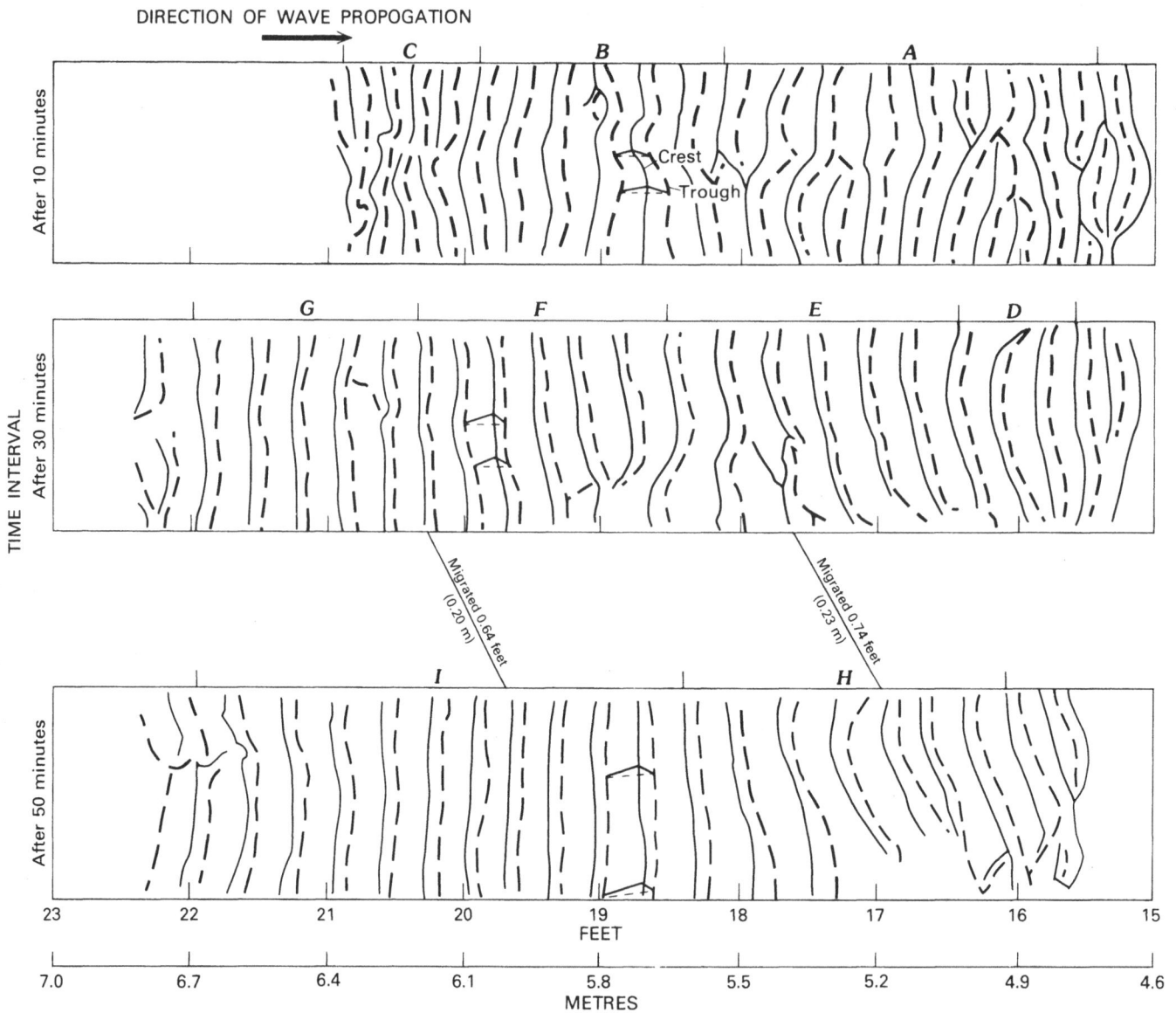

PLAN VIEW OF RIPPLE PATTERNS for experiment 1 after 10, 30, and 50 minutes of wave action. Trough lines dashed. Slanting lines connecting the same ripple demonstrated ripple migration (distance indicated on line). Letters A — I indicate groups of ripples listed in table 20. (Fig. 124.)

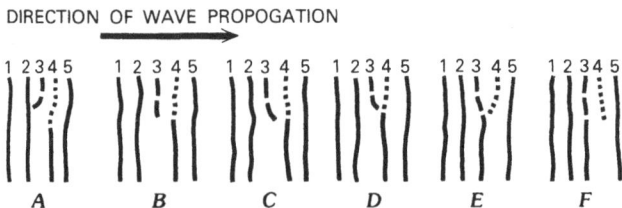

PROCESS OF RIPPLE BRANCHING. Plan view of ripple crests, sets A (earliest time) through F (latest time). (Fig. 125.)

cated between 17.3 and 19.0 feet (5.3 and 5.8 m) were platform and had a parallel-straight to slightly parallel-sinuous pattern. Parallel-straight ripples with distinct crests developed from 19.0 to 20.0 feet (5.8 to 6.1 m). Crests from 20.0 to 21.5 feet (6.1 to 6.6 m) were platform. Between 21.7 and 23.5 feet (6.6 and 7.2 m), the ripples did not completely cross the tank and, therefore, were not included in figure 124 beyond 20.2 feet (6.2 m).

Figures 126 and 127 illustrate the manner in which the ripple parameters change as a function of time. In figures 126A and B, the RI values of 5, 6,

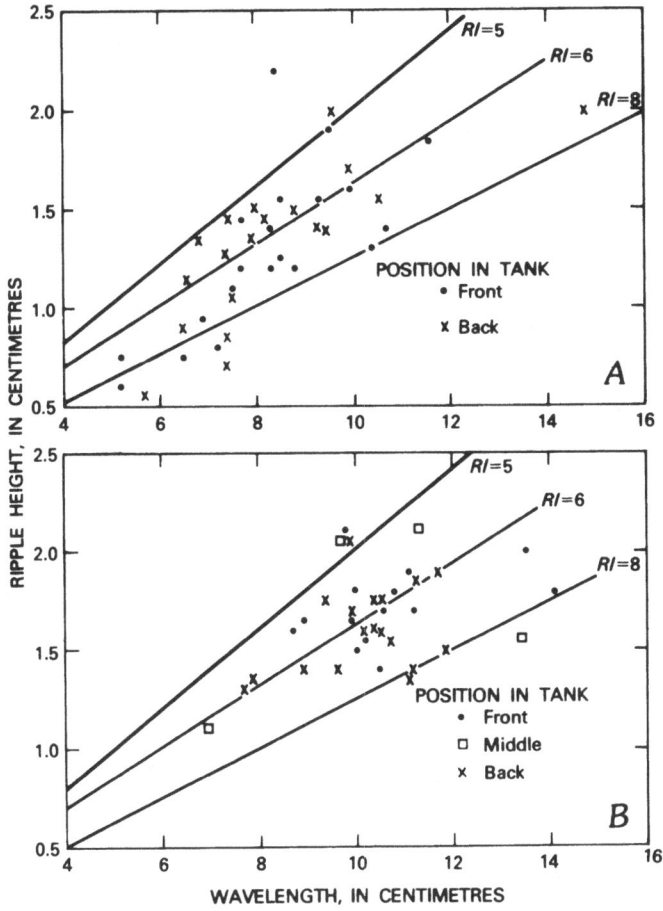

GRAPHS SHOWING RIPPLE HEIGHT AND WAVELENGTH after (A) 10 minutes and (B) 50 minutes of wave action. Ripple index lines (RI = wavelength/height) plotted to reflect their relation to the 5 − 8 range of asymmetrical wave ripples. As shown in A, 74 percent of the ripples are in the 5 to 8 range, and in B, 86 percent are in the 5 to 8 range. Ripple height and wavelength measured approximately 10 cm from front and back of wave tank. 1 centimetre = 0.394 inch. (Fig. 126.)

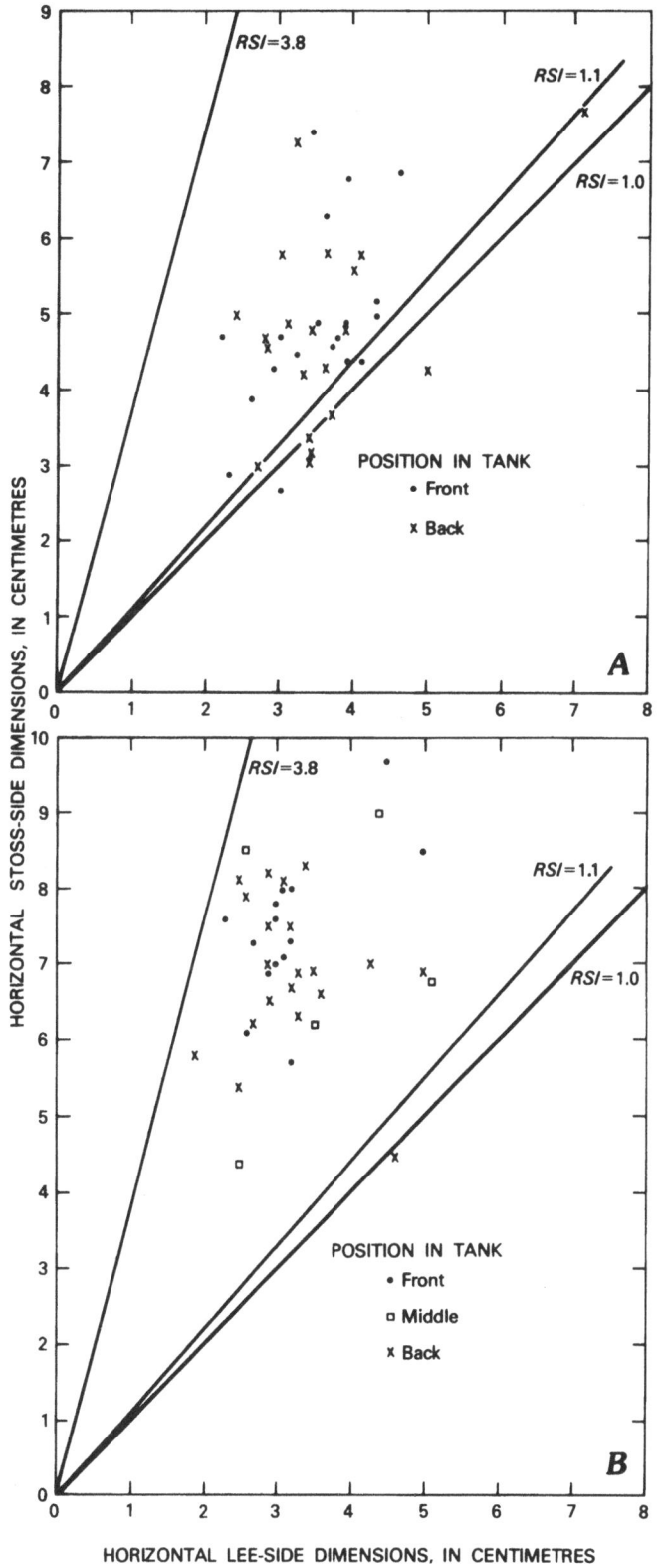

GRAPHS OF STOSS- AND LEE-SIDE horizontal dimensions after (A) 10 minutes and (B) 50 minutes of wave action. Ripple symmetry index lines (RSI = Horizontal projection of stoss side/Horizontal projection of lee side) plotted to reflect their relation to the 1.1 − 3.8 range of asymmetrical wave ripples. As shown in A, 80 percent of the ripples are in the 1.1 − 3.8 range, and in B, 97 percent are in the 1.1 − 3.8 range. Stoss- and lee-side measurements were made approximately 10 cm from the front and back of the tank; measurements were made in the middle of the tank when they could not be made along the front and back. 1 centimetre = 0.394 inch. (Fig. 127.)

and 8 are indicated by solid lines. After 10 minutes, 74 percent of the ripples fell within the 5–8 range, and after 50 minutes 86 percent fell within that range. Lines of constant *RSI* (*RSI* = 1.1, 3.8) are drawn in figures 127 *A* and *B* to correspond to the reported *RSI* range for asymmetrical wave ripples. After 10 and 50 minutes, 80 and 97 percent of the ripples, respectively, fell within this range. Ripples which plot outside the indicated ranges fall into one of four categories: (1) Ripples that are influenced by their position near the bar zone, (2) branching ripples, (3) washed-out ripples, and (4) ripples at the upwave end of the sand bed. The ripples in category 4 are not considered to be fully developed.

Experiments 2 and 3

The starting sand profiles for experiments 2 and 3 are shown in figure 125. For experiment 2, wave break was located at 17 feet (5.2 m), the wave height, *H*, was 4.1 inches (10.4 cm) at 19 feet (5.8 m) and 3.7 inches (9.4 cm) at 23 feet (7 m) (fig. 128). For experiment 3, wave break was located at 17 feet (5.2 m), wave height, *H*, was 3.9 inches (10 cm) at 19 feet (5.8 m) (fig. 129).

In experiment 2, irregular crescentic ripples developed throughout the wave tank after approximately 3 minutes with the best development occurring at about 23 feet (7 m). Parallel-straight to slightly parallel-sinuous ripples with a uniform pattern developed after 10 minutes of wave action (fig. 128; table 21).

In experiment 3, irregular crescentic ripples developed after 2 minutes of continuous wave action. Isolated crescentic ripples coalesced to four large crescentic ripples after 5 minutes at approximately 23 feet (7 m). Parallel-straight, branching bedforms, parallel-sinuous, and irregular crescentic ripples developed after 10 minutes of wave action (fig. 129; table 22).

After 50 minutes, ripple development was less distinct in experiments 2 and 3 (figs. 128, 129; tables 21, 22). The ripples were washed out between approximately 17 and 19 feet (5.2 and 5.8 m) with crests becoming more platform and troughs filling, causing the bottom profile to become more planar.

The *RI* values calculated for experiments 2 and 3 show that after 10 minutes 90 and 77 percent, respectively, fell within the 5–8 range and that

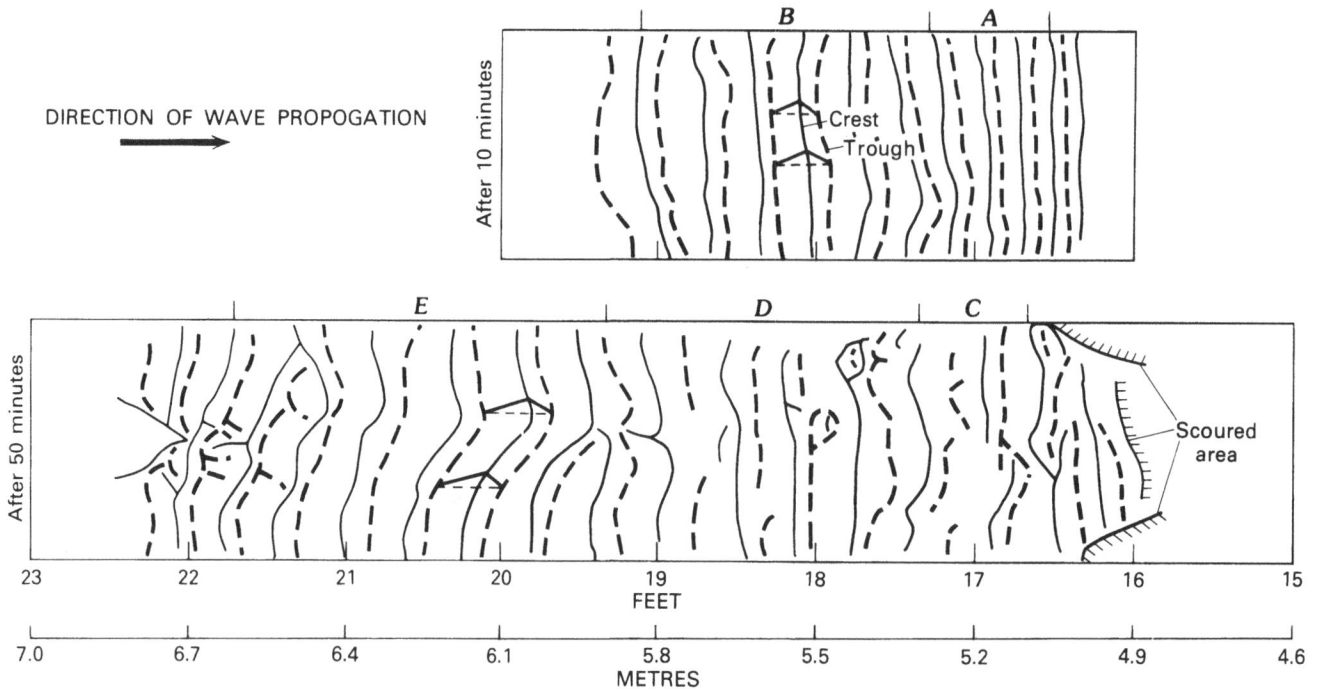

PLAN VIEW OF RIPPLE PATTERNS for experiment 2 after 10 and 50 minutes of wave action. Trough line dashed. Letters *A – E* indicate groups of ripples listed in table 21. (Fig. 128.)

DIRECTION OF WAVE PROPOGATION

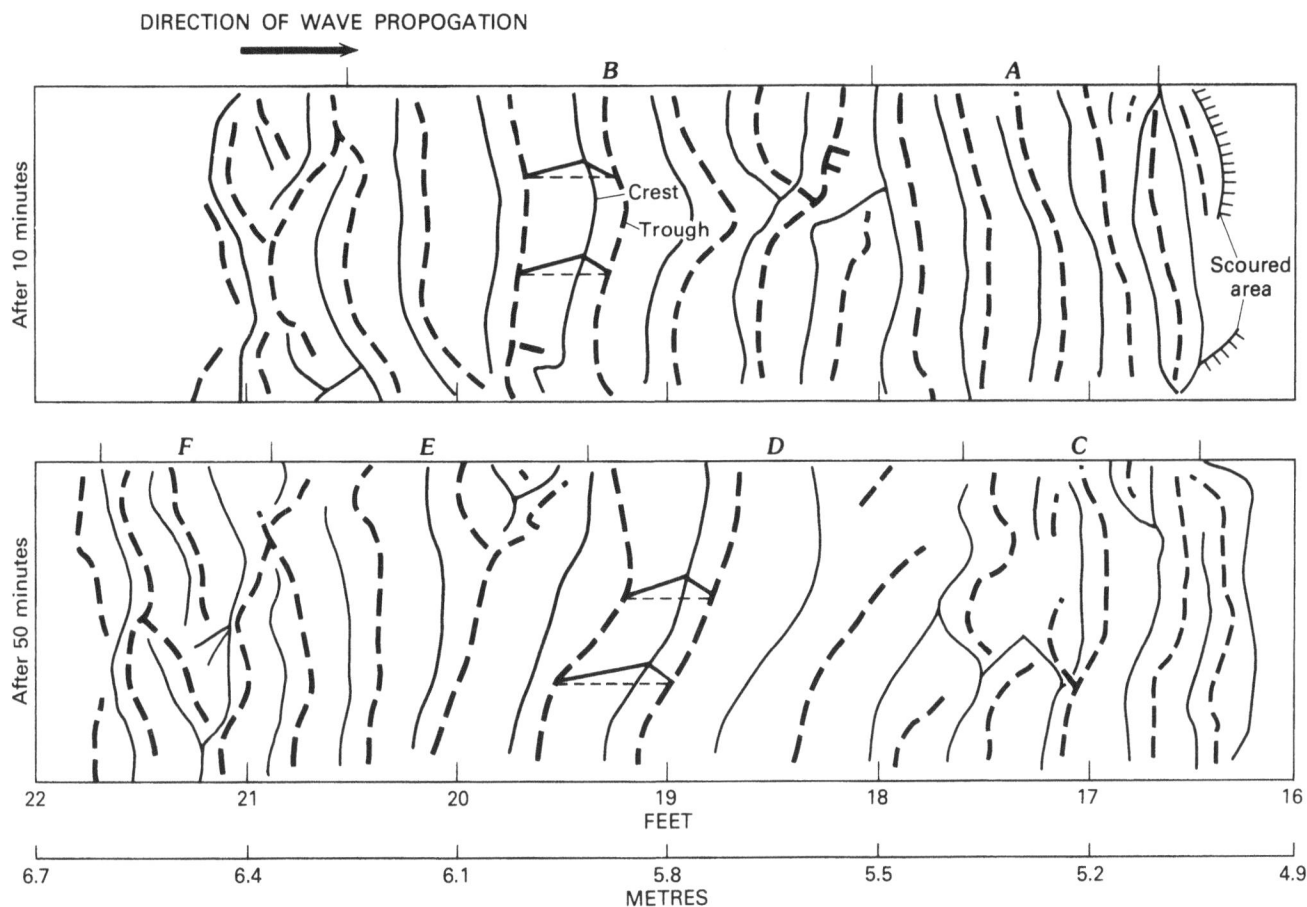

PLAN VIEW OF RIPPLE PATTERNS for experiment 3 after 10 and 50 minutes of wave action. Trough lines dashed. Letters A – F indicate groups of ripples listed in table 22. (Fig. 129.)

TABLE 21. — *Summary of ripple dimensions (in centimetres) and ripple symmetry index for experiment 2 after 10 and 50 minutes*

[Letters designating ripple groups correspond to those in figure 128]

Ripple group	Average wavelength	Average height	Average stoss	Average lee	Ripple symmetry index (RSI)
After 10 minutes					
A	6.7	1.0	4.0	3.2	1.3
B	10.9	1.5	6.8	3.3	2.0
After 50 minutes					
C	12.1	2.1	7.5	4.5	1.7
D	13.2	1.5	8.3	4.9	1.7
E	11.5	1.8	7.6	3.9	2.0

TABLE 22. — *Summary of ripple dimensions (in centimetres) and ripple symmetry index for experiment 3 after 10 and 50 minutes*

[Letters designating ripple groups correspond to those in figure 129]

Ripple group	Average wavelength	Average height	Average stoss	Average lee	Ripple symmetry index (RSI)
After 10 minutes					
A	9.7	1.5	6.4	3.4	1.9
B	10.8	1.8	7.2	3.7	2.0
After 50 minutes					
C	10.2	1.4	6.1	4.1	1.4
D	18.5	2.5	11.0	7.6	1.5
E	18.3	1.7	7.5	4.8	1.6
F	8.1	1.4	5.4	2.7	2.0

after 50 minutes 88 percent of the *RI* values for both experiments fell within this range. *RSI* values calculated for experiments 2 and 3 show that after 10 minutes 95 and 100 percent, respectively, fell within the 1.1 – 3.8 range cited for asymmetrical wave ripples and after 50 minutes 90 and 82 per-

cent, respectively, fell within these limits. The decrease in the *RSI* percentages appears to be caused by ripples that are being washed out in the area between 17 and 19 feet (5.2 and 5.8 m) (figs. 128, 129). This washing out resulted in crests becoming platform and in troughs filling.

Ripple Development Patterns

Figure 130 shows the transition from one form to another that oscillatory ripples make with time. This change in pattern seems to develop as (1) a function of the flow velocity of the water and (or) water depth and (2) time. Ripples during experiments 1, 2, and 3 appeared to develop in the following manner: Sand movement was dominantly by traction and saltation in a downwave direction when the ripples first started to develop. When one ripple became more distinct by its height increasing, sand movement was affected by the vortex or eddy developed in the lee of the ripple. The circular action of the eddy churned up the sediment on the lee surface and in the trough in front of the ripple, and this churning put the finer sand in suspension. The finer material was held in suspension and moved in both a downwave and an upwave direction due to the oscillation of the passing wave. The coarser material dropped out of suspension and caused avalanching on the lee surface of the ripple.

Therefore, sand movement was by traction and saltation on the stoss side and dominantly by suspension and avalanching on the lee side of the ripple. With time, this process of sand movement could change an isolated crescentic ripple into a coalescing crescentic ripple and a coalescing crescentic ripple into a parallel-sinuous or parallel-straight ripple. If U_m (the maximum near-bottom fluid velocity) increased above a certain critical velocity, sheet flow instead of oscillatory wave motion would occur near the bottom. With this change in U_m toward sheet-flow conditions, ripples become washed out, and eventually a planar surface devoid of ripples develops.

Unidirectional current ripples undergo a somewhat similar change in ripple patterns as a function of flow velocity and (or) water depth (Allen, 1968, p. 93) (fig. 131).

Comparison of Aqueous Wave Ripples (Aqueous Dunes) and Eolian Dunes

The similarity in crestal patterns of aqueous ripples (dunes) and eolian dunes is striking (table 23; fig. 132). In an aqueous environment having limited sand supply, crescentic ripples develop (fig. 132, aqueous forms; fig. 121). In an eolian environment barchan dunes occur under the same sand-supply conditions. When sufficient sand is available, parallel-straight and branching aqueous ripples develop, and they appear to be very similar to eolian barchanoid and transverse ridges.

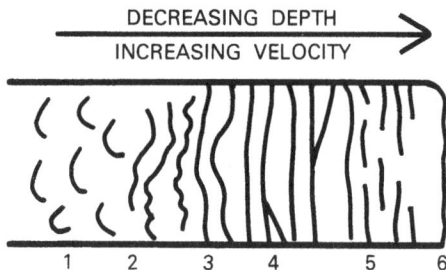

DECREASING DEPTH
INCREASING VELOCITY

SEQUENCE OF SMALL-SCALE RIPPLES (plan view), changing with time and with an increase in flow velocity and (or) water depth. Forms are (1) isolated crescentic, (2) coalescing crescentic, (3) parallel-sinuous, (4) parallel-straight, (5) irregular pattern, and (6) bar zone (zone of ripple destruction). (Fig. 130.)

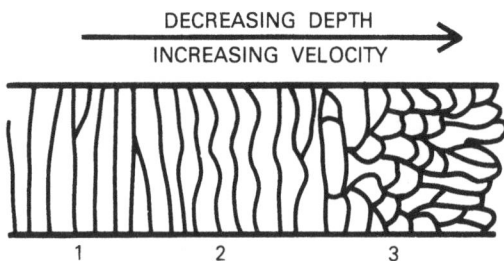

DECREASING DEPTH
INCREASING VELOCITY

SEQUENCE OF SMALL-SCALE RIPPLES (plan view) changing with an increase in flow velocity and (or) water depth. Sketched and modified from J. R. L. Allen (1968, fig. 4.61a). Forms are (1) parallel-straight, (2) parallel-sinuous, and (3) linguoidal. (Fig. 131.)

TABLE 23. — *Bedforms oriented transverse to wave or wind direction*

Oscillatory wave ripples	Eolian dunes
Crescentic solitary form	Barchan form (solitary)
Coalescing crescentic form (parallel-sinuous form)	Barchanoid ridge (coalescing barchanoid form)
Parallel-straight and branching form	Transverse ridge

AQUEOUS BEDFORMS—Laboratory
wave tank examples

1. Crescentic
solitary

2. Coalescing crescentic
(parallel-sinuous)

EOLIAN BEDFORMS—Sketch and
ERTS examples

1. Barchanoid solitary

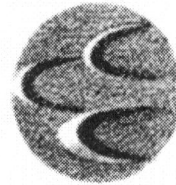

2. Barchanoid ridge
(coalescing barchanoid)

E1183–06194

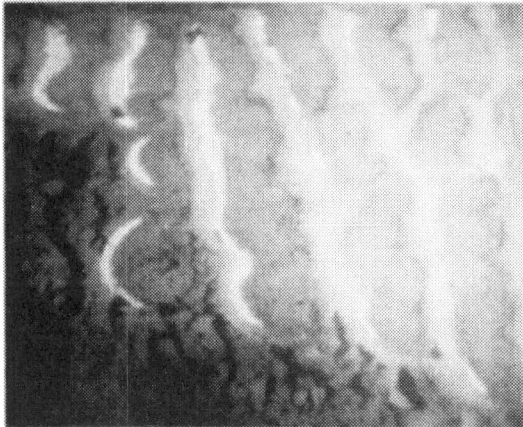

3. Parallel-straight and branching

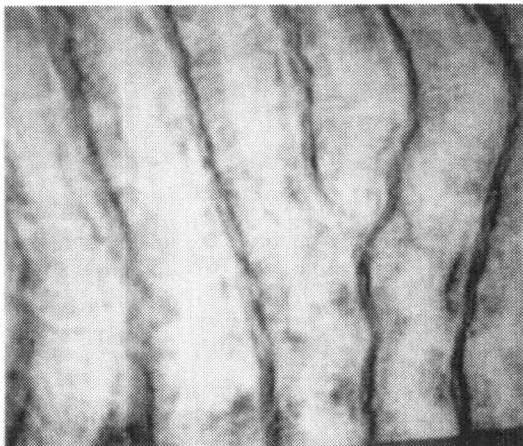

3. Transverse ridge

E1128–04253

COMPARISON OF AQUEOUS AND EOLIAN BEDFORMS (plan views). Direction of wave propogation and of wind is from left to right. White in aqueous bedforms 1 and 2 is bottom of wave tank exposed by erosion due to limited sand supply. Numbers next to round photographs refer to Landsat (ERTS) imagery from which examples were taken. (Fig. 132.)

Effects of Barriers on Ripple Development

Experiments were conducted to study the effects of barriers on wave ripple development. Barriers were placed on a flat sand surface to disrupt the wave motion, and photographs were taken of the resultant ripple development. The barriers used were a steel ball 3.5 inches (9.0 cm) in diameter and clay bricks of various dimensions. Figure 133 illustrates the ball's effect on ripple development. Scour occurred around the edges of the ball, and, with time, crescentic, coalescing crescentic, and transverse ripples propagated in a downwave direction from the ball. The asymmetry of the wave-generated flow prevented upwave ripple formation from the ball during this experiment. Initial small-scale ripple development is similar in an eolian environment (fig. 134). Comparisons of bedforms can also be made on a much larger scale, as evident on Landsat (ERTS) imagery of eolian dunes in the An Nafūd Desert, Arabian Peninsula (figs. 135, 136). The images show questionable barchan dunes developing into barchanoid ridges, and the development of dunes across a geomorphically controlled area.

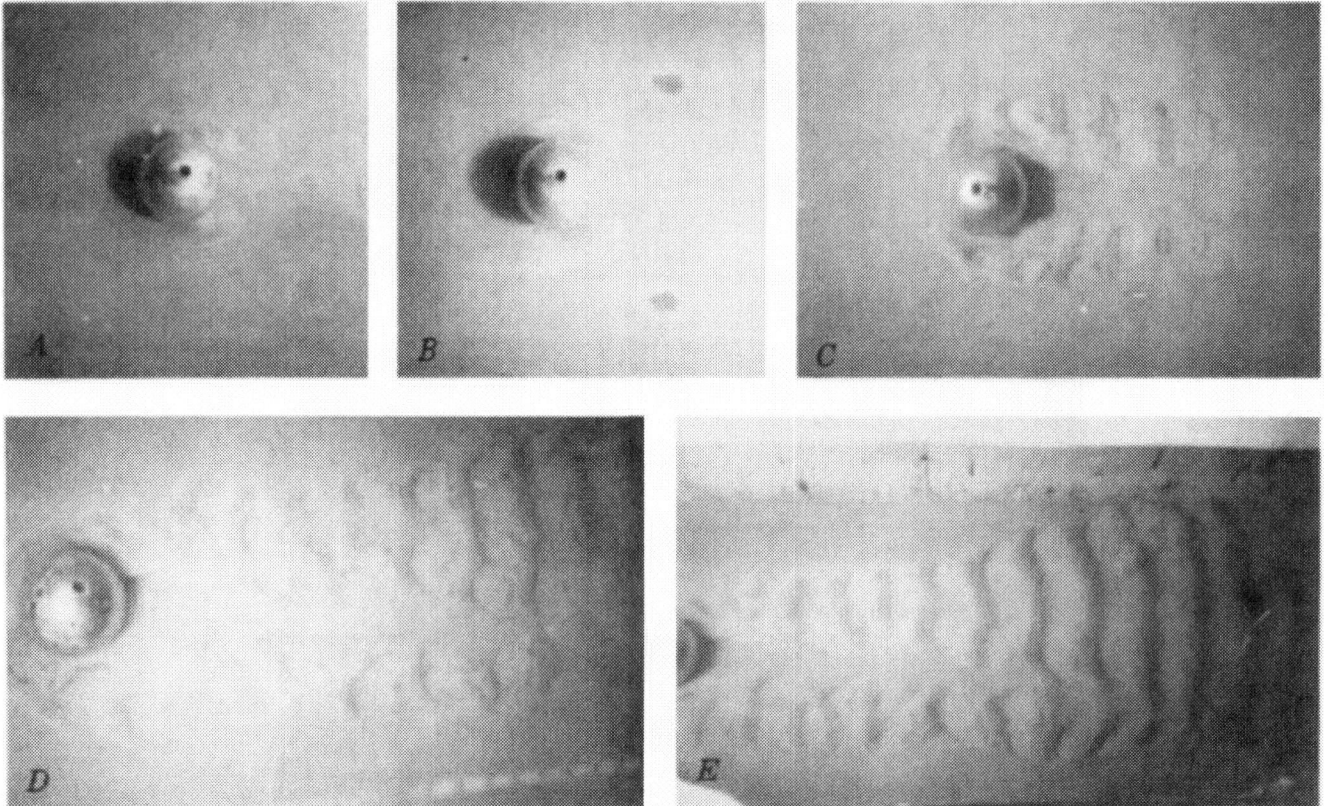

SEQUENTIAL PHOTOGRAPHS SHOWING EFFECTS of a barrier on ripple development in a wave tank. Steel ball 3.5 inches (9 cm) in diameter. Direction of wave propogation is left to right. A, Scour occurring around the edges of the ball after 2 minutes, and, B, after 5 minutes. C, Isolated crescentic form developing after 82 minutes. D, Isolated and coalescing crescentic forms developing after 107 minutes. E, Coalescing crescentic and parallel-straight forms developing after 114 minutes. (Fig. 133.)

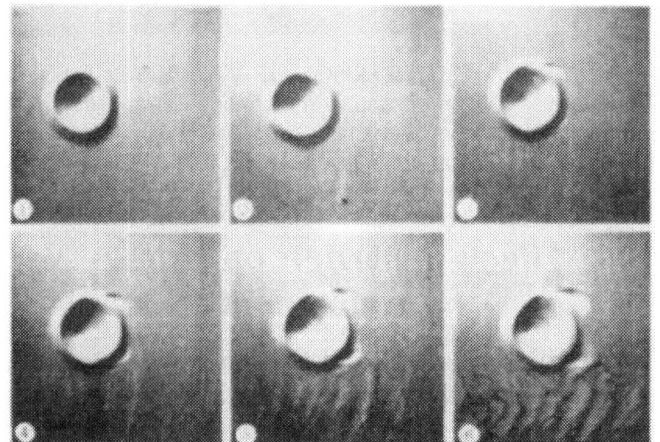

SEQUENTIAL PHOTOGRAPHS OF BEDFORMS (1−6) developed in a wind-tunnel experiment. Wind direction is from left to right. Crater is approximately 7 inches (18 cm) in diameter, modeled in loose sand and placed in the wind tunnel. Wind velocity is 160 inches per second (420 cm/s). Slight rippling of sand in photographs 4−6 is effect of tunnel wall. Photographs from Greeley and others (1974, p. 487). (Fig. 134.)

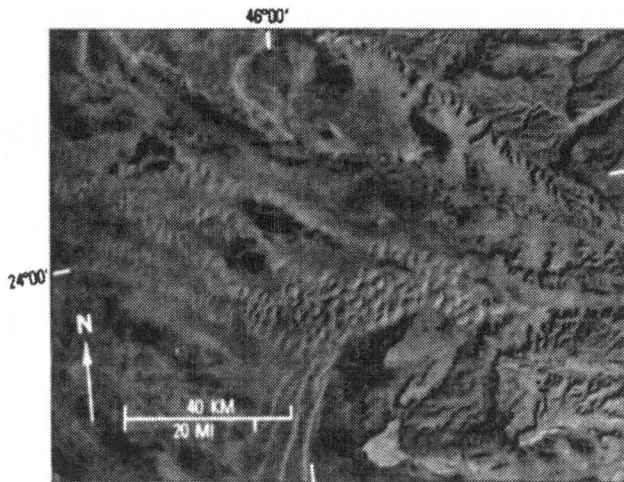

QUESTIONABLE BARCHAN DUNES developing into barchanoid ridges, An Nafūd Desert, Arabian Peninsula. General wind direction from west-northwest. From Landsat (ERTS) imagery E – 1171 – 06530 – 7 01. (Fig. 135.)

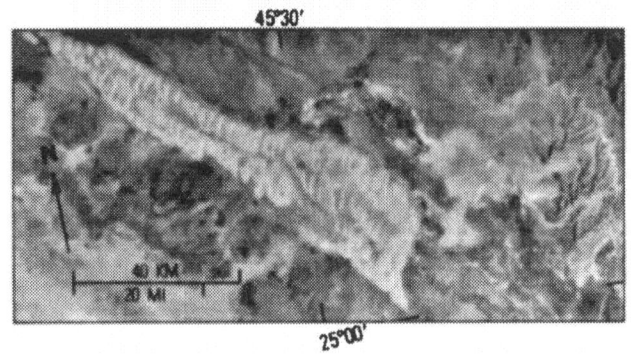

DEVELOPMENT OF DUNES across a geomorphically controlled area, An Nafūd Desert, Arabian Peninsula. General wind direction from northwest. From Landsat (ERTS) imagery E – 1226 – 06593 – 7 01. (Fig. 136.)

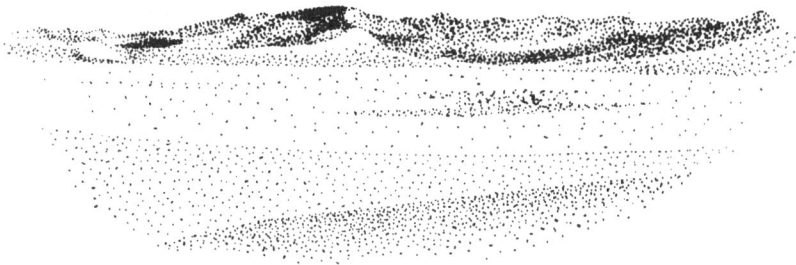

A STUDY OF GLOBAL SAND SEAS

ANCIENT SANDSTONES CONSIDERED TO BE EOLIAN

Chapter H

By EDWIN D. McKEE

With sections by JOÃO J. BIGARELLA[9]

Contents

Illustrations

Table

Summary of Conclusions

ANCIENT SANDSTONES of many ages and in many parts of the world have been attributed to an eolian origin. They are generally recognized by their primary structures, especially cross-strata which characteristically are on a large scale and dip at high angles. Other diagnostic structures are ripple marks with distinctive orientation and indices, unique forms of slump marks, and contorted bedding of types characteristic of dry sand. Commonly, the sand is well sorted and fine to medium grained, and has surficial markings, such as frosting and pitting. Sharply defined animal tracks and trails are also distinctive features of eolian sandstones; associated units of interdune deposits containing nonmarine faunas occur within sequences. Thus, although the genesis of some sandstones of this type is open to question, the origin of most seems sufficiently definite to merit calling them eolian-type sandstones.

As in modern sand seas, the dominant wind or winds were largely responsible for development of individual dune forms preserved in each eolian-type sandstone. Because wind direction commonly can be determined from orientation of slipfaces, dip directions of cross-strata in ancient eolian-type sandstones normally will reveal both the course of the wind and the type of dune represented. Unidirectional, bidirectional, and possibly multidirectional winds can be identified in many ancient sandstones by the clustering and spread of dip-directional measurements in cross-strata. The type of resulting dune — barchan, linear, or star — can also be recognized by the dip-direction spread and by certain other features of the cross-strata. Coastal dunes can be distinguished from interior desert dunes in part by the geometry or shape of the sand body and in part by their intertonguing or other close association with bordering marine deposits.

Introduction

LONG SWEEPING CROSS-STRATA of even-grained sandstone that form the sheer walls of many canyons in southwestern Utah led Huntington in 1907 (p. 381) to ascribe an eolian origin to these rocks (Navajo Sandstone) of Jurassic and Triassic(?) age. Subsequently, many other sandstones of that region, having similar textural and structural characteristics, have been attributed to the accumulation of dune sand. Additional kinds of evidence have been recognized in some of these formations, which make their eolian genesis reasonably certain, but in others the origin has remained debatable and subject to different interpretations.

For the most part, the eolian-type sandstones of the Western United States, over which controversy regarding genesis has continued, are mixtures of two or more facies in which intertonguing relations exist or one rock type grades into another. Where either fluviatile and eolian facies or marine and eolian facies abut, or overlie one another, the diagnostic features of one environment frequently have been used in interpreting both. Furthermore, the surfaces and deposits of interdune areas that normally occur intercalated between sets of dune sands commonly have not been recognized for what they are and have confused interpretations by their noneolian features.

Major advances in understanding the eolian development of certain sandstone bodies resulted from several basic pioneer investigations. Statistical studies on the orientation of cross-strata and their relations to current (wind) directions were initiated in the United States by Knight (1929) in the Casper Formation of Wyoming and by Reiche (1938) in the Coconino Sandstone of Arizona. Analysis of cross-strata patterns and of cross-strata surfaces of erosion and deposition in migrating dunes of the barchanoid and transverse types were elucidated by Shotton (1937) for rocks referred to as the "Lower Bunter Sandstones" in England. Shotton's study served to explain basic relations between wind direction, availability of sand, and dune structure. Still later, work of Bagnold (1941) on sand movement and dune processes in modern deserts clarified many features of wind activity, greatly stimulating the study of eolian deposits in general and providing a better understanding of their development.

Distribution and Age

SANDSTONES OF PROBABLE eolian origin are widespread in both time and space throughout the world. Cross-strata attributed to deposition by wind have been reported from rocks of the Precambrian, Paleozoic, Mesozoic, and Cenozoic Eras; they are described from North America, South America, Europe, Asia, Africa, and Australia. Not all these assignments to eolian genesis have stood the test of critical analysis, but, in general, more sandstones are interpreted as dune deposits today than were a few decades ago. In brief, the well-sorted fine-grained sandstones with large-scale cross-strata that commonly are attributed to eolian processes are numerous in rocks of various ages, and they constitute a considerable amount of the geologic column.

Eolian-type sandstones are especially numerous and widely distributed throughout the Western and Southwestern United States in rocks of late Paleozoic and early to middle Mesozoic age. They have been extensively studied by many geologists, although genesis remains unsolved in a number of formations. A summary of these sandstones of probable eolian origin has been made for the Colorado Plateau by Poole (1962). He grouped them by age and recognized a total of 15 lithic units in 4 major age categories. Some of these sandstones comprise entire formations; some consist of tongues that extend into other lithologically dissimilar formations; others are merely parts or facies of a formation associated with one or more other lithofacies.

The principal eolian-type formations of the Colorado Plateau listed by Poole are as follows:

Late Jurassic time
    Entrada Sandstone
    Cow Springs Sandstone
Jurassic and Triassic(?) time
    Navajo Sandstone
    Aztec Sandstone
Jurassic(?) and Triassic(?) time
    Nugget Sandstone
Triassic time
    Wingate Sandstone
    Moenkopi Formation
Permian time
    Coconino Sandstone
    White Rim Sandstone Member of the Cutler
        Formation
    De Chelly Sandstone
Permian and Pennsylvanian time
    Weber Sandstone

Of the just-listed sandstones in the Western United States, some clearly represent extensive dune deposits of interior deserts, as evidenced not only by the great lateral extent of the sandstone bodies but also by distinctive features of primary structure and by faunal traces. To illustrate desert dune deposits, the Coconino Sandstone and the Navajo Sandstone are selected for detailed discussion later in this chapter. In contrast, coastal-plain sand accumulations are illustrated by the Lyons Sandstone (Permian) and the Casper Formation (Permian and Pennsylvanian), which contain structures typical of a coastal dune environment and intertongue with marine strata. The De Chelly Sandstone seems to be in a position intermediate between the Coconino and various noneolian depositional environments.

Extensive eolianlike deposits are represented in South America (Brazil) by the Botucatú and Sambaiba Sandstones of Mesozoic age and in Africa by the Cave and time-equivalent sandstones which are also of Mesozoic age. Texturally and structurally these formations resemble sandstones of the Western United states and are described by Bigarella later in this chapter. In Africa also are "feldspathic arenites" of Zambia said to be of Eo-Cambrian [late Precambrian] age and attributed by Garlick (1969, p. 114) to an eolian origin. In Australia the Chatsworth Limestone of Cambrian age, with a thickness of 12 m (36 ft) has been attributed to eolian deposition on the basis of structural features (Shergold and others, 1976, p. 10, 33 – 34).

In northwestern Europe the Rotliegendes of Permian age has generally been interpreted as a desert deposit that includes considerable eolian sand. Because of its petroleum potential, it has been extensively studied in England, the Netherlands, Germany, and elsewhere. In England, the "so-called Lower 'Bunter' Sandstone or Lower Bunter Mottled Sandstone, probably Permian in age" and also the "Yellow Sands of Durham and parts of the Keuper Sandstone of probable Permian and Triassic Age" are believed to be eolian (D. B. Thompson, written commun., 1976). For comparative purposes, these stratigraphic units are briefly described in this chapter.

The Barun Goyot Formation of Late Cretaceous age in central Asia probably represents deposits of a great interior desert. Exceptionally thick and widespread interdune sediments occur between sets of large-scale cross-stratified sandstones considered to be eolian. The interdune sediments probably were accumulated both by intermittent streams and in temporary ponds and lakes. Unique in these deposits — both dune and interdune — is their extensive fauna of vertebrate animals.

The Precambrian Makgabeng Formation of the Waterberg Supergroup (northern Transvaal) is considered (Meinster and Tickell, 1976, p. 191) "of aeolian origin and shows many of the characteristics of transverse dunes."

Characteristics of Eolian Sandstones

ANCIENT SANDSTONES believed to be of eolian origin are mostly light colored (white, buff, yellow, and pale red), highly cross-stratified, even grained, and cliff-forming. The criteria by which their eolian genesis may be determined are of two types: (1) features that are distinctive of modern eolian deposits and therefore may be considered diagnostic, or (2) features that are compatible with an eolian environment but furnish no proof of wind deposition.

The geometry of dune sandstone bodies probably depends largely on the general environment of deposition — whether an inland desert basin or a coastal strip. Some ancient sand bodies that are very extensive, such as the Wingate, Navajo, and Coconino Sandstones of western North America, are either tabular- or wedge-shaped on a huge scale and definitely represent sinking basins. Others, such as the Lyons Sandstone and the Casper Formation, farther east, occur as relatively narrow sheets that apparently bordered strandlines and developed along the margins of regressing seas.

Boundaries of eolian sandstones mostly are abrupt, for these sandstones normally interfinger with, rather than grade into, adjacent facies. Examples are numerous of contacts either with marine deposits or with fluviatile and lacustrine deposits. Contrasts in lithology usually make the contacts easily recognizable.

Most ancient dune-type sandstones are composed almost entirely of quartz grains, and many are classed as orthoquartzitic. Some, like the De Chelly, Lyons, and Casper, contain small percentages of feldspar (much authigenic), and others

contain chert grains. Cross-stratified gypsum sands, believed to be wind deposited, occur locally in the Triassic Moenkopi Formation of Arizona (McKee, 1954, pl. 3B).

Textures in the ancient eolianlike sandstones resemble, in most respects, those in modern dune fields; however, many of the same characteristics are also found in modern beach deposits and some other deposits. The sand is typically very fine to medium, well sorted, rounded, and frosted. In many sandstones, these grains represent reworked material (second or third generation); hence, the textural properties may represent inherited features and bear little relation to length or manner of transport in the present regime. Furthermore, frosting may be the result of postdepositional chemical alteration.

In some ancient eolian sandstones the texture has been greatly modified by quartz overgrowths or by iron oxide coatings on the grains, or both. Commonly, there is intergranular silica, calcite, or iron oxide, but matrix is rare because silt and finer particles forming dust were winnowed out by the wind at the time of deposition. Coarse grains, except those developed as lag on the horizontal interdune deposits, are also lacking. Most lag deposits that have been described, as in the Lyons and Casper Formations, are coarse-grained sandstones, but granules may occur on some interdune surfaces, especially near the margins of a dune field.

Whereas composition reflects the source of a sediment and texture is largely a function of transportation, primary structures furnish the record of manner of deposition. The most common and most distinctive structure of eolian deposits is cross-stratification. Other characteristic structures preserved in some eolian-type sandstones are ripple marks, rain pits, and slump marks. All these are diagnostic features of dune deposits, especially when used as part of a critical evaluation of all available data (Selley, 1970, p. 52).

Cross-strata of dune-type sandstones are characteristically, though not entirely, large to medium scale and high angled. They commonly attain lengths of more than 100 feet (30 m) and dip 20°–30°. These dips are somewhat less than the angle of repose for modern dune-sand slipfaces, which is commonly 33° or 34°, but the difference is attributed to subsequent compaction (Glennie,

1972, p. 1058; Walker and Harms, 1972, p. 280) or to differential preservation with higher, steeper parts removed (Poole, 1962, p. 148). Low-angle cross-strata typical of windward-side deposition are occasionally preserved on modern dunes and have been recognized in ancient sandstones by their dip-direc-

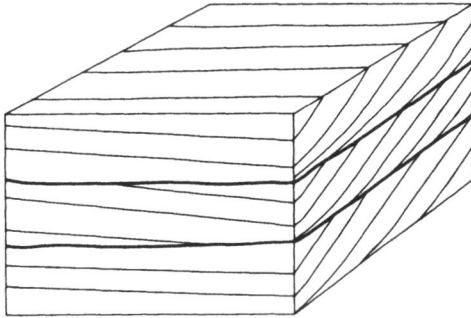

A

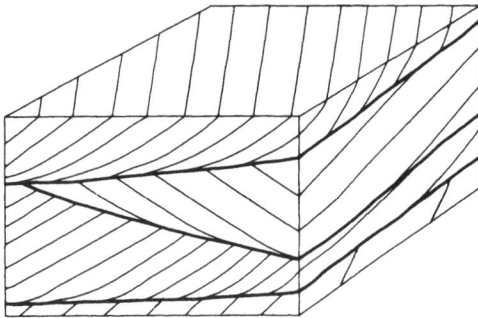

B

C

BASIC FORMS OF CROSS-STRATA recognized in eolian-type sandstones. A, Tabular-planar, abundant; B, wedge-planar, abundant; C, trough type, rare. (Fig. 137.)

tion vector opposite that of the regional current direction. (See description under Coconino Sandstone in this chapter.)

Cross-stratification is mostly of two basic types — tabular planar and wedge planar (fig. 137). In the tabular-planar type, lower bounding surfaces are believed to represent interdune deposits or erosion surfaces; dipping foresets above form tangents at their bases as in most modern barchanoid dunes. In the wedge-planar type, abutting sets of cross-strata probably reflect shifting winds within a single dune sequence. Simple lenticular types are recorded, but the trough type is rare, with the notable exception of the remarkable "festoon structures" in sandstone of the Casper Formation. These structures have not been established unequivocally as eolian but may represent "blowout" dunes.

Ripple marks, not uncommonly preserved in many of the eolian-type sandstones, are distinctive (McKee, 1945, p. 318). They invariably have high ripple indices (above 15, mostly far above), in contrast to the low indices of water-formed ripple marks (Tanner, 1966). In addition, most of them are oriented with their parallel troughs and crests passing up and down the steeply dipping foresets of cross-strata — a feature not recorded for ripple marks under water — and their crests generally are well rounded (fig. 138).

Adhesion ripples, formed where dry sand is blown onto moist surfaces and fixed by surface tension, are recorded (Glennie, 1972, p. 1061) at many horizons in the Rotliegendes of northwest Europe and are viewed as evidence of interdune surfaces.

Raindrop craters with raised rims and reoriented basins formed on the dip slopes of high-angle cross-strata have been recorded from the Coconino Sandstone (McKee, 1934b, p. 102) and the Lyons Sandstone (Walker and Harms, 1972, p. 282). These pits are typical of modern raindrop craters in dry sand and are considered to be good evidence of eolian deposition.

Various forms of avalanche or slump marks are preserved on cross-strata surfaces in several eolian type sandstones (McKee, 1945, p. 321), and contorted strata are locally prominent, although, in general, such structures are surprisingly uncommon. In the Coconino Sandstone the "miniature step" variety seems to be good evidence of slumping at the angle of repose when the sand was wet (Reiche, 1938, p. 918), and other varieties suggest dry sand movement (McKee, Douglass, and Rittenhouse, 1971, p. 368).

RIPPLE MARKS with high ripple indices and rounded crests in slab of eolian-type sandstone, Coconino Sandstone, Kaibab Trail, Grand Canyon National Park, Arizona, U.S.A. (Fig. 138.)

Some of the most conclusive evidence of eolian origin for cross-stratified sandstones is the presence of clearly preserved footprints of four-footed animals on steep dip slopes. Experimental work has shown that tracks with such excellent preservation and with orientation uphill only must have been formed on dry sand at or near the angle of repose. These tracks could not have been formed under water (Peabody, in McKee, 1947, p. 27) or in either wet or damp-crust sand (McKee, 1947, p. 25). Tracks of this type are common and widespread in the Coconino Sandstone and are also present in the De Chelly and Lyons Sandstones.

Selective preservation of certain parts of dune sands doubtless is responsible for absence in the geologic record of many features that occur in modern dunes. Upwind features, for instance, are mostly destroyed by erosion on that side of a dune, and the steeply dipping cross-strata at the top of each set commonly are nearly or entirely removed by planation prior to deposition of the overlying set. Thus, the normally abundant, horizontally oriented ripple marks, as well as many tracks and trails on low-angle surfaces, are very rarely preserved in eolian sandstones.

Vertebrate fossil remains are rarely found in the highly permeable cross-stratified sand of eolian deposits, presumably because conditions for preservation were poor. In the Navajo Sandstone, two articulated skeletons of dinosaurs have been dis-covered, showing clearly that the sand in the area was continental (Brady, 1935; Camp, 1936). Many dinosaur eggs and some dinosaur skeletons have been found in dune-type cross-strata of the Barun Goyot Formation in Mongolia (Gradzinski and Jerzykiewicz, 1974b, p. 140).

Characteristics of Interdune Record

COMPARED TO THE AMOUNT of attention that has been given dune deposits, the record of interdunes has been largely neglected and is as yet only poorly understood. Interdune areas differ greatly from place to place in modern dune fields — ranging from long narrow corridors between barchanoid ridges to small rounded or elliptical depressions in "blowout" areas and from very wide, flat surfaces between linear dunes to the typical radially sloping surfaces within parabolic dunes.

In the geologic record some interdune deposits can be recognized by texture and structure, but many interdunes are recorded merely as erosion surfaces within vertical sequences of dune sands. Whether or not characteristic sediments are accumulated and preserved seems to be at least partly a function of time for detritus to accumulate before being buried by migrating dunes. The process involved in the preservation of interdunes has been described for barchanoid and transverse ridges of the White Sands dune field, New Mexico (McKee and Moiola, 1975).

Some interdune areas within major sand seas develop in an environment controlled by standing bodies of water, either at the surface or as ground water directly below. The term "interdune sabkhas" was applied to such environments by Glennie (1972, p. 1061); he described the deposits that accumulate in them as consisting of "small, irregular, and more or less horizontally bedded adhesion ripples" of sand grains that "range from fine to coarse and generally are poorly sorted." Sediments with similar textures and structures, preserved in the Permian Rotliegendes of northwest Europe, were interpreted by Glennie as interdune deposits. "Steeply dipping eolian sandstones," separated by 9 inches of nearly horizontal adhesion-rippled sandstone are shown (Glennie, 1972, fig. 14) in a drill-core photograph.

Excellent examples of strata believed to have been formed as interdune deposits, without adhesion ripples, are preserved in the type De Chelly Sandstone of Canyon De Chelly, Arizona, where sets of high-angle large-scale cross-strata of eolian type are separated by extensive thin units of horizontal strata composed of poorly sorted red-brown silt and sand (McKee, 1934a).

In many eolian-type sandstones, interdunes are represented by few or no deposits. However, their positions in vertical sections are marked by surfaces of beveling that form the nearly horizontal bounding planes of adjoining cross-strata sets. Many such surfaces are developed in the Coconino, Navajo, and similar sandstones of the Southwestern United States. They are prominent planes and are more irregular than the diagonal erosion surfaces that separate wedge cross sets, probably because they represent greater time spans. In some sandstones, such as the Lyons Sandstone and the Casper Formation, the truncation planes are gently curving, rather than flat, and have coarse grains interpreted as lag scattered on their surfaces. Injection structures in some deposits, like the Lyons Sandstone, have been recorded.

In various ancient eolian-type sandstones, the bounding surfaces overlying large-scale cross-strata commonly preserve the truncated upper traces of those strata. The resulting dune pattern may then etch out on weathering to show on a horizontal plane the character of cross-strata sets. Such structures are observed on many modern interdune surfaces before their burial by fresh sand (chapter E, figs. 45, 70, 71). Likewise, cross-strata sets have been recorded in ancient eolian-type sandstones, including the Lyons Sandstone of Colorado (Walker and Harms, 1972, p. 284) and the Navajo and Coconino Sandstones of Arizona.

Coarse sand or granules formed as lag are introduced in interdune areas by the process of creep resulting from saltation (Bagnold, 1941, p. 102). Probably, they are largely restricted to upwind areas (McKee and Tibbitts, 1964, p. 13) or to interdunes near source materials as along coasts because of the slow particle movement involved. Most examples in ancient dune sandstones containing lag seem to be from coastal dune areas.

Recognition of Dune Types

THE RECOGNITION OF various basic types of dunes in ancient sandstones is possible where certain diagnostic features are well developed and accessible for study. Those differences that result from changes in wind direction, amount of available sand, and the presence or absence of moisture in the sand at the time of deposition are commonly reflected in the form of distinctive structures in the sandstones.

Amount of spread and distribution of cross-strata dip directions are perhaps the most satisfactory features for differentiating types of dunes in ancient rocks. Study of structures in modern analogs has furnished some useful and reliable generalizations. For example, unidirectional winds that are responsible for barchans, barchanoid ridges, and transverse ridges normally have cross-strata dip directions restricted to a single small arc, about one quadrant or less (fig. 139). This type of distribution (A, B) is illustrated by rose diagrams or stereonets for the Navajo Sandstone, Coconino Sandstone, and numerous others in the Colorado Plateau of the Southwestern United States.

In linear dunes, commonly called seif or longitudinal, the distribution of cross-strata dip directions (chapter E, fig. 61) is very different from that in transverse dunes. Even though no general agreement on exact cause of linear dunes has yet been reached, their pattern of cross-strata was predicted (Shotton, 1956, p. 460) and has been recorded in trenches (McKee and Tibbitts, 1964, fig. 6). It is represented on a stereonet by two separate clusters of dip attitudes almost opposite one another (Glennie, 1972, fig. 13b). Ancient eolian sandstones representing linear (seif) dunes have been recognized in the Permian Rotliegendes of the North Sea area by Glennie (1972, p. 1061) and probably occur in many other formations.

A third distinctive pattern represented by the spread in cross-strata dip directions is that of parabolic and other "blowout"-type dunes. These dunes occur wherever vegetation or moisture, or both, tend to stabilize or anchor the sand but are especially characteristic of coastal areas. The feature of the spread in dip directions is its width, encompassing an arc of 180° or more, some being as great as 270°, as shown by Bigarella (chapter E, figs. 78, 79; tables 10, 11). Among ancient rocks, the Casper Formation of Wyoming (Knight, 1929, p. 65, 66; this chapter, fig. 139) and the Lyons Sandstone of Colorado (Walker and Harms, 1972, p. 281) show

the characteristic spread of cross-strata orientation (fig. 139).

Reversing dunes, in which nearly opposite wind directions periodically affect the sand body in alternate direction and so develop two opposed slip-faces, form distinctive patterns of dip directions. Such dunes occur in various modern sand seas (among them, Great Sand Dunes National Monu-

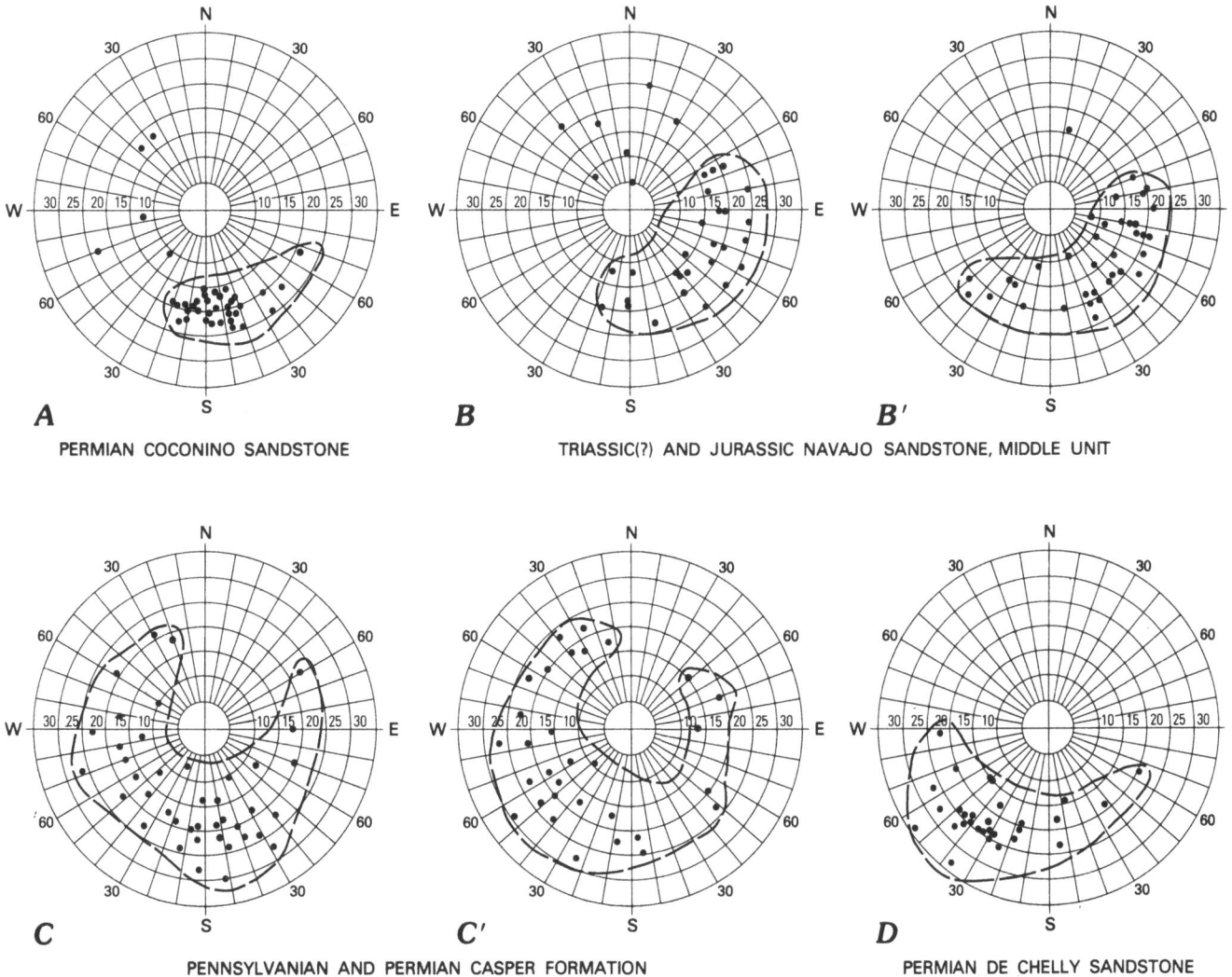

A PERMIAN COCONINO SANDSTONE

B   B'   TRIASSIC(?) AND JURASSIC NAVAJO SANDSTONE, MIDDLE UNIT

C   C'   PENNSYLVANIAN AND PERMIAN CASPER FORMATION

D PERMIAN DE CHELLY SANDSTONE

SELECTED STEREONETS, SHOWING DISTRIBUTION OF CROSS-STRATA dip directions (dots) in typical eolian-type sandstones, Western United States. Azimuth of dot represents direction of dip; distance of dot from center indicates degree of dip. Dashed line enclosing dots represents foreset direction considered the result of downwind-current deposition. A, Bunker Trail, Grand Canyon National Park, Arizona (Reiche, 1938). B, Central Buckhorn Wash, Utah. B', San Rafael River, Utah (Kiersch, 1950). C, Red Buttes area, Wyoming. C', Sand Creek area, Wyoming (adapted from histograms by Knight, 1929). D, White House Trail, Canyon De Chelly, Arizona (Reiche, 1938). (Fig. 139.)

ment, Colorado, U.S.A., and Florinopolis, Brazil) and should be expected in ancient eolian sandstones but as yet have not been recognized or recorded.

Additional structures that may be helpful in determining the type of dune represented by an ancient eolian deposit are the uncommon low-angle dips that sometimes are formed on upwind surfaces of transverse ridge dunes and that contrast with normal high-angle cross-strata of the lee side. These structures were recorded in the Lower Mottled Sandstone by Shotton (1937, fig. 2) and in the Coconino by Reiche (1938, fig. 1, p. 909). Concave downward foresets are distinctive features of many parabolic dunes, formed on the foresets of the dune nose which is vulnerable to undercutting and steepening near its base by crosswinds (McKee, 1966, p. 51; Ahlbrandt, 1973, p. 39).

Other features distinctive of different dune types, discussed elsewhere in this chapter, are the slump marks of wet and dry sand, the contrasts in typical contorted structures, and differences in associated interdune surfaces.

Facies Associated With Dunes

A major difficulty in the recognition of ancient eolian deposits in the geologic record is their intimate association with rocks of other facies and the temptation to use the criteria preserved in one for interpreting the other. Interdune deposits that commonly involve features of an aqueous environment have been discussed, but in addition, all eolian deposits must have on their borders various facies that are fluviatile, lacustrine, or marine and that intertongue with and, under some conditions, grade into them.

Fluviatile deposits in desert areas can generally be categorized as river channel, flood plain, or alluvial fan. The river channels in desert areas commonly are dry streambeds, referred to as arroyos or washes in the Southwestern United States and as wadis in north Africa; their deposits consist primarily of high-energy sediments but may be intermixed with eolian or ponded materials. The flood-plain sediments are mostly overbank deposits and are largely low-energy varieties that contrast with those of the channels. Alluvial fans containing much coarse detritus are common on the margins of most dune areas that are bordered by mountainous terrain.

Deposits of desert lakes and playas, which characteristically are shallow and ephemeral, contain low-energy textures and structures except where they border high-relief areas that contribute alluvial sediments. Most desert lakes and ponds accumulate salt and clay in horizontal beds, various amounts of windblown sand and, in some places, anhydrite or other salts. They typically develop mudcracks and mud curls during dry times when water is absent. Sandstone dikes commonly are associated with these deposits. In inland sebkhas or sinking basins, where the water table rises and periodically reaches the interdune surface, characteristic sedimentary features are adhesion ripples and irregular or horizontal beds, commonly accompanied by crusts of salt or gypsum (Glennie, 1972, p. 1053).

Many dune fields in both humid and arid regions form adjacent to the sea and so are a part of the coastal environment. Whether the wind is offshore or onshore, the beach, bar, or shallow-water marine sands may intertongue with dune sands and provide a complex of environments containing typical dune structures proximate to marine deposits. The Permian Lyons Sandstone of Colorado and the Permian and Pennsylvanian Casper Formation of Wyoming are probable examples of coastal dune deposits that are adjacent to marine sediments.

Selected Examples of Eolian Sandstones

This discussion of ancient eolian sandstones seeks to describe the characteristics of sandstones believed to have been formed by wind action, to indicate those attributes considered evidence of eolian origin, and to show methods of distinguishing between the different basic dune types. Examples of formations thought to have been formed by wind activity are also given. From 30 or more sandstones known to qualify as dunelike types, 12 examples were chosen for analysis.

The sandstones discussed herein (1) are formations that have been studied in detail, (2) represent principal types of deposition and illustrate distinctive features of various dune forms, and (3) are geographically widely distributed. Because the data used came from many different sources and represent various degrees of detailed research, the conclusions presented here differ considerably from area to area in both scope and dependability.

The sandstone formations believed to be examples of interior desert dune-field deposits that were selected for analysis are the Coconino Sandstone of Arizona, U.S.A., of Permian age, and the Lower Mottled Sandstone of England of Permian (?) age; the Navajo Sandstone of Arizona and Utah, U.S.A., the Botucatú Sandstone of Brazil, and the Cave Sandstone of southern Africa, all of Mesozoic age. Sandstones illustrating a coastal dune environment are the Lyons Sandstone of Colorado and the Casper Formation (eolian origin uncertain) of Wyoming, both in the United States and of late Paleozoic age.

The Frodsham Member of the so-called Keuper Sandstone of the Permian-Triassic Cheshire Basin in England, though very limited in extent, illustrates the dome type of dune, and the discussion presents a method of recognizing it through structural characteristics. The De Chelly Sandstone of Arizona typically shows well-developed interdune deposits between sands interpreted as transverse dunes, and it illustrates complex relations along its margins between dune and other continental facies. Finally, evidence from the Rotliegendes of the North Sea area is reviewed because it is believed to include examples of seif dunes as well as transverse types and because typical sebkha structures are recognized in its interdune strata.

Coconino Sandstone (Permian), U.S.A.

THE PERMIAN COCONINO SANDSTONE of northern Arizona is considered by many geologists to represent the deposits of a large sand sea. It consists of large-scale cross-stratified quartz sandstone (fig. 140), flat-bedded interdune deposits are sparse. The general lack of interdunes, with only horizontal bedding planes occurring between sets of cross-strata, suggests an abundant sand source with resulting rapid dune migration and burial of interdune spaces. The large scale of the foresets suggests the growth of high dunes.

The Coconino Sandstone forms a conspicuous cliff in the upper walls of Grand Canyon, along the Mogollon Rim to the south and east, and in many other parts of the region, where it occurs as a prominent white to buff layer, commonly several hundred feet thick. The Coconino is a uniformly fine grained, nearly pure quartz sandstone with siliceous cement, has a massive appearance, and is cross-stratified on a large scale. Cross-strata mostly dip at high angles (fig. 141). Commonly, crossbed sets form large truncated wedges, and numerous individual beds extend as much as 60 or 80 feet (18 or 24 m), curving to form a tangent with the underlying bounding plane.

Although the Coconino characteristically is white, it locally has a brownish cast, as in Marble Canyon, or is bright red, as near Flagstaff. Most of the sloping bedding surfaces are extremely smooth and even, but some contain abundant ripple marks, clearly impressed fossil footprints, and other minor structures. A general uniformity of texture and structure is characteristic.

Distribution and Thickness

The Coconino Sandstone forms a great sedimentary wedge across most of northern Arizona. Along the entire Arizona-Utah boundary, it is either very thin or absent, ranging in thickness from 40 feet (12 m) at Lees Ferry to 60 feet (18 m) at Hacks Canyon, farther west, and 50 feet (15 m) at Grand Gulch mine, near the Nevada State line. To the south 100 miles (160 km) or more, it is more than 500 feet (150 m) thick along the Mogollon Rim of central Arizona and attains a maximum thickness of more than 1,000 feet (300 m) near the town of Pine. The Coconino Sandstone covers about 32,000 square miles (51,500 km²).

Formation Boundaries

The white sandstone of the Coconino rests upon a flat surface of red mudstone of the Hermit Shale across nearly its entire area of deposition. Generally, the contact is sharp and even, but in some places the surface of the red Hermit is trenched with large shrinkage cracks, probably the result of desiccation, which are filled with white sandstone of Coconino type that apparently filtered down from above.

The upper boundary of the Coconino in Grand Canyon sections is marked by sandstone of the Toroweap Formation (Rawson and Turner, 1974). Steeply dipping crossbeds directly below are truncated everywhere to form a flat surface at the contact. Quartz grains probably derived from the Coconino form the basal unit of the Toroweap and suggest that the sand of the Coconino had not yet

PERMIAN COCONINO SANDSTONE, forming cliff on west side of Hermit Basin, Grand Canyon National Park, Arizona, U.S.A. Large-scale tabular-planar cross-strata are prominently displayed. (Fig. 140.)

PERMIAN COCONINO SANDSTONE, showing sets of high-angle cross-strata and tangential bases (arrow) of foresets Dripping Springs Trail, Grand Canyon National Park, Arizona, U.S.A. (Fig. 141.)

consolidated when the formation's top surface was beveled and quartz sand began to cover it.

Texture

The Coconino Sandstone is composed dominantly of well-sorted fine-grained (0.125 – 0.25 mm) sandstone. Nearly all the grains are rounded to subangular, and many are pitted or frosted. The sandstone consists of white quartz with a few scattered feldspar grains and traces of heavy minerals (McKee, 1934b, p. 91). As evident from the pale color, the grains are free of iron oxides in most localities.

Uniformity of grain size and various textural features of the Coconino sand grains have been described in detail by McKee (1934b p. 90 –96). Because sorting and texture are functions of transportation rather than deposition, and, further, may reflect processes of an earlier generation, they cannot be considered to be proof of eolian origin. Grain characteristics of the Coconino are, however, typical in all respects of many modern dune sands.

Primary Structures
Stratification

The most conspicuous primary structure preserved in the Coconino Sandstone is the large-scale cross-stratification. It is composed of long sweeping layers, many of them 30 – 40 feet (9–12 m) long, and some of them as long as 80 feet (24 m). They dip mostly at 25° – 30°, but a few reach a maximum of about 34° (McKee, 1940, p. 823). In general, the dip is southward, but the spread of readings is relatively great, ranging between SW. and SE. (Reiche, 1938, p. 908).

Cross-stratification of the Coconino is uniform throughout wide areas, forming large irregular wedges truncated by overlying layers (fig. 142). Thin laminae form parallel curves, concave upward, that are tangential with underlying bounding planes (McKee, 1934b, p. 98). Contrasting with the common high-angle cross-strata are some low-angle (< 15°) cross-strata that dip in the opposite direction. They occur in limited numbers in a few localities (McKee, 1940, p. 823). Horizontal laminae constitute the basal 3 – 6 feet (1 – 2 m) of the formation at many localities.

A stereographic polar net for representation of crossbedding attitudes was first used by Reiche (1938) in connection with analyses of Coconino cross-strata. His analyses yielded weighted average

WEDGE-PLANAR CROSS-STRATA, a common structure of the Permian Coconino Sandstone, Kaibab Trail, Grand Canyon National Park, Arizona, U.S.A. (Fig. 142.)

directions of dip and expressed numerically measures of the consistency in direction of individual dips. This study was the earliest in America to employ a rigid statistical approach and has been followed by numerous similar investigations made to determine mean current directions within various sedimentary environments. This quantitative and statistical method of study has become an important tool for the analysis of cross-strata.

A distinctive feature of the stratification in both dipping and horizontal strata of the Coconino Sandstone is a characteristic splitting to form flat bedding surfaces. The detachment of one laminae from another in sandstone of such uniform composition suggests either marked differences in grain size or differences in cohesion between individual laminae probably resulting from variations in wetness or dryness of the sand surface at time of burial. Because analyses show few differences in grain size between laminae, the easy detachment along planes of stratification seems to reflect a lack of cohesion in the original deposit (McKee, 1934b, p. 100).

Ripple Marks

Although not abundant throughout the formation, ripple marks are locally numerous and distinctive (fig. 143). They all seem to be of one general form — low, wide, and asymmetrical with ripple indexes (ratio of wavelength to amplitude) greater than 15 (McKee, 1934b, p. 101). Most crests and troughs are oriented parallel to the direction of dip of the foreset slopes.

RIPPLE MARKS ON FORESET SURFACE of cross-strata, Coconino Sandstone, Kaibab Trail, Grand Canyon National Park, Arizona, U.S.A. (Fig. 143.)

A detailed analysis made in 1945 of ripple marks in the Coconino, in which about 50 sets were examined, shows compelling evidence, on the basis of five characteristics, that the ripple marks were wind formed (McKee, 1945, p. 317 – 320). The diagnostic features are as follows:

1. Large ripple index (lowest was 17; most were considerably higher).
2. Ripple crests were definitely rounded in all specimens.
3. A concentration of coarser sand grains occurred on or about the crests in most specimens.
4. All specimens in situ had crests trending up and down surfaces of sloping laminae; none was normal to dip.
5. Crests and troughs were straight and parallel, with little change in direction where exposed for distances of 3 feet or more.

The apparent discrepancy between the abundance of ripple marks on modern dunes and the general scarcity and localized distribution of this structure in the Coconino may be attributed to two factors related to preservation. First, a large majority of ripple marks in modern dunes are formed on the gentle windward slopes of saltation, but in ancient dunes, because of differential preservation, most windward-side deposits have not been preserved. The steep slipfaces of lee sides, where few ripple marks form because deposition is by avalanching, constitute nearly all the sediment preserved in ancient dune sandstones. In barchan dunes or barchanoid ridges, lee-side ripple marks are largely restricted to the sides or wings where crosswinds are effective (McKee, 1934b, p. 101). Second, most ripple marks formed in dry sand are unlikely to be preserved on any surface unless the sand later develops cohesion through moisture received as dew or light rain. These special conditions required for the preservation of ripple marks in dune sand place a strong limitation on the places and times favorable for permanent preservation (McKee, 1945, p. 315 – 316).

Raindrop Impressions

Raindrop impressions in sand form small craterlike pits, some of which have been recognized in the Coconino Sandstone at several localities (McKee, 1934b, p. 102). These pits are especially significant in interpreting genesis, for they give positive evidence of subaerial deposition. They differ in several respects from rain pits in mud, such as have frequently been recorded in the literature.

Rain pits in fine sand form thin, detached layers, or shells, roughly circular but with irregular margins resulting from cohesion of grains by the introduction of moisture. These detached shells of sand, when formed on sloping surfaces like those of the Coconino, have become reoriented with respect to the lamination plane, and each circular pit tends to face upward, or vertically, thus raising the downslope rim (McKee, 1945, p. 323).

Rain pits on sloping surfaces of fine dry sand are readily reproduced in the laboratory; they are very distinctive and are not likely to be confused with craters having other origin. "Only brief showers can leave such a record, for on the surface of a dune that is thoroughly wetted, the shells lose coherence" (McKee, 1945, p. 324, fig. 7). This structure therefore may be considered an additional indication of dry-sand origin.

Slump Marks

Slump marks indicate mass movement or avalanching. In sand dunes they are commonly formed when lee-side slopes near the angle of repose are oversteepened. One form, as described by Bagnold (1937, p. 433), is the result of "successive discontinuous jerks with miniature land-slides." Another (McKee, 1945, p. 321) is a series of variable and irregular lines roughly parallel to the direction of slope and marking the border of the sand mass that has slumped (fig. 144). These and other types of dry-sand slump marks have been recognized on surfaces of the Coconino (McKee, 1945, figs. 5, 6).

SLUMP MARKS ON SLIPFACE of cross-strata, Coconino Sandstone, Grandview Trail, Grand Canyon National Park, Arizona, U.S.A. Scale in inches; 1 inch equals 2.54 cm. (Fig. 144.)

The nature of slump marks and the differences between those formed in dry sand, damp sand, and wetted sand have been studied extensively, both in the laboratory and in the field (McKee, 1945, p. 320; McKee, Douglass, and Rittenhouse, 1971, p. 363–375; McKee and Bigarella, 1972, p. 673–677). Those described from the Coconino Sandstone seem to give further strong evidence of the formation's eolian origin.

Miscellaneous

A minor structure of the Coconino, referred to as "miniature terrace-and-cliff structure" (fig. 145), was said by Reiche (1938, p. 918) to have "ready explanation as a dune feature, but * * * [to be] scarce-

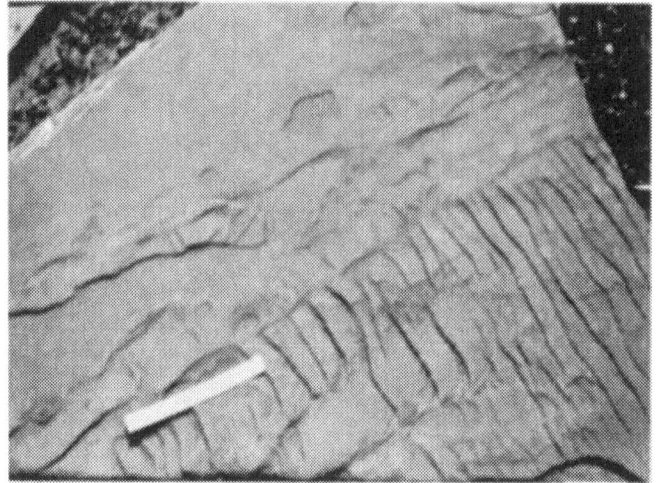

"MINIATURE TERRACE-AND-CLIFF STRUCTURE" (Reiche, 1938, p. 918) or "stair-step fault structure" (McKee, Douglass, and Rittenhouse, 1971, fig. 8b) formed on foreset surface of Coconino Sandstone, Bunker Trail, Grand Canyon National Park, Arizona, U.S.A. White scale is 6 inches (15 cm) long. (Fig. 145.)

ly understandable otherwise." This structure consists of "a succession of abrupt rises or steps of about one-tenth inch each, ascending the slope" (McKee, 1934b, p. 103); otherwise, the surface of the slab is smooth and regular. The genesis of this feature is attributed by Reiche to the local disruption and breaking down, with rapid uphill recession, of successive layers of recently wetted sand on the lee side of a dune. This structure has been noted in the Frodsham Member of the so-called Keuper Sandstone Formation in England (D. B. Thompson, written commun., 1975) and has been reproduced experimentally in the U.S. Geological Survey sedimentary structures laboratory in Denver (McKee, Douglass, and Rittenhouse, 1971, fig. 8b).

Contorted Bedding

Penecontemporaneously deformed cross-strata are not common in the Coconino Sandstone, although they occur in a few places on a large scale. A spectacular example is near Doney Crater, northeast of Flagstaff, Arizona (fig. 146), where dipping foresets many feet in length have been folded locally, but the cross-strata above and below are undisturbed. The sand apparently had some coherence at the time of contortion, but whether the strata were wet or dry is not clear, for this type

CONTORTED BEDDING IN LARGE-SCALE CROSS-STRATA of Coconino Sandstone about 2 miles (3.2 km) northeast of Doney Crater, near Flagstaff, Arizona, U.S.A. (Fig. 146.)

of fold has been developed under both conditions (McKee, Douglass, and Rittenhouse, 1971, p. 372).

Tracks and Trails

The fauna of the Coconino Sandstone consists exclusively of the "footprints of vertebrate animals and the trails of worms and other invertebrates" (McKee, 1934b, p. 106). The invertebrates were not classified by Gilmore (1926, p. 33), but they were compared by Brady (1939, p. 32) with modern arthropods. All the vertebrate tracks were formed by quadrupeds (fig. 147); however, they vary widely in size and character. Fifteen genera and 22 species were named and described by Gilmore (1926, 1927, 1928).

No skeletal parts of the vertebrate animals responsible for Coconino footprints have yet been found. According to Schuchert (1918, p. 350), this lack of organic remains is "probably due to the originally loose and repeatedly reworked sands, a most unfavorable habitat for animals."

The tracks have sharp definition, even margins, and general excellence of preservation. The significance of these features in interpreting the environment of deposition was clearly stated by Matthes (in McKee, 1934b, p. 99), as follows: "the footprints have not the soft, vague outlines which tracks made in loose sand under water commonly have * * * even the deep impressions made by the claws of the animals being well preserved."

FOOTPRINTS OF VERTEBRATE animal, probably reptile, preserved on dip slope (about 30°) of crossbedded Coconino Sandstone. Tracks point uphill. Grand Canyon National Park, Arizona, U.S.A. Photograph by U.S. National Park Service. About one-fourth natural size. (Fig. 147.)

An additional attribute of the Coconino footprints, important to their interpretation, is the fact that these trackways "with only three exceptions out of many hundreds examined, were all leading uphill" (McKee, 1944, p. 62). Explanation of this distinctive feature was obtained through a series of controlled experiments made in 1943 by the writer (McKee, 1944, 1947). On loose sand derived from the Coconino and deposited at high angles (25° – 28°) comparable to that in crossbedding of the Coconino, a chuckwalla lizard was enticed to walk both up and down the surfaces of dry loose sand, damp (sprayed) sand with a crust, and

UPHILL TRACKS OF CHUCKWALLA LIZARD made on foreset slopes at 28° of fine dune sand. A, Dry sand. B, Damp sand with crust. C, Wet sand. Experiments made in laboratory, Museum of Northern Arizona, Flagstaff, Arizona, U.S.A. (Fig. 148.)

wet sand (fig. 148). Only in dry sand, uphill, could footprints be formed similar in size and clarity to those of the Coconino. Damp and wet sands gave very different results, and tracks made downhill were consistently destroyed by avalanching.

Experiments made with modern millipedes lend further support to the belief that the Coconino was deposited as dry sand (McKee, 1944, p. 70). The trails formed by these invertebrates consist of either two or three small furrows between low parallel ridges when made on low-angle slopes (less than 12°) of dry sand (fig. 149). The uphill trails contained two furrows, and the downhill, three. These animals were too light to leave trails in wet or damp sand and were unable to effectively climb slopes steeper than about 20°. The similarity is strong between millipede trails formed under experimental conditions and invertebrate markings in certain low-angle beds of the Coconino.

Environment of Deposition

The Coconino Sandstone is believed to have been deposited largely as dry sand in the form of dunes. Structures are typical of barchan or barchanoid-ridge and transverse types. In some places, the curving shapes of small barchan types are recognizable, but mostly the cross-stratification is on a very large scale with individual lamination planes extending downdip for 50–150 feet (15–30 m) or more. The planes have sweeping curves in their lower parts to form tangents with the underlying bounding surfaces (figs. 141, 142). They clearly are the result of unidirectional winds, as shown by crossbedding vector analyses (fig. 139).

Although most of the minor structures and all the animal tracks and trails probably were formed in dry sand, their permanent preservation required, first, prompt surface dampening by dew or other moisture. Dune surfaces with sand crusts apparently formed in this manner are recorded from many desert areas of today and explain cohesion of surface grains that preserve surface structures (McKee, 1934b, p. 100; Glennie, 1972, p. 1049).

Selective preservation of lee-slope deposits is a second controlling factor in the protection of surface features in dunes. Mostly, the gentle wind-

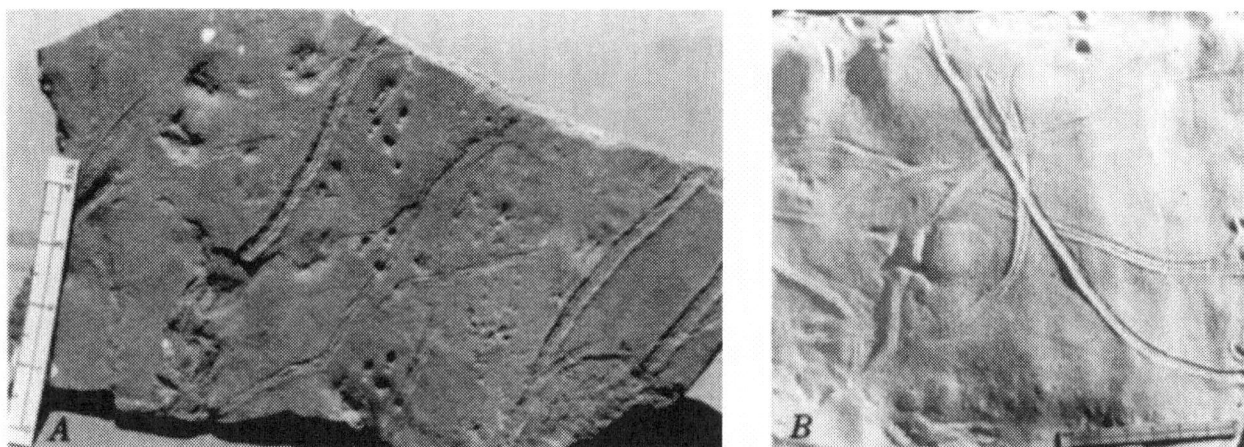

TRAILS OF MILLIPEDES. *A*, In Coconino Sandstone, low-angle foresets near base, Grand Canyon National Park. *B*, In modern dry dune sand on 18° slope. Trails made in experimental trough at Museum of Northern Arizona, Flagstaff, Arizona, U.S.A. Scale in inches; 1 inch equals 2.54 cm. (Fig. 149.)

ward slopes of dunes are continually eroded, but downwind sides are constantly being buried by avalanching sand and protected. Thus, ripple marks and other structures upwind are seldom preserved, yet the relatively rare ripple marks on the slipfaces and the tracks of animals on such slopes have, if buried promptly, a good chance of permanent preservation.

Summary of Criteria for Eolian Origin

The basis for considering the Coconino Sandstone to be of eolian origin involves numerous criteria, some of which are distinctive of an eolian environment and others merely compatible with but not diagnostic of it. No single type of evidence seems entirely conclusive, but, together, the various features present very strong evidence. The principal criteria of dune deposition are as follows:

1. The extent and homogeneity of the sand body.
2. The tabular-planar and wedge-planar type and large scale of cross-stratification. The common high-angle deposits are interpreted as slipfaces on the lee sides of dunes, and the relatively rare low-angle cross-strata that dip toward the opposite quadrant apparently represent deposits of windward slopes.
3. Slump marks of several varieties preserved on the steeply dipping surfaces of lee-side deposits. These are distinctive of dry sand avalanching.

4. Ripple marks which are common on surfaces of high-angle crossbedding suggest eolian deposition both by their high indexes (above 15) and by their orientation with axes parallel to dip slopes.
5. The local preservation of a distinctive type of rain pit. Such pits illustrate the cohesion of sand grains with added moisture and a reorientation of the crater axes with respect to bedding slopes.
6. Successions of miniature rises or steps ascending dip slopes of crossbeds.
7. The preservation in fine sand of reptile footprints and probable millipede trails with sharp definition and clear impression.
8. The consistent orientation of reptilian tracks up (not down) the steep foreset slopes.

De Chelly Sandstone (Permian), U.S.A.

THE DE CHELLY SANDSTONE of northeastern Arizona is generally considered to be the lateral equivalent of the Coconino Sandstone to the west (McKee, 1934a, p. 231; Reiche, 1938, p. 924; Peirce, 1963, p. 137), although Baars (1962) showed it as older in his correlations. Like the Coconino, the De Chelly is believed to be an eolian deposit, at least in part, but is much more complex and includes several distinctive noneolian members and facies.

Contrasts in sedimentary structures, grain-size distribution, and composition distinguish these other facies and indicate various noneolian depositional environments.

Although the contrasting facies just referred to have been recognized as discrete parts of the De Chelly by most geologists (among them, Read, 1951; Allen and Balk, 1954; Peirce, 1963) who have studied them, the following discussion concentrates on the two facies that are considered to be eolian. Facies that occupy the eastern margin of the formation almost certainly represent various subaqueous environments. Farther west in the type locality at Canyon De Chelly, most of the sandstone consists of alternating units believed to be dune and interdune deposits. To the south and the northwest, pure sandstone considered to be eolian closely resembles the Coconino and may indeed be an eastern extension of the main body of that formation.

Distribution and Thickness

Rocks commonly assigned to the De Chelly Sandstone form a relatively narrow belt extending about 125 miles (200 km) from north to south in northeastern Arizona. The formation is about 800 feet (240 m) thick at its type locality at Canyon De Chelly but is progressively thinner both north and south. At Bonito Canyon, Hunter's Point, and Oak Springs Cliffs, between 25 and 40 miles (40 and 65 km) to the southeast, its thickness ranges from 640 feet (200 m) to less than 500 feet (150 m). It thins to approximately 250 feet (75 m) at the south end of the Defiance Plateau (Peirce, 1963, p. 16). Northward from Canyon De Chelly, it terminates in the subsurface, probably somewhere near the Arizona-Utah border.

The eastern margin of the De Chelly is obscure both because of facies complications and because of the lack of surface exposures. In a westward direction, the De Chelly is represented by a facies with large-scale cross-strata essentially identical to those of the Coconino.

A facies consisting of alternating sets of horizontal beds and of large-scale cross-strata is, so far as is known, limited to the upper, or major, cliff member (570 ft (175 m) thick) of the De Chelly in Canyon De Chelly and its environs. The area of outcrop is about 20 miles (30 km) square, but, as suggested by Peirce (1963, p. 59), this facies may be represented in neighboring localities, such as Nazlini and Bonito Canyons, although largely lacking the characteristic sets of horizontal beds. If all areas mentioned are included in this facies, it has a north-south extent of 60 miles (95 km) and definitely merges into normal Coconino-like deposits across a wide belt.

Composition

Compositionally, the De Chelly Sandstone ranges from nearly pure quartz sandstone or orthoquartzite to beds containing enough feldspar to be classified as subarkoses. Specks of what seem to be kaolin are common throughout parts of the formation and may have been developed by the breakdown of feldspars. They are characteristic of the crossbedded sandstones in Canyon De Chelly and canyons to the south (McKee, 1934a, p. 224), and in subsurface samples from the Four Corners area to the north (McKee, Oriel, and others, 1967).

Most of the quartz grains are white, but sparse red quartz grains characteristically are scattered through much of the formation (Gregory, 1917, p. 33; McKee, 1934a, p. 225). They probably result from ferric material that coats the grains or fills surface irregularities. Thin-section study revealed that they have developed partly before and partly after quartz overgrowths (Peirce, 1963, p. 115).

In the thin flat-bedded silty sandstones of probably subaqueous or interdune origin, mica is widespread, but it seems to be rare or absent in the large-scale cross-stratified sandstones. Heavy minerals are rare in all these strata.

Texture

Analyses of 125 samples showed the De Chelly (Peirce, 1963, p. 93) to consist principally of fine sand with a modal size between 0.125 and 0.177 mm, a median diameter of 0.164 mm, and a sorting coefficient (Trask) of 1.25 mm. Therefore, the formation as a whole is considered to be a rather uniform "well-sorted fine-grained sandstone."

Although the size analyses of the De Chelly, as averaged for all its sandstone facies, reveal a general overall similarity, perhaps more significant are comparisons made by Peirce (1963, p. 94) of grains from each of the various structural types (facies) of sandstone. In the upper, or main, sandstone member in Canyon De Chelly, sand grains of the large-scale cross-strata that are believed to be dune sands "average 0.170 mm in median diameter, [which is] slightly coarser than those of the formation as a whole" (Peirce, 1963, fig. 8, D). Horizontally stratified silty sandstones, here interpreted as interdune deposits, that form thin zones repeated throughout this member, have a wider range in grain size, with more coarse grains, as well as a higher silt-clay content (Peirce, 1963, figs. 10B, 10E).

A comparison of sand from the large-scale (eolian) cross-strata of the upper member at Canyon De Chelly with sand from the corresponding unit at six localities to the south and southeast shows similar size properties in all sections (Peirce, 1963, table 2), as follows:

Coarse sand:	Very little or none.
Silt and clay:	Less than 10 percent.
Median grain size (mm):	0.149 – 0.186.
Sorting index (Trask):	1.19 – 1.32.

Comparing the eolian-type sandstone of the upper member with sandstones that Peirce (1963, table 2) considered of aqueous origin in the lower member at Bonito Canyon to the southeast discloses the contrast shown in the following unnumbered table.

	Coarse sand (percent)	Silt-clay (percent)	Median grain size (mm)	Sorting index (Trask)
Canyon De Chelly	Trace	5	0.170	1.21
Bonito Canyon	0.4	9	.126	1.50

Sand grains in the De Chelly, regardless of facies, are nearly all rounded or subrounded. Commonly, they are pitted or grooved, and many are frosted. This frosting, however, is caused by "the light scattering effect of myriads of minute quartz crystals, or incipient silicification" (Peirce, 1963, p. 98).

Primary Structures
Stratification

The upper unit or member of the De Chelly Sandstone in Canyon De Chelly contains two alternating and contrasting types of stratification that together distinguish it from all other facies in the formation. These cross-strata consist of (1) a large-scale high-angle type, interpreted as eolian, and (2) a thin, virtually horizontal type believed to represent interdune deposits. The large-scale cross-strata constitute a major part of the total deposit and occur in thick layers, in terms of feet or metres, whereas the horizontal strata form thin sets, a few of which are more than 3 feet (more than 1 m) thick and which laterally may be represented by bedding (parting) planes. These units of alternating crossbeds and horizontal beds were noted by McKee (1934a, p. 222, 224) and described in detail by Peirce (1963, p. 51 – 67).

The large-scale cross-stratified sandstones that form the main cliff of De Chelly Sandstone in Canyon De Chelly, similar in structure to the Coconino Sandstone and to the Coconino-like sandstones near Ganado to the south and in Monument Valley to the northwest, have been considered extensions (facies) of the De Chelly (Gregory, 1917; Baker, 1936). The large-scale cross-stratified sandstones form two principal structural types — tabular planar and wedge planar. In the tabular-planar type, bounding planes both above and below the foresets are nearly horizontal and truncate the underlying crossbeds (fig. 150).

In the wedge-planar type, oblique erosion surfaces separate sets of cross-strata. The tabular-planar type is interpreted as representing an interdune surface on which no deposits accumulated (McKee and Moiola, 1975); the wedge-planar type is developed within a migrating dune of barchanoid or transverse type and is common in the dunes at White Sands National Monument, New Mexico (McKee, 1966, p. 59).

The thin horizontal silty sandstone beds that characteristically alternate with large cross-strata rest on flat erosion surfaces that truncate cross-strata sets (fig. 151). Followed laterally, some of the horizontal beds are seen to thin and wedge out, but the underlying erosion surface continues as a bounding plane between sets of cross-strata above and below. A detailed description of these structures (Peirce, 1963, p. 56 – 59) suggests that although the flat-bedded silty deposits occupy only restricted areas on truncation planes, they occur as sedimentary structural features throughout an area of many square miles.

The distribution, composition, and structure of the thin lensing flat beds that occur between large-scale cross strata, strongly resemble the interdune deposits now developing in the dune field at White

LARGE-SCALE AND MEDIUM-SCALE CROSS-STRATA of tabular-planar type in Permian De Chelly Sandstone, Canyon De Chelly, Arizona, U.S.A. (Fig. 150.)

Sands National Monument, New Mexico (McKee and Moiola, 1975). If the White Sands analog is valid, the thickness of a flat-bedded deposit (interdune) may be considered a function of the time of exposure before the next dune migrates over and buries it. Likewise, the horizontal bedding planes lacking deposits are the record of surfaces that were buried rapidly by advancing dunes.

Ripple Marks

Ripple marks in several parts of the Monument Valley area are preserved on cross-stratified sandstones assigned to the De Chelly (Peirce, 1963, pls. 25, 26). They are described as "having axes aligned up and down dip" and are therefore considered to be wind-formed. They resemble closely the abundant ripple marks described from the Coconino Sandstone of Grand Canyon and give further indication that the sandstone containing

them constitutes an eastern extension of the Coconino or at least a recurrence of its eolian facies.

Tracks and Trails

Footprints of several species of vertebrate animals and trails of invertebrates, similar to forms represented in the Coconino Sandstone of Grand Canyon, occur in the large-scale cross-stratified sandstone member of the De Chelly at Nazlini Canyon (McKee, 1934a, p. 227) and at Kinlichee (McKee, 1934a, p. 232) south of Canyon De Chelly. They are also reported from western Monument Valley in crossbedded sandstone assigned to the De Chelly (Peirce, 1963, p. 26). All these tracks and trails lead up the steep dip slopes of the cross-strata, as do those in the Grand Canyon.

The significance of these footprints in the history of the sandstone is threefold. First, they suggest a direct correlation between the Coconino Sandstone

HORIZONTALLY BEDDED SILTY SANDSTONE, interdune deposit (arrow) between cross-strata sets of eolian dune type. Permian De Chelly Sandstone, Canyon De Chelly, Arizona, U.S.A. (Fig. 151.)

and the De Chelly Sandstone and furnish evidence of similar environments of deposition. Second, the sharpness of the imprints and the general manner of preservation indicate that the imprints were formed in dry sand. Third, their preferred orientation, up the bedding slope only (never down), is evidence of having been formed in dune sand near the angle of repose. (See the preceding discussion of the Coconino Sandstone.)

Environment of Deposition

Of the three members that form the De Chelly Sandstone in its type locality, only the upper is considered to be dominantly of eolian origin. It forms the major cliff in the walls of Canyon De Chelly where it is nearly 600 feet (180 m) thick. The large-scale cross-strata in this unit show a persistent southwestward dip and are interpreted as indicating a prevailing wind that caused dunes to migrate

in that direction. Between cross-bedding sets, in many places, are thin sets of horizontal beds believed to have developed as interdune deposits.

The other two members in Canyon De Chelly are probably of subaqueous origin. The thin (45 ft, 14m) middle member, composed of micaceous siltstone and silty sandstone, is considered to represent a relatively low-energy deposit and may have formed on the margin of saline basins to the south. The lower member, formed of cross-stratified sand, is believed by Peirce (1963, p. 145) to be "a high energy fluvial" deposit accumulated on "an aggrading delta."

Two additional members (or facies) in the De Chelly Sandstone occur in the eastern part of the Defiance uplift (Peirce, 1963). These members underlie the cliff-forming upper member and are considered by Peirce (1963, p 148—149) to be low-energy lagoon or tidal flat deposits overlying "nearshore marine deposits," together with some cross-stratified sandstone, possibly eolian.

The foregoing interpretations suggest that in the western part of the region the De Chelly Sandstone represents stream deposits followed by extensive dune deposits but that, in a parallel belt to the east, near-shore marine deposits preceded the development of eolian strata.

Summary of Criteria for Eolian Origin

Those facies of the De Chelly Sandstone believed to be composed of windblown deposits are of two principal types. One facies resembles the Coconino in its uniformity of grain size and structure and in large-scale high-angle cross-stratification. Characteristics of the cross-strata indicate deposition as barchans or barchanoid ridges. The orientation of ripple marks and the state of preservation of reptilian tracks are further evidence of eolian origin; these features are common to both the Coconino and the western part of the De Chelly.

The second facies of probable eolian origin in the De Chelly consists of an alternation of (1) large-scale high-angle cross-strata forming tabular-planar and wedge-planar sets and (2) very thin sets of horizontally bedded silty sandstone. From the standpoint of both texture and structure, these two types of sandstone compare well with the alternating dune and interdune deposits of a modern dune field developed under conditions of a prevailing wind.

Navajo Sandstone (Triassic? and Jurassic), U.S.A.

THE NAVAJO SANDSTONE has been selected to illustrate eolian-type sandstone because of its broad geographic extent and great thickness and because of the large-scale and excellent development of its primary structures, permitting it to be analyzed across a wide area and observed in relation to other facies. Preservation in the Navajo Sandstone of small dinosaurs and other forms of continental life makes it unlikely that this sandstone represents accumulations of sand waves in deep marine waters of the North Sea type, at least in areas where those animals are present. Various features of associated facies indicate development in a regional environment of aridity. In brief, considerable evidence is available to suggest that the Navajo represents an extensive sand sea of a great interior desert, mostly or entirely Lower Jurassic (Lias) time.

Obviously, making generalizations concerning the character and genesis of this huge sand body is difficult. Observations made in one area may not apply entirely to another area, yet many of the distinctive features are common to most parts. The local exceptions, doubtless related to subenvironments, have prompted various different interpretations. The really remarkable feature of the Navajo is the high degree of uniformity of the sandstone and its general consistency throughout the region.

Distribution and Thickness

The Navajo Sandstone as currently recognized (McKee, Oriel, Swanson, and others, 1956, table 1) extends across much of Utah, northeastern Arizona, and southwestern Colorado. It is correlated with, and almost certainly is a continuation of, the Aztec Sandstone (Triassic? and Jurassic) to the west in southern Nevada and California, and of the Nugget Sandstone (Triassic? and Jurassic?) to the north and east in northern Utah, southwestern Wyoming, southeastern Idaho, and northwestern Colorado (Baker and others, 1936). As shown by isopachs in figure 152, this great cross-stratified sandstone body extends approximately 600 miles (965 km) from north to south and at least 250 miles (400 km) from east to west. These measurements approximate the present extent of the sandstone; the original sand sea doubtless extended much farther but has been removed by Cenozoic erosion, as in northwestern Arizona and eastern Nevada.

The Navajo Sandstone is more than 1,000 feet (300 m) thick in parts of northeastern Arizona and more than 2,000 feet (600 m) thick in southwestern Utah (fig. 152); its correlative, the Aztec Sandstone, is 3,000 feet (900 m) thick in the Mohave Desert of California. Along its eastern and southeastern margins in Colorado and northern Arizona the Navajo forms a wedge and thins to a vanishing point; elsewhere its margins are less definite.

Along the southern margins of the Navajo Sandstone, prominent tongues of large-scale cross-stratified sandstone extend southward and south-

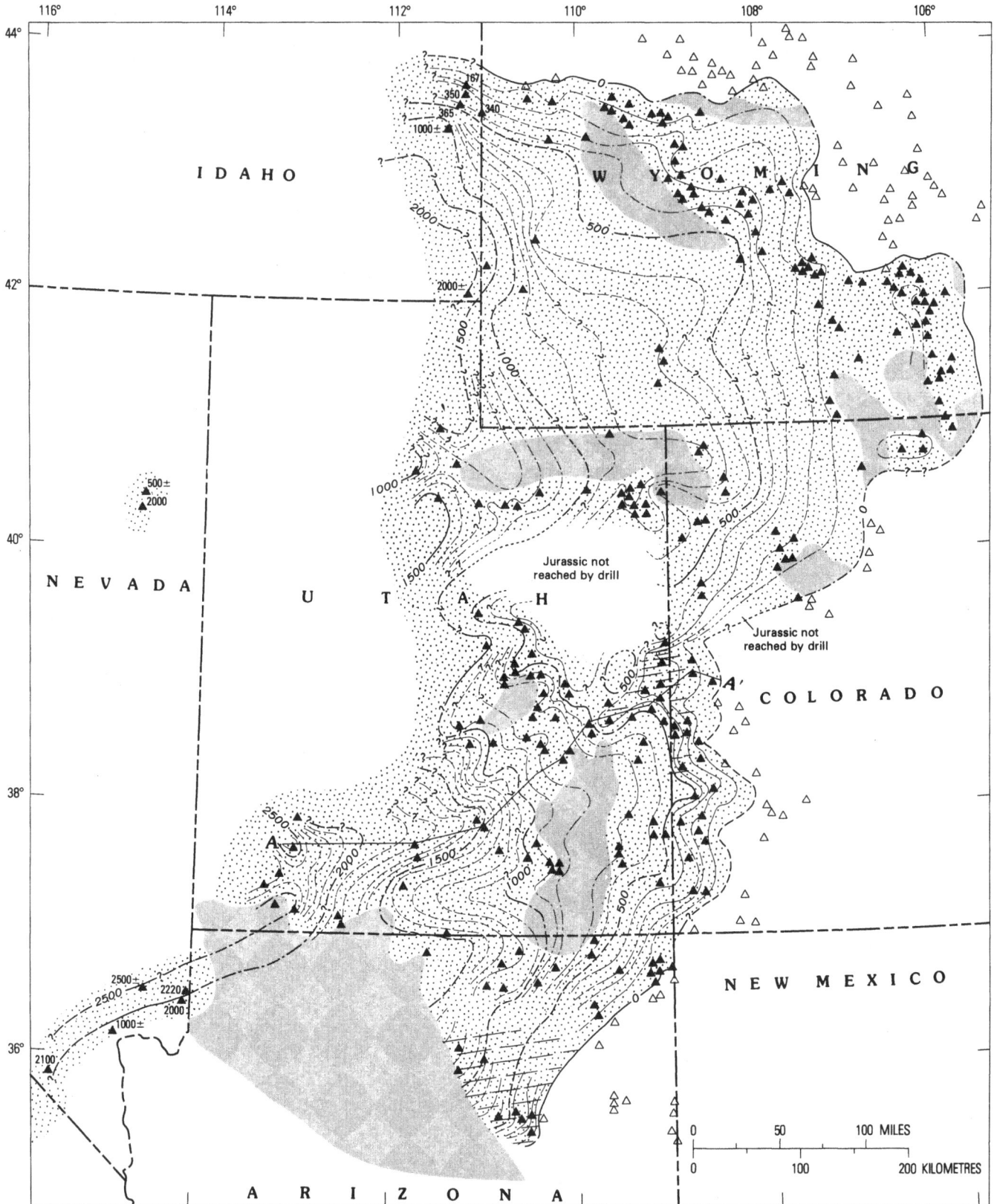

ISOPACH-LITHOFACIES MAP AND CROSS-SECTION showing distribution and thickness of Triassic(?) and Jurassic Navajo Sandstone, Triassic(?) Kayenta Formation, and correlative units. (Modified from McKee, Oriel, and others, 1956, pl. 4). 100 feet equals 30.5 metres. (Fig. 152.)

Navajo, Kayenta, and correlative sandstones
Dashes indicate areas of some shale

Area where Navajo, Kayenta, and correlative sandstones absent within the region of their distribution

▲2000

Outcrop or well where Navajo, Kayenta, and correlative sandstones are present
Thickness, in feet

△

Outcrop or well where Navajo, Kayenta, and correlative sandstones are absent
Older rocks are overlain by younger Jurassic, Cretaceous, or (locally) Tertiary rocks

————————500————————

Isopach in areas of good control
Thickness, in feet

— — — — — 500 — ? — · — · · · · ·

Isopach in areas of poor control
Thickness, in feet; queried where especially questionable. Dotted where inferred

— · — · — · — 500 — · — · — · —

Isopach showing inferred former thickness of Navajo, Kayenta, and correlative sandstones where they are known to be eroded

Thickness, in feet

SECTION OF NAVAJO, KAYENTA, AND CORRELATIVE
SANDSTONES ALONG LINE A-A'

(Fig. 152. — Continued.)

westward into the dominantly flat-bedded sandstones, siltstones, and limestones of the Kayenta Formation (Harshbarger and others, 1957, p. 22; R. F. Wilson, 1958, p. 116). Some of these tongues have appreciable thickness and have been named by workers in various parts of the region — for example, the Triassic (?) Lamb Point (R. F. Wilson, 1958) and Shurtz Sandstone Tongues of the Navajo (Averitt and others, 1955). A correlation of named tongues of the Navajo is included in table 1 of Hatchell (1967, p. 25).

The Kayenta Formation consists of sedimentary rocks believed to be fluvial and probably represents the deposits of southwestward-flowing streams. Thus, the tongues of the Navajo are interpreted (Harshbarger and others, 1957, p. 1) as windblown sands from the north or northwest that "gradually overwhelmed" the streambeds.

Composition and Texture

Studies in several widely separated areas indicate that nearly everywhere the Navajo Sandstone consists largely of fine-grained quartz sand, with very minor amounts of feldspar and a few other accessory minerals. It apparently is homogeneous and well sorted throughout, and in central Utah the Navajo is described as "monotonously uniform" in texture (Kiersch, 1950, p. 927). The quartz particles have been reported as medium- to fine-grained with an average diameter of 0.21 mm in northeastern Arizona (Harshbarger and others, 1957, p. 19) but as fine-grained to very fine grained near Navajo Mountain on the Utah–Arizona border (Hatchell, 1967, p. 30). Sand grains are uniformly about 0.2 mm in diameter (fine grained) in the northern part of the San Rafael Swell area, Utah (Kiersch, 1950, p. 927).

In the Navajo Sandstone, most quartz grains are rounded or subrounded, and relatively few particles are subangular or angular (Kiersch, 1950, p. 927; Harshbarger and others, 1957, p. 22; Hatchell, 1967, p. 30). Among samples from the northeastern Arizona area examined by Harshbarger, Repenning, and Irwin, about 25 percent were determined to be rounded. Frosting has been noted on quartz grains in many areas, and angular overgrowths of silica, forming orthoquartzite, are reported (Kiersch, 1950, p. 927). Cement is mainly silica but locally iron oxide or calcite (Hatchell, 1967, p. 30; Kiersch, 1950, p. 927).

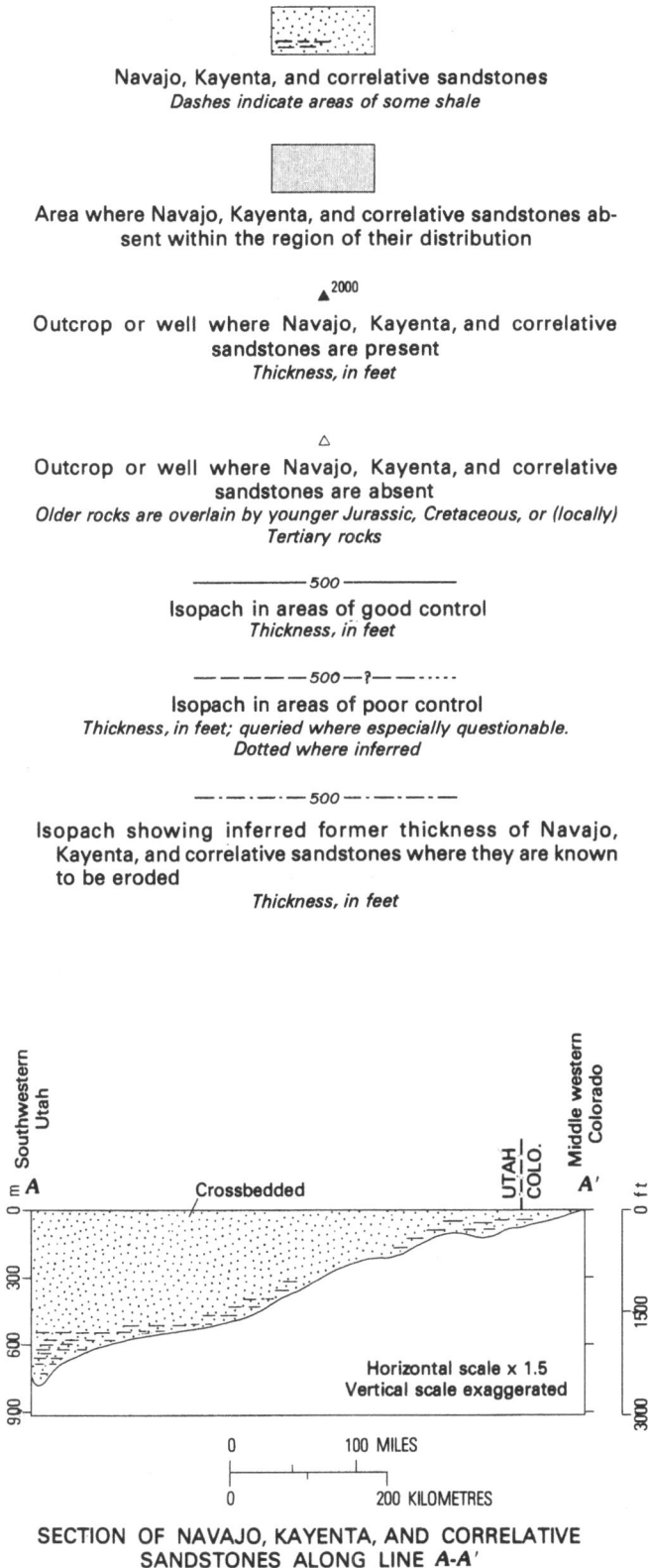

Horizontally interbedded units locally among the massive cosets of sandstone throughout the region are thin layers and lenses of aphanitic limestone and siltstone (Baker and others, 1936, p. 52; Harshbarger and others, 1957, p. 19), that probably represent interdune deposits. Wind-faceted pebbles, ranging in diameter up to more than 2 inches (5 cm), locally occur between units of cross-stratified sandstone in the San Rafael Swell and Green River Desert areas of east-central Utah (Baker and others, 1936, pls. 13, 14). Small pebbles and coarse sand grains of quartz and chalcedony occur along some planes of cross-strata in Buckhorn Wash in east-central Utah (Kiersch, 1950, p. 927).

Primary Structures
Stratification

Primary structures of the Navajo Sandstone of southern Utah consist "of large wedge-shaped planar sets of large-scale, high-angle, tangential cross-strata. The bounding surfaces of individual sets of cross-strata are planar sets that dip gently in a direction opposite to that of the steeply dipping cross-strata" (R. F. Wilson, 1958, p. 112, 141, 145). The formation contains both tabular-planar and wedge-planar cross-strata. These structures are characteristic of the formation in the Zion Canyon area and in the Kanab area farther east (fig. 153).

In the Navajo of northeastern Arizona, a dominance of large-scale wedge-planar and tabular-planar cross-strata has been recorded both by Harshbarger, Repenning, and Irwin (1957, p. 19) and by Hatchell (1967, p. 30). Simple wedge structures (without beveled tops) have also been recorded there by Hatchell and in east-central Utah by Kiersch (1950, p. 930). An absence of trough-type cross-strata was noted by R. F. Wilson (1958, p. 147) and by Hatchell (1967, p. 32), although, according to Wilson, trough structures — some of them large scale — are common in sandstones of the underlying Kayenta Formation of probable fluviatile origin.

In southwestern Utah much of the Navajo cross-stratification is large scale, commonly exceeding 50 feet (15 m) in length and forming sweeping arcs, concave upward. The thickness of sets in east-central Utah is as much as 15 feet (4.5 m) (Kiersch, 1950, p. 930) and in the Navajo Mountain area of Utah and Arizona is up to 100 feet (33 m) (Hatchell,

1967, p. 30). The prevalence of high-angle dips, greater than 20°, in most areas, is noted by many geologists, among them R. F. Wilson, (1958, p. 148, table 6).

Ripple Marks

Ripple marks, though uncommon in the Navajo Sandstone, occur locally in widely scattered parts of the formation. They have been recorded from the east-central Utah area by Kiersch (1950, p. 935) and from the Arizona–Utah border by Hatchell (1967, p. 31), both of whom considered them to be wind-formed types because of orientation with crests parallel to the dip slopes of crossbeds. Ripple marks have also been reported from southwestern Utah and classified as eolian on the basis of ripple indices (Tanner, 1964, p. 432).

CROSS-STRATA IN NAVAJO SANDSTONE, north of Kanab, Utah, U.S.A. A, Wedge-planar type. B, Tabular-planar type. Photographs by E. Tad Nichols. (Fig. 153.)

The relative scarcity of ripple marks in the Navajo, as compared with their great abundance in most modern dune fields, is probably a function of selective preservation, as in the Coconino (McKee, 1934b, p. 101). Only the relatively few ripple marks that form from crosswinds on the lee-side slipface have a chance of burial and preservation. Furthermore, even though wind ripples form in dry sand, permanent preservation requires some degree of cohesion, usually contributed by dew or light rain, a requirement further limiting the chances of ripple-mark survival (McKee, 1945, p. 324).

Contorted Bedding

Zones of contorted bedding in which irregularly folded strata occur between sets of undisturbed strata are fairly numerous in various parts of the Navajo Sandstone (fig. 154). They have been refer-red to by most authors as slump structures. In east-central Utah contorted zones have been described by Kiersch (1950, p. 941) as 35–50 feet (10–15 m) thick, occurring at many widely scattered localities in the upper middle part of the formation. In southwestern Utah, zones of contorted strata (R. F. Wilson, 1958, p. 142) range in thickness from a few feet to 20 feet (6 m) and extend laterally as much as 100 feet (30 m). In northeastern Arizona, they occur in foresets "in many places" (Harshbarger and others, 1957; Hatchell, 1967, p. 31).

Irregular folds, some of them recumbent, in the contorted zones of the Navajo have generally been attributed to oversteepening on the lee sides of dunes or to the weight of overlying sediment (R. F. Wilson, 1958, p. 142). These processes can be distinguished by the type of fold that has developed (McKee, Reynolds, and Baker, 1962a); however, the environmental implications which are recognized

CONTORTED STRUCTURES IN NAVAJO SANDSTONE, Kanab Creek area, Utah, U.S.A. Photograph by E. Tad Nichols, (Fig. 154.)

by the degree of sharpness or "fuzziness" in resulting fold structures are more important. These features are largely a function of the cohesiveness of the sand at the time of deformation and serve to differentiate between saturated, wet, damp, and dry sand (McKee, Douglass, and Rittenhouse, 1971, table 1). Folds resembling many in the Navajo Sandstone are recorded in modern dune deposits, especially where the sand has been wetted (McKee and Bigarella, 1972, p. 674).

Interdune Deposits

Although the term "interdune" has not been mentioned in any of the many publications dealing with the Navajo Sandstone, probably many of the references to thin horizontal beds, siltstone or limestone lenses, and other distinctive deposits among the massive units of cross-stratified sandstones actually refer to interdune deposits. If so, then interdune deposits are well represented in the Navajo by a variety of sedimentary bodies having in common thinness, limited lateral extent, and either horizontal or uneven stratification.

In the Lamb Point Tongue of the Navajo in central southern Utah are "scattered thin lenses of flat-bedded to laminated, pale reddish brown siltstone and claystone" (R. F. Wilson, 1958, p. 141). The lenses are not more than 2 or 3 feet (0.5–1 m) thick and they pinch out along strike. Farther west in Zion Canyon, Utah, R. F. Wilson (1958, p. 150) referred to Navajo cross-strata that alternate with flat-lying beds or with beds that have low-angle dips. The claystone and siltstone lenticular bodies probably formed as interdunes, according to modern analogs (McKee and Moiola, 1975).

Locally, Navajo interbeds include extensive lenses of thin-bedded limestone, presumably formed in ephemeral ponds. Among the sandstones of east-central Utah are "limy and argillaceous beds" (Kiersch, 1950, p. 927). In northeastern Arizona, many lenticular beds of cherty limestone occur within the Navajo (Harshbarger and others, 1957, p. 19), and thin discontinuous beds of gray limestone, calcareous siltstone, sandstone, and limestone conglomerate form platy or flaggy units between sandstones (Hatchell, 1967, p. 2).

Numerous distinctive structures and other features in the Navajo of eastern Utah probably are attributable (Baker and others, 1936, p. 12, 52) to in-

terdune deposits: "sand-filled desiccation cracks in limestone," "a bed of nodular-weathering silty red sandstone," "The occurrence of excellent dreikanter," and "local thin lenses of dense unfossiliferous gray limestone from a few feet to several miles in diameter" interpreted as the product of "ephemeral water-filled basins" and "formed through algal or purely chemical actions." The limestones contain mud cracks.

Fauna

Fossils are uncommon in the Navajo Sandstone, the total recorded forms consisting of two small bipedal dinosaurs (Brady, 1935; Camp, 1936), some dinosaur tracks, a few ostracodes, and some other crustaceans (Harshbarger and others, 1957, p. 1). These animals probably inhabited interdune areas, as evidenced by the dinosaur tracks near Navajo Canyon, Arizona (Gregory, 1917), that are preserved in thin lenses of limestone probably precipitated in ponds by chemical processes. The ostracodes likewise are from limestone lenses (Harshbarger and others, 1957) and presumably lived in shallow ephemeral water bodies between dunes.

Probably more important as evidence of the sandstone genesis than the animals themselves is their manner of preservation. For example, the skeleton of the small dinosaur recorded by Camp (1936, p. 39) from northern Arizona was buried on a surface parallel to the dipping "planes of crossbedding of the Navajo sandstone at that point" with the axis of the body horizontal and the right hind foot pressed into the sand. The creature seems to have been covered "by shifting sand before death, or soon after" (Camp, 1936, p. 41).

The uniform sand that constitutes the Navajo, Nugget, and Aztec Sandstones across a vast area is generally interpreted as an eolian accumulation with a source to the north or northwest (Harshbarger and others, 1957, p. 3). Studies of wind directions as determined from cross-strata dips in these sandstones have been made by numerous geologists in many parts of the region (for example, Poole, 1962, fig. 163.1). Cross-strata orientations, indicated by arrows on Poole's map, show dominant north and northeast winds for the Nugget and Navajo Sandstones in northeastern Utah and northwestern Colorado and for the Aztec

Sandstone in southern Nevada, but show northwest winds for the Navajo Sandstone in most of southeastern Utah and adjacent Arizona. "The consistency of cross-strata dip bearings in a southerly direction suggests [for that period] a broad belt of strong and persistent northerly winds * * *." (Poole, 1962, p. D150).

The sources of the sand have been postulated by various geologists; but no one source sufficiently great to have furnished the entire volume represented by these sandstones can readily be visualized. A major source in western Nevada where mountain uplift probably occurred at that time is suggested by conglomerates in the adjacent region (Harshbarger and others, 1957, p. 3). In northeastern Arizona, where sands of the Navajo apparently advanced eastward and eventually covered the fluvial Kayenta Formation, "Kayenta streams [are believed to have been] the primary source of material for the Navajo dunes * * *." (Harshbarger and others, 1957, p. 25). A source far to the north, certainly for the Nugget sandstone, also has been suggested.

The depositional environment of the Navajo and its correlative sandstones has generally been recognized as a great interior desert basin that sank gradually as the several thousand feet of sand accumulated. Aridity is evident from such features as the wind-faceted pebbles in east-central Utah and desiccation cracks in limestone beds of interdune deposits, and it can readily be inferred from the great size and thickness of the sand sea. That some dunes were very high is shown by the large scale of individual foresets and by the thickness of some cross-strata sets.

Dunes of the Navajo Sandstone are believed to have been transverse or barchanoid types, according to analyses of the cross-strata (R. F. Wilson, 1958, p. 150) and the orientation and average dips of cross-strata. The cross-strata show a very small arc of spread, especially when compared with the spread of underlying fluvial deposits (R. F. Wilson, 1958, p. 148, table 6). Between sets of dune sands in many places are thin beds or lenses thought to be interdune deposits that consist of flat-bedded siltstones, claystones, and limestones (R. F. Wilson, 1958, p. 141) and probably made up the desert floor.

While the Navajo Sandstone was forming in the southern part of the region, the Carmel Sea probably was beginning to transgress across it from the north. This is indicated at the upper contact of the Navajo in east-central Utah, where inundation by the sea is evidenced by reworking of the sand, leaving "swash marks, beach cusps and other water marks" (Kiersch, 1950, p. 927). The gradational sediments locally at the Navajo-Carmel contact indicate that there was no appreciable hiatus but virtually continuous, uninterrupted sedimentation.

Not all geologists agree with the concept of an eolian origin of the Navajo Sandstone. Largely from examination of a thin stratigraphic section near Woodside, Utah, and textural analyses of six sand samples, Visher (1971, p. 1423) postulated a marine high-energy channel sequence. From field measurements and textural studies at seven sections in southern Utah, Marzolf (1969) postulated a marginal marine environment, rather than an arid interior desert.

Summary of Criteria for Eolian Origin

A synthesis of criteria for considering the Navajo Sandstone to be eolian was prepared 40 years ago by Baker, Dane, and Reeside (1936, p. 52) when they suggested five kinds of evidence — three positive and two negative. Positive evidence consisted of (1) the abundance and "gigantic scale" of the cross-strata, (2) the well-rounded grains, mostly well sorted, but with coarse grains scattered locally along bedding planes, and (3) the truncation of cross-strata sets by other sets, dipping "at all angles without system," that is, tangential foresets. Negative evidence was (1) a lack of numerous regular (horizontal) bedding planes and (2) the general scarcity of silt, ripple marks, mudcracks, mud-pellet conglomerates, and other evidences of water action.

As indicated elsewhere in this chapter, textural features and the various types of negative evidence, though compatible with a dune environment, cannot be considered incontrovertible proof. However, the discovery of wind-faceted quartz pebbles in what may be considered interdune deposits of the Navajo in San Rafael Swell and Green River Desert (Baker and others, 1936, pls. 13, 14) probably constitutes important additional evidence. Still other evidence has been accumulated in support of an eolian genesis during later years.

Detailed analysis by R. F. Wilson (1958, p. 143) of cross-strata in both the Navajo Sandstone and the Lamb Point Tongue of the Navajo in southern Utah clarified several points concerning their structure. After comparison with analogs in modern barchan- and transverse-type dunes, the following features seem to be common to both modern and ancient:

1. Type of cross-strata — largely tabular and wedge planar, no large trough types (fig. 155).
2. Numerous large-scale foresets, 50 feet or more (15 m) long.
3. High-angle (>20°) dips of foresets.
4. Cross-strata dip directions with consistent orientation and narrow range of vectors.
5. Cross-strata largely tangential and concave upward.

The numerous zones of penecontemporaneous contorted bedding with sharp outlines and irregular fold patterns in the Navajo are considered evidence of sand that has become wet but not saturated, and those zones are comparable to structures in some modern dune fields.

NAVAJO SANDSTONE, showing large-scale wedge-planar cross-strata (middle) between tabular-planar cross-strata (above and below), Zion National Park, Utah, U.S.A. Photograph by E. Tad Nichols. (Fig. 155.)

The vast extent and considerable thickness of well sorted fine sand with uniform character throughout rarely occur except in eolian environments. A possible exception is in areas of large submarine sand waves that are said (Jordan, 1962, p. 840) to resemble desert dunes in general form; however, nothing is yet known of their internal structure. In any event, the association of articulated skeletons of terrestrial dinosaurs in some parts of the Navajo precludes the possibility of those parts having formed under marine conditions.

Eolian-type ripple marks in the Navajo seem good evidence of a dune environment. Numerous lenses and thin beds of limestone and siltstone between sets of cross-strata in the Navajo are believed to represent deposits of interdunes.

Lyons Sandstone (Permian), U.S.A.

THE LYONS SANDSTONE of Permian age in Colorado was selected for description in this chapter for two principal reasons. First, outcrops at its type locality near Lyons, Colorado, offer excellent opportunities for detailed examination and careful analysis (Walker and Harms, 1972). Second, it apparently represents deposits of a coastal dune area in relation to other facies for which considerable environmental data are available.

In many respects, the Lyons Sandstone closely resembles the Coconino Sandstone of Arizona and many of the features considered to be evidence of the Coconino's eolian origin also occur in the Lyons. The nature of cross-strata, the uniformity of fine sand, the presence of high-index ripple marks, rain pits, and distinctive kinds of slump marks, and the presence of nearly identical tracks of vertebrate animals show a generally similar depositional environment for both formations. In certain other features, the two differ markedly — the size and geometry of the entire cross-stratified sand body, the character of erosion planes considered to be interdunal surfaces, and the abundance and distribution pattern of relatively coarse sand concentrates.

The question presented here is not whether the Lyons Sandstone of the type locality is eolian — this seems to be established on rather firm grounds — rather, what type of dune is represented and why was its depositional environment somewhat different from that of the Coconino.

Distribution and Thickness

The Lyons Sandstone forms a narrow foothills belt, extending about 160 miles (250 km) from its northernmost outcrop near the Wyoming – Colorado line to its southernmost outcrop, south of Colorado Springs (Hubert, 1960, p. 35), Colorado. Its thickness, as determined in a series of measured sections along the outcrop belt between Boulder and Colorado Springs, ranges from 150 to 680 feet (45 – 200 m) (Hubert, 1960, p. 25). At its north end, the formation wedges out between tongues of the Permian Satanka Formation. To the south, it wedges out between red Paleozoic sandstones and siltstones (W. O. Thompson, 1949).

Eastward in the subsurface of eastern Colorado, the Lyons is represented by various facies, probably marine, of sandstone, siltstone and evaporites, as shown on subsurface maps by Maher (1954) and Broin (1957). East of Colorado Springs, the Lyons seems to maintain its identity for nearly 40 miles (65 km) before grading into other facies (Maher, 1954).

The following discussion of the Lyons Sandstone concerns mainly that part which consists of large-scale high-angle cross-strata, believed to be of eolian origin, at the type locality and in a few other limited areas along the outcrop belt. Intimately related noneolian facies of the formation are of interest primarily because they furnished supporting evidence concerning the depositional environment and paleogeographic setting. The eolian-type deposits in this formation are only known over a relatively few miles along the north-south belt and are very different in extent from other wind-formed sandstones, such as the Coconino and the Navajo.

Composition and Texture

Along its outcrop belt the Lyons Sandstone consists of two principal rock types — "arkoses (24 percent) and feldspathic quartzites (75 percent)" (Hubert, 1960, p. 13, 100). The arkoses are mainly conglomerates and pebbly sandstones; intraformational conglomerates, lenticular dolomitic quartzites, and micaceous siltstone and shales constitute about 1 percent. The arkoses and many of the feldspathic quartzites indicate by their textures, structures, and other features noneolian depositional environments. The eolian facies consist of a relatively uniform fine-grained feldspathic sandstone which has conveniently been delimited by Walker and Harms (1972, p. 279) as the "flagstone beds."

Texturally, sandstones of the high-angle large-scale cross-strata in the Lyons Sandstone, as represented at the type locality, are characteristically fine grained and well sorted. Pebbles, granules and beds of fine detritus, such as silt and clay, are absent (Walker and Harms, 1972, p. 284). The grains include both quartz and feldspar which mostly are tightly cemented by quartz; much of the feldspar has been altered by replacement.

Sand grains of medium or coarse size are concentrated on erosion surfaces that truncate sets of cross-strata throughout the formation. Relatively coarse grains uniformly scattered over these surfaces are interpreted as lag deposits of interdunes (Walker and Harms, 1972, p. 284). Such lag deposits occur on serirs and interdune surfaces of various modern dune fields (McKee and Tibbitts, 1964, pl. 1B).

The relatively clay-free clean sand that composes the dune-type sandstone of the Lyons has been called variously a first-generation and a second-generation sand. Its relative purity has been attributed to the abrasion and winnowing of arkosic material shed from the ancestral Front Range highland (Maher, 1954, p. 2237), and to accumulation along a "Permian strand by waves and currents that carried in clean sands probably derived from previously existing sediments" (W. O. Thompson, 1954).

Primary Structures

Stratification

The Lyons Sandstone, as exposed at the type locality near Lyons, Colorado, is characterized by large-scale high-angle cross-strata. These cross-strata are mostly of the tabular-planar type with individual sets as much as 42 feet (13 m) thick and with bounding planes described as "scoop-like erosional surfaces" (Walker and Harms, 1972, p. 279, 281). Many sets of strata extend laterally "for hundreds of yards" with little change in character; laminae within them are straight or gently curved in cross section.

The presence of relatively small subsets of cross-strata within the major sets of high-angle cross-strata is indicated by the traces of laminae exposed on the truncated surfaces that form bounding planes between sets. The nature of these subsets was recorded and analyzed in detail by Walker and Harms (1972, p. 281), who believed them to represent a record of nonuniform accretion along a planar surface resulting from wind drift. Such lateral shifting of cross-strata to form subsets apparently is common in many modern dunes of various types, and can be seen both on the floors of dune trenches and on interdune surfaces before burial (McKee, 1966, pl. VIIA). Shifts in wind direction also result in wedge-planar sets in many dunes. (See chapter E, "Sedimentary Structures in Dunes.")

Cross-strata in the Lyons, as described, resemble structures in many modern dunes but are duplicated in relatively few other environments. The maximum dip angles of the foresets are commonly 4°–5° less than the angles of repose or maximum dips (33°–34°) of many modern dunes. In ancient dune deposits, a maximum angle of dip consistently lower than in modern dry sand has been attributed to compaction (Glennie, 1972, p. 1061; Walker and Harms, 1972, p. 280) and to removal of unstable grains by intrastratal alteration (Walker and Harms, 1972, p. 280). An alternative explanation is differential preservation in cross-strata sets, where the upper parts of foresets that are relatively higher angled commonly are removed by erosion and planation (Poole, 1962, p. D148; Walker and Harms, 1972, p. 280).

Miscellaneous

Several minor structures occur in many places in the ubiquitous cross-stratification of the eolian-type facies of the Lyons Sandstone (Walker and Harms, 1972, p. 279). These structures include asymmetrical ripple marks invariably oriented up and down the dip surfaces of foresets and with very low crests that are parallel and straight for long distances. These ripple marks have the high indices of wind ripples. Other structures are raindrop pits with typical ring shapes and raised rims (fig. 156) and slump marks (called avalanche structures by Walker and Harms), including the distinctive "step type" that is characteristic of many slipface

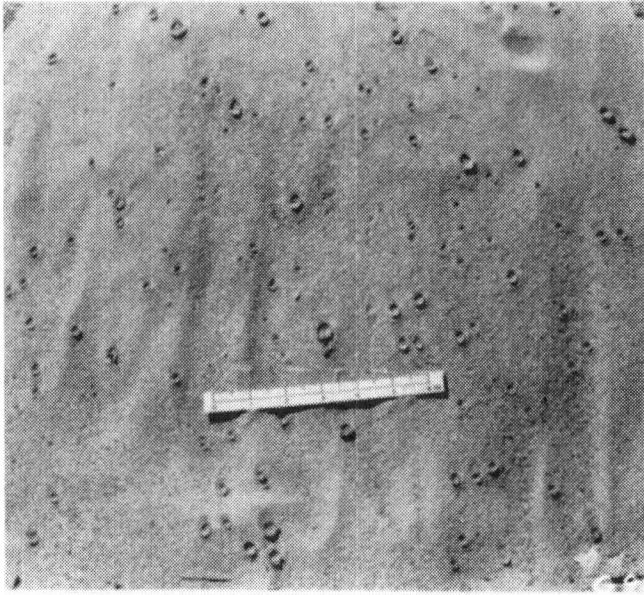

PITS OF RAINDROPS in modern sand of Lyons Sandstone texture. One inch equals 2.54 cm. (Fig. 156.)

avalanches, where the surface is slightly cohesive (McKee, Douglass, and Rittenhouse, 1971, p. 369).

These minor structures seem to be similar in all essential respects to the ripple marks, raindrop pits, and slump marks that are recorded from the Coconino, De Chelly, and some other eolian-type sandstones. Their probable modes of preservation and their interpretation were discussed in the section entitled "Coconino Sandstone." It is sufficient to say that these structures all give strong supporting evidence for deposition in an eolian environment.

Tracks and Trails

The footprints of vertebrate animals, probably reptilian, and the tracks of scorpionlike invertebrates preserved on the steeply dipping foresets of the Lyons Sandstone were described many years ago (Henderson, 1924). These clear and sharp tracks, which preserve delicate details, including the claw marks of the reptiles, probably are identical to forms reported from the Coconino and De Chelly Sandstones. The manner of their preservation and the conditions under which they apparently were preserved indicate deposition in dry dune sand.

The validity of the preceding conclusions is indicated by the following:

1. Experiments with living reptiles on steep slopes of fine sand indicate that such tracks could readily be formed in dry sand but not in wet or damp sand (McKee, 1944, p. 68; 1947, p. 27).
2. Experiments on the forming of tracks by living salamanders showed the virtual impossibility for such animals to leave diagnostic trackways on a sand surface under water (Peabody, in McKee, 1947, p. 27).
3. In extensive observations of modern amphibian trackways on mudflats, no examples of the clear outlines of individual footprints preserved under water were found, even where some trackway patterns remained (Peabody, in McKee, 1947, p. 27).

The environmental conditions under which tracks in dry sand might be permanently preserved were discussed under "Coconino Sandstone."

Environment of Deposition

Two aspects of the regional environment in which the Lyons Sandstone was formed are generally agreed upon. First, the formation was deposited in a long narrow coastal area extending from north to south with a seaway to the east. Second, it was flanked to the west by a granitic terrane referred to as the ancestral Front Range highlands, and this upland contributed arkosic detritus to the formation. Details concerning the subenvironments in which various facies of the formation developed have been interpreted in many ways and ideas about their genesis have differed greatly.

Various investigators have attributed the sedimentary deposits that compose the Lyons Sandstone to accumulation (1) by streams, (2) on beaches and bars, (3) by the wind, and (4) through combinations of these factors. Deposition by streams, at least in part, is advocated by Maher (1954), Vail (1917), W. O. Thompson (1949), and Weimer and Land (1972). Littoral sedimentation, primarily beach and bar accumulation, is referred to by W. O. Thompson (1949), Maher (1954), and Hubert (1960). Eolian deposition, involving the highly cross-stratified facies, is advocated by Vail (1917), Tieje (1923), and Walker and Harms (1972). Most of the early workers gave little specific evidence to support their conclusions, seemingly reached largely on the basis of paleogeographic and other general considerations. Eolian deposition now seems well established for major parts of the Lyons, but numerous other facies, both coastal and alluvial, probably occur in the formation.

The kind of dune that is represented by the large-scale cross-strata of the Lyons Sandstone at and near its type locality has caused much speculation. Several types of pertinent evidence have been discussed, including analysis of cross-strata dip directions, descriptive information on the nature and extent of interdune surfaces, and conclusions based on paleogeographic and environmental considerations.

Systematic directional data based on dip readings have not been obtained for the eolian-type unit of the Lyons Sandstone. The sets of large-scale cross-strata at its type locality were reported by Walker and Harms (1972, p. 281) "to dip in a broad arc from southeast to west." As further stated by them, "the highly variable local dip directions show that accretion occurred along [rock] faces of diverse orientation." Statistically sound data are needed to firmly establish the amount of spread; nevertheless, at least suggestive is the considerable width of spread in dip directions. This spread, as described, forms a pattern similar to that of coastal parabolic dunes in Brazil determined by Bigarella (this chapter). The pattern in the Lyons seems very different from that shown to be typical of barchanoid and transverse dunes (Glennie, 1972, p. 1061) with their narrow spread (clustering of points). It contrasts even more with the nearly opposite double clusters of points on linear (seif, longitudinal) dunes illustrated by Glennie (1972, fig. 13) and by McKee and Tibbitts (1964, fig. 6).

Compelling evidence suggests that the wind-deposited sands of the Lyons represent parabolic dunes, not barchanoid- or linear-type dunes. Interdune surfaces between sets of high-angle cross-strata are "scoop-like erosional surfaces" that "intersect along gently curved nicklines" (Walker and Harms, 1972, p. 281). These surfaces are broadly curved and exposed the beveled tops of earlier cross-strata sets, as is typical of most modern interdunes (McKee and Moiola, 1975). They represent a blowout phenomenon, which marks an early stage of coastal parabolic dunes, and differ from interdunes among barchans and transverse ridges in that they are relatively small and have curving surfaces.

They differ even more from the very broad, flat interdunes between linear (seif) dunes, which commonly have horizontal strata (McKee and Tibbitts, 1964, fig. 8) accumulated on their surfaces. (See section entitled "Characteristics of Interdune Record," this chapter.)

Another aspect of the Lyons dune deposits, lending support to interpretation of their origin as parabolic types, is their geographic position. General agreement among concerned geologists, based on a considerable amount of evidence, indicates that the Lyons Sandstone was formed in a coastal environment, so a variety of coastal dune is inferred. If such an inference is correct, then the evidence already cited — blowout-type interdunes, divergent wind directions, large-scale cross-strata, and common wind ripple marks — all seem compatible with the parabolic dune interpretation.

Summary of Criteria for Eolian Origin

The following features indicate the eolian origin of the highly cross-stratified sandstone facies of the Lyons Sandstone:

1. A composition of clean fine quartzose and feldspathic sand, well sorted, subangular to well rounded, and commonly frosted. Such features, although not incontrovertible proof of genesis, are compatible with the texture of many modern dune deposits.

2. Thin layers of scattered but rather uniformly spaced, coarse grains resting on surfaces of truncation within fine cross-stratified sand. The coarse detritus is interpreted as lag material.

3. Major primary structures consist of large-scale high-angle cross-strata in tabular-planar sets. Such structures are typical of most dunes formed under conditions of unidirectional winds but rarely occur in other environments.

4. Curving scoop-shaped surfaces, exposing the beveled tops of cross-strata that represent sets of earlier development, are characteristic features of interdunes developing in a parabolic-dune environment.

5. Numerous asymmetrical ripple marks oriented parallel to the dip directions of foresets which have high ripple indices characteristic of wind-formed ripple marks.

6. Surfaces of some strata have slump features, both the small steplike variety and the dendritic type that are commonly formed from avalanching of damp, cohesive sand. Surfaces also have raindrop pits.
7. Locally, the sharp, well-preserved footprints of four-footed animals, probably reptiles, occur on the steep foresets. This type of preservation, characteristic of a dry-sand environment, is not possible under water.

Casper Formation (Pennsylvanian and Permian), U.S.A.

THE GENESIS OF the cross-stratified sandstone of the Casper Formation in the southern part of the Laramie Basin, Wyoming, has long been a subject of dispute. Whether this formation is marine or eolian has been difficult to demonstrate largely because composition and texture of the sandstone could represent either environment, and the size and type of cross-strata and contorted beds that are its most distinctive structures have not yet been demonstrated to occur in any modern deposits.

Fortunately, for the understanding of the origin of the sandstone facies in the Casper, an excellent detailed description and analysis of its structure was made many years ago by Knight (1929). This pioneer work, a model for much subsequent work in the study of cross-stratification, called attention to the remarkable and baffling type of structure which Knight termed "festoon cross-lamination." (See discussion of "Festoon Structures" later in this section on the Casper Formation.) Today, it is commonly referred to as large-scale trough cross-stratification.

Festoon structures that are extensive and large, like those in the trough-type cross-strata of the Casper, do not seem to have good analogs in any modern sand deposits that have been examined by trenching; nevertheless, the Casper is believed by many geologists to represent a dune environment, on the basis of other data. Accordingly, the Casper, as a unique form, is described herein. Trough structures of festoon type in some modern noneolian deposits are described, and certain dune environ-

ments which may produce large festoons are discussed on following pages.

Distribution and Thickness

The Casper Formation, including both limestones and sandstones, bounds much of the southern part of the Laramie Basin and flanks the Laramie Range in southern Wyoming, an area 75 miles (120 km) wide and 100 miles (160 km) long. The cross-stratified sandstones, believed to be eolian, occupy the entire thickness of the formation on its southeastern margin encompassing at least 1,000 square miles (1,600 km²) (Knight, 1929, p. 56).

The Casper is more than 700 feet (200 m) thick east of the Laramie Range; near the center of the Laramie Basin about 550 feet (165 km) thick, and in the Sand Creek area to the south (where the cross-stratified sandstones have been most studied) 120 feet (36 m) thick (Steidtmann, 1974). Its facies apparently are lateral extensions of still other and very different facies represented in the Permian and Pennsylvanian Tensleep Sandstone to the northwest. It may also be at least partly contemporaneous with the Permian and Pennsylvanian arkosic Fountain Formation to the south, which is considered to be formed largely of coalescing alluvial fans (Hubert, 1960).

Composition and Texture

The cross-stratified sandstone of the Casper is fine grained and well sorted. Exceptions to this uniformity are laminae at the bottoms of cross-strata troughs that commonly contain evenly distributed, scattered coarse grains of quartz and feldspar. These grains may be lag material (Steidtmann, 1974, p. 1838); if so, the laminae probably represent interdune surfaces developed in an environment of blowout dunes.

Sandstone of the Casper is classed as calcareous subarkose or orthoquartzite, according to Steidtmann (1974, p. 1836). In some areas it is interstratified with beds of marine aphanitic limestone that attain a maximum total thickness of 210 feet (64 m) on the west flank of the Laramie Range. Limestones are absent in the southern part of the area (Knight, 1929, p. 53).

Primary Structures

Stratification

The sandstone facies of the Casper Formation throughout a broad area characteristically consists of large trough cross-stratification, commonly termed "festoon lamination." This type of cross-strata consists of a repetition of two elements — erosion surfaces that form plunging troughs and thin concave-upward laminae that fill the troughs and conform largely to the shape of the trough floors (fig. 157).

The erosion troughs in the Casper differ greatly in size, ranging in width from 5 to 1,000 feet (1.5 to 300 m), in length from 25 feet to a few thousand feet (8 m to a few hundred metres), and in depth from a few feet to 100 feet (30 m) (Knight, 1929, p. 59). The surface of each trough is cut across at one end by one or more overlying troughs, and so each trough has "the general shape of a long, shallow, scoop-like depression." The average inclination of trough surfaces is 10°–15°.

Most of the laminae that fill the erosion troughs form symmetrical sets, only a few being asymmetrical. The angles of inclination of the strata as determined by Knight (1929, p. 64) from about 800 readings were 15°–25° and averaged approximately 20°. Very few sets dipped toward the northeast quadrant; nearly all dip directions fell within the 220° arc from N. 35° W. southwestward to S. 75° E. (Knight, 1929, fig. 30).

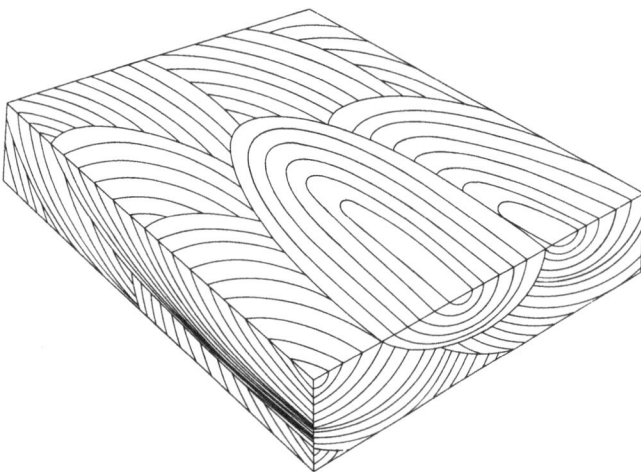

FESTOON CROSS-LAMINATION, as described by Knight (1929), from sandstone of Casper Formation, Wyoming, U.S.A. This structure consists of overlapping, symmetrically filled troughs. (Fig. 157.)

Festoon Structures

The term "festoon cross-lamination" as originally used by Knight (1929) for cross-strata of the Permian and Pennsylvanian Casper Formation was applied to a type of trough cross-stratification filled either symmetrically or asymmetrically and having a large range in size.

The term "festoon-type cross-bedding" has been used by Tanner (1965, p. 564; 1966) for a structure apparently very different from the original "festoon lamination" of Knight. Tanner probably referred to the wedge-planar sets in the Jurassic Entrada Sandstone, which he stated "form additional evidence * * * of a wind origin." These sediment-filled troughs are superimposed one above another and overlap to form the characteristic festoon pattern. The genesis of this structure was attributed by Knight to marine processes. His evidence did not consist of characteristics of the cross-strata, however, and his chief reason was the close association of sandstones of the Casper with other facies that were of undoubted marine origin. Since that pioneer study, several geologists have contended that an eolian origin better explained various features of the sandstone, especially the minor structures.

Today, the origin of the festoon-type cross-stratification in the Casper still remains an enigma. Probably the greatest deterrent to a widespread acceptance of the eolian origin has been the fact that no modern analogs displaying all the Casper's characteristics have been recorded. Point bars in major river systems seem to be the closest modern analog.

Festoon structures in river point bars were first reported and illustrated by Frazier and Osanik (1961) from the Mississippi River Delta. Sections transverse to current direction revealed well-developed festoons forming superimposed semicircular troughs filled with concave-upward strata comparable to those of the Casper. Principal differences were lenses of gravel and other coarse detritus, local accumulations of plant debris, and a lack of any very large channels (up to 1,000 ft (300 m) wide and 100 ft (30 m) deep), such as occur in the Casper. Other examples of festoon structures in river sediments were recorded by Hamblin (1961), Harms, MacKenzie, and McCubbin (1963), Harms and Fahnestock (1965), and Williams (1968), an indication that they are fairly common in a fluvial environment.

How fluviatile large-scale trough cross-strata formed — by a single process of scour-and-fill related to large sand waves, or by two stages involving the filling through megaripples of previously cut troughs — has been controversial. Whatever the process, the resulting structures differ from the Casper-type structures in being considerably smaller, shallower, and less symmetrically filled. Trough structures preserved in dry river channels in South Australia are described (Williams, 1968, p. 138) as having relatively coarse texture — that is, coarse to very coarse sand with some granules and pebbles; locally, they include concentrations of sticks and leaves.

In modern eolian dune deposits, large trough structures seem to be rare, and those that are recorded are mostly shallow and not sufficiently overlapping to form festoon structures. Scattered channels were found at White Sands National Monument, New Mexico, in trenches parallel to the dune crest of a transverse dune (McKee, 1966, fig. 7), but few were found in other types of dunes in that area. Similarly, in the coastal dune ridges of Brazil examined by Bigarella, Becker, and Duarte (1969) uncommon trough structures are shallow and lack well-developed festoon patterns.

In modern dune areas, festoon structure analogous to that of the Casper is most likely among coastal dunes where "blowouts" are abundant. Trenching, followed by detailed analysis of structures, apparently has not been attempted in this environment, but the shape, abundance, and size of blowouts in some regions (fig. 158) may account for large festoon structures.

Contorted Bedding

Numerous sets of trough cross-strata in the Casper Sandstone are crumpled and contorted on a huge scale, with folds having amplitudes up to 25 feet (8 m) locally. This contortion of bedding apparently developed during or immediately after deposition of the sand; examples are numerous of highly deformed strata in one set that is directly adjacent to or overlying uncontorted sets (Knight, 1929, p. 74).

The deformation of strata within the Casper troughs, resulting in large, locally developed folds oriented in many directions, was attributed by

BLOWOUT DUNES ALONG SOUTH SHORE of Lake Michigan, U.S.A. Photograph by R. C. Gutschick. (Fig. 158.)

Knight (1929, p. 27) to asymmetrical filling with sand. This process forms slopes above the angle of repose with a resulting creep down the over-steepened side and a tendency toward overturning the compressional folds. This process was believed by Knight to have required saturated sand; dry sand dune deposits were not considered favorable.

Recent experiments on contorted bedding (McKee, Douglass, and Rittenhouse, 1971, p. 373) showed, however, that a high degree of cohesion is necessary in forming structures of this type and that such cohesion occurs only in wetted sand (such as occurs in dunes that have been exposed to considerable rainfall or other moisture) and not in saturated sand. Coastal dunes of Brazil have been found to contain many contorted structures (Bigarella, Becker, and Duarte, 1969; McKee and Bigarella, 1972, p. 673), although most of these are much smaller than those of the Casper Formation.

Ripple Marks

Low wide ripple marks with parallel crests and troughs, oriented up and down the dip slopes of cross-strata, are relatively numerous in the Casper. These ripple marks are basically like those in the Coconino Sandstone of Arizona which were studied many years ago and described by the author (McKee, 1934b, p. 100; McKee, 1945, p. 323). Similar ripple marks have since been recorded from the De Chelly Sandstone, the Lyons Sandstone, and various others thought to be of eolian origin and discussed in this chapter.

Low-amplitude ripple marks in the Casper Formation were described by Steidtmann (1974, p.

1837), as follows: "The ripple crests are relatively straight, parallel, slightly asymmetric, and spaced 5.5 to 9.0 cm apart. Their heights are from 1.5 to 3 mm; therefore, the ripple indices are 36 to 30." Both orientation of the ripples and the ripple indices are typical of eolian forms.

The ancient ripple marks just described resemble structures developed on the lee faces of various kinds of modern dunes formed when winds normal to the prevailing wind are active. Modern lee-side ripple marks have been observed on both barchanoid ridge and transverse ridge dunes at White Sands National Monument, New Mexico, (Chapter E, figs. 41C, 41D), on reversing dunes at Great Sand Dunes, Colorado, on transverse dunes at Padre Island, Texas, and many others.

Interdune Deposits

Thin lenticular limestones that occur "along low-relief truncation surfaces within the sandstone units" of the Casper Formation were reported by Hanley and Steidtmann (1973). These limestones, of local distribution, do not contain any marine macrofossils, are described as "ostracodal, peloidal, microsparites" and contain shrinkage cracks, gas cavities, and burrow traces.

From the description, these limestone lenses are interpreted herein as interdune deposits, rather than as tongues of normal marine or of lagoonal limestone. They are believed (Hanley and Steidtmann, 1973, p. 428) to have been "precipitated in small, shallow ponds in which environmental conditions were harsh because of fluctuation in water temperature, depth, and salinity."

Trace Fossils

Two varieties of trace fossils in the Casper Formation have been recorded (Hanley and others, 1971). One variety consists of ribbonlike trails on the upper bounding surfaces of cross-strata that fill troughs of the Casper; the narrow trails intersect one another but do not branch. The second variety consists of tubular internal traces. Animals responsible for the first variety of trail have variously been assumed to be insects ("beetles"), arthropods, and annelids (Hanley and others, 1971, p. 1067); however, the main importance of these trails is that they indicate the probable depositional environment of the enclosing trough-type structures. Their

preservation could scarcely have occurred either in saturated sand or in dry loose sand (Hanley and others, 1971, p. 1067); a moist cohesive sand, possibly in a coastal environment, would have been most favorable. Development and burial of these trails probably were like those attributed to the invertebrate tracks of the Coconino (McKee, 1934b, p. 108, pl. 13a).

Environment of Deposition

The broad regional environment in which the Casper Formation was deposited is readily determined. Criteria that seem convincing are available to permit an interpretation, and geologists have reached general agreement on paleogeographic controls (Knight, 1929; Steidtmann, 1974). Apparently, a mountain mass of some relief — the ancestral Front Range — extended from northwest to southeast across south-central Wyoming. Parallel to and northeast of those highlands was the margin of a seaway. Between the mountains and the sea was a broad coastal plain across which rivers transported much arkosic sediment. Evidence for the sea is furnished by limestones of the Casper containing many kinds of marine fossils (Miller and Thomas, 1936; Thompson and Thomas, 1953).

The environment of deposition of the trough-type cross-stratified sandstones in the Casper clearly is related to the environment of the interbedded or closely associated marine limestones. Whether the sandstone is marine or eolian remains uncertain, mainly because no known analog in modern sediments completely matches the sedimentary structures observed in the Casper. Despite some evidence of dune deposition, an interpretation of the remarkably large-scale trough structure cannot be considered to be secure until a modern analog is found.

Summary of Criteria for Eolian Origin

Probably the most compelling evidence that sandstones of the Casper were formed through deposition by the wind is found in the relatively numerous eolian-type ripple marks with crests oriented parallel to the dip of foresets and with high ripple indices.

Further suggestion of eolian origin is furnished by surfaces of scattered coarse sand at the bottoms of cross-strata troughs. These deposits are interpreted as lag sand of interdune areas; thus, the troughs may be considered blowout phenomena, such as are common in many coastal dune fields and the fillings are blowout-type dunes. Supporting this concept is the very wide spread of cross-strata dip directions, characteristic of some coastal dune areas, such as the Lagoa dune field, Brazil (chapter E).

Results of experiments on settling velocities of light and heavy minerals made by Steidtmann (1974, p. 1840) are presented by him as corroborative evidence of an eolian deposition.

The presence of lenticular limestone beds locally distributed along truncation surfaces of cross-stratified sandstones, their ostracodal peloidal texture, and their associated shrinkage cracks, gas cavities, and burrow traces strongly suggest an interdunal origin. Trace fossils, consisting of ribbonlike trails on upper bounding surfaces of cross-strata, indicate by their clear preservation that the sand was neither dry nor saturated at the time of burial but probably was moist and cohesive.

Various features of texture and structure are compatible with the concept of an eolian origin, but a conclusive test of genesis involving dissection of dune types on a modern coastal area, where trough-type structures are exposed, has not yet been made.

Rotliegendes of Northwestern Europe (Permian)

THE PERMIAN RED-BED sequence of northwestern Europe, normally referred to in Germany as the Rotliegendes, represents a great desert area of the past, and so merits discussion here. A careful analysis of these strata and a comparison with deposits of some modern desert areas were made by Glennie (1972, p. 1048). The following brief discussion is largely a synthesis of his work as it pertains to those sandstones in the Rotliegendes believed to be eolian. Additional information on the Rotliegendes and its environment of deposition is derived from Falke (1972) for central and western Europe and from Wills (1948, 1956) for England.

Distribution and Facies

Sedimentary rocks of the Rotliegendes extend over a wide area from Poland to France and western Germany, from Scotland and southern Norway, across England and the Netherlands to central Germany (Kent and Walmsley, 1970). This Permian desert probably approached "in scale the area of the present Sahara, especially if one considers its possible former extension into North America," according to Glennie (1972, p. 1069), who has suggested that it may originally have been "between the paleolatitudes of roughly 10° N. and 30° S.," as indicated by paleomagnetic data. Since then, the landmass containing this low-latitude paleodesert apparently has migrated by continental drift to its present position.

The thickness of the Rotliegendes is about 750 feet (225 m) along its southern margin (Gill, 1967; Kent and Walmsley, 1970) and more than 5,000 feet (1,500 m) in the central part of the depositional basin in northern Germany (Kent and Walmsley, 1970). These figures for distribution and thickness refer to the entire formation or desert environment and include several noneolian facies. The major environments or facies as classified by Glennie (1972, fig. 17) are dune, wadi, sebkha, and desert lake. Sandstones in the southern North Sea area that are attributed to a dune-sand origin have a thickness up to 600 feet (180 m) (Glennie, 1972, p. 1048).

Texture

Quartz grains in those sandstones of the Rotliegendes that are considered to be eolian generally range in size from fine to medium; locally some coarse grains occur. The finer grains are mostly subangular, whereas the coarser ones tend to be subrounded to rounded. "Impact fractures typical of those found in modern aeolian sands have been recognized during electron-microscope investigation of the lower Permian Penrith Sandstone" (D. B. Smith, 1972, p. 13). The cements are principally hematite and authigenic clay, and locally dolomite or anhydrite. Argillaceous matrix is lacking (Glennie, 1972, p. 1058).

Interbedded detrital sediments that comprise thin, roughly horizontal sets between crossbedded units of dune type are interpreted by Glennie (1972, p. 1061) as interdune sebkha deposits that are composed largely of fine to coarse sand. As in many modern interdune deposits, such horizontal beds are characteristically poorly sorted and are silty. They commonly contain many small irregular adhesion ripples but very little argillaceous material.

Primary Structures
Stratification

Cross-stratified sandstones, interpreted as eolian, typically occur as large-scale planar types with steeply dipping (20° – 27°) foresets. Dip directions of laminae are relatively uniform, dips tend to increase upward, and truncated tops of sets are overlain by subhorizontal finely laminated sandstones (Glennie, 1972, p. 1058). In many places these dune-type laminae overlie other sand sequences that are considered to represent a wadi environment. Stereographic polar nets prepared by Glennie (1970, fig. 68) strongly suggest that both barchanoid (transverse) and linear (seif, longitudinal) dunes are represented in the sandstones believed to be eolian. Within barchanoid-type dunes, the dip attitudes of foresets are characteristically concentrated in a single, limited area which is in line with the predominant wind direction. In linear-type dunes, however, foreset dip directions are concentrated in two widely separated, nearly opposite directions (McKee and Tibbitts, 1964, p. 12). Borehole data plotted for the basal Permian "Yellow Sands of Durham," of Great Britain, show a form that, according to Smith and Francis (1967, p. 97 – 98, fig. 18) "strikingly resembles" a seif dune, such as that developed under two alternating wind regimes postulated by Bagnold (1941).

The trend and pattern of cross-strata "in several basins of central and western Britain have shown that the sand [was] accumulated in those places in barchans" under easterly winds as described by D. B. Smith (1972, p. 13), but he further states that in northeastern England "east-north-east trending longitudinal dunes were present." Cross-strata dip directions in the Rotliegendes, according to Glennie (1972, p. 1061), suggest that seif dunes predominated in Britain and West Germany and that the barchanoid-type was prevalent in other areas (that is, away from the Variscan Mountains).

Adhesion Ripples

Interdune deposits on the Rotliegendes are thin zones primarily consisting of "small, irregular, and more or less horizontally-bedded adhesion ripples" (Glennie, 1972, p. 1061). Adhesion ripples are defined as small irregular mounds or ridges formed where dry sand is blown onto a nearly horizontal humid surface and fixed there by surface tension. Laminae are concave upward.

Summary of Criteria for Eolian Origin

Large sequences of cross-stratified sandstone in the Rotliegendes have long been attributed to accumulations of sand by eolian processes, because their textural and structural properties are similar to those of modern dune deposits and because they are closely related to surrounding and interbedded deposits recognized as desert facies. These facies include wadis, desert lakes, and sebkhas. Few, if any, of the properties described for the Rotliegendes can individually be considered proof of a dune origin, yet in aggregate they present strong evidence. On the other hand, recent petrographic studies led Pryor (1971a, b) to question this origin for some of the sandstones and to postulate a genesis in "a shallow marine environment" where sand was reworked by the sea and transported southwestward.

Most eolian sands are characterized by dominantly subrounded frosted well-sorted grains and by lack of argillaceous matter. Such textural features are common in the Rotliegendes. Structures including well-laminated high-angle cross-strata are typical, but not indisputable, evidence. More compelling as evidence are the individual sequences that show upward progressions from subhorizontal to steeply inclined laminae and a record of interdunes shown by intraformational unconformities and by horizontally bedded adhesion ripples. Additional strong evidence furnished by cross-strata is their common occurrence as large-scale planar types.

Statistical studies of dip directions on cross-strata, indicating a small spread in trends of depositional currents, offer evidence of eolian origin and also a means of differentiating between the linear-type dunes (with two nearly opposite clusters of readings) and the barchanoid-types (with a single, larger grouping of directional points).

Strongly suggesting a desert dune genesis is the association of large-scale dune-type cross-strata with evaporite basin deposits of anhydrite, halite, and red clay above and with probable wadi deposits of conglomerate, mudcracked clays, and flat-bedded sands below.

Lower Mottled Sandstone (Permian?), England

ONE OF THE FIRST detailed studies of eolian-type sandstones, critically analyzed and studied in detail, was Shotton's 1937 investigation of the "Lower Bunter Sandstone," now commonly referred to as Lower Mottled Sandstone, of the Birmingham area in England. His work, which formed a model for many later studies, not only presented evidence for ascribing an eolian origin but also developed criteria for interpreting the type of dune, believed to be barchan. His report was a penetrating analysis of the manner in which dune structures are developed and the effect of various factors involved. His statistical study of cross-strata dip directions and discussion of their significance in terms of regional wind currents and dune migration paved the way for an understanding of the environment and history of eolian deposits in many other regions. The classic study by Shotton was published almost simultaneously with Reiche's pioneer work (1938) along similar lines — analysis of the Coconino Sandstone of Arizona.

Both because of its historic significance and because of the excellence of deductive reasoning with regard to a basic dune type, here classed as the barchanoid dune ridge, Shotton's work qualifies the "Bunter" for discussion in this publication. Nearly all data used herein are derived from Shotton's classic paper; additional important references are Wills (1948; 1956, figs. 2416) and Shotton (1956).

Distribution and Thickness

The Lower "Bunter" Mottled sandstone (Shotton, 1937, p. 535) "consists of nothing but sand," with no clay beds and no pebbles in the four areas of outcrop, west and northwest of Birmingham, England, where it was studied. These outcrops extended along a belt roughly 25 miles (40 km) north-south and 10 miles (15 km) wide. A maximum thickness of 850 feet (260 m) for the formation is reported (Shotton, 1937, p. 536) from a borehole in the area studied. In other parts of the southern Cheshire basin, the same formation is known to be 650 feet (217 m) thick.

Composition and Texture

The Lower "Bunter" Mottled Sandstone is formed almost entirely of quartz sand grains but in-cludes some scattered feldspar grains (Shotton, 1937, p. 535). The dull-red coloring of the rock is attributed to thin grain coatings of iron oxide, which is believed to be "the only cementation." The sandstone has "extraordinary uniformity," as confirmed by analyses of samples from 476 exposures. The average grain size was determined to be 0.2 mm, with the median diameter ranging from 0.129 to 0.410 mm; no particles are smaller than 0.05 mm or larger than 1 mm (Shotton, 1937, p. 552). The fineness of grain and high degree of sorting characterize the entire sandstone throughout the area examined.

Primary Structures
Stratification

A systematic investigation of cross-stratification in the "Lower Bunter Sandstone," undertaken by Shotton and others, enabled Shotton (1937) to make a detailed description and careful analysis of its structures. Cross-strata in the formation are recorded as dominantly high angle, ranging between 20° and 30°, and are mostly large scale, forming tabular-planar sets up to nearly 100 feet (30 m) thick. Individual foresets are steepest in their upper parts and are truncated by low-angle erosion surfaces at their tops; they are tangential with the underlying bounding planes. Between sets of cross-strata low-angle erosion surfaces dip in a direction opposite that of the cross-strata. These low-angle planes are interpreted as windward-side dune surfaces.

Statistical measurements of cross-strata dip directions made at 476 exposures and consisting of 1,142 readings furnished data for determining both current-direction vectors and the amount of spread involved in dip directions of each outcrop area studied. Using the mean direction of dip as representing the dominant wind direction, a regional wind from the east was determined. Calculating the size of arc represented by all readings in a particular area disclosed a considerable curvature in dune ridges to be typical. Hence, a barchanoid ridge, rather than a straight transverse-type ridge, is postulated for this formation.

Environment of Deposition

The sand desert with inferred barchanoid-type dunes of the "Bunter" Mottled Sandstone is interpreted by Shotton (1937, p. 550) as occupying an elongate strip west of a highland formed during the

Hercynian orogeny. This highland "fulfilled the double function of abstracting moisture from the prevailing east wind, and of providing the material to build up the sand sea."

Between the postulated highlands to the east and the dune-type cross-strata of the "Bunter" are pebble and sand beds, interpreted as weathering deposits derived from the uplands and spread by intermittent streams across the lower slopes of the hills to the edges of the desert. These alluvial deposits, therefore, are considered the principal source of the dune sands. Despite some uncertainty in correlation of the fluviatile deposits with eolian strata (Shotton, 1937, p. 352) because no transition zone now exists, much evidence supports the relationship.

Summary of Criteria for Eolian Origin

Textural features of the Lower "Bunter" Mottled Sandstone (Shotton, 1937, p. 538) are comparable in all respects with the textures of sands in most modern dunes but show marked differences from the textures developed in many other depositional environments. Distinctive textural features are the fineness of grain, the high degree of sorting, and the absence of both coarse detritus and silt-clay particles.

Most compelling evidences of eolian origin are the size and type of cross-strata represented in the formation — high-angle large-scale tabular-planar, with long curving surfaces tangential to the bounding planes at their bases. The frequency distribution of the cross-strata dips suggests that the dunes were of the barchanoid-ridge type, migrating in a constant direction and accumulating one set upon another.

Frodsham Member of So-Called Keuper Sandstone Formation (Triassic), England

ONE OF THE MOST SUCCESSFUL attempts to interpret ancient deposits of probable eolian origin in terms of specific dune types, is D. B. Thompson's (1969) study of the Frodsham Member of the so-called Keuper Sandstone Formation (Triassic). For his investigation, cross-strata patterns recorded in sandstone of the Frodsham Member in West Cheshire, England, were compared with structures exposed by trenching in modern dunes of New Mexico in the United States (McKee, 1966).

The sandstone study by Thompson represents a detailed description of a relatively small but distinctive eolian-type sandstone and presents a rather convincing case, through cross-strata analysis, for the presence of dome-shaped dunes that had not previously been recognized in ancient rocks. The method of tracing major bounding erosion surfaces, formed at the base and top of dunes, and of recording the patterns of cross-strata sets between these surfaces, is an effective means of establishing the diagnostic structural features of various dune types (D. B. Thompson, 1969). Furthermore, the relationship of these dune sands to other closely associated sedimentary facies illustrates a complex set of depositional environments.

The Frodsham Member of the so-called Keuper Sandstone Formation, as discussed here, is a relatively thin (as great as 55 m or 170 ft) member of red fine-grained sandstone which forms the upper part of a sequence that contains medium- to coarse-grained sandstones and conglomeratic sandstones interbedded with other members composed of soft fine-grained sandstone (table 24; D. B. Thompson, 1970a, b). The Frodsham Member is present in about 2,000 km² (1,240 mi²) of the Cheshire and East Irish Sea Basins, but the Frodsham [facies (sensu stricto)] with dome-shaped and transverse dunes apparently is restricted to a small area near Frodsham and Runcorn in northwest Cheshire, where D. B. Thompson (1969) made his detailed studies.

Composition and Texture

The eolian-type sandstone of the Frodsham Member (D. B. Thompson, 1969, p. 272) consists dominantly of quartz sand with sparse grains of feldspar and little or no mica. The electron microscope enabled the identification of authigenic feldspar (B. Waugh, oral commun. to D. B. Thompson, 1975). In associated beds believed to be fluviatile and lacustrine, feldspars are much more numerous and mica flakes are common; clay minerals are illite and kaolinite.

The roundness of quartz grains in the "Frodsham facies" is described as "subangular to subrounded" and grain-size analyses show (D. B. Thompson,

TABLE 24 — *Age, thickness, composition, and postulated depositional environments of units forming Lower "Keuper" and underlying sandstones in West Cheshire, England*
[Prepared by D. B. Thompson, University of Keele, 1975]

Age	Informal formations	Defined members	Maximum thickness (metres)	Depositional environment
Scythian ?Anisian	"Keuper" Waterstones			Intertidal and ?fluvial.
Scythian	Lower "Keuper" Sandstone (as much as 220 m thick)	Frodsham Soft Sandstone Member	55	Eolian: Sand sea rather than coastal dunes.
		Delamere Pebbly Sandstone Member	85	Fluvial low sinuosity.
		Thurstaston Soft Sandstone Member	45	Eolian and fluvial.
		Alderley Conglomerate Member	34 (not always present)	Fluvial low sinuosity.
		Possible unconformity		
Scythian	Bunter Upper Mottled Sandstone		950	Eolian and fluvial; fluvial mostly near base.
?Scythian	Bunter Pebble beds			

1969, table 3) that in most sections the sand is dominantly fine grained, and in some, mostly medium grained. Very few grains in any sections are above or below these size grades, thus attesting to the high degree of sorting in the sandstone. Most grains have coatings of diagenetic red ferric oxide, deposited presumably by circulating ground waters (D. B. Thompson, 1969, p. 272).

The attributes of composition and texture just described are all consistent with an eolian origin, but they reflect genesis and transportation rather than deposition. The presence of lenses and other concentrations of "millet seed" sand has been viewed by various geologists as evidence of wind deposition; however, according to D. B. Thompson (1969, p. 270) some entire beds in the so-called "Keuper Sandstone Formation," dominated by such sand, also include large pebbles and clay galls.

Primary Structures

Stratification

The "Frodsham facies" — the probable eolian part of the "Keuper Sandstone Formation," as described by D. B. Thompson (1969, p. 273, 285) — consists mostly of large sets of cross-strata with very long laminated foresets, some of which extend downward and laterally into flat-bedded and ripple-bedded bottomsets of siltstone and mudstone. Major bounding planes of erosion mark tops and bottoms of cosets that may include numerous tabular-planar and wedge-planar sets. "Shallow trough structures are present * * * but form only a small proportion of the structures present;" most of them occur directly below major erosion surfaces.

From the cross-stratification patterns analyzed, D. B. Thompson (1969, p. 284) recognized "at least eight vertically and laterally overlapping eolian dunes," six of which were of the dome-shaped type with a tendency toward the transverse type. The beveled tops of cross-strata sets and other structural features as described for dome-shaped dunes at White Sands (McKee, 1966) suggest that the degradation of the transverse dunes by very strong winds is responsible for features described in the "Keuper" dome-shaped dunes.

Penecontemporaneous contorted structures are not uncommon in tabular-planar and wedge-planar sets of cross-strata in the "Frodsham facies of the Keuper Sandstone." "Simple concentrically folded laminae," several feet thick and overlain by undisturbed strata, closely resemble the sand-drag type (McKee, Reynolds, and Baker, 1962b). Other varieties, which are numerous, apparently are like types attributed to dry and to damp sand (D. B. Thompson, 1969, p. 285).

Dip-direction vectors, indicating paleocurrent summaries for 10 dunes analyzed in the "Frodsham facies," were determined by D. B. Thompson (1969,

p. 278). The resulting wind roses show a general westerly wind movement and a spread expectable for dunes, such as dome-shaped and transverse type, formed under a unidirectional wind regime.

Environment of Deposition

The eolian-type sandstone, referred to as the "Frodsham facies" by D. B. Thompson (1969, p. 286), was interpreted by him — largely on the basis of cross-strata analysis — to consist principally of dome-shaped dunes, attributed to strong winds, and, to a lesser extent, of transverse ridge dunes from more moderate breezes. Both dune types result from unidirectional paleowinds that, as determined from cross-strata dips, came from the east. Fluctuating crosswinds were largely from the southeast and southwest.

Various lithofacies associated with the "Frodsham facies" were recognized and interpreted by D. B. Thompson (1969, p. 267): interbedded red mudstones, red-brown shaly mudstones and siltstones, all considered to be of fluvial and lacustrine origin and possibly formed between dunes as interdune ponded deposits. Other deposits, with clay galls, mica concentrates, and argillaceous units, are also undoubtedly of subaqueous origin. Brown shaly flat-bedded and ripplebedded strata among the cross-stratified sandstones resemble modern interdune deposits recorded elsewhere.

The "Keuper" waterstones immediately above the Frodsham Member are partly intertidal, suggesting that the dome-shaped transverse dunes might represent a coastal sand belt. The negative skew of the grain-size distribution of some of the samples suggests derivation from a beach. On the other hand, the basinwide areal extent and geometry of the Frodsham Member suggests the existence of a sand sea followed by a basinwide transgression.

Summary of Criteria for Eolian Origin

The part of the "Keuper Sandstone Formation" called the "Frodsham facies," which is believed to be composed dominantly of dome-shaped and transverse dunes, is intimately associated with facies representing contrasting depositional environments. The closeness of several distinct facies may lead to misinterpretations and confusion where differentiation is not carefully made.

To separate eolian and noneolian deposits, D. B. Thompson (1969, p. 271) used five criteria, which, "when taken as a whole rule out the possibility of fluvial or other environmental interpretations": (1) the recognition of groups and arrangements of sedimentary structures of specific dune types; (2) the presence of very large sets (up to 13 m) of fine-grained cross-stratified sandstone; (3) uniformly fine grained texture of the sandstone; (4) a general absence or virtual absence of feldspar and mica, even though these minerals are common in the associated fluvial lithofacies; and (5) accompanying concentrations of "millet seed" sand.

The first two and the fourth of Thompson's criteria seem to be compelling positive evidence of eolian deposition; the third and fifth, although compatible, may also occur in some other environments. Important additional evidence of dune environment is the recognition in the stratification of bounding surfaces of erosion, formed through interdune processes. Furthermore, the interbedding of eolian- and fluviatile-types is strongly suggestive of blown sands overwhelming the surfaces of river flood plains or of ponded (sebkha) areas.

Barun Goyot Formation (Upper Cretaceous), Mongolia

UPPER CRETACEOUS STRATA called the Barun Goyot Formation, formerly the Upper Nemegt Beds, in the northern part of the Gobi Desert of Mongolia, are reviewed herein because they include exceptionally extensive interdune deposits and contain a large and varied nonmarine fauna for a paleodesert environment. The sandstone of this formation is considered to be eolian not only because of characteristic features of structure and texture attributed to wind deposition in many other deposits but also because of its associated facies and its abundant nonmarine fauna.

Continental deposits of Late Cretaceous age in central Asia have been known for many years and have received considerable fame because of the dinosaur and mammalian remains obtained from them during a series of paleontological expeditions by American, Soviet, and Polish-Mongolian groups in the 1920's, 1940's, and 1960's, respectively. The data on sedimentology reviewed in the following synthesis of the Barun Goyot Formation are largely from the descriptions and interpretations by Gradzinski and Jerzykiewicz (1972, 1974a, b).

Distribution and Thickness

Dune-type deposits of the Barun Goyot Formation are widely exposed in the Nemegt Basin, an east-trending graben in the Gobi and adjoining areas of central Asia. The formation has been studied in detail in two main places — the Nemegt locality, to the west, and the Khulsan locality, to the east, covering an area of "about 100 sq. km [260 mi²]" (Gradzinski and Jerzykiewicz, 1974b, p. 113).

The total thickness of the Barun Goyot Formation has been given by Gradzinski and Jerzykiewicz (1974a, p. 252) as "about 100 meters" (300 ft) and described as grading upward into "fluvial deposits" of the Nemegt Formation "about 400 m [1,200 feet] thick." The Barun Goyot Formation, however, is composed only in part of large-scale cross-strata, considered to be eolian, and much of its thickness comprises interdune deposits of other facies.

Composition and Texture

The results of grain-size distribution studies by Gradzinski and Jerzykiewicz (1974a, p. 263; 1974b, p. 122, table 1, fig. 5) are summarized in their table of statistical parameters and in histograms. From these analyses, sand of the eolian-type cross-strata is shown to be mainly fine- or medium-grained, moderately sorted and slightly skewed on the positive side.

Studies of sphericity, roundness, and surface texture were also made of sand grains from each principal facies. Subrounded and rounded grains predominated in most samples, although some subangular and some well-rounded grains were noted (Gradzinski and Jerzykiewicz, 1974a, p. 257). Mean values for sphericity ranged from 0.66 to 0.72. Surface frosting was recorded for 21 – 45 percent of the grains, and both ridges and pitting were not uncommon on sands of the large-scale cross-strata.

Composition of the sandstone, calculated from eight analyses, was 69 – 82 percent quartz, 18 – 28 percent feldspar, and 1 – 4 percent other minerals, including mica and heavy minerals, principally epidote. The feldspar is mostly fresh but locally weathered, and in some samples fresh and weathered are mixed together (Gradzinski and Jerzykiewicz, 1974a, p. 253).

Primary Structures
Stratification

In the Barun Goyot Formation, five sediment types, based on differences in both lithology and primary structures, were recognized by Gradzinski and Jerzykiewicz (1974b). Of these types, the "mega cross-stratified units" and possibly the "massive, 'structureless sandstone' " are probably of eolian origin. The "flat-bedded sandstones," "diversely stratified sandstones," and "alternating claystones and sandstones" probably are mostly or entirely interdune deposits.

The "mega cross-stratified sandstone," as described (Gradzinski and Jerzykiewicz, 1974a, p. 259), seems to correspond in all essential structural features to the large-scale high-angled cross-strata of other sandstones considered to be of eolian-type that are discussed in this chapter. Its cross-strata dip at high angles (generally 25° – 30°, maximum 36°). They are mostly tabular planar, less commonly wedge planar, and rarely trough shaped. Bounding planes between sets of cross-strata are mostly even and horizontal for long distances and form sets of cross-strata 3 – 10 m (9 – 30 ft) thick, mostly greater than 5 m (15 ft). The length of individual cross-strata commonly is considerable — several tens of metres (90 ft or more). Dip directions are relatively constant, with variations less than 20°.

Because the "mega cross-stratified sandstones" grade both laterally and vertically into "massive, 'structureless' sandstones," some of the latter may also represent wind-deposited sand, but internal structures in them have largely been destroyed. "Indistinct traces of internal structures are visible in some beds * * * [including] horizontal lamination and deformational structures of overturned folds" (Gradzinski and Jerzykiewicz, 1974b, p. 134). Thus, slumping or other types of contortion may have contributed to eradication of primary structure.

Contorted Bedding

Within sets of large-scale cross-strata — between sets that are undeformed — are overturned folds and other varieties of contorted and disrupted bedding indicating a penecontemporaneous origin of the structures (Gradzinski and Jerzykiewicz, 1974a, p. 263, fig. 11a). Deformational structures also are reported from the "structureless" sandstone facies suggesting that processes of contortion may be responsible for the partial or complete obliteration of primary structures in parts of the Barun Goyot Formation (Gradzinski and Jerzykiewicz, 1974a, p. 274; 1974b, p. 134).

Detailed descriptions of the "contorted and disrupted" cross-strata in the Barun Goyot Formation are not available; thus, it cannot yet be determined whether they are structures resulting from loading in loose sand, of the types described from Brazil (McKee and Bigarella, 1972, fig. 4), or whether they were formed by avalanching of sand deposits, as demonstrated in laboratory experiments (McKee, Douglass, and Rittenhouse, 1971, p. 372). If the Barun Goyot sands include overturned folds or overthrusts, the noncohesive or cohesive nature of the sand at the time of deformation might be determined, and, from this information, inferences could be made concerning the dryness or wetness of the environment.

Interdune Deposits

Strata interpreted as interdune deposits are abundantly represented in the Barun Goyot Formation. They consist of facies referred to by Gradzinski and Jerzykiewicz (1974b, p. 136) as "diversely stratified sandstones" and "alternating claystones and sandstones." The first of these types occurs almost exclusively in the lower part of the formation, where it fills erosion channels as much as 3 m (9 or 10 ft) wide, scoured in the large-scale cross-strata of eolian type. The "alternating claystones and sandstones," in contrast, are mostly sheetlike, horizontal or wavy strata that interfinger with the eolian-type large-scale cross-strata and, therefore, were formed contemporaneously with them.

Textural and structural characteristics of the "diversely stratified sandstones" are (1) the common inclusion of small, locally derived pebbles, (2) the incorporation of a few large foreign pebbles and cobbles and numerous skeletal remains of vertebrate animals, (3) poorly sorted sandstones, (4) common trough-type cross-strata, (5) wedging of bedding units within short distances, (6) large variation in the dip directions of cross-strata, and (7) the filling of erosion pockets and scoured sur-

faces (Gradzinski and Jerzykiewicz, 1974b, p. 134). All these features suggest deposition by stream currents.

The facies described as "alternating claystones and sandstones" consists of "quiet water" deposits as indicated by bedding that is commonly flat or wavy, a scarcity of crossbedding, truncation of underlying eolian-type cross-strata, strong deformation of mud layers, and deformational features in sand, such as load casts, disrupted sandstone beds, and isolated sandstone blocks.

An additional facies of the Barun Goyot Formation, referred to as "flat bedded sandstone" by Gradzinski and Jerzykiewicz (1974b, p. 125), occurs mostly in the upper part of the formation and probably represents a major change or shift in environment following eolian deposition. It consists of units a few metres thick composed mostly of sandstone, with some siltstone and sandy claystone. Bedding is largely horizontal or low angle (2°–5°). The low-angle beds occur between horizontal sets and commonly flatten laterally. Erosion channels are rare. This facies is interpreted as playa type (Gradzinski and Jerzykiewicz, 1974b, p. 129–140).

Fauna

No other eolian-type sandstone has produced so rich and varied a fauna as has the Barun Goyot of central Asia. From the sandstones of this formation "several specimens of dinosaurs, very numerous dinosaur eggs, a few crocodiles, several tortoises, a pterosaur, about 80 specimens of lizards, about 50 specimens of mammals .* * * were found within a few weeks in the Khulsan and Nemegt localities" (Gradzinski and Jerzykiewicz, 1974b, p. 140).

Most of these fossils were found in nondune facies — that is, in deposits believed to be interdune consisting of various water-deposited sediments, especially those attributed to intermittent streams and lakes. However, in the large-scale cross-strata of dune-type sandstone, a few dinosaur skeletons and numerous dinosaur eggs have been found. Furthermore, in the massive "structureless" sandstones "in the immediate neighborhood of mega cross-stratified units," nearly complete dinosaur skeletons and many intact dinosaur eggs were found. The distribution of these fossils clearly seems to have been controlled by ecological considerations — the wet or moist vegetated interdune areas being far more favorable, both for the living fauna and for its preservaiton than the bare dune environment (Gradzinski and Jerzykiewicz, 1974b, p. 141).

Environment of Deposition

Interpretation of the depositional environments represented in the Barun Goyot Formation, supported by a considerable amount of evidence, was presented by Gradzinski and Jerzykiewicz (1974a, p. 272). The thick units of large-scale cross-strata (termed "mega cross-stratified units") are considered, principally on the basis of texture and structure, to have formed as wind-deposited dunes. Furthermore, from the direction and narrow spread of cross-strata dip directions, they are believed to have resulted from unidirectional, westerly winds and to have been primarily of the transverse-ridge dune type.

Interdune deposits of at least two varieties are interspersed within the windblown sands. Some of these interdune deposits apparently represent a fluvial variety, formed by flooding of interdune areas; they include both "upper flow regime" sands (largely horizontal beds) and "lower flow regime" types. These deposits — the "diversely stratified sandstones" of Gradzinski and Jerzykiewicz (1974b, p. 139) — include many gravels, channel-fill sediments, wavy laminae, trough bedding, and other indicators of current action.

A second variety of interdune deposits consists of alternating claystones and noncrossbedded sandstones. These strata, deposited synchronously with the eolian sandstones, probably formed in intermittent lakes (Gradzinski and Jerzykiewicz, 1974b, p. 136, 138).

The relatively great thickness, 2–4 m (6–12 ft), of both fluvial-and lacustrine-type interdune units, suggests either that dune migration and burial of interdunes was slow or that interdune sedimentation was much more rapid than for other ancient eolian sandstones described herein. The clear evidence of ponding and the abundance of fossil vertebrate remains suggest considerable vegetation in the interdune area.

Summary of Criteria for Eolian Origin

Evidence that large-scale high-angle cross-strata ("mega cross-stratified units") of the Barun Goyot Formation were wind deposited was summarized by Gradzinski and Jerzykiewicz (1974a, p. 272; 1974b, p. 137) — the considerable thickness of individual sets, the high-angle dips of most cross-strata, the tabular-planar form that predominates among crossbedding sets, and the wide, even form of major bounding surfaces. Furthermore, the relatively constant dip direction of cross-strata and the lateral continuity of sets strongly suggest that the sandstone is formed of dunes of the transverse ridge type developed under the influence of unidirectional winds.

Textural features — fine- to medium-grain size, moderate sorting (good to fair on the Payne scale), and common grain-surface features including frosting and pitting — are all characteristic of most modern dune deposits.

Additional data bearing on the environment of deposition are the associated and intertonguing deposits of flat-bedded and wavy-bedded sands and muds, the thin conglomerate beds, the abundant nonmarine fauna (including dinosaur skeletons and eggs), and other features of interdune deposits that are interspersed among typical dune-type sandstones.

Botucatú and Sambaiba Sandstones of South America (Jurassic and Cretaceous)

By JOÃO J. BIGARELLA

THE BOTUCATÚ SANDSTONE of Brazil constitutes one of the world's largest and most widespread sequences of sandstone of probable eolian origin (fig. 159) — an extent of more than 1.3 million km² (800,000 mi²). This sandstone not only occurs throughout the Paraná Basin (States of São Paulo and Paraná), but also extends beyond this basin's present edge into the States of Mato Grosso and Minas Gerais. Probably, the Botucatú Sandstone originally connected with similar sandstones of the Parnaiba Basin in northern Brazil (States of Maranhão and Piauí). The present distribution of these sandstones is shown in figure 159.

In northern Brazil, the Sambaiba Formation of the Parnaiba Basin is considered to be equivalent to the Botucatú sequence. This formation is composed of large-scale crossbedded sandstones, alternating with probable subaqueous deposits.

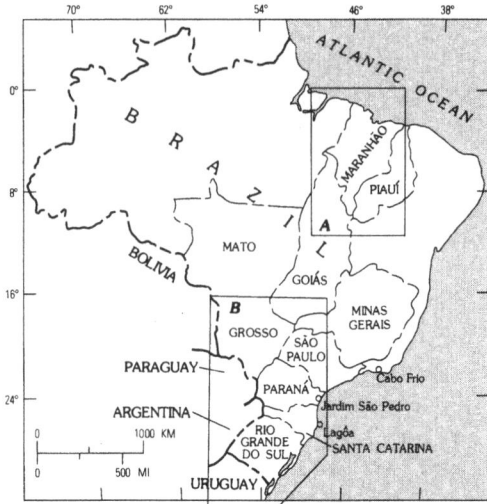

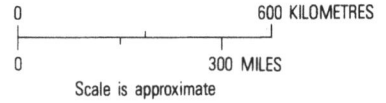

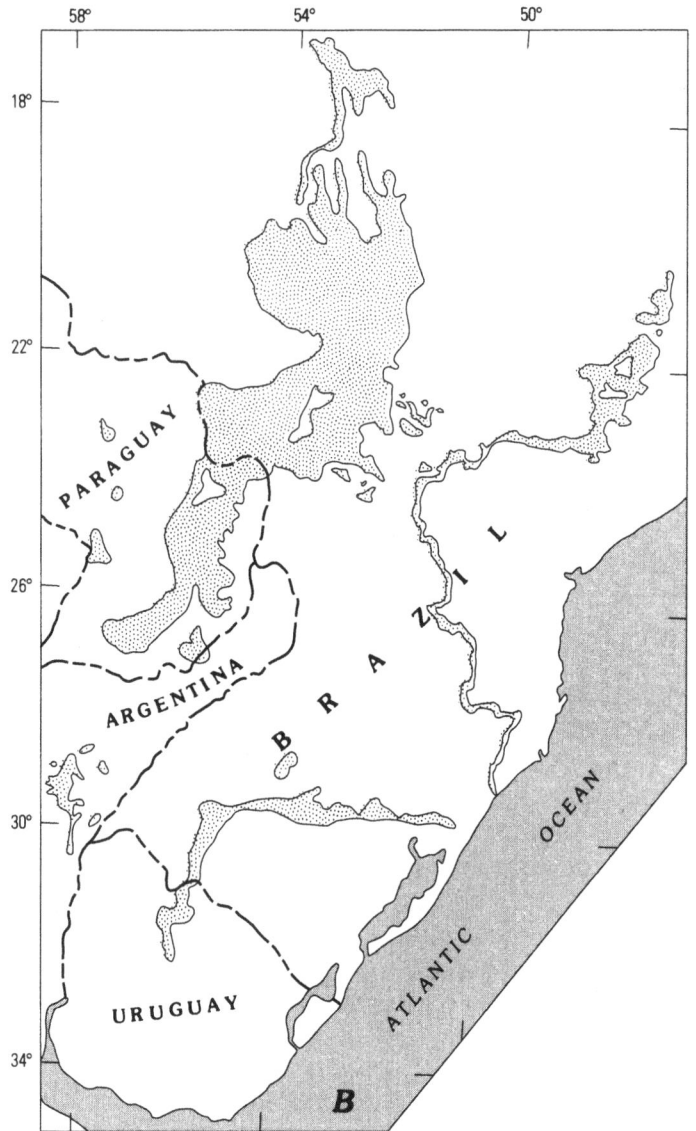

OUTCROPPING AREAS OF SELECTED MESOZOIC eolian sandstones (dark stipple) in South America: A, Sambaiba Sandstone; B, Botucatú Sandstone. (Fig. 159.)

The Botucatú Sandstone, first described by Gonzaga de Campos (1889), includes all large-scale cross-stratified sandstone units intercalated with or underlying the Serra Geral basaltic lava flows. Another sequence of large-scale cross-stratified sandstones of probable eolian origin overlies basaltic flows in the States of Mato Grosso, São Paulo, and Paraná (fig. 159). Some authors have designated it as the Caiuá Sandstone (Washburne, 1930; Maack, 1941), but others (Almeida, 1954) consider it to be a continuation of the Botucatú Sandstone.

The thickness of the Botucatú Sandstone is generally 10–20 m (30–65 ft) but is locally more than 30 m (100 ft).

The Botucatú Sandstone was long considered to be Triassic-Jurassic in age. However, radiometric age determinations of the associated and intercalated basaltic rocks indicate an average age of 120–130 m.y. for the lava flows (Cordani and Vandoros, 1967), and so its age is now considered to be Late Jurassic–Early Cretaceous.

Facies

The sandstone facies of the Botucatú, with large-scale cross-stratification considered to be eolian, is in some places interbedded with another facies, the Piramboia, believed to be fluviatile (Pacheco, 1913). This fluviatile facies generally occurs stratigraphically below eolian sandstones and probably represents intermittent stream deposition in small basins without outlet.

Another supposed fluviolacustrine or playa-lake sequence, which is interbedded with the large-scale crossbedded eolian sandstones of the Botucatú, has been called Santana facies (Almeida and Barbosa, 1953, p. 64). "The Santana facies contains a Conchostraca (branchiopod) assemblage consisting of *Bairdestheria barbosai*; *Palaeolimnadia petrii*; *Euestheria mendesi*; *Pachecoia acuminata*; *P. rodriguesi*; *Candonopsis pyriformis candona?*; *Estheriella sp.*"

Sandstones of probable fluviatile origin occur in the lowermost part of the Botucatú sequence and are interbedded with the large-scale cross-stratified sandstones of probable eolian origin. They crop out in many places in the States of Paraná, São Paulo, and Minas Gerais.

The equivalent of the Botucatú Sandstone is known in Uruguay as the Tacuarembǒ Sandstone and in Paraguay as the Misiones Formation. However, not all the strata mapped as Misiones Formation (Eckel, 1959) are actually equivalent to the Botucatú Sandstone.

Texture

In the Paraná Basin of Brazil, for a distance of 2,500 km (1,550 mi) along the eastern belt of outcrops, 99 samples of eolian-type Botucatú Sandstone were collected for textural studies. Because most samples contained tails of fine-grained material (silt and clay) introduced by weathering or by postdepositional impregnation, grain-size distribution was recalculated, based on quartz grains, to be 0.062 mm. The recalculated samples contain three or four textural classes with more than 1 percent frequency. In the Botucatú Sandstone three classes predominate in much of the formation, but close to source areas in the crystalline shield many of the samples include four textural classes. Sandstones — supposedly fluviatile but intercalated in the dune sequence — contain as many as seven textural classes.

In the Botucatú paleodesert, the modal class, according to recalculated analyses, is mostly fine-grained sand (0.125– 0.25 mm) and includes some medium-grained sand (0.25– 0.5 mm). The medium-grained sand is mainly near the crystalline shield source area; only in a few places is it downwind, toward the central part of the paleodesert.

The average mean diameter of recalculated samples of large-scale cross-stratified Botucatú Sandstone decreases southward from 0.209 mm in Minas Gerais, to 0.183 mm in São Paulo and to 0.178 mm in Paraná; it also decreases northward from 0.233 mm in Rio Grande do Sul to 0.215 mm in Santa Catarina. In both series grain size decreases downwind from crystalline areas. The sandstones of the northern area of the Botucatú paleodesert are generally finer grained (0.186 mm) than those in the southern area (0.218 mm). The northern and southern areas were under the influence of different paleowind regimes. The wind in the southern area was stronger, as revealed by larger cross-strata, larger sand grains, and larger average mean grain-size diameter.

The Botucatú Sandstone is dominantly moderately to well sorted. Sorting mostly is progressively better downwind — that is, away from the crystalline shield areas. Trask sorting coefficients (So) decrease from Minas Gerais (So = 1.44) to São Paulo (So = 1.33) and Paraná (So = 1.32). They decrease also from Rio Grande do Sul (So = 1.46) to Santa Catarina (So = 1.39). The samples from the northern area of the Botucatú paleodesert are slightly better sorted than are those from the southern area. As the mean grain-size diameter increases, the sorting coefficient becomes poorer.

Ventifacts are scattered along some zones of the Botucatú Sandstone.

Primary Structures
Stratification

Cross-stratification typical of eolian slipface deposits is the outstanding sedimentary structure of the Botucatú Sandstone and is abundant in almost all outcrops. The size of sets may range from small to large in the same section. Large sets of cross-strata with lengths of dipping beds of 30 m (100 ft) or more and heights of about 12–15 m (40–50 ft) occur in many localities.

Environment of Deposition

Paleowind circulation in the Botucatú paleodesert was investigated by Almeida (1953, 1954), Bigarella and Salamuni (1961), and Bigarella and Oliveira (1966). Measurements in the correlative Sambaiba Formation of northern Brazil were made by Bigarella, Montenegro, and Coutinho (unpub. data).

Crossbedding dip directions in the southern part of the Botucatú paleodesert (Santa Catarina and Rio Grande do Sul in Brazil, and Uruguay) indicate that paleowinds blew from the west and west-south-west. These winds were deflected toward the north in southern Paraguay. In the northern part of the Botucatú paleodesert (Mato Grosso, Goiás, Minas Gerais, São Paulo, and Paraná in Brazil, and part of Paraguay), crossbedding measurements indicate mean paleowind trends from the north or north-northeast, representing return tradewinds. In Paraná, these northerly paleowinds were then deflected toward the west.

In the Botucatú paleodesert, southern air masses moved northward, leaving their traces mainly in Paraná and São Paulo, and, less frequently, the northern air masses penetrated the southern areas; so, cold air masses and warmer ones probably conflicted. Rainfall was therefore probably greatest north of Santa Catarina. This postulate is supported by deposits in that region of the Piramboia and Santana facies, which seem to represent subaqueous environments. Such deposits, which are at the base of, and interbedded with, supposed eolian deposits, are common in Paraná and São Paulo but decrease in abundance northward. Waterlaid deposits are rare in the southern part of the Botucatú paleodesert (Bigarella and Salamuni, 1961).

Crossbedding measurements in Parana have a very low consistency ratio, probably mostly the result of the conflict between opposing wind directions, westerly and northerly. The area of conflict between these winds extends westward into Paraguay.

Crossbedding measurements of the Sambaiba Formation, which is the Botucatú Sandstone equivalent in northern Brazil, indicates that prevailing winds responsible for sand movement were the southeasterly tradewinds. The Botucatú and Sambaiba eolian sandstones were deposited in a low-latitude paleodesert.

Summary of Criteria for Eolian Origin

Criteria responsible for considering major parts of the Botucatú and Sambaiba Sandstones to be of eolian origin consist of sedimentary structures and grain textures. The most compelling structural evidences are the very large scale and the high angles common to much of the wedge-planar and tabular-planar cross-stratification. Textural features are entirely compatible with this interpretation. Throughout a large part of Brazil, the great extent of moderately sorted to well-sorted sand, which is dominantly of fine to medium size, can be explained by few environments other than eolian. Additional evidence of genesis is furnished by scattered ventifacts.

Cave Sandstone and Similar Sandstones of Southern Africa (Triassic)

By JOÃO J. BIGARELLA

THE SUCCESSION OF GONDWANA BEDS in southern Africa is known as the Karroo System. The basal beds, named the Dwyka Group, are interpreted as glacial and periglacial deposits. The uppermost sedimentary beds are believed to have been deposited under arid or semiarid conditions, although much of the sand was deposited in subaqueous environments. These uppermost strata are known by various names — Cave Sandstone in the Republic of South Africa, Botswana, Lesotho, and Swaziland; the Bushveld Sandstone in the central Transvaal; Etjo Sandstone in South-West Africa; and the Forest and Nyamandlovu Sandstones in Rhodesia. They are mostly Triassic in age but probably represent a range of about 75 m.y.

These sandstones are conformably overlain by or intercalated in their upper parts with lavas of various compositions. Sedimentation continued after the first outpourings of lava. The comparatively thin layers of sandstone between flows indicate the absence of a major unconformity. In places, as in northern Transvaal, the prelava sandstone surface is very uneven, possibly reflecting an uneven dune topography or contemporaneous erosion by running water. The main outcrops of the sandstone in Lesotho were studied by Beukes (1970), who divided the sequence into three parts: lower and middle massive to thick-bedded sandstones, and an upper sandstone containing large-scale cross-stratification. In Rhodesia the upper part is called the Nyamandlovu Sandstone, and its cross-strata are generally considered to be of eolian origin.

Distribution and Thickness

The Cave Sandstone and equivalents in the Republic of South Africa, Lesotho, and Rhodesia conformably overlie a succession of red and purple mudstones, shales, and sandstones. The Cave Sandstone attains a maximum thickness of about 350 m (1,150 ft) in the Orange Free State and in northern Transvaal, but in most other places, it is less than half this thickness. In Rhodesia it is relatively thin; in South-West Africa it generally is 60 – 150 m (200 – 500 ft) thick and attains about 600 m (2,000 ft) in maximum thickness.

Composition and Texture

The composition of the Cave Sandstone differs considerably from place to place. In the upper cross-stratified part, detrital grains include 60 –73 percent quartz and 1.0 – 9.3 percent feldspar. Miscellaneous matrix ranges from 10 to 21 percent (Beukes, 1970). The remainder of the rock consists of rock fragments, silica cement, and heavy minerals.

The mean grain size of the cross-stratified sandstones ranges from 0.170 mm on the northwestern side of Lesotho to 0.080 mm on the southeastern side.

Primary Structures
Stratification

Along the Huab Valley in South-West Africa, and also in parts of northern Transvaal, dunelike structures are very prominent in the Cave Sandstone. However, in most localities in Lesotho, Orange Free State, and along the Lebombo Mountains of eastern South Africa and Swaziland, this sandstone is not clearly stratified. The unstratified layers may represent deposition by flash floods that originated in a semiarid climate. Judged by texture, these layers probably consist of either mudflows or sand flows.

Probably the Cave Sandstone from Lesotho and the Lebombo Mountains should be renamed because it and the Bushveld Sandstone are much older than either the well-cross-stratified eolian sandstone of northern Transvaal near the border with Botswana or the Etjo Sandstone of the Huab Valley.

The Etjo and Cave Sandstones are characterized by eolian-type cross-stratification, mostly large to medium wedge-shaped sets, interpreted as products of transverse and barchan dunes. Some convex-upward strata may be related to parabolic dunes or other types of eolian bedforms which produce convex-upward structures (that is, retention dunes, protuberances on transverse dunes).

Within the same section, the size of the cross-strata sets ranges from small to large. Cross-strata consist of the wedge-planar and tabular-planar types. Dips of the strata very seldom exceed 33°; the maximum dip recorded is 40°. The average dips are:

Cave Sandstone (around Lesotho) . . . . . 16.1°
Cave Sandstone (northern Transvaal) . . 21.9°
Etjo Sandstone . . . . 21.9°

Fauna

Remains of reptiles, crustaceans, fish, insects, and plants indicate an age of Late Triassic or older for the Cave Sandstone in South Africa and Rhodesia (Haughton, 1969, p. 398). The remains of charred trees just below the first lava flow occur in the Orange Free State and reptilian tracks are fairly common in South-West Africa, Lesotho, and the Orange Free State. Sharp, clear tracks of at least five different types of reptiles were recently discovered in the northern Transvaal on dune cross-strata (foresets) having dips of about 27°. At the Kariba Dam in Rhodesia, reptilian remains occur in sandstone intercalated in lava.

Environment of Deposition

Although the climate probably varied considerably, the main bulk of the sandstone apparently was not deposited under extreme desert conditions. In many places, the presence of water is indicated by current ripple marks, erosion channels, and lenses of shale probably deposited in old depressions.

Summary of Criteria for Eolian Origin

Although sandstones that developed in different environments may show similar structural features, the Cave Sandstone interpretation as eolian depends not only on its characteristic cross-strata pattern but also on associated environmental features. As in most eolian sandstones, its sorting is better than for subaqueous sandstones. The Cave is neither associated nor intercalated with sequences of siltstone and claystone, and it does not include gravel beds.

Convex-upward cross-strata are believed to be typical eolian features developed in some types of dunes. Similar types of cross-strata rarely occur in subaqueous sandstones.

A STUDY OF GLOBAL SAND SEAS

SEDIMENTS OF THE ANCIENT EOLIAN ENVIRONMENT—RESERVOIR INHOMOGENEITY

Chapter I

By ROBERT LUPE and THOMAS S. AHLBRANDT

Contents

Illustrations

Summary of Conclusions

𝖀 PPER PALEOZOIC TO MESOZOIC eolian blanket sandstones of the Colorado Plateau and the Rocky Mountains of Colorado and southern Wyoming are texturally complex. As petroleum reservoirs, they commonly have poor performance histories because they represent a depositional system comprised of three closely associated subenvironments — dune, interdune, and extradune. Sediments of each subenvironment have different textural properties which have resulted from different depositional processes. Compared with interdune and extradune sediments, dune sediments are generally more porous and permeable and may be better quality reservoirs. Extradune sediments (a new term) include all deposits adjacent to a dune field and are mainly subaqueous deposits. Interdune sediments are here restricted to those nondune sediments deposited in the relatively flat areas between dunes. A conceptual model is proposed, in which dune sediments may be enveloped by extradune sediments as the depositional system evolves, resulting in a texturally inhomogeneous reservoir that has poor fluid migration properties.

This model of textural inhomogeneity in eolian blanket sandstones was applied to the Weber Sandstone in the Brady field and its equivalent, the Tensleep Sandstone, in the Wertz and Lost Soldier fields, Sweetwater County, Wyoming, U.S.A. Data were obtained from both outcrop and subsurface and included environmental interpretation, textural analysis, and plotting of the distribution of depositional subenvironments. As expected from the model, dune sediments in Brady field were markedly different in texture from interdune and extradune sediments. The predicted geometric distribution of subenvironments was confirmed in Lost Soldier and Wertz fields. However, in these two fields, secondary cementation and fracturing has obscured the original porosity and permeability contrasts. The distribution of porosity and permeability, a characteristic depending partly on depositional processes, could impede fluid migration in the reservoir and significantly reduce recovery of hydrocarbons.

Introduction

MANY OF THE UPPER PALEOZOIC to Mesozoic blanket sandstones of the Colorado Plateau and the Rocky Mountains of Colorado and southern Wyoming were deposited in eolian and closely associated noneolian environments. They are the reservoir rocks in many oil and gas fields. These sandstone units are widespread and of uniform thickness and seem texturally homogeneous. There are, however, complex textural variations within the units that may affect porosity and permeability and be detrimental to their performance as reservoir rocks. These textural differences are related to environments of deposition, and a conceptual model is proposed to explain them. The model will aid the geologist in evaluating porosity and permeability problems and result in significant savings in time and money.

The eolian depositional system contains three subenvironments which are related both in time and by source: dune, interdune, and extradune. The term "dune" is applied in the classic geomorphic sense of Bagnold (1941). The term "interdune" is restricted to the description of the relatively flat area between the dunes of a dune complex. The new term "extradune" describes the area extending beyond the dune field, including some areas previously referred to as interdune. Interdune sediments probably were originally deposited by the same wind activity that built the dune (McKee and Moiola, 1975) and are commonly reworked by water or wind (Glennie, 1970). They are commonly preserved as the flat-lying sediments at the base of the crossbed sets of a dune sequence (fig. 160). Extradune sediments were probably deposited by processes independent of the dune-building process and involve sediments of diverse environments, such as alluvial fan, wadi, serir, stream, sebkha, playa, lake, beach, and tidal flat.

The new term "extradune" fulfills two purposes: (1) It specifically describes a geomorphic area which has often been incorrectly termed "interdune," and (2) it serves as a general term which could include a variety of subenvironments. The term can be used in general reference to various subenvironments or used when the identification of

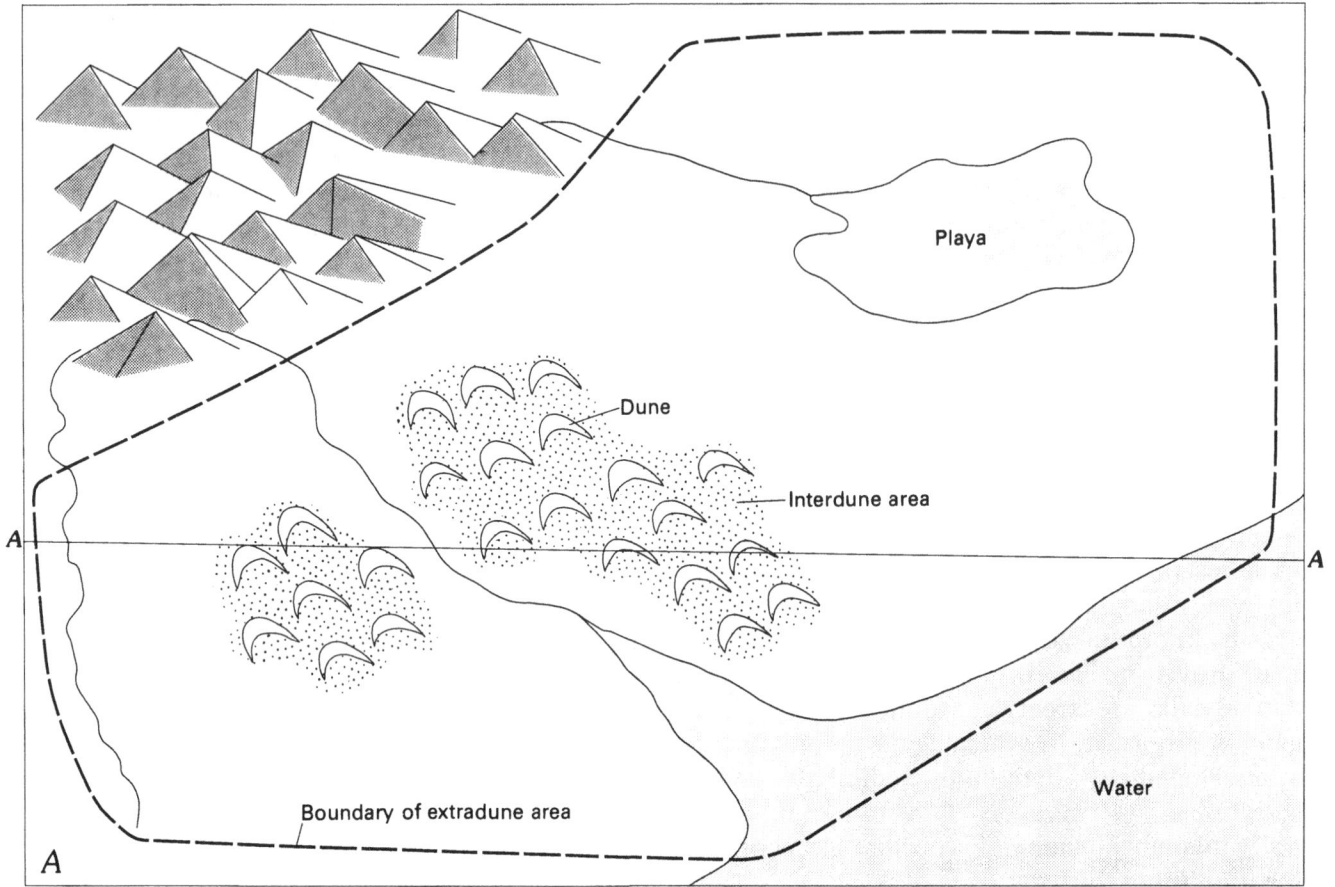

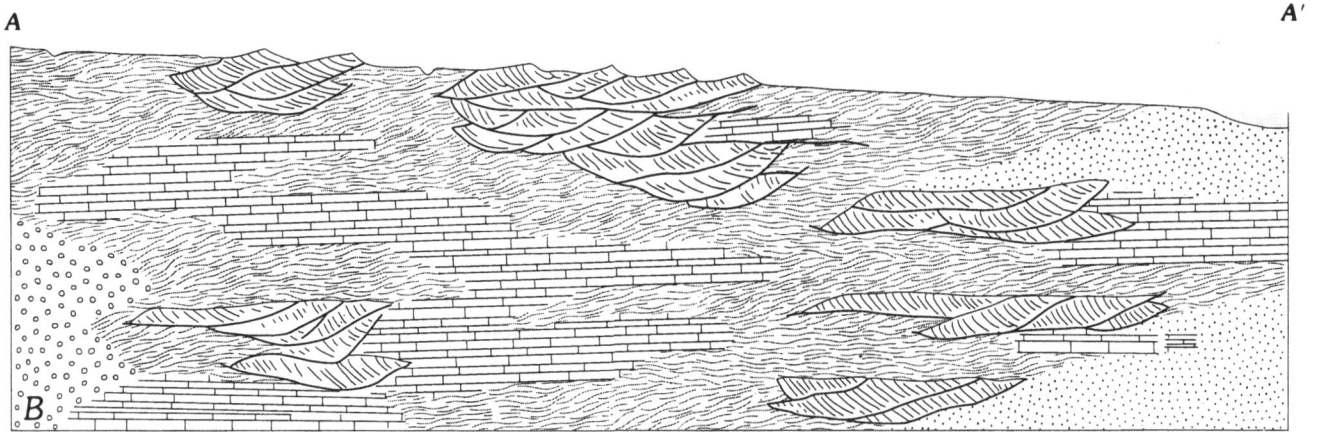

SCHEMATIC DIAGRAM AND CROSS SECTION of dune, interdune, and extradune deposits. *A*, Plan view of typical distribution of dune, interdune, and extradune deposits. *B*, Cross section illustrating evolution of a dune-extradune system, ending with the geographic configuration of *A*. Subenvironments of the system migrate laterally as well as vertically, resulting in isolation of more porous and permeable dune sediments. (Fig. 160.)

specific subenvironments is uncertain but the stratigraphic relation to dune sediments is clear.

Acknowledgments

WE PARTICULARLY THANK Amoco Production Co. for their support and gracious granting of time and expense. We also thank Champlin Petroleum Co., operator of Brady field, and Mountain Fuel Supply Co. and Exxon U.S.A., the other Brady field partners; and Hilliard Oil and Gas Co. for permission to publish data in this report. Pasco, Inc., provided data for the Lost Soldier –Wertz field studies.

The Model

THE DEGREE OF TEXTURAL contrast in dune, interdune, and extradune sediments reflects source-material texture, depositional processes, and the spatial relations of subenvironments within the depositional system.

Dune sediments are relatively better sorted, better rounded, and more compositionally mature than are sediments of most other depositional environments (Bagnold, 1941; Twenhofel, 1945; Kuenen, 1960; Kukal, 1971; Ahlbrandt, chapter B, this paper). And, because porosity and permeability in modern sediments are largely functions of particle size, shape, and sorting (Griffiths, 1967), dune sand is generally more porous and more permeable than are the source, interdune, or extradune sediments.

Extradune sediments are predominantly subaqueous and, therefore, are usually more poorly sorted and less porous and permeable than are dune sediments. Additionally, porosity and permeability in these sediments could be further reduced because the topographically lower interdune and extradune areas are commonly the sites of early cementation (Glennie, 1972). Figure 161 shows the distribution of porosity and permeability in modern dune-interdune sediments; figure 162 illustrates the distribution of porosity and permeability for ancient dune-extradune sediments.

Dune and extradune sediments abut laterally, and successive dune-extradune depositional systems may overlie one another, with dune sands in vertical contact either with other dune sands or with extradune sediments (fig. 160). Dune sand bodies may become isolated from one another, with

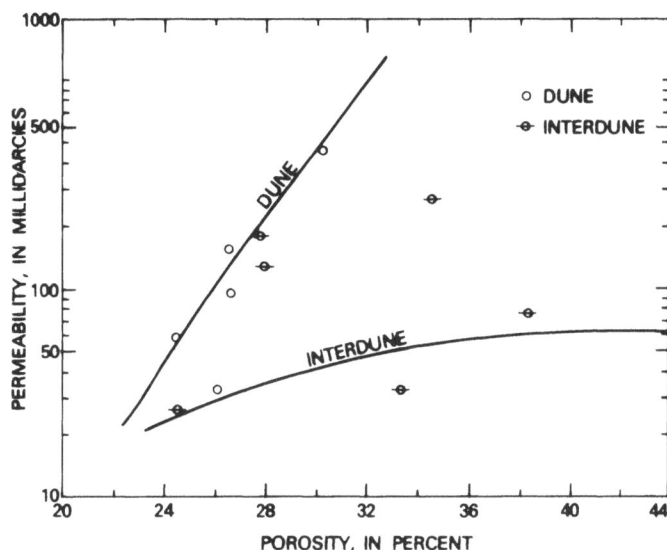

DISTRIBUTION OF POROSITY and permeability in dune and interdune sediments from cores, White Sands National Monument, New Mexico, U.S.A. (Fig. 161.)

the degree of isolation depending on the original distance between dune complexes and on the way in which succeeding depositional systems overlie one another. Movement of fluid would be affected. Furthermore, the porosity and permeability contrast may be enhanced by the effects of burial (Rittenhouse, 1971).

The Model and the Subsurface Situation

EXTENDING THE DUNE-EXTRADUNE model for eolian blanket sands to the subsurface has potentially valuable results for the petroleum industry. Two preliminary steps are required — first, recognition of sedimentary environments from conventional well-bore data, and, second, recognition of the relations between depositional environments and the distribution of porosity and permeability, as just described. Then, the size and distribution of various subenvironments must be determined. In some places the spatial relations of subenvironments can only be approximated from outcrop and (or) subsurface information. Outcrop data may not resemble the actual subsurface situation, and subsurface data may not be geographically dense enough to give a complete picture.

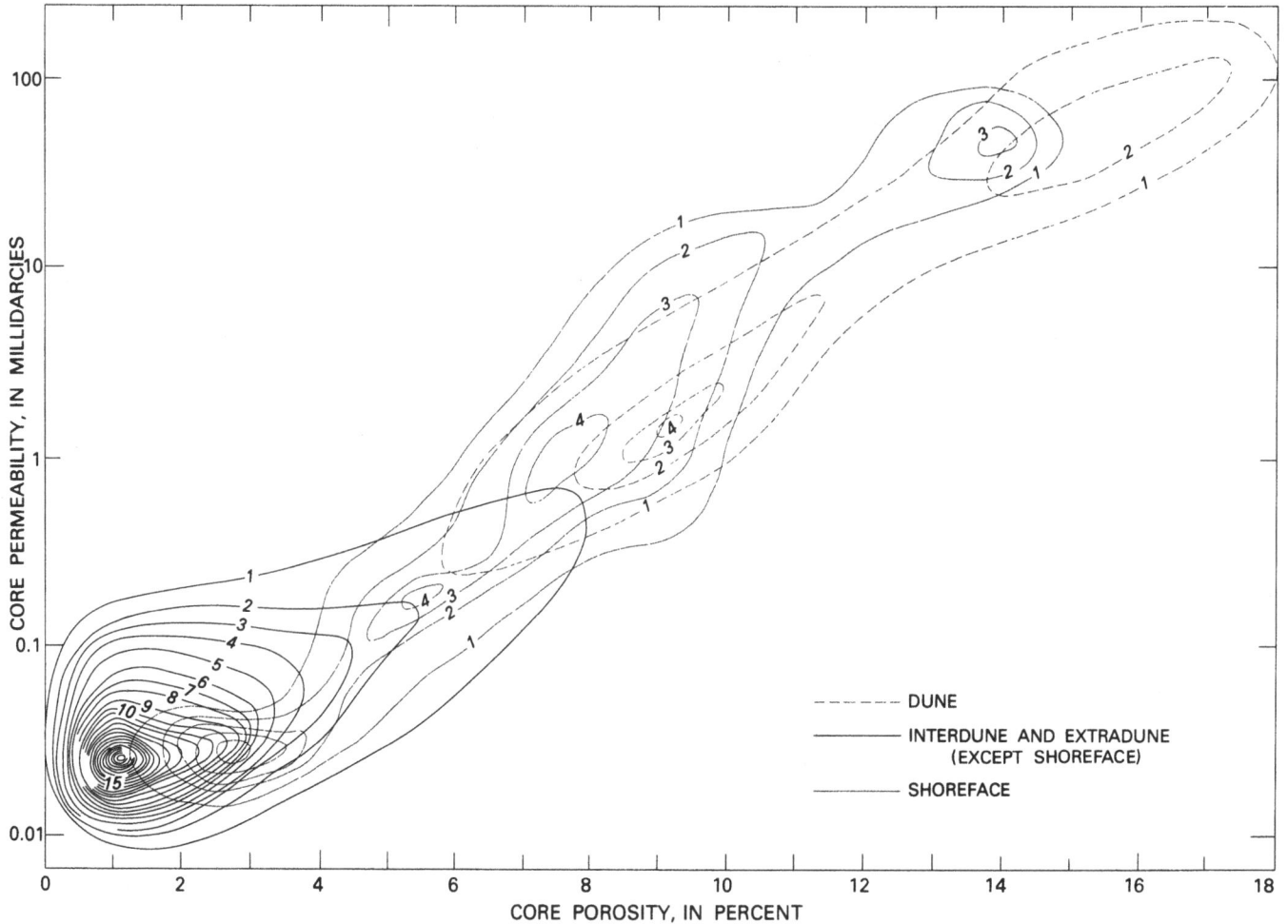

GRAPH OF POROSITY VERSUS PERMEABILITY, showing delineation of three known environments of deposition. Values were obtained each foot from 630 feet (192 m) of core of the Weber Sandstone from three wells in the Brady field, T. 16 N., R. 101 W., Sweetwater County, Wyoming, U.S.A. A grid was superimposed over the plotted data, and the number of points within each square was counted and assigned to the center of each square. Those points are data points for contours (Fig. 162.)

Conversely, subsurface information can provide accurate quantitative data for the model. Routine log analyses include porosity and, commonly, permeability determinations. Cores can provide not only porosity and permeability data but also information on depositional environments. Electric logs, if periodically calibrated with core data, can also provide information on depositional environment.

Figure 163 is an example in which conventional logs were used to interpret environments of deposition. The dipmeter is the most effective tool for describing sedimentary environments (Gilreath and Maricelli, 1964; Campbell, 1968), but the resistivity curve of modern electric logs may also provide the same data. In figure 163 the dipmeter log accurately recorded the high-angle dune crossbed sets and the flat-lying interdune and extradune beds. The resistivity curve of the dual induction laterolog also responded to the same sedimentary structures. The resistivity curve responded inversely to porosity differences — the lower the porosity, the higher the resistivity. In this example, the porosity contrast resulted from porosity differences of sediments deposited in different sedimentary environments — that is, more porous dune sediments versus less porous interdune and extradune sediments. Hence, the most basic suite of logs may not only provide the quantitative data for the construction of a model but also discriminate subenvironments in blanket eolian sandstones, if secondary cementation has not obscured the original porosity distribution. The savings of time and money could be significant.

Application

A STUDY OF THE PERMIAN and Pennsylvanian Weber Sandstone in the Brady field and its correlative 90 miles (150 km) to the northeast, the Tensleep Sandstone in Lost Soldier and Wertz fields, Sweetwater County, Wyoming, tested the application of the dune-extradune model to the prediction of reservoir complexities. The Weber and the Tensleep were selected for three reasons: (1) They were considered to be an eolian blanket sandstone. (2) They had a history elsewhere of requiring an unpredictably large number of closely spaced wells to drain a field. For example, at Rangely field, Colorado, Chevron Oil Co. is developing the field on only 10-acre spacing because of poor fluid migration (Western Oil Reporter, December 10, 1973). The cause of the poor fluid migration was unknown. The 10-acre spacing required to adequately develop the Tensleep of Little Buffalo basin field, northwest Wyoming, described by Emmett, Beaver, and McCaleb (1972), was attributed to reservoir heterogeneity. The possibility of similar poor fluid migration reservoir performance in the Weber Sandstone in the Brady field or other deep fields, where drilling is costly, made the question an economic one. (3) The active exploration underway in 1975 in the Weber in the Brady field and elsewhere permits further testing of the dune-extradune model and resultant porosity and permeability problems.

The study of the Brady, Lost Soldier, and Wertz fields was based largely on core examination. Four thousand five hundred feet of core from 45 wells in Lost Soldier and Wertz fields was described in detail, and depositional environments were differentiated by Reynolds, Ahlbrandt, Fox, and Lambert (1975, 1976) and by Reynolds and Fox (1976). Six hundred fifty feet of core from three wells in Brady field was examined by Lupe. In addition, core porosity and permeability were measured in the cores, and Weber Sandstone outcrops in the Uinta Mountains, 175 miles (280 km) southwest of Brady field, were examined.

Depositional environments were interpreted from cores of selected wells in these fields. The Weber or Tensleep Sandstone is a dune-extradune system composed of mixed nearshore marine and eolian sediments (fig. 164). Dune sandstones were preserved only in Wertz and Brady fields, where they are isolated by foreshore, shoreface, tidal flat, supratidal, or sebkha extradune sediments, and to a lesser degree, interdune deposits. Furthermore, dune sandstone was preserved in upward-thinning transgressive and regressive depositional cycles, which are normally bounded by intraformational unconformities.

Dune sandstones were recognized by their thick bedding, their even and relatively continuous parallel laminae, which commonly dip at high angles, and their lack of holes from burrowing animals. They are very fine grained to fine-grained, well-sorted to very well sorted sandstones containing frosted grains. Interdune deposits separate dune sandstones and are relatively thin (less than 10 ft (3 m)) and horizontally bedded. They contain poorly sorted bimodal sandstones with frosted grains and evaporite cements, commonly anhydrite. Flaser bedding and adhesions ripples were observed in some interdune sediments.

The general criteria used to differentiate extradune subenvironments included many described by Campbell (1971). Supratidal and sebkha deposits contain rip-up clasts, flaser bedding, adhesion ripples, abundant precipitates (such as anhydrite and dolomite), birdseye structures, and desiccation features. Tital flat deposits have discontinuous wavy nonparallel laminae and algal structures, all of which are horizontally bedded and commonly burrowed. Foreshore sandstones contain relatively discontinuous, even-parallel to wavy-parallel laminations, which dip less than 10° and contain rare burrows and scour-and-fill structures. They are moderately to well sorted. Shoreface deposits are extensively burrowed and contain contorted, slumped, and discontinuous wavy nonparallel to parallel relict laminae with maximum dips of 20°. Symmetrical wave ripples were occasionally observed in the relict laminae. The shoreface deposits are also distinctly finer grained than juxtaposed foreshore deposits.

Although the original environments of deposition and textures of the Weber and Tensleep Sandstones in the Brady and Lost Soldier—Wertz fields were virtually identical, the present porosity and permeability of the two areas are significantly different. The Tensleep in the Lost Soldier—Wertz fields has been tightly cemented to an average intergranular porosity of 5 percent, and production is

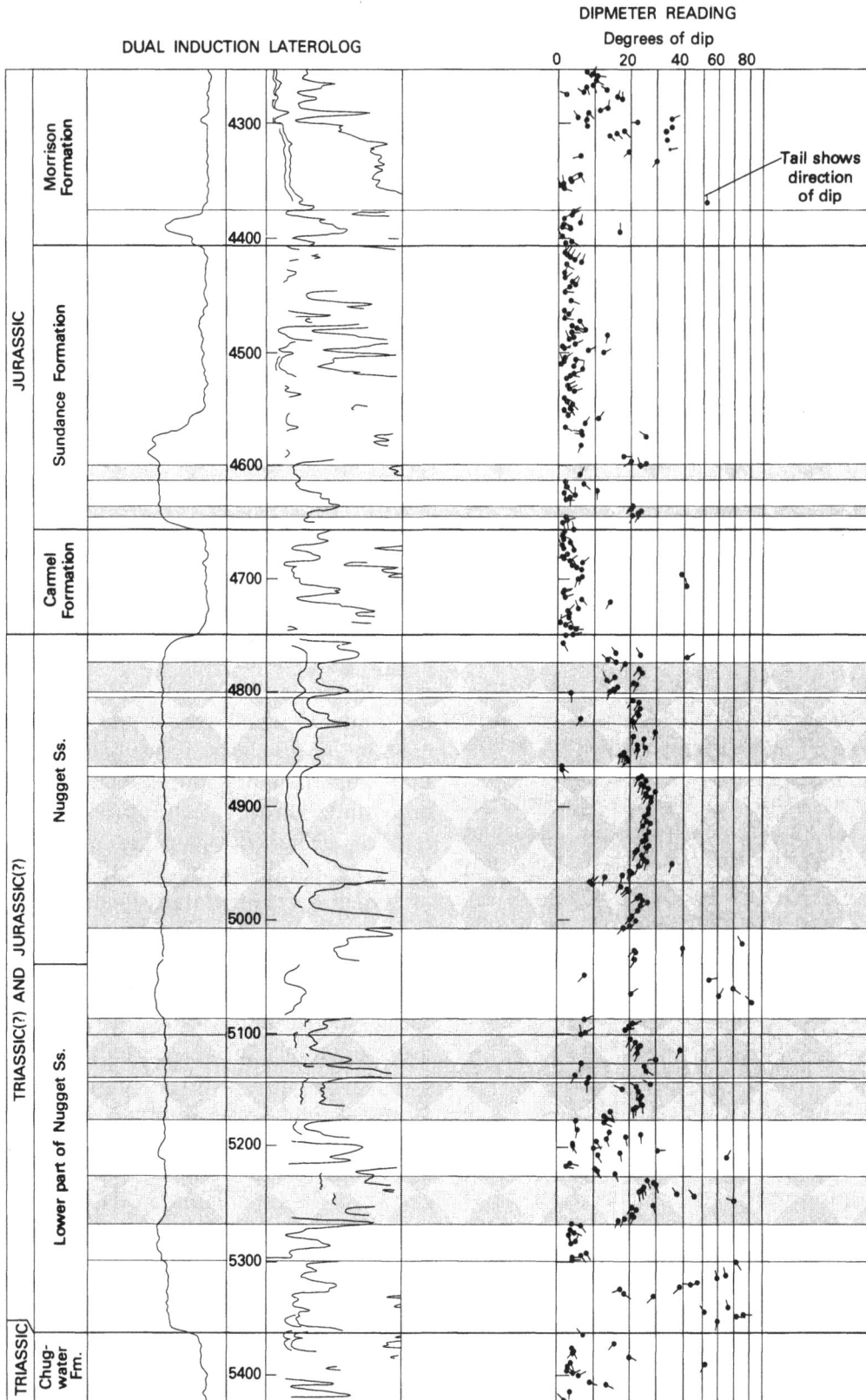

RECOGNITION OF SEDIMENTARY STRUCTURES and depositional environments from electric logs (Hilliard Oil and Gas Co., Joyce dual induction laterolog distinguish between dune, interdune, and extradune sediments. Note correlation of porosity with

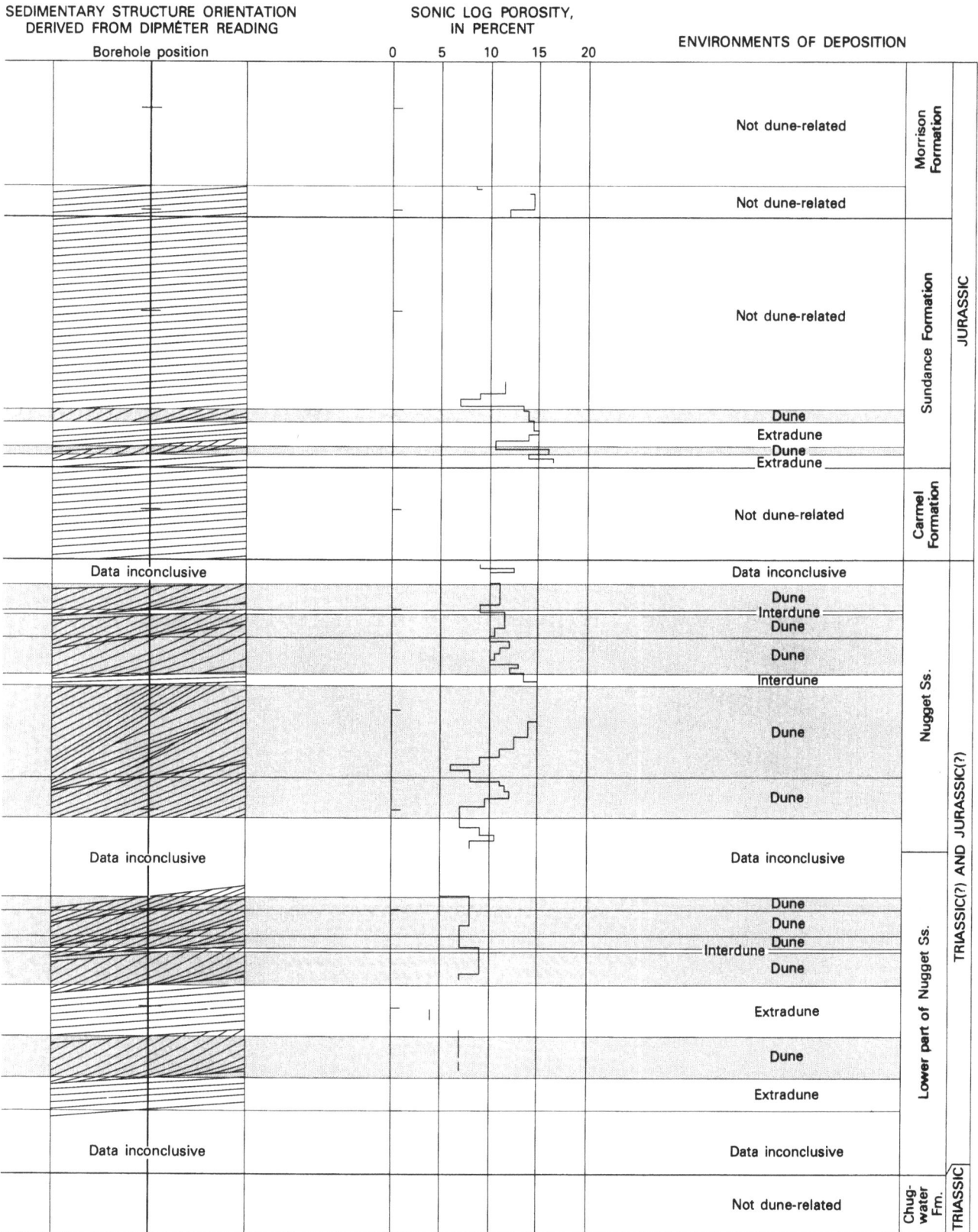

Creek 1, SE¼NW¼ sec. 8, T. 15 N., R. 103 W., Sweetwater County, Wyoming, U.S.A.). Both the dipmeter and resistivity curve of sedimentary structures — that is, with depositional environment. (Fig. 163.)

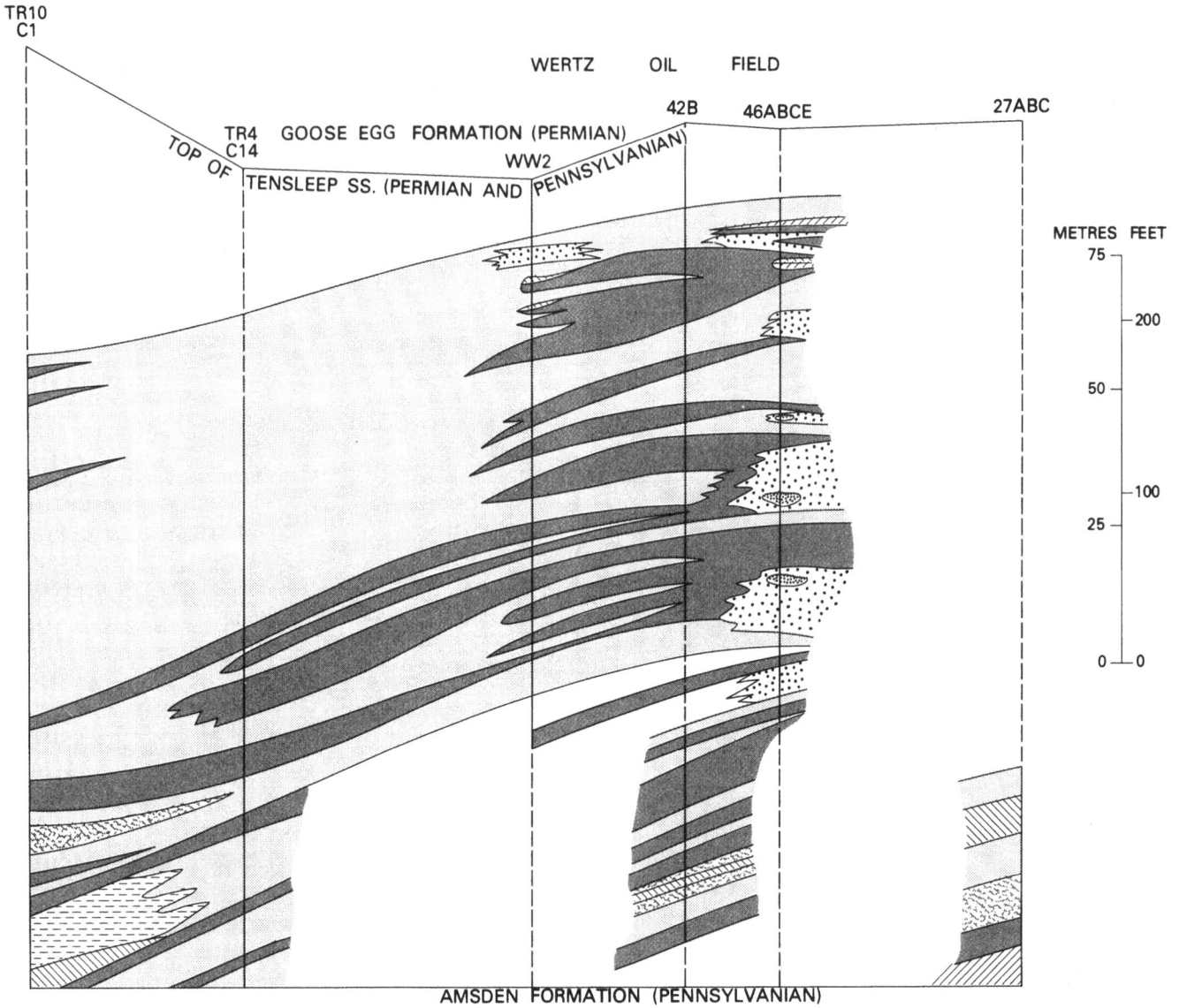

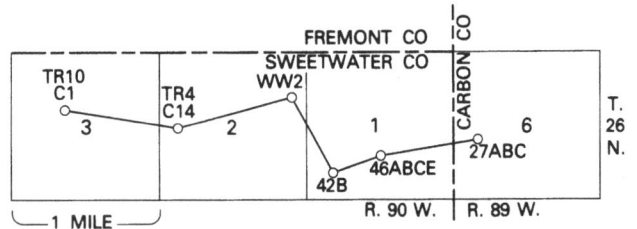

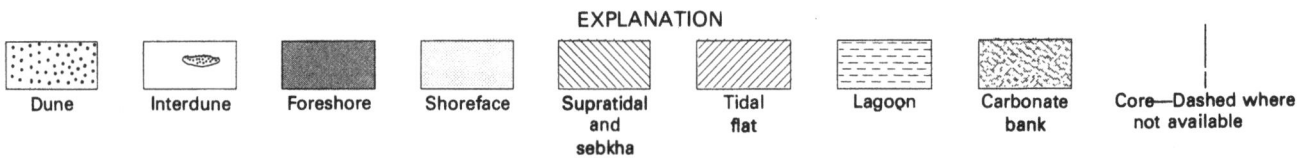

CROSS SECTION OF DEPOSITIONAL ENVIRONMENTS interpreted from cores of the Tensleep Sandstone in the Lost Soldier and Wertz oil fields, Sweetwater County, Wyoming, U.S.A.   From Reynolds, Ahlbrandt, Fox, and Lambert (1975, 1976) and Reynolds and Fox (1976). The top of the Amsden Formation is used as a horizontal datum. (Fig. 164.)

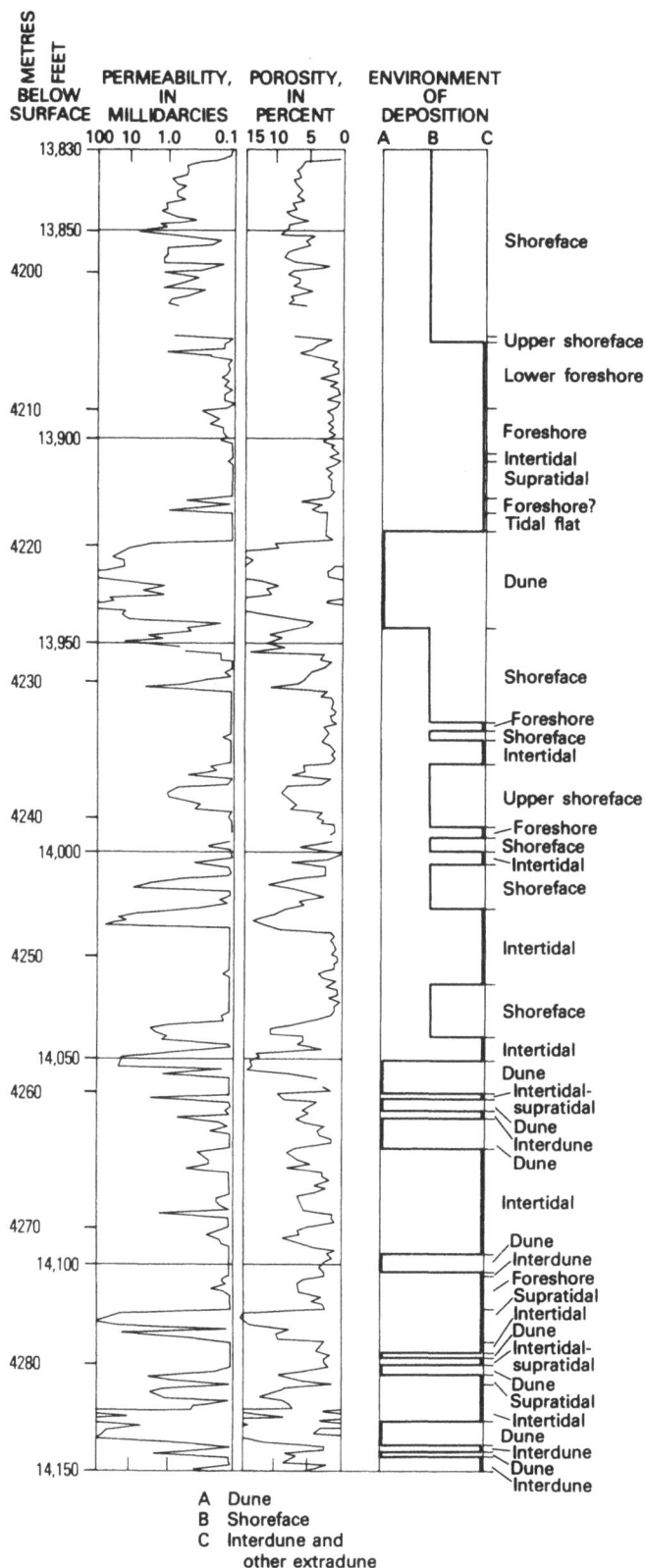

DIAGRAM OF CORE PERMEABILITY and porosity and environment of deposition in the Weber Sandstone, Champlin Petroleum Co. Brady Unit 2, NE¼NE¼ sec. 2, T. 16 N., R. 101 W., Sweetwater County, Wyoming, U.S.A. (Fig. 165.)

from complexly fractured strata. In Brady field, however, much original porosity and permeability remains in the Weber; production is from strata having intergranular porosity of a maximum 15–20 percent.

The combination of the studies of the two fields creates a nearly complete picture of dune-extradune geometry, of the original distribution of porosity and permeability, and of its potential influence on fluid migration. The close control in Lost Soldier–Wertz (fig. 164), combined with the porosity-permeability contrasts resulting from the same depositional processes which are preserved in the Brady field (fig. 165), provides a subsurface example of the dune-extradune model. Relatively more porous and permeable dune sand is isolated by generally less porous and permeable extradune sediment at Brady field, so fluid migration may be influenced.

From the study of the Weber Sandstone and examination of the Nugget Sandstone in southern Wyoming and its correlative, the Navajo Sandstone in the Colorado Plateau, we conclude that dune-extradune systems, such as the Weber and Tensleep Sandstones, present greater reservoir problems than dune-interdune types, such as the Nugget. Dune-interdune systems present less of a problem because (1) interdune sediments are usually thinner than extradune sediments and are therefore less of a porosity and permeability barrier; and (2) interdune sediments are commonly lenticular, a condition which would allow communication between dune bodies. Furthermore, judged from production histories, porosity-permeability problems in dune-interdune reservoirs apparently are not common.

Effective fluid movement between well bores is influenced by the complex permeability barriers that resulted from dune-extradune depositional processes. Where fluid movement between well bores is critical, as in retrograde reservoirs where reservoir pressure must be maintained, the efficient recovery of hydrocarbons requires knowledge of the distribution of depositional environments and the resultant porosity and permeability characteristics.

A STUDY OF GLOBAL SAND SEAS

MORPHOLOGY AND DISTRIBUTION OF DUNES IN SAND SEAS OBSERVED BY REMOTE SENSING

Chapter J

By CAROL S. BREED and TERESA GROW

Contents

Illustrations

Tables

Summary of Conclusions

DUNES OF SAND SEAS are classified in this chapter according to their external shapes and slipface orientations as observed in planimetric view on Landsat imagery and on Skylab and aerial photographs. Types observed (in order of their relative abundance) are linear, crescentic, star, parabolic, and dome-shaped dunes. Within dune types, different stages of slipface development are used in this chapter to distinguish simple, compound, and complex forms. Measurements of dune length, width (or diameter), and wavelength have been made on imagery and photographs of sand seas where dune type and form are uniform.

Dune patterns in sand seas seem regular because the measured values for dune length, width, and wavelength are fairly evenly distributed about mean values that are characteristic of each sand sea. Patterns made by dunes of the same type are similar, despite their occurrence in widely distant sand seas, because the relationships of mean length, width (or diameter), and wavelength are similar among dunes of each type, regardless of differences in size, form, or geographic location. Progression in dune form, from simple to compound or complex, is commonly associated with growth to larger size.

Within dune types and forms, varieties are distinguished in this study by assessing degrees of similarity as shown by correspondence of scale ratios derived from their measured mean lengths, widths (or diameters), and wavelengths.

Introduction

SATELLITE IMAGES OF UNIFORM 1:1,000,000 scale and worldwide coverage provide a base for quantitative and qualitative studies of large-scale eolian landforms in widely distributed sand seas. Desert surfaces recorded on Landsat imagery of Africa, Asia, Australia, and North and South America were examined for evidence of eolian sand. Large-scale eolian landforms, mainly dunes, were measured on approximately 110 selected sample areas on Landsat imagery. Some small dunes, mostly in minor sand seas of North America, were measured in sample areas on aerial and Skylab photographs. Identification numbers for all images and photographs used for sample measurements, and latitude and longitude of the sample areas, are given in the tables at the end of this chapter.

Dune lengths, widths or diameters, and wavelengths (crest-to-crest distances) were measured with an optical comparator on 1:1,000,000 scale bulk-processed Landsat transparencies, according to the conventions illustrated in figure 166. The lower limit of resolution for measurement of sand features was 100 metres (328 ft). Dunes measured on Landsat imagery commonly are as much as 3,000 m (3 km, or about 2 mi) wide. Some individual linear dunes are more than 50 km (30 mi) long and, thus, extend beyond the sample areas. The size of most Landsat sample areas is 2,500 km² (1,550 mi²).

Data resulting from the measurements are arithmetic means of various distance measurements and are two-dimensional. Measured values are analyzed and compared to show distribution of dune sizes and spacing of dune types in various sand seas, and to distinguish varieties among dunes of the same general type. Associations are shown among variables of length, width (or diameter), and spacing (wavelength) of dunes of certain types, regardless of their size and geographic location.

Classification of Eolian Sand Features Based on Landsat Imagery
Genetic and Local Terms

MOST CLASSIFICATIONS OF DUNES are based on local descriptions or on terms which carry genetic implications, such as "longitudinal" or "transverse." (See note by E. D. McKee, *editor*.[10]) The genetic terms imply a certain mode of origin based on slipface orientation relative to the dune body, for many dunes generally move and (or)

---

[10] NOTE — the term "transverse" is perhaps an unfortunate one, for it suggests a genetic connotation and has been used by some geologists to imply a dune with crest transverse to wind direction. However, it has long been used to designate a sand ridge with slip face on one side only, and, therefore, it is related to barchans and barchanoid ridges, differing only in overall shape and some features of structure. In this publication "transverse" is not used in a genetic sense, but the term is retained because of tradition and long usage for a specific type.

E. D. McKee, *editor*.

grow in the direction of avalanching. However, in space imagery where relationships of wind regimes to dune types in a given locality are not yet proved, assumptions that certain dune forms are "longitudinal" or "transverse," in a genetic sense, seem to the authors of this chapter inappropriate. Nevertheless, a few well-established local terms such as "barchans" and "seifs," are retained to describe distinctive types of dunes recognized in most sand seas.

Tables 26–30 list, in chronologic order, terms used by previous authors for dunes classified in this chapter as linear, crescentic, parabolic, star, or dome-shaped. Tables 31–35 provide a similar chronological listing, by author, of local terms for eolian landforms herein classified as complex dunes of all types, sand seas, patterns of eolian sand deposits in sand seas, sand sheets and streaks, and interdune areas. Each entry in tables 26–35 shows the desert region to which the term applies. These tables are at the end of this chapter.

The similarity in name of very large dunes, such as the dome-shaped dunes, compound barchans, and compound linear dunes observed in this study of Landsat imagery, with smaller dunes of comparable shape (simple dome, barchan, and seif or

barchanoid ridge) does not imply that all of the forms within each type category have similar internal structures. The classification based on three-dimensional internal dune structures is given in Chapter A (table 1).

Terms Based on Remote Sensing

Eolian sand features recognized on Landsat imagery, locally supplemented by Skylab or aerial photographs, are dunes, irregularly shaped sand sheets, and elongate sand streaks. A morphological description of eolian sand bodies is primarily a classification of dune types and forms. The term "dune" is used here for all well-defined eolian sand mounds or ridges that exist independently of surrounding topography (Bagnold, 1941, p. 7), whether slipfaces are visible or not. The classification suggested here is based on remote sensing and recognizes types, forms, and varieties of dunes (table 25). (See note by E. D. McKee, *editor.*[11])

Dune Types

Types of dunes as defined here are based on the external shape of the dune and the arrangement of its slipfaces, if any, relative to its shape in plan view.

Linear dunes are straight or slightly sinuous, longitudinally symmetrical sand ridges, much longer than they are wide, that are bounded on both sides by opposite-facing slipfaces. Linear dunes are features commonly referred to as "longitudinal" dunes, "seifs," and "sandridges" (table 26).

Crescentic dunes are mounds or ridges of sand that are crescent-shaped or composed of crescent-shaped segments, each generally as wide or wider than it is long, and each mound or segment bounded on its concave side by a slipface. The category includes sand bodies that are referred to by many geologists as "barchan" and as "transverse" dunes (table 27). In planimetric view on space imagery, even large-scale "transverse" dunes that seem in ground view to be straight

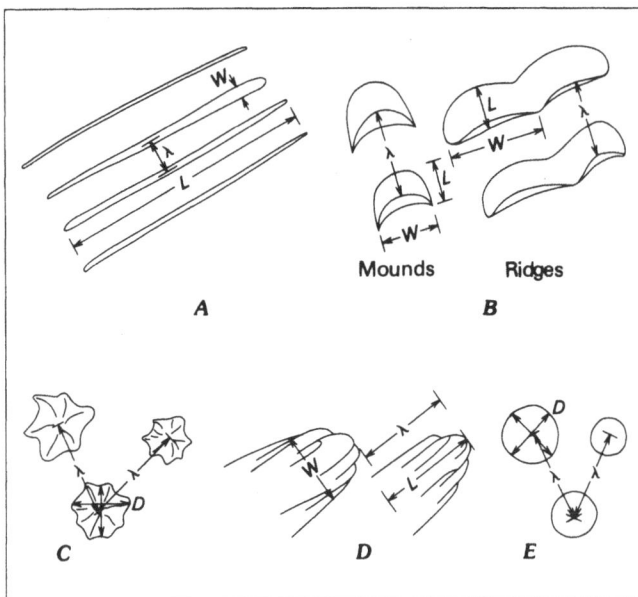

METHODS OF MEASURING ON LANDSAT IMAGERY, Skylab and aerial photographs, the length (*L*), width (*W*), or diameter (*D*), and wavelength (λ) of the following dune types: *A*, Linear; *B*, crescentic; *C*, star; *D*, parabolic compound form; *E*, dome-shaped. (Fig. 166).

[11] NOTE — The classification used here and in the following regional descriptions (chapter K) is derived largely from studies of Landsat imagery supplemented locally by a few Skylab and aerial photographs, and represents an essentially two-dimensional classification. It differs in certain basic respects from the classification based on three-dimensional studies (chapter A), especially structural, used by other authors of this professional paper.

E. D. McKee, *editor.*

TABLE 25. — *Examples of types and forms of dunes, using classification in this chapter, based on remote sensing*
[ERTS (E) and Skylab (SL) numbers are below photographs. NR indicates not recognized on Landsat imagery. Photograph scales 1:1,000,000. Crescentic type includes transverse type as used by other authors in this publication]

Type	Form		
	Simple	Compound	Complex
Linear	E–1179–00593 E–1167–06310	E–1187–06433 E–1117–06551	SL3–RL28 E–1111–09435
Crescentic	E–1128–04253 E–1227–08501	E–1183–06194 E–1183–06194	E–1183–06194 E–1424–03262
Star	NR	E–1183–06203 E–1109–09322	E–1183–06194 E–1111–09435 E–1183–06194
Parabolic	E–1189–17102	E–1177–14255	NR
Dome-like	NR	E–1155–07040 E–1077–04420	

ridges appear to curve slightly in crescent-shaped segments, though each segment may be a kilometre or more wide from horn to horn.

Star dunes are radially symmetrical, pyramidal sand mounds with slipfaces on arms that radiate from the high center part of the mound. They include sand bodies known locally as "oghurd" dunes, pyramidal dunes, and dune "massifs," summarized in table 28. Parabolic dunes are U-shaped or V-shaped mounds of sand, trailed by elongated arms, having slipfaces on the outer, convex sides (table 29), and are typically associated with vegetation. Dome-shaped dunes are semicircular mounds of sand that generally lack surface slipfaces (table 30).

Dune Forms

Forms of dunes are defined by the stage of slipface development and, thus, to some extent by size. Simple dune forms are mounds or ridges with slipfaces, all of which are similar in size or character. Simple linear dunes are the main sand features of the Simpson, Great Sandy, and Great Victoria Deserts in Australia, and of the Kalahari Desert in South Africa. Simple barchans, barchanoid ridges, and simple linear and star dunes are the main dune forms of most North American deserts, but they are minor, inconspicuous components of most major sand seas observed on Landsat imagery.

Many dunes in sand seas of Asia and northern Africa are extremely large and are characterized by extensive development of secondary sets of slipfaces of lesser size than those which define the main mound or ridge. If secondary slipfaces are of the same kind as those that define the main ridge or mound, the dune is a compound form; if they are of a different kind, the dune is a complex form.

Compound dunes are mounds or ridges of sand on which smaller dunes of similar type and slipface orientation are superimposed. For example, a compound crescentic dune consists of a sand mound or ridge having similar but smaller secondary crescent-shaped dunes on its gentle slopes. The arrangement of slipfaces, of similar appearance, gives some dunes of compound form a typically broad, "two-storied" aspect (Kádár, 1966, p. 447). Subsidiary slipfaces on compound crescentic dunes are seldom clearly visible on Landsat imagery, but they can be identified by inspection of aerial or Skylab photographs, or from descriptions based on field observations.

Complex dunes are combinations of two or more dune types. The most common complex forms observed on Landsat imagery are linear or crescentic ridges with superimposed dunes of some other type, such as star dunes, developed along their crests. Generally, complex forms consist of a main dune type and a secondary dune type and can be classed as mainly linear, mainly crescentic, and so forth. However, some complex dunes have approximately equal components of different dune types; where no one dune type is dominant, the dunes are here considered to be complex unclassified.

Dune Varieties

Within the basic classes of type and form are numerous dune varieties. Varieties of simple and compound forms of most dune types have been observed in many sand seas. They are compared by applying a method developed by Strahler (1958, p. 292–294) for assessing similarity of fluvially produced landforms. According to this method of dimensional analysis, landforms are considered to be similar when dimensionless numbers formed by the ratios of their measured values are closely similar (Strahler, 1958, p. 279). Certain varieties of simple and compound dunes are shown by this method to be common to many widely distant sand seas. Common names for these dunes are given in tables 26–30. Varieties of complex dunes, however, tend to be unique to specific localities, so their names usually only have local application. Those names are listed in table 31.

Occurrence in Sand Seas

Large areas of eolian sand constitute sand seas, so-called because the dunes in them tend to be repeated in orderly sequences rather than occurring as scattered solitary features. Sand seas occur from above the Arctic Circle in Alaska to below the Antarctic Circle in Antarctica, and at all latitudes between. Most major sand seas, however, lie within a belt between lats 40° N. and 30° S., in Africa, Asia, and Australia. General distribution of the world's sand seas is discussed by I. G. Wilson (1971b, 1973). Local names for many sand seas are summarized in table 32. Descriptive terms used by various workers to identify orderly patterns of eolian sand deposits in sand seas are summarized in table 33.

In the following parts of this chapter, dunes are described according to type, in order of their relative abundance, as observed on Landsat imagery of major sand seas. Linear dunes are the main dune type in the Southern Hemisphere deserts of Australia and southern Africa. They occupy about half the area of the Rub' al Khali in southwestern Saudi Arabia, and they occur in several sand seas of the Sahara of northern Africa. Crescentic dunes are more common than linear dunes in the Northern Hemisphere. They occupy large areas of the eastern Rub' al Khali in southeastern Saudi Arabia, several sand seas in the Sahara, and most sand seas in northern China, eastern U.S.S.R., western Argentina, and western North America.

Star dunes occupy far less area within sand seas than do either linear or crescentic dunes but, like them, are widely distributed. Major occurrences of star dunes are in sand seas of the northern Sahara in Algeria and Libya, the Gran Desierto of Mexico, and around the margins of sand seas in northern China, the Arabian Peninsula, and the Namib Desert of South-West Africa.

Parabolic dunes of a size visible on Landsat imagery are rare. Observations on these images are mostly limited to the Thar Desert of India and Pakistan, and to low-lying areas along the Colombia-Venezuela border near Buenavista, small dune fields in the Kalahari Desert of South Africa, and White Sands in western North America. Also rare on Landsat imagery are domes. or dome-shaped dunes; they are observed in sand seas of the northern Arabian Peninsula, northern China, and the Sahara of northwestern Algeria.

Sand sheets and streaks (termed "stringers" by many authors) are abundant in most major sand seas, but these features have no common geometric characteristics on Landsat imagery by which they can be compared from desert to desert. Some of them are known to contain small dunes that are below the resolution of Landsat imagery. Common terms for them are given in table 34.

Linear Dunes

General Distribution

SIMPLE, COMPOUND, AND COMPLEX linear dunes are observed on Landsat imagery and on Skylab and aerial photographs (figs. 167, 168). Simple linear dunes are single or bifurcated ridges with narrow crests that may be sharp or subdued but which do not have slipfaces of secondary dunes developed upon them (fig. 167). Compound linear dunes are generally much broader linear ridges that have relatively smaller linear ridges, each with its own slipfaces, on their tops. In the "feathered" variety of compound linear dunes, subsidiary ridges intersect or spread obliquely from the main ridge. Linear dunes of compound form in Arabian sand seas are shown on aerial photographs by Glennie (1970, fig. 74) and by Holm (1960, fig. 7). Complex linear dunes may be narrow or broad linear ridges that have other dune types, such as star, dome-shaped, and barchan (crescentic) dunes, superimposed upon them (fig. 168).

Some linear dunes occur as solitary features (chapter K, fig. 193) but mostly they form parallel ridges separated by sandy, gravelly, or rocky interdune corridors in sand seas. Linear dunes, apparently of simple form, cover most of the Australian deserts and the Kalahari Desert of South Africa, but in many other sand seas linear dunes are mostly compound or complex.

The largest solitary linear dune measured on Landsat imagery is a 100-km-long (26-mi-long), 1-km-wide (0.62-mi-wide), complex dune between Atar and Akjoujt, Mauritania, in the northwest Sahara (chapter K, fig. 193). The largest linear dunes forming parallel ridges in a sand sea are in the Qa'amiyat region of the southwestern Rub' al Khali, Saudi Arabia, where individual dunes can be traced for 190 km (118 mi) and range in width from 0.30 to 2.7 km (0.2 to 1.7 mi).

Morphometry

Dune width, length, and wavelength were measured on sixty-seven 2,500-km^2 (1,550-mi^2) sample areas of Landsat imagery showing linear dunes of simple, compound, and complex form in northern Africa, southern Africa, Australia, and the Arabian Peninsula. Measurements of linear dunes of simple form on the Navajo Indian Reservation, northern Arizona, U.S.A., were made from high-altitude aerial photographs (fig. 167B). Some sample areas on other continents are shown in figure 169. Histograms (figs. 170, 171) show distribution of the measured values in sample areas of various sand seas. In general, measured values for dune width, length, and wavelength (crest-to-crest distance) tend to be grouped about characteristic means (summarized in table 36) that give a regular pattern to each sand sea.

Measured values of dune wavelength for simple linear dunes in sampled areas of the Simpson, Great Sandy, and Kalahari Deserts, and Arizona sand seas are distributed lognormally, unlike the distribution of measured values for simple linear dune wavelength in the sampled area of the Rub' al Khali (fig. 170), and linear dunes of compound and complex forms (fig. 171). The greatly skewed distributions of wavelength[12] indicate that spacing of most

---

[12] Transformations of the data, for statistical treatment, are considered unnecessary because the sample mean of such distributions is considered to be "more than 90 percent efficient" as an estimate of the population mean, "provided that the coefficient of variation is less than 1.2, corresponding to a variance of the logarithms of about 0.9" (Koch and Link, 1970, p. 219–220). No coefficients of variations (C) shown in figure 170 exceed the allowable figure.

simple linear dunes of the variety common to the Kalahari and to Australian and Arizona deserts is fairly even and ranges about a mean value characteristic of each area, but that extremely wide spacing of a few individual dunes is common.

Measured values for widths and lengths of the linear dunes are normally distributed, except in the sampled Kalahari and Arizona localities, for reasons that are not yet clear.

Size and spacing relationships of linear dunes sampled in different sand seas were examined to see whether relationships of these variables are constant regardless of differences in dune size and geographic location. A scatter diagram (fig. 172) shows a linear arithmetic relationship between mean dune width and spacing in groups of linear dunes of all forms and varieties, in 12 sand seas of Africa, Asia, Australia, and North America. Sizes of dunes represented in figure 172 range from an average of 43 m (141 ft) wide, on the Navajo Indian Reservation, Arizona, U.S.A., to an average of 1,480 m (4,860 ft) wide, at the western end of the Rub' al Khali, Saudi Arabia (table 36).

Regression analysis indicates that wavelength, which is presumed to be an independent variable, tends to be about twice the mean width of linear dunes, regardless of their size. The relationship also indicates a progression of dune form, from simple to compound or complex, with increasing size. Increasing size of linear dunes thus seems to be associated with development of subsidiary dunes, which may have slipfaces of the same or different orientations than the slipfaces of the main dune. Growth to great size is believed to be a function of time (I. G. Wilson, 1971a, p. 182–183), as well as of sand supply and effective winds.

Relation to Topographic Barriers and Wind Regimes

In sand seas such as the Australian deserts, the southwest Kalahari Desert of South Africa, and the Qa'amiyat region of the western Rub' al Khali, Saudi Arabia (chapter K), groups of linear dunes extend for hundreds of kilometres. Large groups of linear dunes in the Sahara, on the Arabian Peninsula, in southern Africa, and in Australia generally are alined parallel or subparallel to the resultant direction of effective winds (chapters F and K).

Tests were made of measurements within these deserts to determine whether statistically significant differences in mean dune width, $\overline{W}$, mean dune spacing, $\overline{\lambda}$ (crest-to-crest distance), and mean frequency, $\overline{F}$ (number of dunes per kilometre), are associated with relative distances downwind or with positions in the open parts of sand seas, as opposed to positions near topographic barriers. Several different methods were tried.

First, mean wavelengths of dunes in samples measured near hills, playas, and streambeds in the Australian and Kalahari sand seas were compared with mean wavelengths of dunes in samples measured in open areas without such topograhic barriers. Differences significant at the 0.098 level for the Simpson Desert, 0.047 level for the Great Sandy Desert, and 0.010 level for the Kalahari Desert show that mean spacing of dunes in these sampled areas is significantly wider near

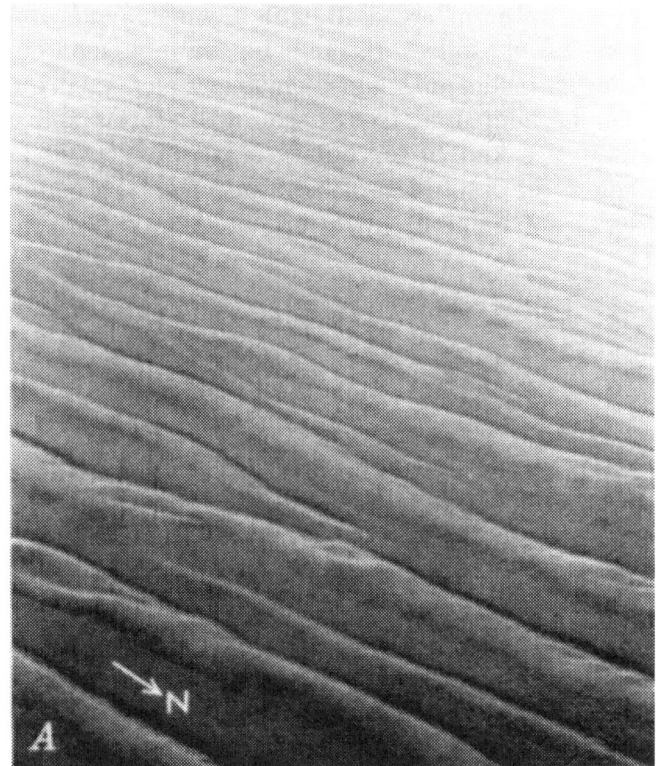

COMMON VARIETIES OF SIMPLE LINEAR DUNES (above and facing page). A, Bifurcating but straight dunes in the northwest Simpson Desert, Australia. Mean width of the dunes is 0.22 km (0.14 mi). Aerial photograph by E. Tad Nichols. B, Linear dunes on the Navajo Indian Reservation, Arizona, U.S.A. Aerial photograph by National Aeronautics and Space Administration (JSC 236 28 0073). (Fig. 167.)

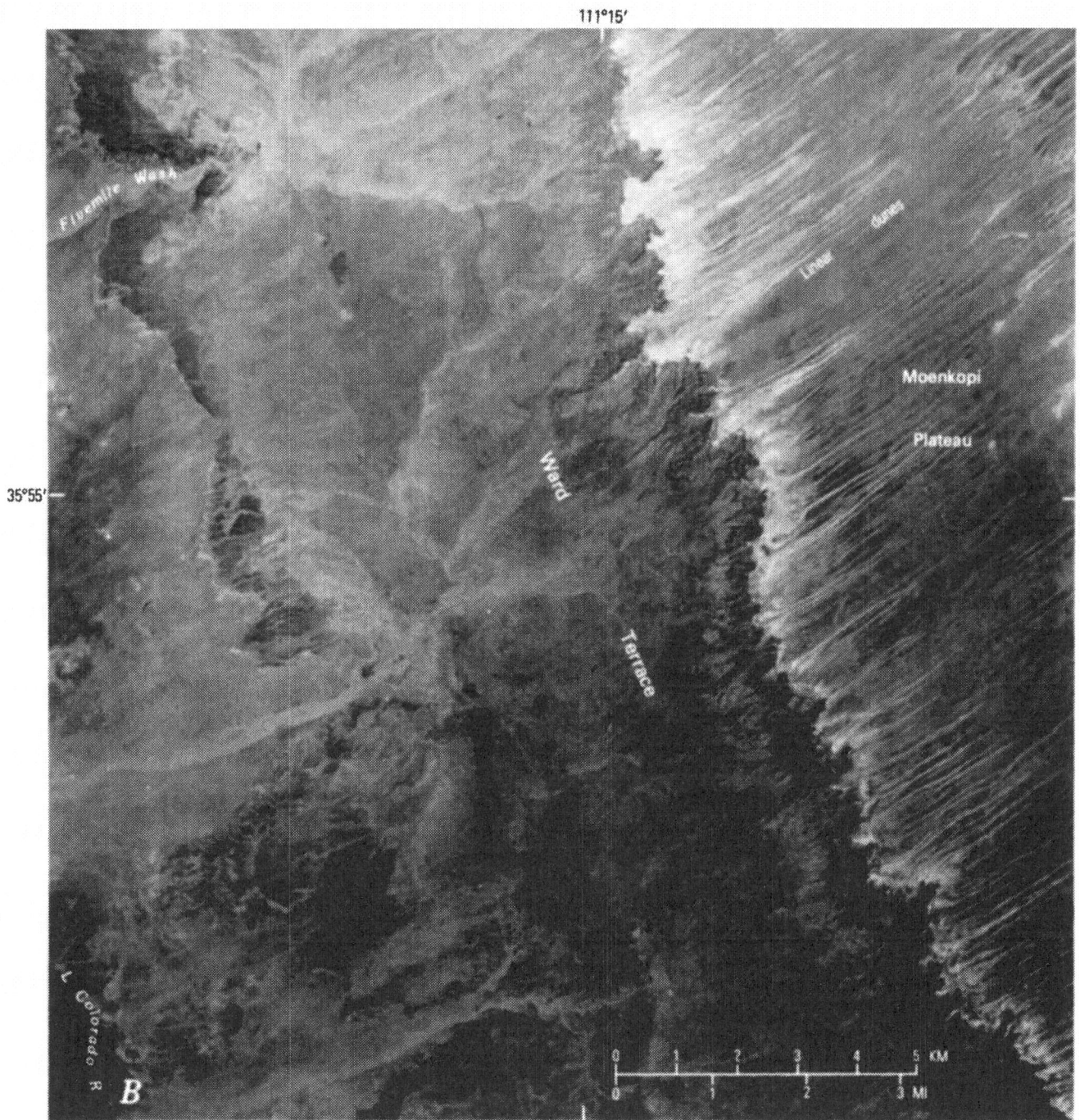

(Fig. 167 — Continued.)

topographic barriers than where such barriers are absent.

Second, regularity of dune spacing (represented by dispersion of wavelength measurements) relative to distance downwind in a sand sea was tested by plotting the coefficient of variation, C (the ratio of the standard deviation to the mean of each sample), versus distance downwind in the Rub' al Khali and the Kalahari, Simpson, and Great Sandy Deserts. No correlation was found.

Third, the numbers of dunes in six sample areas of the Simpson Desert were counted to see whether

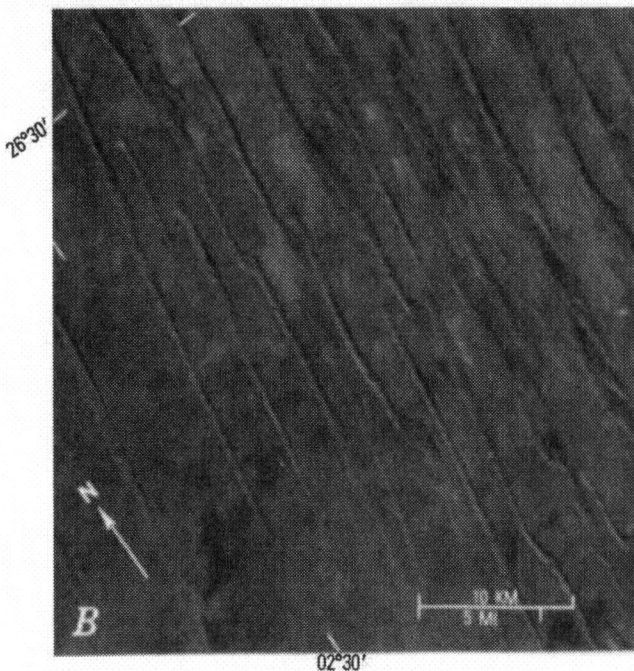

AERIAL AND SPACE PHOTOGRAPHS OF COMPLEX LINEAR DUNES. *A*, Subsidiary star dunes and crescentic dunes on the crests of complex linear ridges in the central Namib Desert, South-West Africa. Dunes average 2.2 km (1.4 mi) from crest to crest. Photograph by H. T. U. Smith. *B*, Skylab EREP (Earth Resources Experiment Package) photograph showing linear dunes with barchanoid ridges on their crests in the northwest Sahara, Algeria, northern Africa. Photograph by National Aeronautics and Space Administration (SL 3–28–365). (Fig. 168.)

significant changes in dune numbers occur with distance downwind (measured in a direction parallel to resultant wind direction). No significant difference in dune numbers was found either between adjacent samples or between extreme upwind (southeast) and downwind (northwest) samples in the Simpson Desert.

The negative results of these tests suggest that simple distance from the upwind end of a sand sea (and, presumably, distance from the sand supply) is not an important factor in determining the distribution patterns of linear dunes.

Identification of Varieties

Ratios derived from measurements of the mean widths ($\overline{W}$), lengths ($\overline{L}$), and wavelengths (λ) of dunes in each sample area define the geometric characteristics of the sampled dunes in each sand sea (table 36). The degree of similarity of simple linear dunes in different sand seas is shown by the degree of correspondence of their ratios (table 37).

At least two varieties of simple linear dunes can be distinguished by the degree of similarity of the sampled dunes:

1. Long narrow linear dunes of the Australia "sandridge" variety, common to the Simpson and Great Sandy Deserts, show excellent to good correspondence to linear dunes of similar size in the Kalahari Desert and good to fair correspondence to much smaller linear dunes on the Navajo Indian Reservation in Arizona.

2. Short linear dunes of the "seif" variety in the Rub' al Khali sand sea of Saudi Arabia, east of Al 'Ubaylah (chapter K, fig. 241) show very poor to good correspondence to the linear dunes in the Navajo Indian Reservation of Arizona in the United States, in southern Africa (fig. 169*A*), and in parts of Australia (fig. 169*B*). Comparison of ratios, derived from measurements, thus seems to support the suggestion by Price (1950, p. 462–463) that two varieties of linear dunes have been described as seifs — one by Bagnold (1941, p. 222–229) in the Sahara of Libya, and the other, called sandridges, by Madigan (1946, p. 45–63) in the Australian deserts.

Comparisons of dimensionless ratios for linear dunes of compound and complex forms are not

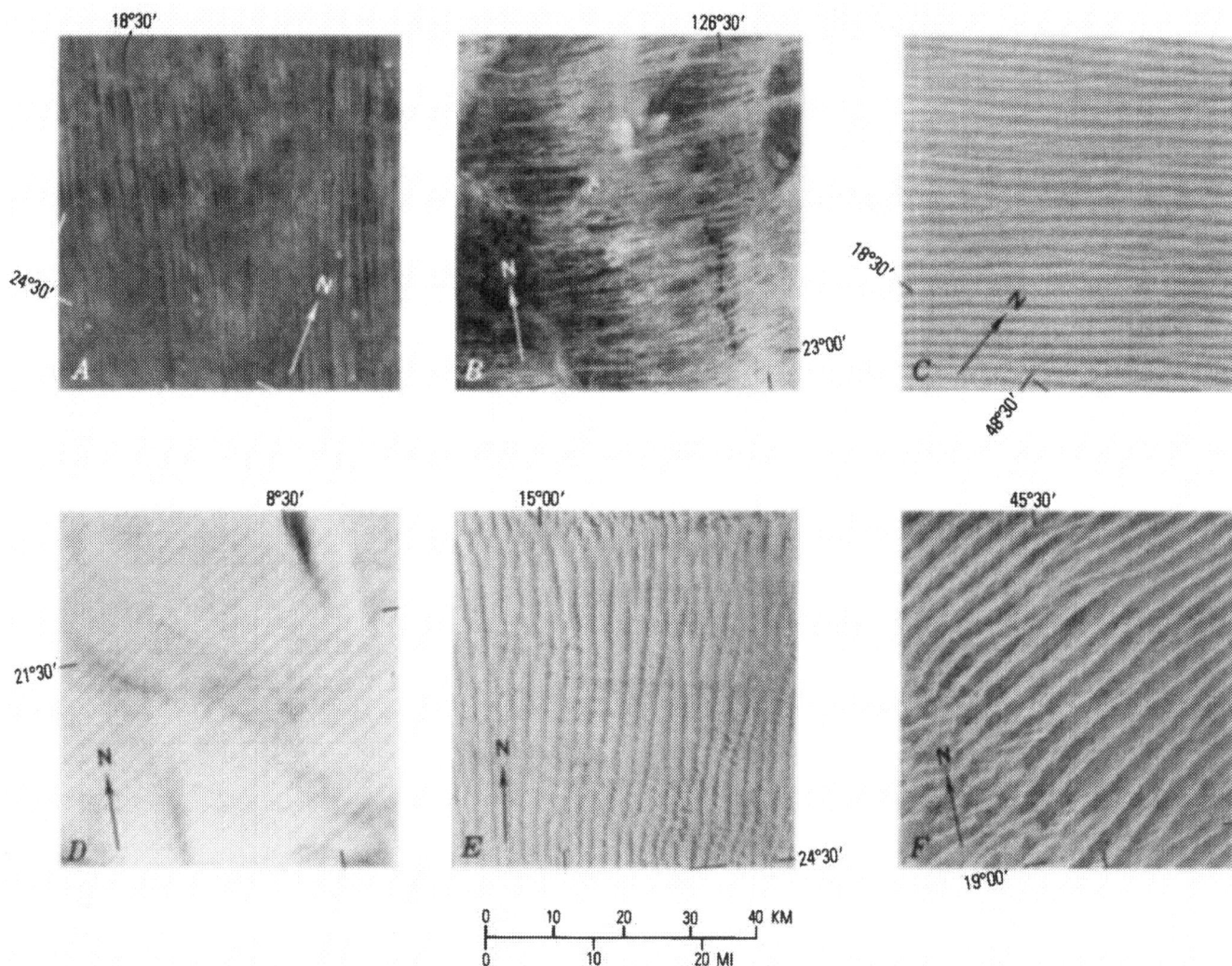

LANDSAT IMAGERY SAMPLE AREAS on which measurements of linear width, length, and wavelength were made. Numbers in parentheses refer to the regions listed in table 36. Each sample is an area 50 km by 50 km, or 2,500 km² (1,550 mi²) in a sand sea. A, Kalahari Desert, southern Africa; simple dunes (4). B, Great Sandy Desert, Australia; simple dunes (3). C, Southwestern Rub' al Khali, Saudi Arabia; compound dunes (8). D, Southwestern Sahara; compound dunes (6). E, Namib Desert, South-West Africa; complex dunes (9). F, Western Rub' al Khali, Saudi Arabia; complex dunes (12). (Fig. 169.)

given because, as indicated in table 36, lengths of these dunes commonly are greater than the sample lengths.

Crescentic Dunes
General Distribution

ON THE BASIS OF SLIPFACE DEVELOPMENT, three basic forms of crescentic dunes — whether mounds or ridges — can be discerned on Landsat imagery and on aerial and Skylab photo-graphs. Simple crescentic dunes are crescent-shaped mounds, commonly referred to as barchan dunes. Crescentic ridges, commonly known either as barchanoid ridges or as transverse ridges, have a single slipface on each arc. In most places they are relatively small (fig. 173). Curvature of the segments of some crescentic ridges is extreme, and the term "barchanoid" has been used (Cooper, 1967, p. 28) to distinguish this variety from the very slightly curved variety of ridge commonly called transverse on the basis of structure (table 27).

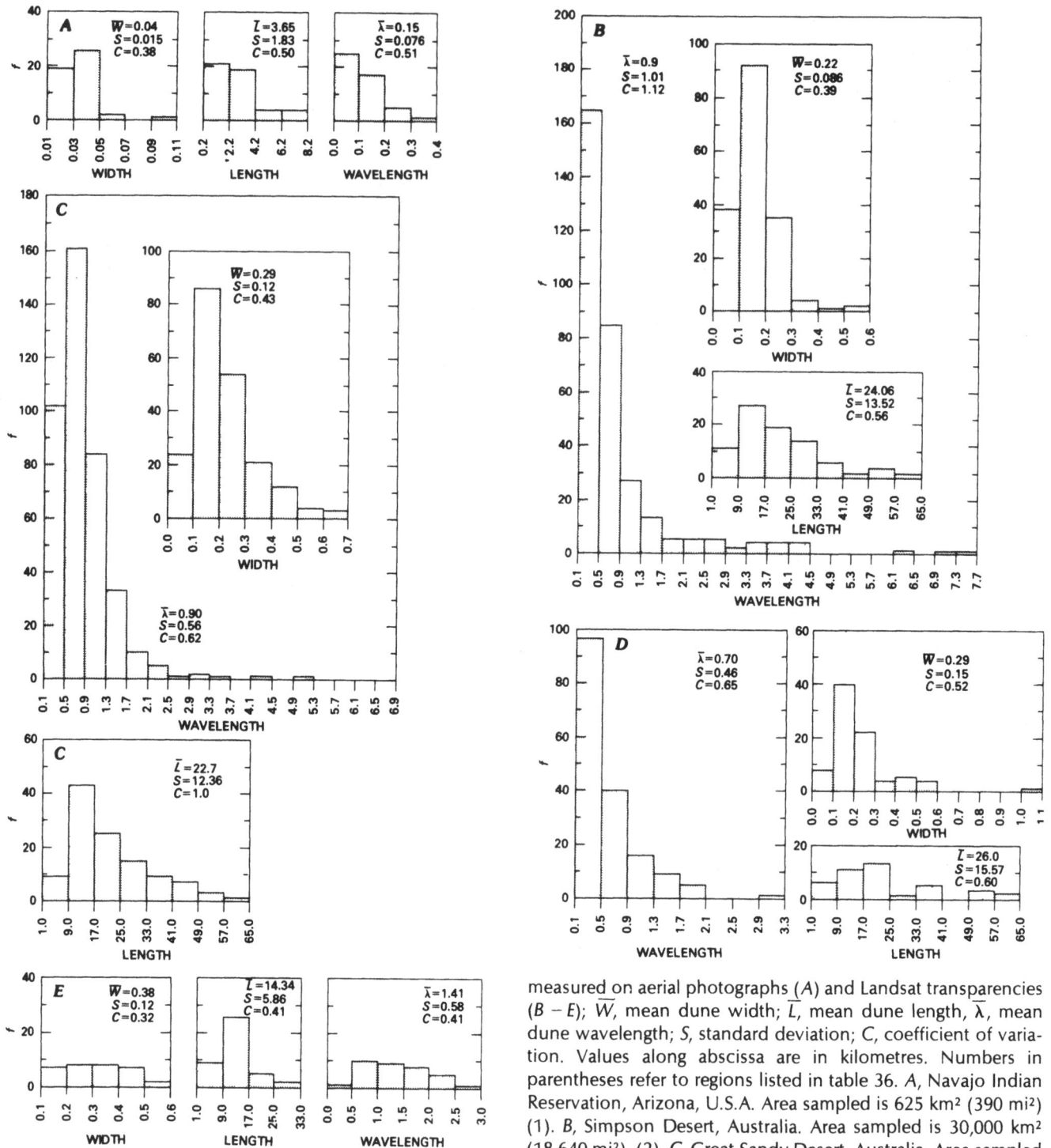

HISTOGRAMS SHOWING FREQUENCY DISTRIBUTION (f) of dune widths (W), dune lengths (L), and dune wavelengths (λ), in kilometres, in sample areas of linear dunes of simple form, measured on aerial photographs (A) and Landsat transparencies (B – E); W, mean dune width; L, mean dune length, λ, mean dune wavelength; S, standard deviation; C, coefficient of variation. Values along abscissa are in kilometres. Numbers in parentheses refer to regions listed in table 36. A, Navajo Indian Reservation, Arizona, U.S.A. Area sampled is 625 km² (390 mi²) (1). B, Simpson Desert, Australia. Area sampled is 30,000 km² (18,640 mi²), (2). C, Great Sandy Desert, Australia. Area sampled is 35,000 km² (21,740 mi²) (3). D, Kalahari Desert, South Africa. Area sampled is 15,000 km² (9,320 mi²), (4). E, Northern Rub' al Khali, Saudi Arabia. Area sampled is 2,500 km² (1,550 mi²) (5). (Fig. 170.)

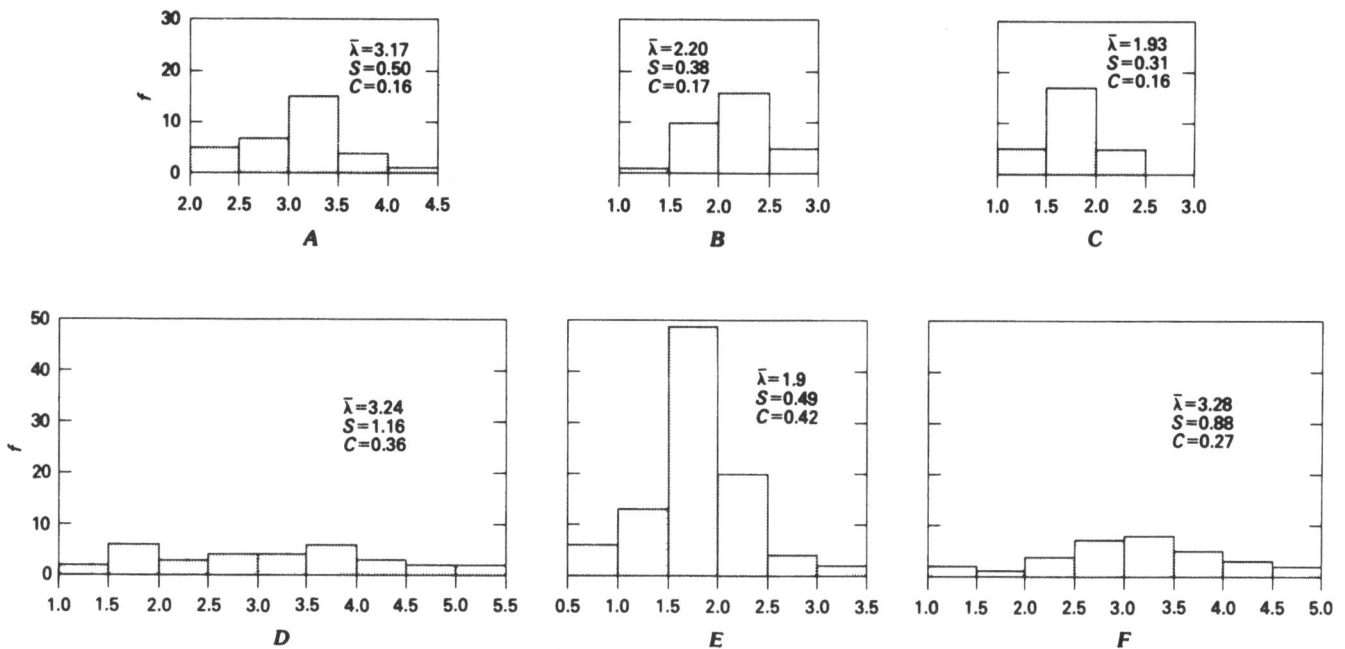

HISTOGRAMS SHOWING FREQUENCY DISTRIBUTION (f) of dune wavelengths (λ), in kilometres, in compound and complex linear dunes measured on Landsat transparencies: $\bar{\lambda}$, mean dune wavelength; S, standard deviation; C, coefficient of variation. Numbers in parentheses refer to regions listed in table 36. Values along abscissa are in kilometres. A, Western Rub' al Khali, Saudi Arabia; complex dunes. Area sampled is 2,500 km² (1,550 mi²) (12). B, Namib Desert, South-West Africa; complex dunes. Area sampled is 2,500 km² (1,550 mi²) (9). C, Southwestern Sahara, Mauritania; compound dunes. Area sampled is 2,500 km² (1,550 mi²) (6). D, Northern Sahara, Algeria; complex dunes. Area sampled is 2,500 km² (1,550 mi²) (10). E, Southern Sahara, Niger; compound dunes. Area sampled is 7,500 km² (4,660 mi²) (7). F, Southern Sahara, Niger; complex dunes. Area sampled is 2,500 km² (1,550 mi²) (11). (Fig. 171.)

Compound crescentic dunes (the "megabarchans" of Norris, 1966, p. 292–296) are generally much larger than simple crescentic dunes and are characterized by the presence of numerous subsidiary crescentic dunes with slipfaces oriented the same as the main slipfaces but developed secondarily on the gentle slope of each major dune mound or ridge. Several examples are shown on aerial photographs in figure 174. Compound crescentic mounds are typified by the Pur-pur dune of Peru (Simons, 1956), which is a large solitary barchan, 0.75–0.85 km (0.46–0.52 mi) wide and 2.1 km (1.3 mi) long, with a major slipface forming its steep slope; numerous minor crescents, each with a slipface on its lee side, occur on the windward slope.

Compound crescentic dune ridges are more common than large solitary barchans and are typified by the Nebraska Sand Hills, described by H. T. U. Smith (1965, p. 564) as "transverse, and probably * * * of the compound variety, with minor second-

ary transverse ridges on the windward slope, which appears to be typical where the basic dune form reaches large size." Similar features are shown on an aerial photograph of the Algodones Dunes, California (fig. 174D), in a close view of that area by Norris and Norris (1961, pl. 3, fig. 1) and in the Peski Karakumy, U.S.S.R., illustrated by Suslov (1961, fig. 14–3). In Saudi Arabia, compound crescentic dune ridges showing pronounced curvature of the segments (fig. 174A) have been described (Holm, 1960, p. 1372) as "giant crescentic massifs" (table 27).

Very large compound crescentic dune ridges were observed on Landsat imagery of sand seas at relatively high latitudes, in the Peski Karakumy (U.S.S.R.), Takla Makan and Ala Shan Deserts (northern China), and in the Nebraska Sand Hills (U.S.A.). Lower latitude sand seas of large compound crescentic dunes occur in the eastern Rub' al Khali (Saudi Arabia), several parts of the Sahara (northern Africa), and in the Algodones and Gran

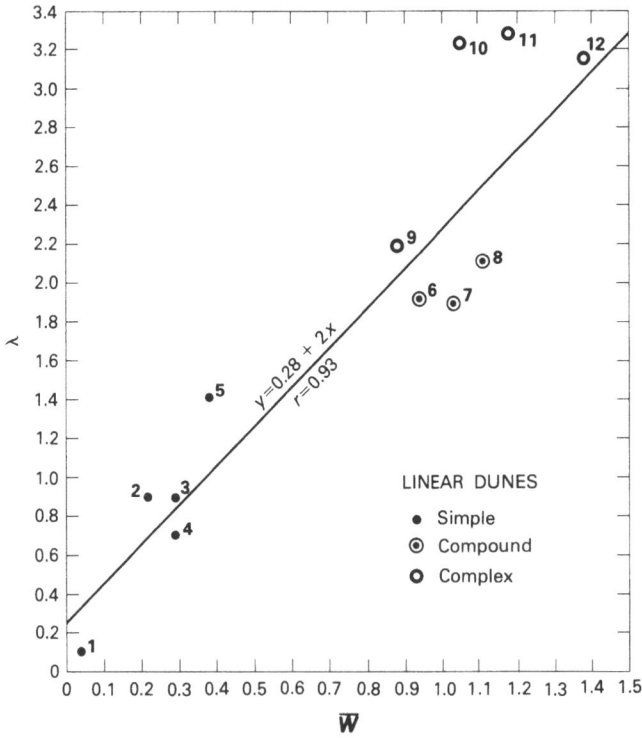

SCATTER DIAGRAM SHOWING CORRELATION of mean dune wavelength $(\bar{\lambda})$ and mean dune width $(\bar{W})$ in linear dunes, in relation to dune form. Measurements in kilometres. The correlation coefficient $(r) = 0.93$, shows excellent correlation. Numbers beside points refer to regions listed in table 36. (Fig. 172.)

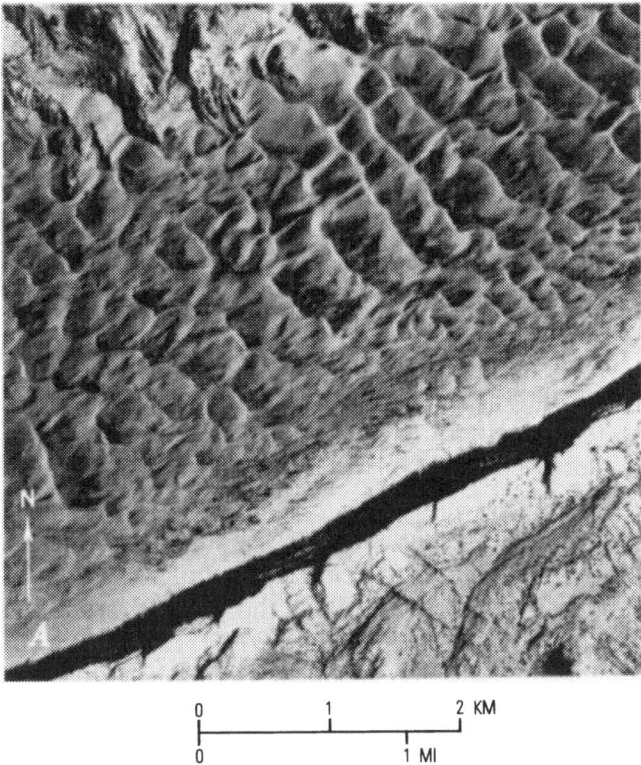

AERIAL PHOTOGRAPHS OF SIMPLE CRESCENTIC DUNES. A, Transverse ridges with slightly crescentic segments north of

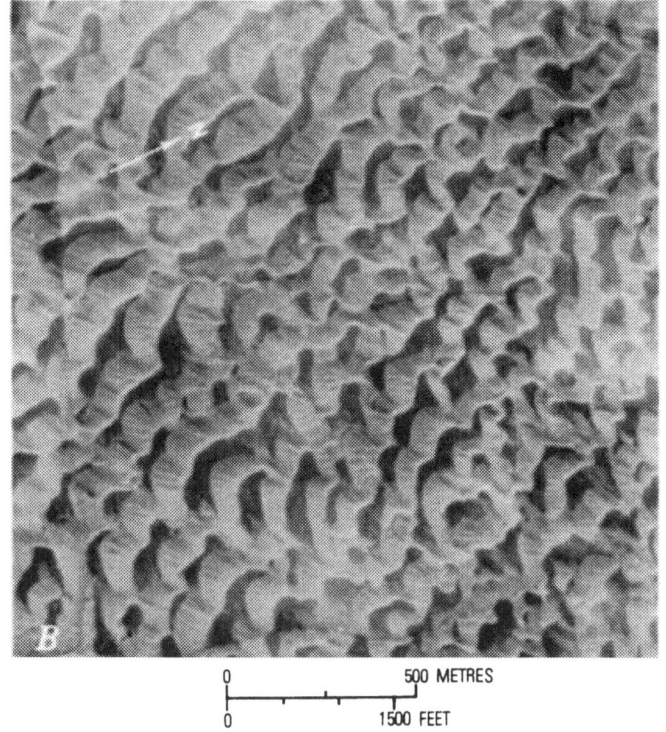

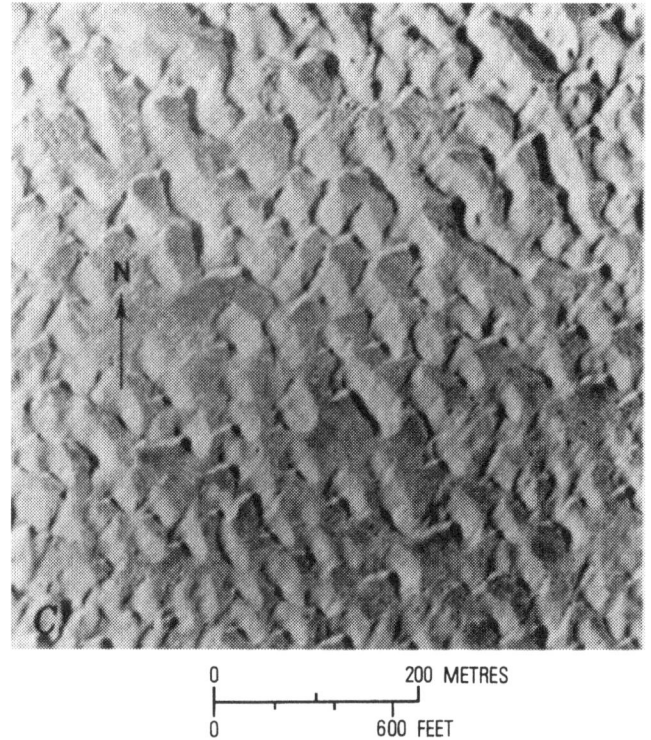

Comb Ridge, near the Arizona-Utah border, U.S.A., at lat 36°51′ N., long 110° W. Photograph from Denny, Warren, Dow, and Dale (1968, p. 14, fig. Ariz. 3). B, Barchanoid ridges, White Sands, New Mexico, U.S.A. (near lat 32°48′ N., long 106°16′ W.). Photograph used by permission of Holloman Air Force Base. C, Transverse ridges grading southward into barchanoid ridges and barchans in the Sahara (near lat 20° N., long 11° W.). Photograph from Monod (1958, pl. XXVII, p. 173, used by permission of the author). (Fig. 173.)

Desierto regions of California and Mexico, respectively. The largest dunes of crescentic type that have been noted on images are in the Takla Makan Desert of China (chapter K), where mean width (horn-to-horn) of the segments is 3.24 km (about 2 mi). Large crescentic dunes, in general, are oriented transverse to resultant directions of effective winds, but present winds in the regions of these large dunes mostly are of relatively low energies (chapter F), compared with energies of present winds in many other desert regions.

Complex crescentic dunes include both large and small mounds or ridges on which subsidiary dunes of a different type and orientation — most commonly star dunes — have formed. Some crescentic dunes, typified by ridges in the Great Sand Dunes National Monument, Colorado (McKee, 1966, p. 61, pl. E; Denny and others, 1968, fig. 10B) are included in this discussion as a variety of complex crescentic dunes because they seem to be combinations of both crescentic and star dune types. Some dunes in the Namib Desert believed to be of the complex crescentic variety are shown on a Skylab photograph in figure 174C.

Groups of relatively small crescentic mounds seen on Landsat images occur mainly in the southern Sahara and Libyan deserts of northern Africa and in the southeastern Rub' al Khali, Saudi Arabia. These barchanlike dunes are 0.20–1.5 km (0.12–0.93 mi) wide and 0.25–1.1 km (0.15–1.7 mi) long; most are probably compound barchans rather than simple barchans. A typical field of these dunes in the western Sahara of Mali is shown on a sample of Landsat imagery (fig. 175A).

Small crescentic dune ridges are characteristic of many dune areas bordering the sea, such as the coastal part of the Namib Desert (chapter K, figs. 222, 223), the Persian Gulf coast (eastern Qatar and the area around Sebkhat Matī) and the southern coast of Mauritania near the Sénégal River (figs. 188, 191). Small crescentic dune ridges also occur inland, where they commonly override much larger dunes of crescentic type, as in the Al Jiwā' area of the eastern Rub' al Khali (chapter K, fig. 240) and on the eastern edge of the Takla Makan sand sea near T'ieh-kan-li-k'o (chapter K, fig. 261A). Size of the small dunes, however, precludes making measurements on Landsat imagery, except for wavelength, so apparent similarities cannot be tested by comparisons of their scale ratios. Small

coastal crescentic dunes commonly are oriented transverse to resultant directions of fairly strong present-day winds (chapters F and K).

Morphometry

Dune width (horn-to-horn distance of each crescent or crescentic segment), dune length (distance from base of windward slope to base of slipface), and dune wavelength (crest-to-crest distance) were measured (fig. 166) on forty-five 2,500-km² (1,500-mi²) sample areas of Landsat imagery, representing large, compound crescentic dunes in both northern and southern Africa, the Arabian Peninsula, Pakistan and India, U.S.S.R., northern China, and western North America. Wavelengths of smaller, possibly simple crescentic ridges were measured on numerous Landsat imagery samples of the Arabian Peninsula, Pakistan, China, and South-West Africa. Some sample areas of Landsat imagery for most of these regions are shown in figure 175. Simple crescentic (barchanoid) dunes at White Sands, New Mexico, and compound crescentic dunes in the Gran Desierto, Mexico, and in the Algodones Dunes, California, were measured on aerial and Skylab photographs.

Measurements in each sampled sand sea are summarized in table 38, and the frequency distributions of measured values are shown by histograms in figures 176 and 177. Measurements of individual dunes show a wide range in many sand seas, but as a group, measurements in each sampled area tend to be distributed close to the mean values. Fairly uniform distribution of dune lengths, widths, and wavelengths defines the apparent regularity of dune patterns in each sand sea.

Size and spacing relationships of crescentic dune ridges sampled in different sand seas were examined to determine whether these relationships are constant despite differences in dune sizes and geographic locations. Sizes of dunes represented in figure 178 range from crescentic segments with a mean width of 111 m (365 ft) at White Sands, New Mexico (fig. 173B), to segments with a mean width of 3.24 km (2 mi) in the Takla Makan Desert of northern China (fig. 175D). Scatter diagrams (fig. 178) show linear arithmetic relationships between the mean widths of crescentic dune segments and their mean wavelengths, between the mean widths of crescentic dune segments and their mean lengths, and between the mean lengths of crescen-

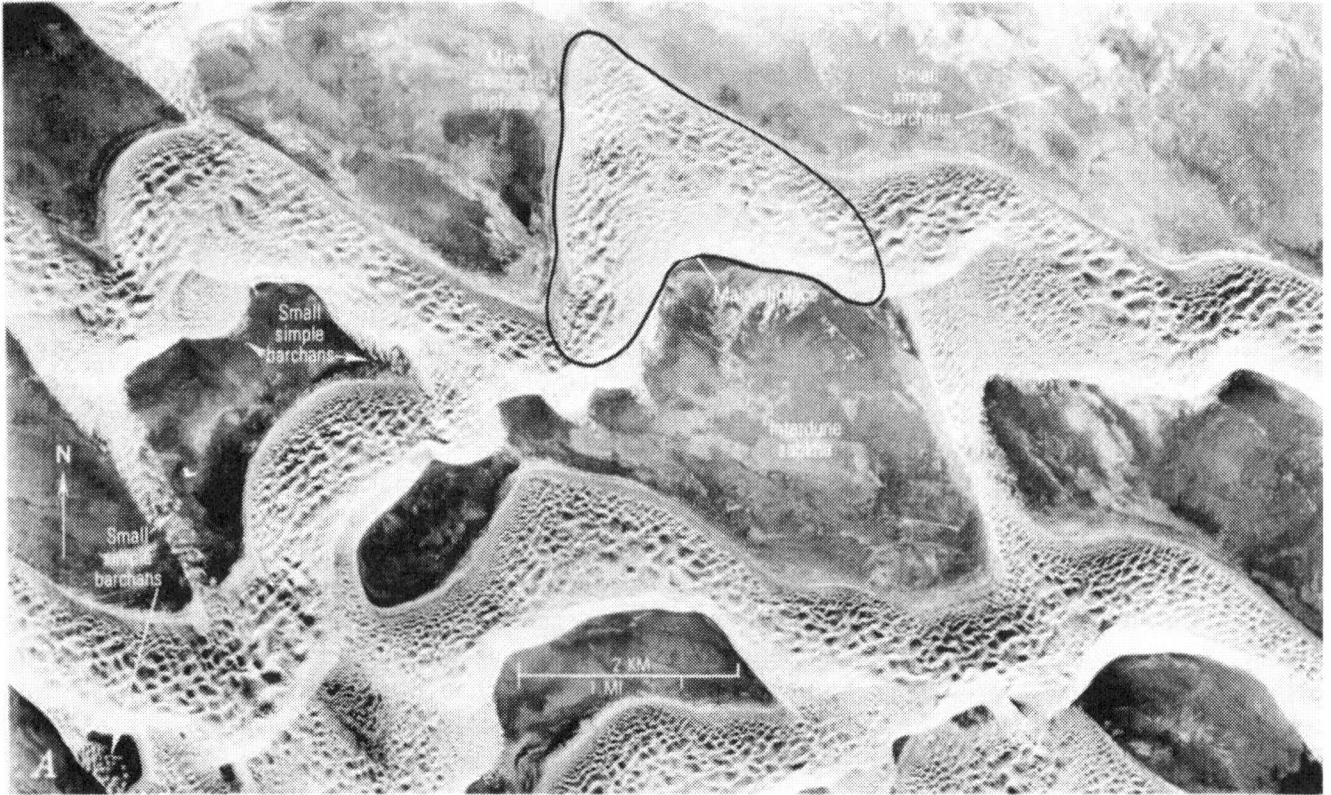

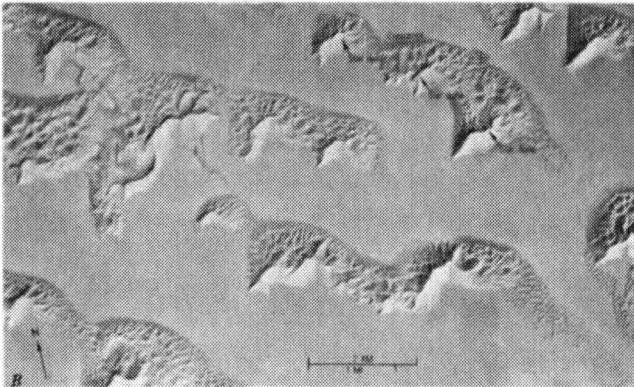

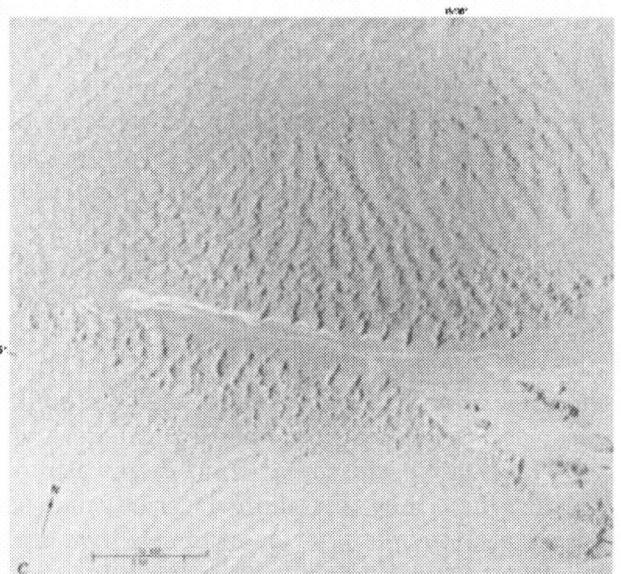

AERIAL AND SPACE PHOTOGRAPHS OF COMPOUND and complex crescentic dunes (above and facing page): *A*, Compound crescentic ridges in the eastern Rub' al Khali (Al'Uruq al Mu'taridah), Saudi Arabia, at approximately lat 23° N., long 54° E. One giant crescentic segment is outlined to show relation to much smaller, simple barchans that cross the interdune areas between the ridges. Segments of these ridges average 2.76 km (about 1.75 mi) from horn to horn and are about 3.1 km (0.3 mi) long, from the base of the gentle slope to base of the main slipface. Photograph by the U.S. Department of Defense. *B*, Complex crescentic dune ridges in the southeastern Rub' al Khali, Saudi Arabia, at approximately lat 21°45' N., long 54°30' E. Star dunes have formed on the crests of the ridges. Photograph by the U.S. Department of Defense. *C*, Complex crescentic dunes along the valley of the Tschaubrivier, South-West Africa. The dunes nearest the river are about 300 m (1,000 ft) high (Barnard, 1973, p. 2, 5). Photograph by National Aeronautics and Space Administration (SL 4–207–8073). *D*, Compound crescentic dune ridges in the Algodones Dunes, California, U.S.A. Major slipfaces are on the south-facing lee slopes, and numerous subsidiary slipfaces are on the gentler slopes. Photograph by National Aeronautics and Space Administration. (Fig. 174.)

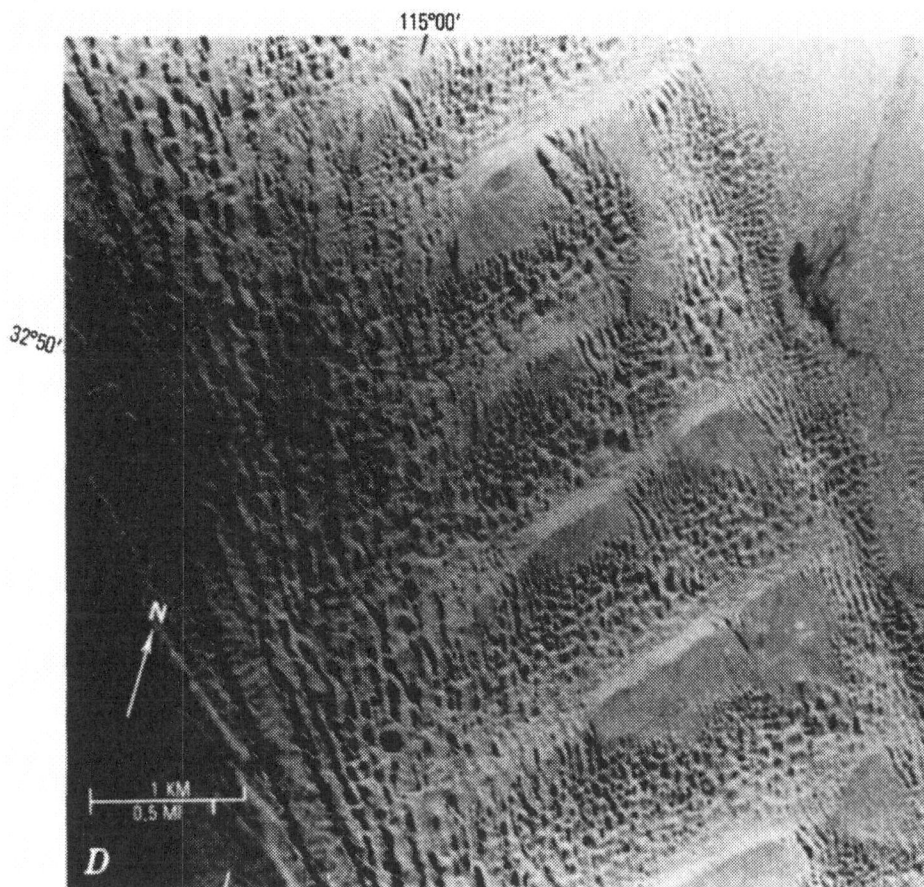

tic dune ridges and their mean wavelengths. The latter association is particularly pronounced (correlation coefficient = 0.90), and this suggests that spacing of crescentic dunes varies directly with dune length, regardless of the size, form, or location of the dunes. Correlation of mean widths of crescentic dune segments with their mean lengths is also strong (r = 0.81) in dunes of all sizes, forms, and locations, which indicates that the correlation of width and length of crescentic mounds (barchans), as recorded by Finkel (1959, p. 628), seems to apply also to segments of crescentic ridges. As in linear dunes (fig. 172) progression in form from simple to compound or complex in crescentic dunes seems to be associated with growth to larger size (fig. 178).

Identification of Varieties

Planimetric similarity among crescentic dune ridges of compound form is shown by correspondence of scale ratios derived from their measurements (table 39). Differences in degrees of correspondence of the ratios indicate that two extreme varieties of compound crescentic dune ridges can be distinguished:

1. Crescentic ridges whose segments are much wider than they are long, and which are not markedly curved; these include the Aoukâr dunes of the southwestern Sahara in Mauritania and the Nebraska Sand Hills, U.S.A. (figs. 175F, E). These ridges are known by most geologists as "transverse" dunes and are referred to by that name in structural studies in this publication.

2. Crescentic ridges whose segments are as wide or slightly wider than they are long and which are markedly curved; these include the compound crescentic dunes of the eastern Rub' al Khali, Saudi Arabia, the Takla Makan and Ala Shan Deserts, China, (figs. 174A; 175B, D, C) and the Thar Desert near Sukkur, Pakistan (chapter K, fig. 256B).

Dunes of the compound crescentic variety show very good to excellent correspondence to each others' scale ratios but mostly poor to fair correspondence to the ratios of dunes of the variety com-

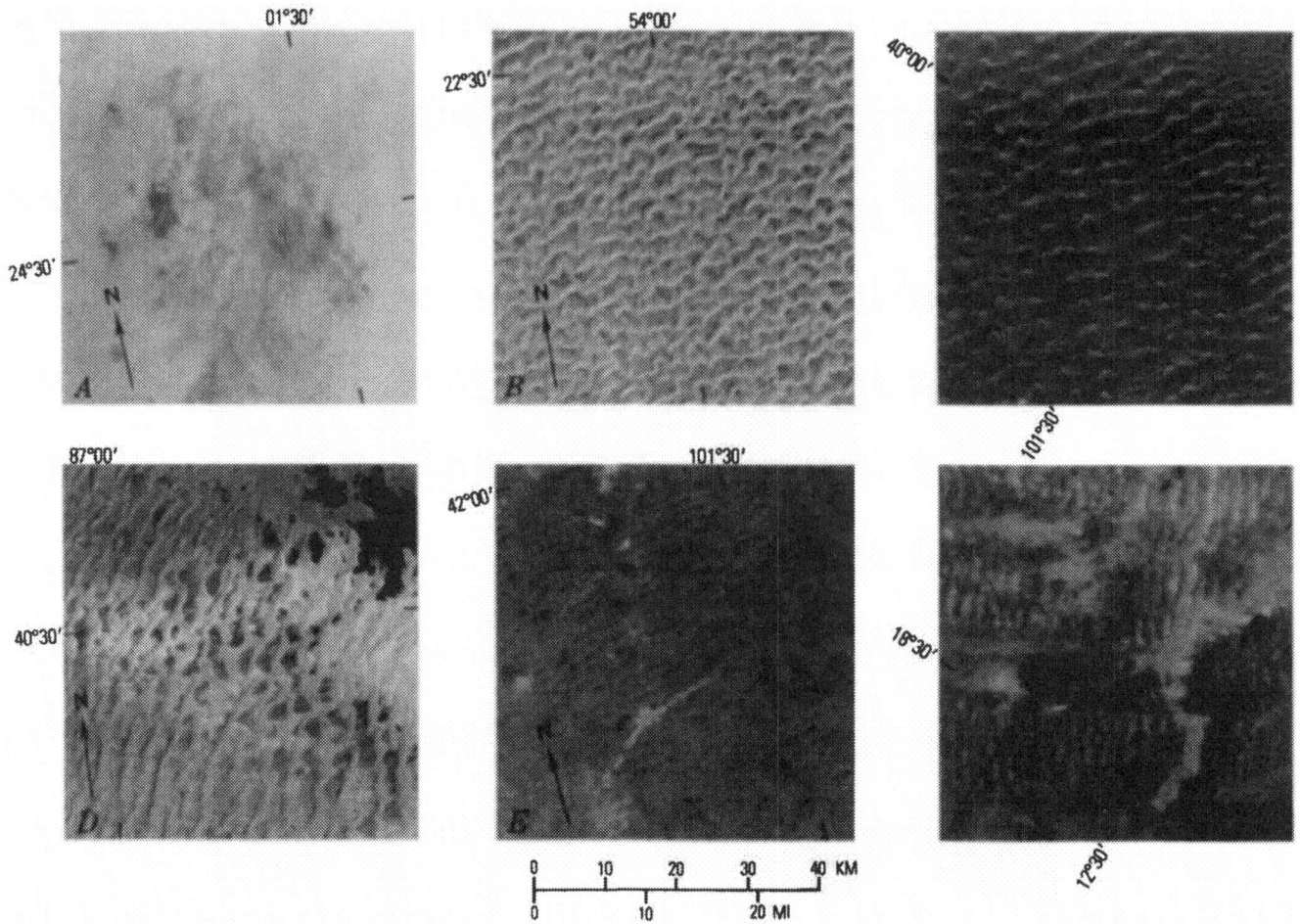

LANDSAT IMAGERY SAMPLE AREAS ON WHICH COMPOUND crescentic dune width, length, and wavelength were measured. Each sample area is 2,500 km² (1,500 mi²). Numbers in parentheses refer to regions listed in table 38. A, Large crescentic dunes, probably compound barchans, in the western Sahara, Mali (8). B, Compound crescentic dunes in the eastern Rub' al Khali, Saudi Arabia (14); these are the dunes seen in closer view in figure 174A. C, Large barchanoid variety of compound crescentic dunes in the central Ala Shan Desert, China (20). D, Compound crescentic dunes in the northeast Takla Makan Desert, China (15). E, Compound crescentic dunes in the Nebraska Sand Hills, U.S.A. (19). F, Compound crescentic dunes in the Aoukâr sand sea, western Sahara, Mauritania (21). (Fig. 175.)

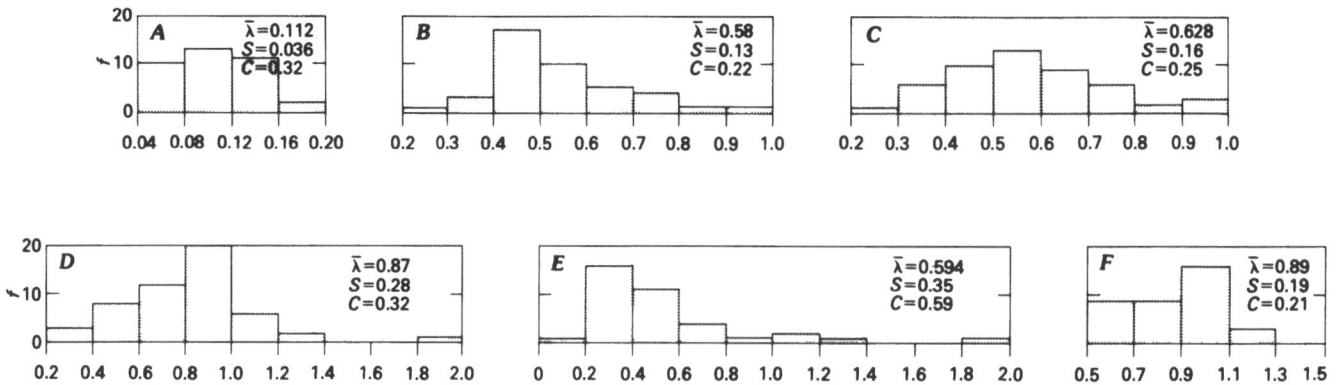

HISTOGRAMS SHOWING FREQUENCY DISTRIBUTION (f) of wavelengths (λ), in kilometres, in small crescentic dune ridges measured on aerial photographs and Landsat imagery. Numbers in parentheses refer to regions listed in table 38. λ̄, mean wavelength, in kilometres; S, standard deviation; C, coefficient of variation. A, White Sands, New Mexico, U.S.A. (1). B, Thar Desert north of Umarkot, Pakistan (3). C, Rub' al Khali, Al Jiwa' area, Saudi Arabia, (5). D, Coastal Namib Desert between Meob Bay and Sandfisch Bay, South-West Africa (4). E, Coastal Arabian peninsula along Persian Gulf between Qatar and Sebkhat Matī, Saudi Arabia (2). F, Eastern Takla Makan Desert near T'ieh-kan-li-k'o, China (6). (Fig. 176.)

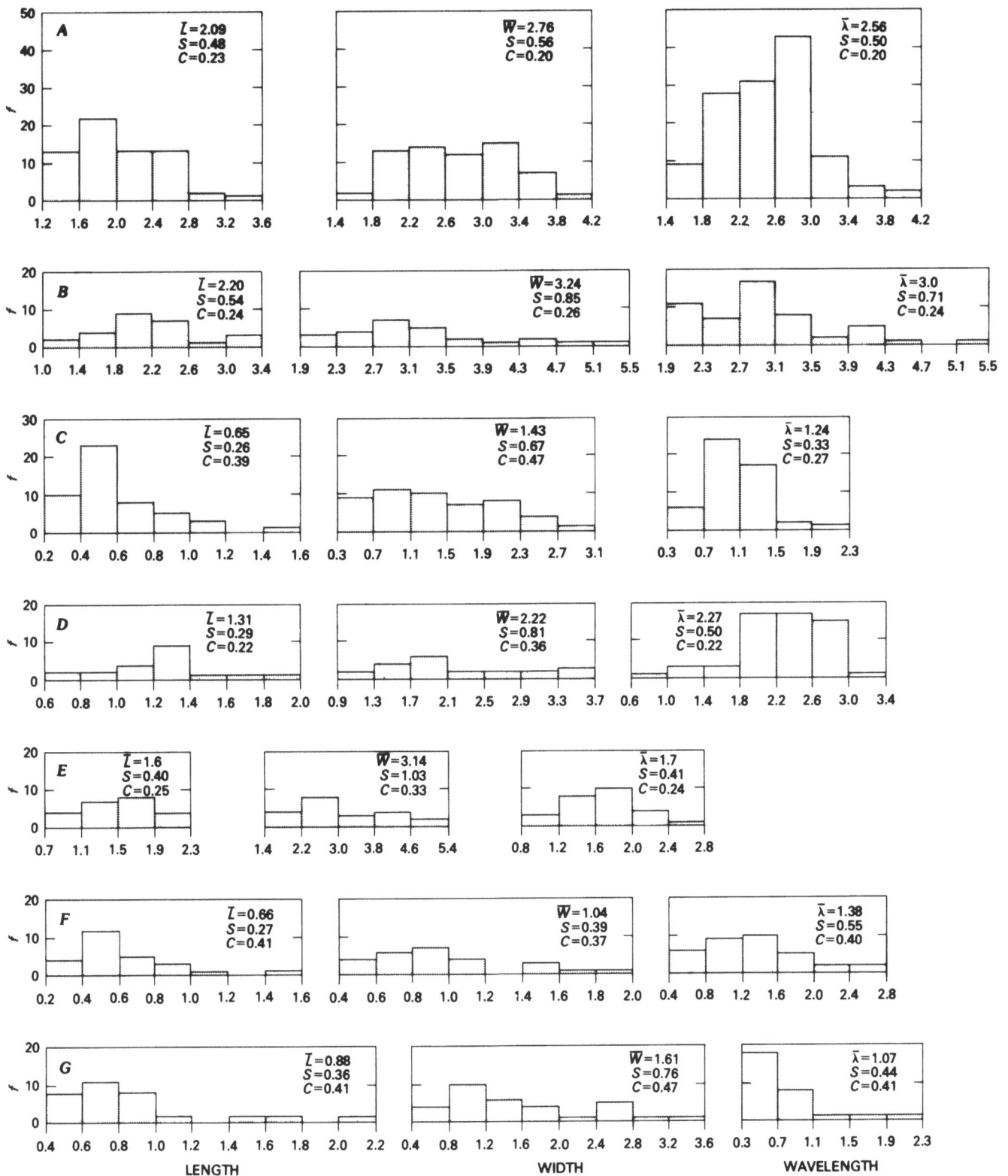

HISTOGRAMS SHOWING FREQUENCY DISTRIBUTION (f) of compound crescentic dune widths (W), lengths (L), and wavelengths (λ), in kilometres, measured on Landsat imagery and Skylab photographs. $\overline{W}$, mean dune width; $\overline{L}$, mean dune length; $\overline{\lambda}$, mean dune wavelength; S, standard deviation; C, coefficient of variation. Numbers in parentheses refer to regions listed in table 38. A, Eastern Rub' al Khali, Saudi Arabia. Area sampled is 10,000 km² (6,200 mi²) (14). B, Takla Makan Desert, China. Area sampled is 5,000 km² (3,100 mi²) (15). C, Northern Sahara, Algeria and Tunisia. Area sampled is 5,000 km² (3,100 mi²) (10). D, Peski Karakumy near Gory Bukantau, U.S.S.R. Area sampled is 2,500 km² (1,550 mi²) (16). E, Nebraska Sand Hills, U.S.A. Area sampled is 3,000 km² (1,860 mi²) (19). F, Gran Desierto, Sonora, Mexico. Area sampled is 650 km² (400 mi²) (12). G, Algodones Dunes, U.S.A. Area sampled is 500 km² (310 mi²) (13). (Fig. 177.)

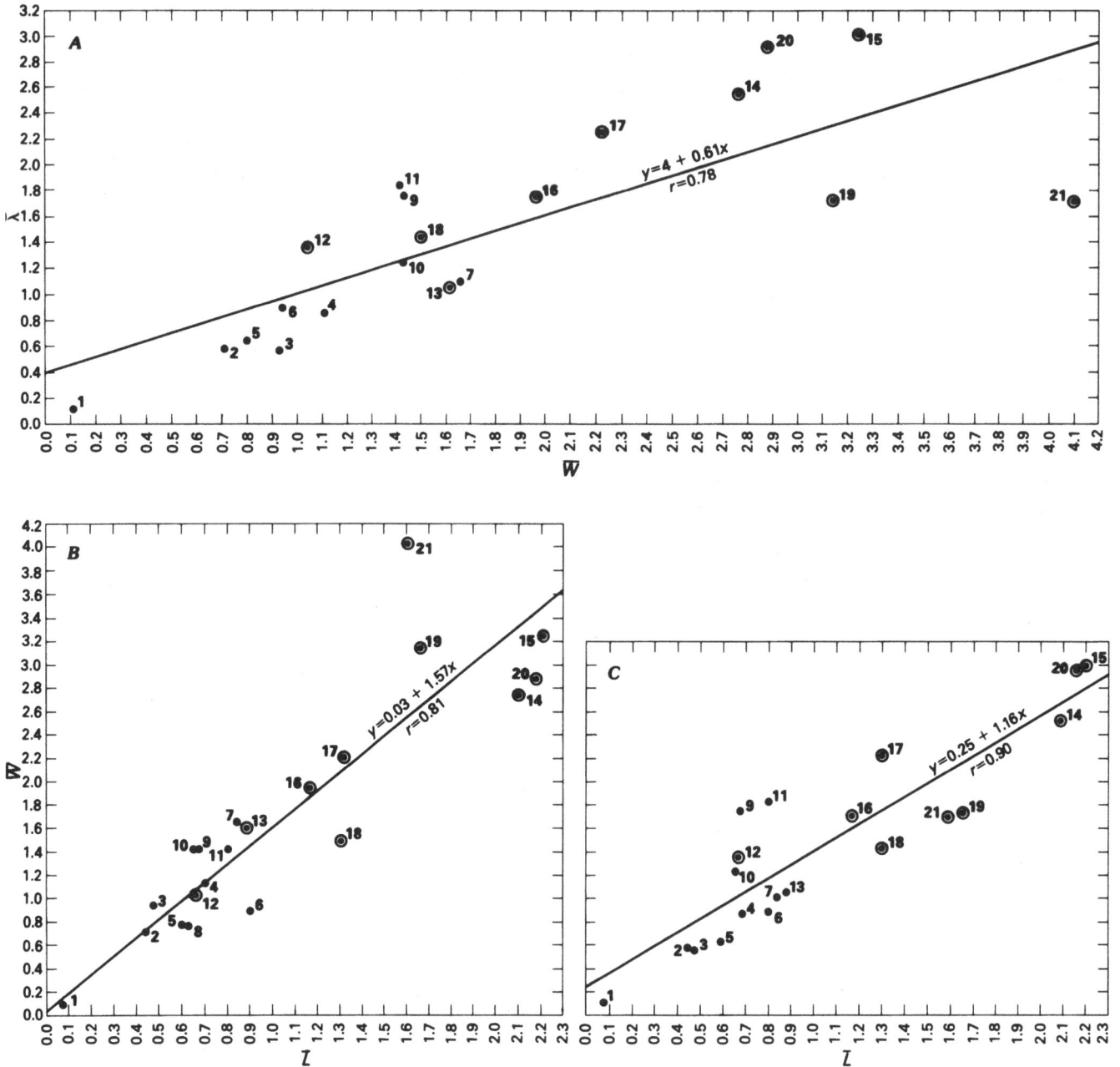

SCATTER DIAGRAMS SHOWING RELATION OF VALUES of crescentic dunes in sand seas measured on Landsat imagery and Skylab and aerial photographs. Circled dots represent dunes identified from aerial photographs and field reports as compound ridges. Solid dots represent small simple(?) crescentic ridges and large barchans. Measurements in kilometres. Num-bers near dots refer to regions listed in table 38. r, correla-tion coefficient. A, Mean dune width ($\overline{W}$) versus mean dune wavelength ($\overline{\lambda}$); B, mean dune length ($\overline{L}$) versus mean dune width ($\overline{W}$); C, mean dune length ($\overline{L}$) versus mean dune wavelength ($\overline{\lambda}$). (Fig. 178.)

monly referred to as "transverse." Some compound crescentic dunes included in the study, such as those of the Algodones Dunes of California, U.S.A. (fig. 174D), the Peski Karakumy near Mary and Gory Bukantau, U.S.S.R., and in the Gran Desierto, Sonora, Mexico, seem to be intermediate in shape between the extremely curved and extremely straight varieties, for their correspondence to each other and to dunes of the two main varieties ranges from very poor to very good, but they are not sufficiently distinct from the others to be identified with separate names.

Only compound dunes, among the crescentic forms, were compared by dimensional analysis. Crescentic dune ridges in several sand seas that are suspected to be compound cannot be definitely classed as such because of poor resolution on Landsat imagery and lack of aerial photographs. Such regions include the An Nafūd in Saudi Arabia and parts of the northeastern Sahara of Algeria and Tunisia. Dunes of equivalent size in the Gran Desierto, Sonora, Mexico, have been determined from Skylab photographs to be compound.

Star Dunes
General Distribution

SIMPLE, COMPOUND, AND COMPLEX star dunes were examined on Landsat imagery and on Skylab and aerial photographs. Simple star dunes have three or four arms, all with slipfaces of about the same size (fig. 179A). Compound star dunes have both large primary arms and small subsidiary arms with slipfaces (fig. 179B). Complex star dunes are combinations of star and other dune types, in which the star element predominates. Examples are star dunes with crescentic curvatures, recorded in the southeast Rub' al Khali, Saudi Arabia (fig. 179C), and Gran Desierto, Mexico (fig. 179D), as well as star dunes in linear chains in the Gran Desierto, Mexico, and the Grand Erg Oriental, Algeria (fig. 180A, D).

Star dunes in simple, compound, and complex forms are the dominant dunes of the Grand Erg Oriental, (fig. 180B), but elsewhere they are limited to rather small areas. In the Ala Shan Desert, China, and the Namib Desert, South-West Africa (chapter K, figs. 264B, 224), star dunes occur around the margins of sand seas, typically near topographic barriers. In southeast Saudi Arabia, Yemen, and Oman, small star dunes occur near the mouths of dry washes along the southeastern edge of the Rub' al Khali, and they grade northward and westward into much larger star dunes (fig. 179C), which in turn merge with very large southward-migrating crescentic dune ridges that have star dunes on their crests (fig. 174B). In the Gran Desierto, Mexico, simple and compound star dunes occur around the Sierra del Rosario (fig. 180A). To the east (toward Cerro Pinacate), star dunes grade into complex dunes that are crescentic in shape but have star dunes on their crests (fig. 179D). Toward the south (Gulf of California), star dunes coalesce in east-west linear chains. The largest star dunes observed in this study are in the Grand Erg Oriental, Algeria (chapter K, fig. 210B).

Analysis of the distribution of star dunes in relationship to distribution of effective winds (chapters F and K) suggests that star dunes occur where effective winds blow with moderate strength from several directions.

Morphometry

Histograms (fig. 181) show frequency distributions of star dune diameters and wavelengths in the sampled sand seas. Measurements in ten 2,500-km² (1,550-mi²) sample areas, and four smaller areas of Landsat imagery and Skylab photographs are summarized in table 40. As in sand seas of other types of dunes, star dune measurements tend to be grouped about characteristic mean values, thus defining the regular pattern for each sand sea. However, the scatter diagram (fig. 182) shows only a very weak association between mean diameters of star dunes and their mean wavelengths in the sampled sand seas. The weak correlation ($r = 0.52$) indicates that spacing of star dunes, unlike that of linear and crescentic dunes (figs. 172, 178) does not vary directly with their size. Differences in the relationship of dune spacing to dune size in different types of dunes may reflect differences in processes of growth. Star dunes, unlike linear and crescentic dunes, grow upward, rather than forward or laterally, in response to effective winds.

Identification of Varieties

Ratios derived from measurements of mean diameters and mean wavelengths of star dunes in sampled sand seas (table 40) are compared in table 41. Reliability of the comparisons is limited by the weak association of diameter and wavelength among star dunes (fig. 182). Also, differences in degrees of correspondence of scale ratios support the suggestion, based on aerial photographs, that three apparently different varieties of star dunes exist:

1. Sharp-pointed radiating pyramidal dunes are typical of parts of the Gran Desierto, Mexico, the Ala Shan Desert, China, parts of the southern Sahara in Niger, and much of the

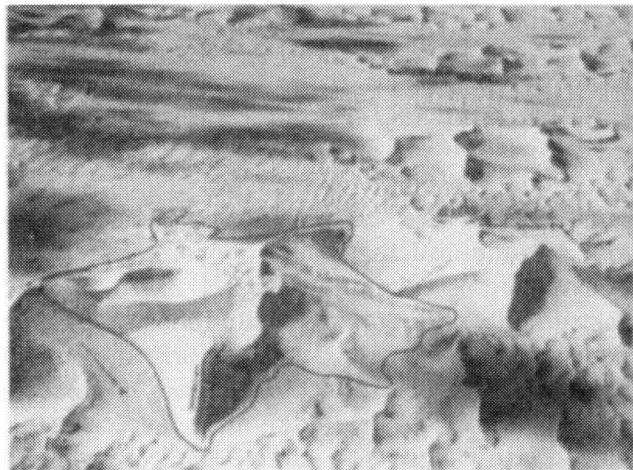

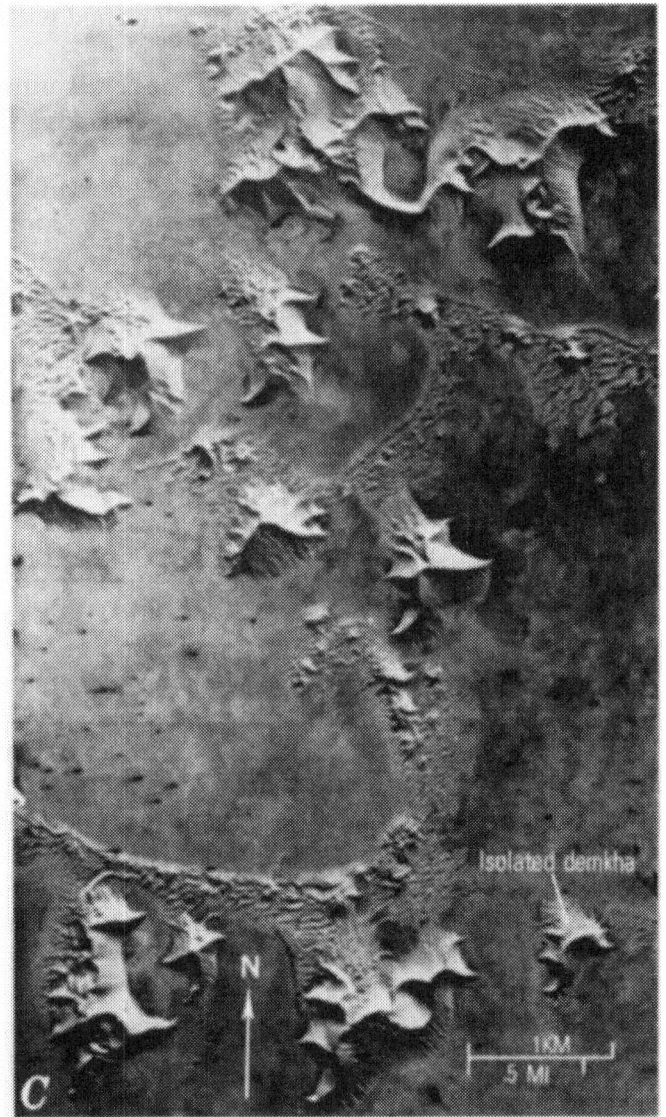

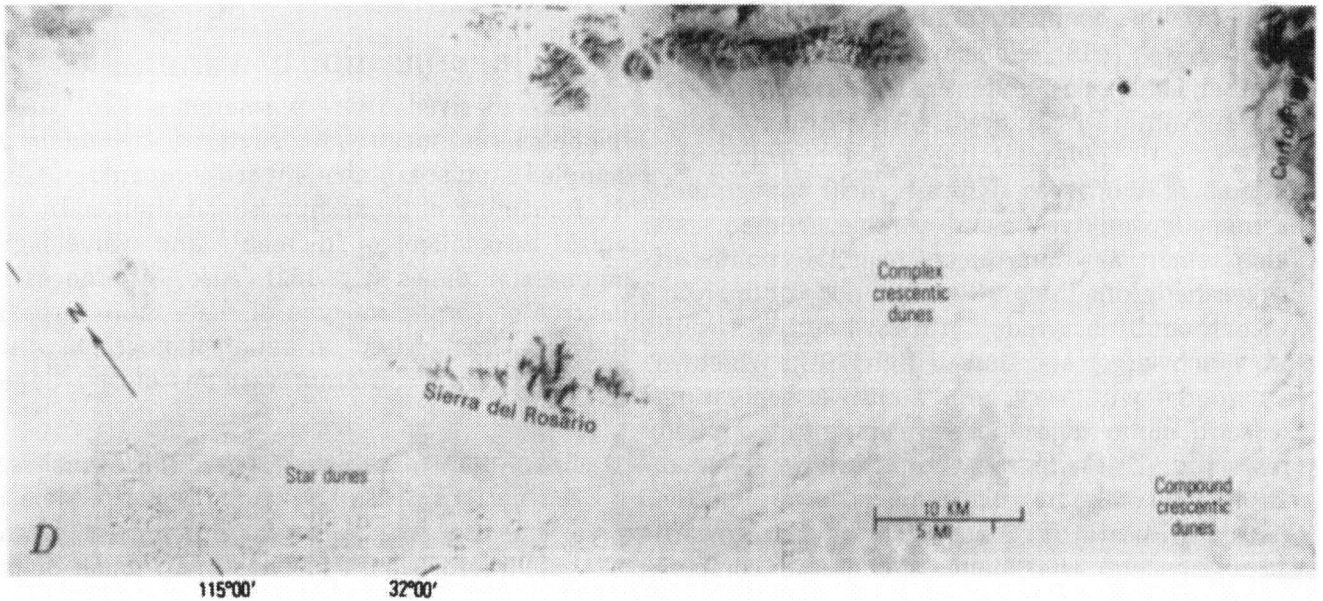

AERIAL PHOTOGRAPHS OF SIMPLE, COMPOUND, and complex star dunes (facing page). *A*, Simple star dunes, Gran Desierto, at approximately lat 32° N., long 115° W. Arms of these dunes are generally less than 100 m (330 ft) long. Photograph by E. Tad Nichols. *B*, Compound star dunes, Gran Desierto, Sonora, Mexico, at approximately lat 32° N., long 114°30' W. Diameter of these dunes averages 700 m (2,300 ft). One dune is outlined for contrast with smaller, simple star dunes. *C*, Complex star dunes, eastern Rub' al Khali, Saudi Arabia, at approximately lat 20°45' N., long 54°45' E. Photograph by the U.S. Department of Defense. *D*, Star dunes in the Gran Desierto, Sonora, Mexico, grade eastward into complex crescentic dunes with stars on their crests, and these grade into compound crescentic dunes near the southern Cerro Pinacate. Photography by National Aeronautics and Space Administration (SL 4–92–358). (Fig. 179.)

Grand Erg Oriental in Algeria. Similarities of dunes in Mexico and in Algeria are illustrated in low-altitude aerial photographs of a part of the Gran Desierto (fig. 179*B*) and a part of the Grand Erg Oriental (Capot-Rey and Capot-Rey, 1948, pl. 2).

2. Rounded compact star dunes with arms that do not project far from the main mass of each dune (H. T. U. Smith, 1963, p. 45) are typical of the southern margin of the Grand Erg Oriental, Algeria (chapter K, fig. 210*B*). Correspondence of the southern rounded variety to the sharp-pointed variety of star dunes farther north in the Grand Erg Oriental is very poor to fair (table 41).

3. Sharp-crested dunes with arms greatly elongated in a preferred direction may be a third variety. Such dunes are observed in the Namib Desert (chapter K, fig. 224), where they grade into complex reversing dunes, in the eastern Rub' al Khali, where they also are associated with complex dunes (fig. 179*C*), and in the northern Grand Erg Oriental, Algeria. These dunes show excellent correspondence to each other but very poor to fair correspondence with dunes of the first two varieties.

Star dunes described by McKee (1966, p. 65–68) in the 'Irq as Subay', Saudi Arabia, are different from all these varieties, for they have extremely long arms that radiate far from the main body of each dune. These dunes are not compared with other star dunes in table 41.

Parabolic Dunes
General Distribution

TWO FORMS OF PARABOLIC DUNES are recognized on Landsat imagery and on aerial photographs. Simple parabolic dunes are U–shaped or V–shaped and consist of a convex mound with two long trailing arms. Dunes of this kind at White Sands National Monument, New Mexico, U.S.A. (McKee, 1966, p. 25, 50–51) are shown on an aerial photograph (chapter E, fig. 50).

Compound parabolic dunes are U–shaped, V–shaped, or rake-shaped and have numerous secondary trailing arms. In the Thar Desert (India and Pakistan, chapter K) rakelike clusters of parabolic dunes cover about 100,000 km² (62,000 mi²). Their planimetric aspect was sketched (Verstappen, 1968a, fig. 3; A. S. Goudie, written commun. to McKee, 1974) from aerial photographs and maps (chapter K, fig. 254). Their aspect on Landsat imagery is shown in figure 183*A*. The largest parabolic dunes observed on Landsat imagery are near the Colombia-Venezuela border at about lat 6° 30' N., long 69° W. (fig. 183*B*). These dunes are U–shaped or V–shaped rather than rakelike and have numerous trailing arms as much as 12 km (7 mi) long.

Parabolic dunes are described from the Midwestern United States (Melton, 1940, p. 126–128; H. T. U. Smith, 1946; Ahlbrandt, 1974b, p. 45), coastal Oregon and California (Cooper, 1967, p. 22–25), White Sands National Monument, New Mexico (McKee, 1966, p. 50–51), localities in coastal Brazil (chapter E) and Europe (Kádár, 1966, p. 446). Parabolic dunes in the Thar Desert and at White Sands are alined parallel to the resultant direction of effective winds from the southwest (dune noses point to the northeast) (McKee, Breed, and others, 1974, figs. 25, 5; this publication, chapters F and K and figs. 252, 253).

Morphometry

Compound parabolic dunes in the sampled part of the Thar Desert (fig. 183*A*) have an average of seven arms, mean lengths of 2.6 km (1.6 mi), and a mean width of 2.4 km (1.5 mi). Their length-width ratio is 1.1. Compound parabolic dunes in South America (fig. 183*B*) have a length-width ratio of 3.7,

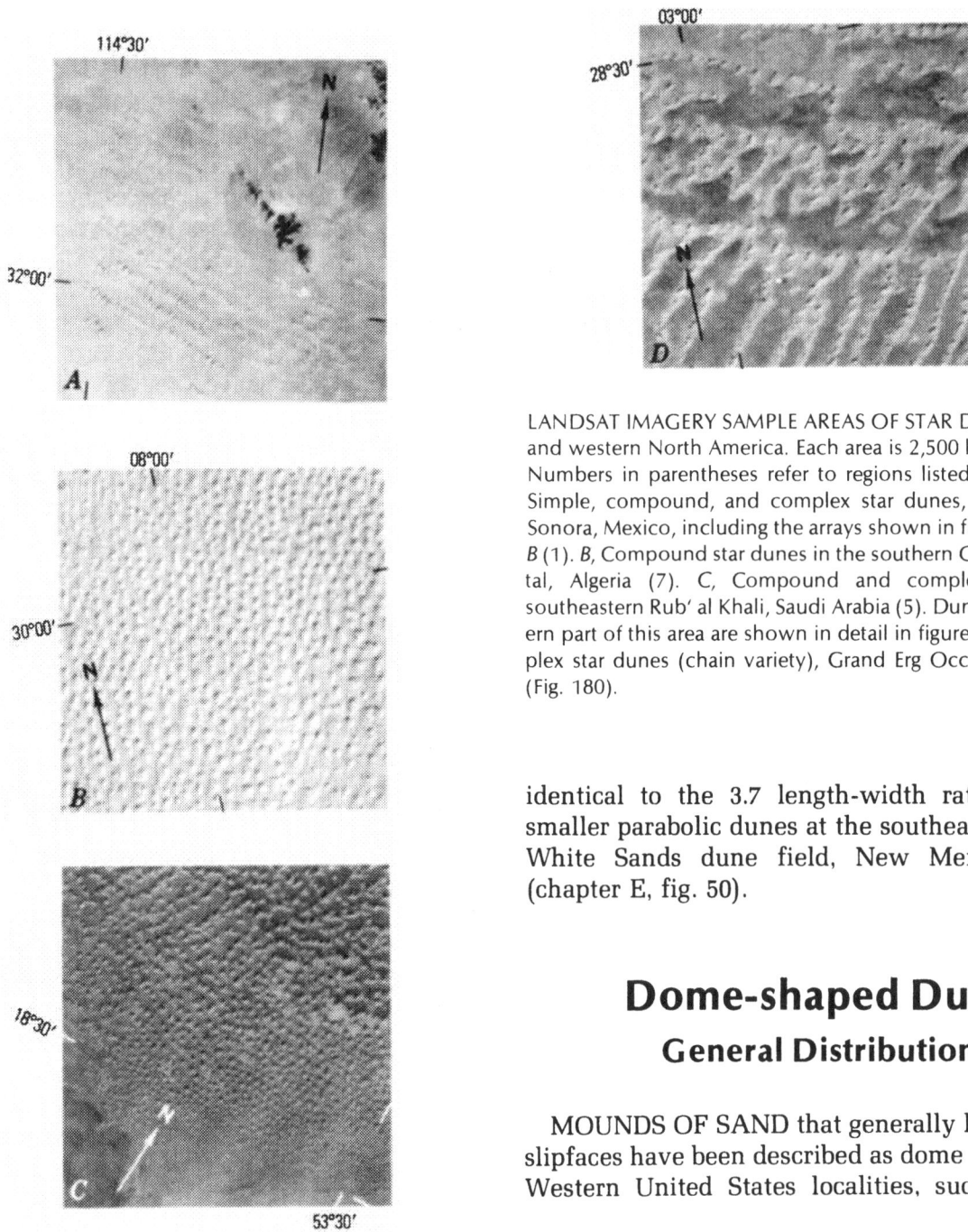

LANDSAT IMAGERY SAMPLE AREAS OF STAR DUNES in Arabia and western North America. Each area is 2,500 km² (1,550 mi²). Numbers in parentheses refer to regions listed in table 40. *A*, Simple, compound, and complex star dunes, Gran Desierto, Sonora, Mexico, including the arrays shown in figures 179*A* and *B* (1). *B*, Compound star dunes in the southern Grand Erg Oriental, Algeria (7). *C*, Compound and complex star dunes, southeastern Rub' al Khali, Saudi Arabia (5). Dunes in the northern part of this area are shown in detail in figure 179*C*. *D*, Complex star dunes (chain variety), Grand Erg Occidental, Algeria. (Fig. 180).

identical to the 3.7 length-width ratio of much smaller parabolic dunes at the southeast end of the White Sands dune field, New Mexico, U.S.A. (chapter E, fig. 50).

Dome-shaped Dunes
General Distribution

MOUNDS OF SAND that generally lack external slipfaces have been described as dome dunes in the Western United States localities, such as White

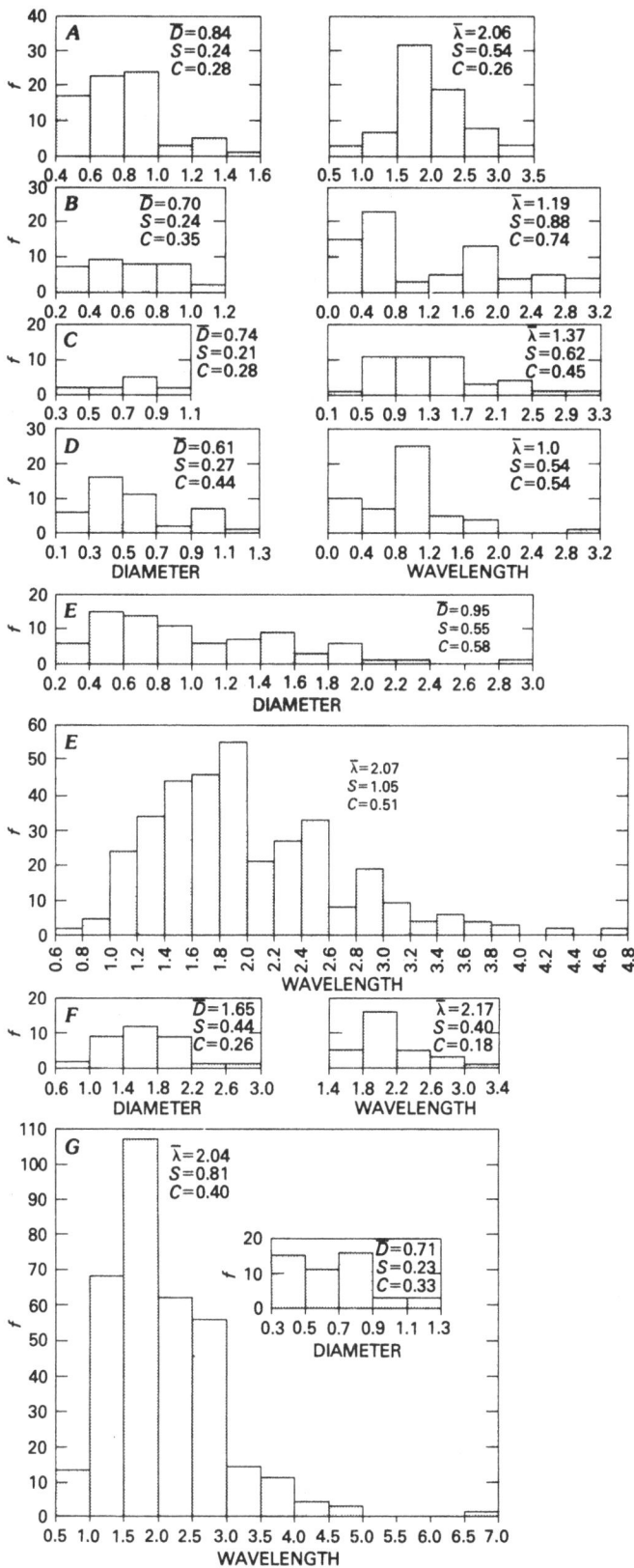

HISTOGRAMS SHOWING FREQUENCY DISTRIBUTION (f) of diameters (D), and wavelengths (λ), in kilometres, of star dunes measured on Landsat imagery and Skylab photographs. $\overline{D}$, mean dune diameter; $\overline{\lambda}$, mean dune wavelength; S, standard deviation; and C, coefficient of variation. Numbers in parentheses refer to regions listed in table 40. A, Southeastern Rub' al Khali, Saudi Arabia. Area sampled is 12,500 km² (7,770 mi²) (5). B, Gran Desierto, Sonora, Mexico. Area sampled is 1,000 km² (620 mi²) (1). C, Ala Shan Desert, China. Area sampled is 2,000 km² (1,240 mi²) (4). D, Southern Sahara, Niger. Area sampled is 1,000 km² (620 mi²) (3). E, Entire Grand Erg Oriental, Algeria. Area sampled is 10,000 km² (6,200 mi²) (6). F, Southern margin of the Grand Erg Oriental, Algeria. Area sampled is 2,500 km² (1,550 mi²) (7). G, Northern part of the Grand Erg Oriental, Algeria. Area sampled is 7,500 km² (4,660 mi²) (8). (Fig. 181.)

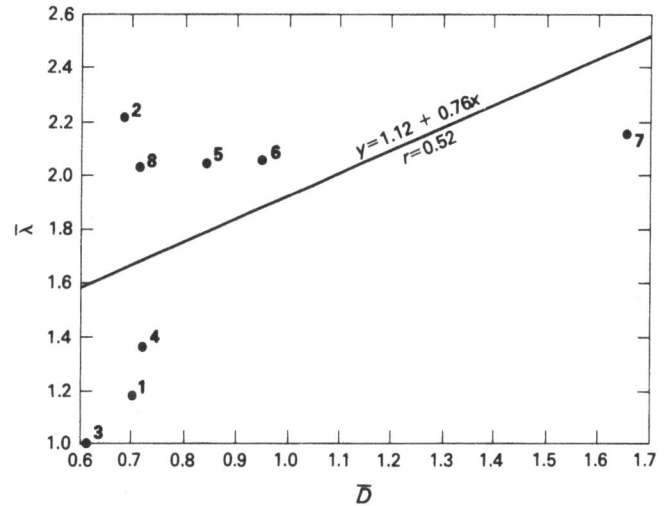

SCATTER DIAGRAM SHOWING RELATIONSHIP of mean dune diameter ($\overline{D}$) to mean dune wavelength ($\overline{\lambda}$) in star dunes sampled on Landsat imagery and Skylab photographs. All dunes are believed to be compound or complex forms. All measurements are in kilometres. r, correlation coefficient. Numbers in parentheses refer to regions listed in table 40. (Fig. 182.)

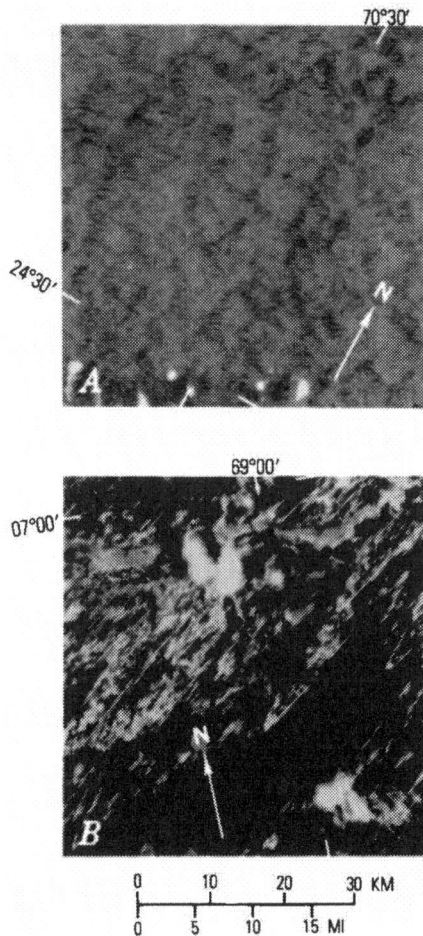

LANDSAT IMAGERY SAMPLE AREAS OF PARABOLIC DUNES of the (A) Thar Desert, India and Pakistan, and (B) Colombia-Venezuela border region. (Fig. 183.)

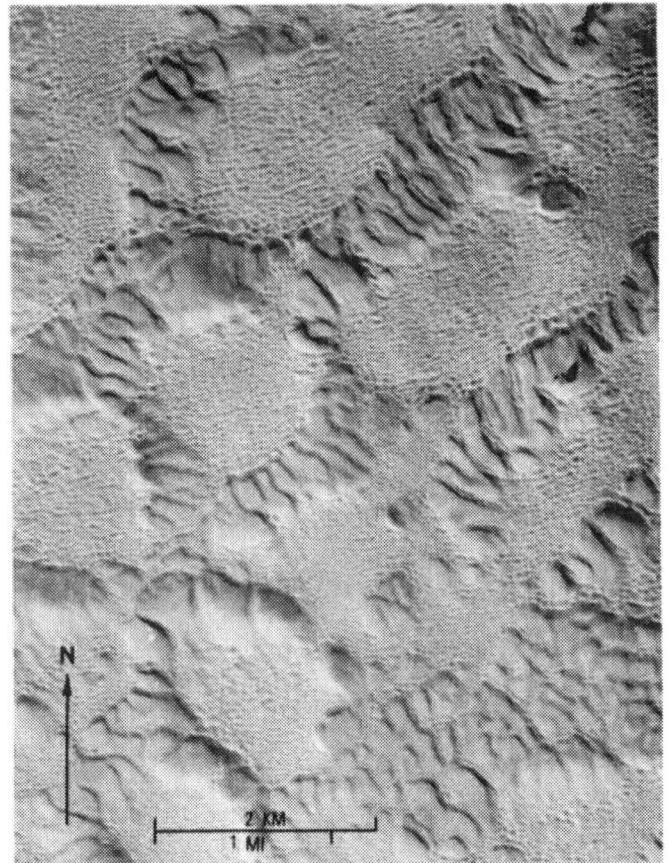

COMPLEX, GIANT DOMES IN THE NAFŪD ATH THUWAYRĀT, northern Saudi Arabia, east of Buraydah at about long 44°30' E., lat 26°45' N. Dark spots in the interdune areas are oases. Diameters of dome-shaped dunes in this sand sea average about 1 km (0.6 mi). The appearance of these dunes on Landsat imagery is shown in chapter K, fig. 244. Aerial photograph by the U.S. Department of Defense. (Fig. 184.)

Sands National Monument, New Mexico (McKee, 1966, p. 26–27), and the Killpecker dunes, Wyoming (Ahlbrandt, 1974b, p. 45). Terminology for dome-shaped dunes is summarized in table 30. Small dome-shaped dunes are not visible on Landsat imagery or on space photographs, but a few groups of very large dome-shaped dunes have been observed. In the northern sand seas of Saudi Arabia, such large features are described (Holm, 1953) as giant domes.

On aerial photographs (fig. 184), the Arabian dome-shaped dunes are interpreted as complex for their rounded crests are covered by rows of crescentic dune ridges. Adjacent domes are separated by low ridges of indeterminate character. The nature of these dome-shaped dunes on Landsat imagery is shown in chapter K, figure 244. Dunes of

similar dome-shaped appearance on Landsat imagery, which have not been studied on aerial photographs nor described in previous work, are observed in the northern Grand Erg Occidental, Algeria (fig. 185A), and in the northern Takla Makan Desert, China (fig. 185B).

In each of the three geographic regions just referred to, large dome-shaped dunes are located at the far upwind margins of sand seas. Small dome dunes are located at the upwind end of both the White Sands dune field (McKee, 1966, p. 26–27) and the Killpecker, Wyoming, dune field (Ahlbrandt, 1974b, p. 45). Locations of dome dunes in both these fields may be related to the presence of fairly strong and unidirectional effective winds, as suggested by analysis of wind patterns (chapters F, K) and by internal structure.

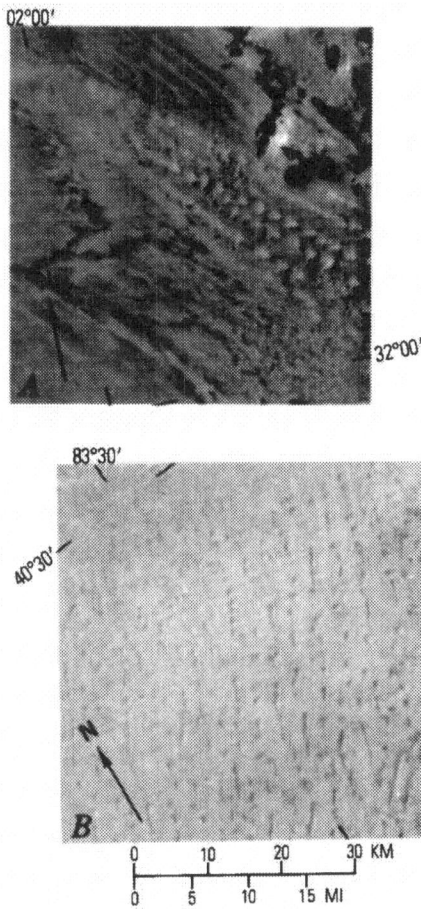

LANDSAT IMAGERY SAMPLE AREAS SHOWING DOME-SHAPED DUNES. A, Giant domes(?) in the northern part of the Grand Erg Occidental, Algeria, north of El Golea. B, Giant domes(?) in the northern Takla Makan Desert, China. Some of these features have linear "tails" 0.9–7.0 km (0.6–4.3 mi) long. (Fig. 185.)

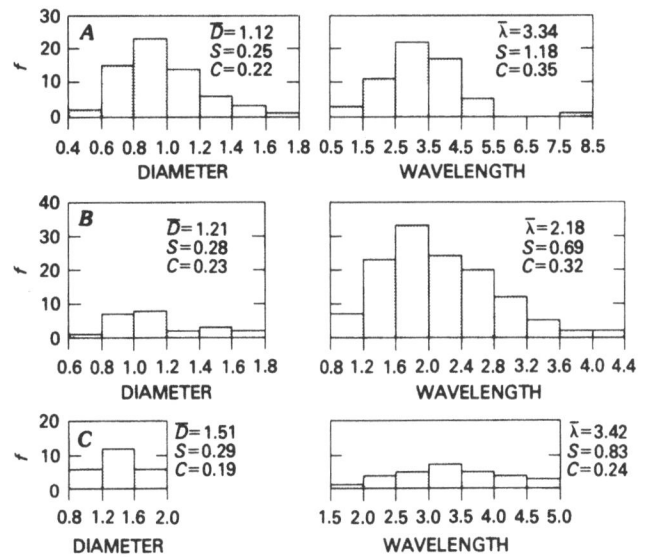

HISTOGRAMS SHOWING FREQUENCY DISTRIBUTION (f) of diameters (D) and wavelengths (λ), in kilometres, of dome-shaped dunes measured on Landsat transparencies. $\overline{D}$, mean dune diameter; $\overline{\lambda}$, mean dune wavelength; S, standard deviation; C, coefficient of variation. Numbers in parentheses refer to regions listed in table 42. A, Takla Makan Desert, China. Area sampled is 2,500 km² (1,550 mi²) (1). B, Sand seas, northern Saudi Arabia. Area sampled is 3,000 km² (1,860 mi²) (3). C, Sahara, Algeria. Area sampled is 2,500 km² (1,550 mi²) (2). (Fig. 186.)

Morphometry

Measurements of domes and dome-shaped dunes were made for five sample areas of Landsat imagery, partly shown in figure 185 and summarized in table 42. Histograms (fig. 186) show that diameters and wavelengths of dome-shaped dunes are broadly distributed about characteristic mean values in each sand sea. Samples are too few, however, to test relationships of mean diameter and mean wavelength between dome-shaped dunes of different sizes and in different geographic locations. Comparisons of dimensionless ratios (table 43) of all observed dome-shaped dunes show close similarity of these features.

Sand Sheets and Streaks

General Distribution

IRREGULAR PATCHES OF EOLIAN SAND with external shapes definable on Landsat imagery in terms of the bounding topography, and without discernible slipfaces, are herein termed "sand sheets." Elongate patches with sharply defined edges as seen on Landsat imagery are herein termed "streaks." (These are the sand stringers mentioned by other geologists in this publication). Other authors' terms for these features are shown in table 34.

Sand sheets are ubiquitous to nearly all sand seas studied, but sand streaks are less common and seem to be associated mainly with sand seas of linear dunes. Figure 187 shows large eolian sand streaks in the northwest Sahara of Mauritania. Figure 167B shows smaller streaks on the Navajo Indian Reser-

13°00'

21°00'

20 KM
10 MI

LANDSAT IMAGERY SHOWING SAND SHEETS and streaks (stringers) associated with linear dunes in the northwestern Sahara, Mauritania. The area shown is about 34,000 km² (31,100 mi²). Effective winds in this region are from the northeast. (Fig. 187.)

vation in northern Arizona, U.S.A. Other areas where streaks are conspicuous are the southern Sahara (chapter K, fig. 201) and the northern part of the Arabian Peninsula (chapter K, fig. 200). Analysis of effective winds in these regions (chapters F and K) suggests that sand streaks are associated with strong, dominantly unimodal effective winds. No measurements were made of sand sheets or streaks.

Sand streaks (stringers) were noted on orbital photographs by Verstappen and van Zuidam (1970, p. 41). Ground investigations were made by S. G. Fryberger of such sand bodies in an area northeast of the Little Colorado River, Arizona, U.S.A. Sand-streak surfaces there had numerous small

barchanoid dunes, some of which are discernible on high-altitude aerial photographs.

Interdune Areas
General Distribution

DUNES IN SAND SEAS are commonly separated from each other by interdune areas whose well-defined shapes are associated with dunes of certain types, regardless of scale of the dunes or their geographic location. For example, interdune areas in sand seas of linear dune type are generally open corridors called "couloirs" in the formerly "French" Sahara, "straate" in the Kalahari and Namib Deserts of southern Africa, and "sandy lanes" in Australia. Some local names distinguish sand-covered interdune corridors from those that are bedrock.

Names for interdune areas in sand areas of crescentic dunes generally reflect the degree of closure of the segments of these dunes and, thus, suggest the ease or difficulty of travel through the sand sea. An early name applied by geographers to fully enclosed interdune areas in sand seas of crescentic dune type are "bajirs." Hedin (1905a), used this term to describe large flat-floored basins between dunes in the eastern Takla Makan Desert of China (fig. 175D). Equivalent terms in the Arabian and Sahara sand seas are "fuljes" and "feidjs." The term "fulje" was applied (Melton, 1940; and H. T. U. Smith, 1946) to interdune hollows in the Midwestern United States.

Examples of small-scale interdune areas of the fulje type are shown in aerial photographs of White Sands and northwest Sahara (figs. 173B, C); large-scale interdune areas of the same type are shown on figures 174A and 175B of the eastern Rub' al Khali, Saudi Arabia. Various local names for interdune areas are listed in table 35.

A density slicer with planimeter can be used to measure relative areas of dune and interdune surfaces in some sand seas, such as the Grand Erg Oriental, Algeria, where Landsat transparences show extreme photographic contrast between dune and interdune surfaces. The technique is only locally successful, however (chapter K), so comparisons from desert to desert have not been made.

Tables 26–39

These tables give chronological lists of local terms for various types of dunes and the morphometry of such dunes derived from measurements of sample areas on Landsat imagery and Skylab and aerial photographs.

TABLE 26. — *Chronological list of terms considered equivalent to dunes classified as linear, based on Landsat imagery*

Reference		Terms (reference page, figure, plate, and table numbers in parentheses)	
Author (and region)	Year	Simple ridge	Compound ridge
Hedin, Sven (Takla Makan Desert, China).	1905a	Thresholds (235-236). Ordinary ridges (323; fig. 267).	[Some] dune accumulations (354-355). Dune-range (357).
Aufrère, Léon (Formerly "French" Sahara, North Africa).	1928	Les dunes longitudinales (833).	[Some] châines de sable (833). Châines longitudinales (833).
	1930	do. (220, 227, 229).	Draa or bras (220). (Some) châines de sable (220, 230).
	1932	Oblique dunes (42, 44).	No term.
	1934	Dunes de conjonction, dunes d'incidence (132, 138).	Châines avec vagues longitudinal (slassels) (137; fig. 2) [Some] châines avec vagues obliques (137; fig. 2). Châines diagonales (139). [Some] dunes de conjonction et dunes d'incidence (132).
Bagnold, R. A. (Libyan desert, North Africa).	1933	Single seif dunes (107-112, 122). Single line or range of seifs (122).	Whaleback ridge (107, 125). Belts of seif dunes (108, 109).
	1941	Longitudinal dune, seif (189, 222-225). Seif dune chain (224-225).	Whalebacks (230-231; pl. 13a).
Madigan, C. L. (Simpson Desert, Australia).	1936	Sandridges (206, 210-227). Longitudinal ridges (224-227).	No term.
	1938	Sandridges (513-515). Longitudinal ridges (513).	Do.
	1946	Sandridges (45-63; pl. 5-8).	[Some] giant sandridges and whaleback sandridges (figs. 1, 2; pl. 8).
		Straight and parallel sandridges (57). Seif (pl. siouf) (50). Crest or seif (61).	Seif ridges crossing sand ridges at a small angle en echelon (50).
Melton, F. A. (Midwestern United States),	1940	Longitudinal dunes (113, 121; fig. 9, 10). [Some] wind-shadow dunes (120). [Some] windrift dunes (129-130). [Some] source-bordering lee dunes (122; fig. 10, 13).	[Some] longitudinal dunes (121-122). Oblique windrift dunes(?) (129, fig. 19, 23). Oblique longitudinal dunes (fig. 27).
Hack, J. T. (Navajo Indian Reservation, U.S.A.).	1941	Longitudinal dune (240-243).	No term.
Capot-Rey, Robert (Formerly "French" Sahara, North Africa).	1945	Sand ridge (400). Longitudinal forms, dune longitudinale (401).	Dunes of quadrature (parallel sif) (401). Dune chain with suif (403; fig. 10). Dunes of incidence (oblique sif) (401).
	1947b	No term.	Cordons (87). Files de siouf allongés (89).
Capot-Rey, Robert, and Capot-Rey, Francoise (Formerly "French" Sahara, North Africa).	1948	Longitudinal forms, dune longitudinale (67). Sif (pl., siouf) (63, 66-73). Elb (pl., aleb) (73).	Do. Dos de baleine (whalebacks) (63-70). [Some] diouf, bras, or draa (72, 75).
Capot-Rey, Robert (Formerly "French" Sahara, North Africa).	1953	No term.	Do. (145).
Smith, H. T. U. (Nebraska, U.S.A.; Sahara, North Africa).	1946	Longitudinal dunes (197).	No term.
	1963	do. (1, 17-18). Lee dune ridges (34; fig. 6).	Fishbone pattern of longitudinal dune ridges (18). Compound longitudinal dunes (20, 41, fig. 13).
	1965	Longitudinal dunes (566-568; pl. 3, 4A). Linear ridges (557).	No term.
	1968	Longitudinal dunes; seif dunes (43).	Do.
Holm, D. A. (Saudi Arabian deserts).	1953	No. term.	Uruq (112).
	1957	[Some] sigmoidal dunes (1746).	Anastomosing linear complexes (1746).
	1960	do. (1371; fig. 4). Linear ridges (1371).	Linear complexes (1373). Uruq (1373, 1377; fig. 7).
	1968	do. (978). [Some] sigmoidal dunes (978).	Linear complexes (978).
McKee, E. D. (White Sands, U.S.A.; Libyan and Saudi Arabian deserts).	1957	Longitudinal dune (1720-1721).	No term.
	1966	do. (16; pl. IIB). Seif dunes (3, 10, 16, 61; pl. IIB).	Do.
Cooper, W. S. (West coast, North America).	1958	Longitudinal dunes (7, 49, 60-64).	[Some] oblique ridge dunes with basal plinth and active crests (49).
		Seif, seif chain (61). [Some] oblique ridge dunes (7, 49, 59).	
	1967	do. (109-110). Longitudinal elements (26).	No term.
Grove, A. T. (Southern Sahara, North Africa; Kalahari Desert, South Africa).	1958	Linear dunes (526). Longitudinal dunes (526-530).	No term.
	1960b	do. (203).	[Some] alab dunes (203).
	1969	do. (195-202, 211; fig. 3). Linear dunes (195). Sand ridges (195-202, 211; fig. 3).	Do. (195-201, 208). [Some] qoz (195).
Monod, Theodore (Formerly "French" Sahara, North Africa).	1958	Le silk (pl., sloûk) (25, 142, 147-150; figs. 10(2), 33). La famille longitudinale; le silk, les cordons vifs (sloûk) et l'elb (pl., alâb) (141, 147-150).	Les alâb longitudinale (89). Alâb: cordons majeures et de sloûk superimposés (71-72; fig. 33). Sloûk oblique sur elb (141, 148).
Brown, G. F. (Arabian deserts).	1960	Longitudinal or seif dunes (156-157).	No term.

TABLE 26. — *Chronological list of terms considered equivalent to dunes classified as linear, based on Landsat imagery —*
Continued

Reference Author (and region)	Year	Terms (reference page, figure, plate, and table numbers in parentheses) Simple ridge	Compound ridge
Norris, R. M., and Norris, K. S. (Algodones Dunes, California and Mexico).	1961	Linear ridges (618; fig. 1; pl. 1).	No term.
Norris, R. M. (Algodones Dunes, California and Mexico).	1966	Sand streamers (294-295, 305). Seif dunes (304).	Do.
Powers, R. W., and others (Saudi Arabian deserts).	1966	[Some] urūq; sand, ridges and dune chains; sayf dunes (D100).	[Some] urūq (D100).
Mabbutt, J. A. (Australia).	1967 1968	Longitudinal dunes (149). do. (143). [Some] parallel ridges (141). Single-crested longitudinal ridges (144). Short dunes (144-146).	No term. [Some] parallel ridges with oblique crests (141). Oblique crestal dunes (143). Climax dune type (147).
Petrov, M. P. (U.S.S.R.).	1967	Linear barkhans and ridges (301; fig. 102(2)).	No term.
Stone, R. O. (All deserts).	1967	Longitudinal dune, ridge, linear dune (232), Seif or sif (246).	Sand levee (245). Whaleback (253).
Glennie, K. W. (Saudi Arabian deserts).	1970	Linear dunes or seif dunes (89-95).	Pleistocene seif, the Wahiba sands (90-93; fig. 74).
Goudie, A. S. (Southern Africa).	1970	Linear forms (94).	Alab ridges (94). Composite forms (95). Clustered dendritic ridges (97; fig. 2a).
Goudie, A. S., Allchin, Bridget, and Hedge, K. T. M. (Thar Desert, India).	1973	Longitudinal dunes (245-246, 248).	No term.
Wilson, I. G. (Algerian Ergs).	1971b 1972 1973	Longitudinal dunes (196). Seif dunes, seif dune chains (186-187; fig. 4). do. (182-183; pl. III B, C). Longitudinal dunes (193, 201; pl. III C). Elemental longitudinal dunes (193; pl. III C). Seif dunes, seif dune chains (84, 103; fig. 21).	Draas (182, 191, 198). Compound (seif) dune ridge (208). No term.
Cooke, R. U., and Warren, Andrew (All deserts).	1973	Longitudinal dunes (295, 301, 304, 312). Oblique dunes (295, 301, 304, 309). Seif dunes (301-304, 309, 317, 327).	Draa (50, 231-232).
Breed, C. S., and McKee, E. D. (All deserts).	1973	Linear dunes (160-171).	No term.
McKee, E. D., and Breed, C. S. (All deserts).	1974a 1974b	do. (9-3; figs. 9-3, 9-4). Parallel straight or linear dunes (665, 668, 674; fig. 3).	Do. Linear and feathered linear megadunes (674; fig. 3, 7)
McKee, E. D., and others (All deserts).	1974	do. (3, 17, 52-55, 62, 64, 84; figs. 2, 9, 13, 17, 20-22; table 1).	Do. (3, 17, 23, 47, 62, 84; figs. 2, 7, 13, 17, 20 table 1).
McKee, E. D., and Breed, C. S. (All deserts).	1976	Parallel straight or linear dunes (81; fig. 55).	No term.

TABLE 27. — *Chronological list of terms considered equivalent to dunes classified as crescentic, based on Landsat imagery*

[Crescentic, as used here, includes barchan, barchanoid ridge, and transverse dunes of other geologists in this publication]

Reference Author (and region)	Year	Terms (reference page, figure, plate, and table number in parentheses) Simple mound	Simple ridge	Compound mound	Compound ridge
Hedin, Sven (Takla Makan Desert, China).	1905a	Ideal crescents (323; fig. 268). Individual dunes (351; fig. 293).	Fused dunes, dune triplets (351-352; figs. 294, 295). Transverse thresholds (316-320, 361-363). Barkhans or sandridges (227, 324, 329, 412). Crescentic dunes (270). Chains of dunes in festoons (320). Chains of connected dunes (351; fig. 296). Compound dune length (357-358; figs. 302-303).	Composite dune (271-273; figs; 238-244).	[Some] dune accumulatio (352-354; figs. 296-298).
	1905b	Crescentic individual dunes (404).	Chains of dunes in festoons (403). Chains of coalescing dunes (416).	No term.	No term.
Aufrère, Léon (Formerly "French" Sahara, North Africa).	1928 1930 1932 1934	Sif (pl. siouf) or barchans (833-835). Barchans (224, 226). No term. do.	Les vagues de sable (833). No term. Les vagues de sable (42). do. (134, 137; fig. 2). Dunes d'incidence normal (132). Les dunes en vague (136).	No term. do. do. Zemlah (pl. zemoul)(?) (133).	No term. Do. Do. Châines transversales (139).

TABLE 27. — *Chronological list of terms considered equivalent to dunes classified as crescentic, based on Landsat imagery* — Continued

Reference			Terms (reference page, figure, plate, and table number in parentheses)			
Author (and region)	Year	Simple mound	Simple ridge	Compound mound	Compound ridge	
Bagnold. R. A. (Libyan desert, North Africa).	1933 1941	Barchan dune (106, 109-112). do. (189, 208-221).	No term. Colony or chain of dunes (188).	No term. Complex barchans (212). Rounded mounds (217).	No term. Do.	
Madigan, C. L. (Simpson Desert, Australia).	1936 1938 1946	Barchans, crescentic dunes (223). Transverse dunes (223-225). Barchans, crescentic dunes (514). do. (50, 56-58).	Transverse sand waves (?23-224). Transverse ridges (227). do. (513). No term.	No term. do. do.	No term. Do. Do.	
Melton, F. A. (Midwestern United States).	1940	Barcan dune (113-117, 140; figs. 1-5, 24, 26).	Peak and fulje (fig. 25). Transverse dune series (113, 117-118; fig. 7). Isolated transverse dune ridges (119; fig. 8).	No term.	Intersecting transverse dune series (fig. 26).	
Hack, J. T. (Navajo Indian Reservation, U.S.A.).	1941	Barchans (241, 250).	Transverse dunes (241, 250). Sand waves (241, 257). Climbing dunes (241). "Free" transverse dunes (250).	No term.	No term.	
Capot-Rey, Robert (Formerly "French" Sahara, North Africa). Capot-Rey, Robert and Capot-Rey Francoise (Formerly "French" Sahara, North Africa).	1945 1947b 1953 1948	No term. Barkhanes (96-97). do. (144). do. (63, 66-69, 71-72). Qôrd (71)	Elb (392). Sif (suif) (400-401). No term. Sif (suif) (144). Dunes transversales (73).	Demkha, amygdaloidal dune massif, mastodon (395; fig. 6). No term. do. Demkha, amygdaloidal dune. massif, mastodon (73; pl. II.1).	No term. Do. Do. Do.	
Smith H. T. U. (Nebraska, U.S.A.: Sahara, North Africa).	1946 1953 1956 1963 1965 1968	Barchans (197). do. (102). No term. Barchans (1, 30; fig. 2). No term. do.	Transverse dune ridges (177). Transverse dunes (102). No term. Transverse ridges (1, 16-17, 36; fig. 8). No term. do.	No term. do. do. Compound transverse dunes (19, 40, fig. 12). do. (557, 564). Compound barchans and barchanoid forms (563; pl. 2A). Compound transverse dunes 35, fig. 3-3).	No term. Do. Giant composite barchans (1735). Compound barchans 16, 19, 39; fig. 11). No term. Composite barchans (39).	
Holm, D. A. (Saudi Arabian deserts).	1953 1960 1968	Barkhan dunes (106). Barchan dunes (1371). Crescentic dunes (1373). Crescentic dunes (975). Barchan dunes (975, 978).	[Some] zibarr (110). [Some] zibar (1372). Barchans in transverse ridge (compound dunes) (977). Transverse ridges (978).	No term. do. Giant crescents (1371). Crescentic hollows(?) (1370, 1373; fig. 3). No term.	No term. Do. Giant crescents, giant crescentic massifs (1371-1372). Giant crescentic dune complex (1375; fig. 5). No term.	
McKee, E. D. (White Sands, Navajo Indian Reservation, U.S.A.; Libyan desert, North Africa).	1957 1966	Barchan (1720-1725). do. (9-15, 17, 25, 39-40, 50, 59-60; pl. IIc). Barchan crescent (9).	No term. Barchanoid forms, gradations from barchans to transverse dunes (9, 17; pl. IID).	No term. do.	No term. Do.	
Cooper, W. S. (West coast, North America).	1958 1967	Barchans (27-28). Compressed isodiametric transverse ridge (28). No term.	Undulating dune ridge (7). Transverse ridges (25, 27-34, 58). Transverse waves (33-34, 43). do. (35). Transverse ridges (21, 31, 109). Barchanoid form (28; fig. 1; pl. 4).	Huge barchan with system of anastomosing transverse ridges on [its] back (28). No term.	No term. Do.	
Grove, A. T. (Southern Sahara, North Africa; Kalahari Desert, southern Africa).	1958 1960b 1969	No term. do. Barchans (195, 205; fig. 10).	Transverse ridges (526-531). Mreyyé or mereie (202). Curved dunes (199).	No term. do. do.	No term. Do. Do.	
Monod, Theodore (Formerly "French" Sahara, North Africa).	1958	Barkhanes (100-141).	Elements transverses mineurs, undulations du type Myreyyé (25, 68, 143, 145, 170; fig. 10; pl. 5). Série de bourrelets transversaux (145). Bourrelets géants (147). Les aklés typiques (92-94, 141; figs. 15-18). Le ghord individuel (141). Les individues ghordiques, engagés dans une série de vagues transversales (143).	Elements transverses majeures-les pseudo-alâb (25, 68, 141-142; fig. 10(2)).	Bourrelets monumentaux(?) (141).	

TABLE 27. — *Chronological list of terms considered equivalent to dunes classified as crescentic, based on Landsat imagery —*
Continued

Reference			Terms (reference page, figure, plate, and table number in parentheses)			
Author (and region)	Year	Simple mound	Simple ridge	Compound mound	Compound ridge
Norris, R. M., and Norris, K. S. (Algodones Dunes, California and Mexico).	1961	Barchan dunes (605).	Transverse dunes (605).	Complex barchans (605, 617). Very large, overlapping barchans (617).	Complex barchans (605, 617). Very large overlapping barchans (617).
Norris, R. M. (Algondones Dunes, California and Mexico).	1966	Barchan dunes (292, 306). Crescentic dunes (292).	No term.	Giant barchans megabarchans (292-296).	Giant barchans, megabarchans (292, 296).
Kádár, Laslo (Central Europe).	1966	No term.	Crescent dunes or barchans, Twin barchans, transversal dunes (438, 447; fig. 28).	Two-storied barchans (447).	No term.
Powers, R. W., and others (Saudi Arabian deserts).	1966	Simple barchan dunes (D100).	Compound barchan dunes, or simple rounded transverse ridge (D100).	Complex barchans, giant barchans, (D100).	Complex barchans giant barchans (D100).
Petrov, M. P. (U.S.S.R.)	1967	Solitary barkhans (295, 300-301; fig. 102(1); table 36).	Barkhan chains (300-301; fig. 102(3); table 36). Ridge-barchan form (307).	No term.	Complex barchan ridges or "davana" (295, 298, 300; fig. 101). Alveolar davanas or alveolar barkhan sands (298, 300; table 36).
Stone, R. O. (All deserts).	1967	Barchan; also, barcan, barchane, barkan, barkhan (215). Crescentic dune (220).	Climbing dune (220). Transverse dune (251).	Megabarchan (232).	No term.
Mabbutt, J. A. (Kalahari Desert, southern Africa; Simpson Desert, Australia).	1968	Barchans (144).	Transverse crescentic links, transverse elements crescentic transverse elements, crescentic transverse dunes or ridges (143-144).	No term.	No term.
Goudie, A. S. (Southern Africa).	1970	Barchans (94-95).	No term.	No term.	No term.
Goudie, A. S., Allchin, Bridget, and Hedge, K. T. M. (Thar Desert, India).	1973	No term.	Transverse dunes (245-246, 249).	do.	Do.
Glennie, K. W. (Saudi Arabian deserts).	1970	Barchan dune or barchan (82-88).	Barchanoid forms (84). Transverse dunes (95-99).	No term.	Giant barchan complex (88; fig. 72). Barchan-like complexes (98-99; fig. 78).
Wilson, I. G. (Algerian Ergs).	1971b 1972	Barchan dunes (185-186; fig. 4). do. (181, 192, 208).	Transverse dunes (196). Crescentic barchanoid or transverse sandy ridge (180-181). Transverse ridges (182, 185, 190, 199-202).	Draas (182, 191, 198). Transverse draas (191). Barchanoid draas (178-179; pl. IID, E).	Draas (182, 191, 198). Transverse draas (191). Barchanoid draas (178-179); pl. IID, E). Giant composite dunes [draas] (181).
	1973	Barchan dunes (83).	No term.	do. (100; fig. 17).	Barchanoid draas (100; (fig. 17).
Cooke, R. U., and Warren, Andrew (All deserts).	1973	Barchans (282, 290, 292-295, 297, 300-301, 304, 306, 310; figs. 4.31, 4.32).	Transverse dune ridges (280, 283, 291, 304, 309). Barchanoid and lingoid transverse ridge (288-290, 297, 311-312). [Some] zibar (309, 327).	Draa-sized barchans (282-283). Mega-barchans (297).	Draa-sized barchans (282-283). Mega-barchans (297).
Breed, C. S., and McKee, E. D. (All deserts).	1973	Barchan (163).	Crescentic dunes (162).	No term.	No term.
McKee, E. D., and Breed, C. S. (All deserts).	1974b	No term.	Parallel wavy or crescentic dunes (665, 668; figs. 9, 10).	Crescentic megadunes, giant crescents, or megabarchans (668; fig. 4, 7).	Crescentic megadunes, giant crescents or or megabarchans (668; (figs. 4, 7, 10).
McKee, E. D. and others. (All deserts).	1974	Barchan (28, 61, 63, 69-71, 81; fig. 28).	do. (3, 24, 62, 64; figs. 17, 25; table 8). Barchanoid dune, fishscale type, peak and fulje, topography (24, 47, 82, 84; figs. 3, 9, 10, 12, 13, 15, 17, 25). Transverse dune (fig. 28). Aklé (21, 81, 84).	Crescentic megadunes, giant crescents, or megabarchans (3, 24, 69, 74, 83; figs. 3, 7, 12, 15, 19; table 1).	Crescentic megadunes, giant crescents or megabarchans (3, 24, 69 74, 83; figs. 3, 7, 12, 14, 15, 19). Dune complexes or draa (28, 69-70, 82). Crescentic, fishscale (fig. 25). Transverse, parallel wavy (62).
McKee, E. D., and Breed, C. S., (All deserts).	1976	No term.	[Some] parallel wavy or crescentic dunes of the simple basic type (81-82).	Megabarchans or giant crescents (82).	[Some] parallel-wavy or crescentic dune complexes including [some] fishscale type and gian crescent or megabarchan type (82; fig. 56).

TABLE 28. — *Chronological list of terms considered equivalent to dunes classified as star, based on Landsat imagery*

Reference		Terms (reference page, figure, plate, and table numbers in parentheses)	
Author (and region)	Year	Simple star	Compound star
Hedin, Sven (Takla Makan Desert, China).	1905a	Pyramidal dunes (330, 363).	No term.
Aufrère, Léon (Formerly "French" Sahara, North Africa).	1930 1932 1934	No term. [Some] rhourds simples (44). Collines de sable (rhourds) (137, 139-140; figs. 2, 4(3)). [Some] rhourds degagés (139).	Oghroud (223). Les grands rhourds polypyramidés (45). Rhourd polypyramidé (141).
Capot-Rey, Robert (Sahara, North Africa).	1945 1947a 1953	[Some] khurds, guerns (393, 397; fig. 7). Oghroud, mastodonte, iguidi (86, 89). No term.	[Some] khurds (397; fig. 7). Pyramides aux sommets pointus (89). Do. (145).
Capot-Rey, Robert, and Capot-Rey, Francoise (Sahara, North Africa).	1948	Qôrd, ghourd (72-73; pl. 113).	Les grandes pyramides (oghroud) 73, 77). Oghroud (pl. II, 2, no. 3).
Smith, H. T. U. (Nebraska, U.S.A.; Sahara, North Africa).	1946 1963	Rhourd (198). Embryonic forms of peaked dunes (21, 44; fig. 16).	No term. Peaked complexes (21, 43; fig. 15). Domal dune complexes (22, 45; fig. 17).
Holm, D. A. (Saudi Arabian deserts).	1953 1960 1968	Sand peaks or star dunes (112). Pyramidal dunes (1371-1373). No term.	No term. Pyramidal dunes (1373). Do. (977).
Monod, Theodore (Formerly "French" Sahara, North Africa).	1958	[Some] ghords (141).	[Some] ghords (141). [Some] châines ghordiques (20).
Brown, G. F. (Saudi Arabian deserts).	1960	Star dunes (159).	No term.
Powers, R. W., and others (Saudi Arabian deserts).	1966	[Some] sand mountains (D100).	Giant sigmoidal and pyramidal sand peaks (D100).
McKee, E. D. (Irq as Subay', Saudi Arabia).	1966	Star dune (3, 10, 62-68; pl. VIII A, B).	No term.
Petrov, M. P. (U.S.S.R.).	1967	Conical dunes (295, 300-301; figs. 101c, 102(4); table 36).	No term.
Stone, R. O. (All deserts).	1967	Rhourd (242). Star dune (249).	Dune massif (225). Guern (230). Rhourd (242). Star dune (249).
Glennie, K. W. (Saudi Arabian deserts).	1970	No term.	Stellate dunes (86-88; figs. 70-71).
Wilson, I. G. (Algerian Ergs).	1971a 1972 1973	[Some] star-shaped pyramids, rhourds (270). No term. do.	Draa peaks, rhourds (268). Star-shaped pyramids, rhourds (270). Stellate rosettes (178-179; pl. II C). Draa peaks (oghourds) (188). Oghourds or star dunes (190-191). Draa peaks, rhourds (99; fig. 16).
Cooke, R. U., and Warren, Andrew (All deserts).	1973	[Some] sand mountains or rhourds (304).	[Some] sand mountains or rhourds (304).
Breed, C. S., and McKee, E. D. (All deserts). McKee, E. D., and Breed, C. S. (All deserts). McKee, E. D., and others (All deserts). McKee, E. D., and Breed, C. S. (All deserts).	1973 1974a 1974b 1974 1976	Star or radial dunes (162; fig. 9). Star dunes (9-3). Star or radial dunes (665, 668). do. (3, 17, 83-85; figs. 4, 8; table 1). Star or radial dunes (81).	Radial (star) megadunes (162; figs. 3, 9). Star dunes (9-3; figs. 9-1, 9-2). Star or radial megadunes (668; fig. 5). Do. (3, 17, 24, 47, 83-85, figs. 4, 8, 10, 12, 17, 19; table 1). Dune complexes or draa (28, 82). Star or radial megadunes (82).

TABLE 29. — *Chronological list of terms considered equivalent to dunes classified as parabolic, based on Landsat imagery*

Reference		Terms (reference page, figure, plate, and table numbers in parentheses)	
Author (and region)	Year	Simple U-shaped dunes	Compound U-shaped dunes
Hedin, Sven (Takla Makan Desert, China).	1905a	Spoon-shaped dunes(?) (331; fig. 270).	No term.
Aufrère, Léon (Formerly "French" Sahara, North Africa).	1928	Les dunes paraboliques et les dunes en U (834-835).	No term.
	1930	Dunes de mousson, dunes en rateau (228).	Do.
	1934	Les dunes paraboliques et les dunes en U (134).	Les Châines bouclées (134).
Madigan, C. L. (Simpson Desert, Australia).	1936	U-shaped dune, fulje-type dunes (224).	No term.
	1946	Blowouts (57). Parabolic bowouts or fuljes (61).	Do.
Capot-Rey, Robert, and Capot-Rey, Francoise (Formerly "French" Sahara, North Africa).	1948	Dunes paraboliques (75).	No term.
McKee, E. D. (White Sands, U.S.A.).	1957	Parabolic dune (1720-1721). Blowouts (1725).	No term.
	1966	do. (9-10). Parabolic dune (3, 9-11, 15, 25, 50-51). U-shaped or V-shaped dune (9).	Do.
Grove, A. T. (Southern Sahara, North Africa; Kalahari Desert, Southern Africa).	1958	Parabolic dunes (529).	No term.
	1969	Blowouts (199).	Do.
Cooper, W. S. (West coast, North America).	1967	Parabola dunes (22-25, 111). [Some] giant parabolas (24).	Compound parabola (31). [Some] giant parabolas (111).
Stone, R. O. (All deserts).	1967	Blowout (217). Chevron dune (219). Elongate blowout dune (226). Hairpin dune (230). Parabolic dune (236). U-shaped dune (251). Upsiloidal dune (252).	No term.
Melton, F. A. (Midwestern United States).	1940	Blowout or parabolic dunes (133, 126-127). Elongate-blowout dunes (127-128; figs. 17, 18).	"Nested" elongate-blowout dune series (128).
Hack, J. T. (Navajo Indian Reservation, U.S.A.).	1941	Parabolic dunes of deflation and accumulation (240, 242-243).	No term.
Smith, H. T. U. (Nebraska, U.S.A.; Sahara, North Africa).	1946	Upsiloidal dune (197-199).	No term.
	1953	Phytogenic dunes (102).	Do.
	1965	Blowouts (557, 569-570).	Do.
	1968	do. (33).	Elongate U-shaped dunes and subparalle ridges (44).
Goudie, A., Allchin, Bridget, and Hedge, K. T. M. (Thar Desert, India).	1973	Parabolic dunes (248).	No term.
Cooke, R. U., and Warren, Andrew (All deserts).	1973	Blowouts, parabolic dune, lunettes (318).	Nested parabolic system (318).
Ahlbrandt, T. S. (Wyoming, U.S.A.).	1974b	Parabolic dunes (45).	No term.
McKee, E. D., and Breed, C. S. (All deserts).	1974b	Parabolic or U-shaped dunes (665, 668; figs. 6, 9).	Parabolic dune array, Thar Desert (668; fig. 6).
McKee, E. D., and others (All deserts).	1974	do. (3, 24, 53, 61-62, 83, 85; figs. 5, 9; table 1).	Do. (24, 62; figs. 5, 18, 25). Dune complexes or draa (28, 82).
McKee, E. D., and Breed, C. S. (All deserts).	1976	do. (81-82).	Parabolic dunes, India (fig. 58).

TABLE 30. — *Chronological list of terms considered equivalent to dunes classified as dome-shaped, based on Landsat imagery*

Reference Author (and region)	Year	Terms (reference page, figure, plate, and table numbers in parentheses)
Bagnold, R. A. (Libyan desert, North Africa).	1941	Rounded mounds (?) (217).
Holm, D. A. (Saudi Arabian deserts).	1953	Dome-shaped dunes (106-112; figs. 2-8).
McKee, E. D. (White Sands, U.S.A.).	1966	Dome-shaped dunes (10, 15, 25-27).
Powers, R. W., and others (Saudi Arabian deserts).	1966	Giant oval sand mounds (D100).
Ahlbrandt, T. S. (Wyoming, U.S.A.).	1974b	Dome dunes (45).

TABLE 31. — *Chronological list of local terms for dunes considered equivalent to those classified as complex, based on Landsat imagery*

Reference Author (and region)	Year	Terms (reference page, figure, plate, and table numbers in parentheses)
Hedin, Sven (Takla Makan Desert, China).	1905a	[Some] dune-accumulations (400).
Auffère, Léon (Formerly "French" Sahara, North Africa).	1930 1932 1934	[Some] draas, bras, or flük (220, 223). Slassels or draa with oghounds (223, 226). Chains with rhourds (44). Les dunes complexes (133). Châines avec vagues obliques et pitons (draas) (fig. 2). Rhourds engagés (1939).
Madigan, C. L. (Simpson Desert, Australia).	1936 1938 1946	Sandridges with minor crescentic dunes on top (212). Do. (514). [Some] plinths with varieties of crest formations (49-51).
Melton, F. A. (Midwestern United States).	1940	Complex dune forms (fig. 28).
Capot-Rey, Robert (Formerly "French" Sahara, North Africa). Capot-Rey, Robert, and Capot-Rey Francoise (Formerly "French" Sahara, North Africa).	1945 1947a 1948	Dunes of conjonction (sif normal to axis of chain) (401). [Some] braas or draa (404; fig. 11). Do. (88-89). Demkhas (73; Pl. II 1).
Holm, D. A. (Saudi Arabian deserts).	1957 1960 1968	Linear belts with V-shaped chevrons; dune complexes; [some] sigmoidal dunes (1746). Hooked dunes (1371-1372). Linear complexes(?) (1373). Complex dunes, dune complexes (977).
McKee, E. D. (Great Sand Dunes National Monument, U.S.A.).	1966	Reversing dune (10, 61, pl. II E).
Monod, Theodore (Formerly "French" Sahara, North Africa).	1958	Grands bnaig du dhra (83). Dunes en rateau, modele transverse sur 'elb; [some] famille des aklés; système complex (141).
Smith, H. T. U. (Nebraska, U.S.A.: Sahara, North Africa).	1963 1968	Complex longitudinal dunes (20-21; figs. 10, 14). Ridged dune complex (22; fig. 18). Do. (36).
Powers, R. W., and others (Saudi Arabian deserts).	1966	[Some] sand mountains [D100].
Petrov, M. P. (China, U.S.S.R.).	1967	Alveolar barkhan sands with conical tops (323). Longitudinal-transverse and interferential dune forms (297; fig. 100).
Stone, R. O. (All deserts).	1967	Fish hook or hooked dune (228). Sigmoidal dune (249).

TABLE 31. — *Chronological list of local terms for dunes considered equivalent to those classified as complex, based on Landsat imagery* — Continued

Reference Author (and region)	Year	Terms (reference page, figure, plate, and table numbers in parentheses)
Mabbutt, J. A. (Australia).	1968	[Some] ridges with multiple or braiding crests (143; Pl. VI). [Some] dune networks with intersecting trends (146). Reticulate dunes (148).
Grove, A. T. (Namib Desert, South-West Africa).	1969	[Some] alab dunes (195-201; pl. VIIIa).
Glennie, K. W. (Saudi Arabian deserts).	1970	Seif dunes with barchan-like slipfaces (94-95; fig. 75). [Some] dunes of the Uruq al Mu'taridah (96-98; fig. 76).
Wilson, I. G. (Algerian Ergs, North Africa).	1971b	[Some] draas (182, 191, 198). Lingoid networks with rhomboidal or hexagonal cells (182). Giant draa ridges (pl. IIIA, F).
Cooke, R. U., and Warren, Andrew (All deserts).	1973	[Some] draa (231-232; fig. 4.27). [Some] draa ridges (235, 296-297). Oblique draa patterns (300).
Breed, C. S., and McKee, E. D. (Namib Desert).	1973	Linear megadunes (fig. 7).
McKee, E. D., and Breed, C. S. (All deserts).	1974a 1974b.	Linear chains of star dunes (9-3; figs. 9-1, 9-2). Do. (668; fig. 5). Chevron or basketweave dunes (668; fig. 4).
McKee, E. D., and others (All deserts).	1974	Star dunes in chains (19; figs. 4, 8; table 1). Linear type (fig. 7). Chevron or basketweave dunes (81; fig. 3; table 1). Reversing dunes (fig. 10).
McKee, E. D., and Breed, C. S. (All deserts).	1976	Chains of star dunes (82; fig. 54).

TABLE 32. — *Chronological list of local terms for various sand seas*

Reference Author (and region)	Year	Terms (reference, page, figure, plate, and table numbers in parentheses)
Hedin, Sven (Takla Makan Desert, China).	1905a	"Sea" of sand (234).
Aufrère, Léon (Formerly "French" Sahara, North Africa.	1928 1930 1934	Pasage dunaire, la mer de sable (834). Erg (225). Un champ de dunes (132). La mer de sable (134, 138-139).
Bagnold, R. A. (Libyan desert, North Africa).	1933	Sand sea (110). Dunefield (108).
Madigan, C. L. (Simpson Desert, Australia).	1936 1946	Mer de sable (224). Sandridge desert (206, 208, 215; fig. 2). Do. (56).
Capot-Rey, Robert (Formerly "French" Sahara, North Africa).	1945 1947a 1953	Erg (391). Sea of sand, la mer de sable (393). Houle de sable (88). La mer de sable (145). Aklé (145-146).
Capot-Rey, Robert and Capot, Rey, Francois, (Formerly "French" Sahara, North Africa).	1948	Houle de sable (79).
Holm, D. A. (Saudi Arabian deserts).	1953 1960	Nefud (106-112; figs. 2-8). Nafud (1369-1374).
Mabbutt, J. A. (Kalahari Desert, southern Africa; Simpson Desert Australia).	1955 1967 1968	Sandveld (24, 26). Sandridge deserts (145). Dunefields (173). Do. (140).
Cooper, W. S. (West coast, North America).	1958 1967	Erg (61). Body of dunes (39). Dune complex (42, 44-47, 57-58; fig. 1; pl. 8-10, 13).
Grove, A. T. (Southern Sahara, North Africa).	1958	Erg (526-533). Qoz (531).
Monod, Theodore (Formerly "French" Sahara, North Africa).	1958	Erg (22, 55, 63).
Norris, R. M., and Norris, K. S. (Algodones Dunes, California and Mexico).	1961	Dune mass (617).
Smith, H. T. U. (Nebraska, U.S.A.: Sahara, North Africa).	1963 1968	Dune field (36, 43; figs. 8, 15). Do. (29).
McKee, E. D. (White Sands, Navajo Indian Reservation, U.S.A.; Libyan desert; Arabian deserts).	1966	Dunefield (10).
Powers, R. W., and others (Saudi Arabian deserts).	1966	Nafud (D100).
Stone, R. O. (All deserts).	1967	Erg (227). Sand sea (245)
Goudie, A. S. (Southern Africa).	1969	Dune country (404).
Goudie, A. S., Allchin, Bridget, and Hedge, K. T. M. (Southern Africa; Thar Desert).	1973	Ergs (245; table 1). Dune fields (247, 249).
Glennie, K. W. (Saudi Arabian deserts).	1970	Dunefield (84).
Wilson, I. G. (Algerian Ergs, North Africa).	1971a 1971b 1972 1973	Erg, sand-sea (264-267). Erg, 180, 184, 190-198; fig. 3). Do. (187-188). Sand-ridge deserts (187). Erg (77-105). Draa ergs, dune ergs (91).
Cooke, R. U., and Warren, Andrew (All deserts).	1973	Sand sea (50, 229). Erg (229. 300, 306, 312, 322-327). Dunefield (311-315, 322).
Breed, C. S., and McKee, E. D. (All deserts).	1973	Sand sea, erg (1960).
McKee, E. D., and others (All deserts).	1974	Sand sea (3-85).
McKee, E. D., and Breed, C. S. (All deserts).	1976	Desert sand seas (81-82).

TABLE 33. — *Chronological list of local terms for patterns of eolian sand deposits in sand seas*

References Author (and region)	Year	Terms (reference page, figure, plate, and table numbers in parentheses)
Hedin, Sven (Takla Makan Desert, China).	1905a	Network of dunes, bajirs, and thresholds (360, 365; figs. 301-303, 317; pl. 52).
Auffere, Léon (Formerly "French" Sahara, North Africa).	1928 1932 1934	Les sable en colonne ou en monome (834). Zemloul (44-45). Dunes espaćees et les dunes serrées (133). Vagues entrecroisées ou reticulées (138-139).
Bagnold, R. A. (Libyan desert, North Africa).	1933 1941	Sand sea of parallel seif dune lines (110). Belt of barchans (106, 123-124; fig. 3). Belt of barchans (218-221). Multiple seif chains (231-235).
Madigan, C. L. (Simpson Desert, Australia).	1936	Sand-ridge grid, grid of parallel ridges (211, 218; fig. 11).
Melton, F. A. (Midwestern United States, U.S.A.)	1940	Array of dunes (118-119). Peak and fulje topography; intersecting transverse dune series (131-132; fig. 25, 26). Trains of barchan dunes (figs. 2, 3).
Smith, H. T. U. (Nebraska, U.S.A.; Sahara, North Africa).	1963 1965	Barchan field (30; fig. 2). Trains of barchans (16). Random pattern (43; fig. 15). Parallel, subparallel, divergent, reticulate group patterns (17-18). Pronounced parallelism like waves of the sea (563). Ridge and furrow topography (566).
Capot-Rey, Robert (Formerly "French" Sahara, North Africa).	1947a 1953	Châinons dunairés (88). Les grandes massifs dunairés (94). Essaims [swarms] de barkhanes (97). Do. (144).
Capot-Rey, Robert, and Capot-Rey, Francoise. (Formerly "French Sahara, North Africa).	1948	Train [of barchans] (66). Massif d'un aspect alvéolé (73).
Cooper, W. S. (West coast, North America).	1958 1967	Reticulate pattern with rhomboidal meshes (31). Oblique ridge pattern (25, 49-60, 62; fig. 2; pl. 11). Longitudinal-ridge pattern (62). Transverse-ridge pattern (25, 27, 58, 64; fig. 1; pl. 7, 11). Dune belts (23; figs. 2, 3; pl. 8, 9). Dune sheets (33-40, 58-60; figs. 1, 2; pl. 8).
Norris, R. M., and Norris, K. S. (Algodones Dunes, California and Mexico).	1961	Peak and hollow topography (608). Chains of dunes, chiflones (617). Barchan chains (618).
Norris, R. M. (Algodones Dunes, California and Mexico).	1966	Barchan swarms (294-295). Dune cluster (292).
Mabbutt, J. A. (Kalahari Desert, southern Africa; Simpson Desert, Australia).	1967 1968	Parallel ridges (58, pl. 18c). Do. (141-143, 146; pl. 1). Barchan trains (114) Widely-spaced parallel ridges (143, 145; pl. 6). Convergence and deflection of dune ridges (145, 148; pl. 7). Alined short dunes (146). Braided, forked, arcuate crests (146). Arcuate belts of dunes (144). Dune networks (145-146; pl. 8). Branching patterns (147).
Monod, Theodore (Formerly "French" Sahara, North Africa.	1958	Régime alâb-sloûk, régime longitudinal typique (59). Regime alâb ayðun (20, 77, 80, 141). Systèmes transverses et systèmes longitudinaux (143). Barchans grégaires (141). Les aklés (58, 92-94, 129, 141, 173; pl. 32). Un aklé reticule (143). Erg en rateau ou en peine (145).

TABLE 33. — *Chronological list of local terms for patterns of eolian sand deposits in sand seas — Continued*

References		Terms (reference page, figure, plate, and table numbers in parentheses)
Author (and region)	Year	
Grove. A. T. (Southern Sahara, North Africa, Kalahari Desert, South Africa).	1958	Ridges and hollows (526-531) Dunefield (527).
Petrov, M. P. (U.S.S.R.).	1967	Crisscrossing barkhan chains (300; table 36).
Stone, R. O. (All deserts).	1967	Complex dunes, peak and depression (fulji) topography (220-221)
Holm, D. A. (Saudi Arabian deserts).	1968	Dune complex, dune terrains, dune belts (976). Waves of dunes (978).
Goudie, A. S. (Southern Africa).	1969	Dune network (404).
Goudie, A. S., Alchin, Bridget, and Hedge, K. T. M. (Thar Desert, India).	1973	Dendritic, reticulate, and parallel ridges (93, 97, 100; fig. 2). Clustered dendritic ridges (97).
Glennie, K. W. (Saudi Arabian deserts).	1970	Dune pattern (101).
Wilson, I. G. (Algerian Ergs, North Africa).	1971a	Network of draa ridges (268). Complex dune nets (270).
	1971b	Two-trend dune networks (186-187; fig. 4). Bedform patterns (182-183).
	1972	Do. (173-192, 195-204). Lunate network, linguoid network with rhomboidal or hexagonal cells (182) [Some] seif dune ridge varieties (194, fig. 4). Gridiron, fishscale, and braided patterns (207-208).
	1973	Bedform patterns (78, 97). Barchan dunes in belts (84). Gridiron, fishscale, and braided patterns (100; fig. 17). Oblique left-handed, en echelon chain (102).
Cooke, R. U., and Warren, Andrew (all deserts).	1973	Aklé (285, 298, 311; fig. 435). Erg patterns (322-327). Transverse patterns (295). Rectangular reticules; complex patterns (296). Dune patterns (229, 283-287, 322-327; figs. 4.25, 4.26). Draa patterns (297; fig. 4.34).
McKee, E. D., and others (All deserts).	1974	Parallel straight or wavy dune complexes (17, 24, 47, 53, 62, 68; figs. 1, 2, 17, 19, 25; table 1). Dune patterns (68). Parabolic dune arrays (fig. 5).
McKee, E. D., and Breed, C. S. (All deserts).	1976	Parallel straight or wavy dune complexes (81-82; figs. 55, 56).

TABLE 34. — *Chronological list of local terms for sand sheets and streaks (stringers)*

[Streaks are considered equivalent to deposits termed "stringers" in preceding chapters of this publication]

Reference		Terms (reference page, figure, plate, and table numbers in parentheses)	
Author (and region)	Year	Sand sheets	Sand streaks
Bagnold, R. A. (Libyan Desert, North Africa).	1933	Sand-sheets (105, 111).	Sand shadows and sand drifts (121; fig. 1).
	1941	Sand sheets (243-245).	Do. (189-195).
Madigan, C. L. (Simpson Desert, Australia).	1938	Sandy plain (514	No term.
	1946	Sand sheets (58).	Do.
Melton, F. A. (Midwestern, United States).	1940	No term.	Umbracer or wind shadow dunes (113, 120).

TABLE 34. — *Chronological list of local terms for sand sheets and streaks* — Continued

[Streaks are considered equivalent to deposits termed "stringers" in preceding chapters of this publication]

Reference		Terms (reference page, figure, plate, and table numbers in parentheses)	
Author (and region)	Year	Sand sheets	Sand streaks
Hack, J. T. (Navajo Indian Reservation, U.S.A.).	1941	No term.	Falling dunes (241).
Smith, H. T. U. (Nebraska, U.S.A.: Sahara, North Africa).	1946 1963	No term. do.	Wind-shadow dunes (197). Do. (35; fig. 7). Sand streamers (32, fig. 4). Sand drifts (33, fig. 5). Climbing dunes, falling dunes, sand glaciers (14-15).
Capot-Rey, Robert and Capot-Rey, Francoise (Formerly "French" Sahara, North Africa).	1948	Nappes de sable, ténéré, reg (63).	No term.
Holm, D. A. (Saudi Arabian deserts).	1953 1960	[Some] zibarr (110). [Some] zibar (1372).	No term. Do.
Mabbutt, J. A. (Kalahari, Desert, southern Africa).	1955 1967 1968	Flat sand spread (27). Sand plain (149). do. (139-140, 145; pl. 1.5).	No term. Do. Do.
Cooper, W. S. (West coast, North America).	1958 1967	[Some] dune sheets (33-34, 64). do. (Fig. 1; pl. 11).	No term. Do.
Grove, A. T. (Southern Sahara, North Africa; Kalahari Desert, southern Africa).	1958	Sandy drift (526-529).	No term.
Norris, R. M. (Algodones Dunes, California and Mexico).	1966	No term.	Sand streamers (294-295, 305).
Powers, R. W. and others (Saudi Arabian deserts).	1966	No term.	Giant elongate sound mounds (D100).
Stone, R. O. (All deserts).	1967	Sand sheet (245)	Windrift dune, windshadow dune (253) Falling dune (228). Sand drift, sand glacier 245). Source--bordering lee dune (249). Umbrafon dune (252).
Glennie, K. W. (Saudi Arabian deserts).	1970	Sheet sands (106-109; figs. 89-90).	No term.
Goudie, A. S. (Southern Africa). Goudie, A. S., Allchin, Bridget, and Hedge, K. T. M. (Thar Desert, India).	1970 1973	Topographically-induced sand sheets (95). do. (246). Sand drifts, windward drifts (246, 249).	No term. Lee drifts (249). Wind drifts (244, 248-249).
Wilson, I. G. (Algerian Ergs, North Africa).	1971a 1972 1973	Sand plain (264). No term. Sand-sheet (103-104; fig. 21). Sand patches (78). Ripple ergs (91). Coarse-sand ergs (104).	No term. Drifts downwind of scarps (191). No term.
Cooke, R. U., and Warren, Andrew (All deserts).	1973	No term.	Lee dunes, trailing dunes, trailing sand patches (314-315).
Sagan, Carl, and others (On the planet, Mars).	1973	No term.	Streaks (179-215).
McKee, E. D., and Breed, C. S. (All deserts). McKee, E. D., and others (All deserts). McKee, E. D., and Breed, C. S. (All deserts).	1974b 1974 1976	Sand sheets (665, 668). do. (3, 52, 61, 73, 85; figs. 17, 19; table 1). Sand sheets (81-82; fig. 59).	Stringers (665, 668). Do. (49, 61; fig. 20). Streaks (3, 52, 55, 61, 85; figs. 6, 17; table 1). Stringers, stringer dunes (81-82; fig. 59).

TABLE 35. — *Chronological list of local terms for interdune areas*

Reference		Terms (reference page, figure, plate, and table numbers in parentheses)
Author (and region)	Year	
Hedin, Sven (Takla Makan Desert, China).	1905a	Bajirs or depressions (234-251, 311-369).
Aufrère, Léon (Formerly "French" Sahara, North Africa).	1928 1930 1934	Feidjs [sandy] or gassis [bare]; bahirs (834). Do. (220). Vallées eoliennes (222). Intervallées interdunaires (226, 230). Les sillons interdunaires (220, 225). Couloirs (136, 138). Haouds (138).
Madigan, C. L. (Simpson Desert, Australia).	1936 1938 1946	Gibber or clay flats (214). Inter-ridge areas (223). Inter-ridge spaces; long, narrow clay pans or gibber plains, flats (514) Sandy lanes (43, 49). Transverse hollows (52).
Melton, F. A. (Midwestern United States, U.S.A.).	1940	Fuljes or enclosed basins (131).
Bagnold, R. A. (Libyan desert, North Africa).	1941	Fuljis or interdune hollows (214).
Hack, J. T. (Navajo Indian Reservation, U.S.A.).	1941	Narrow areas that separate waves of sand (241).
Capot-Rey, Robert (Formerly "French" Sahara, North Africa). Capot-Rey, Robert and Capot-Rey, Francois (Formerly "French Sahara, North Africa).	1945 1947a 1953 1948	Gassi: interdune trough without sand; feij, sand-covered interdune trough (395, 401; fig. 4.5). Sebkhas (392). Interdune passages, couloirs, aftout (404). Do. (89). Cuvettes, une serie d'alveoles (145). Do. (89-90; pl. III.1). Interdune passages, couloirs, aftout (73).
Smith, H. T. U. (Nebraska, U.S.A.; Sahara, North Africa).	1946 1963 1965 1968 1969	Fulji (198). Inter-dune basins (40; fig. 12). Interdune basins, troughs, areas (562). Narrow grooves, network of corridors, broad flats, depressions (32). Interdune depressions (17, fig. 14).
Holm, D. A. (Saudi Arabian deserts).	1953 1960	Interdune valleys (105). Deflation hollows (109). Interdune areas (1372).
Monod, Theodore (Formerly "French" Sahara, North Africa).	1958	Les interdunes (25; fig. 10(2)). Interdunes sableux et interdunes rocheux (142; fig. 33). Ayn (pl., ayoûn), tâyâret (pl., tŷar) (81).
Grove, A. T. (Southern Sahara, North Africa, Kalahari Desert, southern Africa).	1958 1969	Hollows (526, 529). Interdune depressions, interdunal corridors, strate (straat) (199).
Norris, R. M., and Norris, K. S. (Algodones Dunes, California and Mexico). Norris, R. M. (Algodones Dunes, California and Mexico).	1961 1966	Large, flat-floored, sand-free depressions (605, 608-610; figs. 1, 2; pl. 2, 3). Intradune hollows, depressions (292-297, 305).
McKee, E. D. (White Sands, Navajo Indian Reservation, U.S.A.; Libyan desert; Irq as Subay', Saudi Arabia).	1966	Interdune surfaces or areas (10, 51-56, 61).

TABLE 35 — *Chronological list of local terms for interdune areas* — Continued

Reference		Terms (reference page, figure, plate, and table numbers in parentheses)
Author (and region)	Year	
Cooper, W. S. (West coast, North America).	1967	Linear series of hallows between dune waves (34).
Petrov, M. P. (U.S.S.R.).	1967	Interbarkhan depressions (304). Interridge depressions (307).
Stone, R. O. (All deserts).	1967	Dune valley (225). Fulji, or Fulje, depression between barchan dunes (228).
Mabbutt, J. A. (Kalahari Desert, southern Africa; Simpson Desert, Australia).	1968	Fuljes, interdune corridors (143). Enclosed or partly enclosed depressions in dune networks (146).
Glennie, K. W. (Saudi Arabian deserts).	1970	Inland interdune sebkhas, interdune sebkhas, (60-61, 97-98; fig. 77). Interdune deflation plane (77). Sand-covered inter-dune areas (78). Interdune areas (90, 107-108). Feidjes (101).
Goudie, A. S. (Southern Africa).	1970	Inter-dune corridors (95). Inter-cluster corridors, straats (97).
Wilson, I. G. (Algerian Ergs, North Africa).	1971a 1973	Troughs (264) Sand-free hollows (267). Dune hollows (270). Do. (92).
Cooke, R. U., and Warren, Andrew (All deserts).	1973	Interdune corridors (322). Interdraa corridors (327).
Breed, C. S., and McKee, E. D. (All deserts).	1973	Intradune corridors (162).
McKee, E. D., and others (All deserts).	1974	Interdune surfaces or areas (70-71).
McKee, E. D., and Breed, C. S. (All deserts).	1976	Interdune areas (81).

TABLE 36. — *Morphometry of linear dunes in sand seas*

[$\bar{W}$, mean width; $\bar{L}$, mean length (*, some or all dunes extend beyond edge of image); $\bar{\lambda}$, mean wavelength (crest-to-crest distance); $\bar{F}$, mean frequency (number of dunes per kilometre) across sample, normal to trend of dunes. All measurements in kilometres]

Region[1]	Latitude, longitude	Desert or sand sea area	Area sampled (km²)	$\bar{W}$	$\bar{L}$	$\bar{\lambda}$	$\bar{L}/\bar{W}$	$\bar{W}/\bar{\lambda}$	$\bar{L}/\bar{\lambda}$	$\bar{F}$	Figure No.
		Simple(?) linear dunes									
1	35°30'-36° N., 111°-111°15' W.	Navajo Indian Reservation (northern Arizona, U.S.A.)------------------	625	0.043	3.65	0.15	84.9	0.29	24.3	4.92	168B
2	23°-27°30' S., 137°-140°30' E.	Simpson Desert (Australia)-----------------	30,000	.22	24.0	.90	109.1	.24	26.67	1.44	None
3	19°30'-23°30' S., 122°30'-128°30' E.	Great Sandy Desert (Australia)------------	35,000	.29	22.7	.90	78.3	.32	25.22	1.06	169B
4	23°30'-25°45' S., 17°30'-19°30' E.	Kalahari Desert (southern Africa)---------	15,000	.29	26.0	.70	89.7	.41	37.14	1.26	169A
5	20°30'-21°30' N., 52°-52°30' E.	Northeastern Rub' al Khali (Saudi Arabia)-	2,500	.38	14.34	1.41	37.7	.27	10.17	.64	241
		Compound linear dunes									
6	21°-22° N., 08°-10° W.	Southwestern Sahara (Mauritania)----------	2,500	0.94	>100.00*	1.93	106.4*	0.49	51.81*	0.48	169D
7	18°-19° N., 13°-14°30' E.	Southern Sahara (Niger)-------------------	7,500	1.06	>40.*	1.90	37.7*	.56	21.05*	.29	None
8	16°30'-19°15' N., 45°30'-50° E.	Southwestern Rub' al Khali (Saudi Arabia)-	20,000	1.21	>83.*	2.18	68.6*	.55	38.07*	.48	169C
		Complex linear dunes									
9	24°-24°45' S., 14°45'-15°30' E.	Namib Desert (South-West Africa)----------	2,500	0.88	27	2.20	30.7	0.40	12.27	0.47	169E
10	26°30'-28° N., 00°30'-02°30' E.	Northern Sahara (Algeria)-----------------	2,500	1.09	30.75	3.24	28.2	.34	9.49	.33	168B
11	19°-19°30' N., 15°-15°30' E.	Southern Sahara (Niger)-------------------	2,500	1.28	65	3.28	50.8	.39	19.81	.30	None
12	18°45'-19°30' N., 45°15'-46° E.	Western Rub' al Khali (Saudi Arabia)------	2,500	1.48	73	3.17	49.3	.47	23.03	.21	169F

[1]Sources of data--(1) Aerial photographs, National Aeronautics and Space Administration: 236-28-0071, 236-28-0072. Following are numbered Landsat images: (2) E1315-00140, E1227-00264, E1224-00100, E1369-00135, E1369-00141, E1133-00040, E1296-00090, E1367-0025, E1224-00093; (3) E1344-01160, E1127-01111, E1377-00585, E1127-01113, E1415-01091, E1100-01045, E1124-00535, E1100-01051, E1125-00591, E1124-00542, E1344-01165, E1413-00584, E1344-01162; (4) E1147-08160, E1416-08085, E1416-08091, E1145-08052; (5) E1184-06255; (6) E1119-10324; (7) E1192-08553; E1193-09012; (8) E1187-06433, E1132-06380, E1187-06435, E1170-06493; (9) E1383-00270; (10) E1115-10074; (11) E1192-08553; (12) E1189-06550.

TABLE 37. — *Similarity of linear dune varieties in widely distant sand seas shown by comparison of ratios derived from measurements of mean lengths, $\bar{L}$, mean width $\bar{W}$, and mean wavelength, $\bar{\lambda}$*

[Subscript following name of sand sea indicates first or second of the pair. Subscripts to $\bar{L}$, $\bar{W}$, and $\bar{\lambda}$ indicate first or second of sand sea pair]

Pairs of sand seas[1]	$\bar{L}$	$\bar{L}_1/\bar{L}_2$	$\bar{W}$	$\bar{W}_1/\bar{W}_2$	$\bar{\lambda}$	$\bar{\lambda}_1/\bar{\lambda}_2$	$\bar{L}_1/\bar{W}_1$	$\bar{L}_2/\bar{W}_2$	$\bar{L}_1/\bar{\lambda}_1$	$\bar{L}_2/\bar{\lambda}_2$	$\bar{W}_1/\bar{\lambda}_1$	$\bar{W}_2/\bar{\lambda}_2$	Degree of correspondence[2]
Simpson₁	24.0		0.22		0.90								
		6.58		5.12		6.0	109.1	84.9	26.67	24.3	0.24	0.29	Fair.
Northern Arizona₂	3.65		.043		.15								
Great Sandy₁	22.7		.29		.90								
		6.22		6.74		6.0	78.3	84.9	25.22	24.3	.32	.29	Good.
Northern Arizona₂	3.65		.043		.15								
Kalahari₁	26.0		.29		.70								
		7.12		6.74		4.67	89.7	84.9	37.14	24.3	.41	.29	Very poor.
Northern Arizona₂	3.65		.043		.15								
Northern Rub' al Khali₁	14.34		.38		1.41								
		3.93		8.84		9.4	37.7	84.9	10.17	24.3	.27	.29	Very poor.
Northern Arizona₂	3.65		.043		.15								
Kalahari₁	26.0		.29		.70								
		1.15		1.0		.78	89.7	78.3	37.14	25.22	.41	.32	Excellent.
Great Sandy₂	22.7		.29		.90								
Kalahari₁	26.0		.29		.70								
		1.08		1.32		.78	89.7	109.1	37.14	26.67	.41	.24	Good.
Simpson₂	24.0		.22		.90								
Simpson₁	24.0		.22		.90								
		1.06		.76		1.0	109.1	78.3	26.67	25.22	.24	.32	Excellent.
Great Sandy₂	22.7		.29		.90								
Simpson₁	24.0		.22		.90								
		1.67		.58		.64	109.1	37.7	26.67	10.17	.24	.27	Fair.
Northern Rub' al Khali₂	14.34		.38		1.41								
Great Sandy₁	22.7		.29		.90								
		1.58		.76		.64	78.3	37.7	25.22	10.17	.32	.27	Good.
Northern Rub' al Khali₂	14.34		.38		1.41								
Kalahari₁	26.0		.29		.70								
		1.81		.76		.50	89.7	37.7	37.14	10.17	.41	.27	Fair.
Northern Rub' al Khali₂	14.34		.38		1.41								

[1]Geographic location of sand seas: Simpson and Great Sandy Deserts, Australia; northern Arizona, U.S.A.; Kalahari Desert, southern Africa; Rub' al Khali, Saudi Arabia.
[2]Standard for degree of correspondence, based on maximum difference in scale ratios: 0.01-0.50 = Excellent; 0.51-1.0 = Good; 1.01-1.50 = Fair; 1.51-2.0 = Poor; 2.10 or more = Very poor.

TABLE 38. — *Morphometry of crescentic dunes in sand seas*

[$\bar{L}$, mean length; $\bar{W}$, mean width; $\bar{\lambda}$, mean wavelength. All measurements in kilometres. Leaders (---), not determined. Crescentic type includes transverse type as used by other authors in this publication]

Region[1]	Latitude, longitude	Desert or sand sea area	Area sampled (km²)	Range of dune length	$\bar{L}$	Range of dune width	$\bar{W}$	Range of dune wavelength	$\bar{\lambda}$	$\bar{L}/\bar{W}$	$\bar{W}/\bar{\lambda}$	$\bar{L}/\bar{\lambda}$	Figure No.
			Simple and probable simple crescentic dune ridges										
1	32°45' N., 106°15' W.	White Sands (U.S.A.)	7.5	0.03-0.185	0.072	0.06-0.215	0.1112	0.06-0.198	0.1115	0.65	1.0	0.65	173B
2	22°-25° N., 51°30'-52°30' E.	Persian Gulf (United Arab Emirates)	12,500	0.1 -1.1	.44	0.2 -1.8	.71	0.07-2.0	.59	.62	1.2	.75	None.
3	26°-26°45' N., 69°-70° E.	Thar Desert (near Umarkot, Pakistan)	5,000	0.2 -0.6	.47	0.5 -1.5	.93	0.2 -1.0	.58	.51	1.6	.81	255
4	23°30'-24°30' S., 14°30'-14°45' E.,	Namib coast (South-West Africa)	270	0.3 -1.3	.68	0.7 -1.5	1.12	0.2 -1.9	.87	.61	1.29	.78	223
5	23°15'-23°30' N., 53°-54°30' E.	Al Jiwa, Rub' al Khali (Saudi Arabia)	4,500	0.2 -0.9	.59	0.4 -1.3	.80	0.2 -1.0	.63	.74	1.27	.94	240
6	40°15'N., 87°15'-88° E.	Eastern Takla Makan (China)	2,000	0.5 -1.2	.80	0.5 -1.5	.94	0.6 -1.2	.89	.85	1.06	.90	261A
7	39°45' N., 79°30'-80° E.	Western Takla Makan (China)	2,500	0.6 -1.5	.84	1.0 -3.0	1.66	0.7 -1.5	1.1	.51	1.51	.76	None.
			Probable compound crescentic dunes (mounds and ridges)										
8	25°45' N., 01°30'-02° W.	Western Sahara (Mali)	2,500	0.5 -0.8	0.62	0.5 -1.2	0.77	---	---	0.80	---	---	175A
9	20°30' N., 55°15'-56° E.	Southeastern Rub' al Khali (Saudi Arabia)	2,500	0.3 -1.1	.67	0.85-2.2	1.43	0.5 -3.2	1.76	.47	0.81	0.38	None.
10	32°20'-30° N., 08°30'-09°30' E.	Northeastern Sahara (Algeria and Tunisia)	5,000	0.3 -1.5	.65	0.5 -3.0	1.43	0.5 -2.0	1.24	.45	1.15	.52	210A
11	27°30'-29° N., 39°-40° E.	An Nafud (northern Saudi Arabia)	5,000	0.5 -2.0	.80	0.6 -2.7	1.41	0.8 -3.3	1.84	.57	.77	.43	None.

TABLE 38 — Morphometry of crescentic dunes in sand seas — Continued

[$\bar{L}$, mean length; $\bar{W}$, mean width; $\bar{\lambda}$, mean wavelength. All measurements in kilometres. Leaders (---), not determined. Crescentic type includes transverse type as used by other authors in this publication]

Region[1]	Latitude, longitude	Desert or sand sea area	Area sampled (km²)	Range of dune length	$\bar{L}$	Range of dune width	$\bar{W}$	Range of dune wavelength	$\bar{\lambda}$	$\bar{L}/\bar{W}$	$\bar{W}/\bar{\lambda}$	$\bar{L}/\bar{\lambda}$	Figure No.
				Compound crescentic dune ridges									
12	31°45'-32°N., 113°30'-114° W.	Gran Desierto (Sonora, Mexico)------------------	650	0.3 -1.5	0.66	0.5 -2.0	1.04	0.5 -2.5	1.38	0.63	0.75	0.48	179D
13	32°45'-33°10' N., 114°45'-115°15' W.	Algodones (California, U.S.A.)------------------	500	0.5 -2.25	.88	0.5 -3.5	1.61	0.4 -2.5	1.07	.55	1.50	.82	174B
14	21°23' N., 53°30'-55° E.	Al 'Uruq al Mu' taridah, eastern Rub' al Khali (Saudi Arabia)------------	10,000	1.3 -3.3	2.09	1.6 -4.0	2.76	1.5 -4.0	2.56	.76	1.08	.82	174C,175B
15	39°30'-40°30' N., 86°45'-87°30' E.	Cherchen sand sea, Takla Makan (China)-------------	5,000	1.1 -3.4	2.20	2.0 -5.2	3.24	2.0 -5.5	3.0	.68	1.08	.73	175D
16	37°-37°45' N., 62°30'-64°15' E.	Peski Karakumy (near Mary, U.S.S.R.)------------------	2,500	0.8 -1.5	1.16	1.2 -3.5	1.96	1.0 -2.5	1.76	.59	1.11	.66	None.
17	42°30'-43°30' N., 62°30'-65° E.	Peski Karakumy (near Bukantau, U.S.S.R.)-------	10,000	0.8 -2.0	1.31	1.0 -3.1	2.22	0.85-3.1	2.27	.59	.98	.58	Do.
18	26°30'-27°30' N., 68°30'-69°30' E.	Western Thar Desert (near Sukkur, Pakistan)---------	2,500	0.75-2.0	1.30	1.0 -2.5	1.50	0.7 -2.5	1.44	.81	1.04	.90	256
19	41°30'-42°30' N., 101°-102° W.	Nebraska Sand Hills (U.S.A.)	3,000	0.8 -2.3	1.65	1.5 -5.0	3.14	1.0 -2.5	1.73	.52	1.81	.95	175E
20	40°-40°30' N., 101°30'-102°30' E.	Pa-tan-chi-lin Sha-mo, Ala Shan Desert (China)-------	2,500	1.25-3.0	2.16	2.0 -4.5	2.88	1.8 -4.25	2.92	.75	.99	.74	175C
21	17°30'-19° N., 09°-13° W.	Aoukâr sand sea, south-western Sahara (Mauritania)	5,000	1.2 -2.1	1.59	2.2 -9.6	4.10	1.0 -2.5	1.71	.39	2.40	.93	175F

[1] Sources of data--(1) Aerial photograph: Holloman Air Force Base. Following are all numbered Landsat images: (2) E1131-06303, E1131-06310; (3) E1191-05211; (4) E1383-08270; (5) E1148-06251; (6) E1146-04251; (7) E1115-04541; (8) E1115-10083; (9) E1182-06141; (10) E1109-09313; (11) E1231-07270, E1231-07272; (14) E1183-06194; (15) E1146-04251, E1146-04254; (16) E1126-05581; (17) E1126-05563, E1127-0621; (18) E1192-05267, E1120-05267; (19) E1026-17012; (20) E1117-03222; (21) E1227-10335, E1229-10445. SKYLAB photographs: (12) SL4-76-085; (13) SL4-76-083.

TABLE 39. — Similarity of compound crescentic dunes in widely distant sand seas, shown by comparison of ratios derived from measurements of mean length, $\bar{L}$, mean width, $\bar{W}$, and mean wavelength, $\bar{\lambda}$

[Subscript following name of sand sea indicates first or second of the pair. Subscripts to $\bar{L}$, $\bar{W}$, and $\bar{\lambda}$ indicate first or second of sand sea pair]

Pair of sand seas[1]	$\bar{L}$	$\bar{L}_1/\bar{L}_2$	$\bar{W}$	$\bar{W}_1/\bar{W}_2$	$\bar{\lambda}$	$\bar{\lambda}_1/\bar{\lambda}_2$	$\bar{L}_1/\bar{W}_1$	$\bar{L}_2/\bar{W}_2$	$\bar{W}_1/\bar{\lambda}_1$	$\bar{W}_2/\bar{\lambda}_2$	$\bar{L}_1/\bar{\lambda}_1$	$\bar{L}_2/\bar{\lambda}_2$	Degree of correspondence[2]
Ala Shan₁-----------------	2.16		2.88		2.92								
		1.03		1.04		1.14	0.75	0.76	0.99	1.08	0.74	0.82	Excellent.
Eastern Rub' al Khali₂---	2.09		2.76		2.56								
Takla Makan₁-------------	2.20		3.24		3.0								
		1.05		1.18		1.17	.68	.76	1.08	1.08	.73	.82	Excellent.
Eastern Rub' al Khali₂---	2.09		2.74		2.56								
Takla Makan₁-------------	2.20		3.24		3.0								
		1.02		1.13		1.03	.68	.75	1.08	.99	.73	.74	Excellent.
Ala Shan₂-----------------	2.16		2.88		2.92								
Bukantau₁----------------	1.31		2.22		2.27								
		1.13		1.13		1.29	.59	.59	.98	1.11	.58	.66	Excellent.
Mary₂--------------------	1.16		1.96		1.76								
Ala Shan₁----------------	2.16		2.88		2.92								
		1.86		1.47		1.66	.75	.59	.99	1.11	.74	.66	Very good.
Mary₂--------------------	1.16		1.96		1.76								
Takla Makan₁-------------	2.20		3.24		3.0								
		1.90		1.65		1.70	.68	.59	1.08	1.11	.73	.66	Very good.
Mary₂--------------------	1.16		1.96		1.76								
Ala Shan₁----------------	2.16		2.88		2.92								
		1.66		1.92		2.03	.75	.87	.99	1.04	.74	.90	Very good.
Thar₂--------------------	1.30		1.50		1.44								
Takla Makan₁-------------	2.20		3.24		3.0								
		1.68		1.46		1.32	.68	.59	1.08	.98	.73	.58	Very good.
Bukantau₂----------------	1.31		2.22		2.27								
Thar₁--------------------	1.30		1.50		1.44								
		1.97		1.44		1.04	.87	.63	1.04	.75	.94	.48	Poor.
Gran Desierto₂-----------	.66		1.04		1.38								
Takla Makan₁-------------	2.20		3.24		3.0								
		3.33		3.12		2.17	.68	.63	1.08	.75	.73	.48	Very poor.
Gran Desierto₂-----------	.66		1.04		1.38								

TABLE 39 — *Similarity of compound crescentic dunes in widely distant sand seas, shown by comparison of ratios derived from measurements of mean length, $\overline{L}$, mean width, $\overline{W}$, and mean wavelength, $\overline{\lambda}$ — Continued*

[Subscript following name of sand sea indicates first or second of the pair. Subscripts to $\overline{L}$, $\overline{w}$, and $\overline{\lambda}$ indicate first or second of sand sea pair

Pair of sand seas[1]	$\overline{L}$	$\overline{L}_1/\overline{L}_2$	$\overline{w}$	$\overline{w}_1/\overline{w}_2$	$\overline{\lambda}$	$\overline{\lambda}_1/\overline{\lambda}_2$	$\overline{L}_1/\overline{w}_1$	$\overline{L}_2/\overline{w}_2$	$\overline{w}_1/\overline{\lambda}_1$	$\overline{w}_2/\overline{\lambda}_2$	$\overline{L}_1/\overline{\lambda}_1$	$\overline{L}_2/\overline{\lambda}_2$	Degree of corresponden
Algodones₁---------------	.88	1.33	1.61	1.55	1.07	.78	.55	.63	1.50	.75	.82	.48	Fair.
Gran Desierto₂-----------	.66		1.04		1.38								
Takla Makan₁------------	2.20	2.5	3.24	2.01	3.0	2.8	.68	.55	1.08	1.50	.73	.82	Fair.
Algodones₂--------------	.88		1.61		1.07								
Eastern Rub' al Khali₁---	2.09	2.37	2.76	1.71	2.56	2.39	.76	.55	1.08	1.50	.82	.82	Fair
Algodones₂--------------	.88		1.61		1.07								
Nebraska Sand Hills₁-----	1.65	1.88	3.14	2.71	1.73	1.62	.52	.55	1.81	1.50	.95	.82	Very poor.
Algodones₂--------------	.88		1.16		1.07								
Mary₁-------------------	1.16	1.32	1.96	1.22	1.76	1.64	.59	.55	1.11	1.50	.66	.82	Very good.
Algodones₂--------------	.88		1.61		1.07								
Ala Shan₁---------------	2.16	1.31	2.88	.92	2.92	1.69	.75	.52	.99	1.81	.74	.95	Fair.
Nebraska Sand Hills₂-----	1.65		3.14		1.73								
Ala Shan₁---------------	2.16	2.46	2.88	1.79	2.92	2.73	.75	.55	.99	1.50	.74	.82	Poor.
Algodones₂--------------	.88		1.61		1.07								
Thar₁-------------------	1.30	1.48	1.50	.93	1.44	1.35	.87	.55	1.04	1.50	.90	.82	Good.
Algodones₂--------------	.88		1.61		1.07								
Eastern Rub' al Khali₁---	2.09	1.27	2.76	.88	2.56	1.48	.76	.52	1.08	1.81	.82	.95	Fair
Nebraska Sand Hills₂-----	1.65		3.14		1.73								
Eastern Rub' al Khali₁---	2.09	1.31	2.76	.67	2.56	1.50	.76	.39	1.08	2.40	.82	.93	Poor.
Aoukâr₂-----------------	1.59		4.10		1.71								
Aoukâr₁-----------------	1.59	.96	4.10	1.31	1.71	.99	.39	.52	2.40	1.81	.93	.95	Very good
Nebraska Sand Hills₂-----	1.65		3.14		1.73								

[1]Geographic location of sand seas: Ala Shan Desert, China; Rub' al Khali, Saudi Arabia; Takla Makan Desert, China; Peski Karakumy, near Bukantau and Mary, U.S.S.R.; Gran Desierto, Mexico; Thar Desert, India and Pakistan; Algodones Desert and Nebraska Sand Hills, U.S.A.; Aoukâr, Mauritania.

[2]Standard for degree of correspondence, based on maximum difference in scale ratios: 0.01-0.20 = Excellent; 0.21-0.39 = Very good; 0.40-0.59 = Good; 0.60-0.79 = Fair; 0.80-0.99 = Poor; 1.00 or more = Very poor.

TABLE 40. — *Morphometry of star dunes in sand seas*

[$\overline{D}$, mean diameter; $\overline{\lambda}$, mean wavelength. All measurements in kilometres]

Region[1]	Latitude, longitude	Desert or sand sea area	Area sampled (km²)	Range of dune diameter	$\overline{D}$	Range of dune wavelength	$\overline{\lambda}$	$\overline{D}/\overline{\lambda}$	Figure No.
1	31°45'-32°15' N., 114°-114°30' W.	Gran Desierto (Mexico)----------------------	1,000	0.25-1.2	0.70	0.1 -3.0	1.19	0.59	180A
2	25°50' S., 15°40' E.	Namib Desert (South-West Africa)------------	1,000	0.4 -1.0	.68	0.8 -4.5	2.21	.31	None.
3	19°30'-20° N., 16°-17° E.	Southern Sahara (Niger)---------------------	1,000	0.2 -1.2	.61	0.15-3.0	1.0	.61	Do.
4	39°50'-40° N., 99°30'-101°30' E.	Ala Shan Desert (China)---------------------	2,000	0.4 -1.0	.74	0.3 -3.2	1.37	.54	264B
5	19°-20°30' N., 53°30'-54°30' E.	Southeastern Rub' al Khali (Saudi Arabia)---	12,500	0.5 -1.3	.84	0.97-2.86	2.06	.41	180C
6	29°30'-32° N., 007°30'-009° E.	Entire Grand Erg Oriental (Algeria)---------	10,000	0.4 -3.0	.95	0.8 -6.7	2.07	.46	210B
7	29°30'-30°15' N., 007°30'-008°15' E.	Southern margin, Grand Erg Oriental (Algeria)	2,500	0.7 -3.0	1.65	1.5 -3.1	2.17	.76	180B
8	30°30'-32° N., 007°30'-009° E.	Northern Grand Erg Oriental (Algeria)-------	7,500	0.4 -1.3	.71	0.8 -6.7	2.04	.35	210B

[1]Sources of data are numbered Landsat images: (1) E1159-17451; (2) E1382-08214; (3) E1209-08494; (4) E1209-03340; (5) E1417-06191, E1147-06202, E1183-06200. E1183-06203: (6) E1109-09320. E1109-09322. E1110-09374: (7) E1109-09322; (8) E1109-09320. E1110-09374.

TABLE 41. — *Similarity of star dunes in widely distant sand seas, shown by comparison of ratios derived from measurements of mean diameter, $\bar{D}$, and mean wavelength, $\bar{\lambda}$*

[Subscript following name of sand sea indicates first or second of the pair. Subscripts to $\bar{D}$ and $\bar{\lambda}$ indicate first or second of sand sea pair]

Pairs of sand seas[1]	$\bar{D}$	$\bar{D}_1/\bar{D}_2$	$\bar{\lambda}$	$\bar{\lambda}_1/\bar{\lambda}_2$	$\bar{D}_1/\bar{\lambda}_1$	$\bar{D}_2/\bar{\lambda}_2$	Degree of correspondence[2]
Gran Desierto$_1$------------------	0.70		1.19				
		1.03		0.54	0.59	0.31	Fair.
Namib$_2$------------------------	.68		2.21				
Ala Shan$_1$----------------------	.74		1.37				
		1.06		1.15	.54	.59	Excellent.
Gran Desierto$_2$-----------------	.70		1.19				
Eastern Rub' al Khali$_1$---------	.84		2.06				
		1.2		1.73	.41	.59	Fair.
Gran Desierto$_2$-----------------	.70		1.19				
Grand Erg Oriental$_1$------------	.95		2.07				
		1.36		1.74	.46	.59	Good.
Gran Desierto$_2$-----------------	.70		1.19				
Southern Grand Erg Oriental$_1$----	1.65		2.17				
		2.36		1.82	.76	.59	Fair.
Gran Desierto$_2$-----------------	.70		1.19				
Northern Grand Erg Oriental$_1$----	.71		2.04				
		1.01		1.71	.35	.59	Poor.
Gran Desierto$_2$-----------------	.70		1.19				
Gran Desierto$_1$-----------------	.70		1.19				
		1.15		1.19	.59	.61	Excellent.
Southern Sahara$_2$---------------	.61		1.0				
Eastern Rub' al Khali$_1$---------	.84		2.06				
		1.14		1.50	.41	.54	Good.
Ala Shan$_2$----------------------	.74		1.37				
Grand Erg Oriental$_1$------------	.95		2.07				
		1.28		1.51	.46	.54	Very good.
Ala Shan$_2$----------------------	.74		1.37				
Grand Erg Oriental$_1$------------	.95		2.07				
		1.13		1.0	.46	.41	Excellent.
Eastern Rub' al Khali$_2$---------	.84		2.06				
Ala Shan$_1$----------------------	.74		1.37				
		1.21		1.37	.54	.61	Very good.
Southern Sahara$_2$---------------	.61		1.0				
Eastern Rub' al Khali$_1$---------	.84		2.06				
		1.18		1.01	.41	.35	Very good.
Northern Grand Erg Oriental$_2$----	.71		2.04				
Southern Grand Erg Oriental$_1$----	1.65		2.17				
		1.96		1.05	.76	.41	Very poor.
Eastern Rub' al Khali$_2$---------	.84		2.06				

TABLE 41. — *Similarity of star dunes in widely distant sand seas, shown by comparison of ratios derived from measurements of mean diameter, $\bar{D}$, and mean wavelength, $\bar{\lambda}$* — Continued

[Subscript following name of sand sea indicates first or second of the pair. Subscripts to $\bar{D}$ and $\bar{\lambda}$ indicate first or second of sand sea pair]

Pairs of sand seas[1]	$\bar{D}$	$\bar{D}_1/\bar{D}_2$	$\bar{\lambda}$	$\bar{\lambda}_1/\bar{\lambda}_2$	$\bar{D}_1/\bar{\lambda}_1$	$\bar{D}_2/\bar{\lambda}_2$	Degree of correspondence[2]
Southern Grand Erg Oriental$_1$ ----	1.65		2.17				
Alan Shan$_2$ ---------------------	.74	2.23	1.37	1.58	.76	.54	Poor.
Southern Grand Erg Oriental$_1$ ----	1.65		2.17				
Namib$_2$ -----------------------	.68	2.43	2.21	.98	.76	.31	Very poor.
Eastern Rub' al Khali$_1$ ----------	.84		2.06				
Namib$_2$ -----------------------	.68	1.24	2.21	.93	.41	.31	Good.
Grand Erg Oriental$_1$ -------------	.95		2.07				
Namib$_2$ -----------------------	.68	1.4	2.21	.94	.46	.31	Fair.
Grand Erg Oriental$_1$ -------------	.95		2.07				
Southern Sahara$_2$ ---------------	.61	1.56	1.0	2.07	.46	.61	Fair.
Eastern Rub' al Khali$_1$ ----------	.84		2.06				
Southern Sahara$_2$ ---------------	.61	1.38	1.0	2.06	.41	.61	Poor.
Namib$_1$ -----------------------	.68		2.21				
Southern Sahara$_2$ ---------------	.61	1.11	1.0	2.21	.31	.61	Very poor.
Northern Grand Erg Oriental$_1$ ----	.71		2.04				
Southern Sahara$_2$ ---------------	.61	1.16	1.0	2.04	.35	.61	Very poor.

[1]Geographic location of sand seas: Gran Desierto, Mexico; Namib Desert, South-West Africa Ala Shan Desert, China; Rub' al Khali, Saudi Arabia; Grand Erg Oriental, Algeria; Southern Sahara, Niger.

[2]Standard for degree of correspondence, based on maximum difference in scale ratios: 0.01-0.15 = Excellent; 0.16-0.30 - Very good; 0.31-0.45 = Good; 0.46-0.60 = Fair; 0.61-0.75 = Poor; 0.76 or more = Very poor. More stringent criteria are set than for dunes of linear and crescentic types because fewer ratios are compared for dunes of star type.

TABLE 42. — *Morphometry of dome and dome-shaped dunes*

[$\bar{D}$, mean diameter; $\bar{\lambda}$, mean wavelength; *, approximate. All measurements in kilometres]

Region[1]	Latitude, longtitude	Desert or sand sea area	Area sampled (km^2)	Range of dune diameter	$\bar{D}$	Range of dune wavelength	$\bar{\lambda}$	$\bar{D}/\bar{\lambda}$	Figure No.
1	40°-40°30' N., 83°-83°45' E.	Takla Makan (China)----------------------------	2,500	0.6-1.8	1.12	0.8-5.4	3.34	0.34	185B
2	32°-32°30' N., 002°-002°30' E.	Northern Grand Erg Occidental (Algeria)---------	1,000*	0.9-2.0	1.51	1.6-5.0	3.42	.44	185A
3	25°20'-27° N., 44°-45° E.	Nafuds (northern Saudi Arabia)-----------------	3,000*	0.8-1.80	1.21	1.0-4.4	2.18	.56	244

[1]Sources of data are numbered Landsat images: (1) El131-04424; (2) El186-10003; (3) El137-07043.

TABLE 43. — *Similarity of domes and dome-shaped dunes in widely distant sand seas, shown by comparison of ratios derived from mean diameter, $\bar{D}$, and mean wavelength, $\bar{\lambda}$*

[Subscript following name of sand sea indicates first or second of the pair.
Subscripts to $\bar{D}$ and $\bar{\lambda}$ indicate first or second of sand sea pair]

Pairs of sand seas[1]	$\bar{D}$	$\bar{D}_1/\bar{D}_2$	$\bar{\lambda}$	$\bar{\lambda}_1/\bar{\lambda}_2$	$\bar{D}_1/\bar{\lambda}_1$	$\bar{D}_2/\bar{\lambda}_2$	Degree of correspondence[2]
Nafuds$_1$------------------------------	1.21		2.18				
		1.08		0.65	0.56	0.33	Good.
Northern Takla Makan$_2$-----------	1.12		3.34				
Northern Grand Erg Occidental$_1$--	1.51		3.42				
		1.35		1.02	.44	.33	Good.
Northern Takla Makan$_2$-----------	1.12		3.34				
Northern Grand Erg Occidental$_1$--	1.51		3.42				
		1.25		1.57	.44	.55	Good.
Nafuds$_2$------------------------------	1.21		2.18				

[1]Geographic location of sand seas: Nafuds, Saudi Arabia; Takla Makan Desert, China; Grand Erg Occidental, Algera.

[2]Standard for degree of correspondence, based on maximum difference in scale ratios: 0.01-0.15 = Excellent; 0.16-0.30 = Very good; 0.31-0.45 = Good; 0.46-0.60 = Fair; 0.61-0.75 = Poor; 0.76 or more = Very poor. More stringent criteria are set than for dunes of linear and crescentic types because fewer ratios are compared for dunes of dome type.

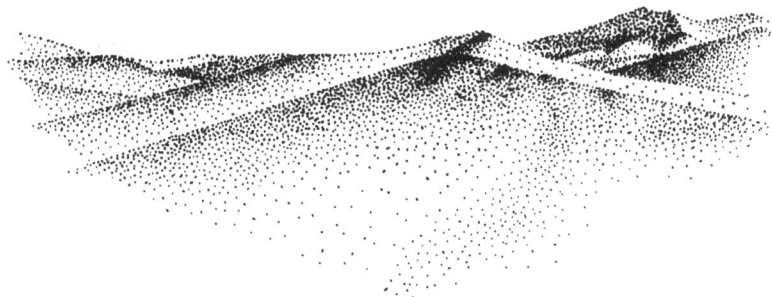

A STUDY OF GLOBAL SAND SEAS

REGIONAL STUDIES OF SAND SEAS, USING LANDSAT (ERTS) IMAGERY

Chapter K

By CAROL S. BREED, STEVEN G. FRYBERGER, SARAH ANDREWS, CAMILLA McCAULEY
FRANCES LENNARTZ, DANA GEBEL, and KEVIN HORSTMAN

Contents

Illustrations

Table

Method of Study

THIS CHAPTER IS A SERIES of regional studies of selected sand seas and is based on available literature, surface wind summaries, and data obtained through remote sensing. Studies of 8 desert regions include sand seas in parts of 17 countries. Mosaics were prepared from Landsat (ERTS) imagery covering each of the regions, and, from each mosaic, a map was made of sand bodies and other features and their distribution. Supplementary information about dunes of each region was obtained from Skylab and aerial photographs, and from the ground studies of other workers.

The Landsat imagery used to map dunes are at a scale of 1:1,000,000. Many features of relatively small dunes, including details of dune shape and slipface development commonly used in dune classifications by field geologists, are not visible at this scale. For this reason, a classification somewhat different from that used in the ground studies and low-altitude aerial photograph studies treated in chapters A – I was found to be expedient here. For example, the term "crescentic dune" is here used to include all dune types that are generally referred to as barchans, barchanoid ridges, or transverse dunes.

Classification of principal dune types according to form and to number of slipfaces is given in chapter E, table 9. We developed an expanded version of this classification (chapter J) for our studies because dunes seen on Landsat imagery may not be directly comparable to those that have been studied on the ground. A comparison of the general terminology for dune classification based on observations of Landsat imagery and that based on field studies is given in chapter A, table 1.

Originally, plans for this study were to include large desert regions in Australia, southern U.S.S.R., and the Sonoran Desert of the United States and Mexico. Although Landsat imagery was obtained and mosaics were prepared for these regions, time limitations precluded compilation of data for use in this chapter.

The areas selected for this study, using Landsat imagery, are as follows: (1) Western Sahara of Mauritania, Senegal, Mali, and Spanish Sahara; (2) Sahel and southern Sahara of Mauritania, Mali, and Niger; (3) northern Sahara of Algeria, western Libya, and Tunisia; (4) Namib Desert of South-West Africa; (5) Kalahari Desert of South-West Africa and Botswana; (6) Arabian sand seas of Saudi Arabia, Oman, Yemen, and adjacent States; (7) Thar Desert of India and Pakistan; and (8) Takla Makan and Ala Shan Deserts of the Peoples Republic of China.

Definitions

The following list of foreign terms used on maps in this chapter are given English definitions. All definitions are adapted from gazetteers prepared by

Term	Definition	Map on which used
Adrar	Mountains	Sahara (western).
Berg	Mountains	Namib, Kalahari.
Cap	Cape, point	Sahara (western).
Dhar	Plateau	Sahara (western).
Djebel	Hills	Sahara (northern).
Erg	Dunes, desert sand region*	Sahara (northern).
Hamada	Plateau, stony plain	Sahara (northern).
Ho	Stream, river	Takla Makan, Ala Shan.
Hu	Lake	Takla Makan, Ala Shan.
Jabal	Hills	Rub' al Khali.
Kum	Sand area	Takla Makan, Ala Shan.
Lac	Lake	Sahara, Sahel.
Massif	Mountains	Sahara.
Oued	Wadi	Sahara, Sahel.
Pan	Playa	Kalahari.
Plato	Plateau	Kalahari.
Po	Lake	Takla Makan, Ala Shan.
Puts	Well, spring	Kalahari.
Rann	Marsh	Thar.
Sebkha	Playa, alkali flat*	Sahara.
Sha-mo	Desert, sand area	Takla Makan, Ala Shan.
Shan	Mountains, range	Takla Makan, Ala Shan.
Shatt	Sebkha	Sahara (northern).
Tassili	Mountains	Sahara (northern).
'Urūq	Sand dunes, dune area	Rub' al Khali.
Vallee	Wadi	Sahel.
Vlei	Pan	Namib.
Wādī	Intermittent or dry stream, stream valley, oasis.	Sahara, Rub' al Khali.

the U.S. Board on Geographic Names. Those definitions marked by an asterisk (*) are adapted from
Webster's New International Dictionary of the
English Language, Second Edition, unabridged,
1960, C & G Merriam Co., Publishers.

Western Sahara of Mauritania, Senegal, Mali, and Spanish Sahara

Summary of Conclusions

Dunes of the western Sahara are limited to two
main types — linear and crescentic — distributed
throughout wide areas. Linear dunes are the most
common type, extending to the coast in both
parallel and en echelon forms from hundreds of
kilometres inland. Compound and simple linear
dunes are the most common forms in northern
Mauritania. Heavily vegetated compound linear
dunes of proportions similar to those in northern
Mauritania occur in southern Mauritania, near
Mederdra and Boutilimit. Crescentic dunes are
common in the southern part of the desert (Aoukâr
sand sea), and in smaller fields along the Atlantic
coast. Sand sheets, generally associated with linear
dunes, occur throughout Mauritania, western Mali,
Spanish Sahara, and northern Senegal.

Rainfall in the western Sahara ranges from less
than 100 mm (4 in.) in the northern regions near
Fort-Gouraud to more than 300 mm (12 in.) in the
southern regions near Rosso. Effective winds in the
western Sahara are mostly from the north and
northeast and result mainly from anticyclonic circulation from the Sahara and Azores high-pressure
cells.

Relatively unvegetated linear dunes and sand
sheets in sand seas of northern Mauritania are in a
high-energy wind environment. These sand seas
typically are elongate and are alined subparallel to
the resultant drift direction of present-day winds.
In some places, however, linear dunes are alined
oblique to trends of sand seas but parallel to directions of some effective winds. Vegetated, or partially vegetated, linear and crescentic dunes in
southern Mauritania are in present-day intermedi-
ate- or low-energy wind environments with considerable rainfall. Large crescentic dunes in the
Aoukar sand seas generally occur downwind of
escarpments or other topographic barriers.

Introduction

Several long, sharply defined sand seas (known
as ergs in northern Africa), extend southwestward
from Algeria across the western Sahara. The
morphology of the dunes and the extent of the sand
seas, as determined from Landsat imagery, is
shown in figure 188. Sand roses representing the
potential sand-moving effect of present-day surface
winds, and isohyets (lines of equal precipitation)
show present climatic factors in relation to sand
distribution (figs. 188, 189). A regional view of the
sand seas is provided by a Skylab photograph
(fig. 190).

Few of the eolian sand areas shown in figure 188
are described in published reports. Sand seas along
the east margin have been described by Monod
(1958). Reconnaissance studies of dune areas near
the Sénégal River and northward to about lat 17° N.
have been made by Tricart and Brochu (1955),
Leprun (1971), Boulet, Guichard, and Vieillefon
(1971), and by others. Grove and Warren (1968)
briefly summarized work in the eastern and
southern parts of the region. Useful maps of the
region are the U.S. Department of Commerce, U.S.
Air Force Operational Navigation Charts H−1, J−1,
K−1, and J−2; and the U.S. Army Map Service
Series 1301, sheets NE 29, NF 29, NF 28, and NE 28;
all maps are at a scale of 1:1,000,000. A regional
landform map by Raisz (1952) is at a scale of 1 inch
equals 60 miles (1:3,801,600).

Review of Previous Work

Dunes in the area bounded by Boutilimit, Saint-
Louis, and Louga (fig. 188) are described (Tricart
and Brochu, 1955, p. 148−149) as belonging to a
zone of old red parallel dunes, 40−50 km (25−30
mi) long, about 1 km (0.6 mi) wide, and 10−20 m
(33−66 ft) high. Near the coast, interdune corridors
are reported to be about 2 km (1 mi) wide and dune
ridges are 20−30 m (66−99 ft) high. Along the
Aftoût'es Saheli (fig. 188), old red dunes have been
truncated by marine erosion and are overlain by recent brown or yellow crescentic coastal dunes (Tricart and Brochu, 1955, p. 148. 150). Sand depths of
20−50 m (60−165 ft) are reported near Mederdra
and Saint-Louis by Tricart and Brochu (1955, p.
146−147).

The mean diameter of dune sand in the southern
area is reported to range from 0.25 to 0.40 mm —

N

EXPLANATION

LINEAR DUNES

Mostly simple, some compound

Feathered compound

CRESCENTIC DUNES

Simple barchanoid ridges

Compound barchanoid ridges with long, straight segments

Complex crescentic ridges with linear elements superimposed

SHEET AND STREAKS

Compound sheets and streaks with linear elements

BEDROCK OUTCROPS OR BARRIERS

ROCK AND GRAVEL DESERT PLAINS WITH NO LARGE DUNES

PLAYAS AND COASTAL SEBKHAS

VEGETATED PLAIN

ISOHYET—Average annual rainfall, in millimetres

BOUNDARY OF SAND SEA

440
700
Bir Mogrein

440
661
Fort-Gouraud

MAKTEIR

ADRAR SOTUF

AZEFAL

SPANISH SAHARA
MAURITANIA

ER RICHAT

283
443
Atar

OURANE

2016
2232
Port-Étienne

AZEFAL

Cap d'Arguin

ATLANTIC

TIJIRIT

AKCHÂR

Orä Maläigdane

Solitary linear dune

238
352
Atjout

Cap-Timiris

OCEAN

AMOUKROUZ

85
133
Tidjikdja

64
118
Nouakchott

AOUKÂR

TAGANT

AOUKÂR

122
259
Boutilimit

AFFOLLÉ

60
85
Kifta

Madedra

Lac-Rkiz

Aleg

AFTOUT ES SAHELI

70
156
Rosso

Senegal

Bogué

Dagena

ASSABA

MAURITANIA/MALI

MAURITANIA

SENEGAL

Kaëdi

River

Saint-Louis

Louga

F E R L O

AFRICA

MAP LOCATION

0   200 KILOMETRES

0   100 MILES

DISTRIBUTION AND MORPHOLOGY OF EOLIAN SAND of the western Sahara of northern Africa. Map based on Landsat imagery. Isohyets from Griffiths and Soliman (1972). (Fig. 188.)

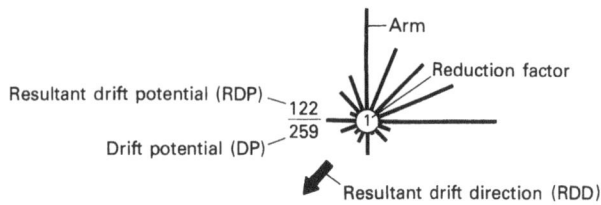

TYPE OF SAND ROSE used in this chapter. See chapter F, figure 93 and caption for explanation of parts of rose. (Fig. 189.)

that is, fine to medium grains (Tricart and Brochu, 1955, p. 156–162). Sands of recent active dunes along the coast and elsewhere are reported to have a maximum diameter of 0.29 mm. Sand of the older red dunes is generally better sorted than that of the recent dunes and ranges in diameter from 0.46 to 0.56 mm (mostly medium grained).

Three groups of immobile dunes are described (Leprun, 1971) in the Ferlo region of northwest Senegal, south of Dagena (fig. 188). Differentiation among the three groups of dunes is on the basis of soil development, texture, color, and vegetative cover.

The sequence of development of dunes in southern Mauritania and northern Senegal, based on paleontologic and pedologic studies, may be summarized as follows:

1. The large red simple to compound linear dunes trending east-northeast in southwestern Mauritania and northern Senegal are believed to have formed during an arid period when sea level was at least 50 m (165 ft) lower than it is at present, about 18,000 years before the present (Grove and Warren, 1968, p. 196; Boulet and others, 1971, p. 104, 113).

2. Reddening of the dune soils and gullying reportedly took place during a pluvial period, when the Sénégal and other rivers cut channels through the linear dune fields to the Atlantic Ocean (Grove and Warren, 1968, p. 196).

3. Sea level rose during the Nouakchottian transgression (5,500 to 5,000 years before the present) and the lower Sénégal River valley was flooded as far as Bogué, 250 km (155 mi) inland. Beaches between the great linear dunes were formed at an elevation 4–6 m (13–18 ft) above the present sea level (Grove and Warren, 1968, p. 196).

ELONGATE SAND SEAS IN THE NORTHWEST SAHARA of Mauritania and Spanish Sahara. Skylab photograph SL4–138–3756. Bright pink areas are interpreted as sand seas; grayish-pink areas are gravelly, sandy, and rocky plains without large dunes; and dark areas include bedrock outcrops and coastal sebkhas. (Fig. 190.)

4. The sea regressed, and sand from the beaches was blown by winds into the coastal "dunes jaunes" of Tricart and Brochu (1955, p. 148).

5. Gullying of old red dunes in the southern part of the Aoûkar sand sea (fig. 188) is continuing at present (Daveau, 1965). Gullying of dunes in the coastal region is interpreted (Tricart and Brochu, 1955, p. 153) as evidence of a transition zone between the arid northern environment and the more humid southern environment.

Interpretation of Landsat Imagery

Analysis of Landsat imagery shows that two main dune types — linear and crescentic — dominate the western Sahara. No dunes of star, dome, or parabolic type were recognized on Landsat imagery of this region. Complex dunes are relatively rare. The simple and compound forms of linear and crescentic dunes are similar to those in other sand seas that have been examined on Landsat imagery (chapter J). Sand sheets and streaks are conspicuous features of the western Sahara, especially in northern Mauritania.

Linear dunes of simple to compound form occur in elongate sand seas that extend from northwestern Mali to the coast of Mauritania (figs. 188, 190, 191). The largest single linear dune seen on Landsat imagery is in central Mauritania.

Measurements of linear dunes near Mederdra and Boutilimit (fig. 191), made on Landsat imagery, yield an average dune width of 0.66 km (0.41 mi). Spacing (wavelength, measured crest to crest) in this area varies widely, but averages 1.58 km (0.98 mi). In some areas, linear dunes as much as 20 km (12 mi) long are alined en echelon, oblique to the trend of the elongate sand seas. Near Atar, at about lat 21°N., long 14°W. (fig. 193), trends of the elongate sand seas are N. 50°E., but linear dunes are oriented N. 40°E.

Compound linear dunes measured on Landsat images of the Makteir sand sea, northeastern Mauritania (fig. 192) commonly are more than 50 km (30 mi) long, have an average width of 0.97 km (0.60 mi), and are very evenly spaced about 1.4 km (0.86 mi) from crest to crest.

A solitary linear dune extends from a gap in an escarpment south of Atar and crosses a sandy plain westward to the base of bedrock outcrops near Akjoujt (fig. 193). The dune (Drâ' Malichigdane) is 100 km (62 mi) long and about 1 km (0.6 mi) wide and is oriented parallel to the resultant drift direction at Atar, as indicated by a sand rose on the map (fig. 188). Near Akjoujt, the dune has small crescentic elements which give it a complex form. Other linear features in the same vicinity resemble the solitary linear dune in size and orientation but are composed of single rows of en echelon short linear dunes.

The Aoûkar sand sea of southern Mauritania and Mali (fig. 194) consists mainly of crescentic dune ridges. Dunes in the western part of this sand sea are compound crescentic. Measurements of simple crescentic dunes were made on images of the Aoûkar sand sea southeast of Tidjikdja. Boundaries of fields of crescentic dunes in the Aoûkar sand sea characteristically match the spurs and reentrants of plateau escarpments.

Small fields of minor, probably simple crescentic dunes occur along the Atlantic coast near the Sénégal River (fig. 188) and among bedrock outcrops south and east of Atar.

Sand sheets dominate the sand seas in the northern part of the western Sahara to a much greater extent than in any other deserts observed on Landsat imagery. North and west of Atar and south of Fort-Gouraud are numerous prominent sand sheets associated with rock outcrops and escarpments (chapter J, fig. 187). Linear dunes occur on many of these sheets, and some sheets have small groups of crescentic dunes. Some sheets are 100 km (62 mi) long, or longer (figs. 190, 194).

Sandstorms which blow southwestward, parallel to the trend of the sand seas, have been observed on Landsat imagery of northern Mauritania, a region characterized by strong winds, sand sheets, and linear dunes.

Surface Wind Flow and Precipitation

Surface wind circulation in the western Sahara (between lat 14° N. and 26° N., and long 6° W. to 19° W.) is controlled by the Azores and Sahara high-pressure cells, and the Intertropical Convergence Zone (I.T.C.Z.). During winter, the Azores high is nearest the African coast, and the Saharan high is strongest (fig. 195A). Outward flow from these

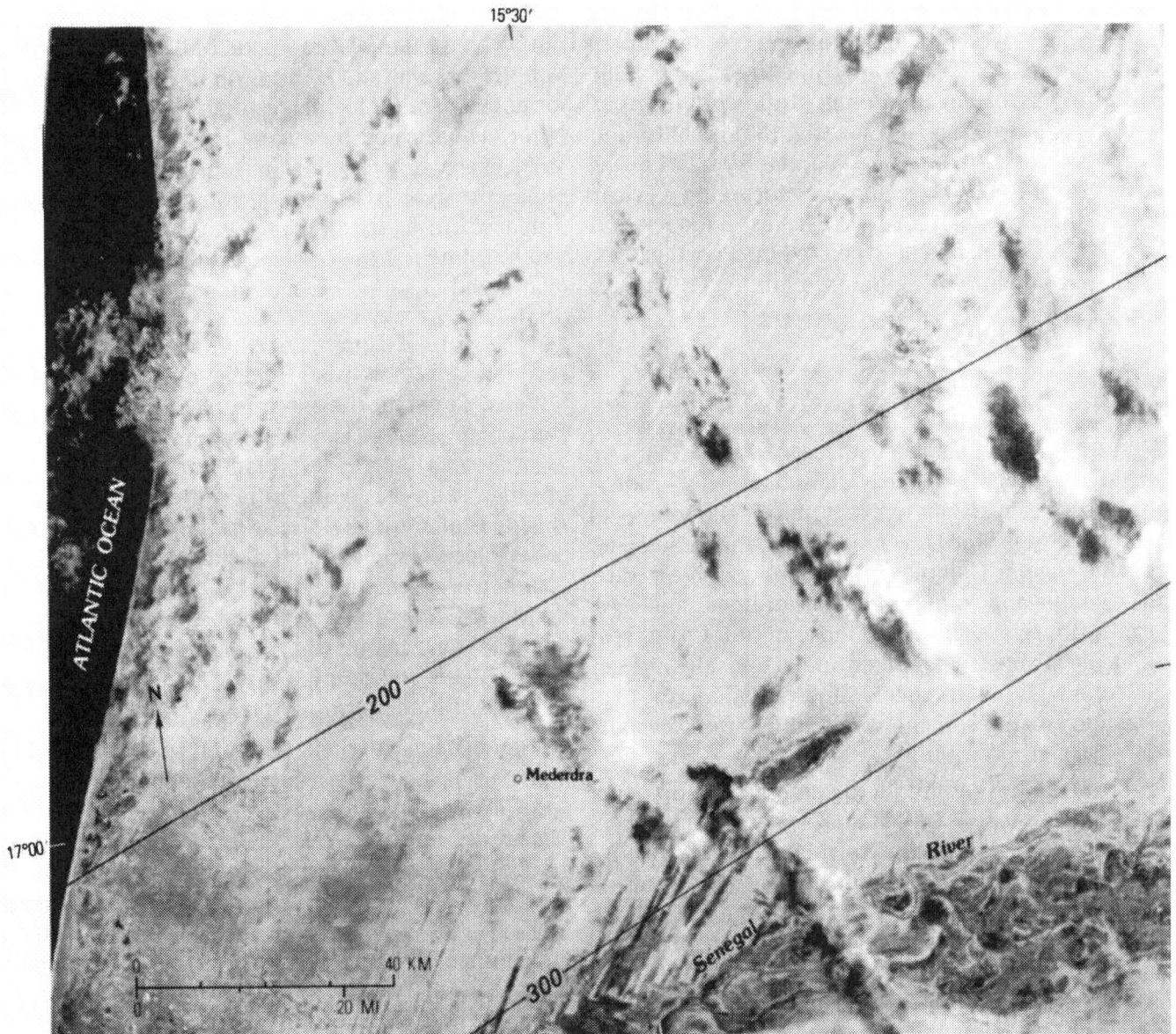

VEGETATED LARGE LINEAR DUNES shown in relation to precipitation isohyets (in millimetres per year) in southern Mauritania. Landsat false-color imagery E1069−10554. Green vegetation appears red, and red dune sands appear yellow on false-color imagery. (Fig. 191.)

pressure systems results in winds from the northeast or east over much of the western Sahara.

By June the Sahara high has been replaced by a thermal low in the central Sahara (fig. 195*B*). At this time the Azores high is strongest and is centered farthest from the coast at approximately lat 30°00′ N., long 32°30′ W., resulting in more northerly wind flow at most stations. Some southern stations near the coast (from Boutilimit westward) experience northwesterly winds during this season. During July−September, when the I.T.C.Z. has moved farthest north, stations south of lat 18° N. experience southwesterly monsoonal winds.

Rainfall in the western Sahara ranges from less than 50 mm/yr (2 in./yr) at Bir Mogreïn (Fort-Trinquet) in the north to more than 300 mm/yr (12 in./yr) at Rosso, in the south (fig. 188). Rain showers and thunderstorms during summer are associated with the I.T.C.Z., which moves northward to about lat 16° N. Thus, rainfall and associated southwest winds begin earlier and end later at southern stations than at northern stations (fig.

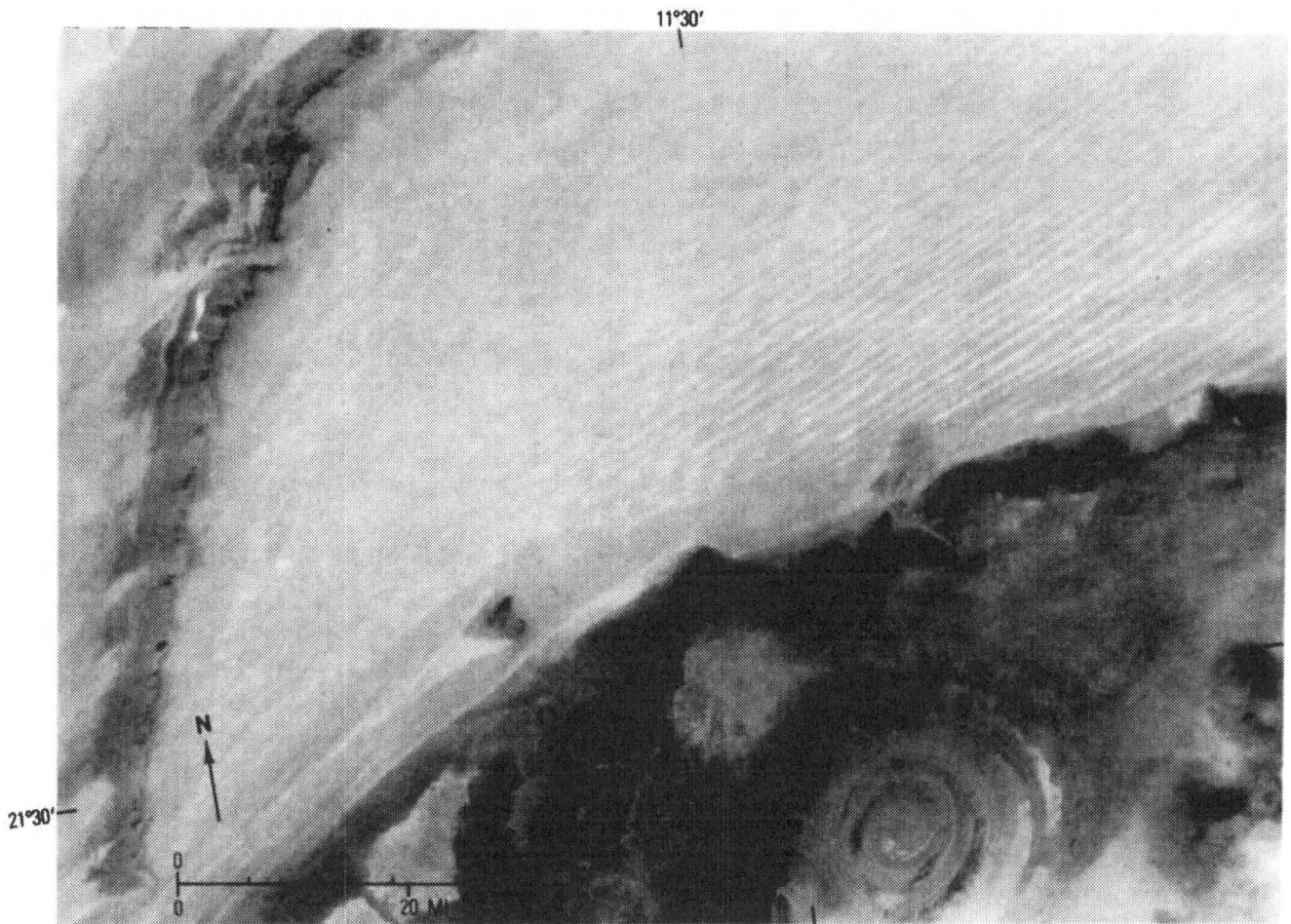

COMPOUND LINEAR DUNES IN THE MAKTEÏR SAND SEA north of the Er Richat structural dome, Mauritania. Landsat imagery E1103–10434. A basin filled with eolian sand is bounded on the west by hogback ridges. Numerous sand streaks and small linear dunes extend downwind (southwestward) from the hogbacks. (Fig. 192.)

196). Areas of active sand as seen on Landsat imagery (fig. 193) are mostly north of the 200-mm (8-in.) isohyet shown in figure 188. Considerable vegetation is visible on imagery of dunes south of this line (fig. 191).

Direction and Amount of Sand Drift

Annual resultant directions of sand drift (RDD's) at most stations in Mauritania are toward the southwest or south(fig. 188). Annual sand roses for these stations indicate considerable directional variability of wind during the year(fig. 188). This variability is cyclic and is reflected in seasonal shifts in resultant drift directions at all stations in the region. These shifts follow trends in the surface wind circulation just described; they produce resultant drift directions generally toward the west and the southwest in the winter (fig. 197A) and toward the south-southwest, south, or southeast during the summer (fig. 197B).

The cyclic swing of resultant drift directions during the year occurs most strongly in the southwestern, heavily vegetated part of the region (fig. 198). Stations south of approximately 18° N. experience reversals in resultant drift directions during July–September that may be caused by the southwest monsoon (fig. 199), but the reversal is usually not of sufficient strength or duration to have much effect on potential sand movement. Moreover, the reversal occurs during the time of maximum rainfall in the south, which tends to further reduce its effect.

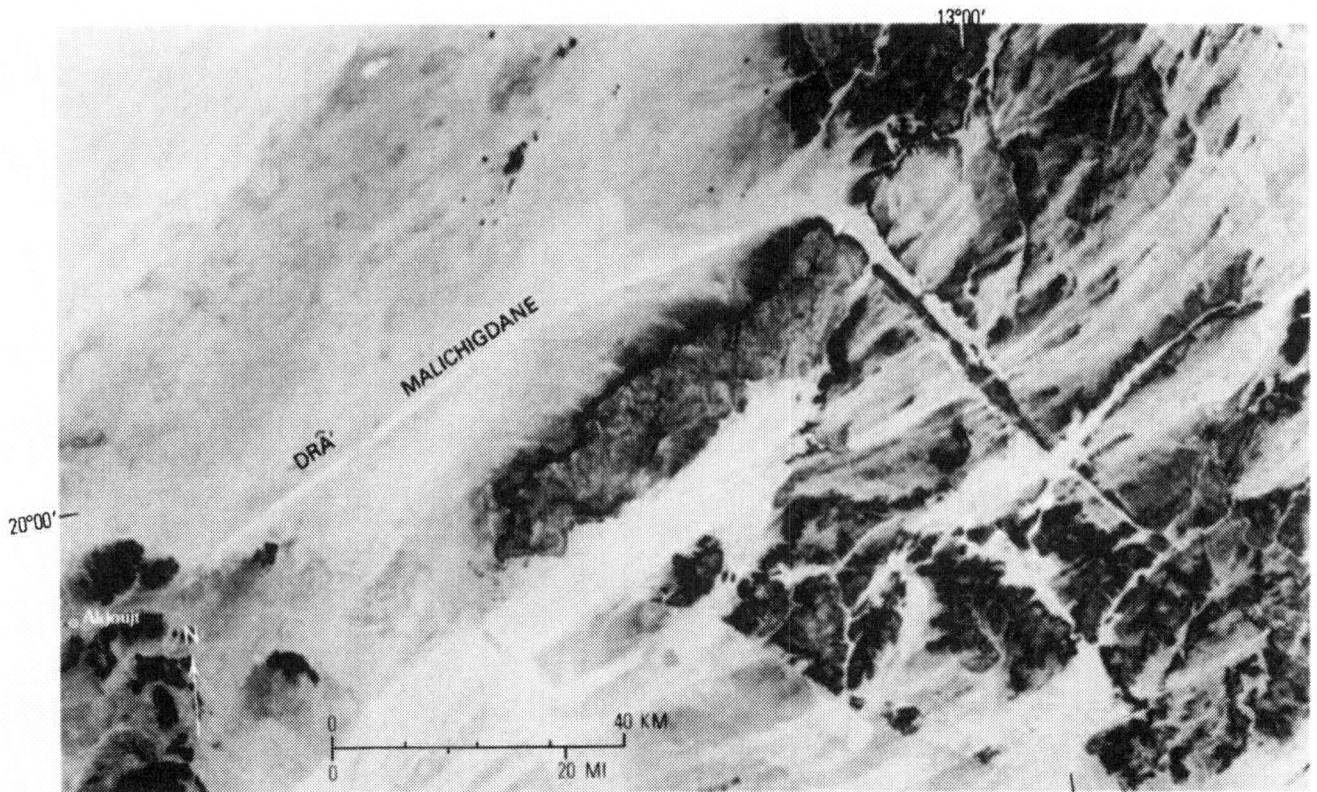

SOLITARY LINEAR DUNE (DRÂ´ MALICHIGDANE) between Atar and Akjoujt, Mauritania. Landsat false-color imagery E1140–10500. Relatively small elongate fields of crescentic dunes occur south of the linear dune. The intense yellow in the false-color imagery indicates that the sand is bright red and lacks vegetative cover (which appears red on false-color imagery); contrast with vegetated (red) linear dunes in figure 191 in southern Mauritania. (Fig. 193.)

Average drift potential of nine stations in the region (excluding Port-Étienne, which has an anomalously high drift potential) is 394 VU (vector units, chapter F), an intermediate value compared to other deserts of the world (chapter F, table 16). If Port-Étienne (2,232 VU) is included in the computations, the average drift potential is 569 VU, and the region as a whole is considered to be a high-energy desert. Total drift potentials and ratios of the resultant drift potential (RDP) to the total drift potential (DP) are highest in the northern areas and lowest in the southern areas (fig. 188). RDP/DP decreases regularly toward the southwestern part of Mauritania, as a result of the greater directional variability of the wind in this region. The strongest potential sand-moving season at some stations, such as Port-Étienne and Bir Mogreïn (Fort-Trinquet), occurs in June, when the western Sahara lies in the zone of strong winds between the Azores high and the thermal low over the Sahara (fig. 195,

200). A second potential sand-moving season in Mauritania occurs at some stations during winter (January and February), when anticyclonicity in north Africa is most strongly developed (fig. 200).

Distribution of the elongate sand seas and linear dunes of the western Sahara agrees in general with the annual resultant drift direction of sand computed from records of present-day winds. The agreement suggests that major changes in wind directions have not occurred since emplacement of the sand seas, thought to be about 18,000 years before the present (Grove and Warren, 1968, p. 196). The anomalous trends of some linear dunes (oblique to the trends of the sand seas in which they occur) may result from the action of certain effective components within the general wind regime, in the manner described by I. G. Wilson (1973, p. 97) and by Warren (1970, 1972) for alinements of dunes in northern and southern parts of the Sahara.

Sahel and Southern Sahara of Mauritania, Mali, and Niger

Summary of Conclusions

LINEAR DUNES, CRESCENTIC DUNES, and elongate sand sheets occur in the Sahel and southern Sahara of northern Africa. Most of these features are at least partially vegetated. Two sets of dunes, with different orientations, occur along the Niger River near Timbuktu. One set of closely spaced, narrow linear dunes with sharp crests is alined at N. 67° E., in approximate agreement with the resultant drift direction of the northeast trade winds. More widely spaced, eroded ridges of indeterminate type form a second set in the same region and are oriented approximately east-west.

Compound crescentic dunes in the Aoukâr sand sea have unusually straight segments with slipfaces toward the west-southwest. Very large barchans with slipfaces toward the southwest or west occur in the Vallée de l'Azaouk in the eastern part of the region. Crescentic ridges there, however, have slipfaces toward the northwest. Broad sand "streaks" extend from east-northeast to west-southwest across the entire middle Niger region, approximately parallel to the resultant drift direction computed from present-day winds.

Most of the region studied averages more than 100 mm (4 in.) of rainfall per year. Isohyets show that much of the region lies in the Sahelian climatic zone, where average yearly rainfall may be as much as 600 mm (24 in.) per year; the rest of the region lies in the sub-Saharan zone, where average annual rainfall ranges from 100 to 250 mm (4 to 10 in.) per year (Hills, 1966, fig. 15, table 15).

Present-day effective winds come mainly from the northeast or southwest throughout the Sahel. Drift potentials vary widely from station to station but are generally lowest in the south. Highest drift potentials at most stations occur in June, but some stations also have a winter sand-moving season. Eolian landform dimensions, orientation, degree of weathering and erosion, and relationship to un-

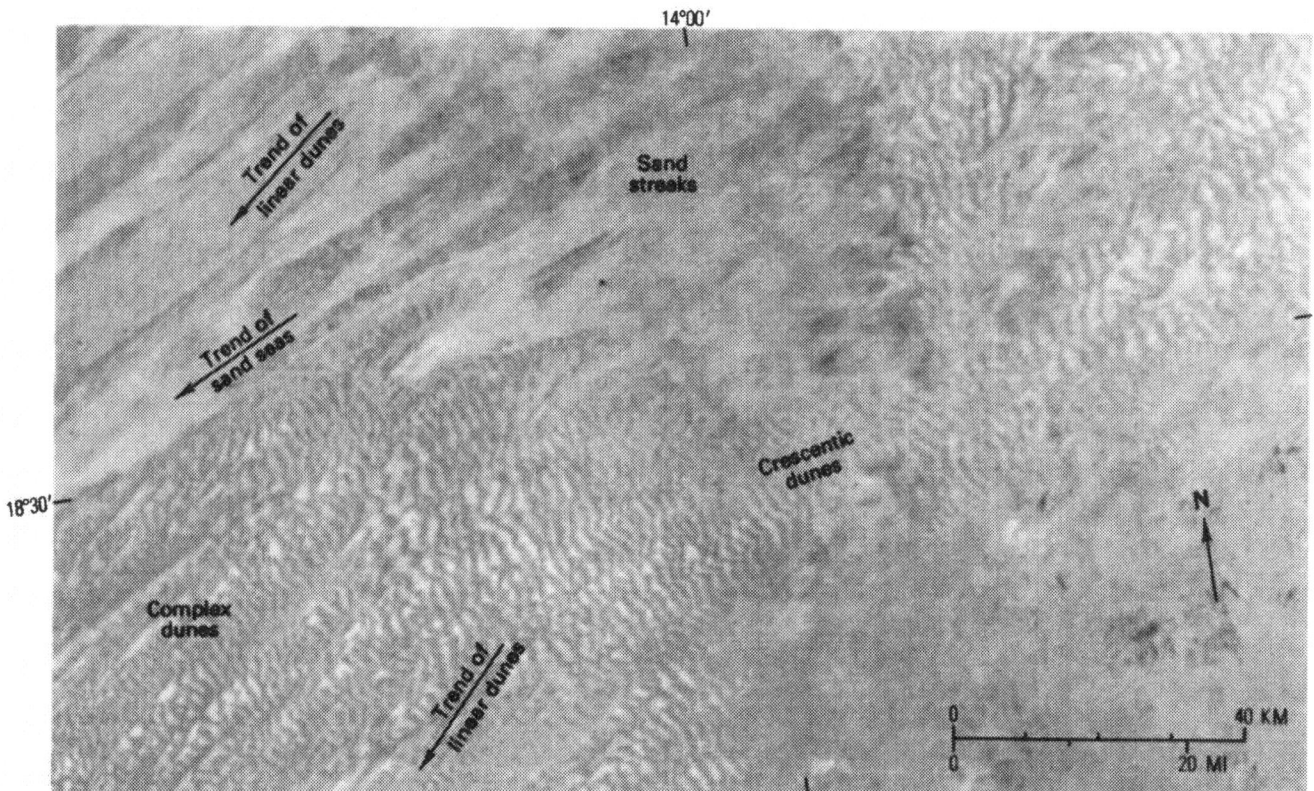

COMPOUND CRESCENTIC DUNES IN THE AOUKÂR SAND SEA, Mauritania. Landsat imagery E1140–10502. This field is directly south of the area shown in figure 193. (Fig. 194.)

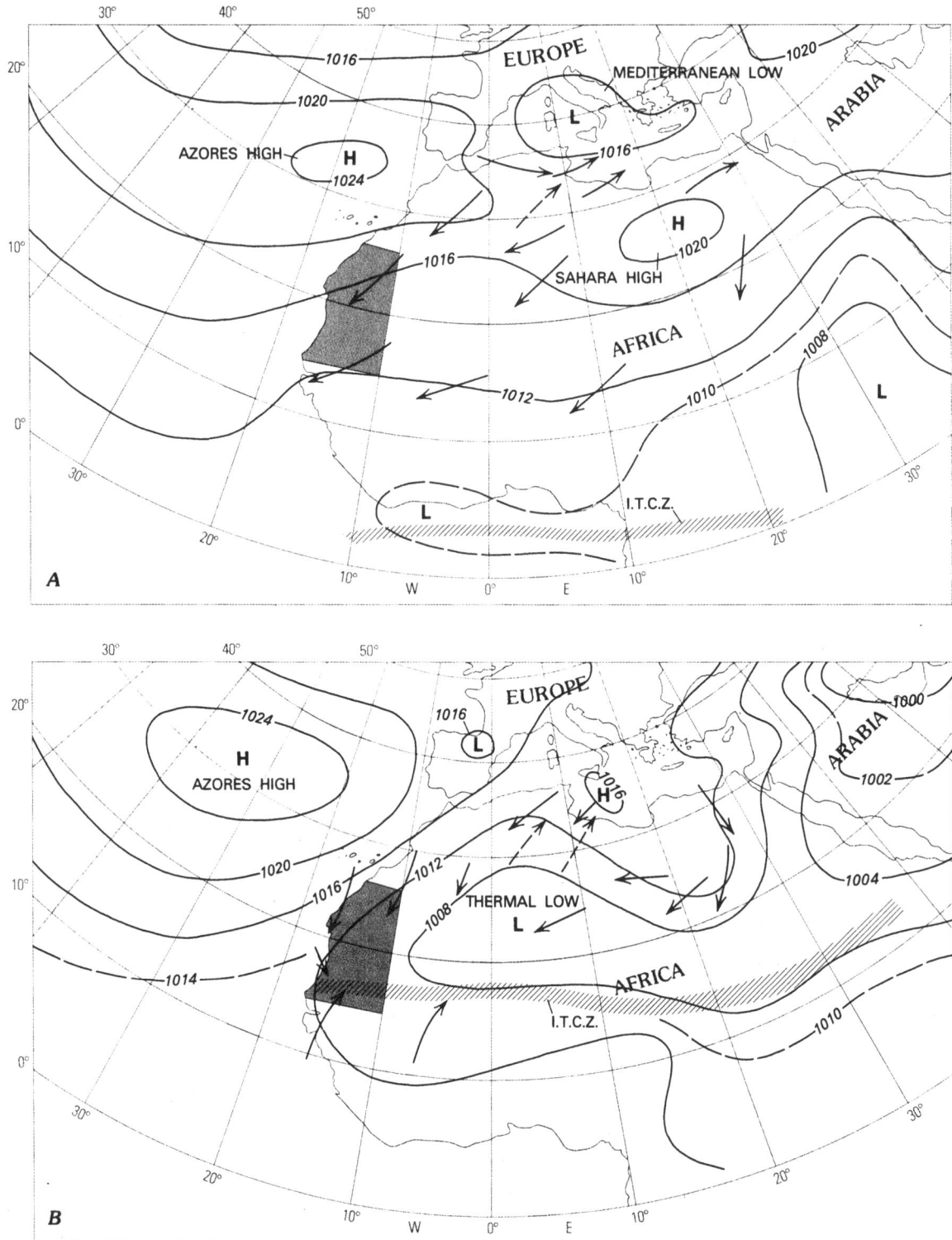

SEA-LEVEL MEAN-PRESSURE CONTOURS, in millibars, of the Sahara and adjacent regions in (A) January and (B) June. Average seasonal trends in surface wind flow indicated by arrows. Region of western Mauritania included in this study is shown by dark pattern. The position of Intertropical Convergence Zone (I.T.C.Z.) indicated by pattern is modified from Crutcher and Meserve (1970). (Fig. 195.)

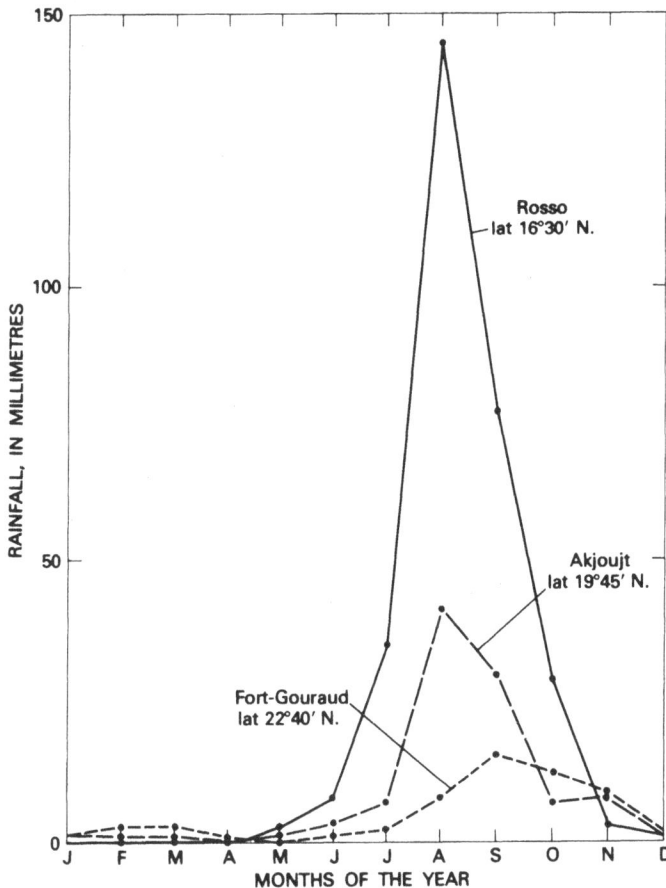

VARIATION IN RAINFALL DURING THE YEAR at Rosso, Akjoujt, and Fort-Gouraud, Mauritania. Rainfall is greater and begins earlier in the summer at the more southern stations. (Fig. 196.)

derlying topography — as observed on Landsat imagery of the sand seas north of the inland delta of the Middle Niger River and in the Vallée de l'Azaouk area — are interpreted as evidence of more than one episode of aridity and eolian activity.

Introduction

Landsat imagery has been used to map and interpret, with the aid of Skylab photographs, the distribution and morphology of eolian sand along the southern border of the Sahara (fig. 201). Severe drought and encroachment of eolian sand in that region are thought by some workers to be caused by poor agricultural and grazing practices but are attributed by others to changing climate.

Published reports (in French) of eolian features in the region are by Urvoy (1935, 1942), Monod

(1958), Tricart (1959, 1965), Dresch and Rougerie (1960), Daveau (1965), Boulet, Guichard, and Vieillefon (1971), Leprun (1971), and others. Other papers (in English) are by Grove (1958), Prescott and White (1960), Grove and Warren (1968), and White (1971). A report on eolian landforms observed on Skylab photographs includes a discussion of the Middle Niger River region (McKee and Breed, 1977). Useful maps of the region are U.S. Department of Commerce, U.S. Air Force Operational Navigation Charts J−2, J−3, K−1, and K−2; and U.S. Army Map Service, Series 1301, NF 30, NE 30, and NE 31; all maps are at a scale of 1:1,000,000.

Review of Previous Work

The Aoukâr sand sea ("Awkar" of Monod, 1958) is an area of compound crescentic dunes in southern Mauritania (the western part was described earlier in this chapter). (See also fig. 188.) Dunes in the eastern part of the Aoukâr sand sea are described (Grove and Warren, 1968, p. 197−198) as a sea of "massed barchans," "unusually rectilinear transverse ridges" 40−50 m (130−165 ft) high, oriented north-northwest to south-southeast (with slipfaces toward the southwest) and covered with bushes, grasses, and trees. Soil development and intermittent gullying of the Aoukâr dunes is interpreted (Daveau, 1965; Grove and Warren, 1968, p. 198) as evidence of climatic oscillations in the Sahel.

Dunes in the inland delta region east of the Aoukâr sand sea, near Timbuktu (fig. 201), are described by Tricart (1959, 1965), by Boulet, Guichard, and Vieillefon (1971, p. 110−111), and by Grove and Warren (1968, p. 198−200). The oldest described set of dunes consists of ridges 10−20 m (about 32−65 ft) high and several kilometres long (Grove and Warren, 1968, p. 199) which formed locally by eolian winnowing of Niger River sands (Tricart, 1959, p. 335; 1965, p. 180−181) during a dry period, probably 20,000−15,000 years before the present (Boulet and others, 1971, p. 111−112). Reddening of the upper 2−3 m (7−10 ft) of the dune sands, and their erosion by the Niger River, occurred during a more humid period 5,500−5,000 years before the present (Grove and Warren, 1968, p. 199). Formation of "grey-brown and yellow dunes which are lower and much less continuous but sharper than the older ones" (Grove and Warren,

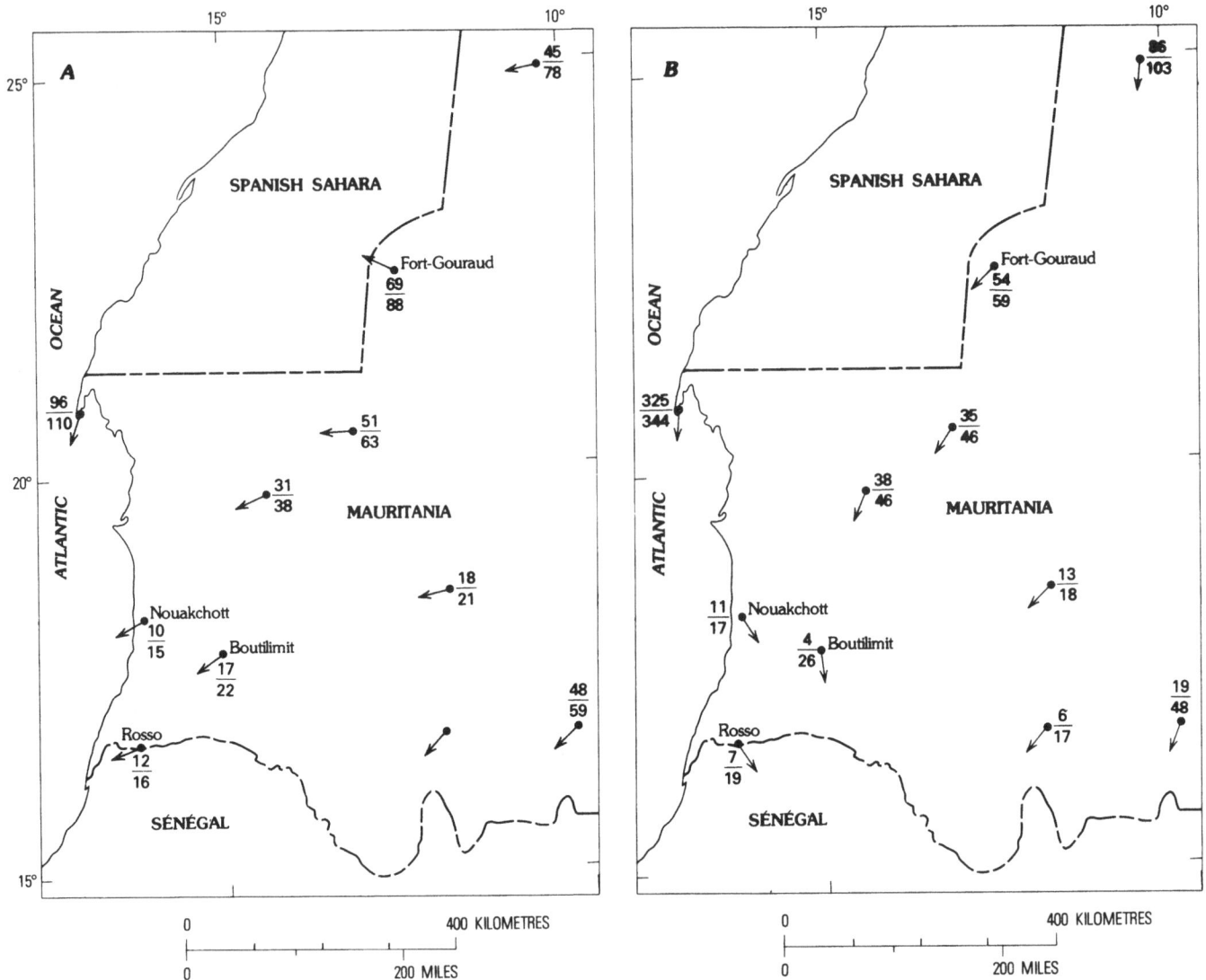

RESULTANT DRIFT DIRECTIONS (arrows), drift potentials (denominators), and resultant drift potentials (numerators) at weather stations in Mauritania during (A) February and (B) July. Terms are defined in chapter F. During February, resultant drift directions are toward the west or southwest over most of the region. During July, resultant drift directions are more toward the south and are roughly toward the southeast in the southern, vegetated parts of the sand sea near Rosso, Nouakchott, and Boutilimit. (Fig. 197.)

1968, p. 199) occurred later, during an arid period.

Early maps of sand seas in the Middle Niger River region between Mopti and Timbuktu (Urvoy, 1942, p. 48; Monod, 1958, p. 27; Grove and Warren, 1968, p. 199) do not distinguish between two dune sets subsequently recognized on Skylab photographs (McKee and Breed, 1977). Relationships observed on space photographs and Landsat imagery are thus difficult to reconcile with work of previous authors.

Crescentic dunes described as "great swellings" several hundred metres wide are reported in the Vallée de l'Azaouk, an abandoned river valley in-

cised into lateritic plateau surfaces east of the great middle bend of the Niger River, near Ménaka (fig. 201). These dunes are assigned to a "pre-Nouakchottian eolian episode," about 7,000–6,500 years befor the present (Boulet and others, 1971, p. 112–113). Other sets of dunes in the area between Tillabéry and Bourem are mainly compound linear dunes (Prescott and White, 1960, p. 200–201). Encroaching eolian sand sheets in the Vallée de l'Azaouk area are described as flat, featureless plains of drift sand from the Sahara that have filled and blocked most river channels (White, 1971, p. 70).

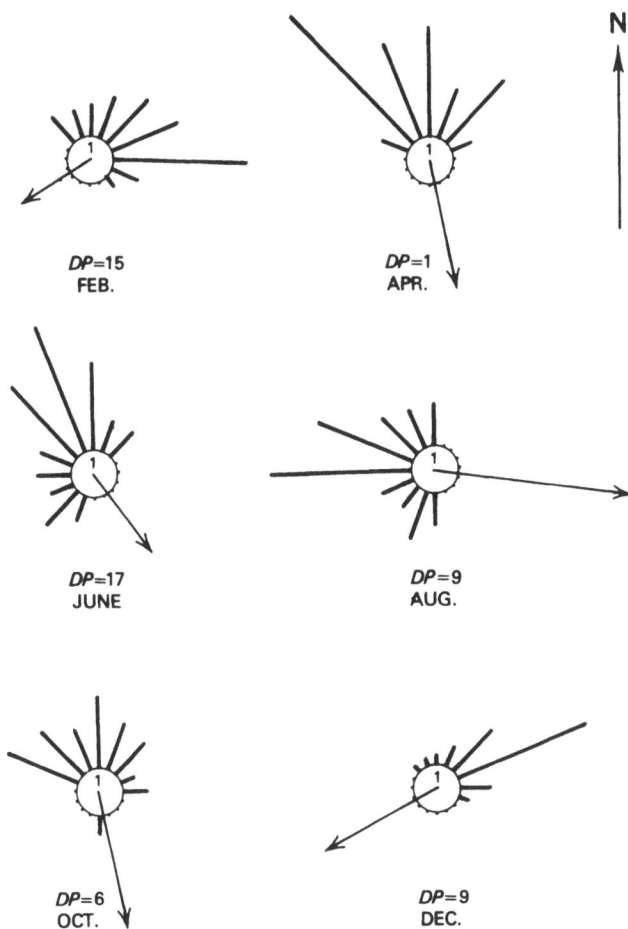

N

DP=15
FEB.

DP=1
APR.

DP=17
JUNE

DP=9
AUG.

DP=6
OCT.

DP=9
DEC.

SAND ROSES FOR 6 MONTHS FOR NOUAKCHOTT, Mauritania (fig. 188). Resultant drift direction (arrow) swings cyclically toward the southwest in February, toward the east-southeast in August, and then back. Southwest winds of the monsoon season do not have much effect at this station (note August sand roses) because Nouakchott is too far north (lat 18°07' N.). Drift potential, in vector units (chapter F), given for each month. (Fig. 198.)

Interpretation of Landsat Imagery

Measurements of Landsat transparencies of compound crescentic dune ridges in the Aoukâr sand sea of Mauritania show that these dunes are composed of slightly crescentic segments that range in horn-to-horn width from 2.2 to 9.6 km (1.4–5.9 mi) (average, 4.1 km or 2.5 mi). The dune ridges are spaced an average 1.71 km (1 mi) apart (crest-to-crest). The Aoukâr dunes have the straightest segments of any ridges of crescentic (transverse) type observed on Landsat imagery of the world's sand seas (chapter J, tables 38,39). The Aoukâr dunes are red (yellow on false-color imagery). Seasonal changes of color on Landsat imagery indicate that these dunes are vegetated.

East of the Aoukâr sand sea, between Mopti and Timbuktu (fig. 201), is a field of much eroded dune ridges of indeterminate type (which appear grayish-pink on imagery) that are oriented nearly east-west across the course of the Niger River. Braided river courses are adjusted to the pattern of these parallel dune ridges, thus forming a trellislike arrangement in which numerous tributary streams and narrow lakes occupy interdune spaces (fig. 202 B).

In contrast to the set of eroded dune ridges are fields of much narrower, sharper, brighter hued, and differently oriented linear dunes. The narrow dunes can be traced on imagery (fig. 202) from compound linear dunes in the Azaouâd sand sea, which occupies the southern Sahara north, west, and east of Timbuktu (Monod, 1958), southward into the Lac Niangay area. These narrower dunes strike west-southwest across parts of the other dunes near Lac Niangay and also across some of the trellis-patterned drainage (fig. 202B). Superposition of the narrow dunes over the east-west set, to which the Niger River had become adjusted is interpreted as evidence of more than one episode of large-scale eolian activity.

East of Bourem and Gao in the Vallée de l'Azaouk region (fig. 201, 202) are numerous individual crescentic dunes, probably barchans, commonly 1 km (0.15 mi) or more from horn to horn, which are clearly visible against the dark laterite plateau surfaces recorded on Landsat transparencies. These are among the largest individual dunes of crescentic type that the authors have observed on Landsat imagery and are probably compound barchans (chapter J). To the south and west in the Vallée de l'Azaouk region are fields of compound crescentic dune ridges similar in form to those of the Aoukâr sand sea farther west but with slipfaces to the northwest, rather than to the southwest.

Broad sheets of pale-red sand strike west-southwest across the entire Middle Niger region (McKee and Breed, 1977) from at least long 5° E. to long 10° W. (fig. 201). Numerous sand-free spaces between the sand "streaks" expose underlying topography, including remnants of abandoned drainages. Triangular-shaped lakes, such as Lac Faguibine and Lac Niangay, occupy depressions in some of the

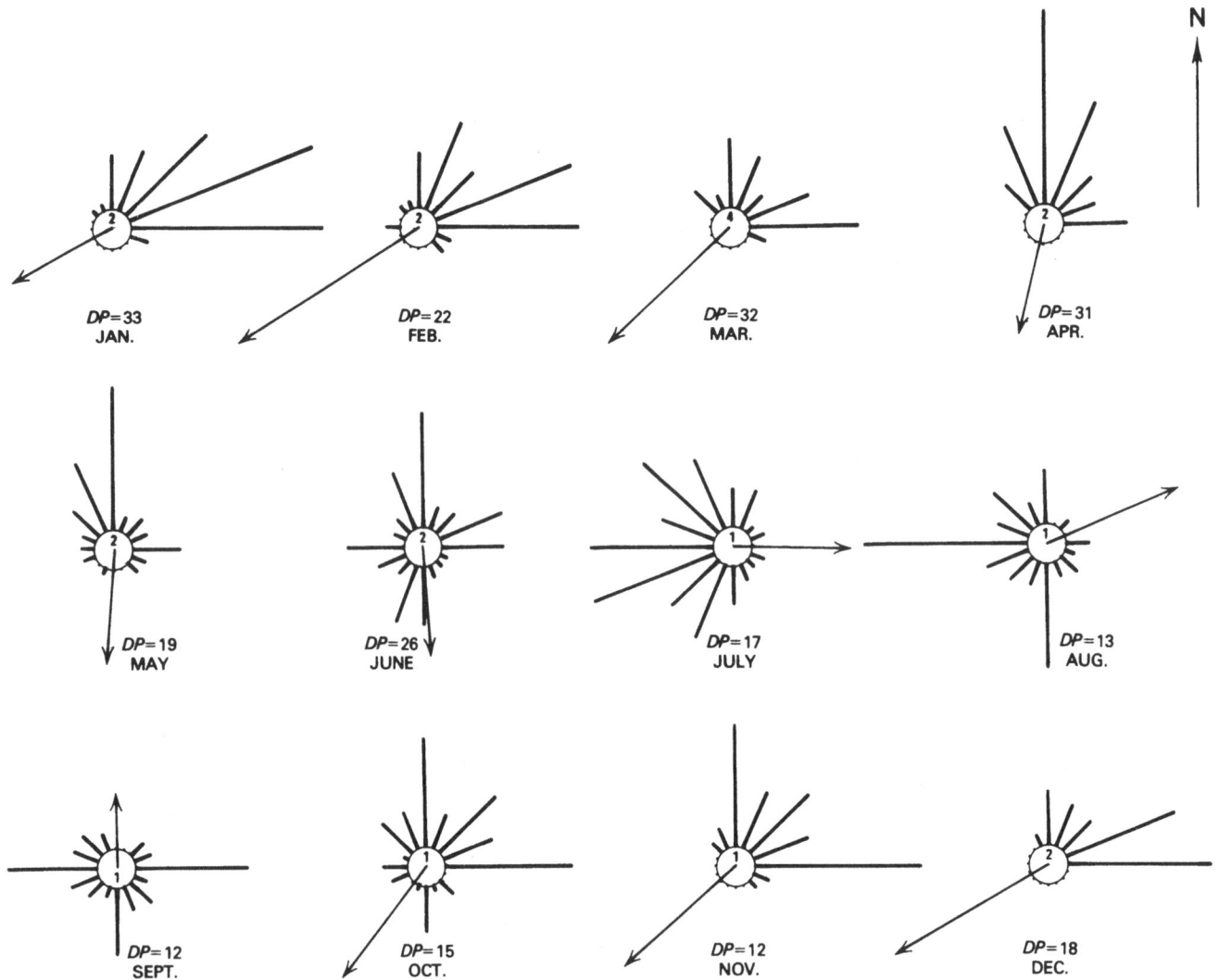

SAND ROSES FOR 12 MONTHS FOR BOUTILIMIT, Mauritania. This station experiences a cyclic swing of the resultant drift direction (arrow) during the year. It also experiences a reversal in effective wind directions during summer. The reversal is a result of westerly through southwesterly winds which occur when the Intertropical Covergance Zone has moved farthest north. Sand roses for Fort-Gouraud, which exemplify the pattern of winds in regions farther north, are shown in chapter F, figure 97B. Drift potential, in vector units (chapter F), is given for each month. (Fig. 199.)

sand-free areas. The sand sheets, like the narrow linear dunes associated with them, extend into and across some of the old eroded dune ridges near the Niger River (fig. 202B).

Surface Wind Flow and Precipitation

Climatic zones of the Sahel and southern Sahel in southern Mauritania, Senegal, Upper Volta, Mali, Niger, and Chad (Hills, 1966, fig. 15–1) are influenced by circulation from the Sahara and Azores highs to the north and equatorial trough (I.T.C.Z.)

to the south. As in the western Sahara (fig. 195), anticyclonic circulation from the Sahara and Azores highs results in flow of surface tradewinds from northeast to southwest across much of the region during winter, but surface wind flow during summer may be from the northeast or southwest. Winds from the southwest (southwest monsoon), resulting from cyclonic circulation associated with the Intertropical Convergence Zone (I.T.C.Z.), are strongest during June through August. Because the monsoon advances from south to north, stations along the southern margin of the Sahara are affected most by the southwest winds. Greater frequency of the southwest winds at the more southern stations is apparent from the annual sand roses shown on figure 201.

Annual wind distributions at most stations are complex, reflecting the interaction of the pressure systems just described. During winter, when the region is dominated by anticyclonic flow from the northeast and east, wind distributions are narrowed, and winds are steadiest (fig. 203 A). During summer, however, interaction of the southwest monsoon and the northeast winds results in com-

plex wind distributions at most stations (fig. 203 B).

All of the Sahel and southern Sahara region, by definition (Hills, 1966, table 15), receives more than 100 mm (4 in.) of rainfall per year, as illustrated by isohyets (fig. 201). Most of this rain occurs during June–September, the period of the southwest monsoon (fig. 204). Rainfall increases from north to south, reflecting the greater influence of the monsoon at the more southern stations. Effects of the increased rainfall are evident on Landsat imagery, which show an increasing amount of vegetation

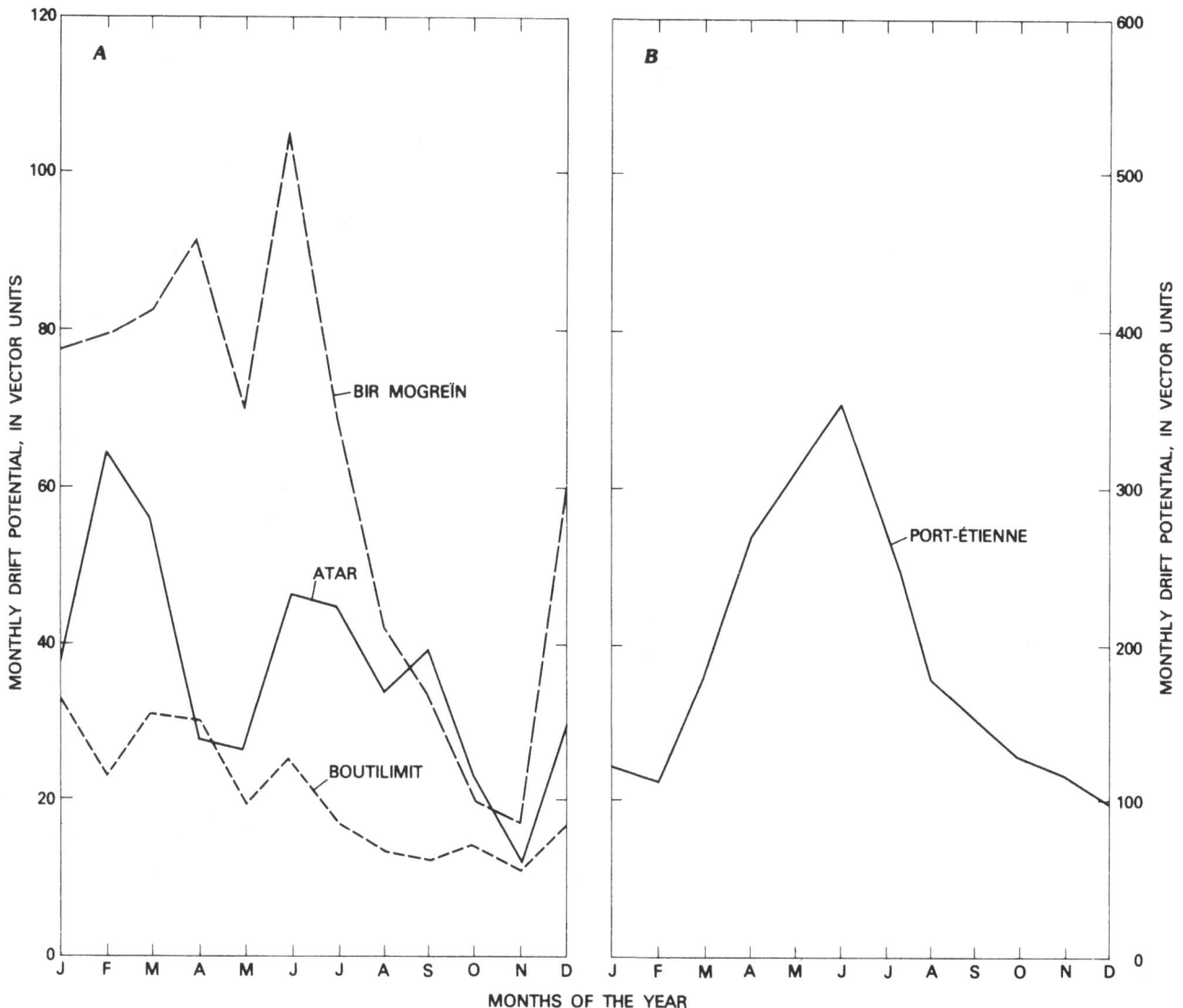

VARIATION IN DRIFT POTENTIAL during the year (A) at Boutilimit, Atar, and Bir Mogreïn (Fort Trinquet) and (B) at Port-Étienne, Mauritania. Bir Mogreïn and Port-Étienne experience strong sand-moving seasons in June, when the northern part of the region is in the zone between the Azores high and the thermal low in the Sahara. Some stations in the north, such as Atar, may, in addition, experience a sand-moving season in January–February, when anticyclonicity is strongest over the central Sahara (Sahara high). (Fig. 200.)

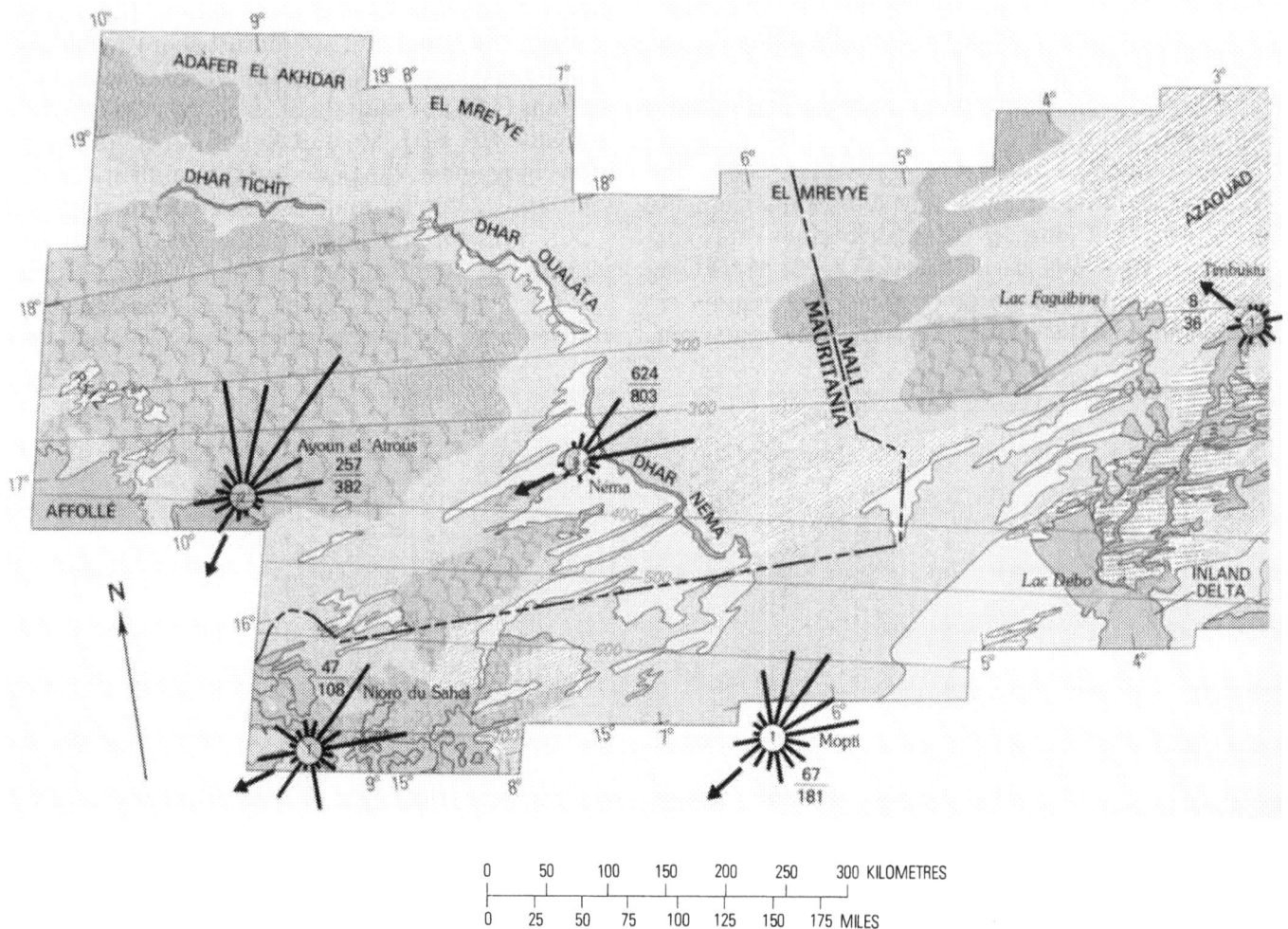

DISTRIBUTION AND MORPHOLOGY OF EOLIAN SAND in the Sahel and southern Sahara of northern Africa (above and facing page). Map based on Landsat imagery. Isohyets from Griffiths and Soliman (1972) and U.S. Naval Weather Service (1968). (Fig. 201.)

southward from areas near the 100-mm (4-in.) isohyet toward the areas of greater rainfall.

Direction and Amount of Sand Drift

Annual resultant drift directions[13] across most of the region shown in figure 201 are generally westward. Dunes and sand sheets observed on Landsat imagery are alined with resultant drift directions at several stations; for example, Kidal (off the map to the north), Tillabéry, Mopti, Nioro du Sahel, and Néma. Near other stations, such as Timbuktu and Gao, resultant drift directions are neither alined with nor transverse to the dunes. The discrepancy may result from the unusual conditions for observation at these stations. An unusually low drift potential at Timbuktu, for example (fig. 201), suggests that the station may be sheltered from the northeast

winds (chapter F, under methods of study, wind data available and limitations). Sand roses are not adjusted for varying effects of vegetation at different stations or for presumed dampening effects of rainfall on sand movement. Potentially effective winds from the southwest, shown on the sand roses for Timbuktu and Gao (fig. 201), blow across heavily vegetated areas and are commonly associated with rainfall. Thus, the true sand-moving effectiveness of the southwest winds, relative to effectiveness of the northeast winds (which blow across the barren Sahara), is probably less than implied by those sand roses.

Compound crescentic dune ridges with northwest-facing slipfaces, in the eastern part of the

---

[13] New terms are defined in chapter F and in the glossary.

EXPLANATION

LINEAR DUNES

Simple, short

Compound

CRESCENTIC DUNES

Simple barchanoid ridges

Compound

Barchanoid ridges

Barchans

SAND SHEETS AND STREAKS

BEDROCK OUTCROP OR BARRIER

ROCK OR GRAVEL DESERT PLAIN
WITH NO LARGE SAND DUNES

VEGETATED FLOODPLAIN OR MARSH

———200——— ISOHYET—Average annual rainfall, in millimetres

———————— BOUNDARY OF SAND SEA

(Fig. 201. —Continued.)

region, may be in agreement with the northward resultant drift directions suggested by the sand roses for Ménaka and Gao (fig. 201). The closely spaced linear dunes north of Timbuktu, oriented N. 67° E. (fig. 202), are in approximate alinement with modern desert tradewinds (northeast winds), the effects of which are reflected in the sand roses for Néma and ʻAyoun el ʻAtroûs farther west.

Drift potentials within the region studied vary widely possibly because vegetation in some localities may shelter stations from sand-moving winds. Some stations north of lat 16° N. record moderate or high drift potentials, for example ʻAyoun el ʻAtroûs, (382 VU), Néma (803 VU), and Gao (388 VU). Drift potentials may be higher in regions north of lat 16° N. than in more southern

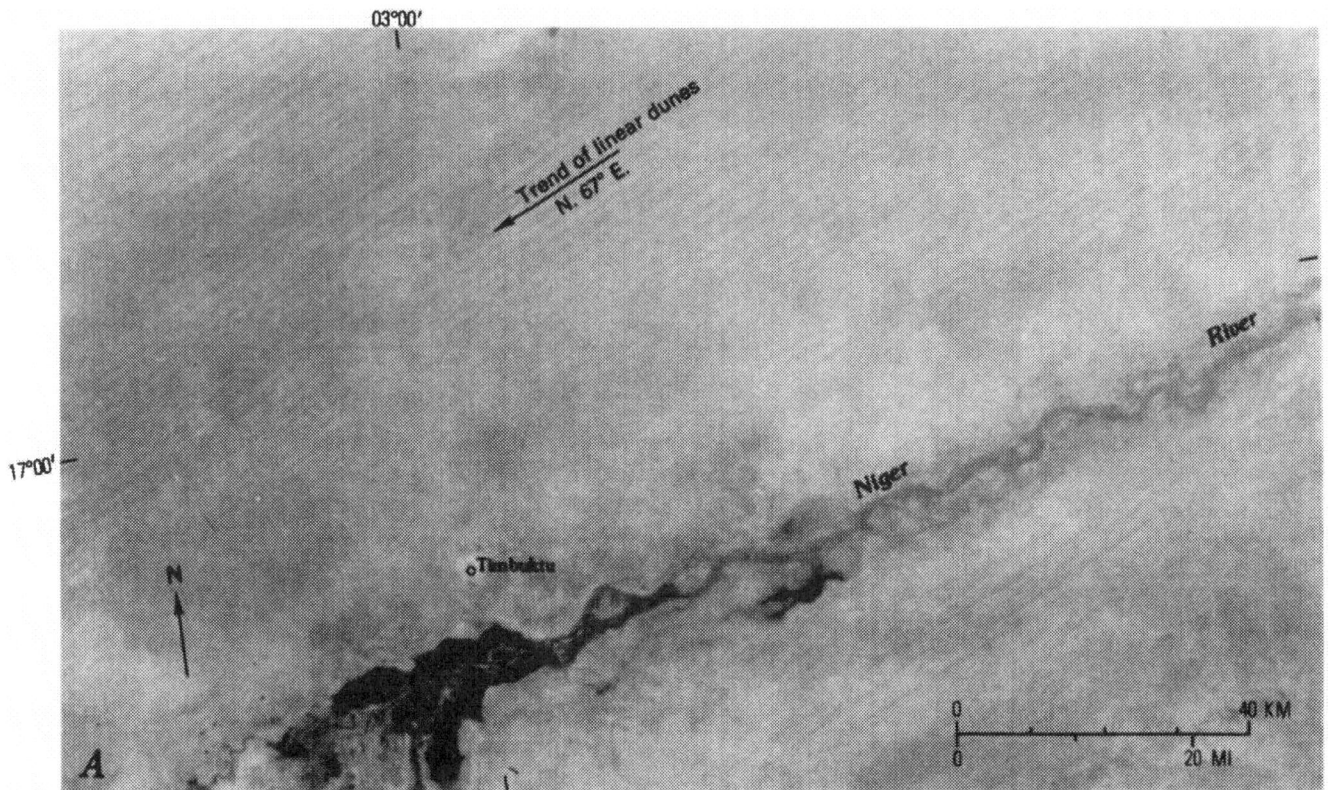

COMPOUND LINEAR DUNES OF THE AZAOUÂD SAND SEA (A) near Timbuktu (above). Landsat imagery E1096–10043. The Niger River runs from west to east. B, Two sets of dunes north of Mopti and south of Timbuktu, directly south of the area shown in A (facing page). Landsat imagery E1132–10052. The Niger River system is adjusted to the presence of large east-west dunes and is overridden on the east by sets of narrow linear dunes and sand sheets oriented like those to the north in A. Spaces between the large east-west dunes and some sand-free areas between the sand sheets are occupied by lakes. (Fig. 202.)

regions because the former are exposed longer to strong northeast winds than to the relatively weak southwest winds, or because more northern regions are less vegetated. Most stations studied have a potential sand-moving season in June, as shown in figure 205. Some stations also have a potentially strong winter sand-moving season; for example, 'Ayoun el 'Atroûs (fig. 205). Portions of the Sahel and southern Sahara, like the Thar Desert of India and Pakistan, generally are exposed to highest drift potentials prior to the onset of the rainy season and after an extended period of dryness.

Northern Sahara of Algeria, Tunisia, and Libya
Summary of Conclusions

SAND SEAS CONTAINING distinctive dune types extend across large parts of the northern Sahara. These sand seas consist mostly of large compound and complex star and crescentic dunes, as in the Grand Erg Occidental, the Grand Erg Oriental of Algeria, and the Ṣaḥrā' Marzūq of Libya. They are mostly confined to basins separated by plateaus and low mountain ranges. Sand seas of linear dunes, such as the Er Raoui, Erg-Iguidi, and 'Erg Chech, commonly extend for long distances across rock and gravel plains.

Large isolated star dunes are a common type in the northern Sahara. In the central and southern parts of the Grand Erg Occidental, star dunes have mean diameters which range from 0.7 to 1.7 km (0.4 to 1.1 mi) and on the downwind margin of the Grand Erg Oriental they have maximum diameters of 2.4 km (1.5 mi). Large star dunes also occur along the southern margin of the Grand Erg Occidental and grade northward into compound crescentic dunes in the interior of the sand sea. Star dunes in the Ṣaḥrā' Awbārī, Libya, and the Grand Erg Orien-

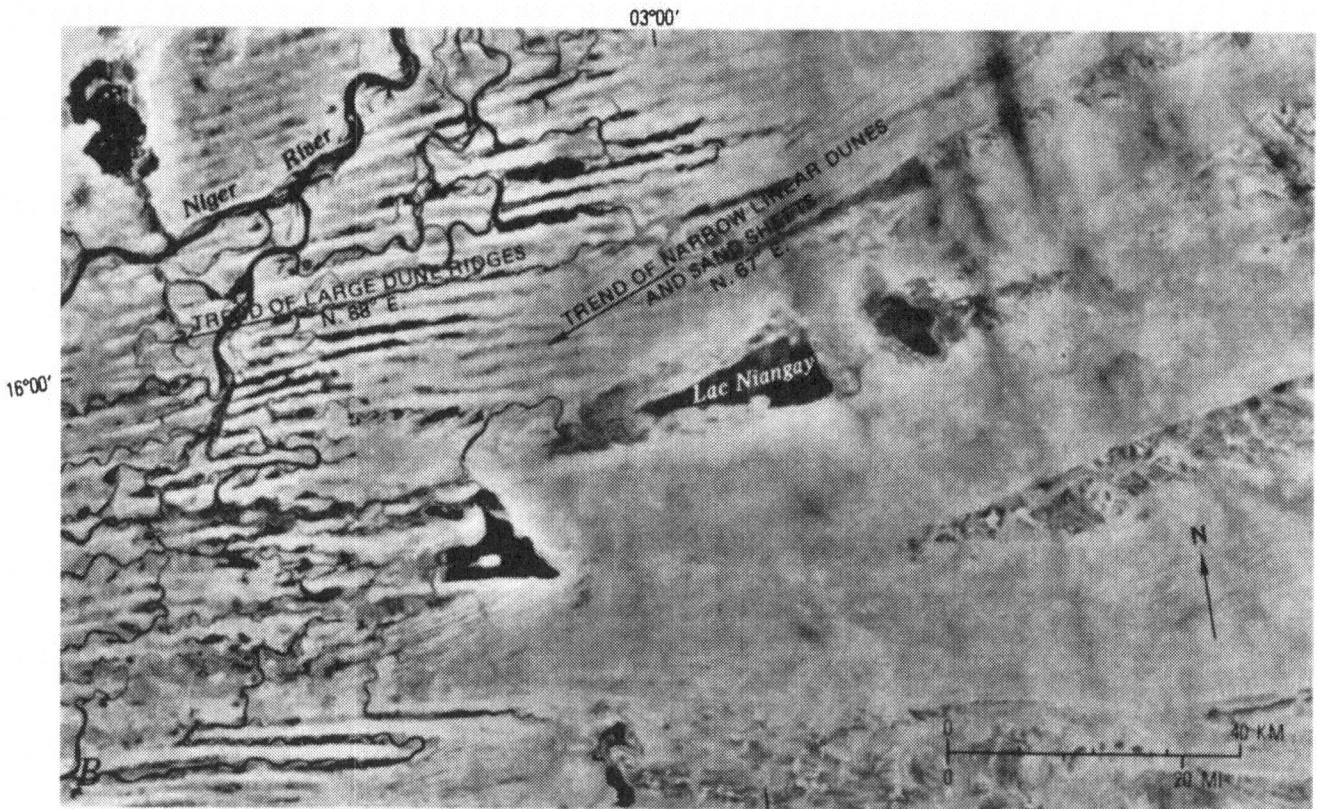

(Fig. 202. —Continued.)

tal, Algeria, are arranged in linear trends which cross.

Widely spaced (5.7 km, or 3.5 mi) compound linear dunes in the Er Raoui, Erg-Iguidi, and 'Erg Chech extend south-southwestward into northern Mali and Mauritania. Barchanoid ridge dunes form a network pattern in the northeastern part of the Grand Erg Occidental, mostly in Tunisia along the border with Algeria and Libya.

Sand seas in the Ramlat Zallāf and Ṣaḥrā' Awbārī of Libya include crescentic dunes, star dunes, and linear dunes. Large crescentic dunes, and isolated star dunes with an average horn-to-horn width (and spacing) of 2 km (3 mi) occur in the Ṣaḥrā' Marzūq, to the south in Libya. Fields of barchans extend in trains from the Ṣaḥrā' Marzūq southwestward across the Plateau du Mangueni to the Great Erg of Bilma in Niger.

Most of the northern Sahara receives less than 100 mm (4 in.) average annual rainfall. Complex intermediate- to high-energy wind regimes characterize the Grand Erg Occidental and the Grand Erg Oriental of Algeria. Very large star dunes occupy hundreds of square kilometres in these deserts.

These complex wind regimes reflect the interaction of winter westerlies, summer northeasterlies, and the southwesterly and northwesterly winds which result from cyclonic circulation and frontal passages. Annual resultant drift directions are toward the southeast at most stations in the Grand Erg Oriental.

Unimodal or bimodal high-energy wind regimes occur in southern Tunisia, a region characterized by a network pattern of barchanoid ridges.

Wind regimes southwards from lat 30° N. in western Algeria are generally unimodal or bimodal. Linear and crescentic dunes are common types in this region. Resultant drift directions are generally toward the southwest in the southern part of the Grand Erg Occidental and the 'Erg Chech.

Complex wind regimes occur in the Ramlat Zallāf and Ṣaḥrā' Awbārī of western Libya. Near Ghudāmis and Nālūt in northwest Libya, annual resultant drift directions are toward the northeast. To the southeast near Sabhā, and farther south near Bordj Omar Driss (Fort Flatters), annual resultant drift directions are toward the west or southwest.

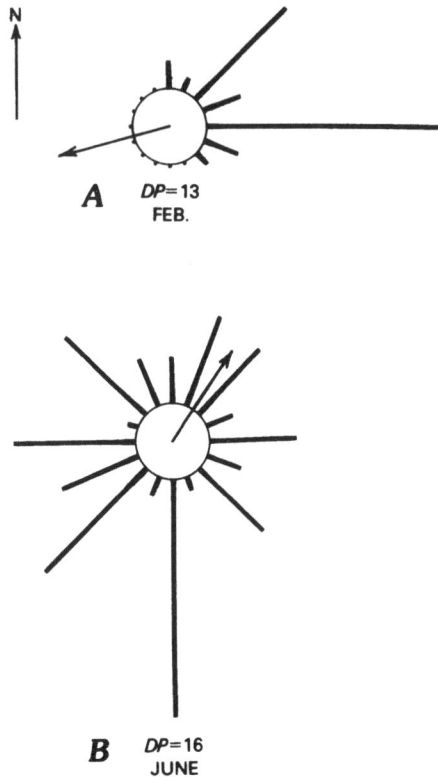

MONTHLY SAND ROSES FOR NIORO DU SAHEL, MALI, for (A) February, during the winter season of steady east-to-northeast winds, and for (B) June, when the southwest monsoon interacts with east-to-northeast circulation to produce a complex wind distribution. Arrowed vector represents resultant drift direction, in vector units. (Fig. 203.)

On the average, intermediate drift potentials occur in the deserts of Algeria, whereas high drift potentials occur in the deserts of northwest Libya. Highest drift potentials occur in March or April at most places in the northern part of the area described but occur in June at some southern stations.

Introduction

Landsat imagery has been used to map and describe the distribution and morphology of eolian sand in several main sand seas of the northern Sahara. The main sand seas in Algeria, Tunisia (fig. 206), and western Libya (fig. 207) are in structural and topographic basins, separated by plateaus and low mountain ranges. Meteorological data were obtained from numerous weather stations in and around the sand seas. Sand roses on the maps (fig. 206, 207) show potential sand-moving power of surface winds, and isohyets are shown for areas that receive 100 mm (4 in.) or more rainfall per year.

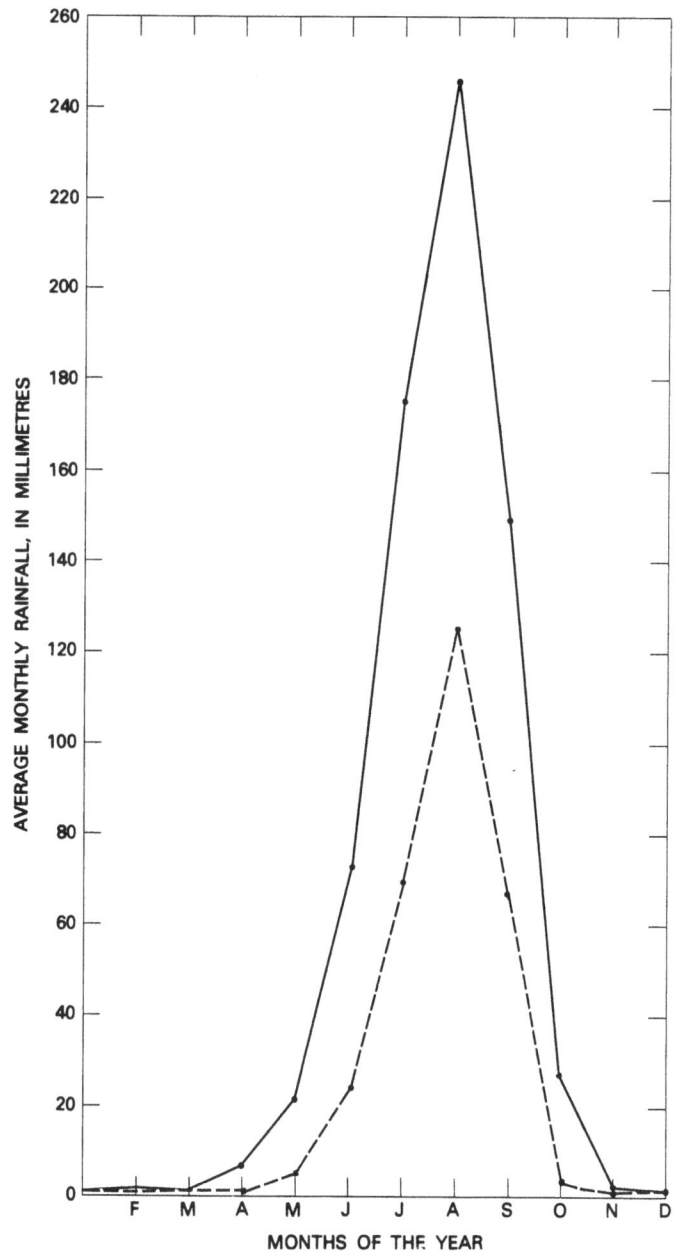

VARIATION IN RAINFALL DURING AN ENTIRE YEAR at Gao, Mali, lat 16°15′ N. (dashed line) and Nioro du Sahel, Mali, lat 15°14′ N. (solid line). The southwest monsoon is farthest north in August, resulting in greatest rainfall at that time. Most southern stations receive greater rainfall than stations in the north because they are subject to the southwest monsoon for a longer period. (Fig. 204.)

The main geologic studies of the sand seas in Algeria and northwestern Libya are by Alimen Doudoux-Fenet, Ferrere, and Palau-Caddoux (1957), Alimen (1965), Capot-Rey (1943, 1945, 1947a, b), and I. G. Wilson (1971a, 1972, 1973). Quaternary stratigraphy of a region that includes

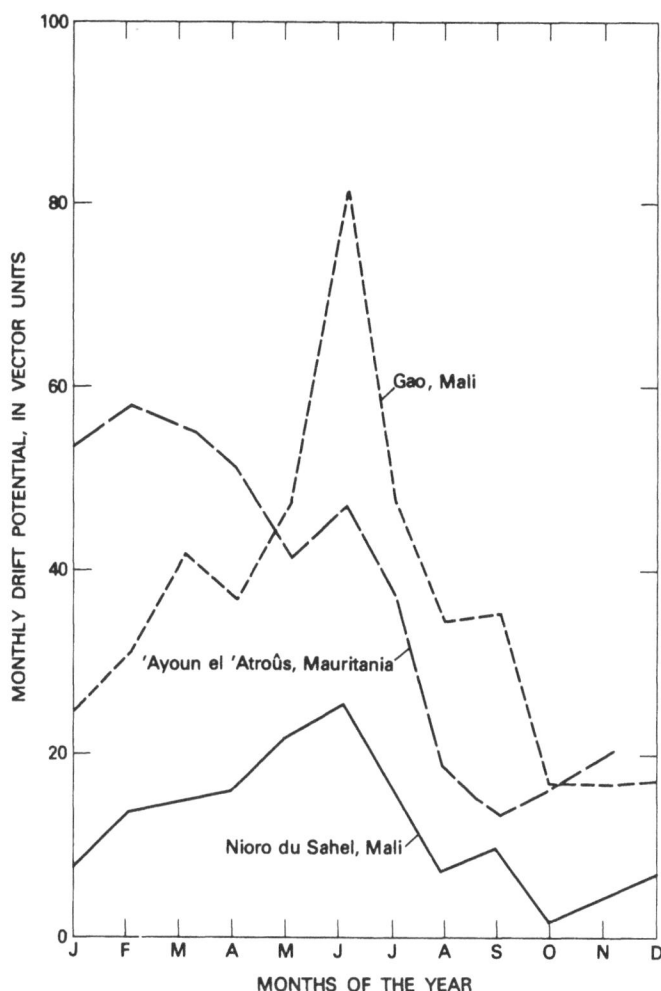

VARIATION IN DRIFT POTENTIAL during the year at representative stations in the Sahel. All stations have a sand-moving season in June; 'Ayoun el 'Atroûs has a sand-moving season in February (winter). (Fig. 205.)

and sand sheets in the northern Ṣaḥrā' Awbārī and Ṣaḥrā' Marzūq are mostly limited to reconnaissance papers (Capot-Rey, 1947b) and brief sections within reports on the geology or hydrogeology of the region (Conant and Goudarzi, 1967, p. 721–728; Jones, 1971).

Field studies of linear dunes in the Ramlat Zallāf are reported by McKee and Tibbitts (1964). Features of Libyan sand seas observed on Gemini space photographs are described by Pesce (1968). A physiographic map of northern Africa by Raisz (1952) shows the sand features at a scale of 1 inch equals 60 miles (1:3,801,600). Other useful maps of the northern Sahara are the U.S. Department of Commerce, U.S. Air Force Operational Navigation Charts H–2, H–3, H–4, G–1, G–2, J–3, and J–4; U.S. Army Map Service, Series 1301, NH 30, NG 29, NH 31, NG 31, NG 30, NF 31, NH 32, NH 33, NG 32, NG 33; and U.S. Department of Commerce, U.S. Air Force World Aeronautical Chart 541. All maps are at a scale of 1:1,000,000.

The Grand Erg Occidental and the 'Erg Chech, Algeria

Review of Previous Work

South of the foothills of the Atlas Saharien and north of the Plateau du Tademaït (fig. 206) is a topographic basin containing the Grand Erg Occidental. The sand sea, which covers about 103,000 km² (64,000 mi²) (Wilson, 1973, p. 86) is bounded on the west by the Oued Saoura, a large intermittent stream. The sand sea is partly separated from the 'Erg Chech (to the southwest) by low mountains, and separated from the Grand Erg Oriental (to the east) by plateaus.

Thirty-two samples of sand from large compound star dunes along the west margin of the sand sea, between Taghit and Kerzaz (fig. 206) were analyzed for grain-size distribution (Alimen and others, 1957). In the large star dunes grain size is unimodal, ranging from 0.20 to 0.22 mm. Sands of minor dunes, sand sheets, and interdune surfaces in the central part of the sand sea are bimodal, with a range in size from 0.18 to 0.26 mm. All dunes of the Grand Erg Occidental, according to Alimen, Doudoux-Fenet, Ferrere, and Palau-Caddoux (1957, p. 108–109, 112, 135–136, 167–169) and Alimen (1965, p. 282), were formed by eolian reworking of

the Grand Erg Occidental is summarized by Chavaillon (1964).

French and Arabic terms for various eolian landforms are shown in tables 26–35 in chapter J. Direct translation of Capot-Rey's nomenclature is complicated by his usage of the tern "sif" as a curved slipface on a dune of any type (Capot-Rey, 1945, p. 392–394, 400, fig. 1; Capot-Rey and Capot-Rey, 1948, pl. 1 (3)), whereas to most authors it means a linear dune.

Studies of dune forms and processes in central and southern Libya (Bagnold, 1933, 1941) have largely been in the Sarīr Kalanshiyū, at Ramlī al Kabīr and Ṣaḥrā' Rabyānah in the Libyan desert, and in the Great Western Desert of Egypt east of the region studied for this chapter. References to dunes

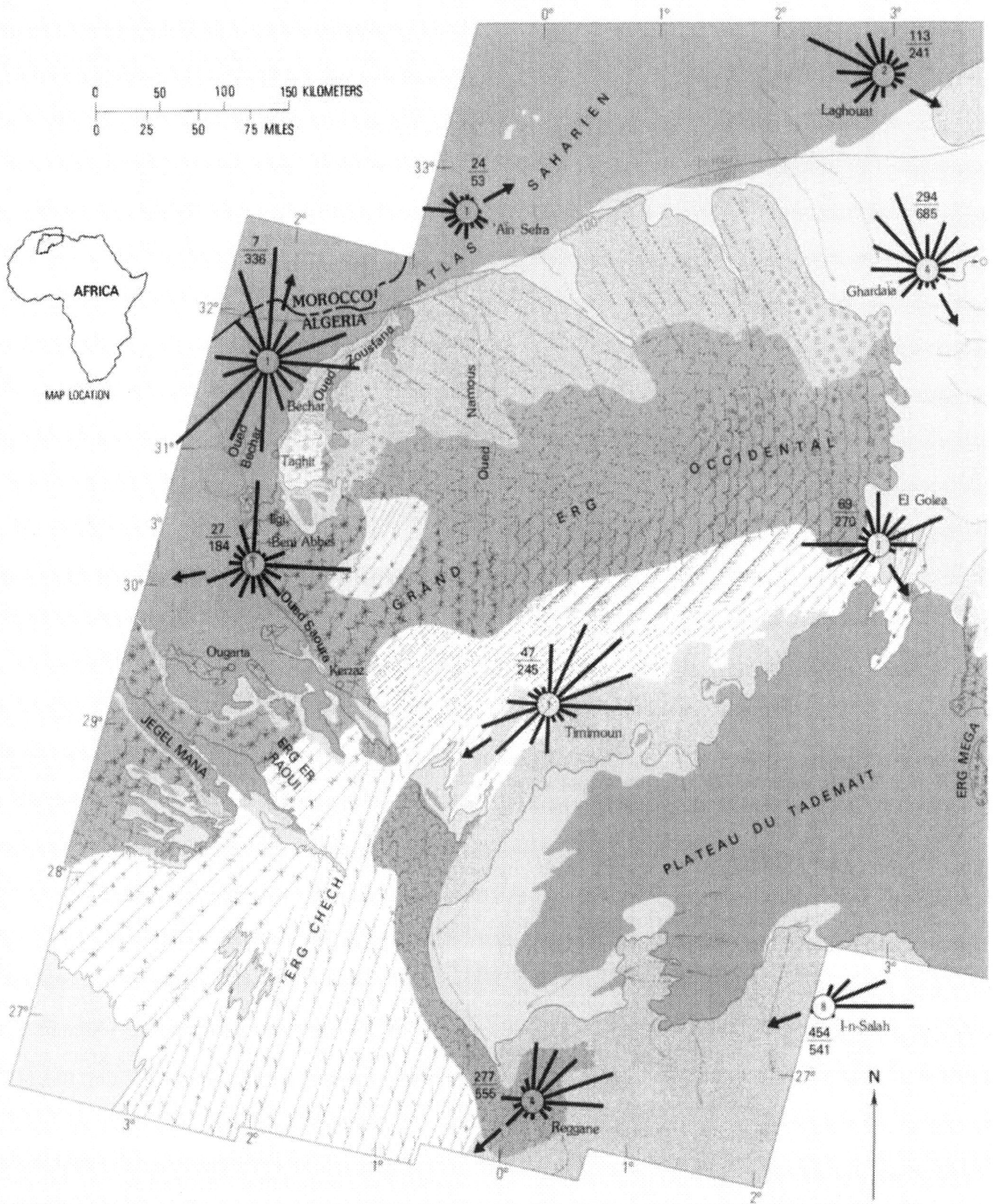

DISTRIBUTION AND MORPHOLOGY OF EOLIAN SAND in Algeria and western Tunisia. Map based on Landsat imagery. Isohyets from Dubief (1952). (Fig. 206.)

Mediterranean Sea

340 / 571

Biskra

Biskra 190 km
north of
Touggourt

113 / 282

44 / 404

El Oued

156 / 312

Qafşah (Gafsa)

Shatt al Jarid

53 / 152

Qābis (Gabes)

Touggourt

56 / 255

Ouargla

62 / 200

Hassi Messaoud

237 / 435

Remādah

TUNISIA
ALGERIA

TUNISIA
LIBYA

O R I E N T A L

E R G

G R A N D

60 / 658

Ghudāmis

LIBYA
ALGERIA

HAMADA   DE   TINRHERT

EXPLANATION

Compound
  Megabarchans
  Barchanoid ridges
  Discontinuous barchanoid ridges
  Barchanoid ridges with closed inter-
    dunes (fishscale pattern)
Complex crescentic ridges with
  Linear elements superimposed
  Dome-shaped dunes superimposed
STAR DUNES
  Simple, and some compound
  Compound, in linear chains
DOME-SHAPED DUNES
  Simple and compound

LINEAR DUNES
  Simple, short
  Compound
    Feathered
  Complex with
    Crescentic dunes super-
      imposed
    Star dunes superimposed
CRESCENTIC DUNES
  Simple barchanoid ridges

Compound, in linear chains
UNDIFFERENTIATED COMPLEX DUNES
SIMPLE SHEETS AND STREAKS
COMPOUND SHEETS AND STREAKS WITH
  LINEAR ELEMENTS
BEDROCK OUTCROP OR BARRIER
ROCK OR GRAVEL DESERT PLAIN WITH
  NO LARGE SAND DUNES
PLAYA
VEGETATED MARSH
ISOHYET—Average annual rainfall, in
  millimetres
BOUNDARY OF SAND SEA
INTERMITTENT DRAINAGE—Showing
  direction of flow

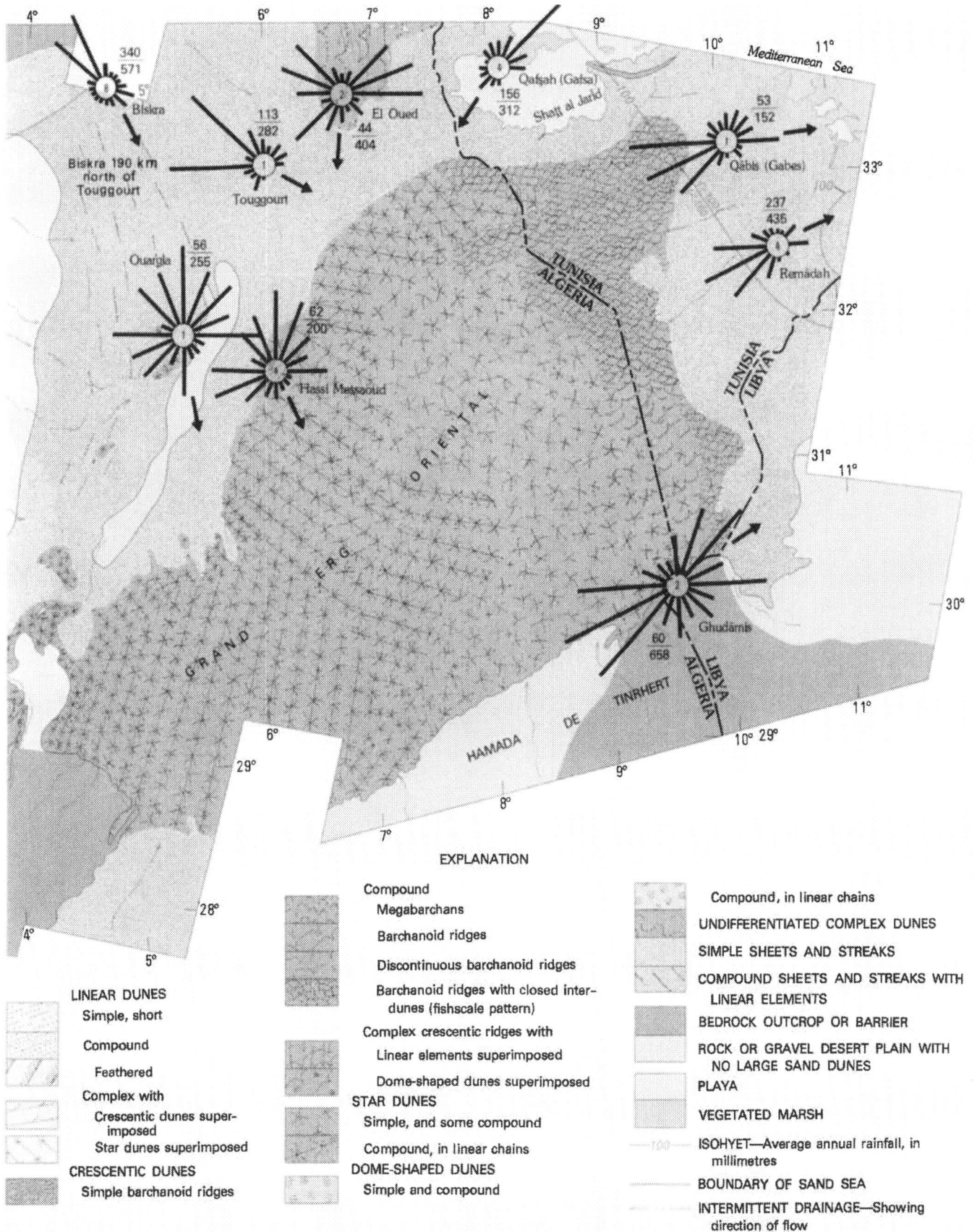

(Fig. 206. — Continued.)

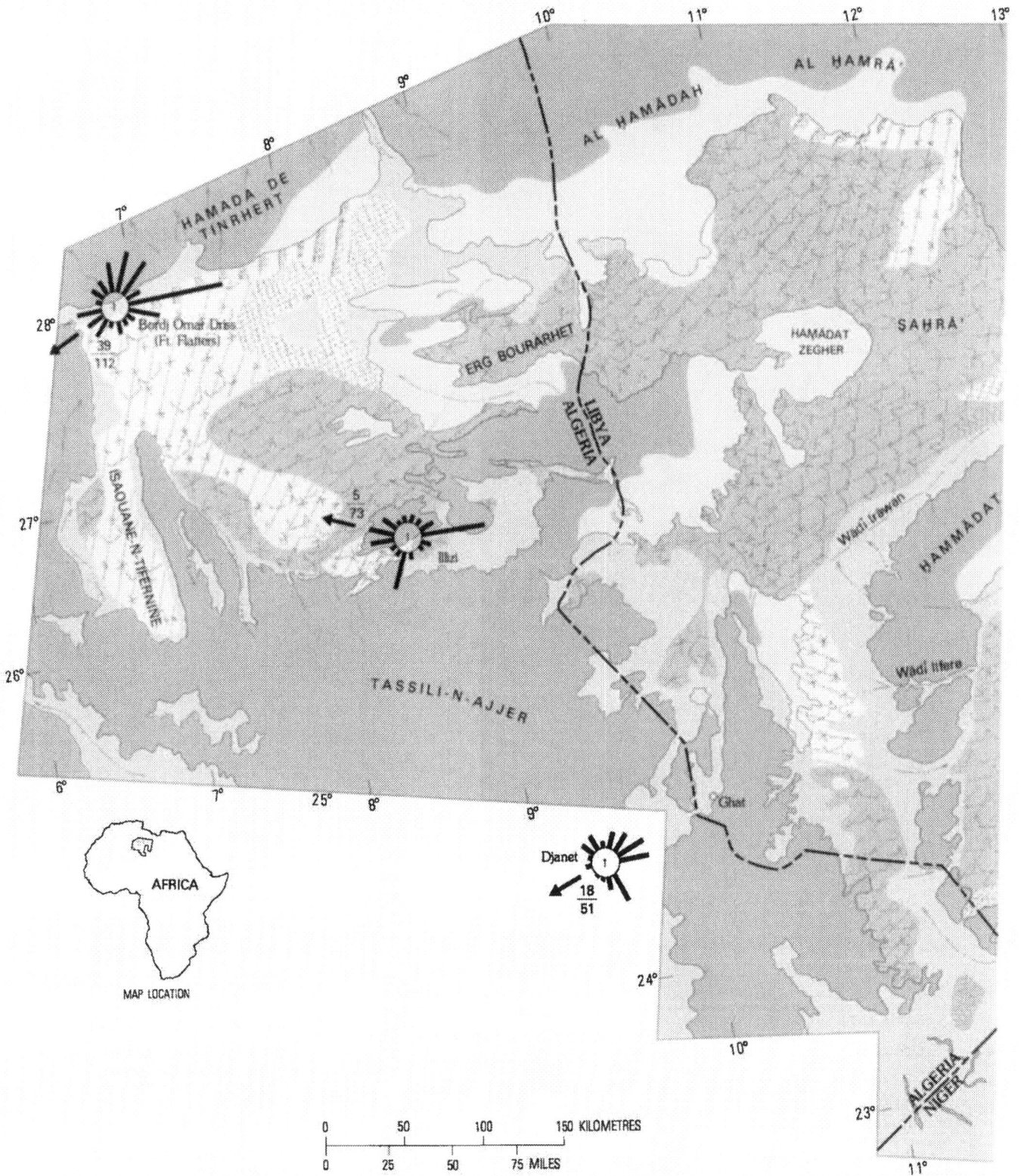

DISTRIBUTION AND MORPHOLOGY OF EOLIAN SAND in sand seas of northwestern Libya and southeastern Algeria. Map based on Landsat imagery. Absence of isohyets indicates that the entire region receives less than 100 mm annual precipitation (Fig. 207.)

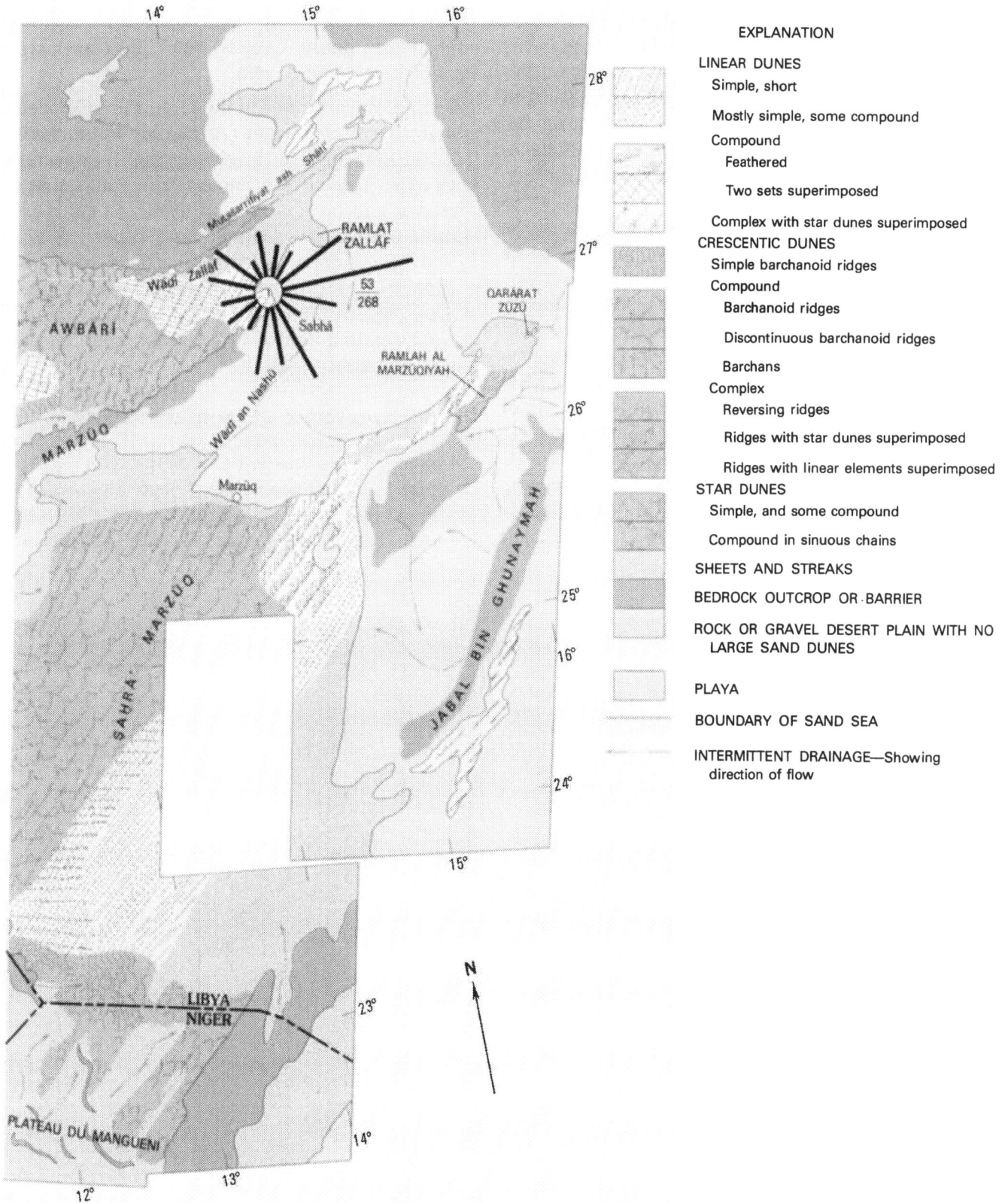

EXPLANATION

LINEAR DUNES
  Simple, short

  Mostly simple, some compound

  Compound
    Feathered

    Two sets superimposed

  Complex with star dunes superimposed
CRESCENTIC DUNES
  Simple barchanoid ridges
  Compound
    Barchanoid ridges

    Discontinuous barchanoid ridges

    Barchans
  Complex
    Reversing ridges

    Ridges with star dunes superimposed

    Ridges with linear elements superimposed
STAR DUNES
  Simple, and some compound

  Compound in sinuous chains

SHEETS AND STREAKS

BEDROCK OUTCROP OR BARRIER

ROCK OR GRAVEL DESERT PLAIN WITH NO
LARGE SAND DUNES

PLAYA

BOUNDARY OF SAND SEA

INTERMITTENT DRAINAGE—Showing
direction of flow

(Fig. 207 — Continued.)

basal early Quaternary (Mazzérien) sediments, mainly alluvium, and she suggested that sand in the big star dunes along the western margin has undergone more eolian reworking than has the sand of the interior. Suggested sources of sand in the Quaternary sediments are outcrops of Miocene-Pliocene sandstones (Alimen and others, 1957, p. 38–40).

Sand of the Grand Erg Occidental is said to be composed almost exclusively of quartz with sparse tourmaline and zircon. This pure composition was interpreted by Capot-Rey (1945, p. 393) as indicating a long sedimentary history. A similar conclusion is reached by Alimen, Doudoux-Fenet, Ferrere, and Palau-Caddoux (1957) on the basis of grain-size distribution of sand and on the variety of sizes, colors, and orientations represented.

Interdune surfaces commonly are silicocalcareous crusts developed on sands similar to those of the dunes but lacking red color (Capot-Rey, 1945). Remnants of Neolithic bog soils, composed of

white calcareous silt and containing gastropod shells, commonly overlie the interdune crusts (Capot-Rey, 1945, p. 398).

Sand of major dunes in the Grand Erg Occidental appears a uniform ocher (Alimen and others, 1957, p. 173–174) but in bright sunlight appears as various shades of red. Recent minor dunes in the sand sea and on Quaternary terraces of the Oued Saoura are yellow (Alimen and others, 1957, p. 173–174). Distinct color differences of various groups of dunes are interpreted (Alimen and others, 1957, p. 175–177, figs. 83, 89) as evidence of dune formation under various climatic conditons and at different times.

Interpretation of Landsat Imagery

Imagery of the western part of the Grand Erg Occidental (fig. 208) shows very large compound star dunes along their downwind (western) margins between Taghit and Kerzaz. Toward the interior of

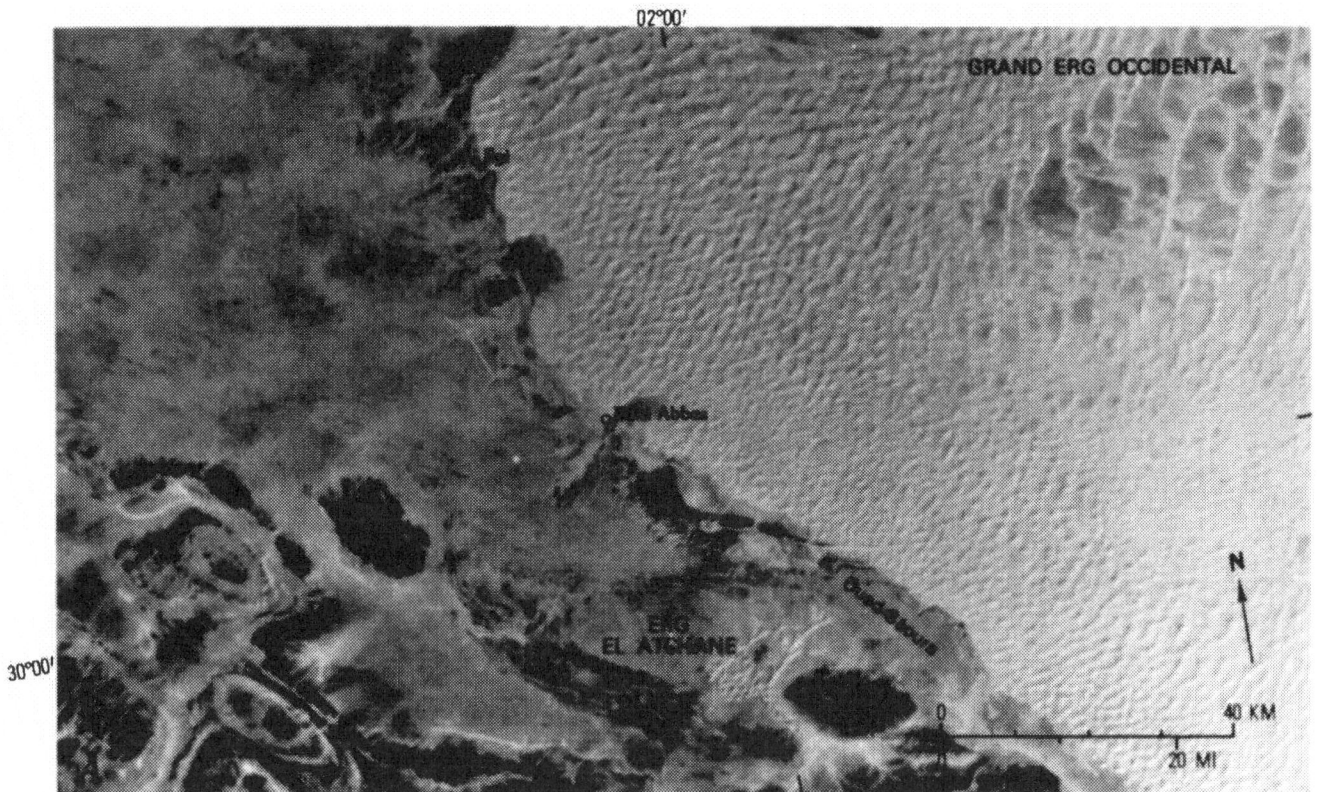

SOUTHWESTERN MARGIN OF THE GRAND ERG OCCIDENTAL, Algeria. A, Complex crescentic and star dunes of the interior of the sand sea grading westward into large compound star dunes along the Oued Saoura. Landsat imagery E1116–10123. B, Southern boundary of the Grand Erg Occidental at Kerzaz and large linear dunes of the Erg er Raoui. Landsat imagery E1115–10071. The area shown in B is just southeast of the area in A. (Fig. 208.)

the sand sea, star dunes grade into complex dunes, crescentic and star (fig. 206). Measurements of Landsat transparencies of star dunes along 50 km (30 mi) of the Oued Saoura, north of Kerzaz, show that the star dunes have a mean diameter of 0.9 km (0.5 mi) and are spaced an average of 1.3 km (0.8 mi) apart. Among this group of dunes is the Great Dune at Kerzaz, illustrated by Alimen, Doudoux-Fenet, Ferrere, and Palau-Caddoux (1957, pl. V, 1), which is more than 200 m (650 ft) high.

The western margin of the Grand Erg Occidental matches the contours of folded mountain ranges (fig. 208), which although of relatively low relief, exert profound topographic control on the downwind boundary of the sand sea.

Beyond the ranges to the southwest are complex linear and star dunes of the Erg er Raoui (fig. 208B). The linear dunes, which have star dunes on their crests, extend southwestward across plains and through gaps in the low-lying ranges to the 'Erg Chech (fig. 209) and Erg Iguidi. The 'Erg Chech consists of very large southwest-trending compound linear and complex linear and star dunes

widely spaced on rock and gravel plains. Dunes measured on Landsat imagery (fig. 209) have a mean width of 1.0 km (0.6 mi) and are spaced an average 5.7 km (3.5 mi) apart. Their wide spacing is unusual, compared with that of linear dunes in other deserts (chapter J, table 36). Dunes of the 'Erg Chech, which are yellow on false-color imagery are actually red (chapter J, fig. 168B).

Linear dunes of the 'Erg Chech can be traced on imagery (partly shown in fig. 209) southwestward into sand seas of northwestern Mali and northeastern Mauritania. Extension of the sand several hundred kilometres southwestward, in the form of linear dunes, seems to support I. G. Wilson's (1971a, p. 196) contention that dunes such as these are "sand-passing" dunes that transport sand great distances from basin to basin across the Sahara.

The Grand Erg Oriental, Algeria and Tunisia
Review of Previous Work

East of Ghardaïa and El Golea and north of the Hamada de Tinrhert lies the Grand Erg Oriental of

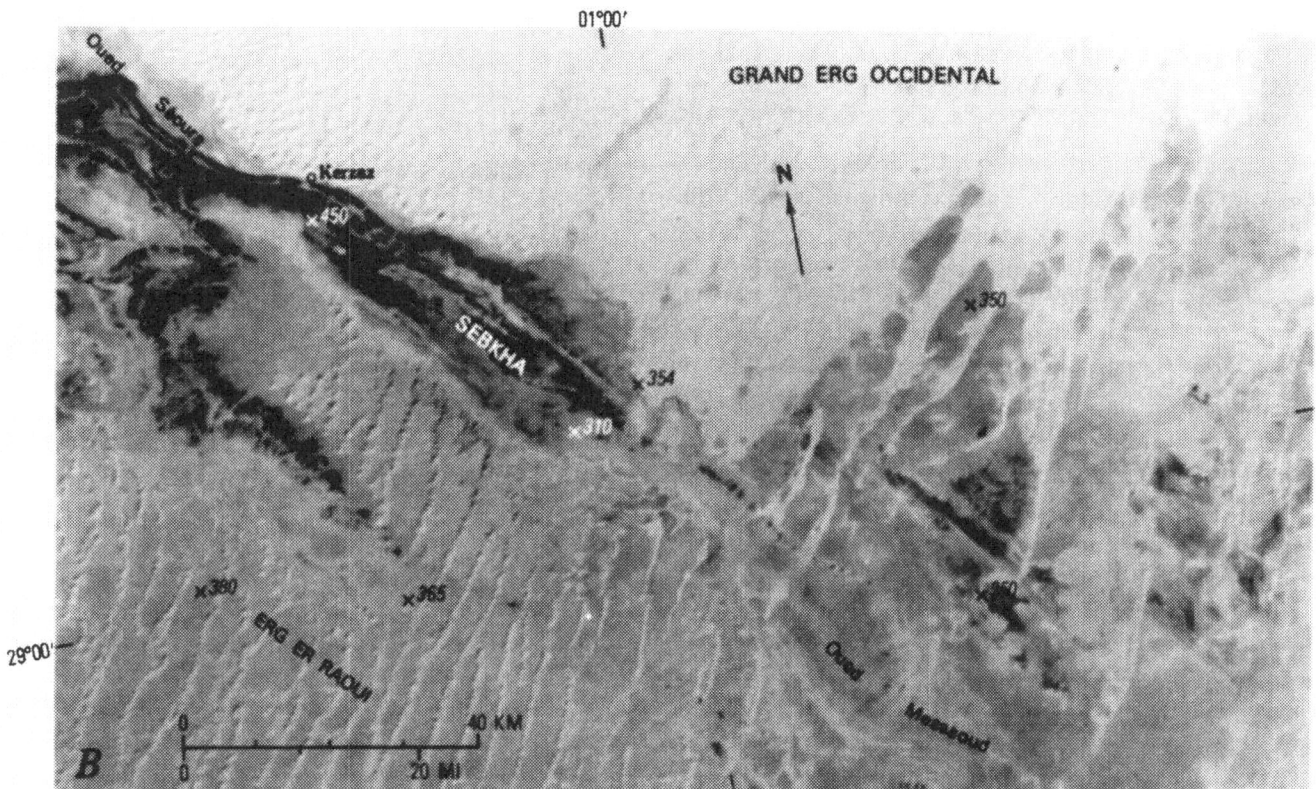

(Fig. 208. —Continued.)

02°30'

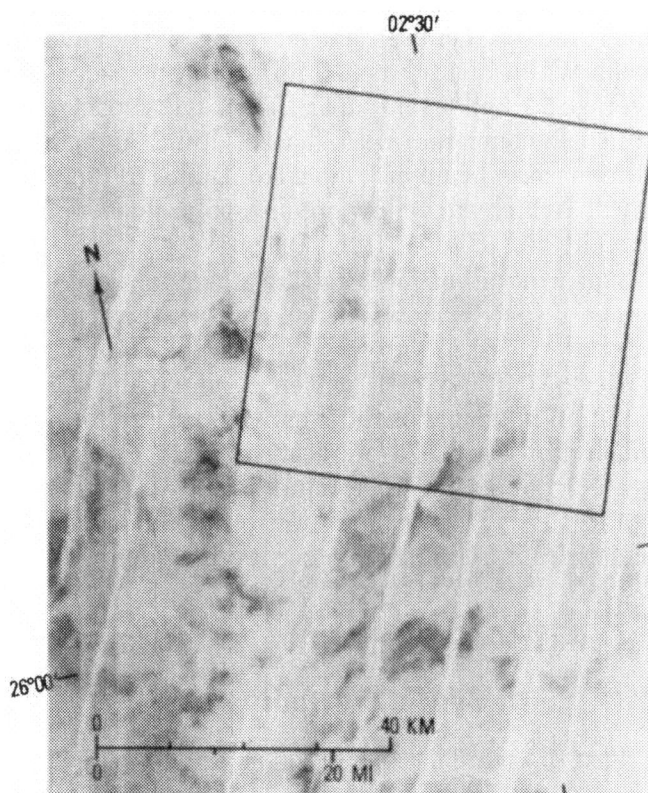

SUBPARALLEL COMPLEX DUNES of linear and star types of the 'Erg Chech, Algeria. Landsat imagery E1116–10135. Widely spaced dunes extend southwestward across gravel and rock plains toward the sand seas of Mali and Mauritania. Outline indicates part of image on which measurements of dune length, width, and wavelength were made. (Fig. 209.)

southeastern Algeria and western Tunisia. The Grand Erg Oriental includes an area of about 192,000 km² (119,000 mi²), and is about 70 percent sand-covered (I. G. Wilson, 1973, p. 91–92). The sand, if spread out, would have a mean thickness of 26 m (85 ft). The mean height of dune ridges in the Grand Erg Oriental is 117 m (385 ft), and they are spaced an average of 1.6 km (1 mi) apart (I. G. Wilson, 1973, p. 91 and fig. 7, p. 88). The center of the Grand Erg Oriental, with about 90 percent sand-cover, is continuous over about 80,000 km² (50,000 mi²) (I. G. Wilson, 1973, p. 92). The rock and gravel basement is nowhere more than a few metres beneath the dunes of the Grand Erg Oriental, even where sand covers 90 percent of the surface (I. G. Wilson, 1973, p. 92). Perennial grass (Aristida sp.) and shrubs grow on dune slopes and in interdune areas (Capot-Rey, 1947b, p. 185).

Fieldwork in the Grand Erg Oriental has not been extensive. A ground traverse across the Erg by I. G. Wilson (1971a) provided sand samples and field observations for that author's (1973) geomorphic analysis, based primarily on aerial photographs and analysis of topographic maps. The regions of high dunes, which I. G. Wilson (1973, p. 104) called "draas," contain mixed sand sizes; the rest of the sand sea, according to him, is composed of sand having a modal size of about 0.16 mm (fine sand).

Whether I. G. Wilson's figures are entirely based on the 620 sand samples collected in the northern part of the Algerian ergs (I. G. Wilson, 1971a, p. 185) or were derived from formulae based on postulated relationships between size, type, and grain-size distribution of dune and draa (I. G. Wilson, 1973, p. 94–97, 101, figs. 18, 19) is not clear. Wilson stated (1973, p. 105) that "in the absence of direct data, the variations of grain size in the ergs can be roughly inferred from the patterns of draa size and orientation, and from aerial photographs by variations in sand color and the distribution of slipfaces on the dunes." Thus, some workers believe that coarse sands develop dunes of size and orientation different from those of fine sands for a given area, presumably because sand of each grain-size is acted upon in various ways by different components of the wind regime (I. G. Wilson, 1973, p. 95–96; Warren, 1972).

Sand in the Grand Erg Oriental is believed (I. G. Wilson, 1973, p. 84–85) to be derived from deflation of alluvium in basins upwind of the sand sea. Estimated mean deflation rates for the bajada surface near Biskra, beyond the northern margin of the Grand Erg Oriental (fig. 206, 219) is 1–4 m (3–13 ft) in 2,000 years (I. G. Wilson, 1971a, p. 188). Estimated age of the Grand Erg Oriental is 1,350,000 years, based on an average annual net flow into the area (by wind) of 6 million tons of sand (I. G. Wilson, 1971a, p. 197).

Interpretation of Landsat Imagery

Landsat imagery of the Grand Erg Oriental (fig. 210) shows that the lowest part of the area is occupied by a large salt playa, the Shaṭṭ al Jarīd. The Grand Erg Oriental lies downwind of the playa, and is bounded on the south by slopes that rise toward the Hamada de Tinrhert and Plateau du Tademait (fig. 206). The abrupt southern boundary of the sand sea seems to be topographically controlled, for its margin matches the 306-m (1,000-ft) contours of the plateau slope (fig. 210). The Grand Erg Oriental tapers off to the north into sand sheets, south of the Shaṭṭ al Jarīd. To the east, in Tunisia, large dunes end in a distinct rampartlike ridge of sand (fig. 210).

Dunes south of the Shaṭṭ al Jarīd and along the border between Algeria and Tunisia are barchanoid and compound barchanoid ridges in a network pattern (Monod, 1958; Cooke and Warren, 1973; chapter J, table 33). Segmented sand ridges (yellow) enclose interdune hollows, which appear bluish- or greenish-gray on the false-color image. Elements of

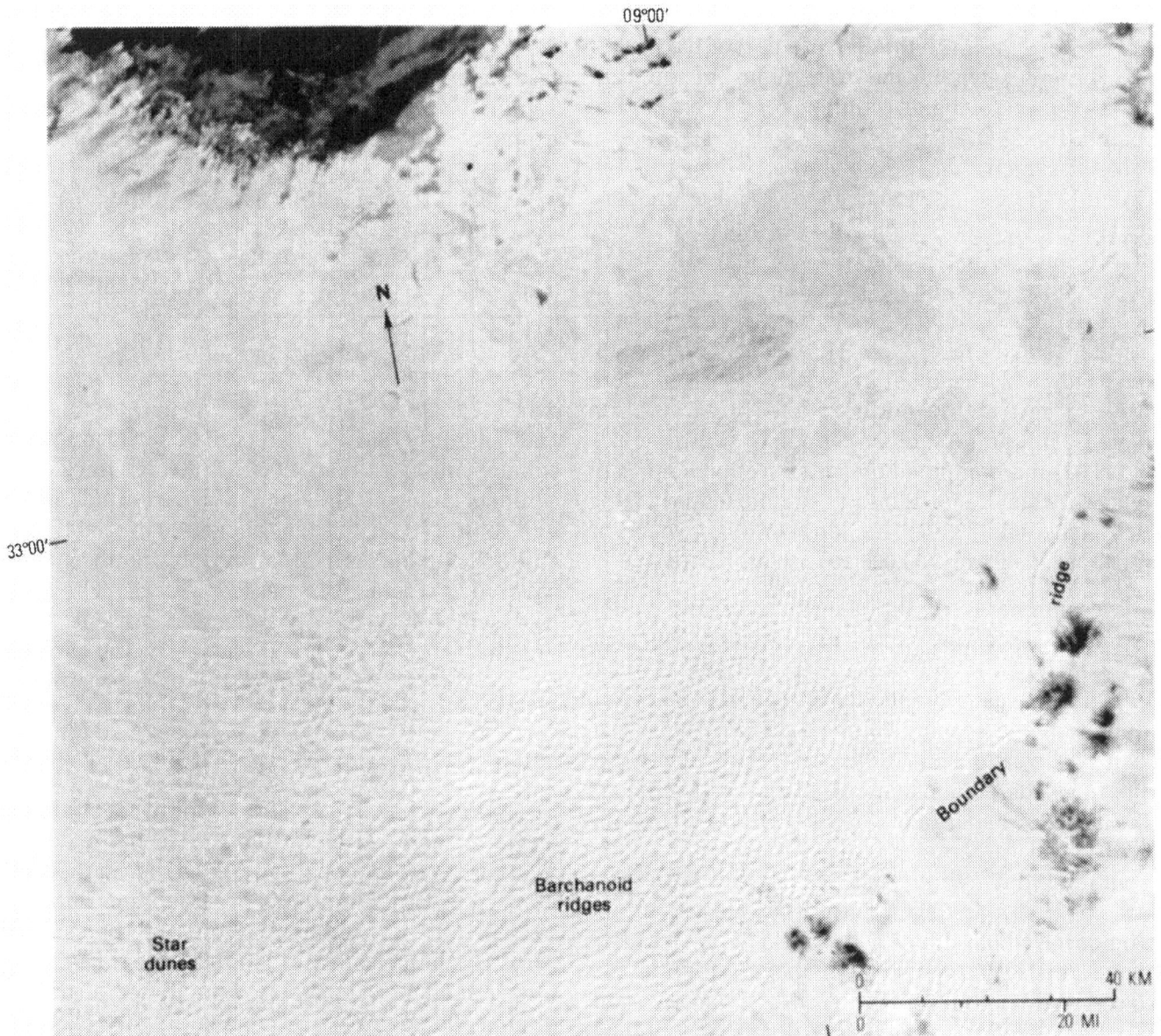

NORTHEASTERN PART OF THE GRAND ERG ORIENTAL, Algeria and Tunisia. Landsat false-color imagery E1109–09313. Size of crescentic ridges gradually increases southward. To the east the dunes, which appear yellow, end in a high, curved boundary ridge of sand. To the south and west, the network pattern of barchanoid ridge dunes (yellow) and fully enclosed interdune areas (bluish- and greenish-gray) merges with a pattern of isolated star dunes. The dark blue area at the top left of the image is a playa, the Shaṭṭ al Jarīd. Red spots are vegetated oases. (Fig. 210.)

the pattern are progressively larger toward the south and west (downwind) margins of the sand sea (fig. 210), where the dune ridges are replaced by star dunes.

The central part of the sand sea and its southern margin west of Ghudāmis consist of isolated star dunes with a distinctive pattern. These star dunes are not distributed at random, but form at the nodes of crossing dune trends, according to I. G. Wilson (1971b, p. 268; 1973, p. 99, fig. 16). The association of star dunes with crossing dune trends can be traced westward into the western part of the Grand Erg Oriental, where the star dunes merge into linear dunes to make complex forms resembling chains (fig. 211).

Measurements of the isolated star dunes in the Grand Erg Oriental were made on Landsat imagery (chapter J, fig. 180B, tables 40, 41). Mean diameter of the dunes ranges from 0.7 km (0.4 mi) at the northern (upwind) end of the Grand Erg Oriental at about lat 32° N., long 8° E., to 1.7 km (1 mi) at the southern margin (fig. 211). Mean wavelength (crest-to-crest distance) ranges from 0.8–6.7 km (0.5–4.1 mi) at the northern end to 1.5–3.1 km (0.9–1.9 mi) at the southern end. I. G. Wilson (1973, p. 92, 98) has computed the cross-sectional areas of relatively small star dunes at the north end of the field (about lat 32°30′ N., long 7°30′ E.) to be as much as 10,000 m² (33,000 ft²) for each dune.

Measurements of the star dune chains in the western part of the Grand Erg Oriental (fig. 211) were made on three sample areas of Landsat imagery. The measurements show that differences in dune size and spacing occur in a downwind direction in the sand sea. At the north (upwind) end of the field, the dune chains are spaced an average of 3.1 km (2 mi) apart and are composed of star dunes with a mean diameter of 0.9 km (0.5 mi). At the southern (downwind) end of the field, the chains are spaced an average 6.3 km (4 mi) apart and are composed of star dunes with a mean diameter of 1.6 km (1 mi). Relative areas of dune/interdune surfaces were measured by Kevin Horstman on Landsat transparencies, using a density-slicer to exploit the color contrast between the two kinds of surfaces. Ratio of dune/interdune areas at the north end of the field is 80/20 and at the south end of the field is 40/60.

Differences in dune size, spacing, and relative area from north to south in the Grand Erg Oriental are measured in a direction parallel to the resultant drift direction in this part of the Sahara. The differences may have resulted from changes in the structure of wind regime with distance across the sand sea.

The Ṣaḥrā' Awbārī and Ramlat Zallāf, Libya

Review of Previous Work

South of the Al Ḥamādah al Ḥamrā' and east of the border with Algeria is the Ṣaḥrā' Awbārī and its eastern extension, the Ramlat Zallāf. Sand in the Ramlat Zallāf is composed mainly of rounded grains with coatings of red or yellow iron oxide (McKee and Tibbitts, 1964, p. 5). Sand samples were taken from linear dunes, interdune areas, and sand and gravel plains near Sabhā (fig. 207), and each type of sand deposit has a distinctive texture (McKee and Tibbitts, 1964, p. 8.–10). Depth of sand is not well known, although, for an area in the southern Ṣaḥrā' Awbārī, Glennie (1970, enclosure 4) showed 120 m (395 ft) of unconsolidated sand overlying Cretaceous rocks. Farther east McKee and Tibbitts (1964, p. 5) recorded well samples of eolian sand more than 1,000 ft (305 m) thick.

Linear dunes in the Ṣaḥrā' Awbārī are described by Glennie (1970, p. 95) as about 100 m (330 ft) high and 100 km (62 mi) or more long, with surfaces covered by small barchanlike dunes. The latter represent eolian activity related to present winds, whereas the large linear dunes were probably formed by stronger winds of the past, according to Glennie, (1970, p. 95).

Interpretation of Landsat Imagery

Landsat imagery of the eastern, central, and western Ṣaḥrā' Awbārī (fig. 212) show a definite relation between topography, sand distribution, and dune type. Crossing trends of large dune ridges are evident on the southern margins of the sand sea (fig. 212); in the center of the field, the ridges merge into a field of star dunes. Star dunes also occur around the edge of the Ḥamādat Zegher (fig. 212). Margins of the sand sea coincide approximately with contours of the surrounding plateaus. Elevation of the Ḥamādat Zegher, although slight (500 m, or 1,650 ft; fig. 207) apparently is sufficient to prevent the large dunes to the east at lat 27°30′ N., long 13° E., which have an elevation of 300 m (985 ft), from invading it.

The Ṣaḥrā' Marzūq, Libya and Niger

Review of Previous Work

The Ṣaḥrā' Marzūq (fig. 207) occupies 58,000 km² (36,000 mi²) in a structural and topographic basin bounded on the north, west and southwest by the Hammādat Marzūq, a 1,450-km-long (900-mi-long) cuesta of post-Tasillian-Nubian continental rocks, mainly sandstones (Capot-Rey, 1947a, p. 71, 81; Conant and Goudarzi, 1967, p. 721–728). Unconsolidated sands 1,200 m (3,940 ft) thick are reported to overlie Paleozoic rocks in the Ṣaḥrā' Marzūq (Glennie, 1970, enclosure 4).

Sand of the Ṣaḥrā' Marzūq is mainly quartz, and, according to Capot-Rey (1947a, p. 85), the median size of the grains is considerably greater than that of dune sand from the Grand Erg Occidental and the Grand Erg Oriental of Algeria. Samples show bimodal distribution of grain sizes, with most of the sand grains larger than 0.35 mm (coarse or larger), although sand less than 0.10 mm (very fine) also occurs. The bimodal character of the Ṣaḥrā' Marzūq sand diminishes from east to west across the sand sea, and at the western margin the sands have a high proportion of fine grains (Capot-Rey, 1947a, p. 85).

A relationship between grain size of the sand and two basic forms of sand topography in the Ṣaḥrā' Marzūq is recorded by Capot-Rey (1947a, p. 86–87). Low, rolling, nearly fixed dunes of the southern part of the sand sea are composed of coarse sands; these are the basal, residual sands of the erg. Above

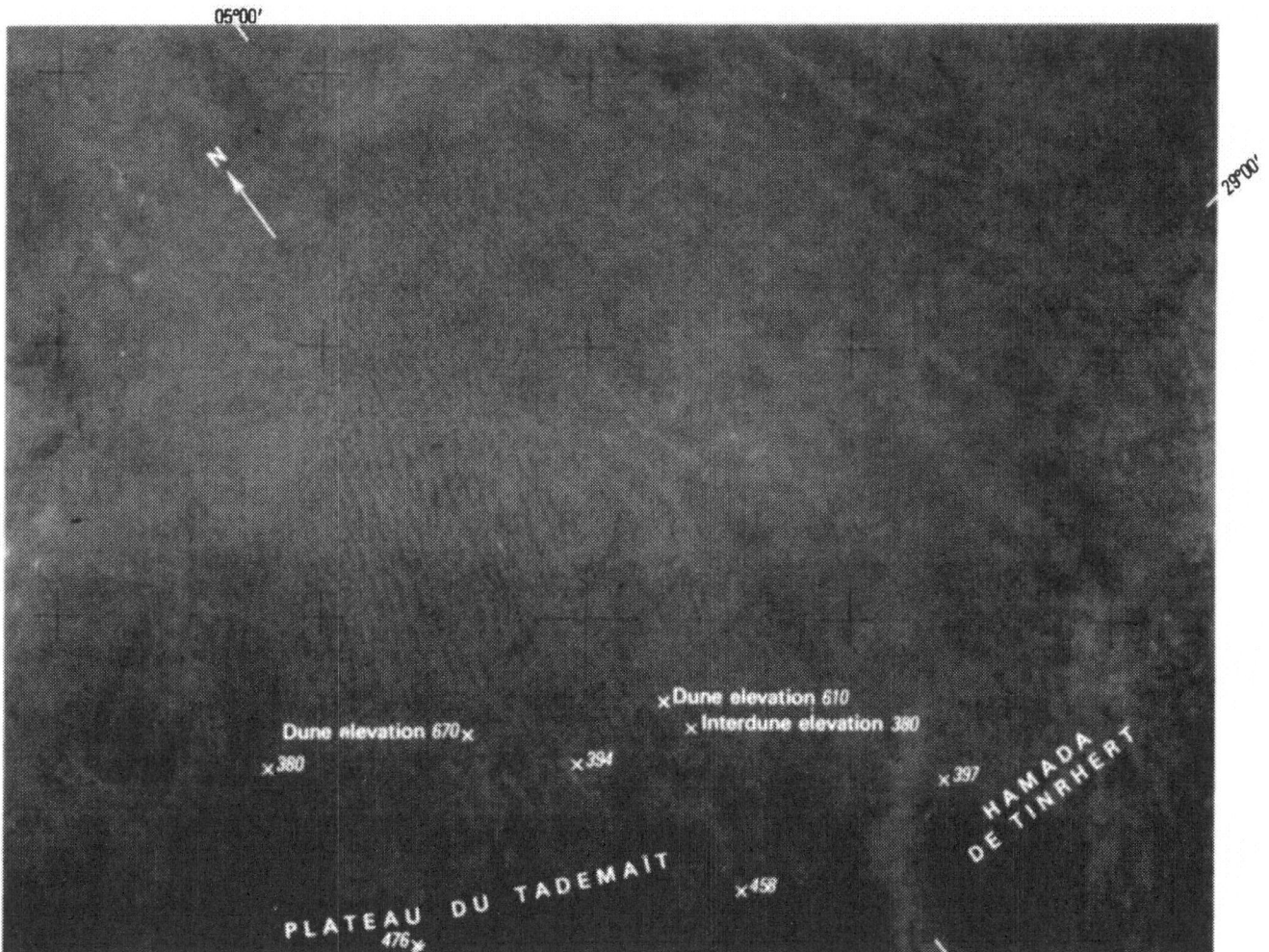

SOUTHWESTERN END OF THE GRAND ERG ORIENTAL, Algeria, showing 200-km-long (120-mi-long) field of star dunes in linear chains. Skylab photograph SL4–138–3885. Southern boundary of the field corresponds approximately to the 400-m (1,300-ft) contour of the Hamada de Tinrhert and Plateau du Tademaït (fig. 206). Height of large star dunes above the rock- and gravel-covered interdune surfaces at the southern margin is about 230 m (750 ft). (Fig. 211.)

the coarse sands — especially in the northern and western parts of the basin — are finer sands which comprise active high dunes with steep slipfaces.

Interdune surfaces in the Ṣaḥrā' Marzūq commonly have calcareous crusts. No trees are reported from the interdune areas, but grasses and shrubs (*Aristida* sp., *Calligonum* sp., *Cornulaca* sp.) are common (Capot-Rey, 1947a, p. 100–103).

Two main types of dunes were recognized during a ground traverse by Capot-Rey (1947a, p. 87–89). In the area north of Marzūq (fig. 207), compound linear dunes are dominant. These are composed, according to Capot-Rey, of dune ridges alined one beside the other, either parallel or oblique to the main axis of the chain. A second main type is the compound crescentic dune ridge associated with isolated large barchans, reported to be as much as 180 m (600 ft) high (Capot-Rey, 1947a, p. 88–89).

Interpretation of Landsat Imagery

A Skylab photograph (fig. 213) shows major dune forms and their distribution at the west end of the Ṣaḥrā' Marzūq. Along the northwestern part of this sand sea, at about lat 25°30' N., long 12°15' E., are numerous isolated large complex dunes, which have a mean horn-to-horn width of 2.0 km (1.3 mi) and a mean length of 1.5 km (1 mi) and are spaced an average 2.0 km (1.3 mi) apart. To the south and east, the isolated dunes merge into compound crescentic dune ridges. Southwest-trending linear elements are part of the dune pattern at the south end of the basin.

Still farther south, fields of barchans extend across escarpments of the Plateau du Mangueni, and sand is apparently being blown southwestward (fig. 213). The crescentic dunes can be traced on Landsat imagery several hundred kilometres south-

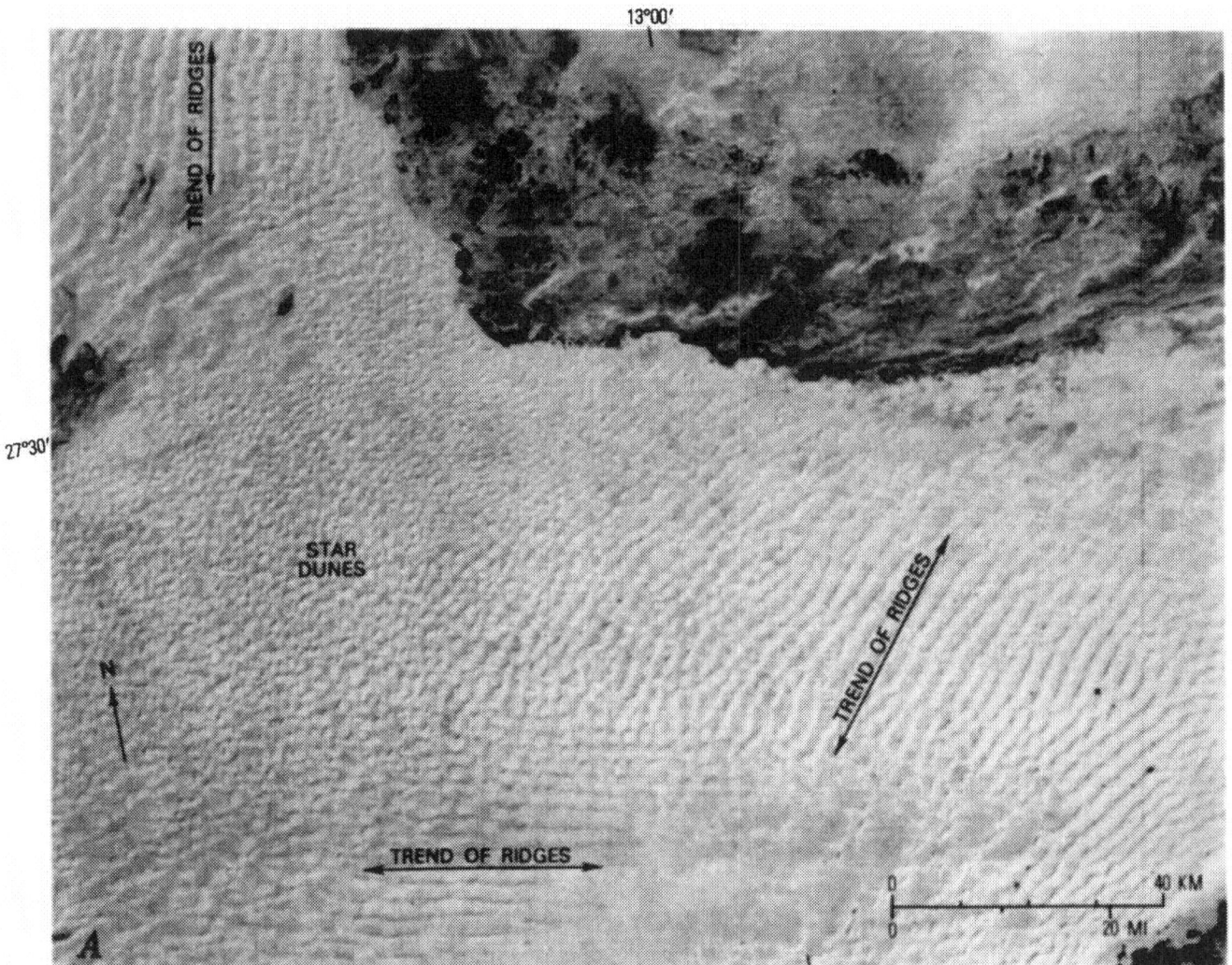

westward, from the edge of the Ṣaḥrā' Marzūg to the Great Erg of Bilma in Niger.

Surface Wind Flow and Precipitation

Surface wind flow in the northern Sahara, including Algeria and Libya, is controlled by several major pressure systems, as shown in figure 195. During the winter, westerly winds prevail in northern Algeria, Tunisia, and northern Libya north of approximately lat 30° N. to the Mediterranean coast. South of this latitude, outward flow from the Azores and Sahara highs results in north to northeast winds. During the summer, a thermal low is centered at approximately lat 26° N. and long 03° E., and the Sahara high has been replaced by a ridge of high pressure extending northwestward into the Mediterranean Sea. Outward flow of air from the Azores high, which is strongest in June, and from high pressure over the Mediterranean Sea into the thermal low results in northeast to east tradewinds over the deserts of Algeria and Libya.

Another important element of surface wind circulation results from the passage of cyclones (low-pressure cells) from east to west along the Mediterranean Sea and from the passage across the desert of cold fronts (fig. 214). Both events result in strong winds from the west and southwest across wide regions of the desert. Northwest winds sometimes occur behind a cold front (fig. 214B). Winter, spring, or late autumn are the principal seasons of cyclonic storms and frontal passages (U.S. Weather Bureau, and U.S. Army Air Forces, 1944, p. 7), and these mainly affect the northern Sahara. Winds resulting from the presumed effects of cyclones and frontal passages are shown by dashed arrows in figure 195.

Except in the highlands and on the northern coast, the northern Sahara receives less than 100 mm (0.4 in.) of rain per year (figs. 206, 207). Average annual rainfall variability for most stations is extreme (Griffiths and Soliman, 1972, p. 97–98). Rainfall over most of the region is insufficient to support vegetation recognizable on Landsat imagery.

Direction of Sand Drift

Stations in Algeria from lat 30°N. to the coast (including El Golea, Ouargla, and Ghardaïa) are characterized by complex wind regimes (fig. 206). These regimes result from the interaction of the westerly effective winds of winter, the northeasterly effective winds of summer, and the southwesterly effective winds resulting from cyclones and cold fronts. Seasonal changes in resultant drift directions occur in the northern Sahara in Algeria at Laghouat (fig. 215), Ghardaïa (fig. 216), and Ouargla (fig. 217). The complex effective wind distributions that characterize northern Algeria may account for the occurrence of star dunes, isolated and in linear chains, throughout the Grand Erg Oriental and parts of the Grand Erg Occidental (fig.

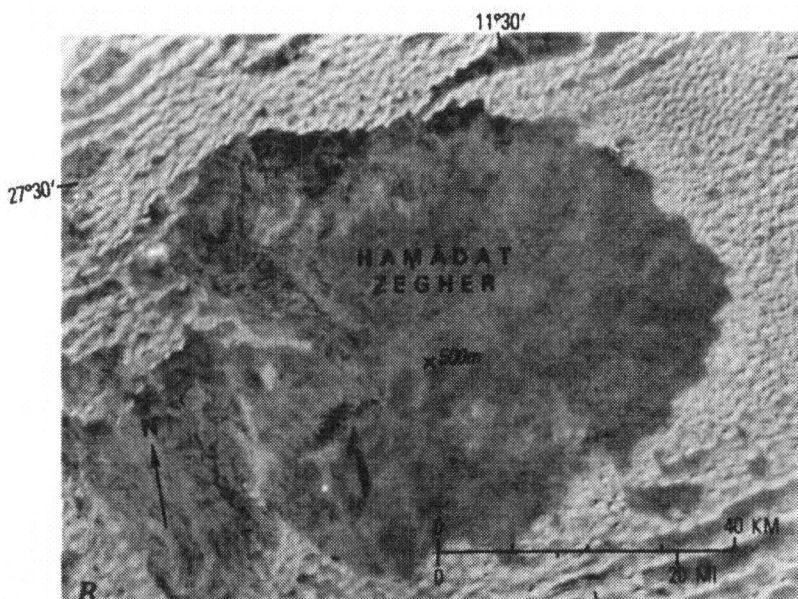

CENTRAL(A) AND WESTERN(B) PARTS OF THE ṢAḤRĀ' AWBĀRĪ, Libya. Landsat imagery E1105–09101; E1106–09160. A, South-trending ridges in the north, southwest-trending ridges in the east, west-trending ridges in the south, and star dunes in the center of the sand sea. B, Star dunes surrounding the dune-free Hamādat Zegher. (Fig. 212.)

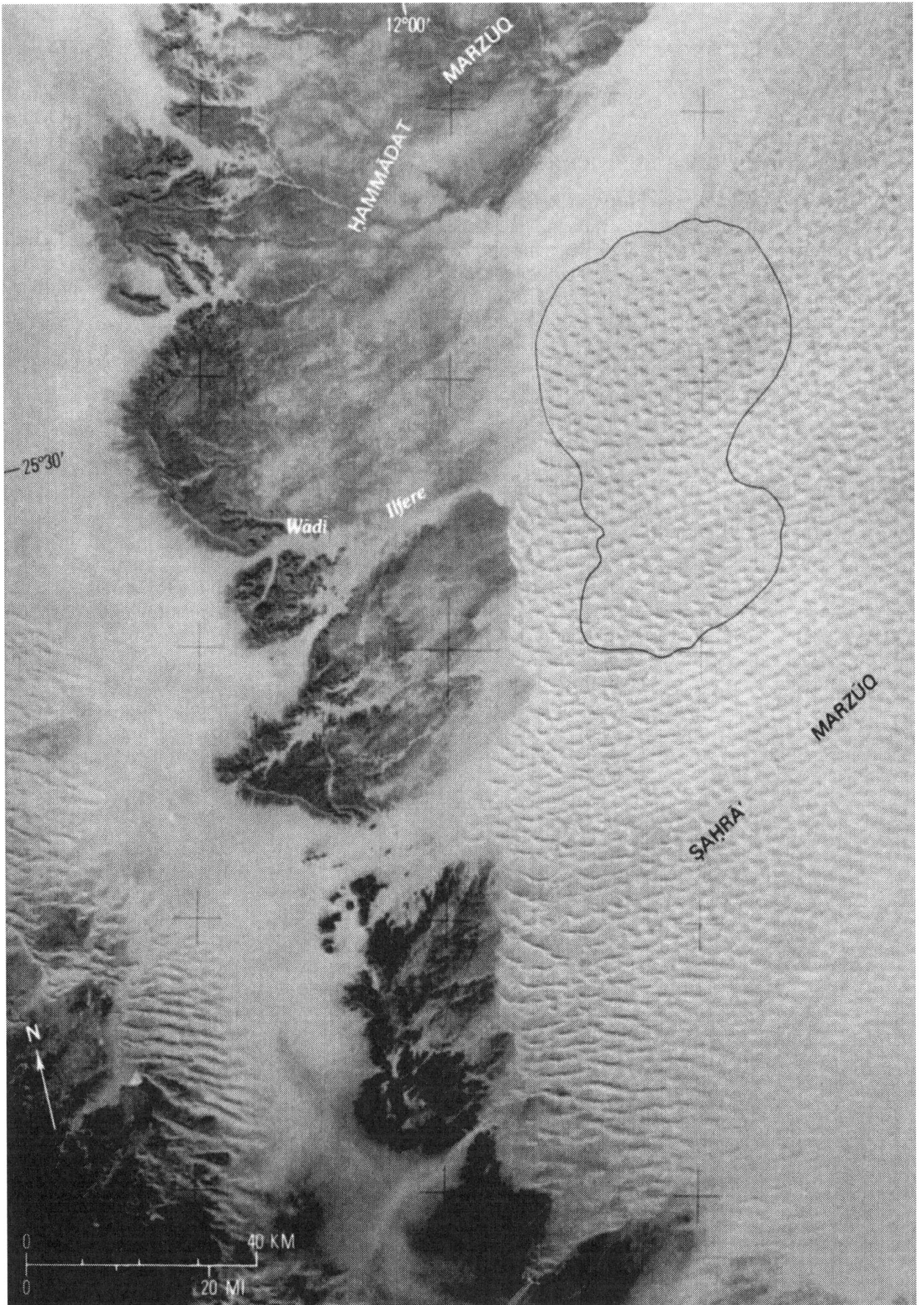

12°00'

MARZŪQ

HAMMĀDAT

— 25°30'

Wādi   Iljere

MARZŪQ

ŞAHRĀ'

N

0   40 KM

0   20 MI

DUNES AT THE WEST EDGE OF THE ṢAḤRĀ' MARZŪQ, Libya (facing page). Skylab photograph. Red eolian sand fills topographically low areas and extends southwestward through gaps in the Ḥammādat Marzūq. Line surrounds a field of isolated combinations of star and crescentic complex dunes. (Fig. 213.)

206). Star dunes are commonly found in areas with complex wind regimes.

Despite complex effective wind distributions, many stations north of lat 30° N. in eastern Algeria have annual resultant drift directions generally toward the southeast (fig. 206). Béchar and 'Aïn Sefra, in western Algeria, along the northwest margins of the Grand Erg Occidental (fig. 206) have annual resultant drift directions to the north and northeast, probably due to influence of the nearby Atlas Saharien on local winds. Annual resultant drift directions at the east margin of the Grand Erg Oriental near Qābis, Tunisia, are mostly toward the northeast (fig. 206, 218). However, weak northeasterly effective winds prevail in this region during the summer, producing a resultant drift direction toward the southwest during that season (figs. 218, 197).

Wind regimes from lat 30° N. southward in western Algeria (including I-n-Salah, Reggane, and Timimoun) are generally not complex (fig. 206). Dune types in the central and southern parts of the Grand Erg Occidental are mostly linear and crescentic ridges, which are associated with less

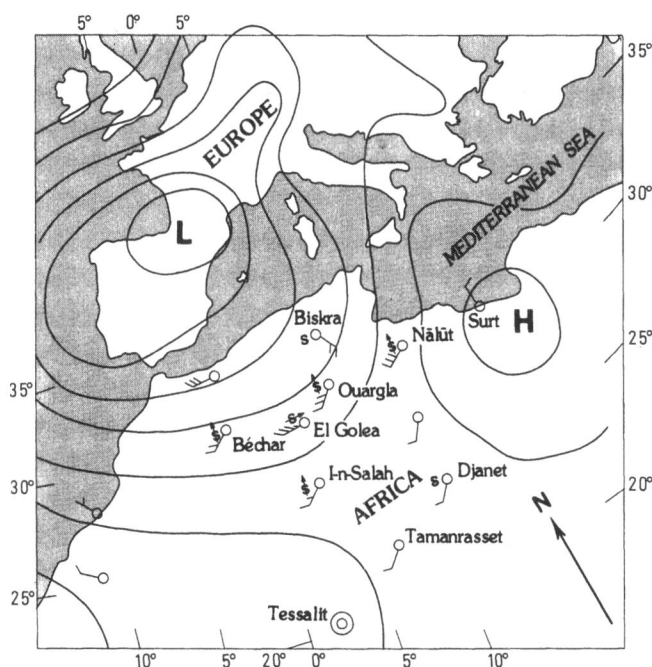

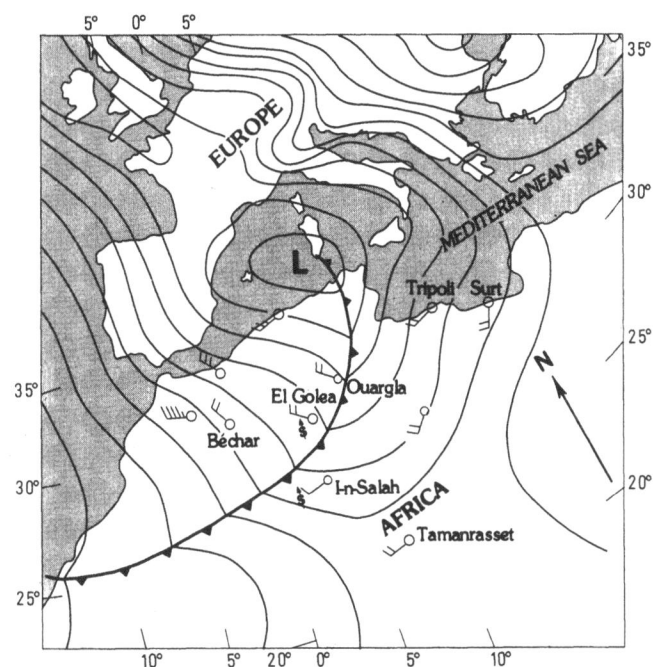

EXPLANATION

H	Center of high-pressure cell	——————	Isobar
L	Center of low-pressure cell	○	Weather station
◎	Calm	∕	Points into wind—large barb, 10 knots; small barb, 3–5 knots
▲—▲—▲	Leading edge of cold front	↕	Direction of sand drift

TWO METEOROLOGICAL EVENTS which produce southwesterly and northwesterly winds over large areas of the northern Sahara. A, Passage of a low-pressure system (cyclone) from west to east to the north of the desert. Counterclockwise circulation around the low results in southwest winds from Tamanrasset to Nālūt, Libya, and in southeast winds at Biskra, northern Algeria. From National Oceanographic and Atmospheric Administration synoptic chart of Northern Hemisphere, April 3, 1971, 1200 G.M.T. B, Passage of a cold front from west to east across the northern desert. Winds in advance of the front are from the southwest, and those behind the front, from the northwest. From National Oceanographic and Atmospheric Administration synoptic chart of the Northern Hemisphere, January 17, 1972, 1200 G.M.T. (Fig. 214.)

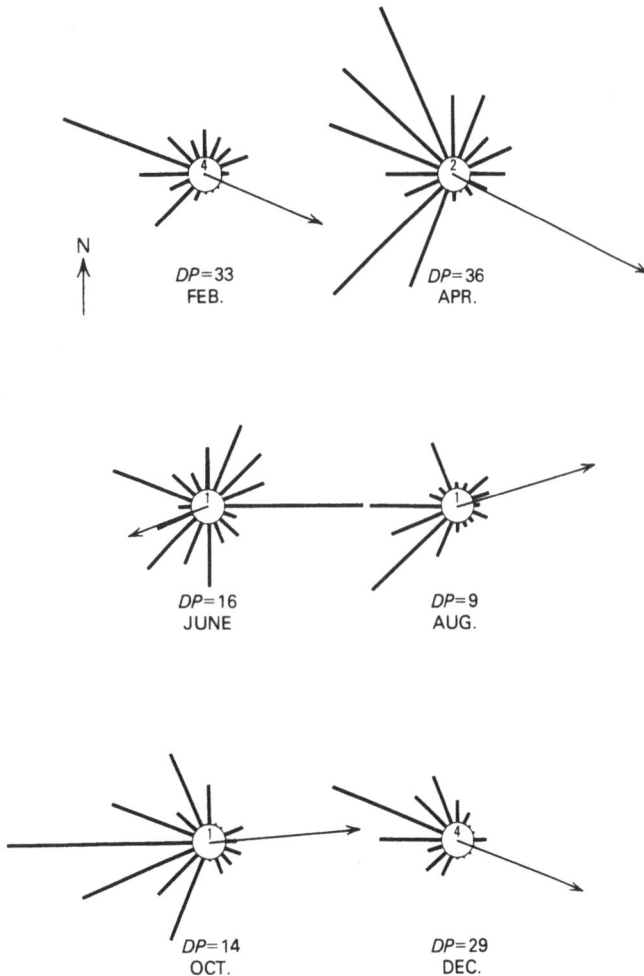

DP=33
FEB.

DP=36
APR.

DP=16
JUNE

DP=9
AUG.

DP=14
OCT.

DP=29
DEC.

SAND ROSES FOR 6 MONTHS FOR LAGHOUAT, north of the Grand Erg Occidental, Algeria. Westerly effective winds in winter (from December to February) result in east-southeastward resultant drift direction. By June, northwest winds have been partially replaced by northeast winds, and resultant drift direction is toward the west-southwest. Southwest winds are most prevalent in August. Drift potential, in vector units, is given for each month. (Fig. 215.)

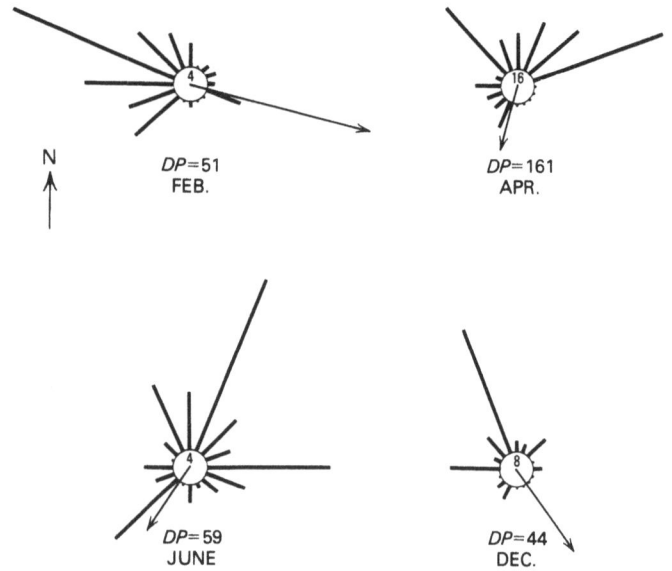

DP=51
FEB.

DP=161
APR.

DP=59
JUNE

DP=44
DEC.

SAND ROSES FOR 4 MONTHS FOR GHARDAÏA, Algeria, between the Grand Erg Occidental and the Grand Erg Oriental (fig. 206). The season of winter westerly winds ends earlier at Ghardaïa than at Laghouat. The sand-moving potential of winds (drift potential) is much greater at this station in the spring at 161 VU (vector units) than in winter (December, 44 VU; February, 51 VU). (Fig. 216.)

Awbārī (lat 27°01' N., long 14°06' E.) (chapter F, fig. 115B) and south-southwestward beyond the map region at lat 29°08' N., long 15°57' E. Surface wind flow in western Libyan sand seas approximates that shown in figure 195, but data are few (fig. 207). Annual sand roses for Bordj Omar Driss (Fort Flatters, lat 28°08' N., long 06°50' E.), Illizi (lat 26°30' N., long 08°38' E.), and Djanet (lat 24°33' N., long 09°38' E.), Algeria (fig. 207), indicate complex wind regimes. All three stations are situated near mountainous terrain which may locally affect wind regimes.

Amount of Sand Drift

The average drift potential of 21 stations around the Algerian sand seas is 239 VU (vector units, chapter F, table 16), and of 7 stations around the Libyan sand seas is 431 VU. These are intermediate and high values, respectively, compared to those of other desert regions around the world (chapter F, table 16). As in other regions, however, a wide range of values occurs at individual stations, from 51 VU at Djanet (fig. 207), to 571 VU at Biskra (fig. 206). In general, northeastern Algeria, southern

variability of wind direction than are star dunes. Resultant drift directions at I-n-Salah, Timimoun, and Reggane (fig. 206) are toward the southwest, approximately parallel to the trend of linear dunes in that region (fig. 209). Monthly resultant drift directions at I-n-Salah are uniformly to the southwest throughout the year (fig. 219) because I-n-Salah is south of the zone of winter westerlies and because frontal and cyclonic activity is greatest in regions farther north.

Annual resultant drift directions are toward the west-northwest at Sabhā in the eastern Ṣaḥrā'

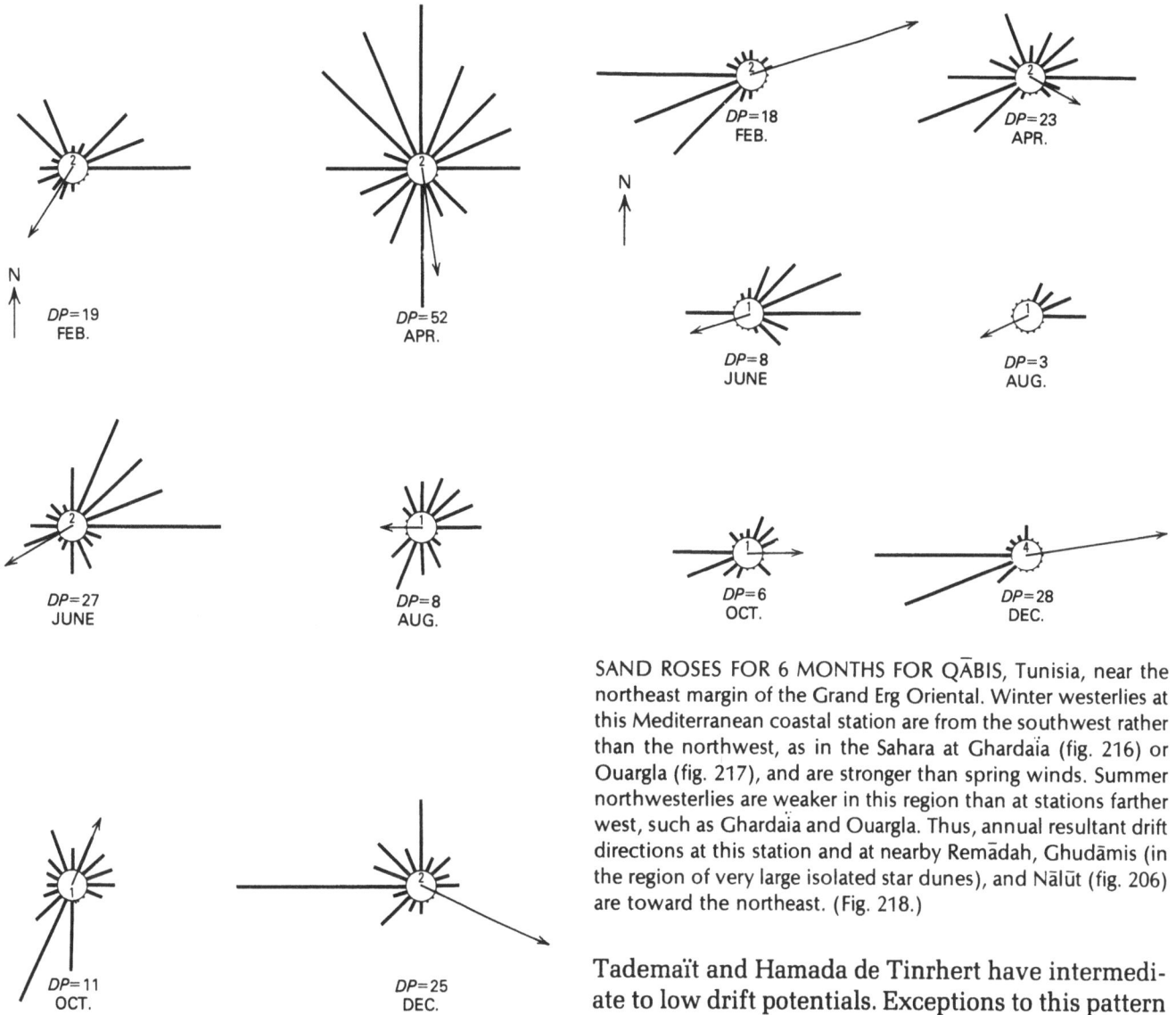

SAND ROSES FOR 6 MONTHS FOR OUARGLA, Algeria, at the northwest edge of the Grand Erg Oriental, illustrating complexity of the wind regime in that sand sea throughout the year. The complexity results from the interaction of the three dominant components of northern Sahara wind regimes; winter westerlies, summer northeasterlies, and the southwesterlies resulting from frontal passages and cyclones. Despite the variation of resultant drift direction during the year at Laghouat (fig. 215), Ghardaïa (fig. 216), and Ouargla, annual resultant drift directions at all of these stations are toward the southeast, approximately parallel to the long dimensions of the chains of star dunes southeast of Hassi Messaoud. (Fig. 217.)

SAND ROSES FOR 6 MONTHS FOR QĀBIS, Tunisia, near the northeast margin of the Grand Erg Oriental. Winter westerlies at this Mediterranean coastal station are from the southwest rather than the northwest, as in the Sahara at Ghardaïa (fig. 216) or Ouargla (fig. 217), and are stronger than spring winds. Summer northwesterlies are weaker in this region than at stations farther west, such as Ghardaïa and Ouargla. Thus, annual resultant drift directions at this station and at nearby Remādah, Ghudāmis (in the region of very large isolated star dunes), and Nālūt (fig. 206) are toward the northeast. (Fig. 218.)

Tademaït and Hamada de Tinrhert have intermediate to low drift potentials. Exceptions to this pattern occur at I-n-Salah (541 VU) and Reggane (555 VU). In Libya, drift potentials decrease southward and eastward toward Jālū (132 VU) and Al Kufrah (81 VU), which are located at long 21°34' E. and long 23°20' E., respectively, near the center of the Sahara high-pressure cell (fig. 195).

The season of highest drift potentials in northern Algeria, Tunisia, and northern Libya usually occurs in March or April (fig. 220). Many stations in the north also have strong winter sand-moving seasons (fig. 220). The range of drift potential at many northern stations, such as Ghardaïa and Biskra, Algeria, and Ghudāmis, Libya, may be extreme (fig. 221). Farther sound towards I-n-Salah and Djanet, Algeria, highest drift potentials can occur in June and July.

Tunisia, and adjacent northwest Libya are in a region of intermediate to high drift potentials (fig. 206). Western Algeria, including the Grand Erg Occidental and areas to the south of the Plateau du

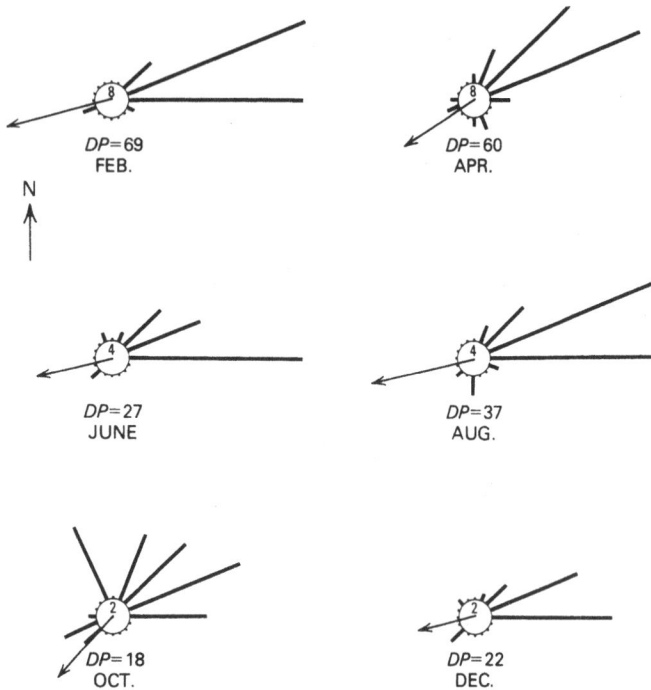

DP=69
FEB.

DP=60
APR.

N

DP=27
JUNE

DP=37
AUG.

DP=18
OCT.

DP=22
DEC.

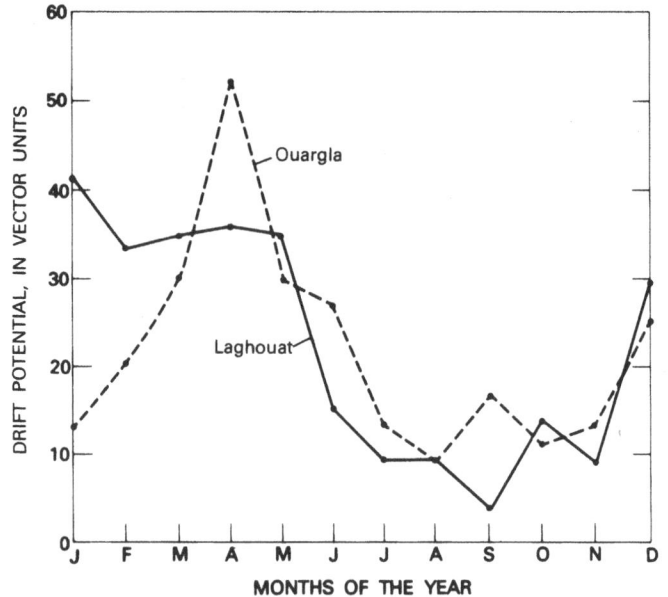

VARIATION IN DRIFT POTENTIAL during the year at Laghouat (solid line) and Ouargla (dashed line), Algeria. These stations typify much of the northern Sahara, including southern Tunisia and northwestern Libya. Laghouat experiences high drift potentials in the winter and spring. Ouargla, which is closer to the interior, experiences highest drift potentials in the spring but is too far south to have equally high drift potentials in the winter. (Fig. 220.)

SAND ROSES FOR 6 MONTHS FOR I-N-SALAH, Algeria, south of the Grand Erg Occidental and east of the 'Erg Chech, illustrating the steadiness of easterly to northeasterly effective winds and resultant drift direction toward the west-southwest and to the southwest. This wind regime, in a region of sand sheets, contrasts markedly with some more complex wind regimes farther north in Algeria in a region of star and complex dunes. (See figs. 216–218.) (Fig. 219.)

Namib Desert of South-West Africa

Summary of Conclusions

CRESCENTIC, LINEAR, AND STAR DUNES occur in the Namib Desert in three elongate zones parallel to the Atlantic coast. Crescentic dunes with an average wavelength of 0.9 km (0.6 mi) occupy a belt 5–30 km (3.1–18.6 mi) wide along the Atlantic coast between Swakopmund and Lüderitz. Compound linear dunes occupy much of the central portion of the desert, from Lüderitz northward to the Kuiseb River. These linear dunes average 2.2 km (1.4 mi) in wavelength and 27 km (16.8 mi) in length. Isolated fields of star dunes, reversing dunes, and complex dunes of indeterminate type occur along the east margin of the Namib Desert. The star dunes have a mean diameter of 0.7 km (0.4 mi) and are spaced an average of 2.2 km (1.4 mi) apart. In general, redness of the dune sands seems to increase inland from the coast.

Average annual rainfall is less than 100 mm (3.9 in.) throughout the Namib Desert, except in some of its extreme eastern parts. Heavy dew along the coast may be the equivalent of as much as 25 mm (1 in.) of annual rainfall. High-energy unimodal winds from the southwest or south occur along the Atlantic coast in a zone which corresponds roughly to the region of crescentic dunes. An intermediate- to low-energy bimodal wind regime may occur in the central zone of linear dunes; a complex low-energy wind regime prevails near the star dunes of the eastern Namib Desert. The observed present-day wind regimes of the Namib Desert are roughly compatible with the observed dune types in the several zones. The rapid decrease in wind energy inland from the coast, however, suggests that linear and star dunes in the interior are considerably less active than are crescentic dunes along the coast.

Introduction

The main sand sea of the central Namib Desert, South-West Africa, is bounded on the north by the Kuiseb River, on the west by the Atlantic Ocean, on the east by the Great Escarpment (Goudie, 1972, p.

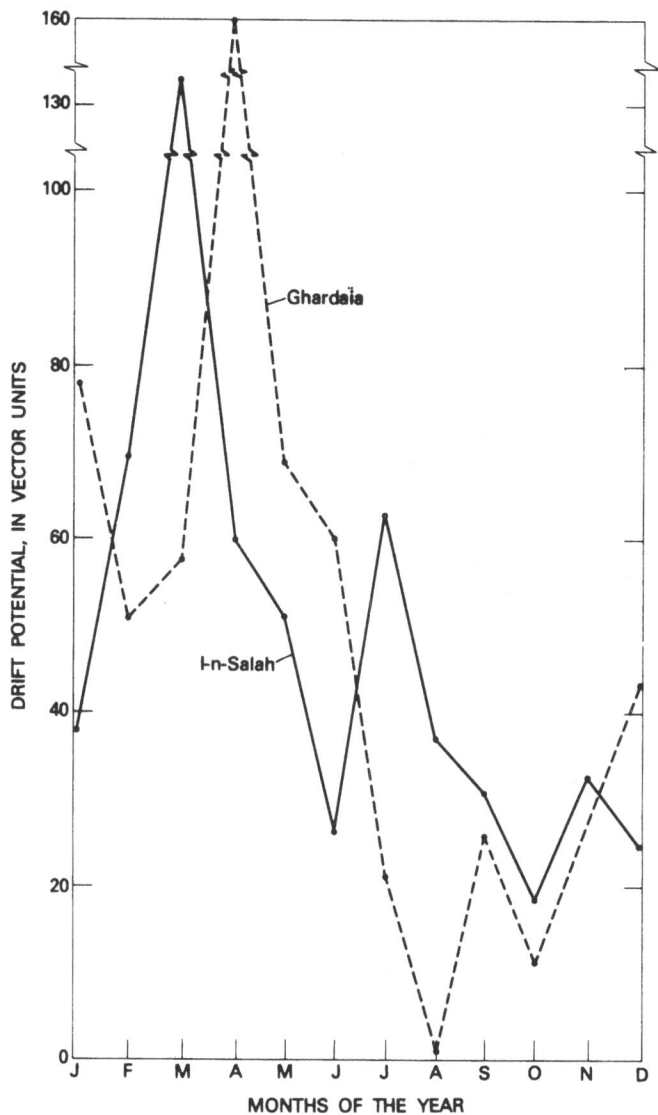

VARIATION IN DRIFT POTENTIAL during the year at Ghardaïa and I-n-Salah, Algeria, illustrating the extreme variation in drift potential which may occur at some stations. Hassi Messaoud, Algeria, and Ghudamis, Libya, also exhibit this variable pattern. I-n-Salah is far enough south to reflect the summer high in drift potentials which occurs at southern stations. (Fig. 221.)

Landsat imagery has been used to map eolian sand distribution and morphology in the main sand sea of the central Namib Desert (fig. 222). Interpretations were made with the aid of Skylab photographs.

Principal published reports (in German) about the Namib sand sea are by Range (1927) and Besler (1972). A major work (in Afrikaans) on dunes of the central Namib is by Barnard (1973). Reports (in English) include a descriptive geomorphic paper by Gevers (1936), general geographic treatises by Wellington (1955) and by Logan (1960), and two papers by Goudie (1970, 1972) that give descriptions of the dunes near Gobabeb. Relationship of dunes to the Tsondab River and Tschaubrivier is described by Seely and Sandelowsky (1974).

Comprehensive aerial photography of the Namib Desert is unavailable and maps of the sand sea, U.S. Department of Commerce, U.S. Air Force, Operational Navigation Charts P–3, P–4, and Q–4, state that "relief data are unavailable."

Review of Previous Work

Sand of the Namib Desert is thought by most workers (J. Rogers, in Seely and Sandelowsky, 1974, p. 61) to be derived mainly from fluvial sediments transported to the Namib platform by the perennial Orange River, in an area off the map (fig. 222) at about lat 28° S., long 17° E. Some sand is also brought in by intermittent streams from the Great Escarpment (fig. 222). Some of the sand near the coast may be of marine origin (Gevers, 1936, p. 71). The sand sea may be as old as Miocene according to Barnard (1973, p. 2). Recent active sands in the Namib are believed to be derived mostly from reworking of Tertiary dunes (Martin, in Seely and Sandelowsky, 1974, p. 61).

Field data from the Namib sand sea consist mainly of descriptions of dune areas near the coast or accessible from Gobabeb (fig. 222). Sand forms near Gobabeb are at the downwind end of the sand sea near topographic barriers and thus, may not be representative of those to the south in the main part of the sand sea. Linear ridges near Gobabeb are described by Goudie (1970, p. 94–95) as "composite forms made up of as many as five subparallel and sinuous segments which sometimes isolate deep hollows within the dune." Interdune valleys near Gobabeb are largely sand free corridors containing a few small barchans (Goudie, 1970, p. 95).

15); it ends at the south in sandy plains north of Luderitz and Aus. The sand sea area is about 34,000 km² (21,000 mi²) (Barnard, 1973, p. 2). It rests upon the Namib platform, a Tertiary erosional surface cut on schists, quartzites, and granite intrusives. Regional slope of the Namib platform is about 2.3 m/km (65 ft/mi) from the base of the escarpment to the Atlantic Ocean. Numerous intermittent streams flow from the escarpment into playas among the dunes, and rock outliers protrude through the sand as inselbergs.

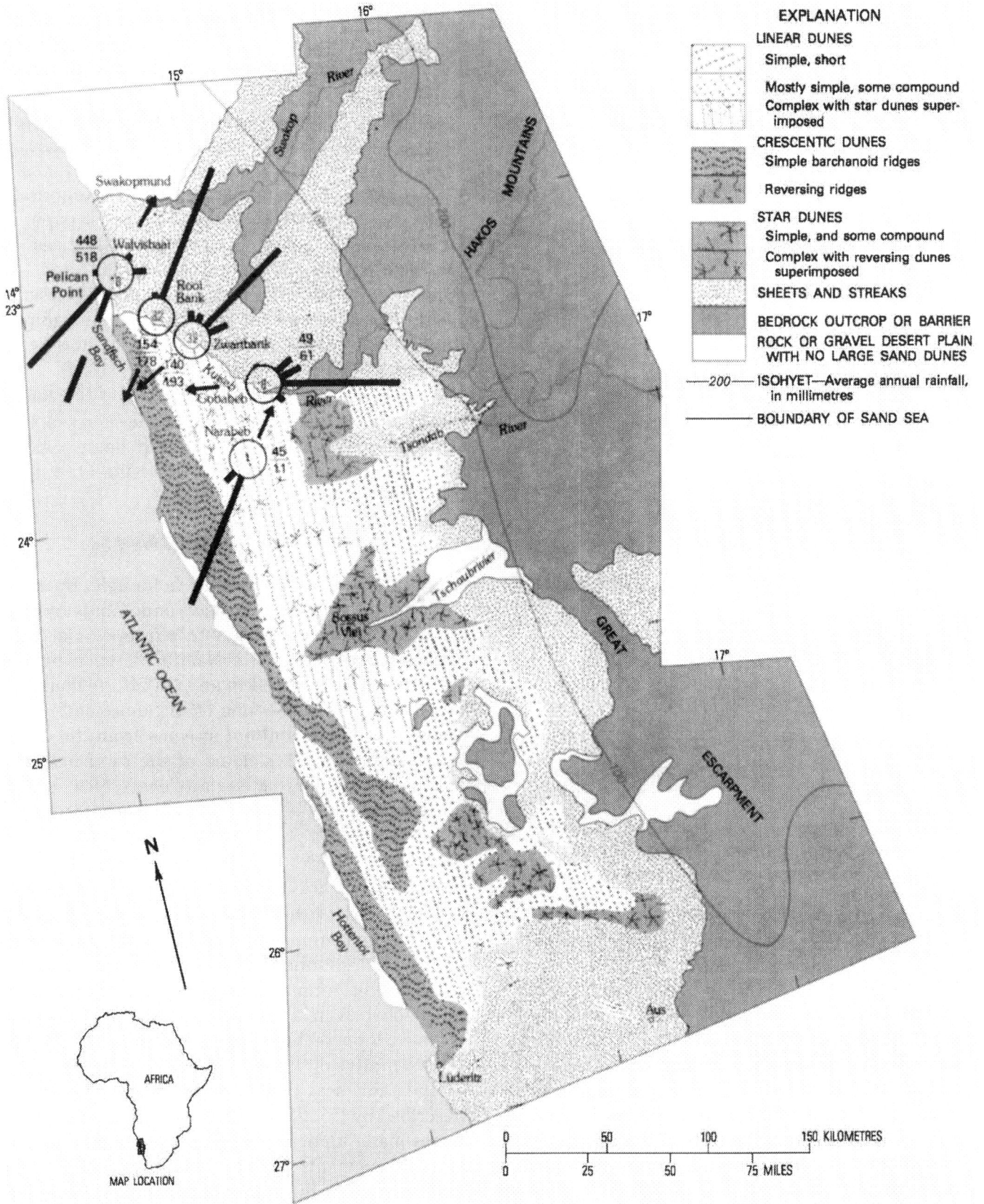

EXPLANATION

LINEAR DUNES

Simple, short

Mostly simple, some compound

Complex with star dunes superimposed

CRESCENTIC DUNES

Simple barchanoid ridges

Reversing ridges

STAR DUNES

Simple, and some compound

Complex with reversing dunes
superimposed

SHEETS AND STREAKS

BEDROCK OUTCROP OR BARRIER

ROCK OR GRAVEL DESERT PLAIN
WITH NO LARGE SAND DUNES

—200—— ISOHYET—Average annual rainfall,
in millimetres

———— BOUNDARY OF SAND SEA

16°

15°

River

Swakop

HAKOS   MOUNTAINS

Swakopmund

448
518

Walvisbaai

Pelican
Point

14°
23°

Rooi
Bank

17°

Zwartbank

49
61

Sandwich
Bay

154
178   140

Kuiseb

A93

Gobabeb

River

Tsondab

River

Narabeb

45
11

24°

ATLANTIC   OCEAN

Tschaubrunen

Sossus
Vlei

GREAT

25°

17°

ESCARPMENT

Tsal

N

26°

Hottentot
Bay

Aus

Luderitz

AFRICA

MAP LOCATION

0   50   100   150 KILOMETRES

0   25   50   75 MILES

27°

DISTRIBUTION AND MORPHOLOGY OF EOLIAN SAND in the central Namib Desert of South-West Africa. Map based on Landsat imagery. Sand roses for Zwartbank, Rooi Bank, and Gobabeb are each for the month during which highest drift potential occurs and are not directly comparable to annual sand roses (annual sand roses for these stations are not available). Drift potentials, resultant drift potentials, and resultant drift directions for Zwartbank and Rooi Bank were estimated from December-to-August, and December-to-July data, respectively. An annual sand rose is shown for Pelican Point. Isohyets are from Grove (1969). (Fig. 222.)

Dune ridges of the central Namib are described by Logan (1960, p. 136) as having a chaotic and disorderly pattern, the result of disruption of their crests by "blowouts," and partial closure of their interdune troughs by "transverse" dunes. A threefold division of dunes is recognized by Barnard (1973, p. 2) in the sand sea of the central Namib. It consists of (1) "a littoral belt of closely packed transverse dunes," (2) an interior zone of "longitudinal" dunes which he believed to have formed by deflation, and (3) an eastern zone of "multicyclic" complex dunes.

Some reported dune heights are as follows: 65 m (210 ft) for dunes in the coastal belt between Sandfisch Bay and Zwartbank (Stapff, 1887, in Goudie, 1970, p. 94); 80–100 m (260–380 ft) for "basically linear dunes" at Gobabeb (Goudie, 1970, p. 94; 1972, p. 24) and 300 m (nearly 1,000 ft) for "high, sharply crested dunes" near the Sossus Vlei along the interior margin of the sand sea (Barnard, 1973, p. 2,5).

Sand of the Namib Desert is generally described as brown (Goudie, 1970, p. 94) or red (Logan, 1960, p. 136). Inland dunes were said by Logan (1960, p. 136) to be "brick red," whereas dunes near the coast are recorded as "yellow-white." He stated that "This difference in coloration [between coastal and inland dunes] is believed to be the result of the greater age of the inland dunes which has allowed greater oxidation of the iron components within the sand" (Logan, 1960, p. 136).

The Namib Desert sand is reported to be 90 percent quartz and of uniform size, except that deviations in median diameter and sorting occur near the coast (Barnard, 1973, p. 2). Textures of Namib sand from dunes near Gobabeb and from "sand glaciers" (sand sheets, including climbing dunes) near Rossing Mountain, east of Swakopmund and north of the Swakop River, have been reported by Goudie (1970, p. 96; 1972, p. 24–25, fig. 16).

The main vegetation among the dunes near Gobabeb consists of sparse clumps of *Stipagrostis sabulicola* (Goudie, 1970, p. 94). Interdune areas support periodic growth of grass following infrequent precipitation (M. K. Seely, oral commun., 1975).

Fresh or brackish water that occurs beneath dunes near the coast, about 85 km (50 mi) west of the playa at the end of the Tsondab River (figs. 222, 223), is interpreted as evidence that the Tsondab River, the Tschaubrivier (fig. 224), and probably other intermittent streams that flow from the Great Escarpment at one time reached the Atlantic but that their courses have been truncated by the northward movement of dunes (Seely and Sandelowsky, 1974). A late Pleistocene age for the earliest occupation by man of the area near Narabeb, west of the Tsondab Vlei (figs. 222, 223), is based on the occurence of Early Stone Age tools, estimated to be 40,000–60,000 years old, in an interdune valley (Seely and Sandelowsky, 1974, p. 64).

Interpretation of Landsat Imagery

Dune types are distributed in three distinct zones across the sand sea — a coastal zone of crescentic dunes, a central zone of compound linear dunes, and an interior zone of star dunes and reversing dunes. Local variations of dune patterns within the zones occur especially near inselbergs and valleys of intermittent streams.

Closely spaced crescentic dune ridges occupy a belt 5–30 km (3–18 mi) wide along the Atlantic coast between Swakopmund and Lüderitz (figs. 222, 223). Measurements of dune width (horn to horn), length, and wavelength (spacing) were made on Landsat imagery (fig. 223, sample area 2) at 10-km (6-mi) intervals between Sandfisch Bay and Meob Bay. Mean dune width is 1.12 km (0.69 mi), mean length is 0.68 km (0.42 mi) and mean dune wavelength, 0.87 km (0.54 mi). These dunes are comparable to crescentic dunes in other areas observed on Landsat imagery (chapter J, table 38).

Compound linear dunes and complex dunes, mainly linear dune ridges with small star dunes and reversing dunes developed along their crests (chapter J, fig. 168), extend from the middle of the desert east of Lüderitz to an abrupt boundary at the south bank of the Kuiseb River (figs. 222, 223).

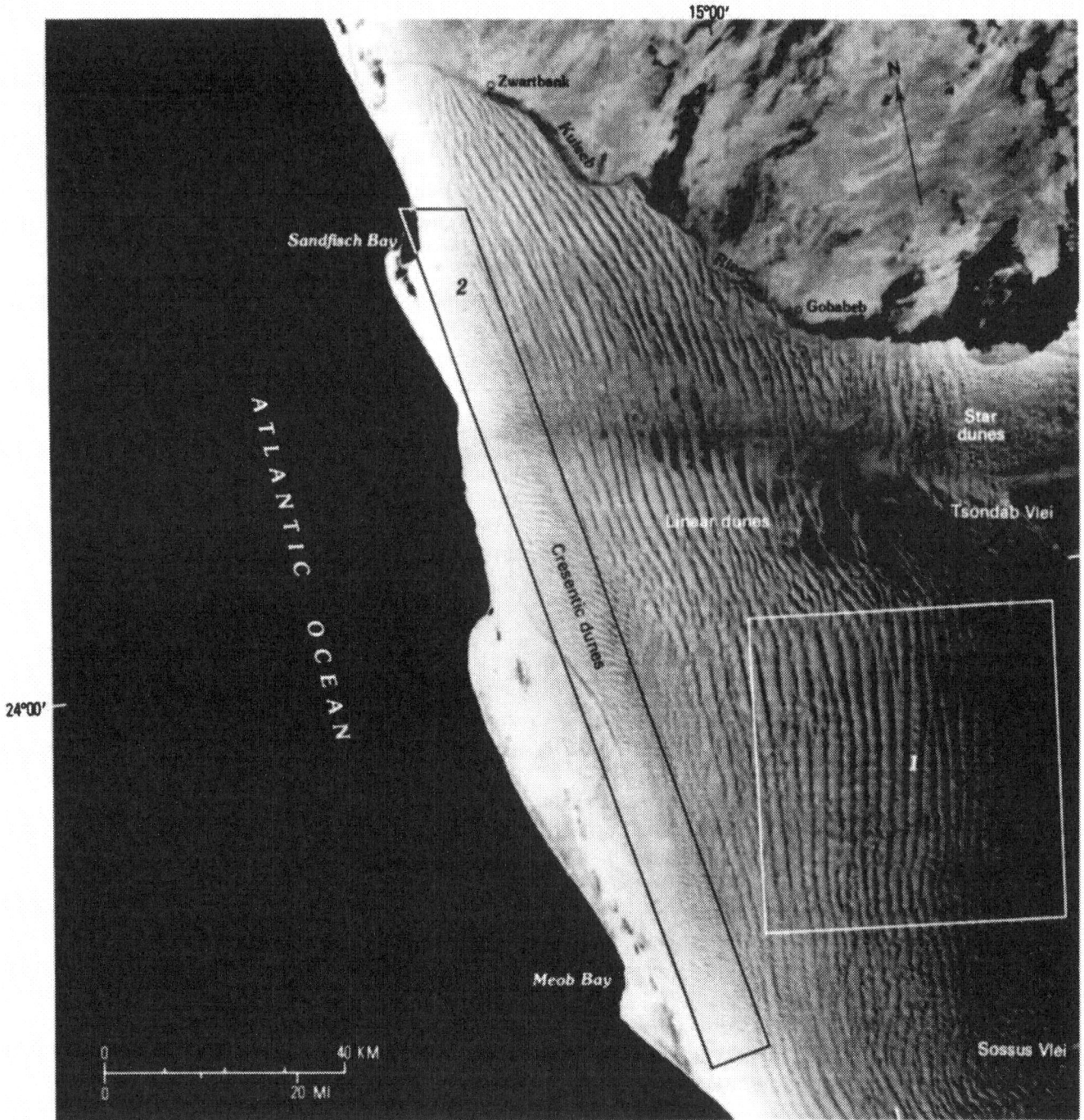

WESTERN PARTS OF THE MAIN SAND SEA in the central Namib Desert. Landsat imagery E1383–08264; E1383–08270. Outlined areas are samples of imagery on which (1) complex linear dunes with superimposed star dunes were measured and (2) crescentic dune ridges were measured. (Fig. 223.)

The linear dunes of the central zone occur in distinct parallel straight patterns (figs. 222, 223, 224). Disruptions of individual dune crests and interdune troughs seem to the authors to create a complexity of dune forms but not a "chaos" as described by Logan (1960, p. 136). Variations from the generally straight pattern of dunes alined N. 9° E. (fig. 223) are most pronounced near valleys of the Tsondab River and Tschaubrivier (figs. 223, 224).

Measurement of the linear dunes in a 2,500 km² (1,550 mi²) sample area in the center of the sand sea (fig. 223, sample area 1) yields the following mean

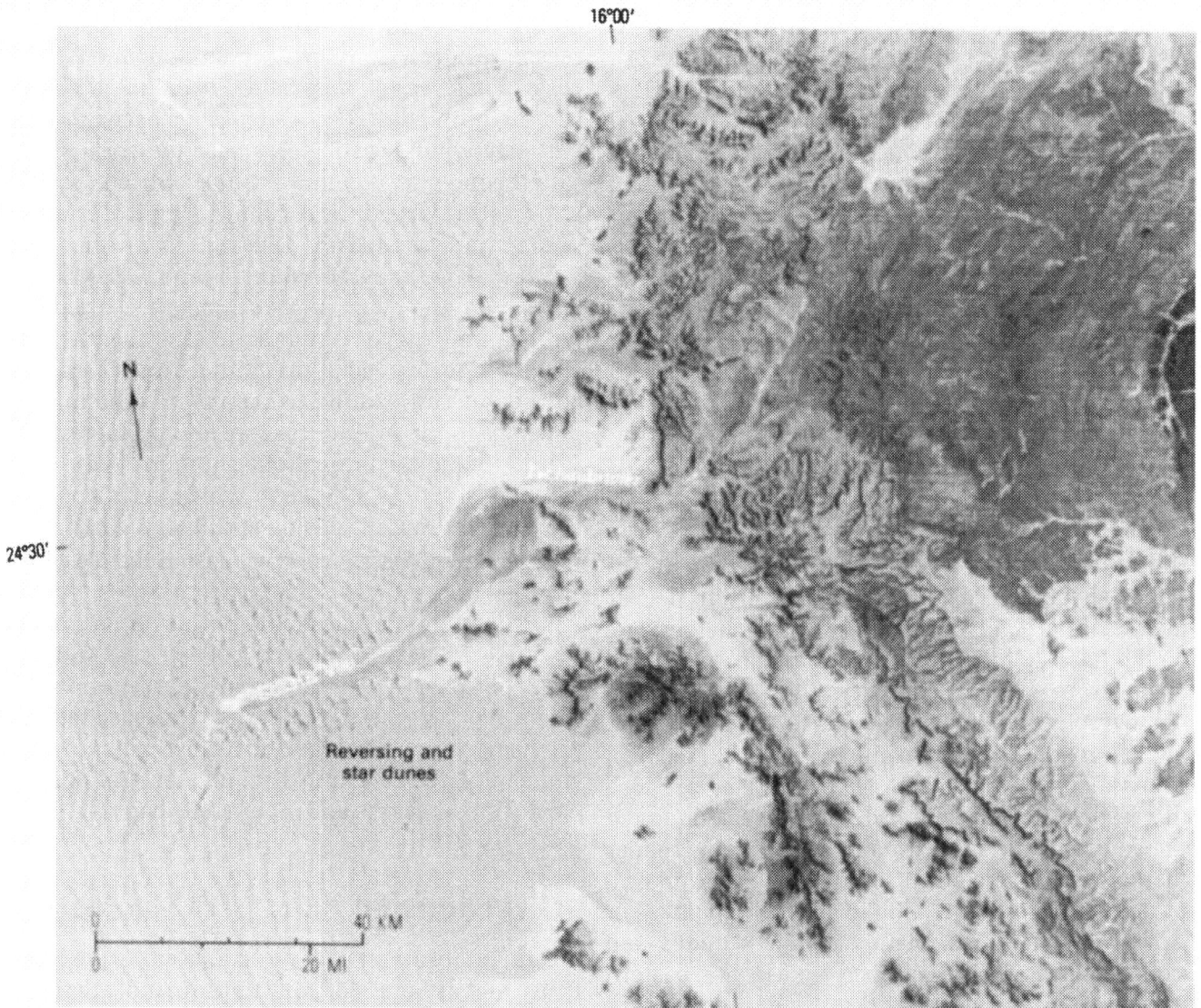

NORTHEAST MARGIN OF THE MAIN SAND SEA near inselbergs of the Great Escarpment (to the east) in the Namib Desert. Landsat false-color imagery E1202 – 08222. Disruption of the dune pattern around the Sossus Vlei (playa) is shown. (Fig. 224.)

values: dune width, 0.83 km (0.51 mi); dune length, 27 km (16 mi); dune wavelength 2.20 km (1.36 mi). The Namib dunes are compared with other dunes of linear type in chapter J (tables 36, 37).

Fields of star dunes, reversing dunes, and dunes of indeterminate (complex?) type occur along the eastern margin of the sand sea, near pediplains and inselbergs of the Great Escarpment (fig. 224). The sharply delineated boundaries of dune fields (fig. 224) coincide approximately with the 1,200-m (4,000-ft) contour line shown on a map of the area (U.S. Department of Commerce, U.S. Air Force Operational Navigation Chart Q–4). Varieties of

dunes in the interior marginal area contrast greatly with the simple crescentic dunes of the coastal belt but merge into the linear dunes of the middle zone (figs. 222, 223, 224).

Star dunes north of Aus (fig. 222) range in diameter from 0.4 km (0.2 mi) to 1 km (0.6 mi), mean diameter, 0.7 km (0.4 mi), and are spaced an average 2.2 km (1.4 mi) apart. The Namib star dunes are of medium size, compared with those of other deserts (chapter J, tables 40, 41). They have a preferred east-west orientation (large arms oriented N. 48° W.).

Landsat imagery shows dune-free surfaces on pediments around outliers of the Great Escarpment and in valleys of intermittent streams (fig. 224). Such dune-free areas around topographic barriers may result from sheetwash by occasional torrential rains along the escarpment. The escarpment is generally parallel to the 100-mm precipitation isohyet on the map (fig. 222). The eastern margin of the sand sea is marked by a high rampartlike ridge of sand (fig. 224), described by Logan (1960, p. 136) and similar in aspect on Landsat imagery to boundaries of sand seas in Tunisia, as described earlier (fig. 210).

Figure 225 shows dunes in the interior zone at the Sossus Vlei. Interbedding of white and tan playa sediments with the red sand of the dunes is common at the ends of the intermittent streams (M. K. Seely, written commun., 1975). Reddish hues of eolian sand translate on Landsat false-color imagery as shades of yellow (fig. 224), and the deeper the red of the sand, the more intense its yellow tone on the false-color imagery. In general, observations made from Landsat imagery (fig. 224), Skylab photographs (McKee and Breed, 1977), and ground photographs (fig. 225) confirm Logan's (1960, p. 136) statement that dunes farthest inland are also the reddest. In general, also, large linear dunes in the central zone of the Namib sand sea are redder than are crescentic dunes both to the west and east of them. Reddest of all are the star dunes and other eolian sand features, mainly complex dunes and sand sheets, around inselbergs near the eastern margin (fig. 224). Relative degrees of redness cannot be attributed simply to age differences, or to different lengths of time in a subaerial environment because numerous factors other than time can also control the degree of redness (Walker, chapter D).

Climatic Regime [14]

Southern Africa lies within a belt of high atmospheric pressure centered at approximately lat 30° S. (Schulze, 1972, p. 503). Surface wind circulation over the Namib Desert during the winter is controlled by the South Atlantic high, the Indian Ocean high, and high pressure over the interior

plateau (fig. 226). Counterclockwise circulation around the South Atlantic high during the winter season results in southerly winds along the Atlantic coast. These winds may be deflected inland by the heating of the desert sands and thus become southwesterly winds, particularly near the coast (Harry Van Loon, oral commun., 1974). Strong easterly winds also occur in the Namib Desert during the winter and are locally referred to as the "Berg Winds." They may be initiated when coastal lows appear on the west coast during the summer or winter and move around to the south coast (Schulze, 1972, p. 506), promoting winds across the adjacent desert. During the summer, a thermal low develops in the interior plateau and promotes landward circulation throughout southern Africa (fig. 226). Onshore (southwest) winds in the Namib Desert are strongest during spring and summer.

The Namib Desert receives less than 100 mm (4 in.) of rain per year, as shown by the isohyets in figure 222, and the variability of this rainfall may be as much as 80 percent from year to year (Schulze, 1972, p. 512, 514). An important source of moisture along the coast is the heavy dew associated with frequent fogs. This moisture may amount to as much as 2.5 cm (1 in.) of equivalent rainfall in a year (Royal Navy and South African Air Force, 1944, p. 37).

Direction and Amount of Sand Drift

Annual resultant drift directions along the Namib coast, and as far as 50 km (31 mi) inland, are toward the north or northeast as shown by annual sand rose and wind roses from Alexander Bay (south of the mapped area), Lüderitz (fig. 227), and Pelican Point, South Africa (fig. 222). This northward and northeastward direction of sand drift along the coast is steady throughout the year, with the exception of May−June, when the east winds ("Berg Winds") increase in strength. (See monthly sand roses for Pelican Point, chapter F, fig. 97A). Resultant drift directions at stations farther inland along the Kuiseb River, at the north end of the sand sea (fig. 222) seem to be toward the southwest. This apparent change in direction of potential sand movement may result from a decrease in the strength of the southwest wind from the coast inland, and a corresponding increase in the strength of the easterly winds ("Berg Winds") in the same direction.

---

[14] Principal sources of meteorological data are National Climatic Center, Asheville, North Carolina, U.S.A., and Desert Ecological Research Unit, Walvisbaai, South-West Africa.

RED SAND DUNES AND TAN PLAYA SEDIMENTS along the interior margin of the Namib Desert. Ground photograph; view at the Sossus Vlei. Photograph by M. K. Seely. (Fig. 225.)

Two lines of evidence suggest that at inland stations the potential sand-moving effect of the east winds is greater than that of southwest winds. First, during June and July, which is the season of east winds, the drift potential of winds from the northeast quadrant (encompasses the "Berg Winds") is highest at inland stations (fig. 228). Second, the high energy of the sea breeze, or southwest wind, seems to diminish rapidly inland, as shown by the decrease in drift potentials from Pelican Point eastward to Gobabeb (fig. 222). At Gobabeb, the south-

west wind is an important sand-moving wind only during October (fig. 229). At Rooi Bank, only 30 km (19 mi) from the coast, the "Berg Wind" is also a very important sand-moving wind (fig. 230).

Available data from stations along the Kuiseb River suggest that the resultant drift direction along the northern interior margin of the sand sea is roughly southwestward but wind data from Narabeb, a station within the sand sea, suggest a northeastward resultant drift direction (fig. 222). All the stations except those at the coast may to

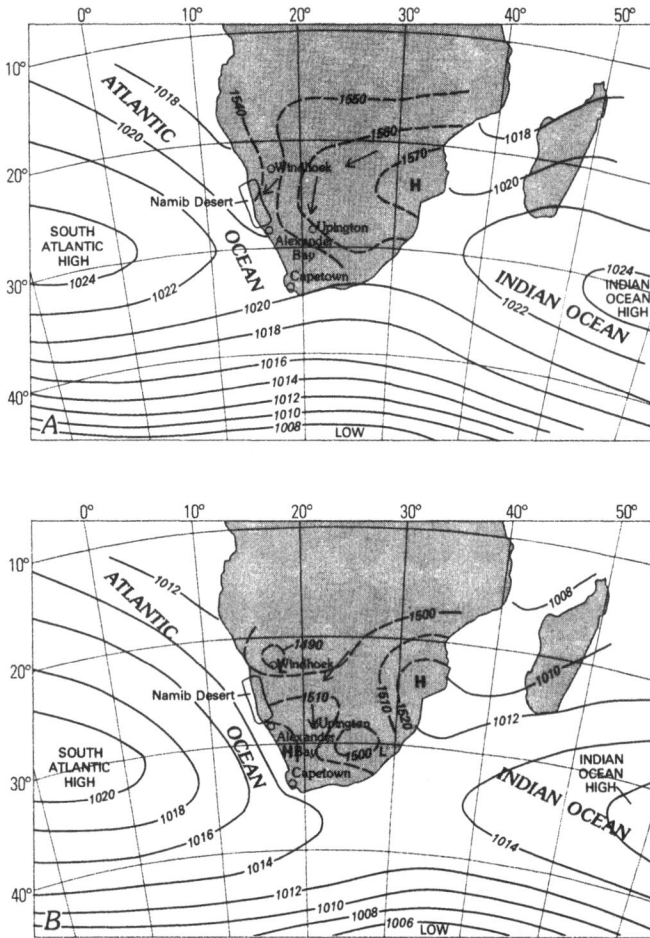

MEAN SEA LEVEL ISOBARS (solid lines) and isobars of the 850-millibar-pressure surface (dashed lines) at 1200 G.M.T. Arrows indicate general wind directions. *A*, June (winter in Southern Hemisphere); *B*, January (summer in Southern Hemisphere). Modified from Schulze (1972, p. 504). (Fig. 226.)

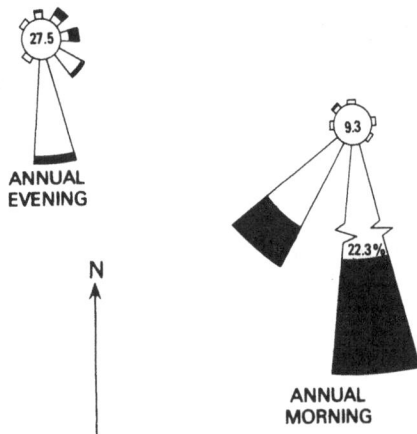

WIND ROSES SHOWING STRONG SOUTHERLY WIND REGIME at the south end of the sand sea in the central Namib Desert at Lüderitz. Number in centers of circles equal percent calm; white, 14–27 knots (25–50 km/hr); black, 28–40 knots (51–75 km/hr). (Fig. 227.)

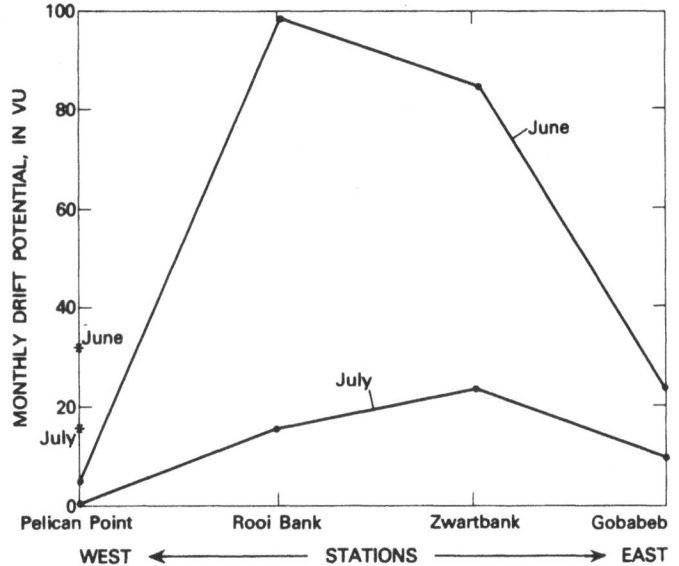

MONTHLY DRIFT POTENTIAL, in vector units, of winds from the northeast quadrant ("Berg Winds") during June and July at four stations from west to east across the northern part of the Namib Desert. During June, the "Berg Wind" increases in strength between Gobabeb and Rooi Bank, then decreases rapidly between Rooi Bank and Pelican Point (fig. 222). Southwest resultant drift directions at Rooi Bank and Zwartbank during June (fig. 222) aline with sand streaks north of the Kuiseb River, partly shown in figure 223. Asterisks indicate drift potential of southwest wind at Pelican Point during June and July, for comparison. (Fig. 228.)

some degree be sheltered from effective winds by high dunes nearby. All wind regimes, however, reflect both the southwest and east winds to some degree and suggest that in the central part of the desert, wind regimes may be roughly bimodal, with the two modes approximately equal in effectiveness. A bimodal wind regime is compatible with linear dunes (chapter F).

A comparison of the low drift potentials at most inland stations with those at stations on the coast suggests that the dunes in the center and along the interior margins of the sand sea are much less active than those near the coast. Wind data from Aus (fig. 231) suggests that some areas along the eastern margin of the Namib Desert may have complex wind regimes. Aus is about 80 km (48 mi) from fields of star dunes, which are generally in areas that have variable wind directions (chapter F, fig. 116).

The average drift potential of the Namib Desert is 237 VU, which places it in a moderate-energy range when compared to other deserts of the world (chapter F, table 15). The coastal stations of Alex-

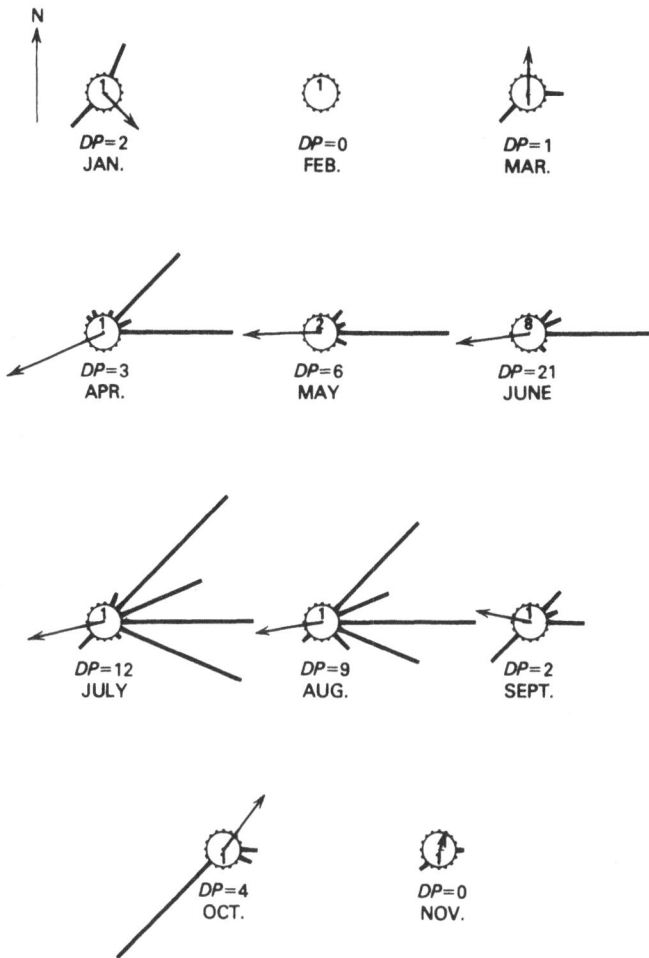

SAND ROSES FOR 11 MONTHS FOR GOBABEB, South-West Africa (location shown in figs. 222, 223). The southwest wind is strongest during October and is reflected in the long southwest arm of the sand rose. East ("Berg") wind is strongest in June, July, and August. December sand rose is unavailable. Drift potential is given for each month. (Fig. 229.)

ander Bay and Pelican Point experience a wide range in drift potential, with highest values occurring at Pelican Point (122 VU) during the spring (fig. 232). Maximum drift potentials at inland stations may occur in June or September (fig. 232).

Kalahari Desert of South-West Africa, South Africa, and Botswana

Summary of Conclusions

Linear dunes are the most common sand features of the western Kalahari Desert. Partially vegetated, mostly simple linear dunes trend northwest-southeast and cover about 100,000 km² (62,000 mi²).

These linear dunes average 0.3 km (0.2 mi) in width and 26.0 km (16 mi) in length, but many individual dunes are much longer. A second type of feature consists of long narrow sand bodies (which may be very large linear dunes or elongate sand sheets) that trend east-west in a region of about 15,000 km² (9,300 mi²) between the Blydeverwachterplato and the Molopo River valley. These sand features have an average width of 2.3 km (1.5 mi), and some extend more than 100 km (62 mi) across hills and valleys. They swing southward at the Molopo River and merge with the sand sea of linear dunes to the east.

Annual rainfall in the southwestern Kalahari Desert is less than 200 mm (8 in.) in most places but increases northward and eastward — from regions of little vegetation into regions of much greater vegetation. At present, the Kalahari Desert as a whole has little wind energy available for sand movement, but drift potentials vary widely within the desert. High drift potentials occur in the less-vegetated, southernmost part of the study area. Low drift potentials occur in the northern parts of the study area in regions of partially vegetated simple linear dunes. The linear dunes and (or) sand sheets in the southwestern Kalahari Desert are alined with the resultant drift directions of present-day winds.

Introduction

The Kalahari physiographic province (Wellington, 1955, p. 52) includes about 1,613,800 km² (1,002,800 mi²) of lowland sandy desert, but only the southwestern part — in South Africa, Botswana, and South-West Africa — has dunes recognizable on Landsat imagery. The sand sea of the southwestern Kalahari Desert is in a basin of internal drainage centered at Abiekwasputs on the Molopo River (fig. 233). The lower parts of this river and its tributaries (the Auob, Nossob, and others) are abandoned drainages that have been choked with eolian sand for at least 1,000 years (Lewis, 1936, p. 30, fig. 6). They are thus cut off from the perennial Orange River, which marks the southern boundary of the Kalahari dune country and drains surrounding territories.

Published reports of the Kalahari region include reconnaissance work (A. W. Rogers, 1934, 1936), field observations of dunes (Lewis, 1936), a general geographic treatise (Wellington, 1955), accounts of geology and geomorphology in the southwestern

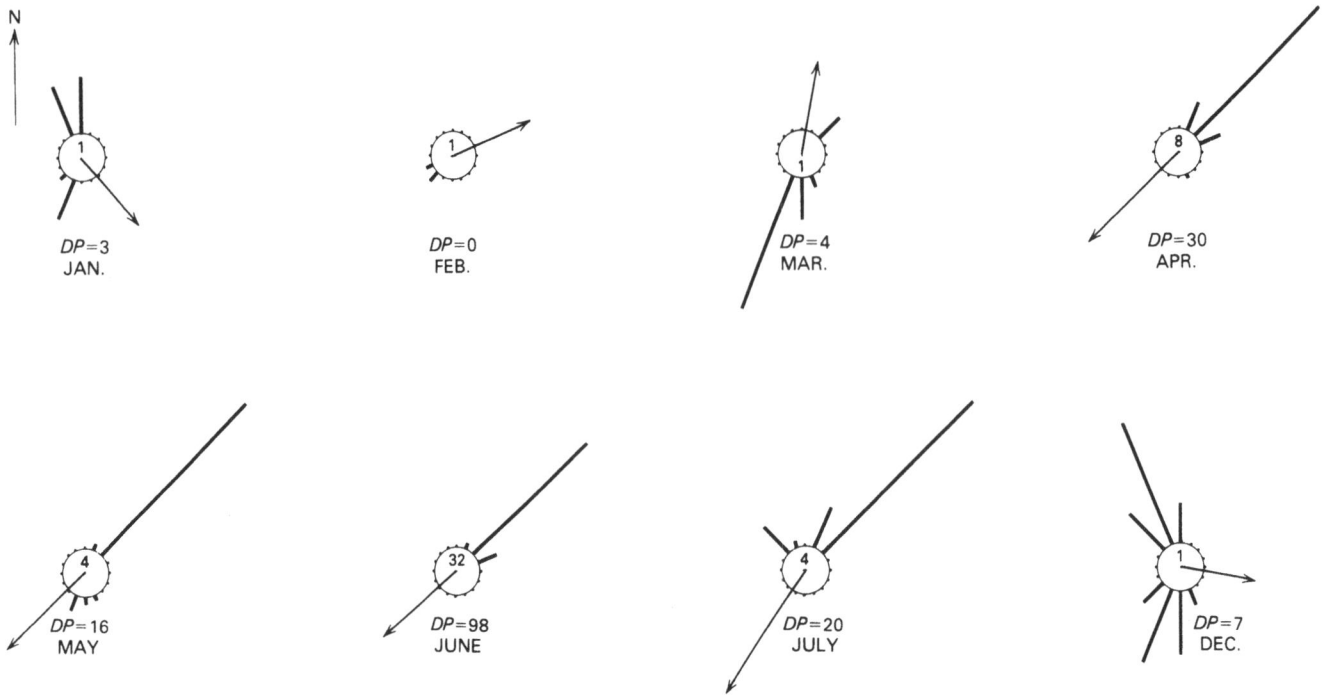

SAND ROSES FOR 8 MONTHS FOR ROOI BANK, South-West Africa, approximately 30 km (18 mi) from the coast. Although this station is very near the coast, the "Berg Wind" (at this station a northeast wind) is the most important sand-moving wind during much of the year. Data for for August–November are unavailable. Compare with sand roses for Pelican Point for the same months, chapter F., figure 97A. Drift potential is given for each month. (Fig. 230.)

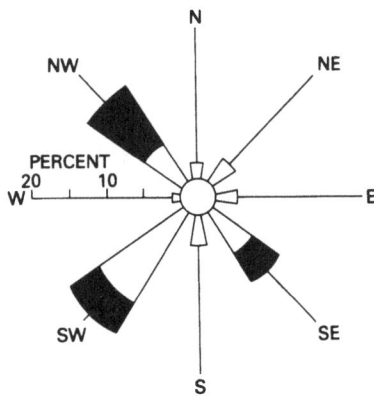

WIND ROSE FOR AUS, South-West Africa, near fields of star dunes in the interior of the Namib Desert. Data were insufficient for computation of sand roses but indicate sand-moving winds from several directions. 1 cm equals 10 percent occurrence. Light areas, winds 14–27 knots (25–50 km/hr); dark areas, winds 28–40 knots (51–75 km/hr). Source of wind data is Royal Navy and South African Air Force (1944). (Fig. 231.)

Kalahari (Mabbutt, 1955) and Botswana (McConnell, 1959; Boocock and van Straten, 1962), a paper describing landforms in relation to climatic change (Grove, 1969), and two papers describing dune morphology and distribution (Goudie, 1969, 1970). Useful maps are the U.S. Department of Commerce, U.S. Air Force Operational Navigation Charts P–4 and Q–4.

Landsat imagery supplemented by the Skylab photographs have been used to map and interpret eolian sand morphology and distribution in the southwestern Kalahari Desert. Data from meteorological stations in southern Africa were used to show relationships of the sand bodies to present wind regimes and precipitation patterns (fig. 233).

Review of Previous Work

About 200 sand samples obtained from dune crests, slopes, and interdune areas near the Aoub and Nossob Rivers, and Kurumanrivier in the southern part of the sand sea are reported (Lewis, 1936) to be composed of partially rounded sand grains that have a composition of 90 percent quartz with accessory zircon, garnet, feldspar, ilmenite, and tourmaline. Samples from dune crests include fine to medium sand, whereas samples from inter-

Eolian and fluviatile sediments of the Kalahari Beds of Tertiary-Pleistocene age occur throughout most of the Kalahari Desert (Boocock and van Straten, 1962, p. 151–154). Suggested sources for some of the younger sands of the Kalahari Beds are the eolian Triassic Cave Sandstone and its equivalents, which crop out in several parts of the region (Boocock and van Straten, 1962, p. 154). Eolian sands of the Upper Kalahari Beds of late Tertiary-Pleistocene age are also known as red sands, Plateau Sands, and Kalahari Sands (Mabbutt, 1955, p. 20, 25–27; table 1), and as sands of Kalahari type (Boocock and van Straten, 1962, p. 151–154).

Samples of some of the near-surface eolian sand in the Kalahari sand sea contain detrital kyanite, epidote, and staurolite, decreasing in amount to the southeast (A. W. Rogers, 1934, p. 11; Poldervaart, 1955, p. 106). This has led several workers to conclude that the latest redistribution of eolian sand of Kalahari type in Botswana has been from the northwest (Boocock and van Straten, 1962, p. 154).

Dune sand in the Kalahari Desert is described (Grove, 1969, p. 199) as yellowish red (5 YR/5/6 to 2.5 YR/4/8; Goddard and others, 1948) without regular differences in hue from dune crests to interdune troughs. Observations in Botswana (Boocock and van Straten, 1962, p. 153) have indicated that sands of Kalahari type occur both as dark-reddish-brown consolidated sand with considerable amounts of clay and silt, and as belts of grayish-white less consolidated sand.

Unconsolidated eolian sand in the Kalahari Desert ranges in thickness from 3 to 33 m (10 to 100 ft) (Grove, 1969, p. 195), but the entire sequence of Tertiary-Pleistocene Kalahari Beds composed of eolian and fluviatile sediments is 100 m (300 ft) or more thick, with maximum thicknesses where these sediments have filled in old drainage channels (Boocock and van Straten, 1962, p. 152, 154). At Sekutani (beyond the map area, in the southeastern part of the basin at about lat 25°40′ S., long 24°20′ E.) a borehole through a buried channel penetrated 157 m (515 ft) of Kalahari Beds (Boocock and van Straten, 1962, p. 152), and at Mafeking (east of Tsabong at approximately lat 26° S., long 25°40′ E.), a borehole penetrated 141 m (463 ft) of "Kalahari deposits" (A. W. Rogers, 1934, p. 10).

Dune crests reportedly are bare except for grasses (Stipagrostis and Eragrostis sp.). Dune slopes sup-

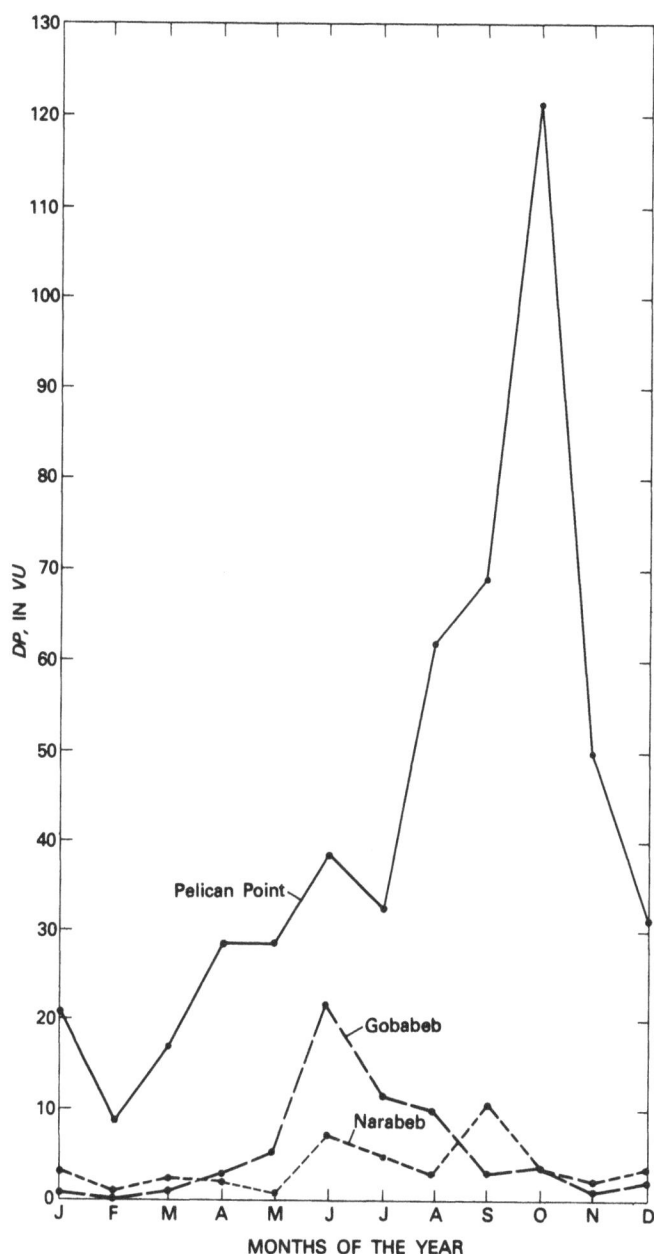

VARIATION IN DRIFT POTENTIAL (DP), in vector units, at Pelican Point, Gobabeb, and Narabeb in the Namib Desert, South-West Africa. All stations experience a sand-moving season in June, a result of the east ("Berg") wind. Pelican Point experiences highest drift potentials during October, when the southwest wind is strongest. (Fig. 232.)

dune areas include very fine to very coarse sand (Lewis, 1936, p. 22–23, 26–28). A "few dozen" samples obtained between Gobabis and the Molopo River in the northwestern part of the sand sea contained moderately rounded sand (A. W. Rogers, 1934, p. 11).

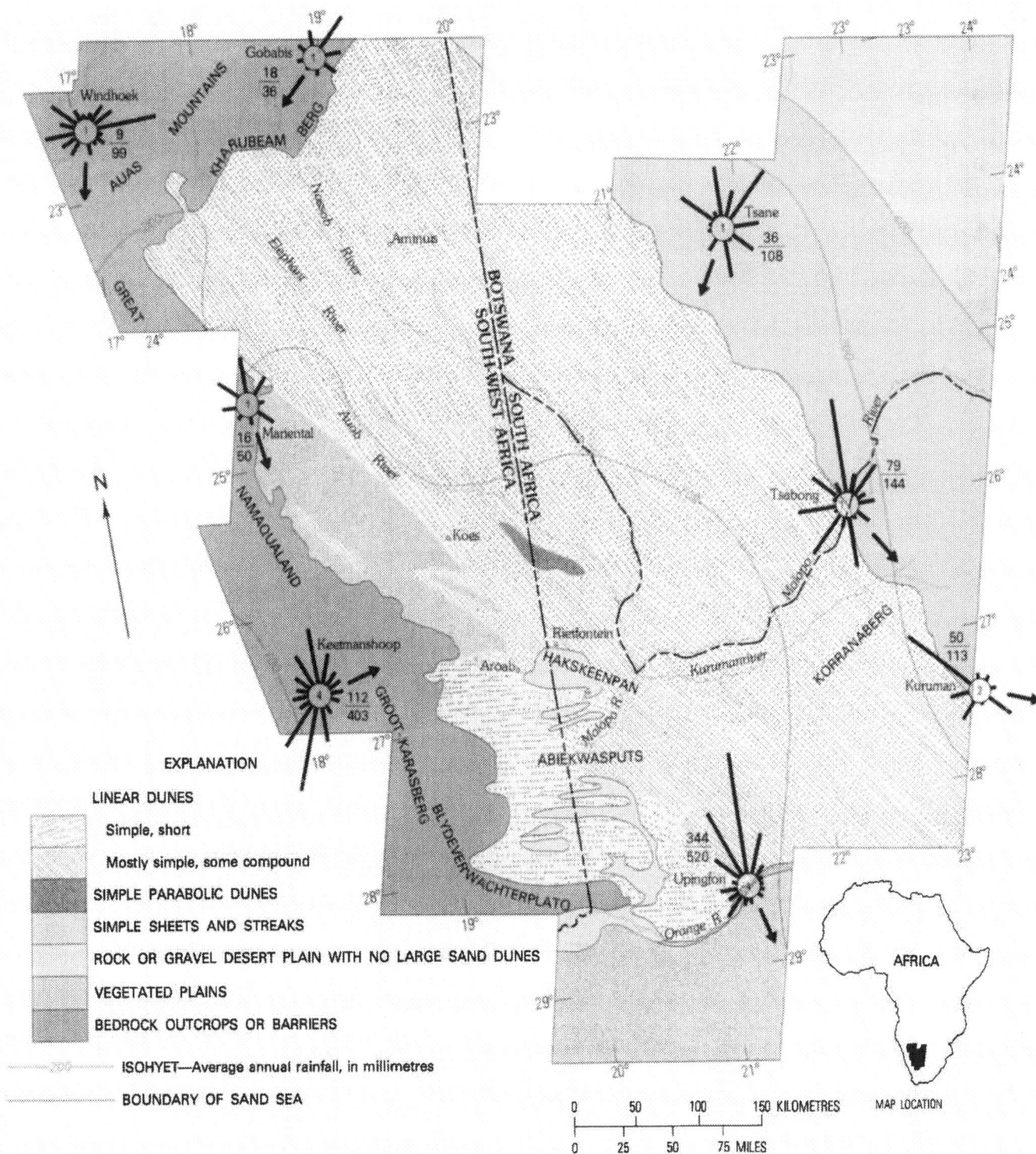

DISTRIBUTION AND MORPHOLOGY OF EOLIAN SAND in the western Kalahari Desert, southern Africa. Map based on Landsat imagery. Isohyets from Grove (1969). (Fig. 233.)

port shrubs and trees as much as 3 m (10 ft) tall, and interdune areas have trees (mainly *Acacia giraffae* and *Acacia haematoxylon*) that grow to heights of 5 m (16 ft) (Grove, 1969, p. 199).

Dunes in the lower Molopo River area above Abiekwasputs (fig. 233) are described (Wellington, 1955, pl. 8A) as longitudinal ridges with Y-junctions opening to the northwest (upwind); they are 10 to 25 m (33 to 82 ft) high and are spaced about 300 m (1,000 ft) apart (Goudie, 1969, p. 404). The distribution of these dunes was measured on aerial photographs by Goudie (1969), who found that their branching is random except where breaks of underlying slope and other topographic influences occur.

Interpretation of Landsat Imagery

Dunes in the southwest Kalahari Desert are mainly of linear (longitudinal) types; barchans and other varieties described by some workers (Lewis, 1936, p. 30; Grove, 1969, p. 199) are mostly too small to be discerned on Landsat imagery and on Skylab photographs . An exception (fig. 233) is a field of parabolic dunes between Reitfontein and the Auob River, which is evident on Landsat transparencies.

Most of the Kalahari sand sea consists of southeast-trending linear dunes of apparently simple form which cover about 100,000 km² (62,100 mi²). Width, length, and wavelength (distance from crest to crest) of these dunes were measured on Landsat transparencies of six 2,500-km² (1,550-mi²) sample areas in and around the regions shown in chapter J, figure 169A. Results are discussed in chapter J. Simple linear dunes of the Kalahari observed on space imagery average 0.3 km (0.2 mi) in width and 26.0 km (16 mi) in length and have an average wavelength of 0.7 km (0.4 mi). Their aspect on imagery made during the rainy season implies that flanks of the dunes are stabilized by vegetation.

Very large, irregularly distributed linear sand features, described previously as "stringer dunes" (McKee and Breed, 1976, fig. 59), cover 15,000 km² (9,300 mi²) and trend eastward from the Blydeverwachterplato to the Molopo River valley. There, they merge with the sand sea of simple linear dunes just described, and the alinement both of dunes and of other sand features swings southward (fig. 234). The extent of vegetative cover on the large sand features is not known. They are shown as areas of "Kalahari sand" on some earlier maps (Mabbutt, 1955, pl. IV).

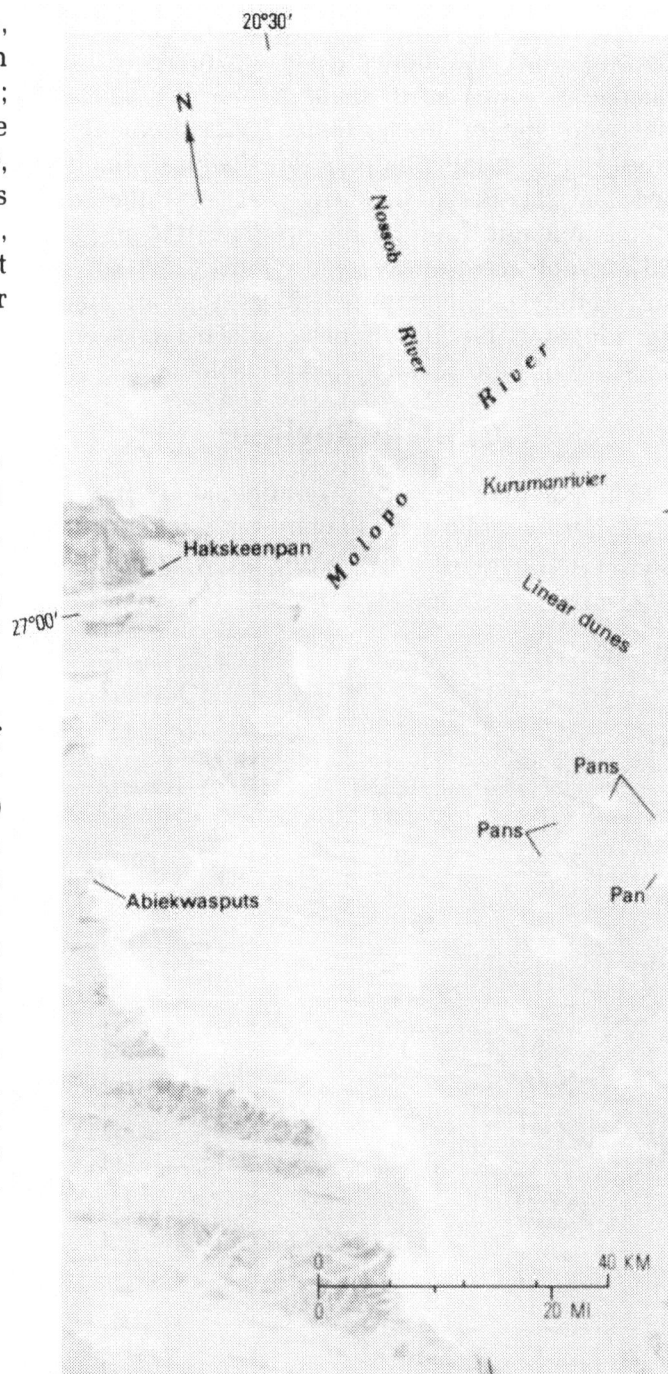

ABANDONED DRAINAGES of the lower Molopo River, Nossob River, and Kurumanrivier (white), playas or pans, (blue and white), and dunes and invasive sand sheets (yellow). Landsat false-color imagery E1198–08000. Elongate sand bodies (sheets or very large linear dunes) extend undeflected across badlands (bluish-green in false-color) in the southwest corner and merge with linear dunes that continue south from the area shown in chapter J, figure 169A. (Fig. 234.)

These large east-trending sand features are observed on Landsat imagery (fig. 234) to cross hilly country and abandoned river channels without change of course or of shape, which is a common characteristic of linear dunes. Measurements of these large sand features on Landsat imagery (McKee and Breed, 1976, fig. 59) show that they range in length from 50 to more than 100 km (31 to 63 mi) and have an average width of 2.3 km (1.4 mi). If they are linear dunes ridges, they are among the longest and widest dunes of linear type seen on imagery of sand seas (chapter J, table 36).

Climatic Regimes

The Kalahari Desert is dominated by high atmospheric pressure in the winter and low pressure in the summer (fig. 226). During the winter, surface wind flow follows a counterclockwise trend, from east-northeast winds at Ghanzi to northerly winds at Upington (Royal Navy and South African Air Force, 1944, fig. 2). During the summer, anticyclonic circulation is weak, and winds are variable. Development of a thermal low over the Kalahari Desert during this season promotes wind flow from the southwest in the region of the desert near Upington and Keetmanshoop (Schulze, 1965, p. 235–238). (See also the annual sand roses for Keetmanshoop and Upington, fig. 233, which reflect this southwest component.)

Annual rainfall in the Kalahari Desert ranges from 144 mm (5.5 in.) at Keetmanshoop and 204 mm (8 in.) at Upington in the west to more than 400 mm (16 in.) at Ghanzi in the northeast. Isohyets are roughly parallel from northwest to southeast, with

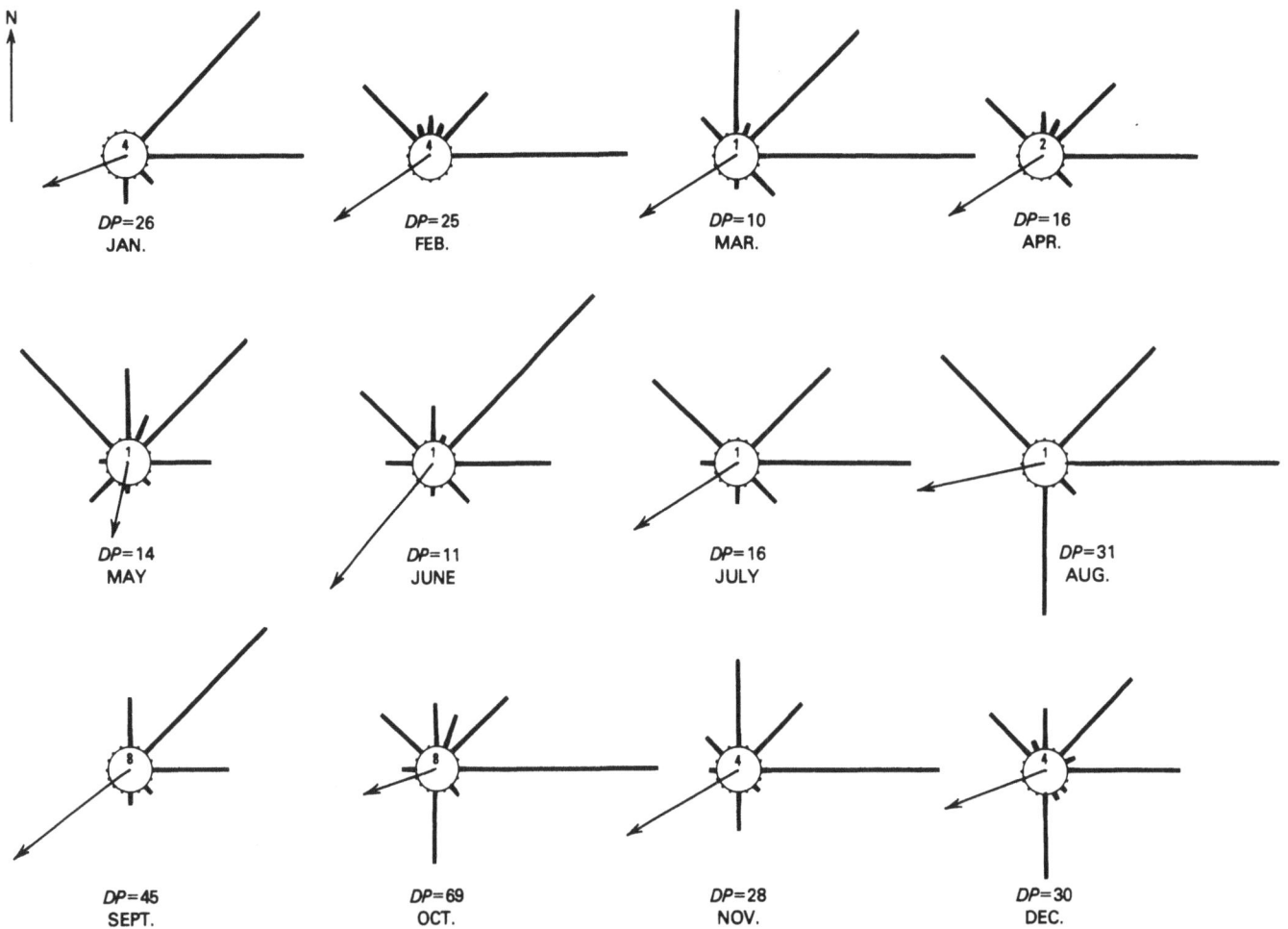

SAND ROSES FOR 12 MONTHS FOR GHANZI, Botswana, in the northeastern part of the Kalahari Desert. Although monthly wind distributions tend to be complex, resultant drift directions from month to month are toward the southwest. (Fig. 235.)

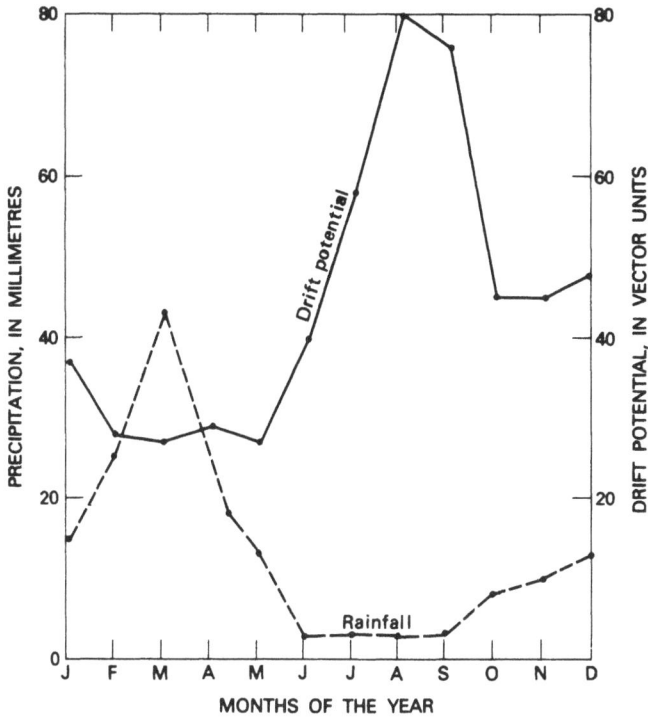

AVERAGE MONTHLY RAINFALL versus average monthly drift potential of Upington, South-West Africa. Season of highest drift potential (June–October) corresponds to season of least rainfall, which increases the probability of sand movement during this period. (Fig. 236.)

rainfall increasing toward the northeast (fig. 233). The driest season in this region is winter, and the wettest season is summer. Summer rainfall occurs mostly in the form of thunderstorms. These begin earliest in the northeastern part of the region (near Ghanzi) and during the summer extend westward across the desert (Schulze, 1972, p. 515, fig. 3).

Direction and Amount of Sand Drift

Evidence of vegetation on Landsat imagery, and data from published sources (Grove, 1969; Flint and Bond, 1968) confirm that most of the Kalahari Desert east of the 200-mm (8-in.) isohyet is a relatively inactive sand sea. However, most of the present-day resultant drift directions are parallel or subparallel to trends of all linear dunes and (or) sand streaks visible on Landsat imagery. Dune alinements seem continuous throughout the southern Kalahari Desert, from the less vegetated southwestern part to the heavily vegetated regions of presumably less active dunes in the northwest and east. Distinct stages of dune development in

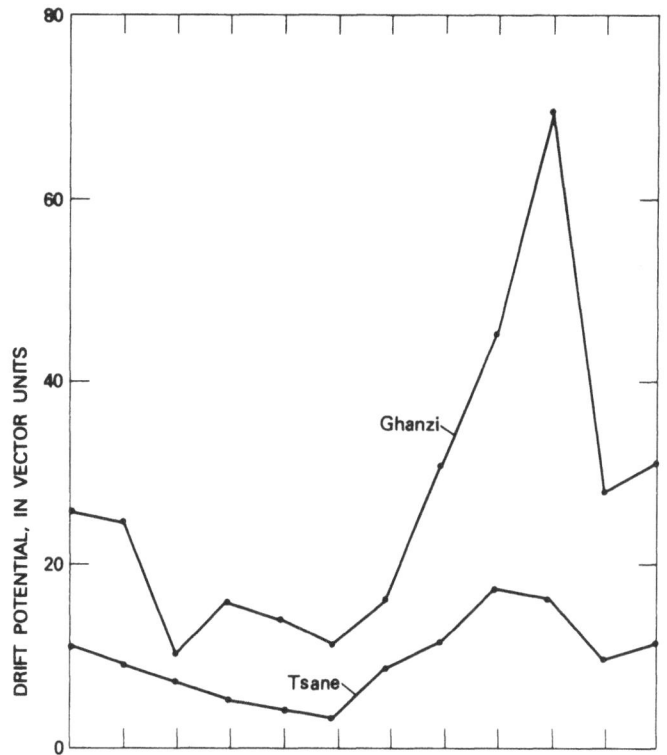

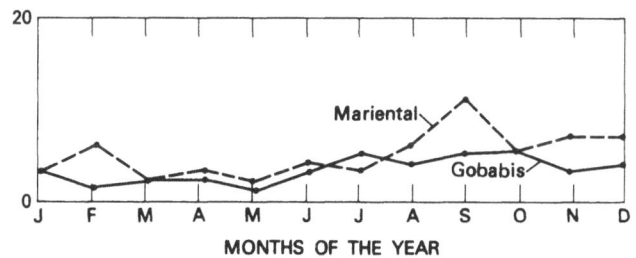

VARIATION IN DRIFT POTENTIAL during the year at four stations in the Kalahari Desert: Ghanzi and Tsane, Botswana (northeast and southeast desert, respectively); and Mariental and Gobabis, South-West Africa (northwest desert). Sand movement at stations with high drift potentials (Ghanzi, Upington) usually occurs during well-defined seasons of high drift potential, with periods of much lower drift potentials the remainder of the year. (Fig. 237.)

this region, thus, cannot be deduced on the basis of morphology alone, as can be done in the Sahel (described earlier in this chapter), where crossed dune trends clearly indicate two periods of dune growth.

Most present-day wind regimes in the Kalahari Desert are complex (fig. 233), although resultant drift directions at the various stations are relatively steady during the year (fig. 235). The region of less vegetated, presumably active sand in the southwestern part of the desert is characterized by obtuse bimodal winds, as shown by the annual sand rose for Upington (fig. 233). The weaker southwest mode occurs principally during the summer. Additionally, the effective wind regime at Upington (fig. 189) is similar on the basis of energy and directional characteristics (chapter F, fig. 105) to present-day effective winds associated with linear dunes in other regions. This relationship indicates that linear dunes visible on Landsat imagery of the Upington region may have developed in response to winds similar to those occuring in the region today.

The average drift potential of nine stations in the Kalahari Desert is 191 *VU*, indicating that the desert as a whole has relatively little energy available for sand movement (chapter F, table 16). However, drift potentials within the desert range widely from 36 *VU* at Gobabis to 520 *VU* at Upington. Most stations have low drift potentials, which, combined with the effects of vegetation in weakening effective winds, may mean that these areas have negligible amounts of sand drift under present-day conditions. The important exception to this pattern is at Upington, where the season of minimum rains coincides with the season of highest drift potential (fig. 236), as in India-Pakistan. The time of highest drift potential at a station depends upon its position in the Kalahari Desert. At Upington (southwest desert), the highest drift potential is in August (fig. 236); at Gobabis and Mariental (west and northwest desert), it is in September (fig. 237); and at Tsane and Ghanzi (eastern desert), it is in September and October. The greatest range of drift potential during the year is observed at the sites with the highest annual drift potentials (Upington, Ghanzi, figs. 236, 237).

Arabian Sand Seas of Saudi Arabia, Oman, Yemen, and Adjacent States

Summary of Conclusions

Large compound and complex linear, crescentic, star, and dome-shaped dunes are observed on Landsat imagery of Arabian sand seas. Dunes of the Rub' al Khali, in the southern part of the Arabian Peninsula, are distributed according to type into three regions — crescentic dunes in the northeast, star dunes along the eastern and southern margins, and linear dunes throughout the western half.

The compound crescentic dunes in Al 'Urūq al Mu'taridah (northeastern Rub' al Khali) are unusually large, with a mean width (horn to horn) of 2.8 km (1.7 mi) and a mean length of 2.1 km (1.3 mi). Presence of a field of smaller south-facing crescentic ridges at the north edge of Al 'Urūq al Mu'taridah might be interpreted as evidence of at least two major episodes of dune building.

Compound linear dunes at the western (downwind) margin of the Rub' al Khali are the largest observed in this study, averaging 1.5 km (1 mi) wide. Compound linear dunes in the southwestern Rub' al Khali have mean widths of 0.7 km (0.4 mi) and lengths of commonly more than 100 km (60 mi). Their southwest trend is parallel to the annual resultant drift direction computed for Al 'Ubaylah. Large compound linear dunes, similar in width and length to linear dunes of the Namib Desert, South-West Africa, occur in the Wahiba Sands, Oman.

Sheets of sand associated with linear, crescentic, and star dunes occur in the Ad Dahnā' and the sand seas in the northern part of the Arabian Peninsula. Giant dome-shaped dunes with a mean diameter of 1.2 km (0.8 mi) fill the Nafūd ath Thuwayrāt, Nafūd as Sirr, and Nafūd Shuqayyiqah.

Average annual precipitation is less than 100 mm per year (0.39 in./yr) throughout most of the sandy regions of the Arabian Peninsula. Surface wind circulation over the Arabian Peninsula is controlled in the winter by the Sahara High of the African continent and in the summer by a large low-pressure cell over the Indian subcontinent. The northern deserts of Arabia experience high wind energies. The most northerly regions, near Badanah, have highest drift potentials in the spring (April); farther southeast, near Dhahran, maximum drift potentials occur during June. Low to intermediate annual drift potentials may occur in the Rub' al Khali.

Annual resultant drift directions in An Nafūd and Ad Dahnā' trend in a broad arc from east through south, exactly paralleling the trend of the sand seas. Annual resultant drift directions in the Rub' al Khali are toward the east near Ash Shariqah, toward the north along the eastern margin, and toward the southwest in the western part of the desert.

Unimodal, bimodal, and complex wind distributions in An Nafūd and Ad Dahnā' are roughly compatible with the variety of dune types in the region. Wind data for the Rub' al Khali are incomplete, but the unimodal effective wind distribution at Al

'Ubaylah is compatible with the crescentic dunes observed near the station.

Introduction

Sand seas with very large dunes extend over 795,000 km² (477,000 mi²) of the Arabian Peninsula (I. G. Wilson, 1973, table 1). Main sand seas are the Rub' al Khali and Wahiba Sands in the south and the An Nafūd, Ad Dahnā', and numerous smaller sand seas in the north. Landsat imagery has been used to map the distribution and morphology of eolian sand in the southern part of the Arabian Peninsula (fig. 238) and to illustrate the varieties of dunes and their distribution in the northern sand seas. Climatic data are shown by precipitation isohyets (fig. 238) and sand roses (figs. 238, 243).

Geologic quadrangle maps of the U.S. Geological Survey published between 1956 and 1963 (Bramkamp and others, 1956; Bramkamp and Ramirez, 1958, 1959a, b, 1960, 1963; Bramkamp, Gierhart, and others, 1963; Bramkamp, Ramirez, and others, 1963; Ramirez and others, 1963a, b; and Brown and others, 1963) show the general distribution and morphology of the sand. Names used herein conform in spelling to usage on the U.S. Geological Survey map of the Arabian Peninsula (U.S. Geological Survey, 1963). Two index maps show distribution of dune areas (Holm, 1960, fig. 1; 1968, fig. 3), and three maps show eolian landforms in the eastern part of the Rub' al Khali (Glennie, 1970, enclosures 1–3). Other useful maps are the U.S. Department of Commerce, U.S. Air Force Operational Navigation Charts J–6, J–7, H–5, and H–6, at a scale of 1:1,000,000.

The Rub' al Khali

Review of Previous Work

Much of southern Arabia is a structural basin with its axis plunging gently northeastward into the Persian Gulf (Powers and others, 1966, p. D106). The Rub' al Khali sand sea (fig. 238) covers about 560,000 km² (216,000 mi²) of the region (I. G. Wilson, 1973, p. 86). It extends from the United Arab Emirates westward almost 1,500 km (930 mi) to the foothills of the Yemen Mountains. Sources of eolian sand in the Rub' al Khali are unknown, although some authors believe that the sand has been transported by wind from alluvial (wadi) sediments that were eroded from Paleozoic and Mesozoic sandstones (Holm, 1960, p. 1370; Brown,

1960, p. 157). The growth of dunes to as much as 300 m (1,000 ft) high in the Arabian sand seas, such as those described by Brown, Layne, Goudarzi, and MacLean (1963), may require thousands of years (I. G. Wilson, 1971a, p. 183). A Pleistocene age for the large dunes of the Al 'Urūq al Mu'taridah, in the eastern Rub' al Khali, is suggested by Glennie (1970, p. 96).

Two forms of crescentic dunes were recognized in the eastern Rub' al Khali (fig. 238) and described by Glennie (1970, p. 88–89) as "predominantly barchan dunes, both simple and complex," and as "giant sand massifs in various forms, often of complex barchanlike structure with their axes transverse to the dominant winds." The latter dunes are termed "giant crescentic massifs" by Holm (1960, p. 1372, fig. 5) and are herein referred to as "compound crescentic dunes," using the classification described in chapter J. (See also chapter J, table 27.)

Compound linear dunes of a "hooked" variety are recognized in the north-central Rub' al Khali (Holm, 1960, p. 1371–1372; 1968, fig. 3). Linear dunes 100 m (330 ft) high, 20–200 km (12–125 mi) long, and 1–2 km (0.6–1.2 mi) wide, separated by flat interdune corridors, are described in the Qa'amiyat region of the southwestern Rub' al Khali by Bunker (1953, p. 428–429) and by Holm (1960, p. 1373). These dunes, which have subsidiary linear ridges oblique and parallel to the main ridge (Holm, 1960, fig. 7), are classified as compound linear dunes (chapter J, tables 25, 26, 36).

Star dunes are recognized in the Ramlat as Sahama at the southeastern margin of the Rub' al Khali by Thesiger (in Bagnold, 1951, fig. 4), Beydoun (1966, p. H5), and Glennie (1970, p. 87, figs. 70–71). "Pyramidal" (star) dunes in the Rub' al Khali and in the northern sand seas are described by Holm (1960, p. 1371, 1373). Star dunes on the edge of the 'Irq as Subay' east of Zalim (about lat 22°30' N., long 43° E.) were sectioned and sampled by McKee (1966).

Interpretation of Landsat Imagery

Landsat imagery and Skylab photographs show that dunes in the Rub' al Khali are distributed by type into three principal areas: (1) the northeast, Al 'Uruq al Mu'taridah area, characterized by crescentic dunes, (2) the eastern and southern margins of the sand sea, characterized by star dunes, and (3) the western half of the sand sea, consisting mainly of linear dunes. The great extent of linear dunes

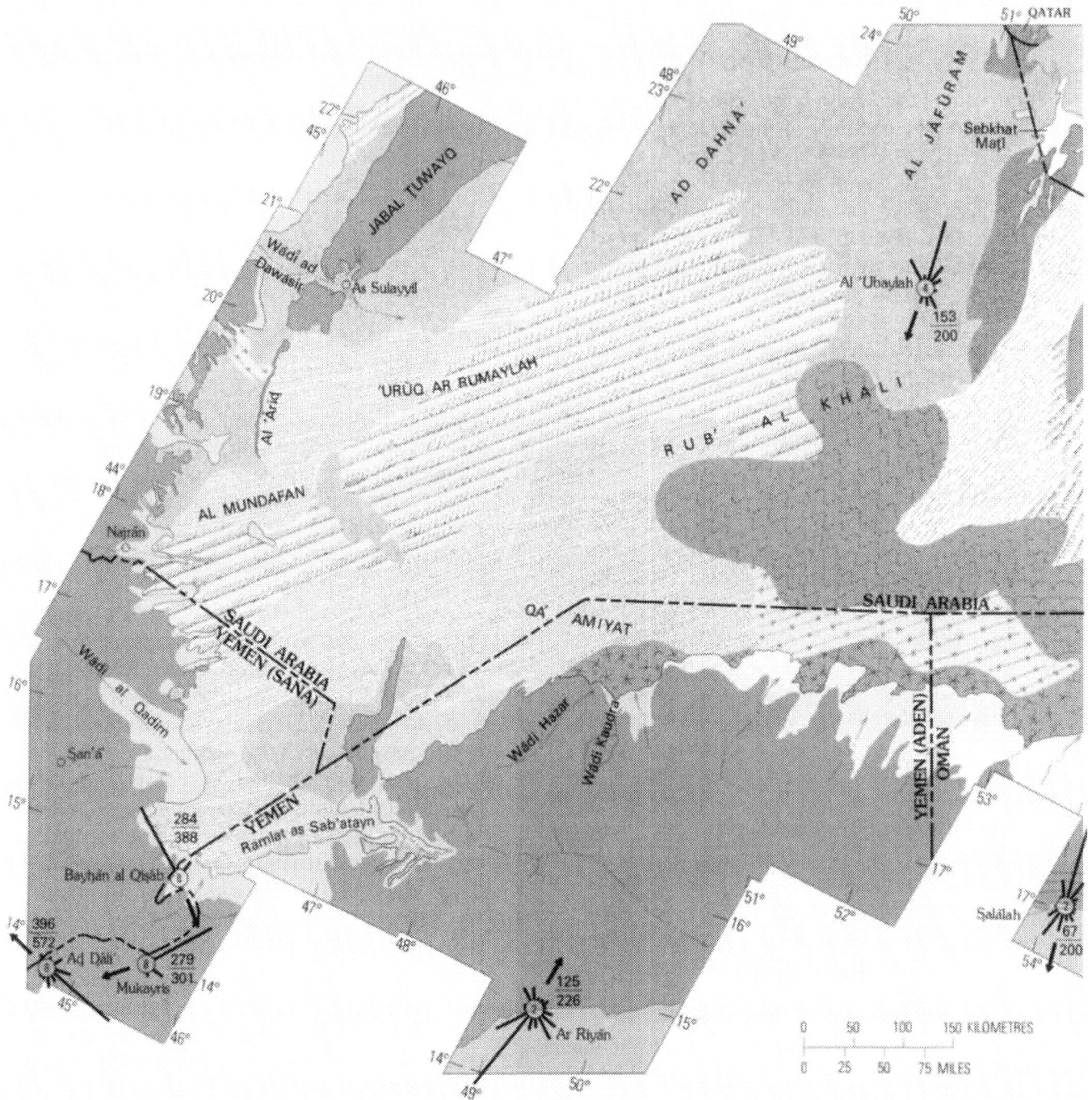

DISTRIBUTION AND MORPHOLOGY OF EOLIAN SAND in Saudi Arabia, Oman, Yemen, and adjacent States of the southern Ara-
higher in elevated

westward — from about long 50° E., to the Jabal Ṭuwayq–Al ʿAriḍ escarpment at about long 45°7′ E., is apparent in the regional oblique view on a Skylab photograph (fig. 239).

The northeastern Rubʿ al Khali includes the Al ʿUrūq al Muʿtaridah sand sea (fig. 238), in which

are some of the largest crescentic dunes observed in this study (chapter J, tables 38, 39). These crescentic dunes are composed of curved segments with slipfaces that have a mean width (horn to horn) of 2.8 km (1.7 mi) (fig. 240). The dunes have a mean length of 2.1 km (1.3 mi) and are spaced an average

EXPLANATION

LINEAR DUNES
Simple, short

Mostly simple, some compound

Compound feathered

Complex
With star dunes

Feathered, with star dunes
superimposed

With crescentic dunes
superimposed

CRESCENTIC DUNES
Simple barchanoid ridges

Compound
Barchanoid and short ridges

Barchanoid ridges

Complex crescentic ridges
with star dunes superimposed

STAR DUNES
Simple and some compound

Complex stars with dome-shaped
dunes overlapping

UNDIFFERENTIATED COMPLEX
DUNES

SHEETS AND STREAKS

BEDROCK OUTCROP OR BARRIER

ROCK OR GRAVEL PLAIN WITH
NO LARGE SAND DUNES

PLAYA

ISOHYET—Average annual rainfall,
in millimetres
BOUNDARY OF SAND SEA

INTERMITTENT DRAINAGE—Showing
direction of flow

bian Peninsula. Map based on Landsat imagery. Isohyets estimated from vegetation signature and station records. Precipitation is
areas. (Fig. 238.)

of 2.6 km (1.6 mi) from crest to crest. Their gentler (stoss) slopes are covered by smaller crescentic dunes, as shown in chapter J, fig. 174B. Thus, they are compound barchanoid or crescentic forms similar in size and pattern to dunes in the Peoples Republic of China, the western Sahara, and

elsewhere (chapter J, fig. 175 and tables 38, 39). Toward the east and south, in Al 'Urūq al Mu'tariḍah, these compound crescentic dune ridges grade into complex ridges with star dunes on their crests (McKee and Breed, 1976, fig. 56; chapter J, fig. 174).

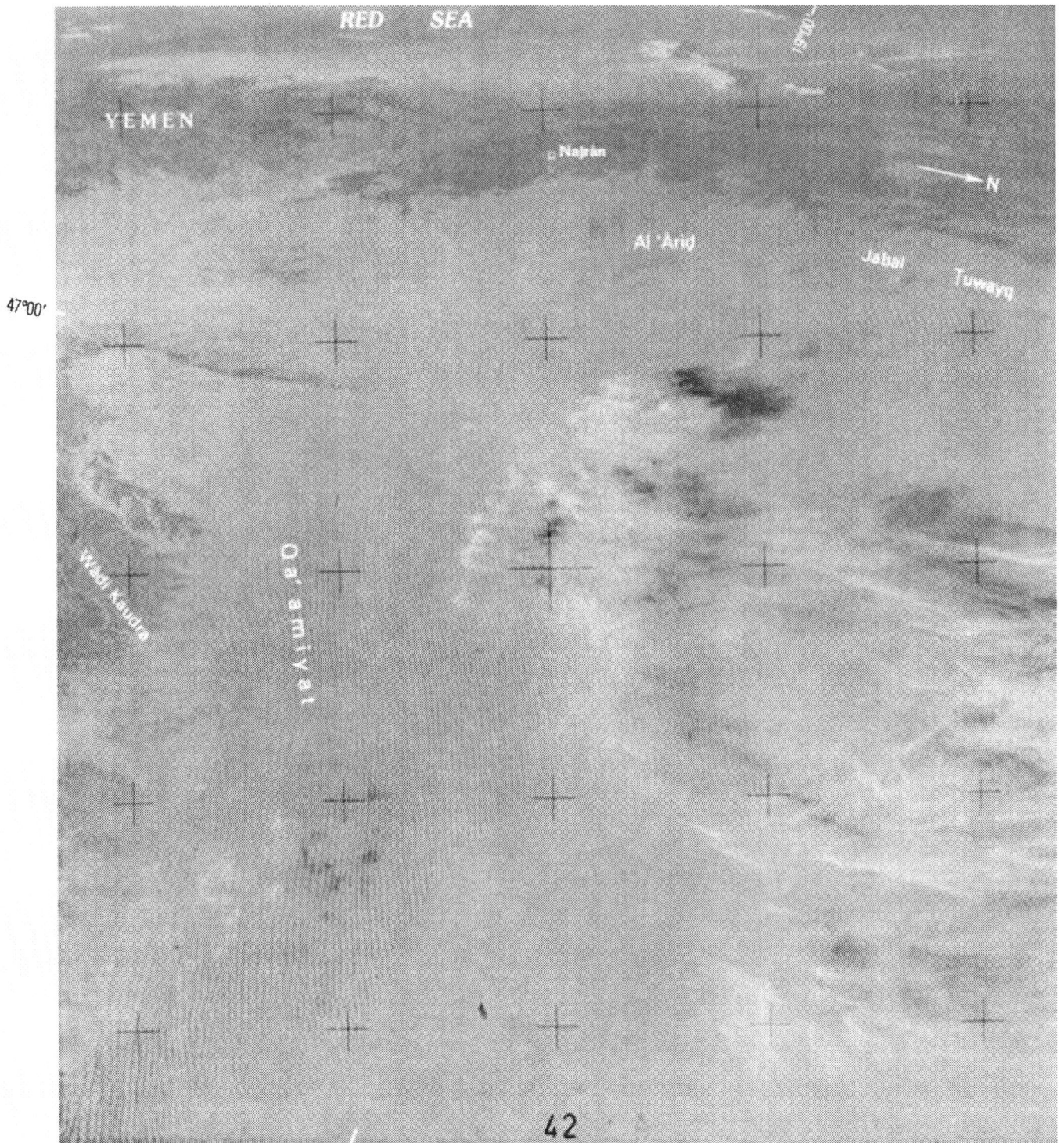

RED LINEAR DUNES EXTENDING 600 km (360 mi) across the western Rub' al Khali to the foothills of Yemen. Skylab oblique photograph SL4–143–4643. Dunes in the Qa' amiyat region have an average crest-to-crest distance of 2.1 km (1.3 mi) and are commonly more than 100 km (60 mi) long. (Fig. 239.)

Dunes at the southeastern margin of the Rub' al Khali, along the border of Saudi Arabia with Oman (fig. 238) are mainly star dunes and isolated crescentic dunes (very wide barchans or very short "transverse" ridges). The isolated crescentic dunes in Oman at about lat 20° N., long 55° E., have slipfaces that face north, opposite those of the south-

ward-facing crescentic slipface segments of barchanoid ridge dunes in Al 'Urūq al Mu'taridah. The southern barchanoid ridge dunes are white on false-color imagery in contrast with the yellow of the dunes farther north. Their different size, orientation, and color suggest that these dunes in the southeastern part of the Rub' al Khali may be ac-

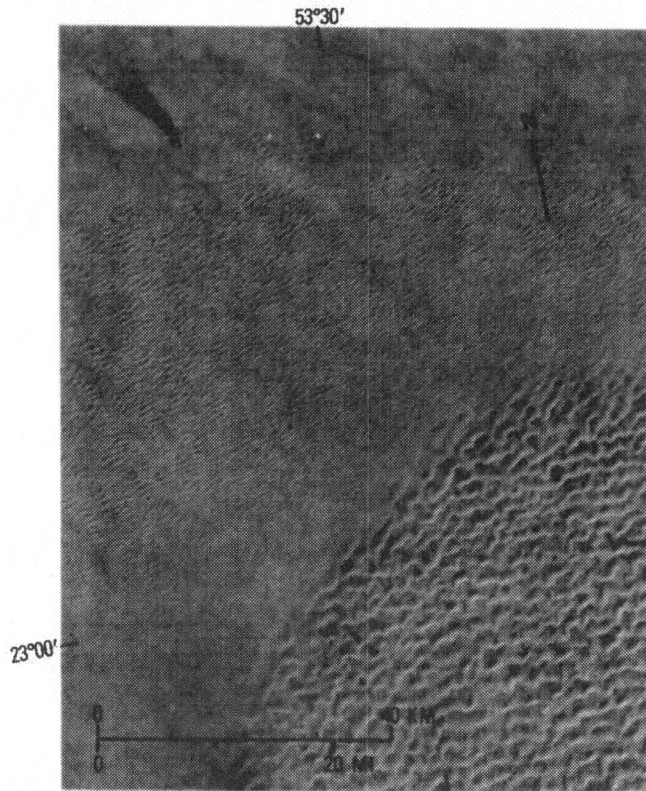

NORTHEASTERN RUB' AL KHALI, Al Jiwā– Al Batīn area. Landsat imagery E1130– 06251. Groups of small crescentic dune ridges, in the area marked by smoke plumes from oil flares, seem to be migrating southward and overriding much larger compound crescentic dunes of the northern Al 'Urūq al Mu'taridah. (Fig. 240.)

tively migrating northward under the influence of strong southwest monsoon winds (fig. 238).

Star dunes along the eastern and southern margins of the Rub' al Khali, mostly in Oman and Yemen, are large isolated compound forms, shown in detail on an aerial photograph (chapter J, fig. 179). The star dunes seem to "fan out" from the mouths of intermittent streams which flow northward from the southern highlands into the Rub' al Khali. Star dunes are larger and redder in a northward direction, where they merge with barchanoid dune ridges, forming complex dunes that have southeast-facing slipfaces (McKee and Breed, 1976, fig. 56). This relationship suggests that some sand in the star dunes along the southern margin of the Rub' al Khali may be derived from the wadi alluvium with which they seem, on the imagery, to be associated.

West of Al 'Urūq al Mu'taridah region of mainly crescentic dunes is a sand sea of dominantly linear dunes. At the boundary between the two major dune types (lat 20° N., long 53° E.) is a region of complex dunes that have both linear and crescentic elements (fig. 241). These include the "hooked" dunes described by Holm (1960, p. 1371 –1372; 1968, fig. 3). Very large compound linear dunes extend west of the region of complex dunes for hundreds of kilometres, through the Qa'amiyat region to foothills in Yemen (fig. 239). At the downwind end

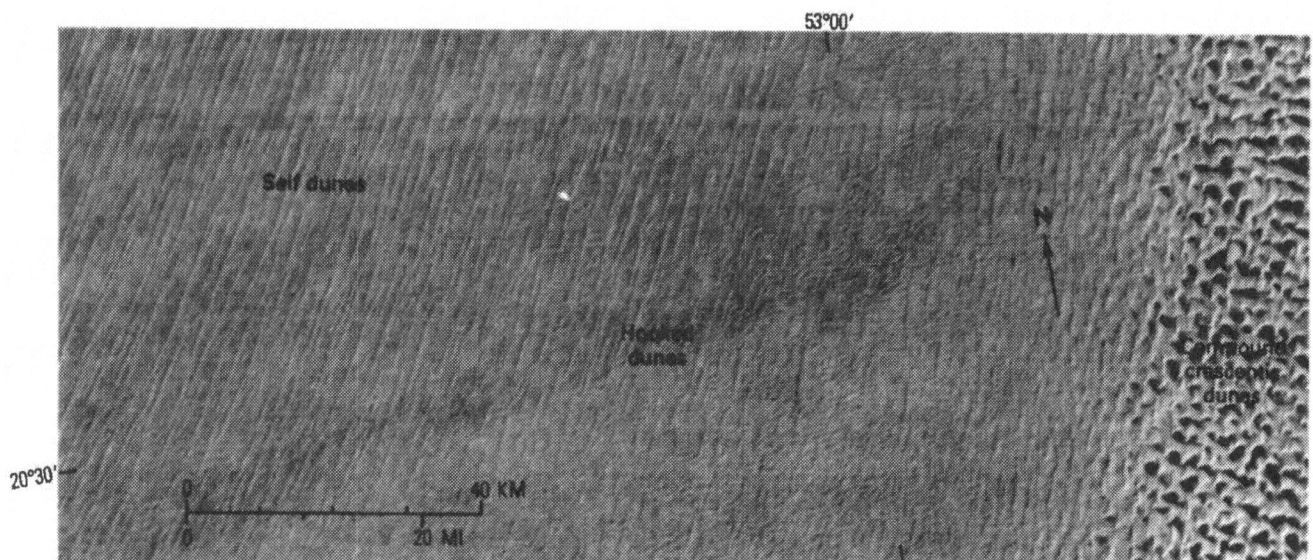

NORTH-CENTRAL PART OF THE RUB' AL KHALI, west of Al 'Urūq al Mu'taridah. Landsat imagery E1166– 06254. Dune patterns change from large crescentic dunes in the east to linear (seif) dunes in the west. In the zone west of the compound crescentic dunes are smaller crescentic ridges. Complex combinations of linear and crescentic dunes ("hooked dunes") occur in the center of the area. (Fig. 241.)

of this sand sea, at the Jabal Tuwayq are the largest linear dunes seen on Landsat imagery during the present study (fig. 239; chapter J, fig. 169F and tables 36, 37). They average 1.5 km (0.9 mi) in width and are spaced an average of 3.2 km (2 mi) apart.

Measurements in the sand sea of linear dunes were made on eight 2.500-km² (1,500-mi²) sample areas of images. At the upwind end of the field, there were 25 ridges per 50-km (30-mi) distance. The widths of the dunes and interdunes corridors are nearly equal, averaging 2 and 2.1 km (1.3 mi), respectively. At the downwind end of the field, dunes are fewer (18 per 50-km (30-mi) distance) and narrower (0.7 km (0.4 mi)), as opposed to wider interdune areas (2.8 km (1.8 mi)). Although the up-wind and downwind distributions of dunes thus differ, a plot of the measurements of all the samples against their distance downwind (expressed as a percentage of the total length of the field) shows no consistent trend.

The Wahiba Sands

Review of Previous Work

The Wahiba Sands is a sand sea of approximately 16,000 km² (10,000 mi²), east of the Rub' al Khali, in Oman (I. G. Wilson, 1973, table 1). The large linear dunes of the Wahiba Sands are Pleistocene features and do not reflect the trends of present-day winds, according to Glennie (1970, p. 90–95). The largest dune ridges are joined by numerous subsidary ridges oblique or parallel to the main trend, as shown on aerial photographs (Glennie, 1970, fig. 74). Thus, the ridges are compound linear dunes (chapter J, table 25).

Interpretation of Landsat Imagery

Compound linear dunes in the Wahiba Sands (fig. 242) have a mean width of 1.1 km (0.7 mi) and are spaced an average of 1.6 km (1 mi) apart. Individual ridges are as much as 100 km (60 mi) long and have an average length of 31.3 km (20 mi). As in the Namib sand sea (fig. 223), crescentic dunes occur along the coast, and linear dunes are farther inland. At the north end of both the Namib sand sea and the Wahiba Sands, linear dunes are truncated along an intermittent stream (Glennie, 1970, fig. 74). Linear dunes in the Wahiba Sands may be alined subparallel to the resultant direction of

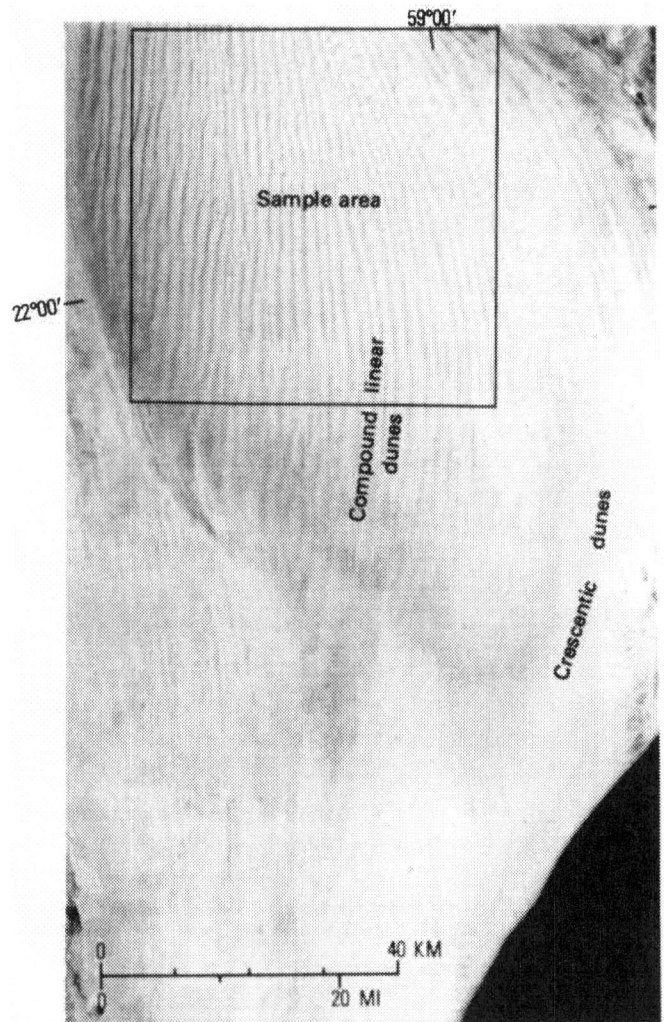

SOUTHERN PART OF THE WAHIBA SANDS, Oman. Landsat imagery E1126–06024. Outlined area is a sample of compound linear dunes on which measurements were made. (Fig. 242.)

effective winds, as suggested by the Al Masīrah sand rose (fig. 238). A good correspondence of dimensionless ratios (Strahler, 1958, p. 292–293) of measured mean widths and mean lengths occurs in the Wahiba Sands and the Namib Desert. Mean wavelengths (spacing) of linear dunes in both deserts is dissimilar (table 44). Spacing is much greater between the linear dunes of the Namib than between the linear dunes of the Wahiba Sands (table 44).

Northern Sand Seas

Review of Previous Work

Sand seas in the northern part of the Arabian desert include An Nafūd, Ad Dahnā', Nafūd ath

TABLE 44 — *Planimetric similarity of linear dunes of the Wahiba Sands, Oman, and linear dunes of the Namib Desert, South-West Africa, shown by comparison of scale ratios derived from their measurements*

[Subscript to sand sea indicates first or second of pair. Subscript to $\bar{L}$, $\bar{w}$, and $\bar{\lambda}$ indicates first or second of sand sea pair]

Sand sea	Mean dune length $\bar{L}$	$\bar{L}_1/\bar{L}_2$	Mean dune width $\bar{w}$	$\bar{w}_1/\bar{w}_2$	Mean dune wavelength $\bar{\lambda}$	$\bar{\lambda}_1/\bar{\lambda}_2$	$\bar{L}_1/\bar{w}_1$	$\bar{L}_2/\bar{w}_2$	$\bar{L}_1/\bar{\lambda}_1$	$\bar{L}_2/\bar{\lambda}_2$	$\bar{w}_1/\bar{\lambda}_1$	$\bar{w}_2/\bar{\lambda}_2$	Degree of correspondence[1]
Wahiba sands$_1$ --	31.25		1.06		1.62								
		1.16		1.20		0.74	29.5	30.7	19.3	12.27	0.65	0.40	Good
Namib Desert$_2$	27.0		.88		2.20								

[1]Degree of correspondence is based on the maximum difference in scale ratios. 0.01-0.20 = Excellent; 0.21-0.39 = Very Good; 0.40-0.59 = Good; 0.60-0.79 = Fair; 0.80-0.99 = Poor.

Thuwayrāt, Nafūd as Sirr, and others southwest of Ad Dahnā' (fig. 243). The largest and least studied is An Nafūd, which covers about 72,000 km² (45,000 mi²) (I. G. Wilson, 1973, table 1: this chapter, fig. 243). Dunes of this region reportedly reach heights of 100 m (330 ft) (Holm, 1960, p.

1373). An Nafūd may furnish some of the sand of Ad Dahnā' and the other northern sand seas (Holm, 1960, p. 1370).

Ad Dahnā' is an arc-shaped sand sea that extends 1,200 km (750 km) southeastward from An Nafūd to the northern Rub' al Khali (Holm, 1960, p. 1370,

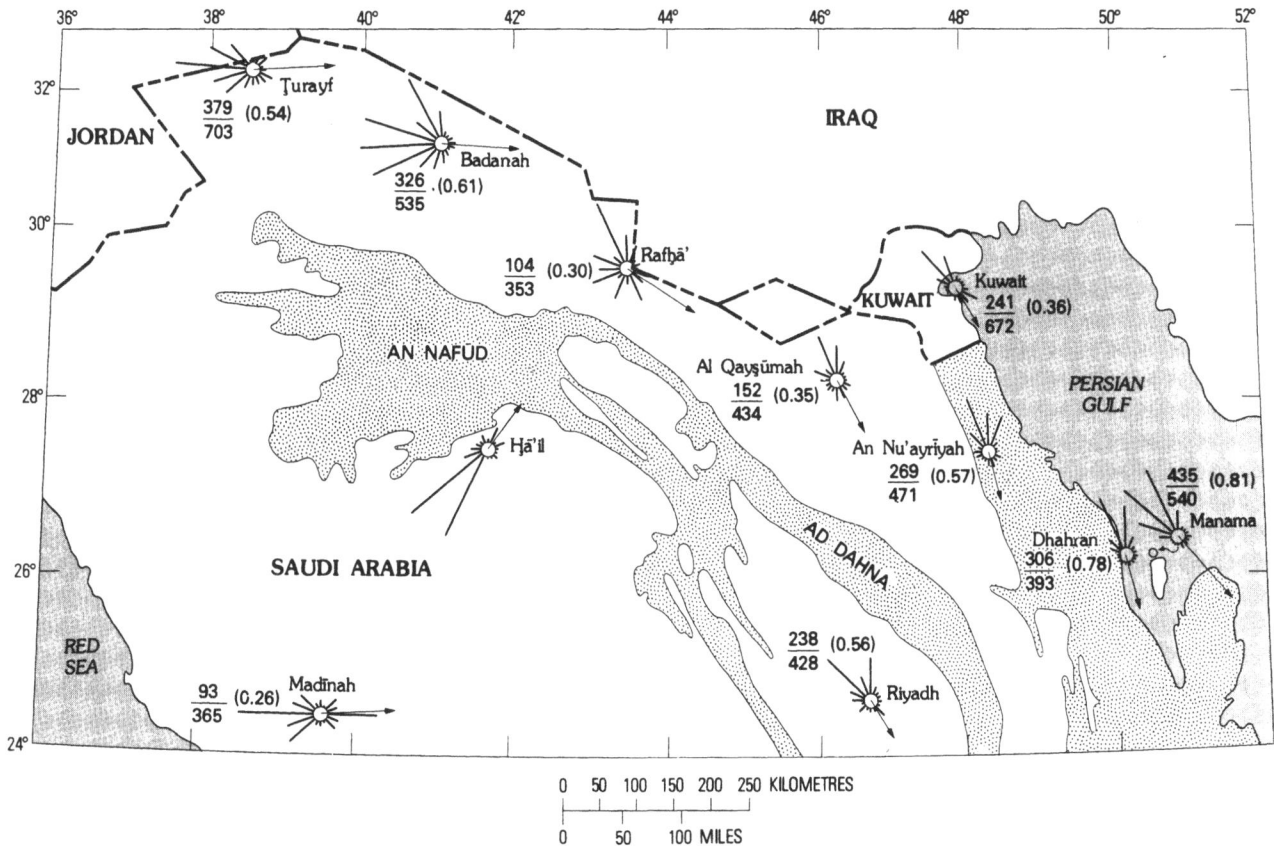

OUTLINE MAP OF NORTHERN ARABIAN PENINSULA. Resultant drift directions are parallel to elongation of An Nafūd, Ad Dahnā', and other northern sand seas, shown by stippled area (sand sea patterns modified from Holm, 1960, fig. 1). Arrows on sand roses indicate resultant drift directions; numbers near sand roses are drift potential, *DP* (on bottom), resultant drift potential, *RDP* (on top), and *RDP/DP* in parentheses (computation of sand roses is described and terms defined in chapter F and in fig. 189). Sand rose for Ḥā'il is based on computations by Bagnold (1953) and is not to scale (*DP* and *RDP* are not available for Ḥā'il). (Fig. 243.)

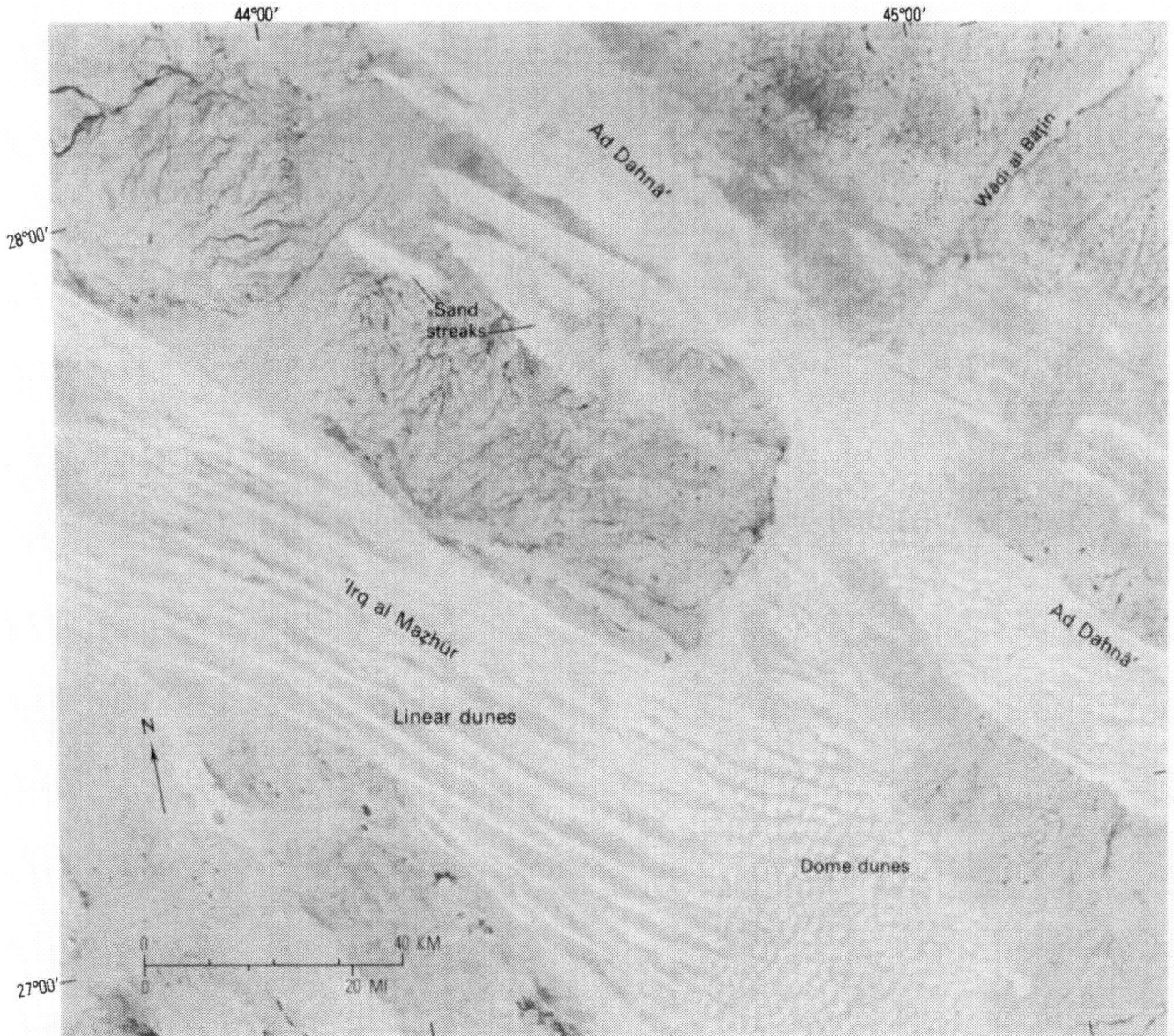

PART OF AD DAHNĀ'OF SAUDI ARABIA. Landsat false-color imagery E1209−07041. Complex dunes composed of linear ridges and giant dome-shaped dunes fill the valleys, illustrating topographic control of sand sea boundaries. Dune sand appears yellow in false-color imagery. Blue-green areas are mostly sandstone and limestone bedrock, with a thin sand cover (yellow). Red areas along stream valleys are vegetated. (Fig. 244.)

fig. 1). Star dunes in Ad Dahnā' are reported to be 150–170 m (500–560 ft) high (Holm, 1960, p. 1370). Southwest of Ad Dahnā' are numerous elongate arcs of eolian sand. Unlike Ad Dahnā', sands of the inner arcs do not extend to the Rub' al Khali. The northern sand seas are dominated by giant dome-shaped dunes (chapter J. fig. 184), each separated from the next by deflation hollows that are com-

monly vegetated and crossed by low sand ridges (Holm, 1953, p. 108–110).

Dome-shaped dunes in the Nafūd ath Thuwayrāt (fig. 243) are commonly 1–1.5 km (0.5–1 mi) in diameter and 100–150 m (330–500 ft) high; compound varieties are as much as 5 km (3 mi) long (Holm, 1953, p. 108–109). The upper surfaces of some dome-shaped dunes have crescentic dune ridges, so they are here classed as complex dunes (chapter J, fig. 184).

Interpretation of Landsat Imagery

Dunes in the northern and southern parts of An Nafūd are complex types, formed of linear ridges, some with star dunes. Compound crescentic dune ridges occur along the west (upwind) margin and in the center of the sand sea. These ridges are spaced so closely that they obscure the interdune areas. The crescentic-shaped hollows observed in An Nafūd by Bagnold (1951, p. 85) and by Holm (1960, p. 1370, fig. 3) may be the slipfaces of these nearly overlapping dunes.

Broad sheets of sand and linear dunes are characteristic of Ad Dahnā' and the northern sand seas, north of the valleys of Wādī Rima' and Wādī al Bāṭin (fig. 244). At the junction of the sand-filled valleys of these intermittent streams, linear dunes grade into dome-shaped dunes of the Nafūd ath Thuwayrāt, Nafūd as Sirr, Nafūd Shuqayyiqah, and the southern part of Ad Dahnā'. Dome-shaped dunes in the Nafūd ath Thuwayrāt, measured on sample areas of Landsat imagery have a mean diameter of 1.2 km (0.8 mi). They are the largest dunes of this type observed during the study (chapter J, table 42). The sand seas in which they occur fill valleys that are bounded on both east and west by sedimentary rock. Dome-shaped dunes in these sand seas increase in size from west to east in a downwind direction across each valley.

Surface Wind Flow and Precipitation

Surface wind circulation on the Arabian Peninsula is mainly controlled during winter by the Sahara High of the African continent, and during summer by a large low-pressure cell on the Indian subcontinent. In winter, clockwise flow of air from the Sahara High and its extension over Arabia produces westward surface wind flow in northern Arabia and south-southwestward flow in the Rub' al Khali (fig. 245A). During the spring and summer, the center of high pressure over eastern North Africa shifts northwestward to approximately lat 35° N., long 15° E. (over the Mediterranean Sea; fig. 245B). Also during this time, low pressure deepens over Asia and, by summer, extends as a trough over

the Arabian Peninsula. This situation results in northwesterly ("shamal") winds over much of the peninsula. On the southeast coast, surface wind flow into the same region of low pressure results in southwesterly winds. From August through November the thermal low over Asia weakens, and the center of high pressure near the eastern part of northern Africa shifts from the Mediterranean Sea southeastward into Libya. This is a season of weak, directionally variable winds over much of the Arabian Peninsula.

Average annual rainfall is less than 100 mm (3.9 in.) at most stations near the An Nafūd, Ad Dahnā', and the Rub' al Khali (figs. 238, 243). Only Ḥa'il (102 mm or 4 in.) and Ṣalālah (106 mm, or 4.1 in.) within the regions shown in figures 238 and 243 have recorded more than 100 mm (3.9 in.) average annual rainfall, based on available data. Vegetation (red on Landsat false-color imagery) is visible only in the sand regions near oases. From north to south across the peninsula, average annual rainfall is 41 mm (1.6 in.) at Badanah, 35 mm (1.3 in.) at Rafḥa', 98 mm (3.8 in.) at Dhahran, 84 mm (3.3 in.) at Ar Riyān, and 58 mm (2.3 in.) at Al Maṣirah (U.S. Naval Weather Service, 1974b).

Direction of Sand Drift

Annual resultant drift directions in the An Nafūd and Ad Dahnā' are toward the east near Ṭurayf and Badanah and toward the south-southeast near Dhahran and Ar Riyān (fig. 243). The curvature of annual resultant drift direction in the northern regions closely parallels trends of the An Nafūd, Ad Dahnā', and other sand seas. The curvature resembles in scale and orientation the clockwise trend of annual resultant drift directions in Algeria (fig. 206).

Monthly resultant drift directions in the An Nafūd are usually eastward during the winter months but shift toward the south in the summer. For example, monthly resultant drift directions at Badanah (fig. 246) are toward the east from December through April but are toward the southeast from June through October. At Dhahran, on the Persian Gulf, the resultant drift direction is toward the southeast most of the year but shifts to the south during March–May (fig. 247).

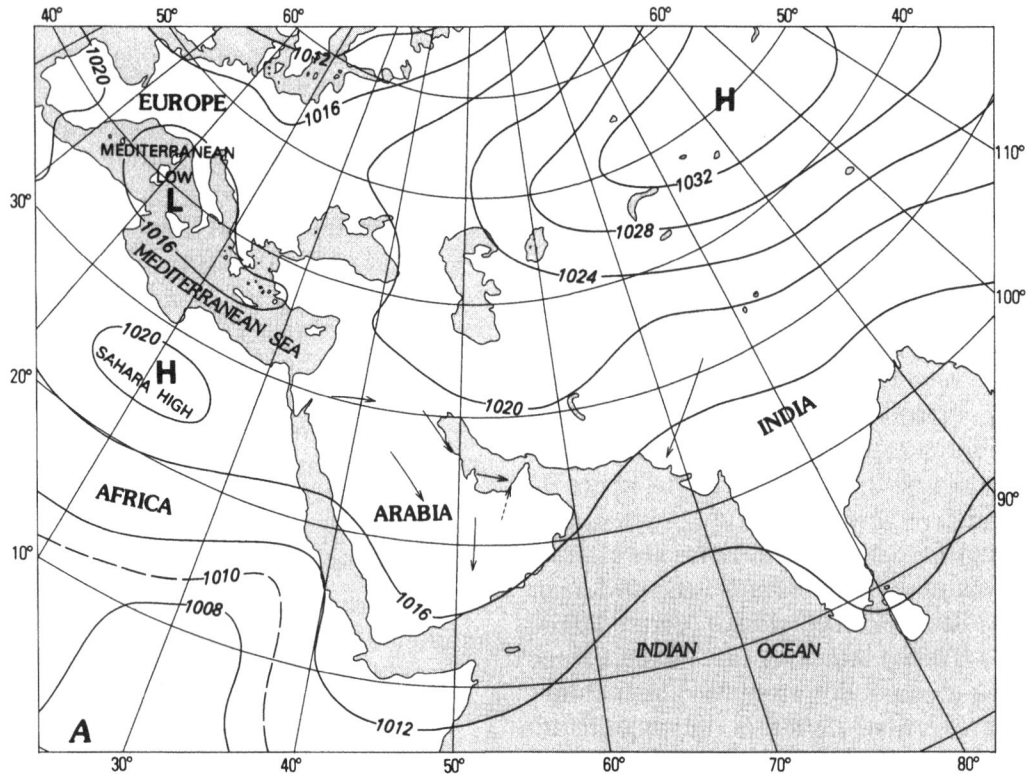

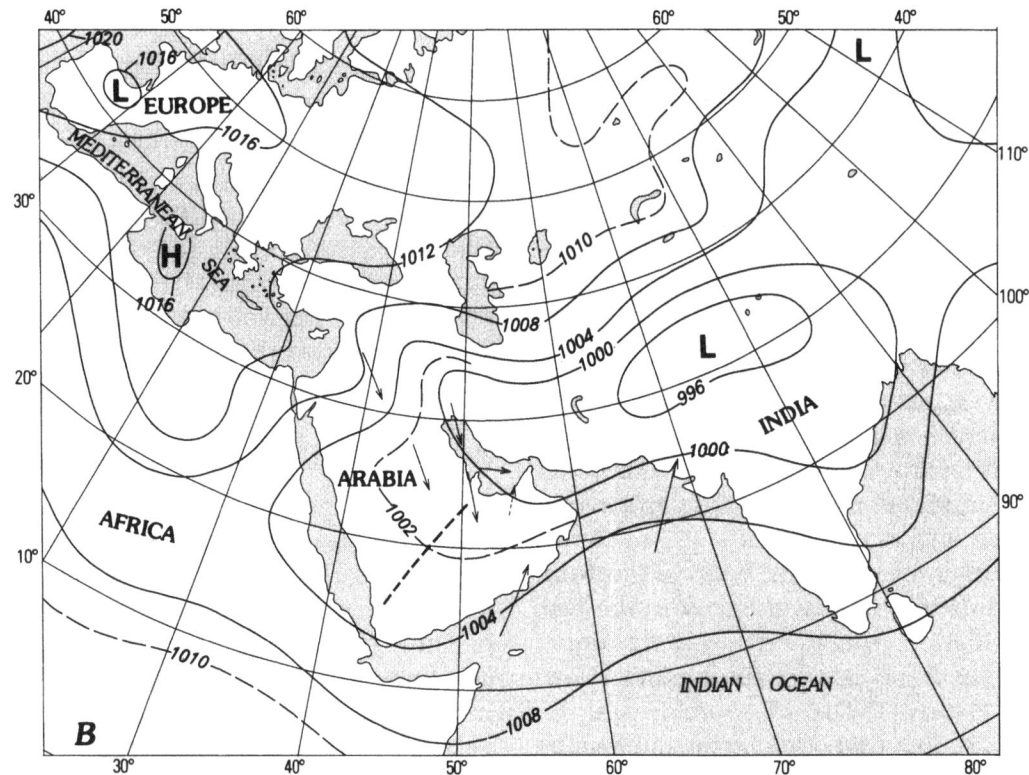

SEA-LEVEL MEAN-PRESSURE CONTOURS (in millibars) on the Arabian Peninsula and adjacent regions for (A) January and (B) June. Average trends in surface wind flow indicated by arrows. L, area of low air pressure; H, area of high pressure. Trough of low pressure over Arabian Peninsula in June indicated by dashed line. Modified from Crutcher and Meserve (1970). (Fig. 245.).

DP=39
FEB.

N

DP=100
APR.

DP=32
JUNE

DP=32
AUG

DP=46
OCT.

DP=22
DEC.

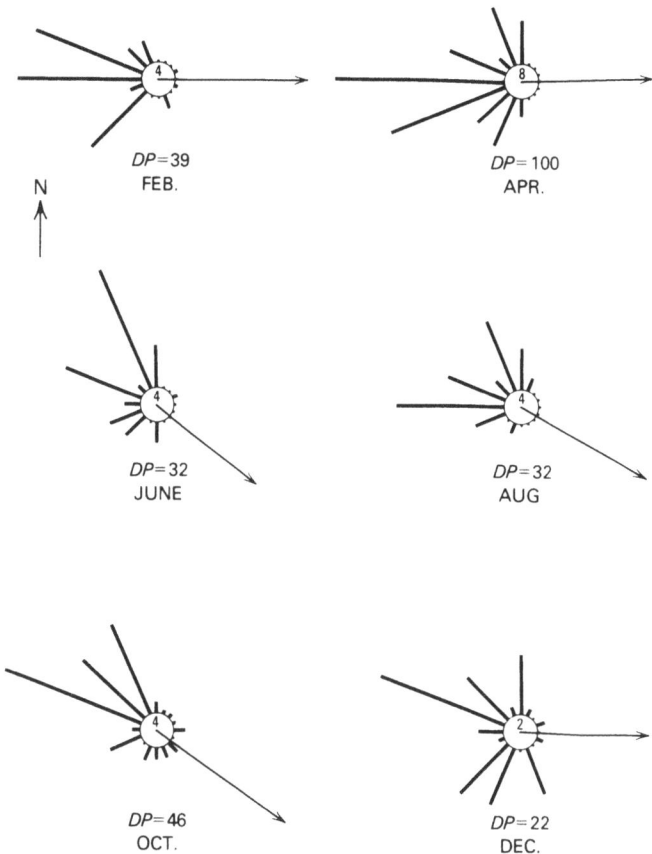

SAND ROSES FOR 6 MONTHS FOR BADANAH, Saudi Arabia. Resultant drift direction (arrow) is generally toward the east from December through April, and toward the southeast from June through October. Drift potential, in vector units, is given for each month. (Fig. 246.)

DP=19
JAN.

N

DP=28
FEB.

DP=33
MAR.

DP=37
APR.

DP=53
MAY

DP=90
JUNE

DP=48
JULY

DP=25
AUG.

DP=21
SEPT.

DP=11
OCT.

DP=12
NOV

DP=19
DEC.

SAND ROSES FOR 12 MONTHS FOR DHAHRAN, Saudi Arabia. Compare with sand roses for Badanah (fig. 246). The wind regime at Dhahran is more directionally steady from month to month than at Badanah; wind distribution is usually unimodal at Dhahran but may be bimodal or complex at Badanah. Resultant drift directions (arrows) are more toward the south at Dhahran than at Badanah. Drift potential, in vector units, is given for each month. (Fig. 247.)

Wind distribution in the An Nafūd –northern Ad Dahnā' region (fig. 243) range from unimodal (Dhahran) to bimodal (Badanah) to complex (Rafḥā'). This range of wind regimes accounts for the wide variety of dune forms in the region (fig. 244).

Annual resultant drift directions in the Rub' al Khali (fig. 238) are toward the south-southwest near Al 'Ubaylah and toward the east along the coast of the Trucial States near Ash Shāriqah. Along the southeastern part of this coast, the resultant drift direction seems to be generally toward the north, based on data from the island station of Al Masirah.

At Ash Shāriqah, on the coast of the Trucial States in the northermost Rub' al Khali, resultant drift directions are toward the east or southeast most of the year (fig. 248). Greatest directional variability of effective winds occurs during the late summer through early fall (August and September, fig. 248) at Ash Shāriqah. Resultant drift directions

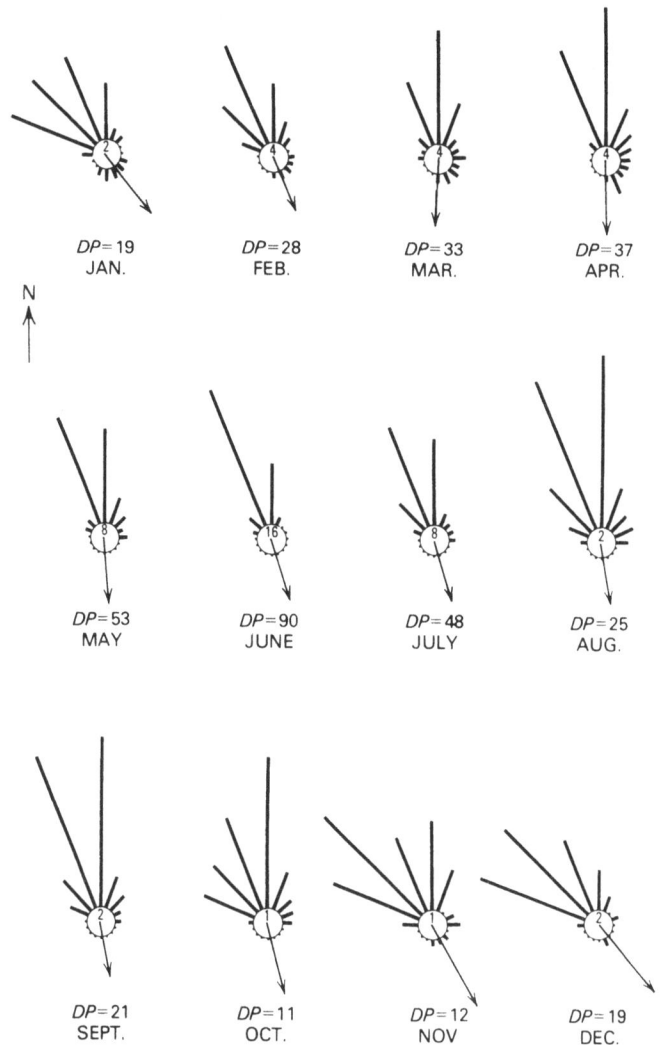

vary little from month to month at Al 'Ubaylah, in the northern Rub' al Khali (fig. 249). This suggests that effective winds farther downwind from Al 'Ubaylah in the western Rub' al Khali, in the region of linear dunes, may also be relatively steady in direction. The eastern and central parts of the Rub' al Khali are characterized by crescentic and linear dunes (fig. 240), which are often associated with narrow wind distributions (chapter F, fig. 105). Star dunes occur along the southern margin of the Rub' al Khali, a region which may be subject to wind flow from the southwest during the summer.

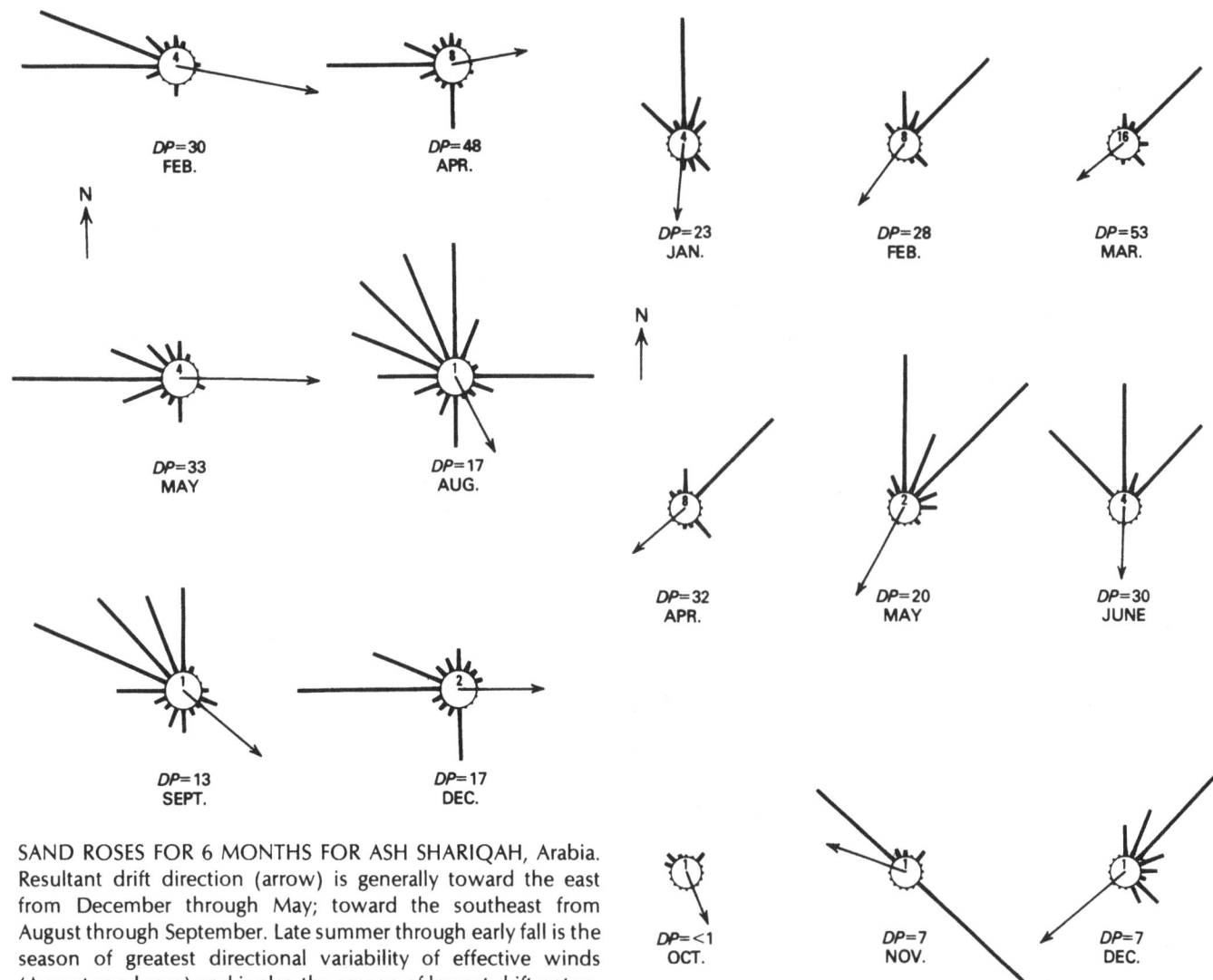

SAND ROSES FOR 6 MONTHS FOR ASH SHARIQAH, Arabia. Resultant drift direction (arrow) is generally toward the east from December through May; toward the southeast from August through September. Late summer through early fall is the season of greatest directional variability of effective winds (August sand rose) and is also the season of lowest drift potentials. Drift potential, in vector units, is given for each month. (Fig. 248.)

Amount of Sand Drift

The average drift potential of the northern sand seas (wind stations shown in figure 243) is 489 VU, a high value compared to that of most other desert regions around the world (chapter F, table 16). Highest drift potentials at stations in An Nafūd, Ad Dahnā', and the Rub' al Khali may occur in the spring or early summer, depending upon location. At many stations along the Persian Gulf, the season of high drift potentials is associated with strong north or northwesterly winds that are locally known as the shamal winds. During that season the wind may blow for 2 or 3 days at 22−26 knots (41−48 km/hr), then diminish to 13−17 knots (24−31 km/hr) for several more days under cloud-

SAND ROSES FOR 9 MONTHS FOR AL 'UBAYLAH, Arabia, in the Rub' al Khali sand sea. Sand roses are not available for July, August, and September. Monthly resultant drift directions (arrows) are relatively steady toward the south or southeast for most of the year. Resultant drift directions at Al 'Ubaylah continue the trend of a broad tradewind arc suggested by data from stations in the An Nafūd farther north (fig. 243). Surface wind data for sand roses published by permission of the Arabian American Oil Co. Drift potential, in vector units, is given for each month. (Fig. 249.)

less skies (ARAMCO Survey, 1974, p. 2). In the northern sand seas (including the An Nafūd), highest drift potentials occur in April at Badanah, in May at Al Qayṣūmah to the southeast, and in June at Dhahran still farther to the southeast (fig. 250). Much of the potential sand movement in the northern sand seas occurs during sharply defined seasons, rather than distributed evenly throughout the year. In the Rub' al Khali, highest drift potentials at Ash Shāriqah and Al 'Ubaylah occur in March (fig. 251). At Al Maṣīrah, the highest drift potentials occur in July (299 VU) at the peak of the southwest monsoon winds. The drift potential at Al 'Ubaylah, approximately 200 VU, suggests that the Rub' al Khali may not be as windy as An Nafūd.

The Thar Desert of India and Pakistan

Summary of Conclusions

Vegetated compound parabolic dunes of a distinctive rakelike shape are the most widespread type of eolian landform in the Thar Desert. Clusters of these dunes average 2.6 km (1.6 mi) long and 2.4 km (1.5 mi) wide. Parabolic dunes extend from the Rann of Kutch northward and eastward to Bīkaner and Jodhpur and occupy an area of at least 100,000 km² (60,000 mi²). Smaller, apparently simple crescentic dunes and short linear dunes occur in small fields surrounded by the parabolic dunes.

Compound crescentic dunes with an average wavelength of 1.3 km (0.8 mi) occur in a broad band along the northwest margin of the Thar Desert, from approximately lat 26°40′ N., to lat 29°40′ N. Eroded, heavily vegetated crescentic dunes occur east of Jodhpur and on the Gujarāt Plain in fields oriented parallel to the eastern shore of the Rann of Kutch. Linear dunes, some of which may be compound forms, are oriented northeast-southwest along both the east and the west margins of the desert. Sand streaks and sheets, similar in appearance to those observed on Landsat imagery of other deserts, are oriented approximately northeast-southwest along the east margin of the sand sea near Jodhpur and Barmer.

Average annual rainfall in the Thar Desert ranges from less than 100 mm (3.9 in.) along the

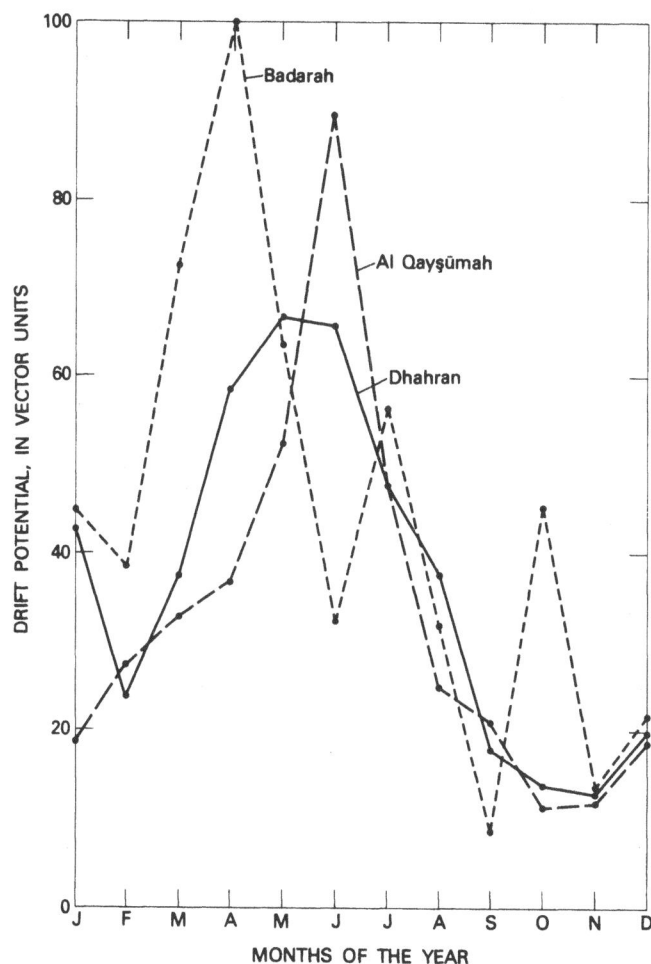

DRIFT POTENTIAL DURING THE YEAR at Badanah, Al Qaysūmah, and Dhahran, Arabia. Location of stations shown in figure 243. Season of highest drift potentials occurs early in the year in the northwestern An Nafūd (Badanah) but later at locatities closer to the Persian Gulf (Dhahran). (Fig. 250.)

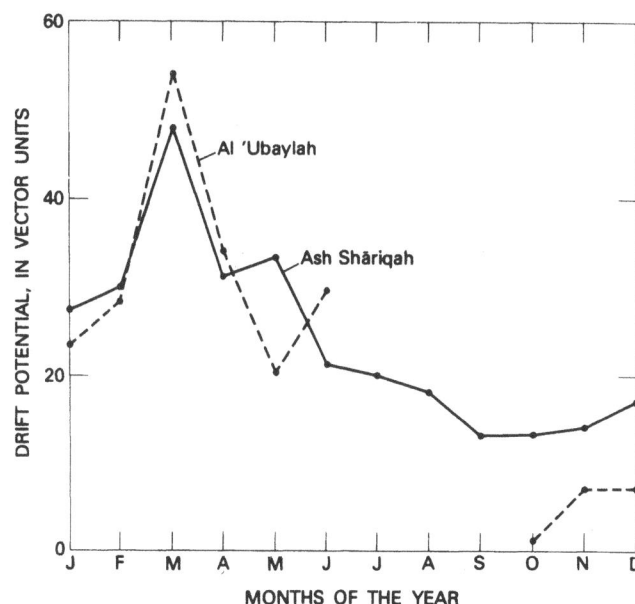

DRIFT POTENTIAL DURING THE YEAR at Ash Shāriqah and Al 'Ubaylah, Arabia. Location of stations shown in figure 238. Both stations probably experience highest drift potentials in March. Monthly drift potentials are not available for Al 'Ubaylah for July through September. Data for Al 'Ubaylah published by permission of the Arabian American Oil Co. (Fig. 251.)

northwest margin of the desert to more than 400 mm (15.8 in.) along the southeast margin. Present-day effective winds in the Thar Desert are mainly from the west-southwest or southwest (southwest monsoon). These winds result from air flow into a thermal low-pressure cell that develops over the Indian subcontinent during the summer. Average annual drift potentials are highest in the south and decrease northward; however, relatively little wind energy occurs at present in the Thar Desert compared to other desert regions. Highest drift potentials at most stations in the Thar Desert occur in June, usually preceding the time of maximum rainfall by 1 or 2 months.

Present-day low-energy unimodal wind regimes are roughly compatible with the parabolic, crescentic, and linear dune morphologies in the Thar Desert. Parabolic dunes and sand streaks aline with resultant drift directions of the present-day winds. The compound parabolic dunes, which are the most widespread and distinctive feature of the sand sea, may have formed several thousand years ago and not in the relatively low-energy wind regime of today. The parabolic dunes occur in a region of approximately 200 mm (7.9 in.) average annual rainfall and are mostly vegetated. Ground studies by other workers suggest that surfaces of these dunes are weakly cemented. However, less vegetated crescentic dunes in numerous places within the sand sea, surrounded by vegetated linear and parabolic dunes, may be active in the present wind regime.

Introduction

The Thar Desert extends westward in northwestern India from the Arāvalli Range to the vegetated flood plain of the Indus River and from the Rann of Kutch northeastward into the State of Punjab, India (fig. 252). The southern and western parts of the desert lie mostly in Pakistan. Physiographically, the region consists of low hills and sand dunes lying on sandy alluvium that is underlain by caliche-cemented concretionary beds. Few topographic barriers interrupt the sea of sand, except for scattered outcrops of limestone and sandstone of Jurassic age near Jaisalmer (fig. 252).

Early work in the Thar Desert was by British geographers (Frere, 1870; Blanford, 1877). The geology of India is described by Wadia (1939), Auden (1950), and Krishnan (1960). Indian and Pakistani studies, mainly for agricultural and sand-control purposes, are numerous. Recent geological fieldwork by Western scientists has been limited to the east margin of the desert, along the Arāvalli Range (Goudie and others, 1973), and in the Rann of Kutch (Glennie and Evans, 1976), and to the Porāli Plain west of the main dune areas, near Karāchi (Verstappen, 1966, 1968a; Snead and Frishman, 1968). Modern geomorphological studies of the main dune area (Verstappen, 1968a, 1970) are based mostly on interpretations of aerial photographs. Useful maps are the U.S. Department of Commerce, U.S. Air Force Operational Navigation Charts H–8 and J–8; and U.S. Army Map Service, Series 1301, Sheets NF 42, NG 42, NH 43, NG 43, and NH 42; all maps are at a scale of 1:1,000,000.

Landsat imagery has been used to map regional distribution and morphology of eolian sand in the Thar Desert. The map (fig. 252) shows eolian sand in relation to precipitation isohyets (in millimetres per year) and potentially effective sand-moving winds.

Review of Previous Work

Thirty-four samples of sand from fixed and active dunes in the eastern part of Rājasthān were collected and analyzed by Goudie, Allchin, and Hedge (1973, p. 247–248). The samples had an average median grain size of 0.148 mm, compared with the median diameters of 0.1–0.15 mm for sand in the central Gujarāt Plain, 0.125 mm in Punjab, and approximately 0.15 mm for sand in western Rājasthān. A mean diameter of 0.14–0.15 mm was determined for sand from the Porāli Plain in southern Pakistan (Goudie and others, 1973, p. 248). The sand in general is well rounded (Wadia, 1939, p. 291).

Sand of the Thar Desert is composed mostly of quartz, with some hornblende and feldspar. It includes calcareous grains, mainly foraminifer tests (Wadia, 1939, p. 291), and salt particles (Qadri, 1957, p. 169). Sand of the Porāli Plain (Verstappen, 1968a, p. 208) also is mainly quartz but has a high $CaCO_3$ content, due in part to the presence of foraminifer and ostracode remains. Fixed dunes of the region are said to commonly have a caliche crust.

The sands are derived from numerous sources including deflating flood plain deposits, especially along the Indus River Valley, weathered materials from the Arāvalli Range, and sand of former shorelines (Seth, 1963, p. 449). Depth of sand in the Thar Desert is several metres (Krishnan, 1960, p. 514). Horizontally bedded fine-grained slightly indurated sandstone which "exactly resembles the windblown sand of the dunes" is reported to occur at depths greater than 100 m (330 ft) in wells along a road connecting Karāchi, Hyderābād, Barmer, and Jodhpur (Seth, 1963, p. 449).

Old, dissected dunes of the Gujarāt Plain, in the southeastern part of the desert, are dark brownish red (7.5 YR 5/8) (Goddard and others, 1948) and they protrude through white, fixed dunes (Goudie and others, 1973, p. 247–248). Sands in all types of dunes on the Porāli Plain in Pakistan are light yellowish brown (10 YR 6/4) (Goddard and others, 1948; Verstappen, 1968a, p. 204). Old fixed dunes in the Thar Desert are vegetated with perennial and annual grasses, shrubs, mainly *Calligonum polygonoides* and *Haloxylon salicornicum*, and low scattered trees, especially *Acacia* sp. and *Prosopis* sp. (Roy, 1969, p. 16–17).

Scientists at the Central Arid Zone Research Institute (CAZRI) at Jodhpur, India have established three categories of eolian sand bodies in the Thar Desert, based on field determinations of relative age and stabilization. These categories are (1) old dissected dunes of indeterminate type and sand "shields," (2) stabilized parabolic, longitudinal, and transverse dunes, and (3) active small-scale barchan, shrub-coppice, and low-longitudinal ridge dunes (Singh and others, 1972).

Old well-vegetated dissected dunes and sand shields (sheets?) of the first category occur in the eastern part of the Thar Desert in the Jodhpur area, along the western slopes of the Arāvalli Range, and on the Gujarāt Plain of India (Singh and others, 1972, p. 51–52). Middle Stone Age artifacts are associated with these dunes which are thought to have formed under conditions of aridity more extensive than those of the present (Goudie and others, 1973, p. 243, 249, 252–255).

Fixed parabolic dunes of the second category northwest of Mandār, India, at about lat 25° N. (fig. 252), are characteristically 35–70 m (115–230 ft) high but may be as much as 150 m (490 ft) high. They average 2.5 km (1.5 mi) long, are variably spaced about 250 m (820 ft) apart, and are oriented N. 50°–55° E. (Verstappen, 1968a, p.210). Farther inland near Jodhpur, Bīkaner, and Jaisalmer, groups

of 15–20 parabolic dunes that range in length from 1 to 3 km (0.6–1.9 mi) and attain heights of 30–70 m (100–230 ft) have been described by Singh, Ghose, and Vats (1972, p. 52). These parabolic dunes extend for great distances (as much as 6 km or 4 mi). Their arms are nearly parallel and more uniformly spaced than in those dunes to the south. At about lat 26° N., remnants of parabolic dunes are only faintly recognizable on aerial photographs, and fixed longitudinal dunes predominate (Verstappen, 1968a, p. 210–211).

The fixed longitudinal dunes of the second CAZRI category in the Thar Desert were observed by Verstappen (1968a, p. 202–203) to have the same N. 55° E. orientation as dunes of similar type north of the Mīanī Hōr Lagoon on the Porāli Plain (fig. 206), which are 20–50 m (65–165 ft) high, 90–350 m (300–1,150 ft) wide, and as much as 5 km (3 mi) long. The fixed longitudinal dunes of the Thar, like those of the Porāli Plain, are believed by Verstappen (1968a, p. 210, 218), to have been derived by blowing out of the noses of parabolic dunes. Both parabolic and longitudinal dunes of the Porāli Plain (and by analogy, dunes of the second CAZRI category in the Thar) are tentatively assigned by Verstappen (1966, p. 7; 1968a, p. 203–204) to the Atlanticum Period, circa 5,000 years before the present.

Areas of active eolian deposition of the third CAZRI category include the desert between Phalodi and Jodhpur, between Barmer and Jaisalmer, between Lohāru and Bhiwāni, and along the courses of the Sutlej and old Ghaggar Rivers (Seth, 1963, p. 449–450). The present northeastward movement of sand and the concurrent development of active small-scale barchans, low longitudinal ridge dunes, and shrub-coppice dunes are attributed by most workers to the erosion of sand from fixed dunes which have been overgrazed or subjected to farming (Khalil, 1957, p. 241; Roy, 1969, p. 16–17; Chaudri, 1957, p. 144; Christian, 1959, p. 7–8). On the Porāli Plain, and in the Thar Desert, the only dunes presently forming, by reactivation of the parabolic dunes, are barchans (Verstappen, 1966, p. 7; 1968a, p. 203, 210–211).

Interpretation of Landsat Imagery

Some of the dunes referred to in the literature have been observed on Landsat imagery of the Thar Desert (fig. 252). Eroded and heavily vegetated

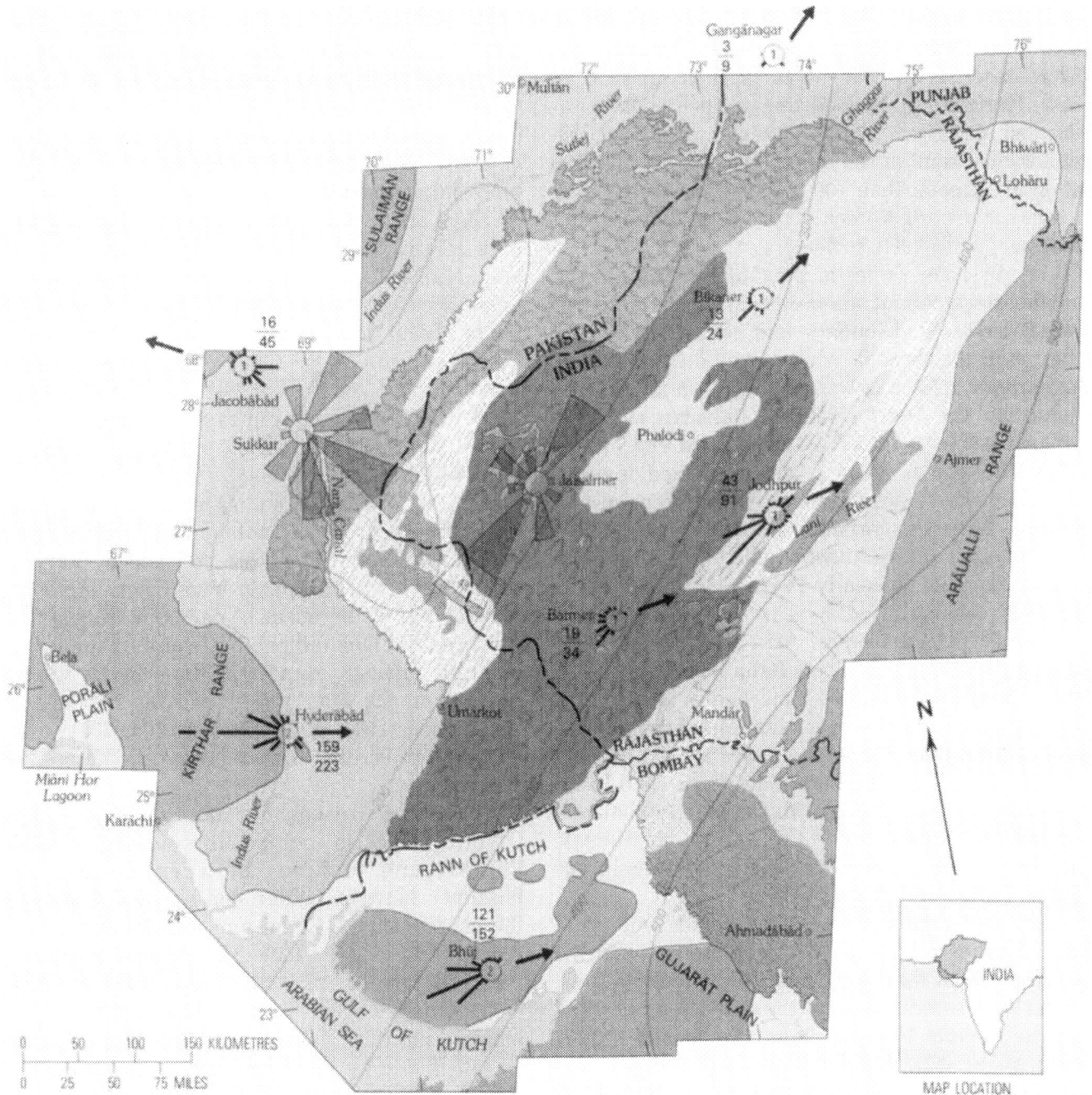

DISTRIBUTION AND MORPHOLOGY OF EOLIAN SAND in the Thar Desert of India and Pakistan (above and facing page). Map based on Landsat (ERTS) imagery. Isohyets are modified from Rao (1958) and from Pramanik (1952). (Fig. 252.)

crescentic dunes visible on images of the Gujarāt Plain may be some of the old, dissected dunes of the first CAZRI category (also described by Goudie and others, 1973). They are in a region that now receives 400–500 mm (16–20 in.) rainfall annually. The old dunes are partly covered by vegetated sand streaks. Clusters of compound parabolic dunes described by Singh, Ghose, and Vats (1972), and by

Verstappen (1968a) are clearly visible in figure 253. Dunes on the Porāli Plain in Pakistan, however, are not visible on Landsat imagery.

Compound parabolic dunes with many arms in rake-shaped clusters have been observed during this study only in the Thar Desert. Parabolic dunes observed in the other sand seas studied are U-shaped, with long trailing arms (chapter J, fig.

EXPLANATION

LINEAR DUNES
Simple, short

Mostly simple, some compound

CRESCENTIC DUNES
Barchanoid ridges

Compound barchanoid ridges

COMPOUND PARABOLIC DUNES

SHEETS AND STREAKS
Simple

Compound, with linear elements

BEDROCK OUTCROP OR BARRIER

DESERT PLAIN WITH NO LARGE DUNES

PLAYA

VEGETATED PLAIN

WIND ROSE—Length of arm is proportional to percentage of wind observations from a given direction. Percentage of observations given where too long for graphic portrayal

ISOHYET—Average annual rainfall, in millimetres

BOUNDARY OF SAND SEA

INTERMITTENT DRAINAGE—Showing direction of flow

(Fig. 252. —Continued.)

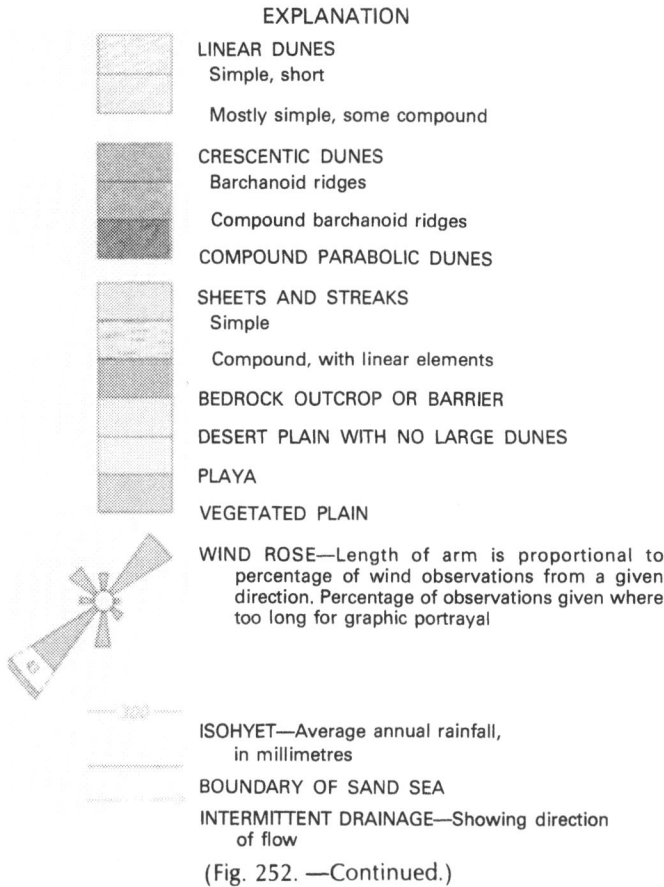

183). The parabolic dunes can be discerned on imagery from the north edge of the Rann of Kutch as far north as Bikaner (fig. 252) and are the dominant dune type of the Thar Desert. In the southernmost Thar, along the Rann of Kutch, dune clusters are massed (fig. 253), but farther inland, the clusters are an average of 2.6 km (1.6 mi) long and 2.4 km (1.5 mi) wide, have an average of seven arms, and, although variably spaced, are an average of 2.3 km (1.4 mi) apart. The rakelike arrangement of the crests of parabolic dune arms is apparent in a sketch by A. S. Goudie (written commun. to McKee, 1974) (fig. 254).

Comparison of imagery made during the dry season with that made after the onset of the monsoon rains confirms that the linear dunes, and the parabolic dunes with which they are associated[15], are seasonally covered by vegetation. However,

PARABOLIC DUNE CLUSTERS along the southern margin of the Thar Desert. Landsat imagery E1137–05214. (Fig. 253.)

fields of crescentic dunes surrounded by the linear and parabolic dunes (fig. 255) showed no evidence of vegetation. The zone of crescentic dunes shown in figure 255 includes an area sketched from aerial photographs by Verstappen (1968a, fig. 3, right) and described by as him containing active barchans. It probably represents a major area of modern eolian activity. The crescentic dunes, whose curved segments are too small to measure on the imagery, have wavelengths of as much as 0.5 km (0.3 mi).

The vegetated linear dunes shown in figure 255 may be some of the "stabilized longitudinal dunes" of the second CAZRI category, believed by Verstappen (1968a, fig. 3, right) to have formed from associated parabolic dunes.

[15] NOTE. — The problem of distinguishing linear dunes from detached arms of parabolic dunes on the basis of morphology alone is great. Parabolic arms have linear form and develop as parallel sand ridges but are structurally very different from linear dunes (Chapter A, table 1; chapter E, table 9). Subsequent correct genetic interpretation of the dune field depends on type of dune identified.

E. D. McKee, editor.

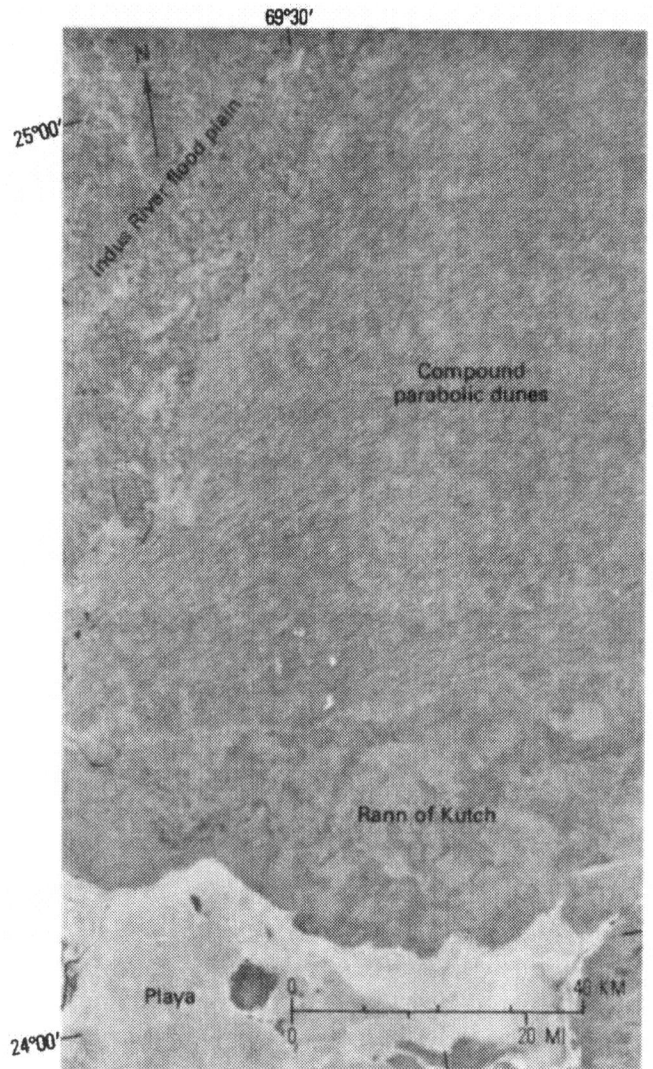

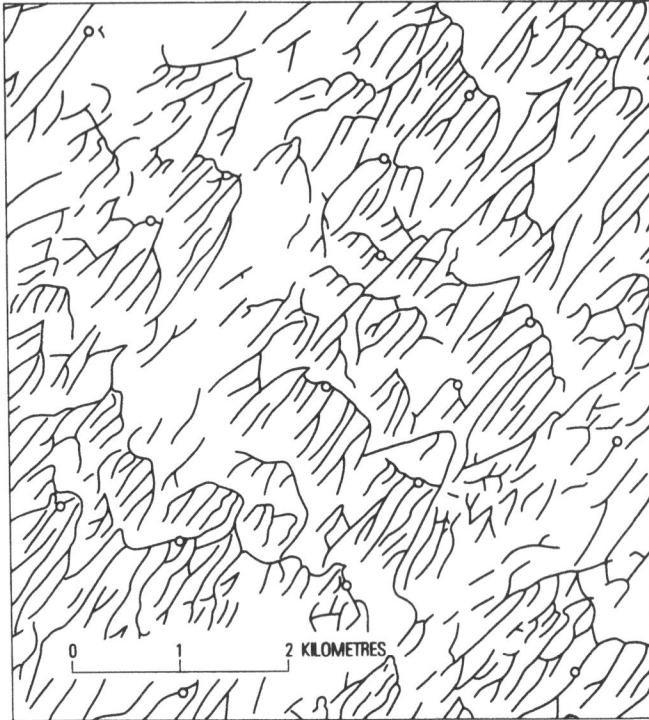

RAKELIKE ARRANGEMENT OF CRESTS of parabolic dune arms comprising compound forms near Jodhpur, India (lat 26°30' N., long 72°5' E.). Sketch traced from a survey map of India based on aerial photography (from A. S. Goudie, written commun. to McKee, 1974). (Fig. 254.)

In the western part of the Thar Desert adjacent to the Indus River flood plain, in a region that generally receives less than 100 mm (3.9 in.) annual rainfall, are fields of large dunes here interpreted as linear and crescentic types (figs. 252, 256). The latter resemble the compound crescentic ridges common to many other sand seas described in this chapter and compared in chapter J (tables 38, 39). Parabolic dunes grade westward into these large dunes, but generally dunes of the parabolic type are not observed on imagery of the more arid, western part of the desert (fig. 252). Segments of compound crescentic dunes, measured on imagery (fig. 256), average 1.5 km (1 mi) from horn to horn, have an average wavelength of 1.3 km (0.8 mi), and their avalanche slopes, which face N. 5°–10° W., have a more westerly trend than do the northeast-facing active crescentic dunes on imagery of the desert near Umarkot (fig. 255). The large linear dunes on the west margin of the sand sea average 2 km (1.2 mi) wide and are commonly 20 km (12.5 mi) long; they are interpreted here to be compound linear dunes (fig. 256).

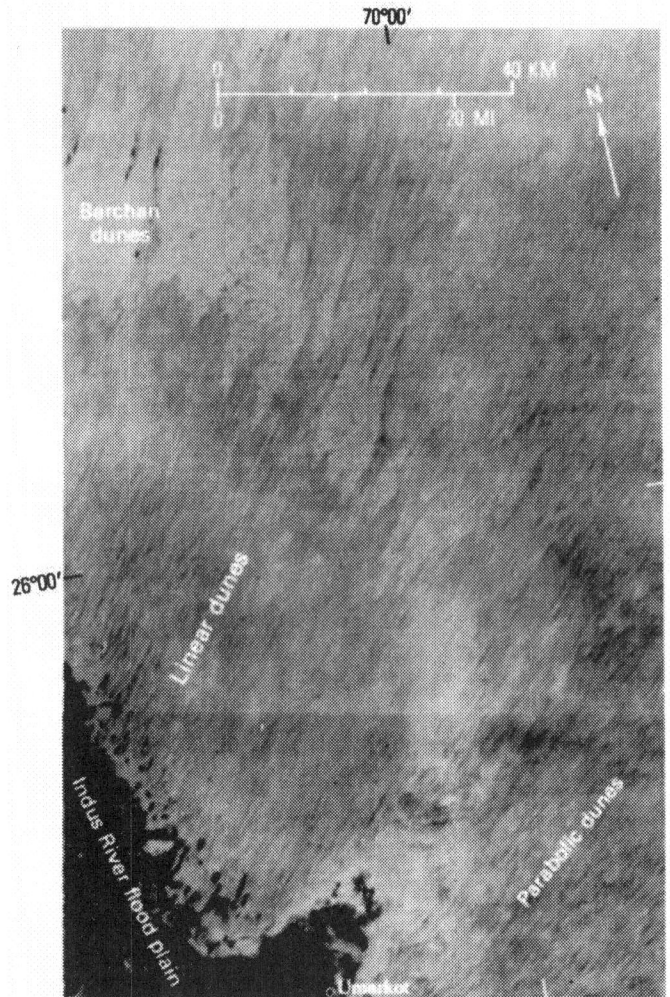

VEGETATED PARABOLIC AND LINEAR DUNES surrounding fields of barren crescentic dunes in the southwestern part of the Thar Desert. Landsat imagery E1137–05212. (Fig. 255.)

In the eastern part of the desert near Jodhpur and Barmer are sheets and streaks of sand which appear as yellow and brown on false-color imagery indicating that the sand is actually red and partly vegetated.

Surface Wind Flow and Precipitation

Surface wind flow in the Thar Desert is from the southwest during the summer and from the northeast during the winter. The summer southwesterly flow occurs when a thermal low-pressure cell develops over the Indian subcontinent due to solar heating (fig. 245B). The winter northeasterly flow is a result of subsidence from a high-pressure cell which develops over central Asia during this season (fig. 245A).

Most of the desert receives more than 100 mm (4 in.) of rain per year. (See isohyets, fig. 252.) This rainfall occurs mainly during the summer southwest monsoon season, figure 257.

Annual resultant drift directions are toward the northeast at most stations in the region, as shown on figure 252. Resultant drift directions trend more northward from Bhūj (N. 73° E.) to Gangānagar (N. 31° E.). Resultant drift directions at all stations aline within 10° with the trends of large parabolic dunes visible on Landsat imagery, except near Jodhpur.

Monthly resultant drift directions show a southward shift at most stations during the winter (fig. 258).

Maximum drift potentials at all Indian stations studied occur during June. Hyderābād and Jacobābād, in Pakistan, however, record maximums in July and May, respectively. The summer maximum of drift potential in the Thar Desert is illustrated in figure 259. Drift potentials are greater during all 12 months at southern stations and are progressively less toward the north (figs. 252, 258).

Takla Makan and Ala Shan Deserts of the Peoples Republic of China

Summary of Conclusions

Major sand seas in northwestern China are in a structural and topographic basin (Tarim Basin) in the Takla Makan Desert and in low areas on stony plains in the Ala Shan Desert farther east. The Takla Makan sand sea, west of the playa Lop Nor, may be divided into four regions characterized by distinctive dune assemblages. Large compound crescentic ridge dunes in the Cherchen Desert, south and west of the Tarim River, grade southward and westward into crossing sets of crescentic dune ridges with distinctive chevron patterns. Dome-shaped and linear dunes occur in the north-central part of the sand sea, and complex dunes occupy the south-central part of the sand sea. Dunes that seem to be simple barchanoid, dome-shaped, and linear characterize the western Takla Makan Desert, west of the Ho-t'ien Ho.

Compound crescentic dunes are the main dune type in approximately 33,000 km² (20,500 mi²) of the Pa-tan-chi-lin Sha-mo of the Ala Shan Desert.

These dunes grade into probable reversing dunes and star dunes near mountain ranges along the eastern and southern margins of the sand sea. To the south and east, in the Ya-pu-lai Sha-mo, Yamalik Dunes, and Wu-lan-pu-ho Sha-mo, dome-shaped dunes and star dunes are in elongate fields that parallel the trends of bounding valley sides. Unusually broad crescentic ridges, not seen elsewhere during this study, occur in the T'eng-ko-li Sha-mo at the southeastern margin of the Ala Shan Desert.

Average annual rainfall in the Takla Makan Desert commonly is less than 100 mm (4 in.). Average annual rainfall in the Ala Shan Desert increases from less than 100 mm (4 in.) in the northwest to more than 200 mm (8 in.) in the southeast. Effective winds in the Takla Makan Desert are mainly from the north and northeast in the eastern part of the Tarim Basin, and from the west and northwest in the western part. Winds in each area tend to be directionally steady throughout the year.

Crescentic and linear dunes, or combinations of these types, commonly occur in areas of the Takla Makan Desert that have both unimodal and bimodal present-day wind regimes. Star dunes and reversing dunes, common in many parts of the Ala Shan Desert, may be attributable to the great directional variability of effective winds there. Little wind energy is available for sand movement in either the Takla Makan or Ala Shan Deserts as compared with other desert regions. Wind energy in both deserts is greatest in the spring.

Introduction

Landsat imagery has been used to map and describe the distribution and morphology of eolian sand in the Taklan Makan and Ala Shan Deserts of the Peoples Republic of China. Sand seas in both deserts are centered at about lat 40° N. and are the highest latitude sand seas included in this study. Meteorological data covering surface winds and precipitation at climatic stations in northern China were obtained mainly from the National Climatic Center, Asheville, North Carolina. Additional data were obtained from reports by U.S. Army Air Forces Weather Division (1945), Dalrymple, Everett, and Wollaston (1970), and others. The sand-moving potential of surface winds is represented by sand roses, and average annual rainfall is shown by precipitation isohyets on the maps (fig. 260, 263).

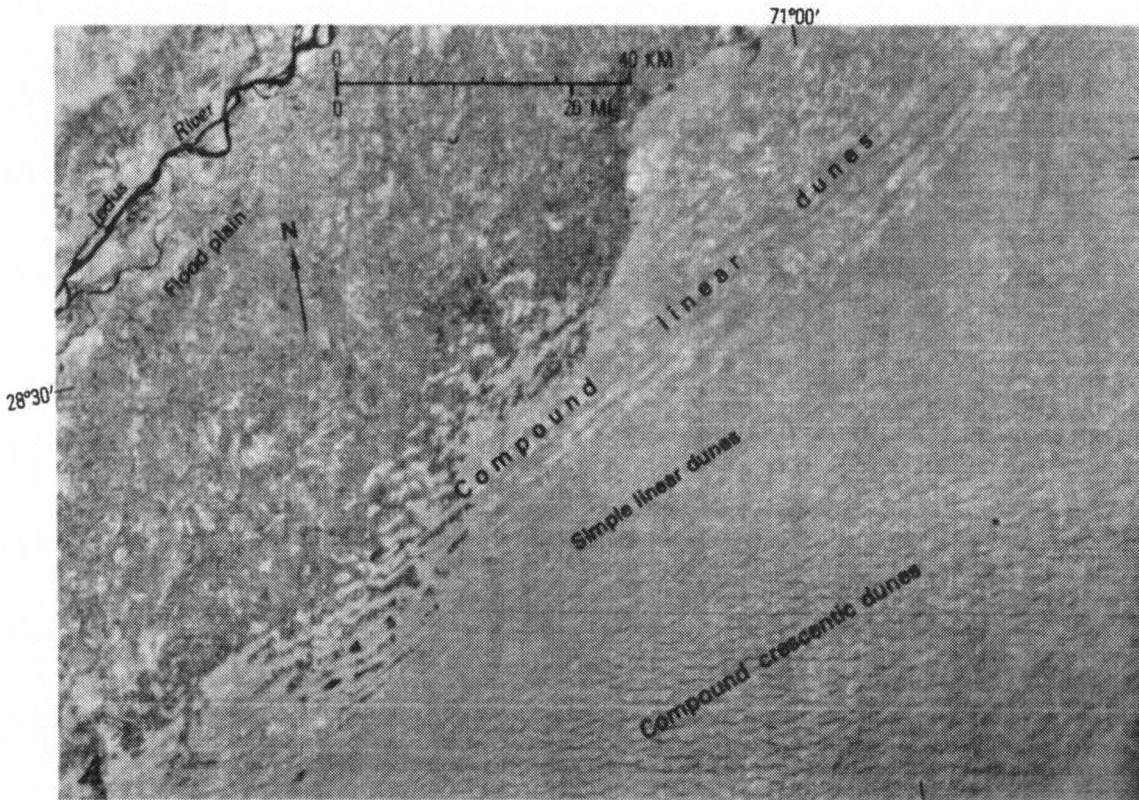

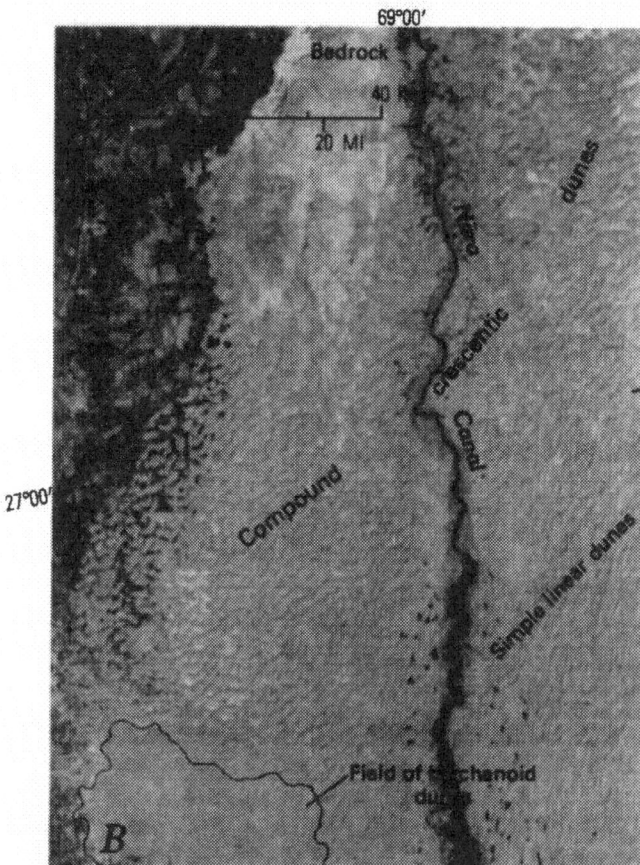

WESTERN THAR DESERT. Landsat imagery E1101–05201; E1120–05263. *A*, Compound linear dunes south of the junction of the Indus and Sutlej Rivers. Vegetation and salt flats occur in interdune areas subject to periodic flooding. *B*, Compound crescentic dunes south of Sukkur, Pakistan. The extreme aridity of the western part of the desert is shown by the sharp demarcation of irrigated, vegetated flood plain from the dune country. (Fig. 256.)

Sand seas of the Takla Makan Desert in the Tarim Basin of northwestern China (fig. 260) were first described by 19th century Russian and Swedish explorers (Bohdanowicz, 1892; Pvetsov (1895), in Hedin, 1905a; Obrutschev, 1895; and Hedin, 1905a, b). Regional geology of the Tarim Basin has been described by Norin (1941). Recent Soviet and Chinese investigations include sampling of eolian sand around the periphery of the Tarim Basin but do not include studies of the main sand seas in the interior of the desert. Useful maps of the Takla Makan and Ala Shan Deserts are the U.S. Army Map Service 1301 Series, sheets NJ 43, NJ 44, NJ 46, NK 46, NK 47, NK 48, NI 49; and the U.S. Department of Commerce, U.S. Air Force Opera-

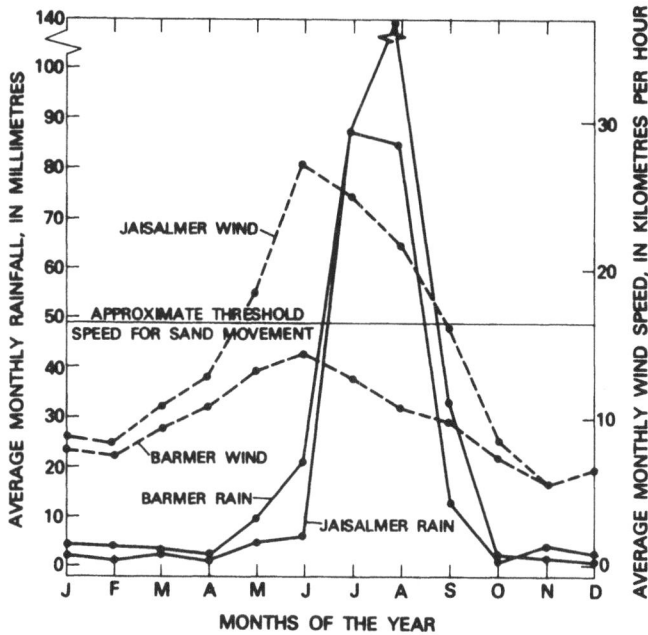

RELATION OF RAINFALL PATTERNS to effective winds at Jaisalmer and at Barmer, Thar Desert. Maximum rainfall precedes time of maximum winds by about 1 month at each station. Modified from McKee, Breed, and others (1974, fig. 27). (Fig. 257.)

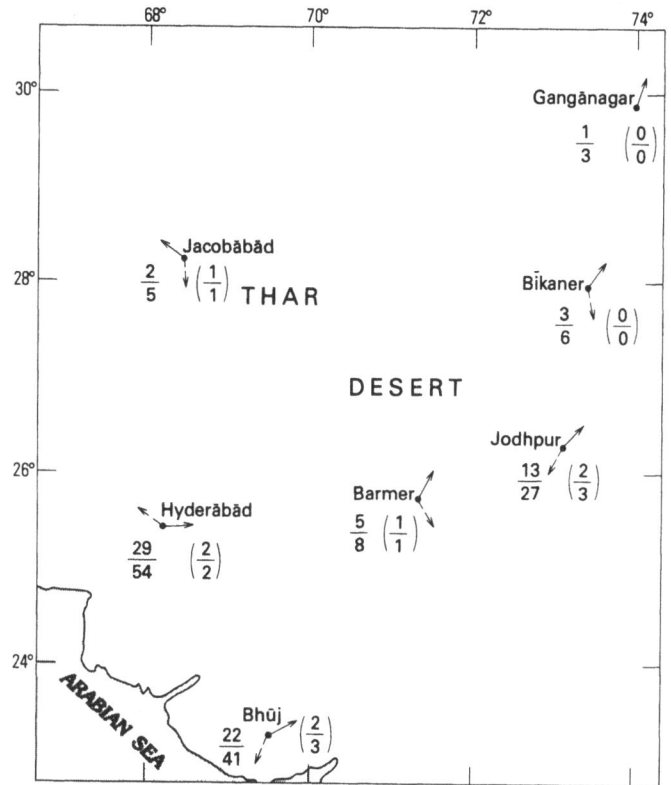

OUTLINE MAP OF THE THAR DESERT, India and Pakistan, with resultant drift directions for June (solid arrows) and January (dashed arrows). Also shown, drift potential (lower number) and resultant drift potential (upper number) for June and January. January values are in parentheses. (Fig. 258.)

tional Navigation Charts F−6, F−7, F−8, F−9, G−7, G−8, and G−9.

Recent publications about the Takla Makan Desert are mostly in Russian or Chinese (listed in Murzayev, 1967, p. 580−615), but some have been translated into French (Petrov, 1962) and English (Petrov, 1966, 1967; Murzayev, 1967). A glossary of geographic names used in central Asia is included by Murzayev in his publication (1967, p. 533−579). A book on the physical geography of China (Zaychikov and others, 1965, p. 533−579) includes information about the Takla Makan sand sea gathered by the Sinkiang Multi-Field Expedition of 1957−59.

Early work in the Ala Shan Desert was by Scandinavian geologists (Hörner, 1936; Norin, 1941). In 1957, Chinese and Soviet members of the Central Huang Ho Complex Expedition studied eolian processes in the Ala Shan and O-erh-to-ssu sand seas, followed in 1958 by the Tsinghai-Hansui Expedition of the Chinese Academy of Sciences, which explored the eastern and southern parts of the Ala Shan, including the T'eng-ko-li Sha-mo. In 1959, 17

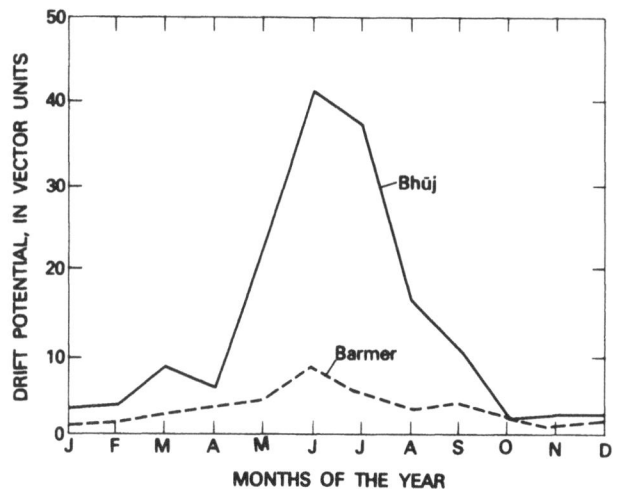

GRAPH OF DRIFT POTENTIAL at two representative stations in the Thar Desert. Both Bhüj and Barmer have highest drift potentials in June. Lower drift potentials at Barmer illustrate northward decrease of drift potential within the desert. (Fig. 259.)

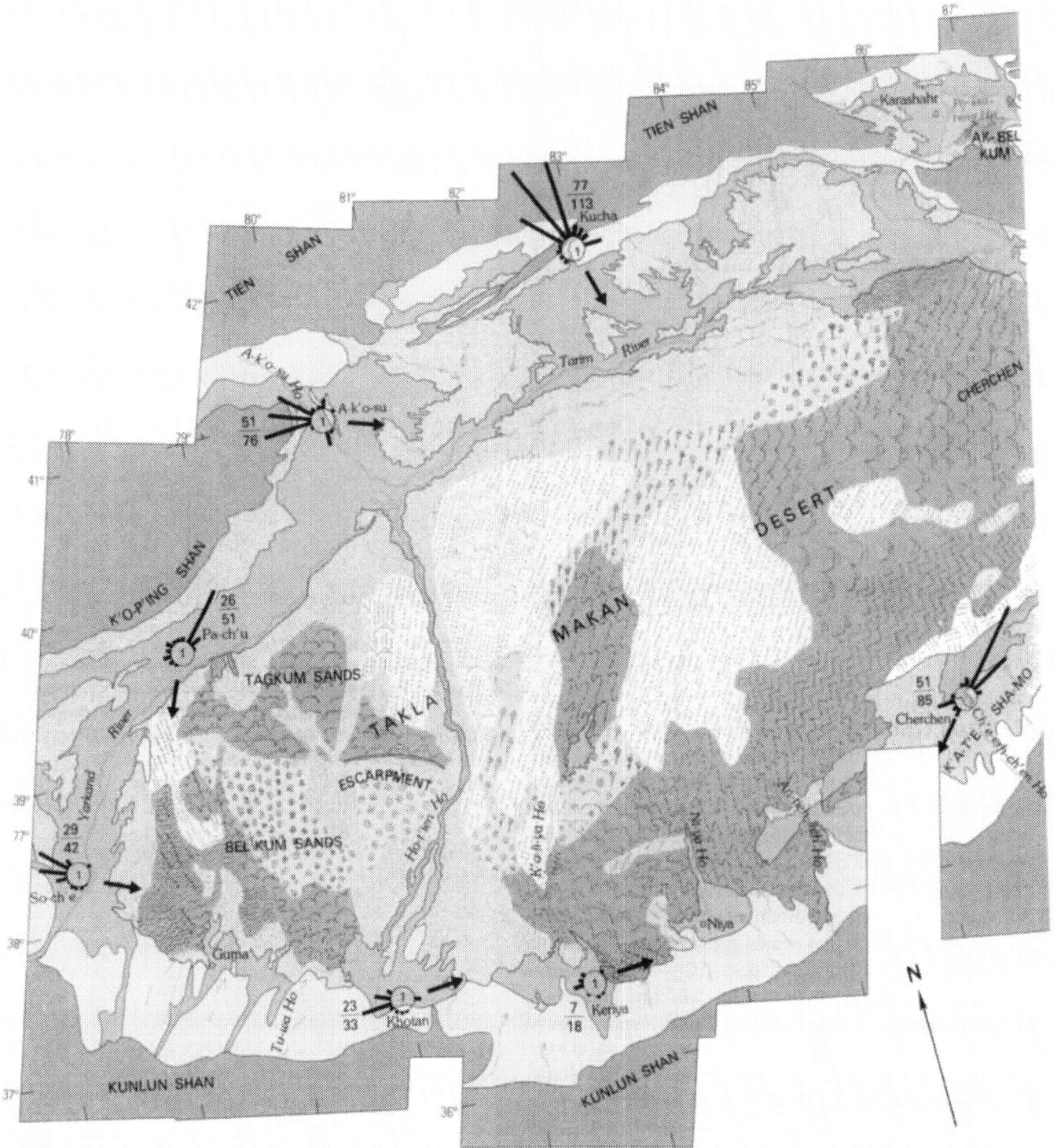

DISTRIBUTION AND MORPHOLOGY OF EOLIAN SAND in the Takla Makan Desert, Peoples Republic of China. Map based on Landsat imagery. Isohyets are not recorded. (Fig. 260.)

field teams of Chinese and Soviet investigators began a 3-year Complex Sand Expedition into the Ala Shan and O-erh-to-ssu Deserts, the Ch'ai-ta-mu P'en-ti, southern Dzungaria, and the Takla Makan Desert. Field stations were established for studies of sand stabilization and reforestation, vegetation

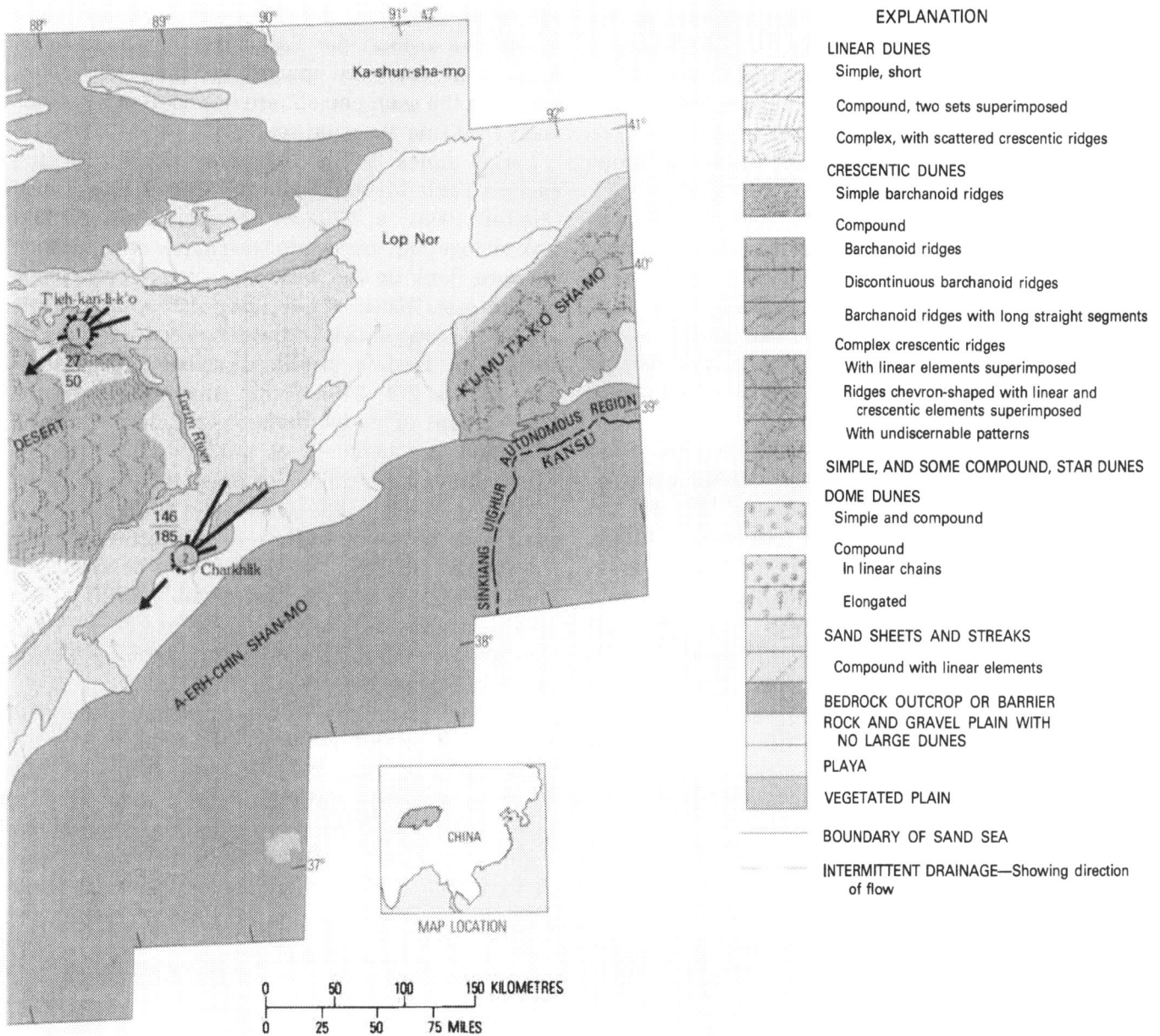

EXPLANATION

LINEAR DUNES

Simple, short

Compound, two sets superimposed

Complex, with scattered crescentic ridges

CRESCENTIC DUNES

Simple barchanoid ridges

Compound
    Barchanoid ridges

    Discontinuous barchanoid ridges

    Barchanoid ridges with long straight segments

Complex crescentic ridges
    With linear elements superimposed

    Ridges chevron-shaped with linear and
        crescentic elements superimposed

    With undiscernable patterns

SIMPLE, AND SOME COMPOUND, STAR DUNES

DOME DUNES

Simple and compound

Compound
    In linear chains

    Elongated

SAND SHEETS AND STREAKS

Compound with linear elements

BEDROCK OUTCROP OR BARRIER

ROCK AND GRAVEL PLAIN WITH
    NO LARGE DUNES

PLAYA

VEGETATED PLAIN

BOUNDARY OF SAND SEA

INTERMITTENT DRAINAGE—Showing direction
    of flow

(Fig. 260. — Continued.)

was surveyed, and about 300 eolian sand samples were collected. Results of those surveys were reported by Petrov (1961, 1966, 1967), but locations of dune areas described in the Soviet literature are difficult to relate to dunes of the Ala Shan Desert as observed on Landsat imagery.

The Takla Makan Desert
Review of Previous Work

Sand seas of the Takla Makan Desert (fig. 260) occupy about 261,000 km² (162,000 mi²) in the center of the Tarim Basin (I. G. Wilson, 1973, table 1), which is one of the largest basins with no exterior drainage (530,000 km², or 330,000 mi²) on Earth (Norin, 1941, p. 1). This basin is of late Tertiary tectonic origin and is nearly ringed by high glaciated mountains composed of Paleozoic, Mesozoic, and Tertiary strata (Norin, 1941, p. 4; Suslov, 1961, p. 533–534). The surface of the Tarim Basin slopes gently eastward, from an elevation of 1,400 m (4,600 ft) at So-ch'e to 780 m (2,560 ft) at Lop Nor, a 2,000 km² (1,240 mi²) playa into which all drainage of the basin seeps (fig. 260). The

Yarkand-Tarim-Ho'tien Ho drainage system carries water across the desert to Lop Nor during flood periods in July and August, when snow melts in the surrounding Tien Shan and Kunlun Mountains. Other streams, such as the Ni-ya Ho and K'o-li-ya Ho, flow only a short distance across the piedmont and into the sands.

The sand sea in the middle of the basin is ringed by a border of alluvial fans, braided river channels, and swampy salt marshes (fig. 260). In historic time, cultivated areas on the periphery of the sand sea have been inundated by eolian sands, as the flow of the Ni-ya Ho and other rivers fed by glaciers in the Kunlun Mountains have diminished (Norin, 1941, p. 30).

Lowered water tables around the margins of the sand sea have contributed to the restriction of poplar forests which were extensive until about 1,500 years ago (V. M. Sinitsyn, in Zaychikov and others, 1965, p. 588; Suslov, 1961, p. 557). Poplars, tamarisk, Russian thistle, dogbane, and reeds are now restricted to old stream valleys and interdune marshes. Except for thin covers of grass (*Aristida pinnata, Agriophyllum arenarium, Corispermum* sp. (Petrov, 1962, p. 147–150) and uncommon saltbushes (*Haloxylon* sp.), the sand seas are "almost completely devoid of vegetation" (Petrov, 1962, p. 147, 150; 1967, p. 306–328; Suslov, 1961, p. 558).

The Soviet-Chinese Complex Sand Expedition samples sands from peripheral areas of the Takla Makan Desert, including the K'a-t'e Sha-mo and dune fields near Guma (Hotien) and A-k'o-su (fig. 260). All the sands were from relatively small dune fields, many of which formed during histroic time and are of "insignificant thickness" (Petrov, 1967, p. 278–283, 295, 300–301, 308–335, table 33). Peripheral eolian sands of the Tarim Basin are generally fine and include large amounts of silt. They are typically grayish yellow and are believed to have formed in place from the winnowing of nearby alluvial sands (Petrov, 1967, p. 283, 301, 307, 331, table 33).

The Soviet and Chinese scientists recorded no sand samples from the large dunes in the main sand seas of the Takla Makan Desert, but Petrov (1967, p. 327) stated that the sands there are similar to the muddy-yellow peripheral sands in texture and composition. Dunes in the main sand sea may have formed by processes described by Hedin (1905a, p.

458–459), in which detrital material eroded from mountains around the Tarim Basin is transported by streams eastward toward the playa, Lop Nor, and then the sand and silt are transported by winds westward into the sand sea.

Large dunes in the Cherchen Desert in the eastern Takla Makan region (fig. 260) are described (Hedin, 1905a, p. 270) as "regularly formed and linked together, crescentic in shape, with a steep leeward flank on the west and a long gentle slope on the east." Rows of subsidiary crescentic ridges are on the long slopes of these crescentic ridges, as shown on Hedin's profile sketches (1905a, figs. 238–257, p. 271–275). Hedin apparently was the first student of desert dunes to describe the form classified in chapter J as compound crescentic. These dunes were called "complex barkhan ridges" by Petrov (1967, p. 295–298), who recognized that they were of the same form as dunes in the Peski Karakumy, U.S.S.R. The dunes were described by Hedin (1905b, p. 400) as "remarkably uniform" in height, ranging only within 10–15 m (33–49 ft) of a measured 89.5 m (293.6 ft) and reaching a maximum of 100 m (330 ft).

Between the dunes of the Cherchen Desert are flat-floored interdune basins (the bahirs of Hedin, 1905a, p. 234–243, 269–276, 311–328, 360–369; 1905b, p. 408–409). Interdune areas in the interior of the Cherchen Desert are commonly covered to a depth of 1–2 m (3–7 ft) by silt or horizontal layers of extremely fine yellow clay that is eroded into yardangs (Hedin, 1905a, p. 317–318, 321–324).

Interpretation of Landsat Imagery

Takla Makan sand seas west of Lop Nor can be divided, on the basis of dune types observed on Landsat imagery, into four sections (fig. 260): (1) the Cherchen Desert in the eastern part of the Takla Makan Desert, (2) the north-central part of the Takla Makan south of Kucha, (3) the south-central part of the Takla Makan, near K'o-li-ya Ho and Ni-ya Ho, and (4) western sand seas between the Ho-t'ien Ho and the Yarkand River.

The main part of the eastern Takla Makan Desert, west of Lop Nor, is known as the Cherchen Desert. Very large compound crescentic dune ridges in the northern part of the Cherchen Desert are separated by interdune basins commonly 1.7 km (1 mi) long and 1.2 km (0.8 mi) wide. Interdune areas near the Tarim River are subjected to periodic

COMPOUND CRESCENTIC DUNES and closed elongate interdune areas in the Cherchen Desert, northern Takla Makan Desert south of the Tarim River, Peoples Republic of China. Landsat imagery E1128-04253. Dunes and interdune basins were measured in the outlined sample area. At the east (upwind) edge of the sand sea (adjacent to the sample area) a 700- km² (453- mi²) field of smaller crescentic (barchanoid) dune ridges encroaches westward over the larger compound barchanoid ridges. Patterns of both sets of dunes are enhanced by snow. To the south, the pattern is a complex chevronlike combination of linear and crescentic elements. (Fig. 261.)

flooding. Several that are adjacent to the river valley contain lakes, whereas others contain salt flats (fig. 261). Interdune areas observed on Landsat imagery of the Cherchen Desert closely resemble interdune areas in other sand seas, where they are classed by Glennie (1970, p. 60–61, 98) as "interdune inland sabkhas."

Measurements of the large barchanoid dunes in the Cherchen Desert (fig. 261) show that the ridges are composed of crescentic segments with a mean horn-to-horn width of 3.2 km (2 mi) and a mean length of 2.2 km (1.4 mi) and are spaced an average of 3 km (1.9 mi) apart. They are the largest dunes of compound crescentic form to be noted on Landsat imagery (chapter J, table 38). The large dunes are overridden south of T'ieh-kan-li-k'o by a field of much smaller barchanoid dune ridges (fig. 261)

composed of segments with a mean width of 0.9 km (0.6 mi) and a mean length of 0.8 km (0.5 mi) and spaced an average of 0.9 km (0.6 mi) apart. The smaller ridges are oriented toward the southwest, as are the big dunes, in agreement with the resultant drift direction of present-day winds (fig. 260). Difference in size of the two sets of dunes might be interpreted as evidence of two different episodes of dune-building. A similar development of two sets of crescentic dunes is described earlier in the section on the eastern Rub' al Khali, Saudi Arabia (fig. 240).

The compound crescentic dune ridges in the Cherchen Desert grade westward through a zone of complex ridges into dome-shaped and linear dunes directly south of the Tarim River. Measurement of the dome-shaped dunes in this area shows that they

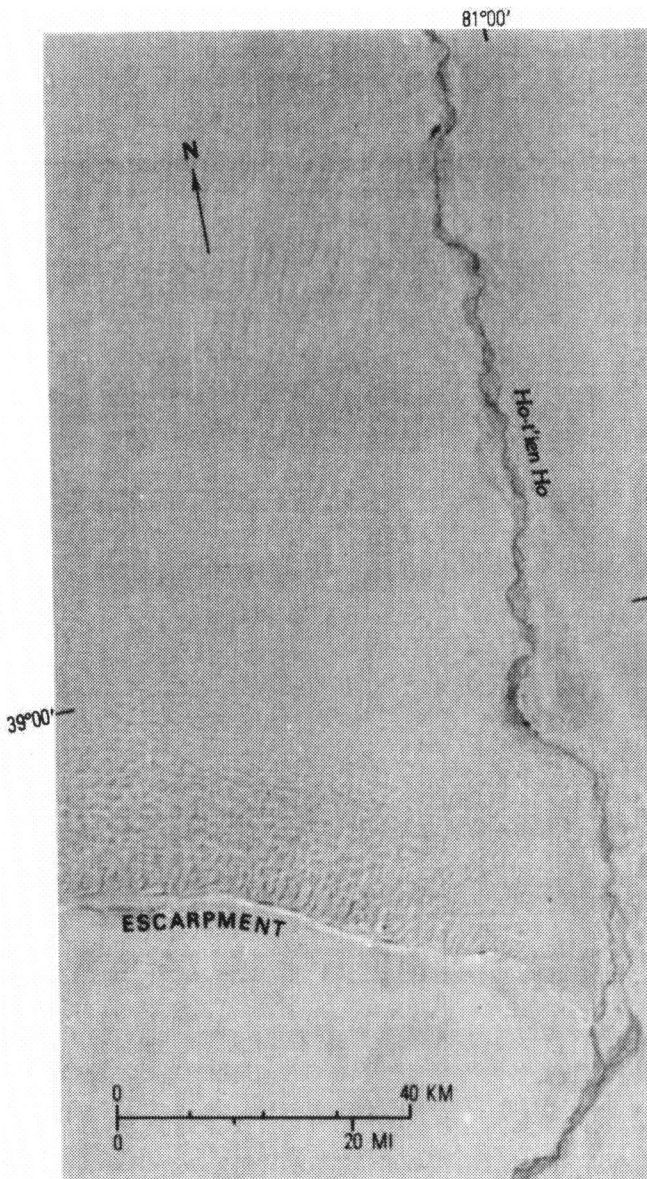

PROBABLE TOPOGRAPHIC CONTROL of position and height of dunes by an escarpment along the Ho-t'ien Ho, an intermittent stream that crosses the western Takla Makan Desert. Landsat imagery E1078–04481. (Fig. 262.)

Dunes in the south-central Takla Makan Desert between Cherchen and Nochiang are mainly complex ridges in which two trends (south and west), can be observed on Landsat imagery (figs. 260, 261). Linear elements predominate in some areas, and crescentic elements predominate in others in the southern part of the desert. The two trends result in a chevronlike dune pattern first described by Hedin (1905a, p. 364). The linear elements may represent full development of the dunes that were called "thresholds" by Hedin, for they connect and lead up to adjacent crescentic dune ridges in at least one of the regions he investigated during four crossings of the desert (Hedin, 1905a, p. 315–327, 360–362, figs. 263–264; 1905b, p. 409).

Sand seas west of the Ho-t'ien Ho are called the Bel'kum and Tagkum Sands, divided at lat 38°30' N. by a large escarpment (Norin, 1941, p. 3). The escarpment is the only major rock outcrop visible on imagery of the interior of the main Takla Makan sand sea and is composed of folded Tertiary strata (Norin, 1941. p. 3). It seems to be an important topographic barrier to sand movement. Compound crescentic dune ridges are piled up along the north (upwind) side, but such massed features are absent on the south side, where only scattered linear ridges, dome-shaped dunes, and simple barchanoid ridges occur (fig. 262). N. A. Belyayevskiy (in Murzayev, 1967, p. 56) reported that dunes near the escarpment rise to 300 m (1,000 ft) above the surrounding plain. The general size of dunes in the western part of the Takla Makan Desert is given as 60 m (200 ft) (Norin, 1941, p. 3). The western boundary of the Takla Makan sand seas is at the Yarkand River.

The Ala Shan Desert
Review of Previous Work

The Ala Shan Desert is a hilly plain about 700 km (435 mi) east of the Takla Makan Desert, at the east end of the Ho-hsi-tsou-lang (fig. 263). Sand seas occupy about 65 percent of the Ala Shan Desert, between the intermittent Jo Shui on the west, and the perennial Huang Ho (Yellow River) on the east. To the south, eolian sands are bounded by foothills of the Nan Shan Range, and to the east, by the Ho-lan-shan Range and the valley of the Huang Ho. To the north, scattered dune fields occur in numerous depressions on the stony plains of northern China

have a mean diameter of 1.1 km (0.7 mi). A sample of these dunes is compared with dome-shaped dunes in other deserts (chapter J, tables 42, 43). Some of the dome-shaped dunes in the Cherchen Desert have linear "tails" that extend downwind (southward), subparallel to the resultant drift direction at Kucha (fig. 260). The dome-shaped dunes grade southward and westward into fields of short linear dunes which occupy the central portion of the Takla Makan sand sea northeast of Ho-t'ien Ho (fig. 260).

and Mongolia. Main sand seas in the Ala Shan Desert are known as the Pa-tan-chi-lin Sha-mo, Yamalik Dunes, Ya-pu-lai Sha-mo, T'eng-ko-li Sha-mo, and Wu-lan-pu-ho Sha-mo (fig. 263).

The Pa-tan-chi-lin Sha-mo supports shrubs (*Hedysarum mongolicum, Calligonum* sp., *Haloxylon* sp., *Cargana microphylla*), sages (*Artemisia sphaerocephala*), gramen (*Psammochlea villosa*), annual grasses (*Agriophyllum gobicum* and *Pugionium cornutum*), and clumps of *Ephedra* sp. In some interdune areas and especially around interdune lakes are reeds (*Phragmites communis*), dogbane (*Apocynum hendersonii*), and tamarisk (*Tamarix ramosissima*) (Zaychikov and others, 1965, p. 345, 547; Petrov, 1966, p. 252–253; 1967, p. 26). Vegetation is reported to be sparse in the Yamalik Dunes and is generally restricted to interdune areas in the T'eng-ko-li Sha-mo. In the Ya-pu-lai Sha-mo, vegetation is mostly the shrub (*Hedysarum scoparium*), sagebrush *(Artemisia sphaerocephala)*, and grass (*Agriophyllum arenarium*). In the Wu-lan-pu-ho Sha-mo, adjacent to the Huang Ho (fig. 260), Ordos sage (*Artemisia ordosica*), gramen, and *Hedysarum scoparium* grow on the dunes, and *Nitraria schoberi* and even reeds (*Phragmites communis*) grow in interdune areas (Petrov, 1966, p. 252).

The Pa-tan-chi-lin Sha-mo in the central part of the Ala Shan Desert (fig. 263) are fine grained and red, (Petrov, 1966, p. 243). Sand of the T'eng-ko-li Sha-mo, in the southern part of the desert, is "muddy yellow, fine grained, and well winnowed"; very fine grains (0.25-0.1 mm) account for 98.3 percent of the sand in some areas, whereas lag gravel and coarse sand occur in the interdune areas (Petrov, 1966, p. 247). The Yamalik Dunes, in the northeastern part of the desert, are composed of red sand deflated from the stony plains and gray sand winnowed from alluvium (Petrov, 1966, p. 251). The Ya-pu-lai Sha-mo, in the south-central part of the Ala Shan Desert, is made up of fine sand (Petrov, 1966, p. 253). The Wu-lan-pu-ho Sha-mo, at the east margin of the desert, is made up of fine sand with some silt, according to Petrov, who stated that it is derived partly from local alluvium but also comes from the interior of the Ala Shan Desert. Sand and silt of the Wu-lan-pu-ho Sha-mo commonly migrate into the Huang Ho (Petrov, 1966, p. 252).

Suggested sources of sand in the Ala Shan Desert are outcrops of Cretaceous sandstone (Petrov, 1966,

p. 243) and extensive Quaternary and modern alluvial-lacustrine sediments of the Jo Shui drainage system (Norin, 1941). The sand-carrying capacity of the Jo Shui is not known, but the amount of mud carried in suspension at normal high water was measured at Hei-ch'eng (Hörner, 1936, p. 729–730, fig. 7) at more than 40,000 tons per day, and most of the sediment is deposited upstream (south) from the Ka-shun-no-erh and So-kuo-no-erh. Dominant west and northwest winds are reported to carry, in single day, several tons of sand eastward from the lower Jo Shui across the gravel plains toward the sand seas of the Ala Shan Desert (Hörner, 1936, p. 735).

Thickest sand deposits in the Ala Shan Desert occur in its western, driest part (Petrov, 1962, p. 141). Sand in the Pa-tan-chi-lin Sha-mo, in the center of the Ala Shan Desert, fills depressions and covers rock outcrops to an estimated depth of 150 m (490 ft) or more (Petrov, 1966, p. 243). Main sand ridges in the Pa-tan-chi-lin Sha-mo move very slowly southeastward under the influence of dominant northwest winds (D. Fedorovich, 1961, in Zaychikov and others, 1965, p. 53), but small barchans and barchanoid ridges in the interdune areas move northeastward as much as 14 m (46 ft) per year because orientation of the main dune ridges affects wind directions in the interdune areas (Petrov, 1966, p. 244).

Height of dunes in the Ala Shan Desert are reported (Petrov, 1962, p. 141; 1966, p. 243–244, 252), as follows: Large crescentic dunes in the main body of the Pa-tan-chi-lin Sha-mo average 300 m (990 ft), with a maximum height of about 400 m (1,310 ft); isolated star and reversing dunes in the northern part of the field are as much as 90 m (295 ft) high. The Yamalik Dunes are as much as 20–25 m (65–80 ft) high, and dunes in the Wu-lan-pu-ho Sha-mo are as much as 25–30 m (80–100 ft) high (Petrov, 1966, p. 251, 252).

Interpretation of Landsat Imagery

Dunes of the Pa-tan-chi-lin Sha-mo cover about 33,000 km² (20,500 mi²) in the center of the Ala Shan Desert (fig. 264). Northwest (upwind) gentle slopes of large crescentic dune ridges are covered with subsidiary crescentic dunes (fig. 264), so the dunes are compound crescentic (large barchanoid) ridges in the classification system used for Landsat imagery. They are the "complex barchan hill-

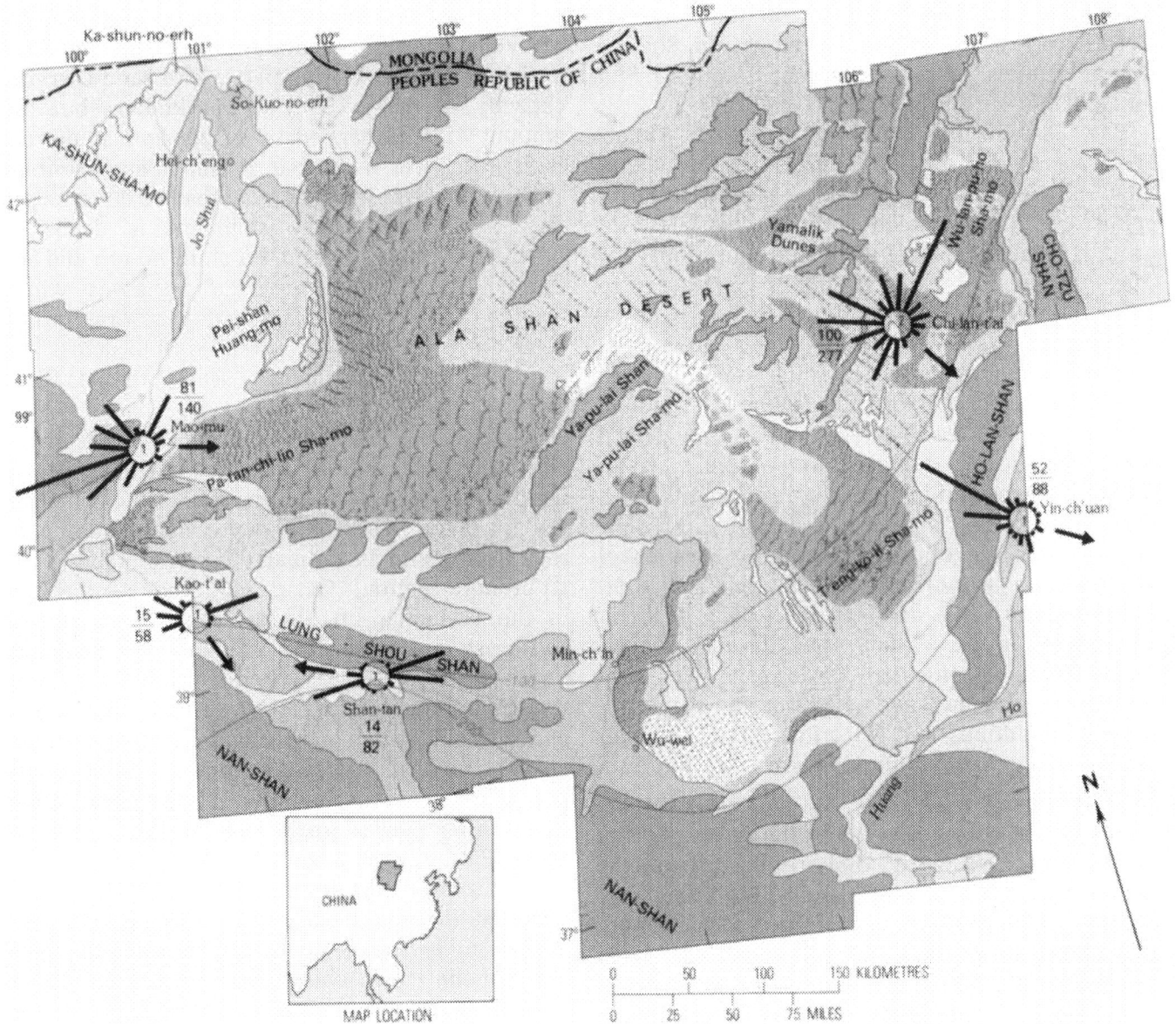

DISTRIBUTION AND MORPHOLOGY OF EOLIAN SAND in the Ala Shan Desert, Peoples Republic of China (above and facing page). Map based on Landsat imagery. Isohyets from Petrov (1967) and World Meteorological Organization station data. (Fig. 263.)

shaped ridges" of Petrov (1966, p. 243). Measurements of the crescentic segments on Landsat imagery (fig. 264) show that they have a mean width (horn to horn) of 2.9 km (1.8 mi) and a mean length of 2.2 km (1.3 mi). The ridges are spaced an average of 2.9 km (1.8 mi) apart. Numerous lakes occur in interdune areas. The crescentic dunes of the Pa-tan-chi-lin Sha-mo differ in pattern from those in the Cherchen Desert described earlier (fig. 261) but are closely similar to those of the Ak-bel Kum on the south and east shores of Po-ssu-t'eng Hu in the

Tarim Basin (fig. 260). Differences in dune patterns (as opposed to differences in dune types) in the Pa-tan-chi-lin Sha-mo and Cherchen Desert seem to be due mainly to differences in relative size and spacing of the dune and interdune elements (chapter J). Compound crescentic dune ridges of the Pa-tan-chi-lin Sha-mo are compared in chapter J (tables 38, 39) with similar dunes in other deserts.

Toward the southeast margin of the Pa-tan-chi-lin Sha-mo (fig. 264B) dune ridges are shorter, distinctly S-shaped, and sharp crested. They are in-

EXPLANATION

MOSTLY SIMPLE, SOME COMPOUND, LINEAR DUNES

CRESCENTIC DUNES
  Simple barchanoid ridges

  Compound barchanoid ridges

  Complex with
    Bulbous ridges

    Feathered ridges

    Feathered ridges with linear elements

    Reversing ridges

    Ridges with linear elements superimposed

SIMPLE, AND SOME COMPOUND, STAR DUNES

SIMPLE AND COMPOUND DOME-SHAPED DUNES

SHEETS AND STREAKS

  Compound, with linear elements

BEDROCK OUTCROP OR BARRIER

ROCK OR GRAVEL PLAIN WITH NO LARGE DUNES

PLAYA OR SALT MARSH

VEGETATED PLAIN OR MARSH

---200--- ISOHYET—Average annual rainfall, in millimetres

············· BOUNDARY OF SAND SEA

···‹···› INTERMITTENT DRAINAGE—Showing direction of flow

(Fig. 263. —Continued.)

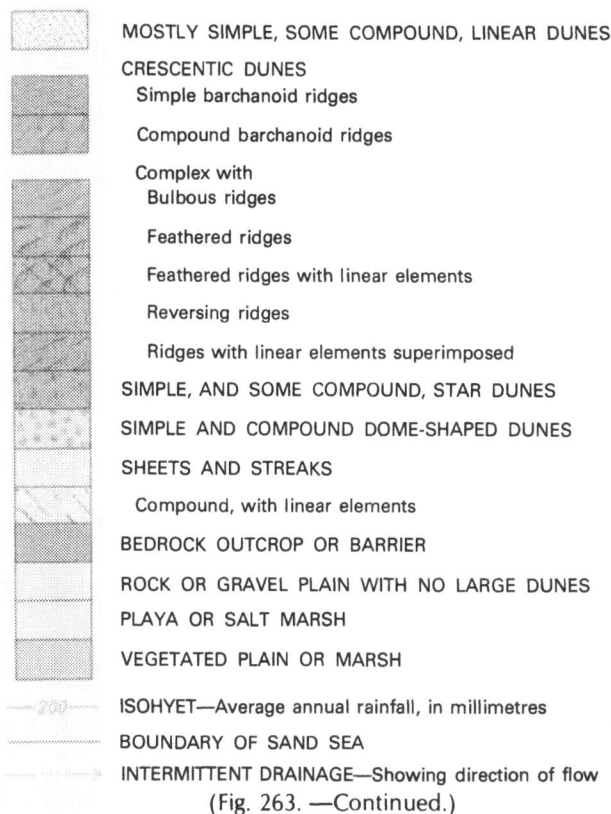

terpreted here as reversing dunes. Isolated dunes at the extreme eastern and southern margins of this sand sea are interpreted here as star dunes.

Small crescentic dune ridges, dome-shaped dunes, and small star dunes of the Ya-pu-lai Sha-mo, Yamalik Dunes, and Wa-lan-pu-ho Sha-mo fill abandoned stream valleys and other low areas on the stony plains of the eastern Ala Shan Desert. Dome-shaped dunes and star dunes in the Yamalik Dunes tend to be in linear patterns that parallel the trends of bounding valley sides (fig. 265A).

Dunes in the T'eng-ko-li Sha-mo at the downwind (southeastern) margin of the Ala Shan Desert are broad ridges of a complex variety not seen elsewhere. The broad crescentic ridges seem to lack main slipfaces, but the tops of the ridges have clusters of star dunes at irregular intervals (fig. 265B). Southwest of the dunes are elongate salt flats here interpreted as deflation hollows. Dunes in the T'eng-ko-li Sha-mo are reported (Petrov, 1966, p. 247) to be more than 100 m (328 ft) high.

Surface Wind Flow and Precipitation

Surface wind circulation over the Takla Makan and Ala Shan Deserts is controlled principally by the central Asiatic high-pressure cell and a thermal low-pressure cell which develops over the Indian subcontinent in the summer. During the winter, subsidence from the central Asiatic high tends to produce northerly to northeasterly winds across the Takla Makan and Ala Shan Deserts (fig. 266A). During the spring the central Asiatic high weakens, its center shifts 10° westward, and a thermal low begins to develop over the Indian subcontinent (fig. 266B). At this time, both the Takla Makan and Ala Shan Deserts lie in a zone between the high- and low-pressure cells, and also during this time the highest drift potentials occur, as discussed later. By summer time (fig. 266C) low pressure dominates much of Asia, especially India, and wind regimes in the deserts are weaker and more directionally unstable than at other times of the year.

Mountains that ring the Takla Makan Desert and those in and around the Ala Shan Desert may deflect wind flow, resulting in variations from the trends of wind direction and strength as just discussed and as shown in figure 266. In addition, cold frontal passages, associated with midlatitude cyclones, may produce the westerly winds observed at some stations (figs. 267, 268), particularly during the spring.

The Takla Makan Desert receives less than 100 mm (4 in.) precipitation per year (fig. 260). Average annual precipitation is only 30 mm (1.2 in.) at Charkhlik and Cherchen, and in the interior of the Tarim Basin is commonly zero (Zaychikov and others, 1965, p. 532, 555). In the Lop Nor area, average annual precipitation is about 10 mm/yr (0.4 in./yr) (Petrov, 1967, p. 306).

The climate of the Ala Shan Desert is transitional from the desert climate of the Tarim Basin to the steppe climate of Inner Mongolia (Petrov, 1967, p. 22). Average annual rainfall increases from northwest to southeast in the Ala Shan Desert (fig. 263). The southern and eastern margins of the desert may receive approximately 200 mm (8 in.) of rainfall per year (fig. 263).

Direction of Sand Drift

Annual resultant drift directions differ widely from place to place within the Takla Makan Desert, but resultant drift directions at each station are mostly steady throughout the year (fig. 260). At Kucha, in the northern part of the Tarim Basin, annual resultant drift directions are south or

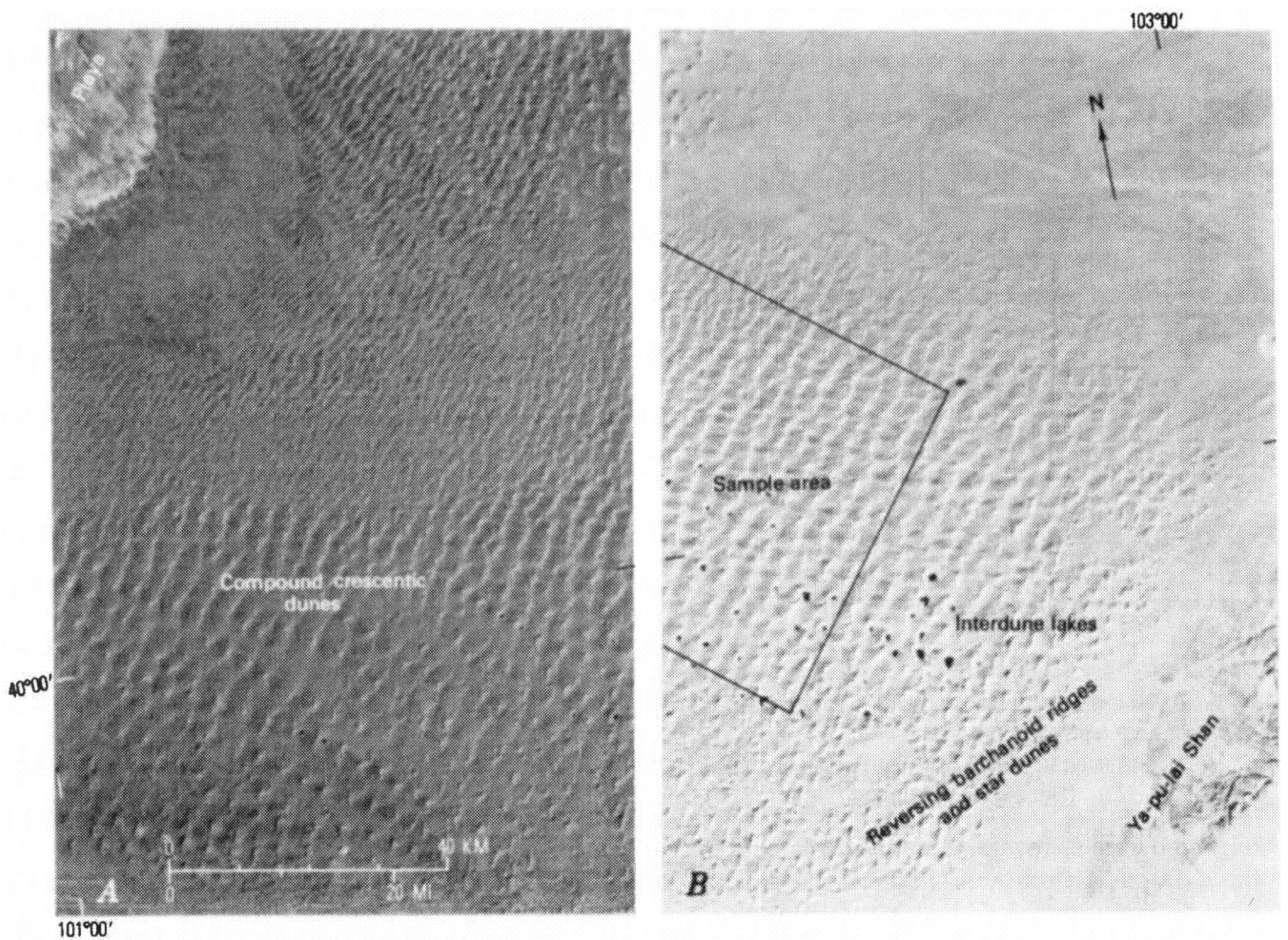

PA-TAN-CHI-LIN SHA-MO, Ala Shan Desert, Peoples Republic of China. Landsat imagery E1424 – 03262; E1117 – 03222. A, The main mass of sands, showing compound crescentic dune ridges, many with interdune lakes. To the northwest are playas of the Jo Shui intermittent drainage system. B, East edge of the sands, where they are bounded by the Ya-pu-lai Shan; sharply recurved S-shaped dunes are interpreted as reversing barchanoid ridges. Star dunes are also evident. (Fig. 264.)

southeastward (fig. 269), subparallel to the trend of the dome-shaped and linear dunes of the northern and central parts of the sand sea. In the western Tarim Basin —including the Bel'kum and Tagkum Sands —resultant drift directions may be toward the southwest as at Pa-ch'u or toward the southeast as at So-ch'e or east as at Khotan (fig. 260). Resultant drift directions in the Cherchen Desert near T'ieh-kan-li-k'o show that the large transverse (crescentic) dunes normally are associated with unimodal or acute bimodal wind distributions, such as that at T'ieh-kan-li-k'o. Dunes in the southern part of the Cherchen Desert near Nochiang, however, are complex combinations of linear and crescentic ridges in a chevronlike pattern (figs. 260, 261). This complexity may be a result of intersect-

ing wind patterns as suggested by Hedin (1905a, p. 152, 360–362); however, available wind data for Nochiang, about 60 km (37 mi) from these dunes (fig. 260 and chapter F) do not confirm Hedin's interpretation.

Annual resultant drift directions in the Ala Shan Desert (fig. 263) are toward the southeast except at Shan-tan, which is in an intermontane valley (fig. 263). Resultant drift directions are relatively steady throughout the year, but some wind regimes are complex from month to month. For example, monthly wind regimes tend to be complex at Chi-lan-t'ai (fig. 268). The average RDP/DP (fig. 189) is 0.40 for stations in the Ala Shan Desert versus an average RDP/DP of 0.60 for stations near the Takla Makan Desert. This difference reflects the greater directional variability of wind regimes in the Ala Shan Desert and may account for the common occurrence there of reversing dunes and star dunes (fig. 264). Star dunes are associated with complex annual wind distributions.

Amount of Sand Drift

The average drift potential of stations in the Ala Shan Desert (129 *VU*) is slightly greater than the average drift potential of stations in the Takla Makan Desert (82 *VU*), but both deserts have low drift potentials compared with other deserts around the world (chapter F, table 16). These low values may occur because the deserts are positioned relatively near the center of the central Asiatic high-pressure cell, where winds tend to be weak, and because they are to some degree sheltered by mountains.

Drift potentials in the Takla Makan Desert are highest in the northeast, near Charkhlik, and west at Ai-ssu-la-k'o-chan, off the map at lat 39°47′ N., long 75°47′ E. Minimum values occur at stations along the southern desert margin (fig. 260). Springtime is the season of highest drift potentials in both the Takla Makan and Ala Shan Deserts (figs. 270, 271), because they are in the zone between the high-pressure cell in China and the low-pressure cell in India (fig. 266B). Some stations also have a winter season of relatively high drift potentials (fig. 270B).

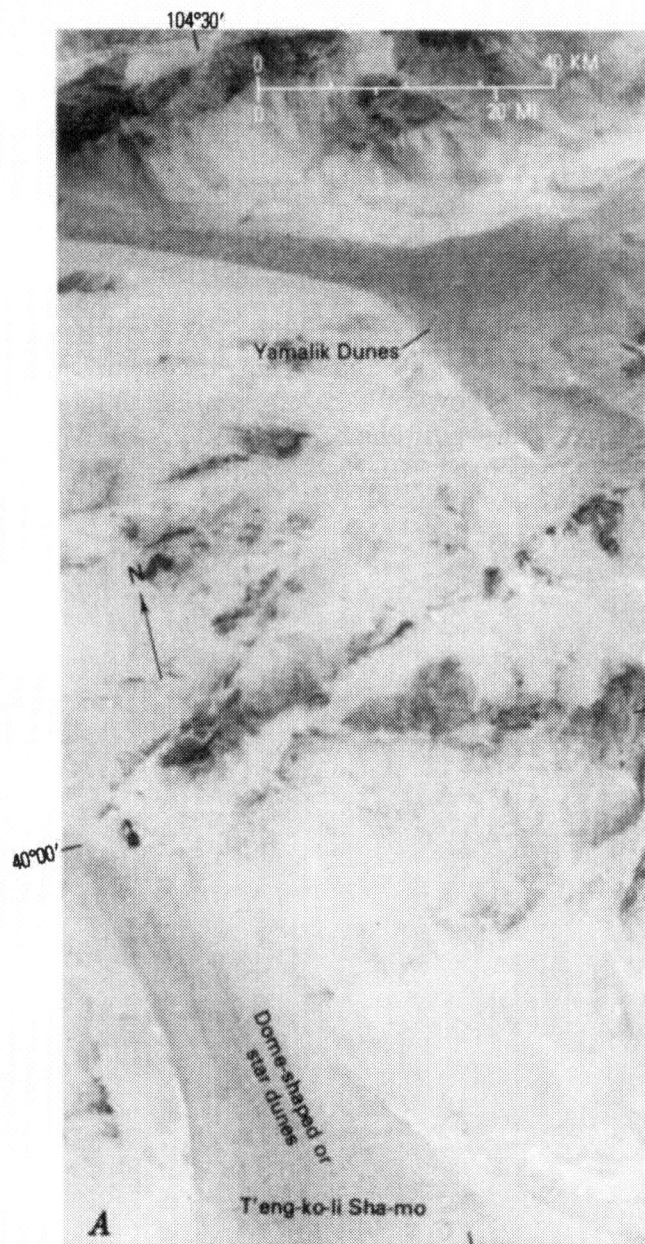

DUNE FIELDS IN THE EASTERN ALA SHAN DESERT (above and facing page). Landsat imagery E1224 – 03165; E1133 – 03112. A, Yamalik Dunes fill valleys on the stony plains. B, Complex dune ridges in the T'eng-ko-li Sha-mo, continuous to the southeast with the area shown in A. To the east is the partly snow-covered Ho-lan-shan Range. East of Ho-lan-shan is part of the Huang Ho valley. (Fig. 265.)

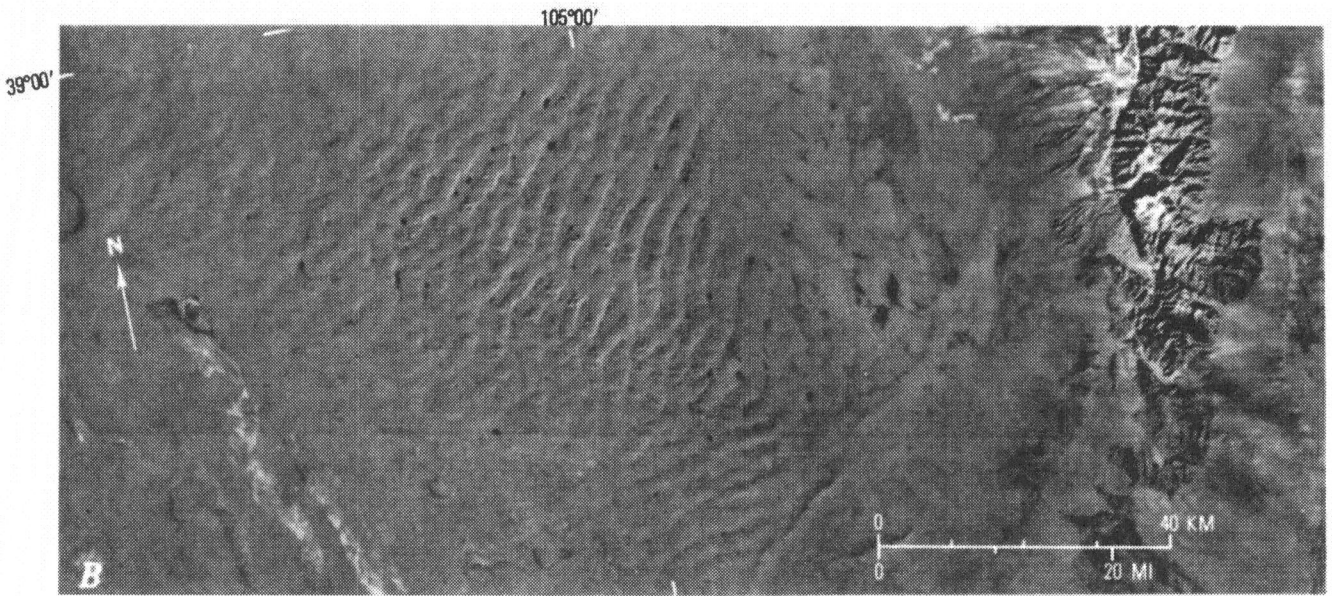

(Fig. 265. —Continued.)

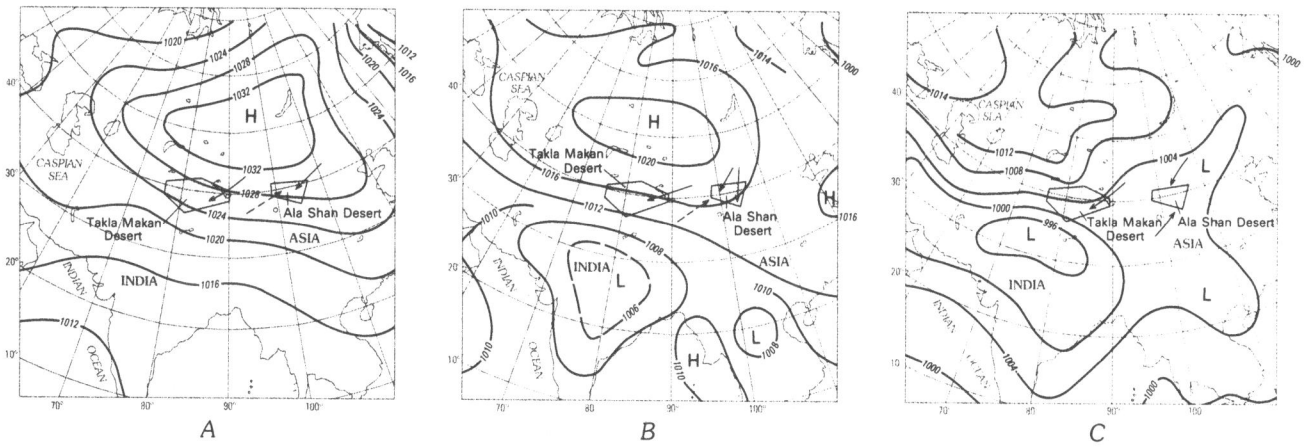

MEAN SEA-LEVEL PRESSURE (in millibars of mercury) distribution over the Takla Makan and Ala Shan Deserts of China for 3 months of the year. H, high-pressure cell; L, low-pressure cell. Trends of surface wind flow are shown by solid arrows; dashed arrows indicate winds which may result from frontal passages or cyclones, or from deflection of prevailing winds by mountains. A, January; B, April; C, June. (Fig. 266.)

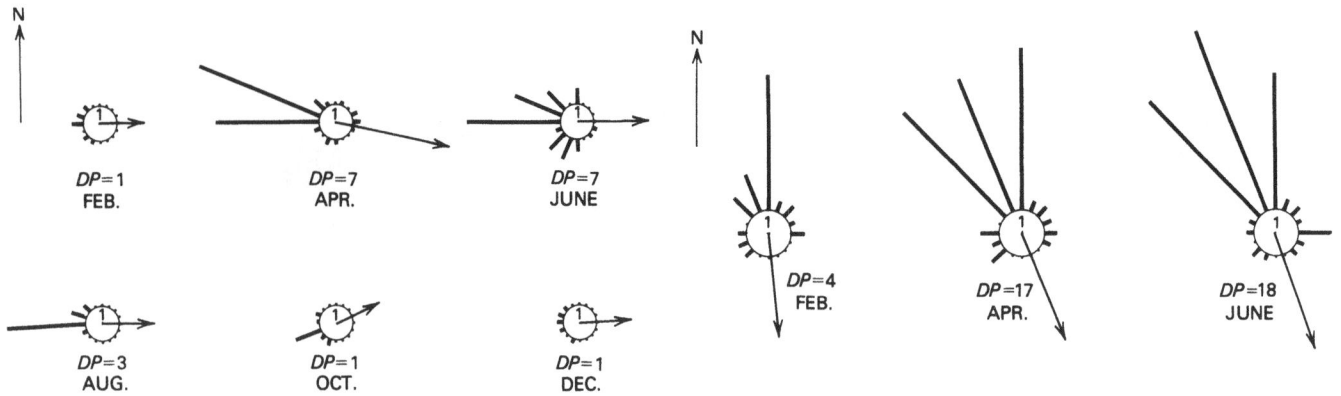

SAND ROSES FOR 6 MONTHS FOR KHOTAN, Peoples Republic of China, illustrating the eastward resultant drift direction along the southwestern margin of the Takla Makan Desert, west of the region of complex dunes (fig. 260). As at So-ch'e (fig. 260) effective winds are directionally steady throughout the year. Drift potential, in vector units, is given for each month. (Fig. 267.)

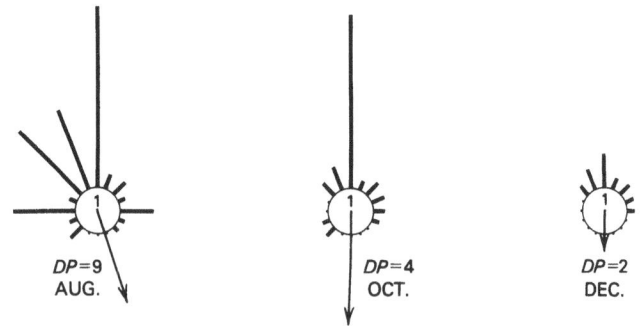

SAND ROSES FOR 6 MONTHS FOR KUCHA, Peoples Republic of China, illustrating southward resultant drift directions in the north-central Tarim Basin. Monthly resultant drift directions vary little because of directional steadiness of effective winds. This station also exhibits low drift potentials characteristic of stations near the Takla Makan sand seas. Drift potential, in vector units, is given for each month. (Fig. 269.)

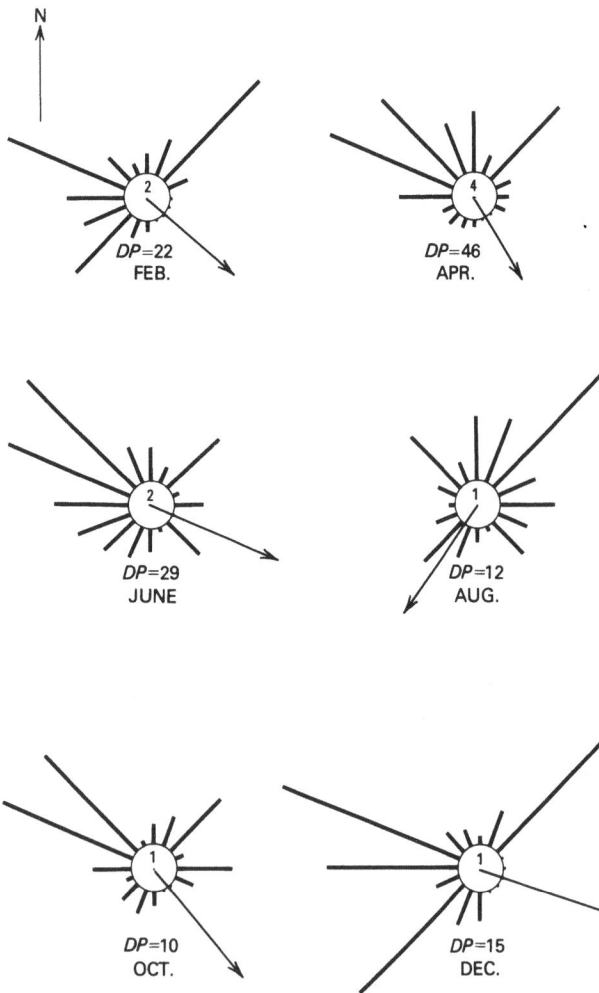

SAND ROSES FOR 6 MONTHS FOR CHI-LAN-T'AI, Peoples Republic of China, illustrating the complex wind regimes which may occur from month to month in the mountainous, eastern Ala Shan Desert. Resultant drift directions are relatively steady (southeasterly, with the exception of August) during the year. Drift potential, in vector units, is given for each month. (Fig.

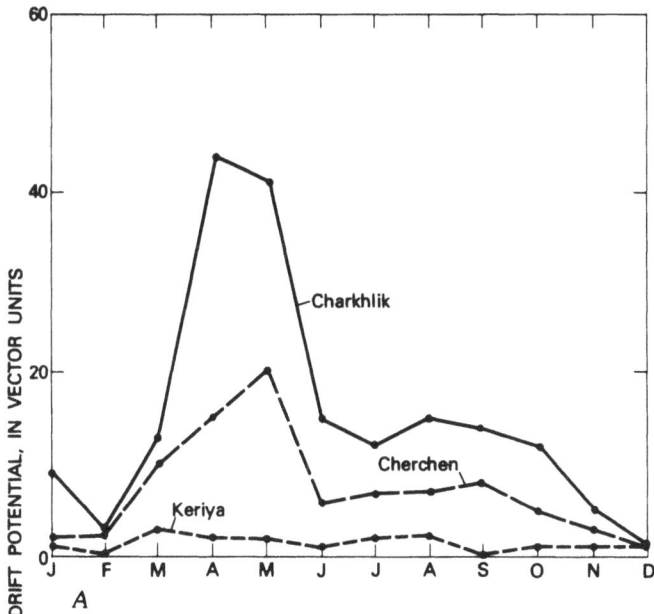

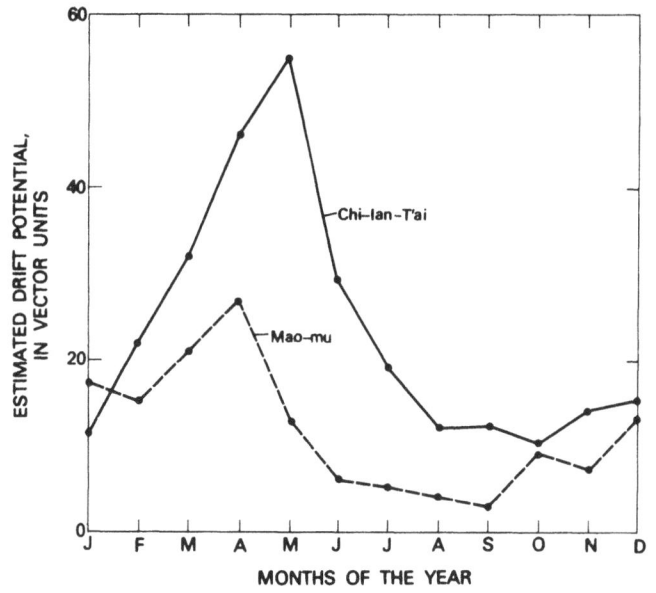

VARIATION IN DRIFT POTENTIAL during the year at Chi-lan-t'ai and Mao-mu on the east and west margins of the Ala Shan Desert, Peoples Republic of China. These stations have higher average monthly and annual drift potentials than most stations in the Takla Makan Desert but, in common with them, experience highest drift potentials in the spring. (Fig. 271.)

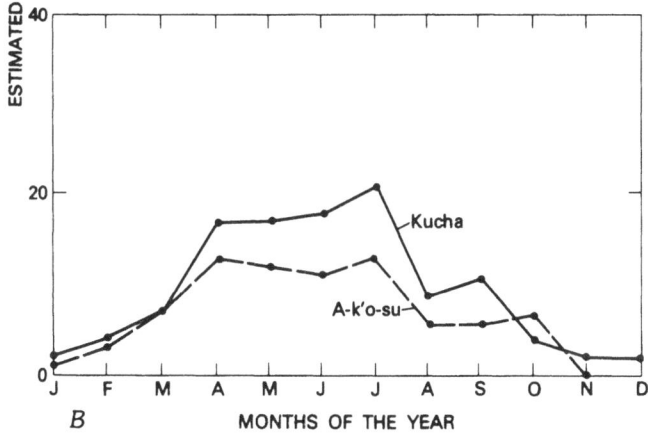

VARIATION IN DRIFT POTENTIAL during the year at representative stations within the Takla Makan Desert. A, Charkhlik, Cherchen, and Keriya. B, Kucha, A-k'o-su. Most stations have highest drift potentials in the spring (A). Northern stations also have relatively high drift potentials in the summer. Part B also illustrates the decrease in average drift potential from the east end of the Takla Makan Desert along the south edge toward Keriya. (Fig. 270.)

A STUDY OF GLOBAL SAND SEAS

GLOSSARY

Prepared by MARGUERITE GLENN

T HIS GLOSSARY defines many of the terms used in this Professional Paper; it is for the convenience of the reader who may not be a specialist on deserts or eolian deposits. Many of the common physiographic, geologic, meteorological, and mathematical terms used are not defined here because their meanings are well established.

Entries and their synonyms are printed in boldface. The most commonly used term for any one meaning is the one that is defined, and the synonymous entries are referred to it. When more than one definition is accepted for an entry, the definitions will be numbered to show the currently more common or preferred usage. Geographic areas of usage are shown in parentheses following some terms.

Many sources were used in preparation of this glossary. Especially helpful were "A Desert Glossary," by R. O. Stone (1967), and "Glossary of Geology," published by American Geological Institute (1972).

Some entries are terms used for the first time in this paper, and they are so identified; for example: (McKee, this paper).

Acute Bimodal wind distribution. Wind distribution with two modes whose peak directions itnersect at an angle of 90° or less (Fryberger, this paper).

Adhesion ripple or **Antiripplet**. Irregular sand ridge transverse to wind direction, formed when dry sand is blown across a smooth moist surface. It may be 30–40 cm (12–16 in.) long and a few centimetres (about 1 in.)

high. The crest is asymmetrical and migrates upwind. The stoss side is steeper than the lee side.

Adhesion wart. A small irregular wartlike sand accumulation made by wind that changes direction rapidly while blowing on a moist sand surface.

Akle (North Africa). See **Fishscale dune pattern**.

Anchored dune. See **Fixed dune**.

Angle of repose. Maximum angle at which loose detritus is stable (commonly 32°–34° for dry dune sand).

Antiripplet. See **Adhesion ripple**.

Arm (of dune). The trailing outer extension of a parabolic or star dune.

Arroyo (Southwestern United States), **Dry wash**, or **Wadi** (North Africa). Channel of an ephemeral or intermittent stream. It commonly has vertical walls of unconsolidated detritus 60 cm (2 ft) or more high.

Avalanche bedding. Steeply inclined bedding in dunes produced by avalanche of sand down the slipface of the dune.

Avalanche slope. See **Slipface**.

Backshore. The part of a beach that is landward of a berm or beach crest. In opposition to foreshore.

Bahr (Lake Chad). Seasonally flooded interdune area.

Bajada (Southwestern United States). A broad alluvial slope at the base of a mountain range. Fan forms may or may not be present.

Bajir (China). See **Interdune corridor**.

Barchan, Barcan, Barchane, Barkan, or **Barkhan**. Crescent-shaped sand dune. Windward slope is gentle and lee slope or slipface is at the angle of repose of dry sand. Horns of the crescent extend downwind.

Barchanoid dune. Any dune of the barchan or barchanoid-ridge type.

Barchanoid ridge. Asymmetrical wavy dune ridge, oriented transverse to wind direction with gently dipping windward slopes and steeply dipping (32°–34°) lee slopes, commonly separated from downwind wavy dune ridges by open interdune corridors. It consists of coalesced barchans in rows.

Basin. A depressed area without a surface outlet, into which runoff from the surrounding terrain drains.

Basketweave dune pattern. Type of parallel wavy dune complex in which the dune ridges are offset, forming a chevron or herringbone pattern of ridges and interdune hollows.

Berg wind (Namib Desert, South-West Africa). A local name for an east wind.

Blowout or **Reactivated dune.** Saucer-, cup-, or trough-shaped hollow formed by wind erosion on a preexisting sand deposit. The adjoining accumulation of sand derived from the depression, where readily recognizable, is commonly considered part of the blowout.

Brink. The top of the slipface of a dune. It may correspond to the crest.

Caliche. An essentially calcium-carbonate deposit remaining in or on soil following loss of water through evaporation or other processes.

Calcrete. Hard, dense layer of caliche in soil.

Captation. The trapping of irregular eolian sand masses against cliffs or mountains, thus forming climbing or hanging dunes.

Chevron dune. A V-shaped variety of a parabolic dune.

Clay dune. Eolian accumulation of sand-sized clay aggregates.

Claypan. Playa or dry lake formed in a shallow, undrained depression that has a hard, sunbaked surface of clay.

Climbing dune. Sand deposited by wind against a cliff or mountain slope.

Coastal dune. Sand dunes on low-lying land abandoned or built up by the sea.

Cold desert. Arid region in which the mean annual temperature is less than 18°C (64.4°F), such as the Gobi Desert or some Arctic and Antarctic regions, where plant and animal life is restricted by low temperature.

Colluvium. The sediment on a slope or at the base of a slope formed by mass-wasting or unconcentrated sheet runoff.

Complex dune. A combination of two or more different dune types in a single dune, in contrast to compound dune, which is a combination of two or more dunes of the same type.

Complex wind distribution. Wind distribution with more than two modes or with poorly defined modes (Fryberger, this paper).

Compound linear dune. A major longitudinal dune ridge with small linear ridges formed on it. The minor ridges may be parallel to the main ridge or diverge from it, as in feathered seif dunes.

Contorted bedding. All disturbed strata, regardless of cause, where deformation occurred after deposition but before lithification. Structures are restricted mostly to simple sets of strata or cross-strata with unaffected sets of strata above and below.

Couloir (North Africa). Open interdune corridor.

Crescentic dune. (1) See **Barchan**; (2) in chapters J and K used in classification of Landsat (ERTS) imagery patterns for all barchans, barchanoid ridge, and transverse type dunes.

Crescentic dune ridge. (1) See Barchanoid ridge; (2) used in chapters J and K as a sand ridge or elongate dune with arclike or barchanoid segments.

Crescent ripple. A small ripple in the shape of a moon in its first or last quarter. The horns or arms of the ripple extend downcurrent.

Crest. A dune summit. It is the highest natural projection of a dune.

Cumulative frequency distribution curve or **Cumulative curve.** A curve drawn to represent the percentage of occurrences of a number of observations of a variable less than or greater than any given value for an entire sample.

Deflation. Removal or "blowing out" of fine-grained sediment by the wind.

Deflation basin. Hollow surface formed by removal of sand and dust by eolian action. It commonly has a rim of resistant material surrounding the depression.

Deflation ripple. See **Granule ripple**.

Desert. An arid or semiarid region characterized by an excess of evaporation over precipitation.

Desert pavement. A sheetlike residual concentration of wind-polished closely packed gravels, or rock fragments, mantling a desert surface where wind has removed fine material. Gravels or fragments commonly are cemented by mineralized solutions (AGI).

Desert polish. Smooth and polished surfaces on rocks caused by the action of windblown sand and dust.

Desert varnish, Desert lacquer, or **Desert patina.** Surface stain or crust of manganese or iron oxide which characterizes many exposed rock surfaces in deserts. It coats outcropping rocks, boulders, and pebbles, imparting a brown or black color and a bright luster to their surfaces.

Desiccation crack, Mud crack, Shrinkage crack, or **Sun crack.** Small fissure or crack formed by shrinkage of clay or clayey beds in drying by the sun.

Dikaka. Accumulation of dune sand covered by scrub or grass vegetation; extended to include plant-root cavities in dune sediments (calcified root tubules).

Discriminant analysis. A statistical procedure for classifying subsequent samples into categories previously defined and differentiated on the basis of samples from known populations.

Dissipation structure. Wavy, irregular layers in sediments, resulting from the modification of original stratification by the concentration of colloidal materials (clay minerals, hydrous iron oxides, and humic compounds) (J. J. Bigarella, written commun., 1975).

Dome dune. Low circular or oval mound formed where dune height is inhibited by unobstructed strong winds. It generally lacks a slipface.

Draa (North Africa). Large-scale accumulation of eolian sand. Used by I. G. Wilson (1973).

Drift potential (DP). A measure, in vector units, of the sand-moving power of the wind. It is derived from reduction of surface wind data through a weighting equation and usually represents 1 year (Fryberger, this paper).

Dund (India). Interdune area filled by a lake.

Dune. Accumulation into a mound or ridge of windblown sediment, commonly having a gentle upwind slope and one or more steep lee slopes (slipfaces).

Dune-extradune depositional system. A system containing all the sediments of a dune field and marginal area. It includes dune and interdune sediments and associated extradune sediments, such as alluvial-fan, wadi, serir, stream, sabkha, playa, lake, beach, and tidal-flat deposits (Lupe and Ahlbrandt, this paper).

Dune-interdune depositional system. A system containing only dune and interdune sediments (Lupe and Ahlbrandt, this paper).

Dune massif. A compound star dune.

Dune spacing index. A measure of the average number of dune ridges crossing a 50-km-long (31.1-mi-long) line normal to the trend of dune ridges in a given area.

Dune valley. See **Interdune**.

Elb (Algeria). A transverse dune (Capot-Rey, 1945, p. 392).

Elephant-head dune (Colorado Desert, Calif., U.S.A.). Small sand shadow whose windward face is covered with vegetation and with a long tapering snout of sand extending to the lee.

Elongate blowout dune. Wind-excavated basin with a crescent-shaped rim of sand about the lee side. It differs from blowout dune in that it has an elongate form, and there is slight migration of the basin and rim in the direction of the prevailing wind.

Eolianite. All consolidated sedimentary rock formed from wind-deposited sand, commonly having large-scale eolian-type cross-strata.

Eolian sand. Wind blown sand.

Eolian sandstone. Sedimentary rock formed of consolidated windblown sand. See **Eolianite**.

Erg (North Africa). See **Sand sea**.

Established dune. See **Fixed dune**.

Extra dune. See **Satellite dune**.

Extradune. Area marginal to dune field which contains sediments of the same age and source as the dunes (Lupe and Ahlbrandt, this paper).

Falling dune. Sand accumulation (sloping at the angle of repose for dry sand) to the lee of a cliff or mountain side as sand is blown off the top.

Fan (alluvial). A sedimentary deposit having a distinctive convex upward surface that radiates downslope from a point where a stream emerges from a canyon floor onto a plain with more gentle gradient.

Fanglomerate. Indurated alluvial-fan gravel.

Feathered dune. A variety of compound linear dune in which subsidiary ridges diverge from a main ridge.

Festoon structure or **Festoon-type cross bedding**. A cross-lamination pattern resulting from repetition of a two-step process: (1) Erosion of plunging troughs and (2) filling of the troughs by thin concave upward layers generally conforming to the shape of the trough floors (fig. 157). Festoon pattern is most conspicuous on vertical faces cut at right angles to longitudinal axes in a series of superimposed troughs plunging in the same direction (Knight, 1929, p. 58). The troughs may be filled symmetrically or asymmetrically.

Fishhook dune. See **Hooked dune**.

Fishscale dune pattern, Akle, or **Peak and Fulji topography**. Type of parallel wavy dune pattern in which the interdune areas are enclosed by crescentic elements of the dune ridges.

Fixed dune. Nonmigratory dune fixed by vegetation or by cementation.

Fluid threshold shear velocity. That shear velocity at which sand movement starts as a result of the shearing stresses imparted to the sand by the wind. (See Bagnold, 1941, p. 88.)

Frosted sand grains. Typical sand grain of eolian desert sediments, whose etched and pitted surface is created by impact with other wind-transported grains and causes a scattering of light. It is the opposite of a glassy or polished sand grain.

Fulji or **Fulje**. (1) A depression (interdune) between barchans or barchanoid ridges, especially where dunes are pressing closely on one another (Arabia); (2) blowout or small parabolic dune (Australia).

Giant crescent dune pattern. Type of parallel wavy dune pattern in which ridges are composed of coalesced megabarchans and are separated by open or closed interdune hollows.

Gibber (Australia). Gravel plain.

Grain flowage. A type of sand avalanching in which individual grains move independently of each other, in

opposition to avalanching by mass movement of slumping.

Granule ripple. Unusually large wind ripples composed in part
of detritus approaching or attaining granule size, 2–4
mm (0.08–0.16 in.). They are commonly associated with
local concentrations of lag material.

Granulometric control. The ability of various grain sizes to
control the shape or dimension of eolian bedforms, such
as ripples or dunes (I. G. Wilson, 1973).

Graphical measures. Graphical method of obtaining statistical
values of grain size in which selected values from points
along a cumulative frequency curve are used in formulas to calculate the desired statistical data.

Guern. See **Khurd.**

Hairpin dune. A greatly elongated parabolic dune that has
migrated downwind, its horns drawn out parallel to
each other, formed where wind is in conflict with
vegetation and where winds are strong and of constant
directions.

Hammada or **Hamada.** A bare rock surface in a desert.

Hanging dune. See **Falling dune.**

Hardpan. Cemented or indurated layer of soil usually composed of calcium carbonate, silica, iron oxide, or clay, at
or some distance beneath the ground surface.

High-energy wind environments. A wind environment with
an annual drift potential of 400 vector units or more.

Histogram. Multiple-bar diagram showing relative abun-
dances of specimens, materials, or other quantitative
determinations divided into a number of regularly arranged classes.

Hooked dune or **Fishhook dune.** Dune consisting of a long
sinuous ridge forming the shaft and a well-defined crescent forming the hook.

Horns or Arms. The pointed ends of a dune, especially the forward extensions of a barchan dune.

Hot desert. Arid region in which the mean annual temperature
is greater than 18°C (64.6°F) (Stone, 1967, p. 230).

Impact threshold shear velocity. The minimum shear
velocity required to maintain sand in saltation. (See
Bagnold, 1941, p. 88.)

Impact threshold wind velocity. The minimum wind velocity
required to maintain sand in saltation. (See Bagnold,
1941, p. 32.)

Inselberg. Prominent residual steep-sided rocky hills rising
abruptly from a plain or desert.

Interdune. Relatively flat area between dunes.

Interdune corridor, Interdune hollow, or **Interdune area.**
Continuous flat or gently sloping surface between dune
ridges or rows of dunes.

Interior drainage, Centripetal drainage, or **Internal drainage.**
Drainage toward the lowermost part of an enclosed
basin.

Intermediate-energy wind environments. Wind environments with an annual drift potential of 200–399 vector
units.

Irq. (1) (Arabia) Elongate dune or belt of elongate dunes
(Stone, 1967, p. 255); (2) (North Africa) narrow
longitudinal dune or a seif (Stone, 1967, p. 263).

Isohyet. Line on a map connecting points which receive equal
amounts of average rainfall.

Khurd (Algeria), **Guern.** A high (80–100 m (262–328 ft))
pyramid-shaped sand dune with curved slopes, formed
by the intersection of seifs (Capot-Rey, 1945, p. 393).

Kurtosis. The quality, state, or condition of peakedness or flat-
ness of the graphic representation of a statistical distribution.

Lag gravel. Residual accumulation of coarse detrital particles
(larger than sand size) from which the fine material has
been deflated.

Laguna (Southwestern United States). Shallow ephemeral lake
in the lower part of a bolson, fed by intermittent
streams. See also **Playa.**

Lee dune. Dune formed to the lee of an obstruction.

Lee slope, lee side, leeward side. The downwind side of a
dune or other protuberance in the path of a wind.

Leptokurtic (1) A frequency distribution that has a concentration of values about its mean, greater than for the corresponding normal distribution; (2) a narrow frequency-
distribution curve that is more peaked than the corresponding normal distribution curve.

Lettau equation for rate of sand drift. The equation suggested
by Heintz Lettau (University of Wisconsin, Madison,
written commun., 1975) for the rate of sand drift. This
rate is a function of grain size and drag velocity (shear
velocity) of the wind. See chapter F, fig. 92.

Linear dune, Longitudinal dune, or **Seif.** Parallel, straight
dune with slipfaces on both sides and with its length
many times greater than its width.

Loess. Soft, porous, yellow- or buff-colored accumulation of
wind-laid particles that are of silt size or smaller.

Longitudinal dune. See **Linear dune.**

Low-energy wind environment. A wind environment with an
annual drift potential of less than 200 vector units.

Low latitude desert. Hot and dry desert located between lats
15° and 35° N. or lats 15° and 35° S. in the region of
subtropical high-pressure air masses and the trade
winds.

Mean grain size. The diameter equivalent of the arithmetic
mean or average grain size.

Medano. Sand dune (Spanish term).

Megabarchan. Giant barchan, 100 m (328 ft) or more in height.
These dunes commonly coalesced into ridges to form
parallel wavy dune patterns. Each crescent in the ridge
normally has one major slipface.

Mesokurtic. Closely resembling a normal frequency distribution — A distribution curve that is neither leptokurtic (very peaked) nor platykurtic (flat across the top).

Middle latitude desert. An arid region between lats 35° and 50° N. or lats 35° and 50° S. that owes its existence either to its location in the deep interior of a continent or to high mountains across the path of the moisture-bearing prevailing wind.

Migration (dune). Movement of a dune due to transfer of sand from windward to the leeward side.

Miniature terrace-and-cliff structure. A succession of abrupt rises or steps of about 2.54 mm (0.1 in.), each ascending an inclined bedding plane.

Moderate-energy wind environment. A wind environment with an annual drift potential of 200 –399 vector units.

Modified lee slope or **surface.** A dune slipface or slipfaces altered by slumping, grain flowage, or deposition of suspended load.

Moment measures. The expected values of each of the powers of a random variable that has a given distribution; a weighted measure of central tendency; in sedimentology they are related to the center of gravity of the particle-size-distribution curve and are defined about the mean value of the variable.

Monsoon. Type of wind system in which there is a complete or almost complete reversal of prevailing direction from season to season. It is especially prominent within tropics on eastern sides of great landmasses.

Morphology. The shape or physiography of land-surface features.

Mud polygon, Desiccation polygon, or **Mud crack polygon.** Block of fine-grained sediment, bounded by mud cracks or shrinkage cracks, that commonly has three to five sides, although in some examples it may have as many as eight sides.

Narrow unimodal wind distribution. A wind regime that produces a distribution of drift potential on a sand rose in which 90 percent or more of the drift potential lies within a 45° arc of the compass.

Nefud. A high sand dune. See also **Sand sea.**

Observer bias. Tendency for observers to record wind directions from prime directions instead of from intermediate directions of the compass.

Obtuse bimodal wind distribution. A wind regime that produces a distribution of drift potential on a sand rose that has two modes whose peaks intersect at an angle greater than 90°.

Oghurd (North Africa). A massive summit above the general sand dune level, or a massive mountainous dune resting on an underlying rocky topographic feature (Stone, 1967, p. 263).

Omiribi (Kalahari Desert, Africa). See **Arroyo.**

Onshore wind. Winds that blow from the sea onto a land surface.

Orographic desert. See **Rain shadow desert.**

Oscillation ripples. Symmetrical ripple marks consisting of sharp, narrow, relatively straight crests between broadly rounded troughs.

Oued. See **Arroyo.**

Packing. The manner of arrangement or spacing of the solid particles in a sediment or sedimentary rock.

Pan. (1) Natural basin or depression that intermittently contains water that may leave a saline deposit; (2) a dry lake or playa.

Parallel ripples. Ripple marks with relatively straight crests, oriented transverse to wave or current direction and having a symmetrical or asymmetrical profile.

Parabolic dune. A U-shaped or V-shaped dune representing a type of blowout in which the middle part has moved forward with respect to the sides or arms. Open end of U or V faces upwind and the arms commonly are anchored by vegetation. One to three slipfaces occur on the convex outer part of the nose and arms of the dune.

Parallel sinuous ripples. Ripple marks in which the horizontal distance, normal to flow, is many times the wavelength, and the crests swing from side to side as viewed in plan.

Parallel straight dune, Seif, or **Longitudinal dune.** A linear ridge of windblown sand whose length is much greater than its width, and which commonly has slipfaces along both sides. It moves in a direction parallel to its length.

Parallel wavy dune complex. A group of crescentic dune segments that have coalesced into wavy ridges. Each crescentic segment of the ridge commonly is as wide as it is long, and the segments are asymmetrical. All segments of a ridge have their major slipfaces oriented normal to wind directions.

Patina. A thin film of **Desert varnish.**

Peak and Fulji topography. See **Fishscale dune pattern.**

Pediment. A broad surface or plain formed in a desert terrain by erosion at a mountain base. It may be bare bedrock or veneered by a thin layer of alluvium.

Piedmont. A surface lying or formed at the base of a mountain. In desert regions it is a plain or slope and may include a pediment, bajada, or both.

Platykurtic. (1) A frequency distribution that has a concentration of values about its mean less than for the corresponding normal distribution; (2) a broad, flat-topped frequency distribution curve that is less peaked than the corresponding normal distribution curve.

Playa (Southwestern United States), **Clay pan, Dry lake, Salt flat,** or **Salt pan.** An extremely flat vegetation-free area of clay, silt, and (or) evaporite minerals in the lowermost part of a basin having interior drainage in an arid or semiarid region. It is formed by a temporary lake which has evaporated, leaving behind fine sediment.

The term is also applied to the temporary lake. (Spanish term meaning beach, shore, or strand.)

Playa lake. Temporary or ephemeral lake in the desert.

Plinth. Lower and outer parts of seif, beyond the slipface boundaries. This part of the dune has never been subjected to sand avalanches (Bagnold, 1941, p. 229).

Polypyramid. A star dune.

Precipitation ridge (Cooper, 1958, p. 55). See **Retention ridge**.

Prevailing or predominant wind. A wind that blows most of the time from one particular direction.

Procedure bias. Error introduced into a wind summary when observations originally recorded from 36 points of the compass are summarized into 16 directions. It results in disproportionate representation of observations from the prime directions of the compass at the expense of the intermediate directions.

Produne. The part of the dune lying beyond the dune front and gently sloping down to the desert floor across which the dune is advancing.

Pyramidal dune or **Polypyramid**. See **Star dune**.

Raindrop sand crater. A thin detached layer or shell of sand, roughly circular with thin margin, formed by raindrops striking dry sand. Center of the shell is depressed and margin is raised slightly above the general sand level. After drying, the shell remains firm and retains the same shape; on a tilted sand surface it has a raised lower rim.

Rain shadow desert or **Orographic desert**. Arid land on the lee side of a mountain or a mountain range which intercepts moisture-laden air.

Recurved dune. A dune in which a forward extension is curved back toward main part of dune as by deflection of wind or by opposing action of two or more wind durections. See also **Reversing dune** and **Sigmoidal dune**.

Reg (Algeria). A large gravel deposit on a desert surface. It differs from serir in that most of the fine detritus has not been removed (Gautier, 1935, p. 62).

Resultant drift direction (RDD). The direction of the resultant drift potential.

Resultant drift potential (RDP). A measure in vector units, of the net sand-moving power of the wind at a station. Vector unit totals from various directions on a sand rose are treated as vector quantities and resolved trigonometrically to a resultant, the magnitude of which is the resultant drift potential, RDP, the direction of which is the resultant drift direction, RDD.

Resultant drift potential/drift potential (RDP/DP). A dimensionless parameter which is a measure of the variability of a wind regime without regard to specific directional components or total energy of environment.

Retention ridge or **Precipitation ridge**. A ridge formed parallel to shore by deposition of sand on the landward margin of the backshore. It accumulates against a vegetation barrier and is a nearly stabilized dune.

Reversing dune. A dune that tends to grow upwards but migrates only a limited distance because seasonal shifts in direction of dominant wind cause it to move alternately in nearly opposite directions.

Rhourd. A pyramid-shaped sand dune, formed by the intersection of other dunes (Aufrere, 1934, p. 139). See also **Khurd**.

Ripple. Any topographic deviation on the bedding surface of a sedimentary deposit that resembles a ripple of water.

Ripple height. The difference in elevation between the lowest point of a trough and the highest point of an adjoining crest in a rippled sand surface.

Ripple index. Wavelength of a ripple divided by height of a ripple (or ratio of wavelength to height).

Ripple symmetry index. The length of the horizontal projection of the stoss side divided by length of the horizontal projection of the lee side of a ripple (or ratio of stoss side to the lee side of a ripple).

Ripple wavelength. The horizontal distance between the deepest points of two troughs or the horizontal distance between two crests in a rippled sand surface.

Rising dune. See **Climbing dune**.

Roughness factor during sand driving (K'). The focus during sand driving, or that height, above the surface, at which lines depicting logarithmic height versus wind velocity for several wind velocities converge. See also Bagnold, 1941, p. 59.

Sabkha. See **Sebkha**.

Salina. A place where crystalline salt deposits are formed, such as a salt flat, a salt pan, a salada, especially a salt-incrusted playa or a wet playa.

Saltation. Bounding movement of rock particles, usually sand size, when carried by wind.

Sand avalanche. Movement of large masses of sand down a dune face when the angle of repose is exceeded, or when the dune is disturbed. It may consist of mass movement (slumping) or grain flowage or both.

Sand sea, Erg (North Africa). Vast region covered with sand and occupied by dunes.

Sand drift. (1) A deposit of eolian sand formed as the result of a decrease in wind velocity. It is commonly developed on the leeward side of windbreaks, such as vegetation, boulders, rock outcrops, or other irregular surface relief. (2) Used in chapter F (this paper) as a general term for movement of sand by wind occurring in a desert or along the shore.

Sand dune. Ridge or pile of windblown sand. See **Dune**.

Sand fall. Sand swept over a cliff or escarpment. See also **Falling dune.**

Sand glacier. Sand accumulation shaped as a broad fanlike plain that develops where sand is blown up the sides of a mountain, through passes and saddles, to spread out on the opposite side.

Sand levee. See **Whaleback.**

Sand ridge. (1) A parallel straight or linear dune; (2) a relatively coarse-grained high-crested widely spaced ripple of sand.

Sand ripple. A ripple mark formed by wind in dry sand; the wavelength depends on wind strength and remains constant with time. See **Ripple.**

Sand rose. A circular histogram depicting the amount of sand potentially moved by winds from various compass directions at a given geographic locality.

Sand shadow. (1) Small mound or dune that forms around clumps of vegetation, behind rocks, or around other obstructions; (2) eolian deposit formed in the lee of a fixed obstruction. See also **Sand drift.**

Sand sheet. Accumulation of sand in essentially flat laminae forming a sheetlike or blanketlike deposit. It has no slipfaces, but has distinct geographic boundaries.

Sand streaks. See **Stringers.**

Sand strip. Long, narrow strip of sand that extends for a considerable distance downwind from each horn of a dune.

Satellite dune or **Extra dune.** Auxiliary or smaller dune associated with main dune.

Sebkha, Sabkha, Sebkra, and many others (especially North Africa and Arabia). A flat area of clay, silt, or sand commonly with saline incrustations; a surface of deflation in which dry, loose particles are removed down to the level of the groundwater or to the zone of capillary concentration (Friedman and Sanders, 1978); includes: (1) **Coastal sebkha** — Coastal flat that occurs just above level of normal high tide in hot, arid, desert climate. (2) **Inland sebkha** — Flat inland areas of clay, silt, or sand, often with saline incrustations. See also **Playa, Clay pan.** (3) **Interdune sebkha.**

Seif, Sif, Sief, Er (North Africa and Arabia), **Linear dune**, or **Longitudinal dune.** Linear dune which, in groups, commonly makes a parallel straight pattern oriented in the direction of prevailing wind or winds. It characteristically has two slipfaces.

Semiarid. Partially arid. It refers to region in which the mean annual rainfall is 30–40 cm (12–16 in.) according to some geographers, or between 25–50 cm (10 and 20 in.) according to others.

Serir (Egypt and Libya). A rough desert surface of fairly large rounded pebbles with no fine material remaining between pebbles (Gautier, 1935, p. 62).

Shear sorting. Sorting of sediments, as in a sand dune, whereby smaller grains tend to move toward the zone of greatest shear strain, the shear plane, and the larger grains toward the zone of least shear. In some dunes it is characterized by a lamination of fine-grained minerals and occurs in a zone between moving and residual sand (Stone, 1967, p. 246).

Shrinkage crack. See **Desiccation crack.**

Shrub-coppice dune. Small streamlined dune that forms to the lee of vegetation on a smooth very shallow sand surface.

Sigmoidal dune. S-shaped steep-sided sharp-crested dune formed under the influence of alternating and opposing winds of roughly equal velocities (Holm, 1957). See also **Reversing dune** and **Recurved dune.**

Silcrete. A secondary deposit of siliceous material in soils of hot, arid, and semiarid regions.

Silk (North Africa). See **Seif.**

Singing sand, or **Acoustical, Barking, Booming, Musical, Roaring, Sonorous, Sounding, Whispering,** or **Whistling sand.** Terms used for sand that, when sliding down a slipface, emits a musical tone or a humming or booming sound.

Skewness. The state of a frequency distribution which bunches on one side of the average and tails out on the other side. Skewness results from lack of coincidence of the mode, median, and arithmetic mean of the distribution. It is measured by the quotient of the difference between the arithmetic mean and mode divided by the standard deviation.

Slipface. Steep face on the lee side of a dune, usually at the angle of repose of dry sand (commonly 30°–34°).

Slumping. A mass movement in which grains normally remain relatively cohesive within sand sheets, even though individual groups of laminae may be warped, stretched, or otherwise contorted. This movement is in contrast to sand-flow avalanching during which laminae are being progressively destroyed by the separation and dispersion of the component grains.

Slump mark. A surface marking formed on a foreset deposit by slumping or mass avalanching. It may consist of a crack or step fault resulting from tensional stresses near the avalanche summit or of wrinkles and other compressional structures in the zone of compression below. Dendritic and other varieties of slump marks also may form.

Southeast trades. Winds persistently from the southeast in the Southern Hemisphere, caused by counterclockwise circulation about permanent oceanic high-pressure cells, centered about lat 20° S.

Stabilized dune. See **Fixed dune.**

Stabilized free dune or **active form.** A dune stabilized by vegetation but still retaining its initial form.

Star dune. A pyramidal dune, roughly star-shaped or resembling a pinwheel, with three or more radial buttresses or arms extending in various directions from a high central cone. Slipfaces dip in at least three directions.

Stellate rose. A star dune.

Stoss side. Side of dune facing dominant wind direction and direction from which dune has migrated. It is opposite the lee side.

Straat (South Africa). See **Interdune corridor**.

Street. See **Interdune corridor**.

Stringers. Elongate sand bodies that have no slipfaces. They are varieties of sand sheets that have stringlike shape and that commonly, though not exclusively, develop along the margins or at the downwind extremity of a sheet. The term "streak" is used instead of "stringer" in chapters J and K, which are based on observations of Landsat images.

Surface creep. The slow movement of large detrital grains as a result of being pushed by the impact of smaller saltating grains.

Surface wind. Wind near ground-surface level. It is usually measured at a height of 10 m (32.8 ft) or less.

Suspended load. The part of the total wind load that is carried for a considerable period of time in suspension, free from contact with the desert floor. It consists mainly of silt and finer detritus.

Suspension. A mode of sediment transport in which the upward currents in eddies of turbulent flow are capable of supporting weight of particles and keeping them indefinitely held in the air.

Swash zone. Area of a beach where water rushes up onto it following the breaking of a wave.

Tabular-planar cross-strata. Set of cross-strata bounded by planar, essentially parallel surfaces.

Thematic map. Interpretive drawing, based on ERTS –1 color mosaic, which expresses with patterns the relationships of features observed on the imagery.

Thermal low. An area of low atmospheric pressure, caused by heating of the land surface by solar radiation, resulting in expansion of adjacent air.

Traction. A mode of sediment transport in which particles are swept along, or are immediately above and parallel to, the ground surface. Movement may be by rolling, sliding, dragging, pushing, or saltation.

Trade wind desert. See **Low latitude desert**.

Transverse dune. Asymmetric sand ridge (leeward slope, steep; windward, gentle) in which long dimension is normal to dominant wind direction. It has one slipface. It shows largely horizontal strata (apparent dip) in cross sections parallel to crest, in contrast to curving strata of barchanoid ridges viewed in similar sections.

Tropical desert. See **Low Latitude desert**.

Trough cross strata. Set of cross-strata whose lower bounding surface is curved surface of erosion. It results from channeling and subsequent deposition. Overlapping sets may show festoon pattern.

Two-storied barchan. A compound barchan.

Umbracer dune. Type of longitudinal dune that tapers to a point downwind, formed under constant wind direction in the lee of an obstacle. It may form behind prominent bedrock obstacles, or more commonly behind clumps of bushes (**Shrub-coppice dune**).

Unidirectional dune field. Dune field in which dunes migrate in only one principal direction.

Upsiloidal dune. See **Parabolic dune**.

Uruq. (1) (Arabia) Elongate or seif dune (Stone, 1967, p. 256); (2) (North Africa) chain of elongate dunes (Stone, 1967, p. 264).

U-shaped dune. See **Parabolic dune**.

Vector unit (VU). Number arbitrarily used in computation and derived from application of a weighting equation for sand movement by wind to surface wind data. The meaning of a vector unit in terms of sand transport kg/m²/yr (lb/ft²/yr) will differ as different weighting equations or other variables are used, but its value is directly proportional to the effective wind energy of the environment.

Ventifact. Wind-polished or abraded pebble.

Wadi. See **Arroyo**.

Wash. (1) See **Arroyo**; (2) alluvium, especially coarse alluvium.

Wedge-planar cross-strata. Set of cross-strata bounded by oblique erosion surfaces that separate it from other sets of cross-strata.

Weighting equation. A function (used in this report) to weight surface winds of different velocities according to their relative ability to move sand. See chapter F.

Weighting factor for surface winds. A numerical expression of the rate at which wind of a certain velocity moves sand compared to winds of other velocities.

Whaleback or Sand levee. Coarse residue or platform built up and left behind by the passage of a long continued succession of seif dunes along the same path (Bagnold, 1941, p. 242).

Wide unimodal wind distribution. A wind regime that produces a distribution of drift potential on a sand rose that has a single peak or mode on the compass and an arc of more than 45°.

Winddrift dune or **Windrift dune.** Linear dune deposited on the lee slide of blowout areas.

Wind regime. The pattern of winds characteristic of a particular region.

Wind rose. A circular histogram depicting the percentage occurrence of wind speed groups from a given set of compass directions, at a certain geographic locality. It has arms proportional in length to the number (or percentage) of observations of wind from a given direction.

Wind shadow. The area adjacent to a slope, cliff, or escarpment which is protected from the direct action of the wind blowing over it.

Wind shadow dune. See **Umbracer dune** and **Shrub-coppice dune**.

Wind sweep. Windward slope of an advancing dune, or that portion of the windward slope of a dune up which the main wind currents pass.

Windward slope. The upwind side of a dune.

Xerophyte. Plant adapted to an arid environment.

Yardang or Yarding. (1) Sharp salient ridge of incompetent sedimentary material oriented parallel to the direction of the prevailing effective wind and presumably produced by the abrasion of wind-driven sand; (2) irregular ridge bounded by rounded troughs formed by eolian erosion; (3) landscape form produced in limestone or sandstone regions by infrequent rains combined with wind action resulting in a surface covered with a fine and compact lacework of sharp ridges pitted by corrosion.

A STUDY OF GLOBAL SAND SEAS

SELECTED REFERENCES

Ahlbrandt, T. S., 1973, Sand dunes, geomorphology and geology, Killpecker Creek area, northern Sweetwater County, Wyoming: Wyoming Univ. Ph. D. thesis, 174 p.; available from Xerox University Microfilms, Ann Arbor, Mich. [U.S.A.], no. 73–25,539, 212 p.

———1974a, Dune stratigraphy, archaeology, and the chronology of the Killpecker dune field, in Michael Wilson, ed., Applied geology and archaeology — The Holocene history of Wyoming: Wyoming Geol. Survey Rept. Inv. 10, p. 51–60.

——— 1974b, The source of sand for the Killpecker sand-dune field, southwestern Wyoming: Sed. Geology, v. 11, no. 1, p. 39–57.

——— 1975, Comparison of textures and structures to distinguish eolian environments, Killpecker dune field, Wyoming: Mtn. Geologist, v. 12, no. 2, p. 61–63.

Airy, G. B., 1845, Tides and waves: [London] Encyclopedia Metropolitans, v. 5, p. 241–396.

Alimen, Henriette, 1965, The Quaternary era in the northwest Sahara, in H. E. Wright, Jr., and D. G. Frey, eds., International studies on the Quaternary: Geol. Soc. America Spec. Paper 84, p. 273–291.

Alimen, Marie-Henriette, Doudoux-Fenet, Danièle, Ferrere, Jacqueline, and Palau-Caddoux, Marguerite, 1957, Sables quaternaires du Sahara nord-occidental (Saoura-Ougarta): Algeria Service de la Carte Géologique Bull. 15, 207 p. [1958].

Allen, J. E., and Balk, Robert, 1954, Mineral resources of Fort Defiance and Tohatchi quadrangles, Arizona and New Mexico: New Mexico Bur. Mines and Mineral Resources Bull. 36, 192 p.

Allen, J. R. L., 1968, Current ripples — Their relations to patterns of water and sediment motion: Amsterdam, Netherlands, North-Holland Publishing Co., 433 p.

Almeida, F. F. M. de, 1953, Botucatú, a Triassic desert of South America: Internat. Geol. Cong., 19th, Algiers 1952, Comptes Rendus, sec. 7, pt. 7, p. 9–24.

———1954, Botucatú, a Traissic desert of South America: [Brazil] Div. Geologia e Mineralogia Notas Prelim. e Estudos 86, 21 p.

Almeida, F. F. N. de, and Barbosa, Octavio, 1953, Geologia das quadrículas de Piracicaba e Rio Claro estado de São Paulo [Geology of the Piracicaba and Rio Claro quadrangles, São Paulo]: [Brazil] Div. Geologia e mineralogia Boletím 143, 96 p.

American Geological Institute, (Margaret Gary, Robert McAfee, Jr., and C. L. Wolf, eds.), 1972, Glossary of geology, with a foreword by Ian Campbell: Washington, 857 p.

ARAMCO Survey, Projects Construction Services (Shaik Mahboob, compiler), 1974, AER –716, Southern area climate, estimated climatic information for Shaybah EPI 151–192 and EPI 169–29: Arabian American Oil Co., Dhahran, Saudi Arabia, 35 p. and appendix [34] p.

Auden, J. B., 1950, Introductory report on the ground-water resources of western Rajasthan: New Delhi, India Geol. Survey Bull., ser. B, no. 1, 59 p.

Aufrère, Léon, 1928, L'orientation des dunes et la direction des vents: [Paris] Acad. Sci. Comptes Rendus, v. 187, p. 833–835.

———1930, L'orientation des dunes continentales: Internat. Geog. Cong., 12th, Cambridge, England, 1928, Proc., p. 220–231.

———1932, Morphologie dunaire et météorologie saharienne: Assoc. Géographes Français Bull. 56, p. 34–47.

——— 1934, Les dunes du Sahara algérien: Assoc. Géographes Français Bull. 83, p. 130–142.

Averitt, Paul, Detterman, J. S., Harshbarger, J. W., Repenning, C. A., and Wilson, R. F., 1955, Revisions in correlation and nomenclature of Triassic and Jurassic formations in southwestern Utah and northern Arizona: Am. Assoc. Petroleum Geologists Bull., v. 39, no. 12, p. 2515–2524.

Baars, D. L., 1961, Permian blanket sandstones of the Colorado Plateau, in J. A. Peterson and J. C. Osmond, eds., Geometry of sandstone bodies — A symposium, 45th Ann. Mtg., Atlantic City, N.J., 1960: Tulsa, Okla., Am. Assoc. Petroleum Geologists, p. 179–207.

———1962, Permian System of Colorado Plateau: Am. Assoc. Petroleum Geologists Bull. v. 46, no. 2, p. 149–218.

———1972, Southwestern area, in Permian System, in Geologic Atlas of the Rocky Mountain region, United States of America: Rocky Mtn. Assoc. Geologists, p. 148–165.

Bagnold, R. A., 1933, A further journey through the Libyan desert: Geog. Jour., v. 82, p. 103–129.

———1935, The movement of desert sand: Geog. Jour., v. 85, p. 342–369.

——— 1937, The transport of sand by wind: Geog. Jour., v. 89, p. 409–438.

———1941. The physics of blown sand and desert dunes: London, Methuen & Co., Ltd., 265 p.

——— 1951, Sand formations in southern Arabia: Geog. Jour., v. 117, pt. 1, p. 78–86.

——— 1953, The surface movement of blown sand in relation to meterology, in Desert Research — Internat. Symposium, Jerusalem, 1952, Proc.: Israel Research Council Spec. Pub. 2, p. 89–96.

Baker, A. A., 1936, Geology of the Monument Valley –Navajo Mountain region, San Juan County, Utah: U.S. Geol. Survey Bull. 865, 106 p.

Baker, A. A., Dane, C. H., and Reeside, J. B., 1936, Correlation of the Jurassic formations of parts of Utah, Arizona, New Mexico, and Colorado: U.S. Geol. Survey Prof. Paper 183, 66 p.

Barnard, W. S., 1973, Duinformasies in die Sentrale Namib: Tegnikon [South Africa, Stallenbosch Univ.], December, p. 2 –13.

Bartenstein, Helmut, 1968, Present status of the Paleozoic paleogeography of northern Germany and adjacent parts of northwest Europe, in D. T. Donovan, ed., Geology of shelf seas — Inter-Univ. Geol. Cong., 14th, Proc.: Edinburgh, London, Oliver and Boyd, p. 31 –54.

Beard, D. C., and Weyl, P. K., 1973, Influence of texture on porosity and permeability of unconsolidated sand: Am. Assoc. Petroleum Geologists Bull., v. 57, no. 2, p. 349 –369.

Beheiry, S. A., 1967, Sand forms in the Coachella Valley, southern California: Assoc. Am. Geographers Annals, v. 57, no. 1, p. 25 –48.

Belly, Pierre-Yves, 1964, Sand movement by wind, with Addendum 2, by Abdel-Latif Kadib: U.S. Army Corps of Engineers, Coastal Eng. Research Center Tech. Memo. 1, [85 p].

Berner, R. A., 1969, Goethite stability and the origin of red beds: Geochim. et Cosmochim. Acta, v. 33, no. 2, p. 267 –273.

Besler, Helga, 1972, Geomorphologie der Dunen: Namib und Meer, [South-West Africa, Swakopmund], v. 3, p. 25 –35.

Bettenay, Eric, 1962, The salt lake systems and their associated aeolian features in the semi-arid regions of Western Australia: Jour. Soil Sci., v. 13, p. 10 –17.

Beukes, N. J., 1970, Stratigraphy and sedimentology of the Cave Sandstone Stage, Karroo System, in Internat. Union Geol. Sci. Symposium, 2d, South Africa, 1970, Proc. and Papers — Gondwana stratigraphy and paleontology: Marshalltown, Transvaal, South Africa, Geol. Soc. South Africa, p. 321 –341.

Beydoun, Z. R., 1966, Geology of the Arabian Peninsula — Eastern Aden Protectorate and part of Dhufar: U.S. Geol. Survey Prof. Paper 560 –H, 49 p.

Bigarella, J. J., 1965, Sand-ridge structures from Paraná coastal plain [Brazil]: Marine Geology, v. 3, p. 269 –278.

——— 1972, Eolian environments — their characteristics, recognition, and importance, in J. K. Rigby and W. K. Hamblin, eds., Recognition of ancient sedimentary environments: Soc. Econ. Paleontologists and Mineralogists Spec. Pub. 16, p. 12 –62.

——— 1973a, Paleocurrents and the problem of continental drift: Geol. Rundschau, v. 62, no. 2, p. 447 – 477.

——— 1973b, Textural characteristics of the Botucatú Sandstone: Boletím Paranaense Geociências, no. 31, p. 85 –94.

——— 1975a, Structures developed by dissipation of dune and beach ridge deposits: Catena, v. 2, no. 1 –2, p. 107 –152.

——— 1975b, Lagoa dune field, State of Santa Catarina, Brazil — A model of eolian and pluvial activity, in International symposium on the Quaternary, southern Brazil, July 15 –31, 1975: Boletím Paranaense Geociências, no. 33, p. 133 –167.

Bigarella, J. J., Alessi, A. H., Becker, R. D., and Duarte, G. M., 1969, Textural characteristics of the coastal dune, sand ridge, and beach sediments: Boletím Paranaense Geociências, no. 27, p. 15 –80.

Bigarella, J. J., Becker, R. D., and Duarte, G. M., 1969, Coastal dune structures from Paraná [Brazil]: Marine Geology, v. 7, p. 5 –55.

Bigarella, J. J., and Oliveira, M. A. M., 1966, Preliminary note about the directions of sand grain transportation at Furnos and Botucatú in the seventieth part of Paraná Basin [Brazil]: Boletím Paranaense Geografia, nos. 18 –20, p. 247 –256. [In Portuguese.]

Bigarella, J. J., and Popp, J. H., 1966, Contributions to the study of recent shoreline sediments; IV — Shore and dunes of Barra do Sul (SC) [Santa Catarina, Brazil]: Boletím Paranaense Geografia, nos. 18 –20, p. 133 –149. [In Portuguese; English abs.]

Bigarella, J. J., and Salamuni, Riad, 1961, Early Mesozoic wind patterns as suggested by dune bedding in the Botucatú sandstone of Brazil and Uruguay: Geol. Soc. America Bull., v. 72, no. 7, p. 1089 –1106.

——— 1967, Some paleogeographic and paleotectonic features of the Paraná Basin, in J. J. Bigarella, R. D. Becker, and I. D. Pinto, eds., Problems in Brazilian Gondwana geology: Internat. Symposium on the Gondwana Stratigraphy and Paleontology, 1st, Mar del Plata, 1967; p. 235 –301.

Blanford, W. T., 1877, Geological notes on the Great Indian Desert between Sind and Rájpútána: India Geol. Survey Records, v. 10, p. 10 –21.

Bohdanowicz, Karol, 1892, Geologicheskiia izsliedovaniia v Vostochnam Turkestanie—Geograficheskoe obshchestvo SSSR Tibetskaia ekspeditsii 1889 –1890, Trudy no. 2 [Geological research in eastern Turkestan — Works of 1889 –1890 Tibet expedition]: St. Peterburg, Tipografiia M. M. Stasiulevicha, 167 p. [In Russian.]

Boocock, C., and van Straaten, O. J., 1962, Notes on the geology and hydrogeology of the central Kalahari region, Bechuanaland Protectorate: Geol. Soc. South Africa Trans., v. 65, pt. 1, p. 125 –176.

Boulet, R., Guichard, E., and Vieillefon, J., 1971, Observations pédologiques et leurs relations avec les faits géomorphologiques dans le delta su Sénégal — Comparison avec les observations effectuées au Niger [Pedology and its relation to geomorphology in the Senegal delta — Comparison with observations made on the Niger]: Assoc. Sénégal. Étude Quaternaire Ouest African, Bull. 29, p. 103 –114.

Brady, L. F., 1935, Preliminary note on the occurrence of a primitive theropod in the Navajo: Am. Jour. Sci., ser. 5, v. 30, no. 177, p. 210 –215.

——— 1939, Tracks in the Coconino sandstone compared with those of small living arthropods: Plateau, v. 12, no. 2, p. 32 –34.

Bramkamp, R. A., Gierhart, R. D., Brown, G. F., and Jackson, R. O., 1956, Geology of the southern Tuwayq quadrangle, Kingdom of Saudi Arabia: U.S. Geol. Survey Misc. Geol. Inv. Map I –212A, scale 1:500,000.

Bramkamp, R. A., Gierhart, R. D., Owens, L. D., and Ramirez, L. F., 1963, Geologic map of the western Rub' al Khali quadrangle, Kingdom of Saudi Arabia: U.S. Geol. Survey Misc. Geol. Inv. Map I–218A, scale 1:500,000.

Bramkamp, R. A., and Ramirez, L. F., 1958, Geology of the northern Tuwayq quadrangle, Kingdom of Saudi Arabia: U.S. Geol. Survey Misc. Geol. Inv. Map I–207A, scale, 1:500,000.

———1959a, Geology of the northwestern Rub' al Khali quadrangle, Kingdom of Saudi Arabia: U.S. Geol. Survey Misc. Geol. Inv. Map I–213A, scale 1:500,000.

———1959b, Geology of the Wadi al Batin quadrangle, Kingdom of Saudi Arabia: U.S. Geol. Survey Misc. Geol. Inv. Map I–203A, scale 1:500,000 [1960].

———1960, Geographic map of the Jawf-Sakakah quadrangle, Kingdom of Saudi Arabia: U.S. Geol. Survey Misc. Geol. Inv. Map I–201B, scale 1:500,000. [1961].

———1963, Geologic map of the Darb Zubaydah quadrangle, Kingdom of Saudi Arabia: U.S. Geol. Survey Misc. Geol. Inv. Map I–202A, scale 1:500,000.

Bramkamp, R. A., Ramirez, L. F., Brown, G. F., and Pocock, A. E., 1963, Geologic map of the Wadi ar Rimah quadrangle, Kingdom of Saudi Arabia: U.S. Geol. Survey Misc. Geol. Inv. Map I–206A, scale 1:500,000.

Breed, C. S., and McKee, E. D., 1973, Sand dunes in desert areas: Tucson, Ariz., Remote sensing in arid lands, 4th Ann. Conf., Arizona Univ., Proc., p. 160–171.

Broin, T. L., 1957, Stratigraphy of the Lykins Formation of eastern Colorado: Colorado Univ. Ph. D. thesis, 256 p.; available through Univ. Microfilms, Ann Arbor, Mich.

Brookfield, Muriel, 1970, Dune trends and wind regime in central Australia, in Piedmont plains and sand formations in arid and humid tropic and subtropic regions: Zeitschr. Geomorphologie, Supp. 10, p. 121–153.

Brown, G. F., 1960, Geomorphology of western and central Saudi Arabia: Internat. Geol. Cong., 21st, Copenhagen, Rept. 21, pt. 21, p. 150–159.

Brown, G. F., Layne, Newton, Goudarzi, G. H., and MacLean, W. H., 1963, Geologic map of the northeastern Hijaz quadrangle, Kingdom of Saudi Arabia: U.S. Geol. Survey Misc. Geol. Inv. Map I–205A, scale 1:500,000.

Bryson, R. A., 1974, A perspective on climatic change: Science, v. 184, no. 4138, p. 753–760.

Bucher, W. H., 1919a, On ripples and related sedimentary surface forms and their paleogeographic interpretation: Am. Jour. Sci., 4th ser., v. 47, no. 279, p. 149–210.

———1919b, Fossil ripples and their paleogeographic interpretation, pt. 2 of On ripples and related sedimentary surface forms and their paleogeographic interpretation: Am. Jour. Sci., 4th ser., v. 47, no. 280, p. 241–269.

Bunker, D. G., 1953, The south-west borderlands of the Rub' al Khali: Geog. Jour., v. 119, pt. 4, p. 420–430.

Camp, C. L., 1936, A new type of small bipedal dinosaur from the Navajo sandstone of Arizona: California Univ. Pubs. Geol. Sci. Bull., v. 24, no. 2, p. 39–55.

Campbell, C. V., 1971, Depositional model — Upper Cretaceous Gallup beach shoreline, Ship Rock area, northwestern New Mexico: Jour. Sed. Petrology, v. 41, no. 2, p. 395–409.

Campbell, R. L., Jr., 1968, Stratigraphic applications of dipmeter data in mid-continent: Am. Assoc. Petroleum Geologists Bull., v. 52, no. 9, p. 1700–1719.

Campos, L. F. Gonzaga de, 1889, Secção geologica: [Brazil] São Paulo Commissão Geog. Geol. da Provincia Relatorio, 1889, Annexo, p. 21–34.

Capot-Rey, Robert, 1943, La morphologie de L'Erg Occidental: Algiers Univ. Inst. Recherches Sahariennes Travaux, v. 2, p. 69–103.

———1945, Dry and humid morphology in the western Erg: Geog. Rev., v. 35, no. 3, p. 391–407.

——— 1947a, L'Edeyen de Mourzouk: Algiers Univ. Inst. Recherches Sahariennes Travaux, v. 4, p. 67–109.

——— 1947b, Chronique — Exploration de L'Erg Oriental: Algiers Univ. Inst. Recherches Sahariennes Travaux, v. 4, p. 181–187.

——— 1953, Le Sahara français: Paris [France] Presses Univ., 564 p.

Capot-Rey, Robert, and Capot-Rey, Francoise, 1948, le déplacement des sables éoliens et la formation des dunes désertiques, d'après R. A. Bagnold: Algiers Univ. Inst. Recherches Sahariennes Travaux, v. 5, p. 47–80.

Carroll, Dorothy, 1958, Role of clay minerals in the transportation of iron: Geochim. et Cosmochim. Acta, v. 14, nos. 1–2, p. 1–27.

Chaudhri, I. I., 1957, Succession of vegetation in the arid regions of West Pakistan Plains, in Food and Agricultural Council of Pakistan and UNESCO, Symposium on soil erosion and its control in the arid and semiarid zones, Karachi, Nov. 1957: p. 141–156.

Chavaillon, Jean, 1964, Étude stratigraphique des formations Quaternaires du Sahara nord-occidental (Colomb-Béchar à Reggane): [France] Centre Recherches sur les Zones Arides, Série Géologie no. 5, 393 p.

Chepil, W. S., 1945, The transport capacity of the wind, Pt. 3 of Dynamics of wind erosion: Soil Sci., v. 60, no. 6, p. 475–480.

Chepil, W. S., and Woodruff, N. P., 1963, The physics of wind erosion and its control, in A. G. Norman, ed., Advances in agronomy, v. 15: New York, London, Academic Press, p. 211–302.

Christian, C. S., 1959, An arid zone research institute for India: Arid Zone [UNESCO Newsletter], no. 5, p. 6–9.

Chu, Chen-jen, Lieu, Hua-hsun, Ch'en, En-chiu, Wu, Kung-ch'eng, Mi, Kuo-yuan, Hsiao, Yu-ch'uan, Meng, Te-cheng, and Hsu, Chen-fu, 1962, The natural characteristics and the melioration of the southwestern section of the Taklamakan Desert of Sinkiang, in Academia Sinica Sand Control Group, eds., Research on sand control: U.S. Joint Publications Research Service JPRS 19993, Tech. Trans. 63–31183, p. 1–46. [Figs. omitted.]

Conant, L. C., and Goudarzi, G. H., compilers, 1964, Geologic map of the Kingdom of Libya: U.S. Geol. Survey Misc. Geol. Inv. Map I–350A, scale 1:2,000,000.

———1967, Stratigraphic and tectonic framework of Libya: Am. Assoc. Petroleum Geologists Bull., v. 51, no. 5, p. 719–730.

Cooke, R. U., and Warren, Andrew, 1973, Geomorphology in deserts: Los Angeles, California Univ. Press, 374 p.

Cooper, W. S., 1958, Coastal sand dunes of Oregon and Washington: Geol. Soc. America Mem. 72, 169 p.

———1967, Coastal dunes of California: Geol. Soc. America Mem. 104, 131 p.

Cordani, U. G., and Vandoros, Paul, 1967, Basaltic rocks of the Paraná Basin, in J. J. Bigarella, R. D. Becker, and I. D. Pinto, eds., Problems in Brazilian Gondwana geology: Internat. Symposium on the Gondwana Stratigraphy and Paleontology, 1st, Mar del Plata, 1967, p. 207–231.

Cornish, Vaughan, 1897, On the formation of sand dunes: Geog. Jour., v. 9, p. 278–309.

Cressey, G. B., 1928, The Indiana sand dunes and shore lines of the Lake Michigan Basin: Geog. Soc. Chicago, Bull. 8 [Chicago Univ. Press], 80 p.

Crone, A. J., 1974, Experimental studies of mechanically infiltrated clay matrix in sand: Geol. Soc. America Abs. with Programs, v. 6, no. 7, p. 701.

———1975, Laboratory and field studies of mechanically infiltrated matrix clay in arid fluvial sediments: Boulder, Colo., Colorado Univ. Ph. D. thesis, 162 p.: available from Xerox Univ. Microfilms, Ann Arbor, Mich., no. 76–3897, 174 p.

Crutcher, H. L., 1957, On the standard vector deviation wind rose: Jour. Meteorology, v. 14, p. 28–33.

———1966, Components of 1,000-mb winds (or surface winds) of Northern Hemisphere: Washington, U.S. Govt. Printing Office, [NAVAIR 50 –1C –51], 74 p.

Crutcher, H. L., and Baer, Ledolph, 1962, Computations from elliptical wind distribution statistics: Jour. Applied Meteorology, v. 1, p. 522–530.

Crutcher, H. L., and Meserve, J. M., 1970, Selected level heights, temperatures and dew points for Northern Hemisphere: Washington, U.S. Govt. Printing Office, [NAVAIR 50–1C –52, Sept. 1966], 410 p.

Cuvillier, J., 1930, Révision du nummulitique égyptien: Institut d'Égypte Mém., v. 16, 371 p. [In French.]

Dake, C. L., 1921, The problem of St. Peter sandstone: Missouri Univ. School Mines and Metallurgy Bull., Tech. Ser., v. 6, no. 1, 225 p.

Dalrymple, P. C., Everett, K. R., and Wollaston, Sarah, 1970, Environments of Central Asian Highlands: U.S. Army Natick Labs. Tech. Rept. 71–19 –ES, Earth Sci. Lab. Ser. ES –62, 58 p.

Daveau, Suzanne, 1965, Dunes ravinees et dépôts du Quaternaire Récent dans le Sahel Mauritanien: Rev. Géog. Afrique Occidentale, v. 1, no. 2, p. 7–47.

Denny, C. S., Warren, C. R., Dow, D. H., and Dale, W. J., 1968, A descriptive catalog of selected aerial photographs of geologic features in the United States: U.S. Geol. Survey Prof. Paper 590, 79 p.

Dingler, J. R., 1975, Wave-formed ripples in nearshore sands: San Diego, Calif., California Univ. Ph. D. thesis, 136 p.; available from Xerox Univ. Microfilms, Ann Arbor, Mich., no. 75–11,227, 152 p.

Dresch, Jean, and Rougerie, G., 1960, Observations morphologiques dans le Sahel du Niger: Rev. Géomorphologie Dynamique, v. 11, no. 4 –6, p. 49–58. [In French; English abs.]

Dubief, Jean, 1952, Le vent et le déplacement du sable au Sahara: Algiers Univ. Inst. Recherches Sahariennes Travaux, v. 8, p. 123–164.

Eckel, E. B., 1959, Geology and mineral resources of Paraguay — A reconnaissance, with sections on Igneous and metamorphic rocks, by Charles Milton and E. B. Eckel, and Soils, by P. T. Sulsona: U.S. Geol. Survey Prof. Paper 327, 110 p.

Emmett, W. R., Beaver, K. W., and McCaleb, J. A., 1972, Pennsylvanian Tensleep reservoir, Little Buffalo Basin oil field, Big Horn Basin, Wyoming: Mtn. Geologist, v. 9, no. 1, p. 21–31.

Engel, C. G., and Sharp, R. P., 1958, Chemical data on desert varnish: Geol. Soc. America Bull., v. 69, no. 5, p. 487–518.

Falke, Horst, ed., 1972, Rotliegend — Essays on European Lower Permian: Leiden, Netherlands, E. J. Brill & Co., Internat. Sed. Petrographical Ser., v. 15, 299 p.

Finkel, H. J., 1959, The barchans of southern Peru: Jour. Geology, v. 67, no. 6, p. 614–647.

———1961, The movement of barchan dunes measured by aerial photogrammetry: Photogrammetric Eng., v. 27, no. 3, p. 439–444.

Fisher, R. A., 1936, The use of multiple measurements in taxonomic problems: Annals of Eugenics, v. 7, p. 179–188.

Flint, R. F., and Bond, Geoffrey, 1968, Pleistocene sand ridges and pans in western Rhodesia: Geol. Soc. America Bull., v. 79, no. 3, p. 299–314.

Folk, R. L., 1962, Of skewnesses and sands: Jour. Sed. Petrology, v. 32, no. 1, p. 145–146.

——— 1968a, Bimodal supermature sandstones — Product of the desert floor: Internat. Geol. Cong., 23d, Prague, 1968, Proc., Genesis and classification of sedimentary rocks, sec. 8, p. 9–32.

——— 1968b, Petrology of sedimentary rocks: Austin, Tex., Hemphill, 170 p.

———.1969, Grain shape and diagenesis in the Simpson Desert, Northern Territory, Australia: Geol. Soc. America Abs. with Programs for 1969, pt. 7, p. 68–69.

———1971a, Longitudinal dunes of the northwestern edge of the Simpson Desert, Northern Territory, Australia, [pt.] 1, Geomorphology and grain size relationships: Sedimentology, v. 16, p. 5–54.

———1971b, Genesis of longitudinal and oghurd dunes elucidated by rolling upon grease: Geol. Soc. America Bull., v. 82, no. 12, p. 3461–3468.

——— 1976, Reddening of desert sands, Simpson Desert, Northern Territory, Australia: Jour. Sed. Petrology, v. 46, no. 3, p. 604–615.

Folk, R. L., and Ward, W. C., 1957, Brazos River bar — A study in the significance of grain-size parameters: Jour. Sed. Petrology, v. 27, no. 1, p. 3–27.

Fourtau, R., 1902, Sur le grès nubien: [Paris] Acad. Sci. Comptes Rendus, v. 135, p. 803–804.

Frazier, D. E., and Osanik, A., 1961, Point-bar deposits, Old River Locksite, Louisiana: Gulf Coast Assoc. Geol. Socs. Trans., v. 11, p. 121–137.

Frere, H. B. E., 1870, Notes on the Runn [Rann] of Cutch and neighboring region: Royal Geog. Soc. [London] Jour., v. 40, p. 181–207.

Friedman, G. M., 1961, Distinction between dune, beach, and river sands from their textural characteristics: Jour. Sed. Petrology, v. 31, no. 4, p. 514–529.

_____ 1967, Dynamic processes and statistical parameters compared for size frequency distribution of beach and river sands: Jour. Sed. Petrology, v. 37, no. 2, p. 327 –354.

_____ 1966, Textural parameters of beach and dune sands, *in* Abstracts for 1965: Geol. Soc. America Spec. Papers 87, p. 60.

_____ 1973, Textural parameters of sands — Useful or useless?: Geol. Soc. America Abs. with Programs, v. 5, no. 7, p. 626 –627.

Friedman, G. M., and Sanders, J. E., 1978, Principles of sedimentology: New York, John Wiley & Sons.

Fürst, Manfred, 1970, Beobachtungen an quartären Buntsedimenten der zentralen Sahara [Observations on variegated sediments in the central Sahara]: Abhandlungen des Heissischen Landesamtes für Bodenforschung, no. 56, p. 129 –150.

Garlick, W. G., 1969, Special features and sedimentary facies of stratiform sulphide deposits in arenites [with discussion], *in* C. H. James, ed., Sedimentary ores, ancient and modern [revised]: Inter-Univ. Geol. Cong., 15th, Univ. Leicester [England], Dec. 1967, Proc., Leicester Univ. Dept. Geology Spec. Pub. 1, p. 107 –169.

Garrels, R. M., and Christ, C. L., 1965, Solutions, minerals, and equilibria: New York, Harper and Row, 450 p.

Gautier, E. F., 1935, Sahara — The Great Desert: New York, Columbia Univ. Press, 280 p. [A translation by D. F. Mayhew.]

Gees, R. A., 1965, Moment measures in relation to the depositional environments of sands: Eclogae Geol. Helvetiae, v. 58, no. 1, p. 209 –213.

Gevers, L. W., 1936, The morphology of western Damaraland and the adjoining Namib Desert of South-West Africa: South African Geog. Jour., v. 19, p. 61 –79.

Gill, W. D., 1967, The North Sea basin: World Petroleum Cong., 7th, Mexico, 1967, Proc., Origin of oil, geology, and geophysics, v. 2, p. 211 –219.

Gilmore, C. W., 1926, Fossil footprints from the Grand Canyon: Smithsonian Misc. Colln., v. 77, no. 9, 41 p.

_____ 1927, Fossil footprints from the Grand Canyon; second contribution: Smithsonian Misc. Colln., v. 80, no. 3, 78 p.

_____ 1928, Fossil footprints from the Grand Canyon; third contribution: Smithsonian Misc. Colln., v. 80, no. 8, 16 p.

Gilreath, J. A., and Maricelli, J. J., 1964, Detailed stratigraphic control through dip computations: Am. Assoc. Petroleum Geologists Bull., v. 48, no. 12, p. 1902 –1910.

Glennie, K. W., 1970, Desert sedimentary environments, *in* Developments in sedimentology [v.] 14: Amsterdam, London, New York, Elsevier Publishing Co., 222 p.

_____ 1972, Permian Rotliegendes of northwest Europe interpreted in light of modern desert sedimentation studies: Am. Assoc. Petroleum Geologists Bull., v. 56, no. 6, p. 1048 –1071.

Glennie, K. W., and Evans, G., 1976, A reconnaissance of the recent sediments of the Ranns of Kutch, India: Sedimentology, v. 23, p. 625 –647.

Goddard, E. N., Chm., and others, 1948, Rock-color chart: Natl. Research Council (reprinted by Geol. Soc. America, 1951, 1970), 6 p.

Goudie, A. S., 1969, Statistical laws and dune ridges in southern Africa: Geog. Jour., v. 135, pt. 3, p. 404 –406.

_____ 1970, Notes on some major dune types in southern Africa: South African Geog. Jour., v. 52, p. 93 –101.

_____ 1972, Climate, weathering, crust formation, dunes and fluvial features of the Central Namib Desert, near Gobabeb, South-West Africa: Madoqua, Series II, v. 1, no. 54 –64, p. 15 –31.

Goudie, A. S., Allchin, Bridget, and Hedge, K. T. M., 1973, The former extensions of the Great India Sand Desert: Geog. Jour., v. 139, pt. 2, p. 243 –257.

Gradzinski, Ryszard, and Jerzykiewicz, Tomasz, 1972, Additional geographical and geological data from the Polish-Mongolian palaeontological expeditions, *in* Results of the Polish-Mongolian palaeontological expeditions, Pt. 4: Palaeontologia Polonica, no. 27, p. 17 –30.

_____ 1974a, Dinosaur- and mammal-bearing aeolian and associated deposits of the upper Cretaceous in the Gobi Desert (Mongolia): Sed. Geology, v. 12, p. 249 –278.

_____ 1974b, Sedimentation of the Barun Goyot Formation, *in* Results of the Polish-Mongolian palaeontological expeditions, Pt. 5: Palaeontologia Polonica, no. 30, p. 111 –146.

Greeley, Ronald, Iversen, J. D., Pollack, J. B., Udovich, Nancy, and White, Bruce, 1974, Wind tunnel simulations of light and dark streaks on Mars: Science, v. 183, no. 4127, p. 847 –849.

Greenwood, Brian, 1969, Sediment parameters and environment discrimination — An application of multivariate statistics: Canadian Jour. Earth Sci., v. 6, p. 1347 –1358.

Gregory, H. E., 1917, Geology of the Navajo country — A reconnaissance of parts of Arizona, New Mexico, and Utah: U.S. Geol. Survey Prof. Paper 93, 161 p.

Griffiths, J. C., 1967, Scientific method in analysis of sediments: New York, McGraw-Hill Book Co., 508 p.

Griffiths, J. F., and Soliman, K. H., 1972, The northern desert (Sahara), Chap. 3, *in* H. E. Lansberg, ed., World survey of climatology, climates of Africa, v. 10: Amsterdam, New York, London, Elsevier Publishing Co., p. 75 –132.

Grove, A. T., 1958, The ancient erg of Hausaland, and similar formations on the south side of the Sahara: Geog. Jour., v. 124, pt. 4, p. 526 –533.

_____ 1960a, Geomorphology of the Tibesti region, with special reference to western Tibesti: Geog. Jour., v. 126, pt. 1, p. 18 –31.

_____ 1960b, [A note] *in* J. R. V. Prescott and H. P. White, Sand formations in the Niger Valley between Niamey and Bourem: Geog. Jour., v. 126, pt. 2, p. 202 –203.

_____ 1969, Landforms and climatic change in the Kalahari and Ngamiland: Geog. Jour., v. 135, pt. 2, p. 191 –212.

Grove, A. T., and Warren, Andrew, 1968, Quaternary landforms and climate on the south side of the Sahara: Geog. Jour., v. 134, pt. 2, p. 194 –208.

Hack, J. T., 1941, Dunes of the western Navajo country: Geog. Rev., v. 31, no. 2, p. 240 –263.

Hamblin, W. K., 1961, Micro-cross-Lamination in upper Keweenawan sediments of northern Michigan: Jour. Sed. Petrology, v. 31, no. 3, p. 390 –401.

Hand, B. M., 1967, Differentiation of beach and dune sand, using settling velocities of light and heavy minerals: Jour. Sed. Petrology, v. 37, no. 2, p. 514 –520.

Hanley, J. H., and Steidtmann, J. R., 1973, Petrology of limestone lenses in the Casper Formation, southernmost Laramie Basin, Wyoming and Colorado: Jour. Sed. Petrology, v. 43, no. 2, p. 428–434.

Hanley, J. H., Steidtmann, J. R., and Toots, Heinrich, 1971, Trace fossils from the Casper Sandstone (Permian), southern Laramie Basin, Wyoming and Colorado: Jour. Sed. Petrology, v. 41, no. 4, p. 1065–1068.

Harms, J. C., and Fahnestock, R. K., 1965, Stratification, bed forms, and flow phenomena (with an example from the Rio Grande), in Primary sedimentary structures and their hydrodynamic interpretation — A symposium: Soc. Econ. Geologists and Mineralogists Spec. Pub. 12, p. 84–115.

Harms, J. C., MacKenzie, D. B., and McCubbin, D. G., 1963, Stratification in modern sands of the Red River, Louisiana: Jour. Geology, v. 71, no. 5, p. 566–580.

Harshbarger, J. W., Repenning, C. A., and Irwin, J. H., 1957, Stratigraphy of the uppermost Triassic and the Jurassic rocks of the Navajo country [Colorado Plateau]: U.S. Geol. Survey Prof. Paper 291, 74 p.

Hatchell, W. O., 1967, A stratigraphic study of the Navajo Sandstone, (Upper Triassic(?) –Jurassic), Navajo Mountain, Utah and Arizona: Albuquerque, N. Mex., New Mexico Univ., unpub. M.S. thesis, 121 p.

Haughton, S. H., 1969, Geological history of southern Africa: Cape Town, Geol. Soc. South Africa, 535 p.

Haun, J. D., and Kent, H. C., 1965, Geologic history of Rocky Mountain region: Am. Assoc. Petroleum Geologists Bull., v. 49, no. 11, p. 1781–1800.

Hedin, Sven, 1896, A journey through the Takla-Makan Desert, Chinese Turkistan: Geog. Jour., v. 8, July to December, p. 264–278.

_____ 1899, Through Asia: New York, London, Harper & Bros., v. 1, p. 1–649; v. 2, p. 653–1255.

_____1905a, The Tarim River, in Scientific results of a journey in central Asia, 1889–1902, v. 1: Stockholm, Swedish Army Lithographic Institute, 523 p.

_____ 1905b, The central Asian deserts, sand-dunes, and sands, Pt. 2, in Scientific results of a journey in central Asia, 1899–1902, v. 2: Stockholm, Swedish Army Lithographic Institute, p. 379–718.

Henderson, Junius, 1924, Footprints in Pennsylvanian sandstones of Colorado: Jour. Geology, v. 32, no. 3, p. 226–229.

Hesselberg, T., and Björkdal, E., 1929, Über das Verteilungsgesetz der Windunruhe: Beitraege zur Physik der Frein Atmosphäre, v. 15, p. 121–133. [In German.]

Hills, E. S., ed., 1966, Arid Lands — A geographical appraisal: London, Methuen & Co., Ltd., and Paris, UNESCO, 461 p.

Hogue, Mominul, 1975, An analysis of cross-stratification of Gargaresh calcarenite (Tripoli, Libya) and Pleistocene Paleowinds: Geol. Mag., v. 112, no. 4, p. 393–401.

Holm, D. A., 1953, Dome-shaped dunes of the central Nejd, Sauda Arabia: Internat. Geol. Cong., 19th, Algiers, 1952, Comptes Rendus, sec. 7, pt. 7, p. 107–112.

_____ 1957, Sigmoidal dunes — A transitional form [abs]: Geol. Soc. America Bull., v. 68, no. 12, pt. 2, p. 1746.

_____1960, Desert geomorphology in the Arabian Peninsula: Science, v. 132, no. 3437, p. 1369–1379.

_____ 1968, Sand dunes, in R. W. Fairbridge, ed., The encyclopedia of geomorphology: New York, Reinhold Book Corp., p. 973–979.

Hörner, N. G., 1936, Geomorphic processes in continental basins of central Asia: Internat. Geol. Cong., 16th, Washington, D.C., 1933, Rept., v. 2, p. 721–735.

Hubert, J. F., 1960, Petrology of the Fountain and Lyons Formations, Front Range, Colorado: Colorado School Mines Quart., v. 55, no. 1, p. 1–242.

Huntington, Ellsworth, 1907, Some characteristics of the glacial period in nonglaciated regions: Geol. Soc. America Bull., v. 18, p. 351–388.

Jones, J. R., 1971, Ground-water provinces of Libyan Arab Republic, in G. Carlyle, ed., Symposium on the geology of Libya, Tripoli, 1969: Univ. Libya Fac. Sci. Press, p. 449–457.

Jordan, G. F., 1962, Large submarine sand waves: Science, v. 136, no. 3519, p. 839–848.

Kádár, Lázló, 1966, Az eolikus felszíni formák természetes rendszere [Natural systems of eolian landforms]: Foldrajzi Értesíto, v. 15, no. 4, p. 413–448.

Kawamura, R., 1951, Study on sand movement by wind: Tokyo Univ. Inst. Sci. and Technology Rept., v. 5, no. 314.

Kent, P. E., and Walmsley, P. J., 1970, North Sea progress: Am. Assoc. Petroleum Geologists Bull., v. 54, no. 1, p. 168–181.

Khalil, M. S., 1957, The role of vegetation in checking erosion in arid and semiarid tracts, in Food and Agricultural Council of Pakistan and UNESCO, Symposium on soil erosion and its control in the arid and semiarid zones, Karachi, Nov. 1957: p. 237–260.

Kiersch, G. A., 1950, Small-scale structures and other features of Navajo sandstone, northern part of San Rafael Swell, Utah: Am. Assoc. Petroleum Geologists Bull., v. 34, no. 5, p. 923–942.

King, D., 1960, The sand ridge deserts of South Australia and related aeolian landforms of Quaternary arid cycles: Royal Soc. South Australia Trans., v. 83, p- 99–108.

Knight, S. H., 1929, The Fountain and Casper formations of the Laramie Basin — A study on genesis of sediments: Wyoming Univ. Sci. Pub., Geology, v. 1, no. 1, p. 1–82.

Koch, G. S., Jr., and Link, R. F., 1970, Statistical analysis of geological data: New York, John Wiley & Sons, 375 p.

Krishnan, M. S., 1960, Geology of India and Burma [4th ed.]: Madras, India, Higginbothams, Ltd., 604 p.

Krumbein, W. C., and Graybill, F. A., 1965, An introduction to statistical models in geology: New York, McGraw-Hill Book Co., 475 p.

Kuenen, P. H., 1960, Experimental abrasion; [pt.] 4, Eolian action: Jour. Geology, v. 68, no. 4, p. 427–449.

Kuenen, P. H., and Perdok, W. G., 1962, Experimental abrasion; [pt.] 5, Frosting and defrosting of quartz grains: Jour. Geology, v. 70, no. 6, p. 648–658.

Kukal, Zdeněk, 1971, Geology of recent sediments: London, New York, Academic Press, 490 p.

Laming, D. J. C., 1966, Imbrication, paleocurrents, and other sedimentary features in the lower New Red Sandstone, Devonshire, England: Jour. Sed. Petrology, v. 36, no. 4, p. 940–959.

Langmuir, Donald, 1971, Particle-size effect on the reaction goethite = hematite + water: Am. Jour. Sci., v. 271, no. 2, p. 147–156.

Le Ribault, Loïc, 1974, L'eroscopie, méthode de détermination de l'histoire géologique des quartz détritiques [Exoscopy, method to determine the geologic history of detrital quartz]: Rev. Géographie Phys. et Géologie Dynam., v. 16, no. 1, p. 119–130. [In French.]

Lewis, A. D., 1936, Sand dunes of the Kalahari within the borders of the Union: South African Geog. Jour., v. 19, p. 22–32.

Leprun, J. C., 1971, Nouvelles observations sur les formations dunaires sableuses fixées du Ferlo nord occidental (Senegal): Assoc. Senegal. Étude Quarternaire Quest Africain, Bull. 31–32, p. 69–78.

Lindsay, J. F., 1973, Reversing barchan dunes in Lower Victoria Valley, Antarctica: Geol. Soc. America Bull., v. 84, no. 5, p. 1799–1806.

Logan, R. F., 1960, The central Namib Desert, South-West Africa: Natl. Acad. Sci. – Natl. Research Council Pub. 758, 162 p.

Maack, Reinhard, 1941, Algumas observacões à respeito da existência e da extensão do arenito superior São Bento ou Caiuá no estado do Paraná [Some observations regarding the existence and enlargement of the Superior São Bento or Caiuá Desert, Paraná]: Curitiba, Paraná, Brazil Mus. Arquivos Paranaense, v. 1, p. 107–129.

Mabbutt, J. A., 1955, Erosion surfaces in Namaqualand and the ages of surface deposits in the south-western Kalahari: Geol. Soc. South Africa Trans. and Proc., v. 58, p. 13–30.

_____ 1967, Denudation chronology in central Australia; structure, climate, and landform inheritance in the Alice Springs area, in J. N. Jennings and J. A. Mabbutt, eds., Landform studies from Australia and New Guinea, with a foreword by Prof. E. S. Hills: Canberra, Australian Natl. Univ. Press, p. 144–181.

_____ 1968, Aeolian landforms in central Australia: Australian Geog. Studies, v. 6, no. 2, p. 139–150.

Mabesoone, J. M., 1963, Coastal sediments and coastal development near Cádiz (Spain): Geológie en Mijnbouw, no. 2, p. 29–43.

McConnell, R. B., 1959, Notes on the geology and geomorphology of the Bechuanaland protectorate: Internat. Geol. Cong., 20th, Mexico, 1956, Assoc. de Servicios Géologicos Africanos, Actas y Trabajos de las reuniones celebradas en Mexico en 1956, p. 175–186.

MacGregor, P. A., 1955, Development of the theories of the circulation of the atmosphere over South Africa: South African Geog. Jour., v. 49, p. 41–52.

McKee, E. D., 1934a, An investigation of the light-colored cross-bedded sandstones of Canyon de Chelly, Arizona: Am. Jour. Sci., 5th ser., v. 28, no. 165, p. 219–233.

_____ 1934b, The Coconino sandstone — Its history and origin: Carnegie Inst. Washington Pub. 440, p. 77–115.

_____ 1938, Structures in modern sediments aid in interpreting ancient rocks: Carnegie Inst. Washington Pub. 501, p. 683–694.

_____ 1940, Three types of cross-lamination in Paleozoic rocks of northern Arizona: Am. Jour. Sci., v. 238, no. 11, p. 811–824.

_____ 1944, Tracks that go uphill [in Coconino sandstone, Grand Canyon, Arizona]: Plateau, v. 16, no. 4, p. 61–72.

_____ 1945, Small-scale structures in Coconino sandstone of northern Arizona: Jour. Geology, v. 53, no. 5, p. 313–325.

_____ 1947, Experiments on the development of tracks in fine cross-bedded sand: Jour. Sed. Petrology, v. 17, no. 1, p. 23–28.

_____ 1952, Uppermost Paleozoic strata of northwestern Arizona and southwestern Utah, in Utah Geol. Soc. Guidebook to Geology of Utah: no. 7, p. 52–55.

_____ 1954, Stratigraphy and history of the Moenkopi Formation of Triassic age: Geol. Soc. America Mem. 61, 133 p.

_____ 1957, Primary structures in some Recent sediments: Am. Assoc. Petroleum Geologists Bull., v. 41, no. 8, p. 1704–1747.

_____ 1966, Structures of dunes at White Sands National Monument, New Mexico (and a comparison with structures of dunes from other selected areas): Sedimentology, v. 7, no. 1, Spec. Issue, p. 3–69.

McKee, E. D., and Bigarella, J. J., 1972, Deformational structures in Brazilian coastal dunes: Jour. Sed. Petrology, v. 42, no. 3, p. 670–681.

McKee, E. D., and Breed, C. S., 1974a, An investigation of major sand seas in desert areas throughout the world, in Third Earth Resources Technology Satellite–1 Symposium, Washington, D. C., Dec. 10–14, 1973: Natl. Aeronautics and Space Admin. Spec. Pub. NASA SP–351, p. 665–679.

_____ 1974b, Preliminary report on dunes, in Skylab 4 visual observations project report, JSC–09053, June 1974, NASA TM X–58142, p. 9–1 to 9–9.

_____ 1976, Sand seas of the world, in R. S. Williams, Jr., and W. D. Carter, eds., ERTS–1, A new window on our planet: U.S. Geol. Survey Prof. Paper 929, p. 81–88.

_____ 1977, Desert sand seas, chap. 2, in G. P. Carr and others, eds., Skylab explores the Earth: Natl. Aeronautics and Space Admin. Spec. Paper 380. [NASA SP–380], p. 5–48.

McKee, E. D., Breed, C. S., Fryberger, S. G., Gebel, Dana, and McCauley, C. K., 1974, A synthesis of sand seas throughout the world — Type III, Final Report for 1 July 1972–31 March 1974: NASA Earth Resources Survey Program, NASA–CR–139266, 95 p. [Goddard Space Flight Center, Greenbelt, Md.]

McKee, E. D., and Douglass, J. R., 1971, Growth and movement of dunes at White Sands National Monument, New Mexico, in Geological Survey research 1971: U.S. Geol. Survey Prof. Paper 750–D, p. D108–D114.

McKee, E. D., Douglass, J. R., and Rittenhouse, Suzanne, 1971, Deformation of lee-side laminae in eolian dunes: Geol. Soc. America Bull., v. 82, no. 2, p. 359–378.

McKee, E. D., and Moiola, R. J., 1975, Geometry and growth of the White Sands dune field, New Mexico: U.S. Geol. Survey Jour. Research, v. 3, no. 1, p. 59–66.

McKee, E. D., Oriel, S. S., Berryhill, H. L., Jr., Cheney, T. M., Cressman, E. R., Crosby, E. J., Dixon, G. H., Fix, C. E., Hallgarth, W. E., Ketner, K. B., MacLachlan, M. E., Maughan, E. K., McKelvey, V. E., Mudge, M. R., Myers, D. A., and Sheldon, R. P., 1967, Paleotectonic maps of the Permian System: U.S. Geol. Survey Misc. Geol. Inv. Map I–450, 164 p.

McKee, E. D., Oriel, S. S., Swanson, V. E., MacLachlan, M. E., MacLachlan, J. C., Ketner, K. B., Goldsmith, J. W., Bell, R. Y., and Jameson, D. J., 1956, Paleotectonic maps of the Jurassic system, with a separate section on paleogeography, by R. W. Imlay: U.S. Geol. Survey Misc. Geol. Inv. Map I–175.

McKee, E. D., Reynolds, M. A., and Baker, C. H., Jr., 1962a, Laboratory studies on deformation in unconsolidated sediment, in Short papers in geology, hydrology, and topography: U.S. Geol. Survey Prof. Paper 450 – D, p. D151 – D155.

_____ 1962b, Experiments on intraformational recumbent folds in crossbedded sand, in Short papers in geology, hydrology, and topography: U.S. Geol. Survey Prof. Paper 450 –D, p. D155 –D160.

McKee, E. D., and Tibbitts, G. C., Jr., 1964, Primary structures of a seif dune and associated deposits in Libya: Jour. Sed. Petrology, v. 34, no. 1, p. 5 –17.

Madigan, C. T., 1936, The Australian sand-ridge deserts: Geog. Rev., v. 26, no. 2, p. 205 –227.

_____ 1938, The Simpson Desert and its borders: Royal Soc. New South Wales Jour. and Proc. , v. 71, pt. 2, p. 503 –535.

_____ 1946, The Simpson Desert Expedition, 1939, Scientific reports; No. 6, Geology — The sand formations: Royal Soc. South Australia Trans., v. 70, p. 45 –63, 205 –227.

Maher, J. C., 1954, Lithofacies and suggested depositional environment of Lyons sandstone and Lykins formation in southeastern Colorado: Am. Assoc. Petroleum Geologists Bull., v. 38, no. 10, p. 2233 – 2239.

Manton, W. I., 1968, The origin of associated basic and acid rocks in the Lebombo-Nuanetsi igneous province, Southern Africa, as implied by strontium isotopes: Jour. Petrology, v. 9, pt. 1, p. 23 –39.

Margolis, S. V., and Krinsley, D. H., 1971, Submicroscopic frosting on eolian and subaqueous quartz sand grains: Geol. Soc. America Bull., v. 82, no. 12, p. 3395 –3406.

Martins, L. R., 1967, Aspectos texturais e deposicionais dos sedimentos praiais e éolicas da planície costeira do Rio Grande do Sul [Brazil]: Escola de Geologia, Rio Grande do Sul. Fed. Univ., Spec. Pub. 13, 102 p.

Marzolf, J. E., 1969, Regional stratigraphic variations in primary features of the Navajo Sandstone, Utah: Geol. Soc. America Abs. with Programs for 1969, pt. 5, p. 50 –51.

Mason, C. C., and Folk, R. L., 1958, Differentation of beach, dune, and aeolian flat environments by size analysis, Mustang Island, Texas: Jour. Sed. Petrology, v. 28, no. 2, p. 211 –226.

Maughan, E. K., and Wilson, R. F., 1960, Pennsylvanian and Permian strata in southern Wyoming and northern Colorado, in Guide to the geology of Colorado: Geol. Soc. America, Rocky Mtn. Assoc. Geologists, and Colorado Sci. Soc., p. 34 –42.

Meigs, Peveril, 1953, World distribution of arid and semi-arid homoclimates, in Reviews of research on arid zone hydrology: Paris, Arid Zone Programme –1, UNESCO, p. 203 –209.

Meinster, B., and Tickell, S. J., 1976, Precambrian aeolian deposits in the Waterberg Supergroup: South Africa Geol. Soc. Trans., p. 191 –199.

Melton, F. A., 1940, A tentative classification of sand dunes — Its application to dune history in the southern High Plains: Jour. Geology, v. 48, no. 2, p. 113 –174.

Merk, G. P., 1960, Great sand dunes of Colorado, in Guide to the geology of Colorado: Geol. Soc. America, Rocky Mtn. Assoc. Geologists, and Colorado Sci. Soc., p. 127 –129.

Merriam, Richard, 1969, Source of sand dunes of southeastern California and northwestern Sonora, Mexico: Geol. Soc. America Bull., v. 80, no. 3, p. 531 –533.

Miller, A. K., and Thomas, H. D., 1936, The Casper Formation (Pennsylvanian) of Wyoming and its cephalopod fauna: Jour. Paleontology, v. 10, no. 8, p. 715 –738.

Miller, R. L., and Kahn, J. S., 1962, Statistical analysis in the geological sciences: New York, John Wiley & Sons, 483 p.

Moiola, R. J., Spencer, A. B., and McKee, E. D., 1973, Linear discriminant analysis — A technique for analyzing sand bodies: Geol. Soc. America Abs. with Programs, 1973 Ann. Mtgs., v. 5, no. 7, p. 741 –742.

Moiola, R. J., Spencer, A. B., and Weiser, Daniel, 1974, Differentiation of modern sand bodies by linear discriminant analysis: Gulf Coast Assoc. Geol. Socs. Trans., v. 24, p. 321 –326.

Moiola, R. J., and Weiser, Daniel, 1968, Textural parameters — An evaluation: Jour. Sed. Petrology, v. 38, no. 1, p. 45 –53.

Monod, Théodore, 1958, Majâbat al-Koubrâ; Contribution à l'étude de l'"empty quarter" ouest-saharien: Français d'Afrique Noire Inst. Mém. 52, 406 p.

_____ 1961, Majâbat al-Koubrâ (supplément): Français d'Afrique Noire Inst. Bull. 23, no. 3, p. 591 –637.

Munsell Color Co., 1954, Munsell soil color charts: Baltimore, Md.

Murzayev, E. M., 1967, Nature of Sinkiang and formation of the deserts of central Asia: U.S. Joint Publications Research Service JPRS 40,229, TT 67 –30944, 617 p. [Technical translation from Russian — Priroda sin'tszyana I formirovaniye pustyn' tsentral'nay Azii: Akad. Nauk SSSR Inst. Geograffi, 1966.]

Nairn, A. E. M., ed., 1961, Descriptive palaeoclimatology: New York, Interscience Publishers, 380 p.

_____ 1964, Problems in palaeoclimatology, in NATO palaeoclimates conf., Univ. Newcastle-upon-Tyne and Durham, England, 1963, Proc.: New York, Interscience Publishers, 705 p.

Newell, N. D., and Rigby, J. K., 1957, Geological studies on the Great Bahama Bank, in Regional aspects of carbonate deposition: Soc. Econ. Paleontologists and Mineralogists Spec. Pub. 5, p. 15 –72.

Newell, N. D., Rigby, J. K., Whiteman, A. J., and Bradley, J. S., 1951, Shoal-water geology and environments, Eastern Andros Island, Bahamas: Am. Mus. Nat. History Bull., v. 97, art. 1, p. 1 –29.

Norin, Eric, 1941, The Tarim Basin and its border regions: Leipzig Akad. Verlagsgesellschaft, Regionale Geologie der Erde, v. 2, pt. 4b, 40 p.

Norris, R. M., 1966, Barchan dunes of Imperial Valley, California: Jour. Geology, v. 74, no. 3, p. 292 –306.

_____ 1969, Dune reddening and time: Jour. Sed. Petrology, v. 39, no. 1, p. 7 –11.

Norris, R. M., and Norris, K. S., 1961, Algodones dunes of southeastern California: Geol. Soc. America Bull., v. 72, no. 4, p. 605 –620.

Obrutschev, V. A., 1895, [Uber] Ueber die verwitterungs und deflationsprocesse in Central-Asien: St. Petersburg, Russisch-Kaiserlich Mineralogische Gesell. Verh., ser. 2, no. 33, p. 229 –272.

Opdyke, N. D., 1961, The palaeoclimatological significance of desert sandstone, chap. 3, in A. E. M. Nairn, ed., Descriptive Palaeoclimatology: New York, Interscience Publishers, p. 45 –60.

Opdyke, N. D., and Runcorn, S. K., 1960, Wind direction in the Western United States in the Late Paleozoic: Geol. Soc. America Bull., v. 71, no. 7, p. 959–971.

Pacheco, J. A. do Amaral, 1913, Notas sobre a geologia do valle do Rio Grande a partir da fóz do Rio Pardo até a sua confluencia com o Rio Paranahyba; exploração do Rio Grande e de seus affluentes: [Brazil] São Paulo Commissão Geog. e Geol., p. 33–38.

Page, W. D., 1972, The geological setting of the archaeological site at Oued el Akarit and the paleoclimatic significance of gypsum soils, southern Tunisia: Boulder, Colo., Colorado Univ. unpub. Ph. D. thesis, 111 p.; available from Xerox Univ. Microfilms, Ann Arbor, Mich., no. 73–1812, 121 p.

Peabody, F. E., 1940, Trackways of Pliocene and Recent salamandroids of the Pacific Coast of North America: Berkeley, Calif., California Univ., unpub. M.A. thesis.

Pederson, S. L., 1953, The stratigraphy of the (Pennsylvanian and Permian) Casper and Fountain formations of southeastern Wyoming and north-central Colorado: Laramie, Wyo., Wyoming Univ., unpub. M.A. thesis, 87 p.

Peirce, H. W., 1963, Stratigraphy of the De Chelly Sandstone of Arizona and Utah: Tucson, Ariz., Arizona Univ., Ph. D. thesis, 206 p.; available from Univ. Microfilms, Ann Arbor, Mich., no. 63–3650, 291 p.

Pesce, Angelo, 1968, Gemini space photographs of Libya and Tibesti — A geological and geographical analysis: Tripoli, Petroleum Exploration Soc. Libya, 81 p.

Peterson, J. A., 1972, Jurassic System, in Geologic atlas of the Rocky Mountain region, United States of America: Rocky Mtn. Assoc. Geologists, p. 177–189.

Peterson, J. A., and Osmund, J. C., eds., 1961, Geometry of sandstone bodies — A symposium, 45th Ann. Mtg., Atlantic City, N. J., 1960: Tulsa, Okla., Am. Assoc. Petroleum Geologists, 240 p.

Petrov, M. P., 1961, Geographic explorations in the deserts of Central Asia: U.S. Joint Publications Research Service JPRS 9221, 31 p. [Technical translation from Russian.]

_____ 1962, Types de deserts de l'Asie centrale: Annales de Geographie, no. 384, v. 71, p. 131–155.

_____ 1966, The Ordos, Alashan and Peishan, v. 1 of The deserts of Central Asia: U.S. Joint Publications Research Service JPRS 39,145, p. 241–260. [Translated from Russian.]

_____ 1967, The Hoshi Corridor, Tsaidam and Tarim Basin, v. 2 of The deserts of Central Asia: U.S. Joint Publications Research Service JPRS 42,772, 335 p. [Translated from Russian.]

Pettijohn, F. J., Potter, P. E., and Siever, Raymond, 1972, Sand and sandstone: Berlin, New York, Springer-Verlag, 618 p.

Poldervaart, Arie, 1955, Kalahari sands: Pan-African Cong. on Prehistory, 3rd, Livingstone, Northern Rhodesia, 1955, p. 106–114.

Poole, F. G., 1962, Wind directions in late Paleozoic to middle Mesozoic time on the Colorado Plateau, in Short papers in geology, hydrology, and topography: U.S. Geol. Survey Prof. Paper 450–D, p. D147–D151.

_____ 1964, Palaeowinds in the Western United States, in A. E. M. Nairn, ed., Problems in palaeoclimatology: NATO Palaeoclimates Conf., Univ. Newcastle-upon-Tyne and Durham, England, 1963, Proc.; New York, Interscience Publishers, p. 394–405 [1965].

Powers, R. W., Ramirez, L. F., Redmond, C. D., and Elberg, E. L., Jr., 1966, Geology of the Arabian Peninsula — Sedimentary geology of Saudi Arabia: U.S. Geol. Survey Prof. Paper 560–D, 147 p.

Pramanik, S. K., 1952, Hydrology of the Rajasthan Desert — Rainfall, humidity, and evaporation, in Symposium on the Rajputana Desert, India, 1952, Proc.: India Natl. Inst. Sci. Bull. 1, pt. 4, Meteorology and Hydrology, p. 183–197.

Prescott, J. R. V., and White, H. P., 1960, Sand formations in the Niger Valley between Niamey and Bourem: Geog. Jour., v. 126, pt. 2, p. 200–203.

Price, W. A., 1950, Saharan sand dunes and the origin of the longitudinal dunes — A review: Geog. Rev., v. 40, no. 3, p. 462–465.

Pryor, W. A., 1971a, Petrology of the Permian Yellow Sands of northeastern England and their North Sea basin equivalents: Sed. Geology, v. 6, p. 221–254.

_____ 1971b, Petrology of the Weissliegendes sandstones in the Harz and Werra-Fulda areas, Germany: Geol. Rundschau, v. 60, no. 2, p. 524–551.

_____ 1973, Permeability-porosity patterns and variations in some Holocene sand bodies: Am. Assoc. Petroleum Geologists Bull., v. 57, no. 1, p. 162–189.

Qadri, S. M. A., 1957, Wind erosion and its control in Thar, in Food and Agricultural Council of Pakistan and UNESCO, Symposium on soil erosion and its control in the arid and semiarid zones, Karachi, Nov. 1957: p. 169–184.

Raisz, Erwin, 1952, Landform map of North Africa: Boston [Mass.], 02114, 130 Charles St., scale 1:3,801,600.

Ramirez, L. F., Elberg, E. L., Jr., and Helley, H. H., 1963a, Geologic map of the south central Rub' al Khali quadrangle, Kingdom of Saudi Arabia: U.S. Geol. Survey Misc. Geol. Inv. Map I–219A, scale 1:500,000.

_____ 1963b, Geologic map of the southeastern Rub' al Khali quadrangle, Kingdom of Saudi Arabia: U.S. Geol. Survey Misc. Geol. Inv. Map I–220A, scale 1:500,000.

Range, Paul, 1927, Geologie der Küstenwüste Südwestafrikas zwichen dem Kuiseb und der Lüderitzbucht — Eisenbahn: Internat. Geol. Cong., 14th, Madrid, 1926, Comptes Rendus, v. 3, p. 901–905.

Ratner, Benjamin, 1950, A method for eliminating directional bias in wind roses: Monthly Weather Review, v. 78, no. 10, p. 185–188.

Rao, K. N., 1958, Some studies on rainfall of Rajasthan with particular reference to trends: Indian Jour. Meteorology and Geophysics, v. 9, no. 2, p. 97–116.

Rawson, R. R., and Turner, C. E., 1974, The Toroweap Formation — A new look, in T. N. V. Karlstrom, G. A. Swann, and R. L. Eastwood, eds., Geology of northern Arizona, with notes on archaeology and paleoclimate, pt. 1 of Regional studies: Geol. Soc. America Rocky Mtn. Sec. [Guidebook 27, pt. 1], p. 155–190.

Read, C. B., 1951, Stratigraphy of the outcropping Permian rocks around the San Juan Basin, in New Mexico Geol. Soc. Guidebook 2d Field Conf., The south and west sides of the San Juan Basin, New Mexico and Arizona, 1951: p. 80–84.

Read, C. B., and Wanek, A. A., 1961, Stratigraphy of outcropping Permian rocks in parts of northeastern Arizona and adjacent areas: U.S. Geol. Survey Prof. Paper 374 –H, 10 p.

Reiche, Parry, 1938, An analysis of cross-lamination — The Coconino sandstone: Jour. Geology, v. 46, no. 7, p. 905 –932.

Reineck, H. E., and Singh, I. B., 1973, Depositional sedimentary environments with reference to terrigenous clastics: New York, Heidelberg, Berlin, Springer-Verlag, 439 p.

Reynolds, M. W., Ahlbrandt, T. S., Fox, J. E., and Lambert, P. W., 1975, Description of selected drill cores from Paleozoic rocks, Lost Soldier oil field, south-central Wyoming; pt. I, Wells 114A; tract 13, C –128; tract 4, C –14; tract 10, T –1; tract 10, C –2: U.S. Geol. Survey Open-File Rept. 75 –662, 73 p.

_____1976, Description of selected drill cores from Paleozoic rocks, Wertz oil field, south-central Wyoming; pt. I, Wells 27 ABC; 46 ABC&E; 42 B; West Wertz 2: U.S. Geol. Survey Open-File Rept. 76 –377, 56 p.

_____ 1977, Preliminary description of cores from selected wells, Wertz oil field, south-central Wyoming; pt. 2: U.S. Geol. Survey Open-File Rept.

Reynolds, M. W., and Fox, J. E., 1973, Description of selected drill cores from Paleozoic and Precambrian rocks, Lost Soldier oil field, south-central Wyoming; pt. II, wells tract 11, T –1; 109A; tract 13, C –115; 103; tract 13, C –127; tract 14, T –11; tract 9, M –1; U.S. Geol. Survey Open-File Rept. 76 –64, 50 p.

Rice, C. M., 1951, Dictionary of geological terms (exclusive of stratigraphic formations and paleontologic genera and species): Ann Arbor, Mich., Edwards Bros., 465 p.

Ries, Heinrich, and Conant, G. D., 1931, The character of sand grains: Am. Foundrymen's Assoc. Trans., v. 2, no. 10, p. 353 –392.

Rigby, J. K. and Hamblin, W. K., eds., 1972, Recognition of ancient sedimentary environments: Soc. Econ. Paleontologists and Mineralogists Spec. Pub. 16, 340 p.

Rittenhouse, Gordon, 1971, Pore-space reduction by solution and cementation: Am. Assoc. Petroleum Geologists Bull., v. 55, no. 1, p. 80 –91.

Rogers, A. W., 1934, The build of the Kalahari: South African Geog. Jour., v. 17, p. 3 –12.

_____ 1936, The surface geology of the Kalahari: Royal Soc. South Africa Trans., v. 24, p. 57 –80.

Rogers, John, and Tankard, A. J., 1974, Surface textures of some quartz grains from the west coast of southern Africa: Southern Africa Electron Microscopy Soc. Proc., v. 4, p. 55 –56.

Roy, B. B., 1969, Technical report — Progress of research, in Central Arid Zone Research Inst., Jodhpur, Ann. Rept.: p. 12 –20.

Royal Navy and South African Air Force, 1944, Introduction; The west coast of Africa from River Congo to Olifants River, with an appendix on the climates of St. Helena and Tristan da Cunha, pt. 1, in Weather on the coasts of southern Africa, v. 2: p. iii –viii; pt. 1, p. 1 –61.

Rumney, G. R., 1968, Climatology and the world's climates: New York, Macmillan Co., 656 p.

Sagan, C., Veverka, J., Fox, P., and others, 1973, Variable features on Mars, 2, Mariner 9 global results: Jour. Geophys. Research, v. 78, no. 20, p. 4163 –4196.

Sahu, B. K., 1964, Depositional mechanisms from the size analysis of clastic sediments: Jour. Sed. Petrology, v. 34, no. 1, p. 73 –83.

Salamuni, Riad, and Bigarella, J. J., 1967, The Botucatú Formation, in J. J. Bigarella, R. D., Becker, and I. D., Pinto, eds., Problems in Brazilian Gondwana geology: Internat. Symposium on the Gondwana Stratigraphy and Paleontology, 1st, Mar del Plata, 1967, p. 197 –206.

Schiffers, Heinrich, 1971, Die Sahara und ihre Randgebiete; Darstellung eines Naturgrossraumes Gesamtbearbeitung, Bd. 1, Physiogeographie: Munich, Weltforum Verlag, 674 p.

Schlee, John, Uchupi, Elazar, and Trumbull, J. V. A., 1964, Statistical parameters of Cape Code beach and eolian sands, in Geological Survey research 1964: U.S. Geol. Survey Prof. Paper 501 –D, p. D118 –D122.

Schuchert, Charles, 1918, On the Carboniferous of the Grand Canyon of Arizona: Am. Jour. Sci., 4th ser., v. 45, p. 347 –361.

Schulze, R. R., 1965, General Survey, pt. 8 of Climate of South Africa: South Africa Weather Bur. Pub. WB 28, 322 p.

_____ 1972, South Africa, chap. 15, in H. E. Lansburg, ed., World survey of climatology, climates of Africa, v. 10: Amsterdam, New York, London, Elsevier Publishing Co., p. 501 –586.

Seely, M. K., and Sandelowsky, B. H., 1974, Dating the regression of a river's end point: South Africa Archeological Soc. Bull., Goodwin Series no. 2, p. 61 –64.

Selley, R. C., 1968, A classification of paleocurrent models: Jour. Geology, v. 76, no. 1, p. 99 –110.

_____ 1970, Ancient sedimentary environments — A brief survey: Ithaca, N.Y., Cornell Univ. Press, 237 p.

Seth, S. K., 1963, A review of evidence concerning changes of climate in India during the protohistorical and historical periods, in Changes of climate: UNESCO and World Meteorol. Organization Symposium, Rome, 1961, Proc., Arid Zone Research [v.] 20, p. 443 –454.

Sevon, W. D., 1966, Distinctions of New Zealand beach, dune, and river sands by their grain-size distribution characteristics: New Zealand Jour. Geology and Geophysics, v. 9, p. 212 –223.

Sharp, R. P., 1963, Wind ripples: Jour. Geology, v. 71, no. 5, p. 617 –636.

_____ 1966, Kelso dunes, Mohave Desert, California: Geol. Soc. America Bull., v. 77, no. 10, p. 1045 –1073.

Shea, J. H., 1974, Deficiencies of clastic particles of certain sizes: Jour. Sed. Petrology, v. 44, no. 4, p. 985 –1003.

Shepard, F. P., and Young, Ruth, 1961, Distinguishing between beach and dune sands: Jour. Sed. Petrology, v. 31, no. 2, p. 196 –214.

Shergold, J. H., Druce, E. C., Radke, B. M., and Draper, J. J., 1976, Cambrian and Ordovician stratigraphy of the eastern portion of the Georgina Basin, Queensland and eastern Northern Territory: Internat. Geol. Cong., 25th, Australia, 1976, Field Excursion 4C Guidebook, 54 p.

Sherzer, W. H., and Grabau, A. W., 1909, The Sylvania Sandstone; its distribution, nature and origin, chap. III, in

A. W. Grabau and W. H. Sherzer, The Monroe formation of southern Michigan and adjoining regions: Michigan Geol. and Biol. Survey Pub. 2, p. 61 – 86.

Shotton, F. W., 1937, The Lower Bunter sandstones of north Worcestershire and east Shropshire [England]: Geol. Mag. no. 882, v. 74, no. 12, p. 534 –553.

_____ 1956, Some aspects of the New Red desert in Britain: Liverpool and Manchester Geol. Jour., v. 1, pt. 5, p. 450 –465.

Simons, F. S., 1956, A note on Pur-Pur dune, Virú Valley, Peru: Jour. Geology, v. 64, no. 5, p. 517 –521.

Singh, S., Ghose, B., and Vats, P. C., 1972, Genesis, orientation, and distribution of sand dunes in arid and semi-arid regions of India: Central Arid Zone Research Inst., Jodhpur, Ann. Rept., p. 50 –53.

Sinitsyn, V. M., 1959, Tsentralnaya Aziya: Moscow, Gosudar. Izd. Geog. Lit., 456 p.

Smith, D. B., 1972, The Lower Permian in the British Isles, in Horst Falke, ed., Rotliegend — Essays on European Lower Permian: Leiden, Netherlands, E. J. Brill & Co., p. 1 –33.

Smith, D. B., and Francis, E. A., 1967, Geology of the country between Durham and West Hartlepool [England]: Great Britain Geol. Survey Mem., 354 p.

Smith, H. T. U., [1943], Aerial photographs and their applications: New York, D. Appleton-Century Co., 372 p.

_____ 1946, Sand dunes: New York Acad. Sci. Trans., ser. 2, v. 8, no. 6, p. 197 –199.

_____ 1953, Classification of sand dunes [abs.], in Déserts actuels et anciens: Internat. Geol. Cong., 19th, Algiers, 1952, Comptes Rendus, sec. 7, p. 105. [In English.]

_____ 1956, Giant composite barchans of the northern Peruvian deserts [abs.]: Geol. Soc. America Bull., v. 67, no. 12, pt. 2, p. 1735.

_____ 1963, Eolian geomorphology, wind direction, and climatic change in North Africa — final report: U.S. Air Force Cambridge Research Lab., Geophysics Research Directorate [Contract AF 19(628) –298], AFCRL –63 –443, 49 p.

_____ 1965, Dune morphology and chronology in central and western Nebraska: Jour. Geology, v. 73, no. 4, p. 557 –578.

_____ 1968, Nebraska dunes compared with those of North Africa and other regions, in C. B. Schultz and J. C. Frye, eds., Loess and related eolian deposits of the world: Internat. Assoc. Quaternary Resarch Cong., 7th, Colorado, U.S.A., 1965, Proc., v. 12, p. 29 –47.

_____ 1969, Photo-interpretation studies of desert basins in northern Africa — Final report (pt. 1), March 1963 –July 1968: U.S. Air Force Cambridge Research Lab., AFCRL –68 –0590, 60 p.

Snead, R. E., and Frishman, S. A., 1968, Origin of sands on the east side of the Las Bela valley, West Pakistan: Geol. Soc. America Bull., v. 79, no. 11, p. 1671 –1676.

South African Weather Bureau, 1960, Climate of South Africa; pt. 6, Surface winds: Pretoria, 202 p.

Stanley, K. O., Jordan, W. M., and Dott, R. H., Jr., 1971, New hypothesis of Early Jurassic paleogeography and sediment dispersal for Western United States: Am. Assoc. Petroleum Geologists Bull., v. 55, no. 1, p. 10 –19.

Steidtmann, J. R., 1974, Evidence for eolian origin of cross-stratification in sandstone of the Casper Formation, southernmost Laramie Basin, Wyoming: Geol. Soc. America Bull., v. 85, no. 12, p. 1835 –1842.

Stokes, W. L., 1961, Fluvial and eolian sandstone bodies in Colorado Plateau, in J. A. Peterson and J. C. Osmond, eds., Geometry of sandstone bodies — A symposium, 45th Ann. Mtg., Atlantic City, N.J., April 25 –28, 1960: Tulsa, Okla., Am. Assoc. Petroleum Geologists, p. 151 –178.

Stone, R. O., 1967, A desert glossary: Earth-Sci. Rev., v. 3, no. 4, p. 211 –268.

Strahler, A. N., 1958, Dimensional analysis applied to fluvially eroded landforms: Geol. Soc. America Bull., v. 69, no. 3, p. 279 –299.

Stride, A. H., 1963, Current-swept sea floors near the southern half of Great Britain: Geol. Soc. London Quart. Jour., v. 119, pt. 2, no. 474, p. 175 – 199.

Suslov, S. P., 1961, Physical geography of Asiatic Russia [English translation by N. D. Gershevsky]: San Francisco, London, W. H. Freeman & Co., 594 p. [First published in Russian in Leningrad and Moscow, 1947.]

Tanner, W. F., 1964, Eolian ripple marks in sandstone: Jour. Sed. Petrology, v. 34, no. 2, p. 432 –433.

_____ 1965, Upper Jurassic paleogeography of the Four Corners region: Jour. Sed. Petrology, v. 35, no. 3, p. 564 –574.

_____ 1966, Numerous eolian ripple marks from Entrada Formation: Mtn. Geologist, v. 3, no. 3, p. 133 –134.

_____ 1967, Ripple mark indicies and their uses: Sedimentology, v. 9, no. 2, p. 89 –104.

Terwindt, J. H. J., 1971, Sand waves in the southern bight of the North Sea: Marine Geology, v. 10, no. 1, p. 51 –67.

Thesiger, W., 1949, A further journey across the Empty Quarter: Geog. Jour., v. 113, p. 21 –46.

Thompson, D. B., 1969, Dome-shaped aeolian dunes in the Frodsham Member of the so-called "Keuper" Sandstone Formation (Scythian-?Anisian; Triassic) at Frodsham, Cheshire (England): Sed. Geology, v. 3, no. 4, p. 263 –289.

_____ 1970a, Sedimentation of the Triassic (Scythian) red pebbly sandstones in the Cheshire Basin and its margins: Geol. Jour., v. 7, pt. 1, p. 183 –261.

_____ 1970b, The stratigraphy of the so-called "Keuper" Sandstone Formation (Scythian-?Anisian) in the Permo-Triassic Cheshire Basin: Geol. Soc. London Quart. Jour., v. 126, p. 151 –181.

Thompson, M. L., and Thomas, H. D., 1953, Systematic paleontology of fusulinids from the Casper Formation: Wyoming Geol. Survey Bull. 46, pt. 2, p. 15 –56.

Thompson, W. O., 1949, Lyons sandstone of Colorado Front Range: Am. Assoc. Petroleum Geologists Bull., v. 33, no. 1, p. 52 –72.

_____ 1954, Lyons sandstone in Denver basin [Colo.] [abs.]: Am. Assoc. Petroleum Geologists Bull., v. 38, no. 6, p. 1311.

Tieje, A. J., 1923, The redbeds of the Front Range in Colorado — A study in sedimentation: Jour. Geology, v. 31, no. 3, p. 192 –207.

Tillman, R. W., 1971, Multiple group discriminant analysis of grain size data as an aid in recognizing environments of deposition: Internat. Sedimentological Cong., 8th, Heidelberg, Program with Abs., p. 102.

Trask, P. D., 1931, Compaction of sediments: Am. Assoc. Petroleum Geologists Bull., v. 15, no. 3, p. 271–276.

Tricart, Jean, 1959, Géomorphologie dynamique de la moyenne vallée du Niger (Soudan): Annales Géographie, no. 368, p. 333–343.

_____ 1965, Rapport de la mission de reconnaissance géomorphologique de la vallée moyenne du Niger: Institut Français d'Afrique Noire Mém. 72, 196 p.

Tricart, Jean, and Brochu, Michel, 1955, Le grand erg ancien du Trarza et du Cayor (Sud-Ouest de la Mauritanie et Nord du Sénégal): Rev. Géomorphologie Dynam., v. 4, p. 145–176.

Twenhofel, W. H., 1926, Treatise on sedimentation: Baltimore, Md., The Williams & Wilkins Co., 661 p.

_____1945, The rounding of sand grains: Jour. Sed. Petrology, v. 15, no. 2, p. 59–71.

Twidale, C. R., 1972, Evolution of sand dunes in the Simpson Desert, central Australia: Inst. British Geographers Pub. 56, [reprinted from Trans., July 1972], p. 77–109.

Udden, J. A., 1898, The mechanical compositions of wind deposits: Augustana Library Pub., no. 1, 69 p.

_____ 1914, Mechanical composition of clastic sediments: Geol. Soc. America Bull., v. 25, p. 655–744.

U.S. Air Force [MATS], 1956, Aeronautical chart 541, Sebha Oasis, Algeria-Libya.

U.S. Army Air Forces Weather Division, 1945, Weather and climate of China, Pts. A and B: Rept. 890; U.S. Air Weather Service Tech. Rept. 105–34.

U.S. Geological Survey, 1963, Map of the Arabian Peninsula: U.S. Geol. Survey Misc. Geol. Inv Map I–270 B–2, scale 1:200,000.

U.S. Naval Weather Service, 1967, World-wide airfield summaries — Middle East: Asheville, N.C., U.S. Natl. Weather Records Center, v. 2, pt. 1, p. 1–334; v. 2, pt. 2, p. 335–698.

_____1968, World-wide airfield summaries — Africa (northern half): Asheville, N.C., U.S. Natl. Weather Records Center, v. 9, pt. 1, 559 p.

_____1974a, World-wide airfield summaries — China, North Korea, Mongolia: Asheville, N.C., U.S. Natl. Weather Records Center, v. 12, pt. 1, 371 p.; v. 12, pt. 2, 322 p.

_____1974b, World-wide airfield summaries — Middle East: Asheville, N.C., U.S. Natl. Weather Records Center, v. 2, pt. 2, 338 p.

U.S. Weather Bureau and U.S. Army Air Forces, 1944, Northern Africa: U.S. Weather Bureau Prelim. Rept. 41, p. 1–23.

Urvoy, Y., 1935, Terasses et changements de climat quaternaires à l'est du Niger: Annales de Géographie, v. 44, no. 249, p. 254–263.

_____1942, Les bassins du Niger; étude de géographie physique et de paléogéographie: Institute Français de L'Afrique Noire, Mém. no. 4, 141 p.

Vail, C. E., 1917, Lithologic evidence of climatic pulsations: Science, New Ser., v. 46, p. 90–93.

Van de Kamp, P. C., 1973, Holocene continental sedimentation in the Salton basin, California — A reconnaissance: Geol. Soc. America Bull., v. 84, no. 3, 827–848.

Verstappen, H. T., 1966, Landforms, water, and land use west of the Indus Plain: Nature and Resources, v. 2, no. 3, p. 6–8.

_____1968a, On the origin of longitudinal (seif) dunes: Zeitschr. Geomorphologie, v. 12, no. 2, p. 200–220.

_____ 1968b, Geomorphology at the Central Arid Zone Research Institute, India: Nature and Resources, v. 4, no. 3, p. 6–9.

_____1970, Aeolian geomorphology of the Thar Desert and paleo-climates, in Piedmont plains and sand-formations in arid and humid tropic and subtropic regions: Zeitschrift für Geomorphologie, supp. v. 10, p. 104–120.

Verstappen, H. T., and van Zuidam, R. A., 1970, Orbital photography and the geosciences — A geomorphological example from the Central Sahara: Geoforum, v. 2, p. 33–47.

Visher, G. S., 1971, Depositional processes and the Navajo Sandstone: Geol. Soc. America Bull., v. 82, no. 5, p. 1421–1423.

Wadia, D. N., 1939, Geology of India: London, Macmillan and Co., 460 p.

_____1953, Geology of India [3d ed.]: London, Macmillan and Co., 531 p.

Walker, T. R., 1967, Formation of red beds in modern and ancient deserts: Geol. Soc. America Bull., v. 78, no. 3, p. 353–368.

_____1976, Diagenetic origin of continental red beds, in Horst Falke, ed., The continental Permian in west, central and south Europe: NATO Advanced Study Inst., Mainz, Germany, 1975, Proc., p. 240–282.

Walker, T. R., and Crone, A. J., 1974, Mechanically-infiltrated clay matrix in desert alluvium: Geol. Soc. America Abs. with Programs, v. 6, no. 7, p. 998.

Walker, T. R., and Harms, J. C., 1972, Eolian origin of flagstone beds, Lyons Sandstone (Permian), type area, Boulder County, Colorado, in Environments of sandstone, carbonate, and evaporite deposition: Mtn. Geologist, v. 9, no. 2–3, p. 279–288.

Walker, T. R., and Honea, R. M., 1969, Iron content of modern deposits in the Sonoran desert — A contribution to the origin of red beds: Geol. Soc. America Bull., v. 80, no. 3, p. 535–543.

Wallington, C. E., 1968, A method of reducing observing and procedure bias in wind-direction frequencies: Meteorological Mag., v. 97, no. 1155, p. 293–302.

Walther, Johannes, 1888, [Über] Ueber Ergebnisse einer Forschungsreise auf der Sinaihalbinsel und in der Arabischen Wüste: Berlin Gesellschaft für Erdkunde Verh., 15, p. 244–255.

Warren, Andrew, 1969, A bibliography of desert dunes and associated phenomena, in W. G. McGinnies and B. J. Goldman, eds., Arid lands in perspective: Am. Assoc. Adv. Sci. and Arizona Univ. Press, p. 75–99.

_____1970, Dune trends and their implications in the central Sudan, in Piedmont plains and sand-formations in arid and humid and subtropic regions: Zeitschrift für Geomorphologie, supp. v. 10, p. 154–180.

_____1972, Observations on dunes and bi-modal sands in the Tenere Desert: Sedimentology, v. 19, no. 1–2, p. 37–44.

Washburne, C. W., 1930, Petroleum geology of the State of São Paulo, [Brazil]: São Paulo Commisão Geog. e Geol. Boletím 22, 272 p.

Weimer, R. J., and Land, C. B., Jr., 1972, Lyons Formation (Permian), Jefferson County, Colorado — A fluvial deposit, *in* Environments of sandstone, carbonate, and evaporite deposition: Mtn. Geologist, v. 9, no. 2–3, p. 289–297.

Wellington, J. H., 1955, Southern Africa — A geographical study: v. 1, Physical geography: London, Cambridge Univ. Press, 528 p.

Wentworth, C. K., 1932, The mechanical composition of sediments in graphic form: Iowa Univ. Studies in Natural History, v. 14, no. 3, 127 p.

White, L. P., 1971, The ancient erg of Hausaland in southwestern Niger: Geog. Jour., v. 137, pt. 1, p. 69–73.

Wilks, S. S., 1962, Mathematical statistics: New York, John Wiley & Sons, 644 p.

Williams, G. E., 1968, Formation of large-scale trough cross-stratification in a fluvial environment: Jour. Sed. Petrology, v. 38, p. 136–140.

Wills, L. J., 1948, The paleogeography of the Midlands: London, Hodder and Stoughton, Ltd., 144 p.

———— 1956, Concealed coalfields — A paleogeographical study of the stratigraphy and tectonics of mid-England in relation to coal reserves: London, Glasgow, Blackie and Son, 208 p.

Wilson, I. G., 1971a, Desert sandflow basins and a model for the development of ergs: Geog. Jour., v. 137, pt. 2, p. 180–199.

———— 1971b, Journey across the Grand Erg Oriental: Geog. Mag., v. 43, p. 264–270.

———— 1972, Aeolian bedforms — Their development and origins: Sedimentology, v. 19, no. 3–4, p. 173–210.

———— 1973, Ergs: Sed. Geology, v. 10, no. 2, p. 77–106.

Wilson, R. F., 1958, The stratigraphy and sedimentology of the (Jurassic) Kayenta and (?Triassic) Moenave formations, Vermilion Cliffs region, Utah and Arizona: Stanford Univ. unpub. Ph. D. thesis, 167 p.

Wopfner, H., and Twidale, C. R., 1967, Geomorphological history of the Lake Eyre basin, *in* J. N. Jennings and J. A. Mabbutt, eds., Landform studies from Australia and New Guinea: Cambridge [England] Univ. Press, p. 119–143.

Yaalon, D. H., and Ganor, E., 1973, The influence of dust on soils during the Quaternary: Soil Sci., v. 116, p. 146–155.

Yakubov, T. F., and Bespalova, R. Y., 1961, Soil-formation processes during the invasion of sands by plants in the northern deserts of the Caspian region: Soviet Soil Sci., no. 6, p. 651–658.

Yu, Shou-chung, Li, Po, Ts'ai, Wei-ch'i and T'an, Chien-an, 1962, A survey of the Gobi of western Inner Mongolia and Pa Tan Chi Lin desert, *in* Academia Sinica Sand Control Group, eds., Research on sand control: U.S. Joint Publications Research Service JPRS 19993, Tech. Trans. 63–31183, p. 244–276.

Zaychikov, V. T., [ed.], and others, 1965, The physical geography of China: U.S. Joint Publications Research Service JPRS 32,119, TT 65–32612, 634 p. [Technical translation from Russian.]

Zingg, A. W., 1953, Wind tunnel studies of the movement of sedimentary material, *in* J. S. McNown and M. C. Boyer, eds., Hydraulics Conf., 5th, Iowa [U.S.A.], 1952, Proc.: Iowa Univ., p. 111–135.

INDEX

[Italic page numbers indicate major references]

Lightning Source UK Ltd.
Milton Keynes UK
UKOW07f0912131015

260422UK00005B/135/P